Quantitative Organic Analysis via Functional Groups

Quantitative Organic Analysis via Functional Groups

FOURTH EDITION

Sidney Siggia, Ph.D.

UNIVERSITY OF MASSACHUSETTS

AMHERST, MASSACHUSETTS

J. Gordon Hanna

THE CONNECTICUT AGRICULTURAL EXPERIMENT STATION

NEW HAVEN, CONNECTICUT

A Wiley-Interscience Publication

JOHN WILEY AND SONS

New York · Chichester · Brisbane · Toronto

Library of Congress Cataloging in Publication Data:

Siggia, Sidney.
Quantitative organic analysis via functional groups.

"A Wiley-Interscience publication."
Bibliography: p.
Includes index.
1. Chemistry, Analytic—Quantitative.
2. Chemistry, Organic. I. Hanna, J. Gordon, joint author. II. Title.
QD271.7.S53 1978 547'.35 78-5940
ISBN 0-471-03273-5

Printed in the United States of America

10 9 8 7 6 5 4 3 2

Preface

The field of organic functional group determination by chemical means is still developing very rapidly. This is readily reflected by the number of new methods that appear in this edition. This may seem surprising with the advent and growth of the instrumental methods of analysis. However, we have seen that in the past 10 years all existing analytical technologies have grown quickly, along with new ones being born.

This revision is extensive, as can be judged by the increase in the size of this book over the third edition. This is due in part to the addition of new methods and to an increase in the scope of the discussions. Included is a discussion of the chemistry involved in each system; the historical development that led to the method in the form in which it is presented; the merits and limitations of each approach; and a description of how the spectrum of analytical situations for each functional group is covered by the series of methods presented. The text is also completely cross-referenced, showing the interrelationships between the methods for the different functional groups.

Thus this edition is meant to be a convenient handbook of methods of analysis. It is meant also to present a coordinated picture of the chemical methods of analysis of each functional group and of the field of chemical organic functional group analysis as a whole. The chemical analytical approaches for each group are corelated, and a thread is drawn through each item discussed to tie the pieces together into a whole.

The format of each chapter is as follows. The chemical methods for each functional group are subdivided into the various reactions used to determine that group. The chemical and historical backgrounds are given for the application of each reaction; then a method or series of working methods is presented using the particular reaction. The working procedures are presented in full detail and can be used directly without reference to the original sources. In our opinion these methods are the most up to date and/or best in the area being described. The methods were chosen to meet the following standards: general applicability for the purpose intended, simplicity, accuracy, and precision. For the analysis of some functional groups there were so few methods available that there was little choice. There may be better procedures for the determination of certain groups, but those described are, in our opinion, the best and the simplest. In some procedures, slight changes have been made from the description in the original source. These changes—in the nature of sample sizes, simpler equipment, reaction time, and solvent—are included because, in the use of these procedures, they were found to improve the method slightly in simplicity, time, general applicability, accuracy, or precision. Thus the term "adopted from" is often used. In the case of reprinted material where we have made some changes, the variation from the original material is specifically noted.

For the interest of the reader and the education of the student, the historical evolution of each approach is described to show how the present methods were derived. For example, in the section on carbonyl group analysis, the evolution of the sulfite-sulfuric acid reagent from the original sodium sulfite or bisulfite is discussed. Also, the use of unsymmetrical dimethylhydrazine rather than the hydrazines originally used for carbonyl groups is discussed.

The discussions of instrumental methods and methods of separations are not included in this edition. These areas are now so large that to include them as they should be included would be impossible in a book of this type; moreover, they are discussed thoroughly elsewhere.*

In addition to the newer methods is a chapter on the use of reaction rates to chemically analyze mixtures of compounds containing the same functional groups. Methods are now available that will differentiate between homologs and between isomers of a species, and even between two of the same functional group on the same molecule.

I thank the management of General Aniline and Film for their encouragement in the writing of the earlier editions of this book and the management of Olin Mathieson Chemical Corporation for their encouragement in writing the third edition. I also thank again all the people who helped with the first and second editions as well as those who helped with the preparation of the third edition: E. Kuchar, A. Gray, H. Nadeau, P. Thomas, R. Stoessel, F. Reidinger, R. Rittner, H. Agahigian, A. Krivis, B. Starrs, Thomas Palmer, and H. Ackermann.

Special thanks go to Sylvia Maraskauskas and Mary Anne Kusmit who handled the typing, correspondence, and related activities in the earlier editions.

In the case of this fourth edition, we thank Alan Carpenter, Satish Mehta, and Thomas Mourey who read the manuscript and proofs and Charlotte Peet who did the secretarial work.

SIDNEY SIGGIA
J. GORDON HANNA

April 1978
Amherst, Massachusetts
New Haven, Connecticut

* *Instrumental Methods of Organic Functional Group Analysis*, edited by Sidney Siggia, Wiley-Interscience, New York, 1972.

Contents

Introduction

This book deals with the chemical analysis of organic compounds based on reactions of the functional groups on these compounds. Thus the method of measurement is based on the determination of the moieties characteristic of the organic compounds being examined.

With the advent of instrumental approaches such as infrared absorption, gas chromatography, and nuclear magnetic resonance, it may seem to some that the wet chemical methods are outmoded. One practicing in the field of analysis, however, knows that this is not the case. In fact in the analytical laboratories of most manufacturers of organic chemicals, the wet chemical analysts usually outnumber the gas chromatographers and the spectroscopists. It is true that the instrumental approaches make possible a great many things that could not be accomplished previously. They also make possible, in some cases, faster analysis than was possible with the wet methods. The chemical field has grown so fast, however, that even with these new, powerful approaches, the chemical methods not only still persist but have continued to flourish; witness the number of new methods in this text over the previous edition. One need only look at elemental analysis (carbon, hydrogen, etc.), which is one of the first quantitative organic analytical approaches to be developed. This approach is still very much used and is continually being developed beyond its present scope with the new, automated furnaces and the new methods for simultaneous determination of multiple elements. Thus the new analytical instruments serve to enlarge the analytical tool kit but do not displace the older analytical approaches. Just as in carpentry, power tools have enlarged the carpenter's tool kit, but the older, hand tools still have a distinct indispensability.

The reason for the persistence of the chemical types of analysis can be summarized as follows:

1. There are many chemical situations that are better handled by chemical, rather than instrumental methods. The broad spectrum of reactions available gives the wet analyst quite a versatility. Hence, for example, we find that the analysis of complex systems relies heavily on wet

analysis, since specific reactions are generally available for classes of organic compounds. In addition, the area of trace analysis relies heavily on chemical methods to develop specific colors for the materials in question. The foregoing are broad generalizations; instrumental methods figure to some degree in these types of analysis, and also, these types of analysis are not the only ones where wet methods can be used. This paragraph is meant to indicate that there is a "spectrum" of analytical approaches available to the analyst to help solve his diverse problems, and the wet chemical methods occupy a very definite portion of this spectrum.

2. Another advantage to "wet" chemical analysis can be stated as follows. Most instrumental analyses are dependent on calibration curves or calibration data. To obtain the necessary calibrations, pure samples of the compounds in question must be available or preparable. The "wet" chemical methods generally do not require such calibrations. Hence when an analytical laboratory is faced with a rather short series of samples for analyses, it is generally more practical to use the wet methods than to go through the calibration of an instrument and the attendant preparation of standards. In most organic research laboratories and in some plant laboratories, an analyst sees many single samples or small groups of samples that do not recur. It is in these cases that "wet" chemical analyses are the most practical. If a long series of samples is expected, it pays to set up calibrations for an instrument.

3. The cost of equipment for chemical analysis is generally quite low, since such standard laboratory equipment is used as balances, burets, pipets, beakers, and flasks. This aspect makes analysis possible for individuals and groups of limited means.

The general principle behind "wet" chemical methods is the use of a characteristic reaction for the group being measured. This reaction must not only be as specific as possible for the functional group, but it must also be rapid, and it must involve a reactant or product that can be easily measured. Hence fast, specific reactions are used in which the following types of reagent or product are used or produced: acids, bases, oxidants, reductants, gases, water, metallic ions, precipitates, or colored compounds or complexes. The following reactions are typical of those used for functional group analysis. The material being measured is marked by an asterisk, and the group being determined is on the first compound shown in the equation.

ACID CONSUMED

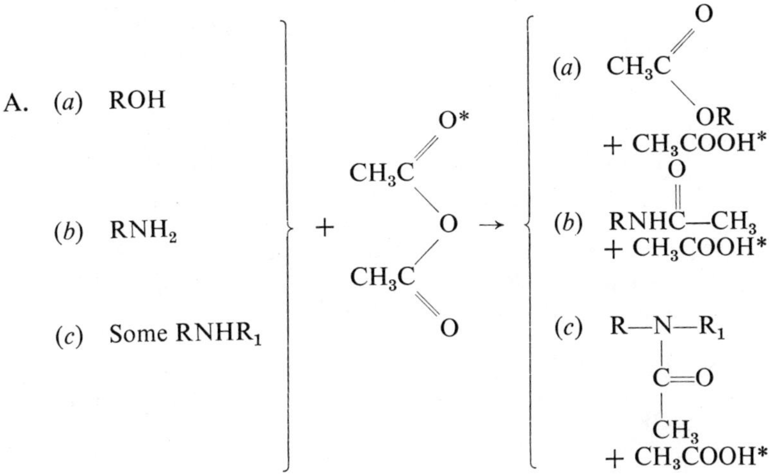

A. (*a*) ROH; (*b*) RNH_2; (*c*) Some $RNHR_1$

$$\left.\begin{array}{l}(a)\ ROH\\(b)\ RNH_2\\(c)\ \text{Some } RNHR_1\end{array}\right\} + (CH_3C(=O^*))_2O \rightarrow \left\{\begin{array}{l}(a)\ CH_3C(=O)OR + CH_3COOH^*\\(b)\ RNHC(=O)-CH_3 + CH_3COOH^*\\(c)\ R-N(C(=O)CH_3)-R_1 + CH_3COOH^*\end{array}\right.$$

B. $$2RCHO + 2Na_2SO_3 + H_2SO_4^* \rightarrow 2RCH(SO_3Na)OH + Na_2SO_4$$

$$\left.\begin{array}{l}(a)\ RCH(OR_1)_2\\(b)\ ROCH{=}CH_2\end{array}\right\} + H_2O \xrightarrow{H^+} \left\{\begin{array}{l}(a)\ RCHO\\(b)\ CH_3CHO\end{array}\right. + \begin{array}{l}(a)\ 2R_1OH\\(b)\ ROH\end{array}$$

Aldehydes determined as in previous equation.

C. Titration of basic materials such as amines (primary, secondary, tertiary), pyridine, quinoline, and carboxylic acids salts with standard acid solutions.

D. 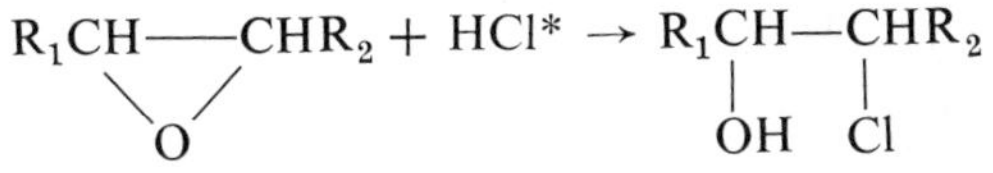 $$R_1CH\underset{O}{\diagdown\!\diagup}CHR_2 + HCl^* \rightarrow R_1CH(OH)-CHR_2(Cl)$$

ACID PRODUCED

A. 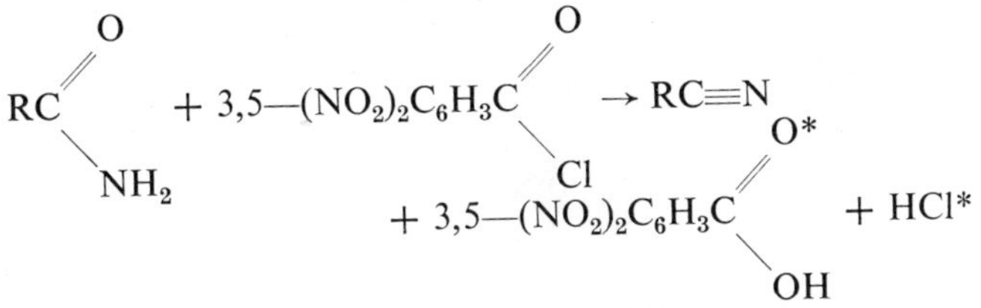 $$RC(=O)NH_2 + 3,5\text{—}(NO_2)_2C_6H_3C(=O)Cl \rightarrow RC{\equiv}N + 3,5\text{—}(NO_2)_2C_6H_3C(=O^*)OH + HCl^*$$

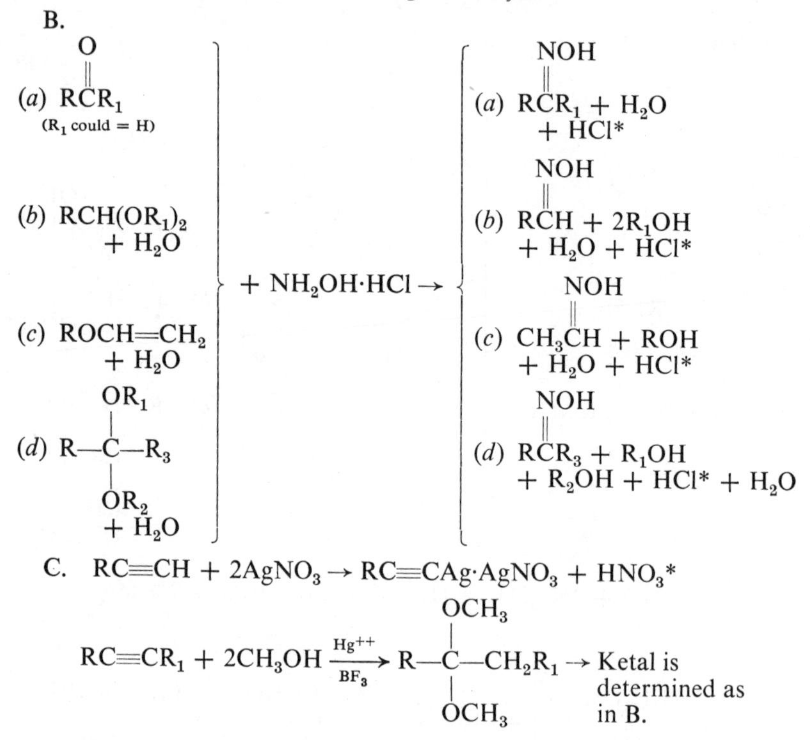

D. Dehydrohalogenation (applicable only to some halogen compounds so that a general equation cannot be written).

BASE CONSUMED

A.

$$\left.\begin{array}{l}(a)\ \text{some } RC(=O)NH_2 \\ (b)\ \text{some } RC{\equiv}N + H_2O \\ (c)\ RC(=O)OR_1\end{array}\right\} + NaOH^* \rightarrow \left\{\begin{array}{ll}(a)\ RC(=O)ONa & + NH_3 \\ (b)\ RC(=O)ONa & + NH_3 \\ (c)\ RC(=O)ONa & + R_1OH\end{array}\right.$$

$$\text{B.}\quad \left.\begin{array}{l}(a)\ RN{=}C{=}O \\ \\ (b)\ RN{=}C{=}S\end{array}\right\} + \underset{(R_1 \text{ is aliphatic})}{R_1NH_2^*} \rightarrow \left\{\begin{array}{l}(a)\ RNH\overset{\overset{\displaystyle O}{\|}}{C}NHR_1 \\ \\ (b)\ RNH\overset{\overset{\displaystyle S}{\|}}{C}NHR_1\end{array}\right.$$

C. Titration of acids with standard alkali.

BASE PRODUCED

$$\text{A.}\quad RC\begin{array}{l}\nearrow O \\ \searrow NH_2\end{array} + NaOH \rightarrow RC\begin{array}{l}\nearrow O \\ \searrow ONa\end{array} + NH_3^*$$

$$\text{B.}\quad RCOOX \xrightarrow{\text{Ignition}} X_2CO_3$$

(X = Na, K, Ca, or Ba with proper valences taken into account)

OXIDANT CONSUMED

A. $RSR_1 + Br_2^* + H_2O \rightarrow R\overset{\overset{\displaystyle O}{\uparrow}}{S}R_1 + 2HBr$

B. $RSSR_1 + 5Br_2^* + 4H_2O \rightarrow 2RSO_2Br + 8HBr$

C. $2RSH + I_2^* \rightarrow RSSR + 2HI$

D. $\underset{\displaystyle OH}{\underset{|}{RCH}}-\underset{\displaystyle OH}{\underset{|}{CH}}-R_1 + HIO_4^* \rightarrow RCHO + R_1CHO + HIO_3 + H_2O$

OXIDANT PRODUCED

Iodine liberated from iodides by peroxides is measured (not applicable to all organic peroxides).

REDUCTANT CONSUMED

A. $TiCl_3$ reductions of —NO_2; —NHNH—; —N=N—; diazonium salts. ($TiCl_3^*$ is measured.)

B. As_2O_3 consumed on reaction with peroxides is measured.

Reactions in which easily determinable materials such as water, silver ions, and sodium nitrite are used or produced are also applicable to determining functional groups. Water takes part in, or is formed in, the reaction of many functional groups—a system of analysis has sprung up around this basis of measurement and is described in *Aquametry* 2nd ed., by John Mitchell (Wiley-Interscience, New York, 1977). Silver enters into reactions with mercaptans and acetylenic hydrogen compounds and it is easily determined by standard methods.

$$RSH + AgNO_3{}^* \rightarrow RSAg + HNO_3$$

$$RC{\equiv}CH + 2AgNO_3{}^* \rightarrow RC{\equiv}CAg{\cdot}AgNO_3 + HNO_3$$

Sodium nitrite takes part in many organic reactions as nitrous acid and can be used to determine primary aromatic amines by diazotization and some secondary aromatic amines and active methylene group compounds by nitrosation.

$$RNH_2{\cdot}HCl + HONO^* \rightarrow RN{\equiv}NCl + 2H_2O$$

$$RNHR_1 + HONO^* \rightarrow \underset{\displaystyle |\atop NO}{RNR_1} + H_2O$$

$$\begin{matrix} O & \phi \\ \| & | \\ C & \!\!-\!\! N \\ / & | \\ CH_2 & | \\ \backslash & | \\ C & \!\!=\!\! N \\ | & \\ CH_3 & \end{matrix} + HONO^* \rightarrow \begin{matrix} O & \phi \\ \| & | \\ C & \!\!-\!\! N \\ / & | \\ ONCH & | \\ \backslash & | \\ C & \!\!=\!\! N \\ | & \\ CH_3 & \end{matrix} + H_2O$$

GAS PRODUCED

$$RN{\equiv}NCl + H_2O \xrightarrow{\Delta} ROH + N_2\uparrow^* + HCl$$

$$RNHNH_2 \xrightarrow[HCl]{Cu^{++}} RN{\equiv}NCl \xrightarrow[H_2O]{\Delta} ROH + N_2\uparrow^* + HCl$$

$$\text{Active hydrogen compound} + CH_3MgX \rightarrow CH_4\uparrow^*$$

GAS CONSUMED

$$\left.\begin{matrix}\text{Unsaturated compounds}\\ \text{Some nitro compounds}\\ \text{Some aldehydes}\end{matrix}\right\} + H_2{}^* \rightarrow \left\{\begin{matrix}\text{Saturated compounds}\\ \text{Amino compounds}\\ \text{Alcohols}\end{matrix}\right.$$

Although there are methods described in this book for determining trace quantities of the various functional groups, practically all the methods in this text can be very simply reduced to a micro scale if such becomes necessary when small amounts of sample are involved. In his textbook *Introduction to the Microtechnique of Inorganic Analysis* (Wiley, New York, 1942), A. A. Benedetti-Pichler describes techniques for micro volumetric and gravimetric analysis. He describes these for inorganic systems, but the techniques are just as applicable to organic systems. Dr. Benedetti-Pichler's approach is to keep the analytical method intact, including concentrations, time of reaction, and so on, but to reduce the scale of the apparatus to permit handling of the small quantities. He uses micro reaction vessels, microburets, and micro gravimetric devices for handling micro samples.

1

Hydroxyl Groups

The behavior of the hydroxyl group varies depending on the molecule to which it is attached. On an aliphatic chain (ROH), the hydroxyl group contributes to the molecule the chemical characteristics attributed to alcohols. A compound with a hydroxyl group on a carbon containing a double bond $RCH{=}\underset{\displaystyle OH}{\underset{|}{C}}R$ is classed as an enol. If there are hydroxyl groups on adjacent carbon atoms, $R\underset{\displaystyle OH}{\underset{|}{C}}H{-}\underset{\displaystyle OH}{\underset{|}{C}}HR$, these are classified as glycols. Hydroxyl groups on aromatic rings yield the properties characteristic of phenols. Thus we can have hydroxyl groups in different situations, each exhibiting a different set of characteristics.

The most general method for determining hydroxyl groups is esterification. This approach will operate for all the types of hydroxyl groups mentioned previously. The only exceptions are tertiary alcohols (R_3COH), which esterify with difficulty, as do trisubstituted phenols and other sterically hindered hydroxyl groups. The infrared methods for following hydroxy groups are recommended in these latter cases. Also, active hydrogen methods can be used (p. 478), but these are not characteristic of hydroxyl groups alone.

Hydroxyl groups on adjacent carbon atoms (glycols) have a characteristic reaction, namely the oxidation with periodic acid to the corresponding aldehydes, which can be used for measurement (see p. 42).

Enolic and phenolic hydroxyl groups exhibit sufficient acidity that they can be titrated directly as acids in certain nonaqueous media (see p. 46). Other hydroxyl groups, although acidic, generally are not sufficiently acidic to be titrated directly. It should be kept in mind, however, that certain other substituents on the molecule such as nitro groups can enhance the acidity of the hydroxyl group such that even some aliphatic alcohols can be titrated as acids (i.e., dinitropropanol).

Phenols can be coupled with diazonium compounds (see p. 62), and also they often can substitute some of the hydrogen on the ring with bromine (see p. 57). These reactions can also be used to determine this variety of hydroxyl compound.

Esterification Procedures

A convenient method for the determination of hydroxyl groups is esterification. The reaction is general for hydroxyl in most situations; the reaction is rather fast and relatively specific; also the reagent is easily, accurately, and precisely measured.

The use of an organic acid for esterification,

$$RCOOH + R_1OH \rightleftharpoons RC(=O)OR_1 + H_2O$$

is not desirable, since the reaction is an equilibrium system. A quantitative reaction is thus not possible unless the water is either removed or circumvented. To achieve the desired results the acid chlorides or acid anhydrides are used for the esterifications. In this way the formation of water is avoided, yet all the desirable aspects of esterification are maintained.

Carboxylic acid anhydrides are the most commonly used esterification reagents. The carboxylic acid halides are used (1, 2) but are more reactive than the anhydrides and hence more difficult to handle. At present there are three anhydrides used for hydroxyl group determination; acetic anhydride, phthalic anhydride, and pyromellitic dianhydride.

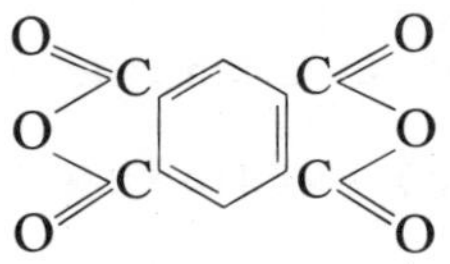

Acetic anhydride and phthalic anhydride were the most popularly used anhydrides until the introduction of pyromellitic dianhydride. The latter is very new, but it appears to embody the advantages of both the acetic and the phthalic systems.

The acetic anhydride system is the most widely used of the older methods. The acetic system reacts faster than the phthalic system and also could be applied to a wider range of alcohols, since steric effects are not as significant with the introduction of the acetyl group onto a molecule as

1. D. M. Smith and W. M. D. Bryant, *J. Am. Chem. Soc.*, **57,** 61 (1935).
2. W. T. Robinson, R. H. Cundiff, and P. C. Markunas, *Anal. Chem.*, **33,** 1030–4 (1961) (see pp. 29–37).

they are with the introduction of the phthaloyl group. For example, acetic anhydride reacts readily with 1,2-glycols, whereas phthalic anhydride reacts quantitatively with difficulty, if at all. Even when the hydroxyl groups are more widely separated, as in 2-butyne-1,4-diol ($HOCH_2$-$C{\equiv}CCH_2OH$), phthalic anhydride reacts so slowly that a 10-mole excess of anhydride over hydroxyl is needed for complete reaction in a given time interval. In the same time interval, the reaction between acetic anhydride and the butynediol, only 3 moles of anhydride per hydroxyl is required. The acetic anhydride also reacts quantitatively with phenols even when disubstituted (does not react quantitatively with trisubstituted phenols, however), but phthalic anhydride does not react with phenols at all.

A disadvantage of the acetic anhydride system is that aldehydes interfere in the determination of hydroxyl groups. No stoichiometric relationship has been found between aldehydes and anhydrides, although it is quite evident that the reaction proceeds more rapidly in the case of the lower molecular weight aldehydes. Formaldehyde presents the most serious interference, acetaldehyde next, and so on. The interfering reaction has not been elucidated, but some possibilities have been suggested.

The following reaction (3) is consistent with the fact that ketones do not undergo the reaction and are known not to interfere in the acetic anhydride determination of hydroxyls.

$$RCHO + (CH_3CO)_2O \rightarrow RCH(O\overset{\overset{\displaystyle O}{\|}}{C}CH_3)_2$$

Another suggestion is that a Perkin reaction is involved (4).

$$RCHO + (CH_3CO)_2O \rightarrow RCH{=}CHCOOH + CH_3COOH$$

However this reaction proceeds in known cases only with aromatic aldehydes, slowly and at high temperatures (>150°C).

Aliphatic aldehydes usually self-condense more rapidly than they undergo this type of reaction with anhydride.

$$2RCH_2CHO \rightarrow RCH_2\underset{\underset{\displaystyle OH}{|}}{C}H{-}\underset{\underset{\displaystyle R}{|}}{C}HCHO$$

3. A. F. Holleman, *Organic Chemistry*, Elsevier, New York, 1951, p. 105.
4. K. A. Connors, *Reaction Mechanisms in Organic Analytical Chemistry*, John Wiley & Sons, New York, 1973, p. 503.

The anhydride could then act on the resultant hydroxyl group of the condensation product. However the reaction should occur also in connection with the aromatic anhydrides, but it is not observed. Also some aldehydes, formaldehyde, for example, do not condense as shown, yet they interfere with the acetic anhydride reagent. Therefore this condensation reaction is also not a likely explanation of the interference.

Since aldehydes do not interfere in the reactions involving phthalic anhydride or pyromellitic dianhydride, these anhydrides have the advantage of specificity. Alcohol samples may contain some aldehyde as a result of oxidation of the hydroxyl group. The authors have noted that hydroxyl group values obtained on the same samples with the phthalic and pyromellitic anhydrides are often lower than those obtained with acetic anhydride. Aldehyde in the sample is the most probable explanation of this behavior. It is well to note that aldehydes combined in the form of acetals do not interfere with any of the foregoing reagents. In fact, small amounts of alcohols in the presence of gross quantities of acetals can be easily determined even with acetic anhydride.

Pyromellitic dianhydride (PMDA) can be considered as a fast-reacting phthalic type. As has been mentioned, aldehydes do not interfere. Also, phenols do not react at all with PMDA. In fact, alcohols are determinable in the presence of phenols using this reagent. In terms of speed of reaction, the PMDA is much faster reacting than phthalic and faster than acetic anhydride when reacted under comparable conditions.

The PMDA and phthalic anhydride both have the advantage of being solids, hence are nonvolatile. For this reason they make ideal reagent solutions. Acetic anhydride is volatile; hence precautions must be taken to avoid volatilization during the esterification of the sample. The phthalic anhydride reacts so slowly, however, that even though it is relatively nonvolatile, the sample often is volatile; thus precautions must be taken to avoid loss of sample. The PMDA, on the other hand, reacts so rapidly that the alcohol is consumed before it can be volatilized. Even methanol and ethanol, which are the lowest boiling alcohols, can be determined with PMDA using open Erlenmeyer flasks.

Pyridine is used in all the anhydride methods. It serves not only as solvent but also to accelerate the reaction by tying up the carboxylic acids formed during the reaction. In addition the pyridine is a weak enough base that the carboxylic acids can be titrated away from the pyridine by using a strong base such as sodium hydroxide.

The anhydride reactions can be accelerated through the use of perchloric acid (see p. 14). This has been thoroughly studied in the case of acetic anhydride. Perchloric acid, however, occasionally causes difficulty by oxidatively altering the hydroxyl group or otherwise oxidatively altering the sample. The polyglycol ethers

$$\left(H(OCH_2\overset{\overset{\displaystyle R}{|}}{C}H)_xOH \right)$$

are the most outstanding examples of cases where perchloric acid catalysis cannot be applied. These materials are easily oxidized, and high, erratic results are obtained by using perchloric acid catalyzed esterification methods.

Acetylation

UNCATALYZED ACETIC ANHYDRIDE METHODS

Determination of Hydroxyl Content of Organic Compounds Adapted from the Acetic Anhydride Method of C. L. Ogg, W. L. Porter, and C. O. Willits

[*Ind. Eng. Chem., Anal. Ed.*, **17,** *394–7* (*1945*)]

$$(CH_3CO)_2O + ROH \rightarrow CH_3COOR + CH_3COOH$$

REAGENTS

Acetylating reagent consisting of 1 volume of ACS grade acetic anhydride and 3 volumes of reagent pyridine. Reagent should be prepared fresh each day.

n-Butanol, technical grade.

Mixed indicator solution. One part of 0.1% aqueous cresol red neutralized with sodium hydroxide and 3 parts of 0.1% thymol blue neutralized with sodium hydroxide.

Standard alcoholic sodium hydroxide, approximately $0.5N$. Sodium hydroxide, $0.1N$, can be used for semimicro samples, but the clarity of the end point is poorer than when the $0.5N$ reagent is used. Alcoholic sodium hydroxide is best prepared by mixing the required amount of saturated aqueous sodium hydroxide (approximately 18*N*) with aldehyde-free ethanol or with CP methanol. The alcoholic alkali is standardized against potassium acid phthalate or against standard acid by use of the mixed indicator.

PROCEDURE

Introduce a weighed sample containing about 0.010 to 0.016 mole of hydroxyl into a glass-stoppered iodine flask together with 10.00 ml of the acetic anhydride–pyridine reagent. Measure the acetylating solution accurately, using a pipet. Moisten the glass stopper well with pyridine and seat loosely in the flask. Place the flask on a steam bath for 45 minutes. Then add 10 ml of water by way of the well on the top of the flask, and swirl the flask to bring the water in contact with all the reagent. After 2 minutes, cool the flask in ice or under running water, with the stopper partly open to prevent a partial vacuum from forming inside the flask. Rinse the sides of the flask and the stopper with 10 ml of *n*-butanol, add a few drops of indicator, and titrate the contents with 0.5*N* sodium hydroxide. If the sample contains 0.001 mole of hydroxyl, it is advisable to titrate with 0.1*N* sodium hydroxide even though the end point may not be as sharp as with the 0.5*N* reagent.

Samples that yield highly colored solutions, making the indicator useless, can be titrated potentiometrically by use of a potentiometer or pH meter with glass and calomel electrodes.

Any free acid or alkali in the sample should be determined on a separate sample by dissolving the sample in 5 ml of pyridine and titrating with standard alkali or acid by using the mixed indicator.

CALCULATIONS

Milliliters of NaOH used for blank minus milliliters of NaOH used for sample (corrections being applied for any free acid or alkali that may be present) = A

$$\frac{A \times N\,\text{NaOH} \times \text{OH} \times 100}{\text{Grams of sample} \times 100_0} = \%\,\text{OH}$$

If the sample contains considerable amounts of water (more than 0.002 mole), additional acetic anhydride should be added, so that enough will be present to acetylate the alcohol readily. Water does not enter into the calculations, but it does destroy the reagent by hydrolyzing it to acetic acid. If too much reagent is destroyed, there will not be enough present to esterify the alcohol completely.

Primary and secondary amines will interfere in this analysis. In fact, they acetylate so readily that this procedure can be used to determine them quantitatively (see Amino Groups).

Aldehydes of low molecular weight also interfere by reacting with the anhydride. The reaction between the anhydride and the aldehyde is not altogether clear (see p. 10).

It is well (keeping the equation on p. 10 in mind) to remember that when all the anhydride is consumed there will still be a titration equal to one-half the blank. If, when unknown samples are analyzed, the titration should be in the vicinity of one-half the blank, it is best to repeat the analysis with a smaller sample to ensure having enough reagent for all the hydroxyl present.

Ogg, Porter, and Willits have successfully determined the following hydroxyl compounds: dihydroxystearic acid, monohydroxystearic acid, oleyl alcohol, cyclohexanol, and benzyl alcohol.

The authors have found that the foregoing procedure works successfully for alcohols from methanol to octadecanol, ethylene glycol, glycerol, glycerol monoacetate, phenol, octylphenol, decyl- and dodecylphenol, ditertiary butylphenol, 2-butyne-1,4-diol, propargyl alcohol, and 3-methoxybutanol. With proper choice of sample size (0.010–0.016 mole of hydroxyl), an average accuracy and precision within 1.0% can be achieved.

As can be seen from the materials used to test the procedure, the procedure will not differentiate between hydroxyl groups on primary and secondary carbon atoms. Hydroxyl groups on tertiary carbon atoms and hydroxyls of 2,4,6-trisubstituted phenols do react with acetic anhydride but only very slightly. This type of hydroxyl group cannot be determined by this method (there may be a few exceptions, but they are very few). Hydroxyl groups in less highly substituted phenols react readily with acetic anhydride, although these hydroxyl groups might be considered to be attached to tertiary carbon atoms.

PERCHLORIC ACID CATALYZED METHOD

Method of J. S. Fritz and G. H. Schenk

[*Reprinted in Part from Anal. Chem.* ***31,*** *1808* (*1959*)]

With the present method, primary and secondary alcohols are determined by acetylation in ethyl acetate or pyridine solution using perchloric acid to catalyze the reaction. Soluble alcohols in ethyl acetate are completely acetylated within 5 minutes at room temperature. In pyridine, a somewhat longer reaction period is required for secondary or hindered alcohols. The amount of alcohol present is calculated from the difference between the blank and sample titrations with sodium hydroxide.

REAGENTS AND SOLUTIONS

2*M* ACETIC ANHYDRIDE IN ETHYL ACETATE. Add 4 grams (2.35 ml) of 72% perchloric acid to 150 ml of ACS grade ethyl acetate in a clean 250-ml glass-stoppered flask. Pipet 8 ml of ACS grade acetic anhydride into the flask and allow it to stand at room temperature for at least 30 minutes. Cool the contents of the flask to 5°C and add 42 ml of cold acetic anhydride. Keep the flask at 5°C for an hour, then allow the reagent to come to room temperature. Some yellow color will develop, but the color and anhydride content of the reagent remain at satisfactory levels for at least 2 weeks at room temperature.

2*M* ACETIC ANHYDRIDE IN PYRIDINE. Cautiously add 0.8 gram (0.47 ml) of 72% perchloric acid dropwise to 30 ml of reagent grade pyridine in a 50-ml flask. Pipet 10 ml of acetic anhydride into the flask with magnetic stirring. Because this reagent discolors and decreases in anhydride content after a few hours, it should be prepared fresh daily. For acetylation of sugars at 50°C, use 1.2 grams of *p*-toluenesulfonic acid instead of the perchloric acid.

3*M* ACETIC ANHYDRIDE IN PYRIDINE. Follow the directions above, but use 40 ml of pyridine, 20 ml of acetic anhydride, and 0.94 ml of 72% perchloric acid.

0.55*M* SODIUM HYDROXIDE. To 185 ml of saturated aqueous sodium hydroxide (carbonate free), add 430 ml of water and 5400 ml of Methyl Cellosolve (Union Carbide Chemicals Co.) or absolute methanol. Use only unopened cans of Methyl Cellosolve, because a solvent that has been exposed to the air for some time develops a yellow color in the sodium hydroxide titrant.

MIXED INDICATOR. Mix 1 part 0.1% neutralized aqueous cresol red with 3 parts 0.1% neutralized thymol blue.

POTASSIUM ACID PHTHALATE. Primary standard grade.

ALCOHOL SAMPLES. Most liquid samples were fractionally distilled through a 24-in Podbielniak partial reflux fractional distillation column. Many of the solids were vacuum sublimed. The estimated purity of the purified samples is in the range 98 to 100%.

PROCEDURE

Weigh accurately a sample containing from 3 to 4 mM of hydroxyl into a 125-ml glass-stoppered flask and pipet into it exactly 5 ml of 2*M* acetic anhydride in ethyl acetate or pyridine. Stir the solids or immiscible liquids until they are dissolved. Allow the reaction to proceed for at least 5 minutes at room temperature; some alcohols require a somewhat longer reaction period if pyridine is used as the solvent. Add 1 to 2 ml of water, shake the mixture, then add 10 ml of 3:1 pyridine-water solution and allow the flask to stand for 5 minutes. Titrate with 0.55*M* sodium hydroxide using the mixed indicator, and take the change from yellow to violet as the end point. Titrate dark-colored samples to an apparent pH of 9.8 using glass-calomel electrodes and a pH meter.

Run a reagent blank by pipetting exactly 5 ml of acetylating reagent into a 125-ml flask containing 1 to 2 ml of water. Add 10 ml of 3:1 pyridine-water solution, allow to stand 5 minutes, and titrate as above. Use the difference between the blank V_b and the sample titration V_s to calculate the percentage of hydroxyl compound in the sample. CAUTION! Dilute solutions of perchloric acid in various organic solvents have been widely used in nonaqueous titrations. There is no hazard under the conditions given in the procedure above. Solution acetylated with perchloric acid present, however, should not be heated, and the sample and blank solutions should be disposed of promptly after the determination is completed.

Determine sugars that dissolve slowly in the reagents above by heating them 5 to 10 minutes at 50°C with 5 ml of a pyridine reagent that is 0.15*M* in *p*-toluenesulfonic acid instead of perchloric acid. Moisten the glass stopper with pyridine and seat loosely in the flask. After heating, cool the flask and hydrolyze the anhydride with the 3:1 pyridine-water mixture at room temperature. Treat the blank similarly. Dry sugar samples only if analyzed at room temperature.

ALTERNATE PROCEDURE FOR WATER-FREE SAMPLES. Use acetic anhydride in ethyl acetate for the acetylation. After the acetylation period, add 10 ml of a 1.5*M* solution of distilled *N*-methylaniline solution in chlorobenzene to the flask instead of the water and water-pyridine solution. After 15 minutes, titrate the excess *N*-methylaniline potentiometrically with 0.2*M* perchloric acid in glacial acetic acid. For this titration use a glass indicator electrode and a sleeve-type calomel reference electrode containing lithium chloride in glacial acetic acid as the electrolyte solution. Determine the blank by reacting exactly 5 ml of the acetic anhydride reagent with 10 ml of *N*-methylaniline solution as above and titrate potentiometrically with 0.2*M* perchloric acid in glacial acetic acid. This procedure is necessary for ethylsulfonylethyl alcohol.

CONCENTRATION OF REAGENTS

The anhydride reagent used contained 3 volumes of pyridine or other solvent to 1 volume of acetic anhydride. This is a sufficiently high concentration of anhydride to ensure rapid acetylation, yet dilute enough that the reagent can be accurately measured with a 5-ml pipet, and the blank titration with $0.55M$ sodium hydroxide will be within the limits of a 50-ml buret.

Preliminary work established the value of perchloric acid in catalyzing the reaction; however it was necessary to establish a proper concentration of perchloric acid. Table 1 shows the effect of varying concentrations of perchloric acid on the acetylation of 2-ethylhexanol in 1:1 pyridine–acetic anhydride. The time of acetylation was 10 minutes at room temperature.

Table 1. Effect of Perchloric Acid on the Acetylation of 2-Ethylhexanol

$HClO_4$ Molarity	% Reaction
0.0	81
0.025	84
0.05	88
0.10	98.5
0.15	99.8

In 3:1 ratios of solvent to acetic anhydride, $0.15M$ perchloric acid is also satisfactory. A $0.30M$ perchloric acid solution is unsatisfactory, since the results are somewhat erratic and the water in the perchloric acid reduces the anhydride concentration of the reagent.

SOLVENTS

After testing several organic solvents, ethyl acetate was selected as the solvent of choice. Acetylations in this solvent proceed rapidly and quantitatively. Acetic anhydride has reasonably good stability in ethyl acetate. The titration of acetic acid after hydrolysis gives essentially constant values for a period of 2 weeks, and the anhydride content as determined by reaction with aniline decreased only about 5% in 2 weeks. After 2 weeks the yellow color of the reagent darkens to orange, and the end point of the titration is no longer as sharp as it should be. The method of preparing the reagent described previously minimizes color formation.

Solution of acetic anhydride in pyridine are less stable than in ethyl acetate and must be prepared fresh daily. Also, acetylation of some compounds requires a longer reaction period in pyridine. Nevertheless, pyridine is an important solvent and supplements the use of ethyl acetate for acetylation of alcohols.

Ethyl benzoate, diethyl malonate, and acetonitrile are suitable solvents for acetylation reactions, but acetic anhydride dissolved in any of these solvents develops an unsatisfactory color rather quickly. Freshly distilled dimethoxyethane is satisfactory, but the peroxides in undistilled dimethoxyethane cause a dark brown color to develop when acetic anhydride is added. Chloroform and triethyl phosphate show very good solvent characteristics and may be regarded as possible alternatives to ethyl acetate. Acetylation of 2-*tert*-butyl-cyclohexanol in the latter three solvents is quantitative in 5 minutes.

To show the effect of acid catalysis in different solvents, several alcohols were acetylated for 5 minutes at room temperature (Table 2).

Table 2. Acetylation of 4 mmole of Alcohols with 3:1 Solvent–Acetic Anhydride

	Reaction, No Acid		% Reaction, 0.15M $HClO_4$	
Alcohol	Ethyl acetate	Pyridine	Pyridine	Ethyl acetate
Methanol	66	87	100	100
Ethyl alcohol	25	45	100	100
2,2-Dimethyl-1-propanol	17	38	95	100
2-Propanol	5	10	80	100
Diisobutyl carbinol	2	7	64	100
Cyclohexanol	0	0	75	100
2-Methylcyclohexanol	0	0	60	100
2-*tert*-Butylcyclohexanol	0	0	7	100
2-Methyl-2-propanol	0	0	0	70

These data confirm the fact that acetylation is catalyzed by the presence of a basic solvent such as pyridine. However, the rate of acetylation is much faster when catalyzed by perchloric acid. Acid catalysis occurs to a significant extent even in pyridine, where the basic solvent is present in large excess over the perchloric acid added.

Table 3 summarizes results on diverse hydroxy compounds.

CALCULATIONS

Same as acetic anhydride method shown above.

In addition to the alcohols used by Fritz and Schenk shown in Table 3

Table 3. Determination of Pure Alcohols

(Average of 3 or 4 determinations, 5- to 7-minute reactions, 4-mM. samples)

	% Purity Found
Primary	
Ethyl Acetate Reagent	
Ethyl alcohol	99.1 ± 0.5
2-Ethylhexanol	99.4 ± 0.2
Ethylsulfonylethyl alcohol[a]	98.9 ± 0.6
2-Methyl-4-butanol	98.5 ± 0.4
Methanol	99.9 ± 0.3
2,2-Dimethyl-1-propanol	98.0 ± 0.3
2-Propyn-1-ol	100.1 ± 0.3
2,2,2-Trifluoroethyl alcohol	98.9 ± 0.6
Secondary	
Benzhydrol	100.3 ± 0.1
Benzoin	99.3 ± 0.3
2-*tert*-Butylcyclohexanol	99.5 ± 0.6
Cyclohexanol	100.1 ± 0.3
2-Cyclohexylcyclohexanol	98.6 ± 0.3
Diisobutyl carbinol	100.5 ± 0.1
2-Propanol	100.3 ± 0.3
2-Methylcyclohexanol	102.9 ± 0.3
2-Phenylcyclohexanol	102.2 ± 0.4
Glycols	
Glycerol	97.1 ± 0.5
2,2,4-Trimethyl-pentane-1,3-diol	100.5 ± 0.1
Sugars	
Cellobiose[b]	100.3 ± 0.8
Glucose[b]	99.4 ± 0.2
Lactose	100.1 ± 0.5
Maltose	100.0 ± 0.1
Mannose	100.8 ± 0.3
Pyridine Reagent	
Alcohols	
3-Phenylpropenol	98.1 ± 0.2
α-Furancarbinol	99.2 ± 0.2
2-Propyn-1-ol[c]	99.8 ± 0.5
α-Tetrahydrofurancarbinol	97.5 ± 0.5
α-Benzoin oxime[c,d]	99.8 ± 0.1

[a] *N*-Methylaniline and perchloric acid used to determine anhydride reacted.

[b] Cellobiose, 35 minutes; glucose, 45 minutes.

[c] 10-minute reaction time.

[d] 3*M* acetic anhydride in pyridine; oxime and amino groups acetylated quantitatively.

Table 3 (*Continued*)

(Average of 3 or 4 determinations, 5- to 7-minute reactions, 4-mM. samples)

Primary	% Purity Found
Pyridine Reagent	
Glycols	
cis-Butene-1,4-diol	98.8 ± 0.4
Tris(hydroxymethyl) aminomethane[d]	98.5 ± 0.1
tert-Hydroperoxides	
tert-Butyl hydroperoxide[c]	88.7 ± 0.3
2.5-Dimethyl-2,5-d hydroperoxy hexane[c]	94.9 ± 0.4
Sugars	
Fructose(4 OH)[e]	100.5 ± 0.2
Glucose[e]	100.7 ± 0.3
Glucose[f]	99.5 ± 0.3
Sucrose[f]	100.4 ± 0.2

[e] Fructose, 20 minutes; glucose, 40 minutes.

[f] Heated 5 to 10 minutes with 0.15*M* *p*-toluenesulfonic acid instead of perchloric acid.

to test this procedure, the authors of this book have successfully used ethylene and propylene glycols, 1,4-butanediol, stearyl, lauryl, butyl, octyl, isooacyl, and allyl alcohols. The method was found not to apply to polyethylene and polypropylene glycol ethers of the type $H(\underset{\substack{| \\ R}}{O}CHCH_2)_xOH$, where R is a CH_3 group or a hydrogen atom as the case may be; also $R_1(OCH_2CH_2)_xOH$, where R_1 is a fatty chain or an alkyl phenol. In these cases high, erratic results were obtained; possibly because of oxidation of the chain by the perchloric acid, yielding hydroxyl or aldehydic groups. These ethers oxidize readily via a peroxide mechanism. Even the uncatalyzed acetic anhydride analysis yields high values with these glycol ethers, probably because of aldehydic interference. The uncatalyzed system gives only slightly high values, however, and these are reproducible. For the polyglycol ethers, the pyromellitic dianhydride method is recommended. The phthalic anhydride method also is operable but requires 2 hours for complete reaction; the PMDA method requires 30 minutes or less.

1,2-DICHLOROETHANE AS SOLVENT

Adapted from the Method of J. A. Magnuson and R. J. Cerri

[*Anal. Chem.*, ***38,*** *1088* (*1966*)]

1,2-Dichloroethane is superior to ethyl acetate as a solvent in many ways. The acetylating agent can be prepared in 1,2-dichloroethane without cooling. Although the newly prepared reagent does become somewhat warm, it may be used within an hour. Cooling to 5°C is necessary with ethyl acetate during one step of the acetic acid–perchloric acid mixing.

This acetylating mixture is virtually colorless, or a very light yellow at most. The reagent in ethyl acetate is yellow to yellow-brown. Indicator end points are sharper when the yellow is absent.

The acetylating agent has a useful life, at least 2 months, which is 2 to 3 times longer than the reagent in ethyl acetate.

Presently, 1,2-dichloroethane is the only solvent known besides ethyl acetate in which alkoxysilanes quantitatively acetylate within 10 minutes.

REAGENTS

Acetic anhydride, 1*M*, in 1,2-dichloroethane (0.15*N* perchloric acid). Pour 420 ml of 1,2-dichloroethane into a 500-ml glass-stoppered flask and add 6.2 ml of 72% perchloric acid and slowly, with stirring, add 55 ml of acetic anhydride.

PROCEDURE

Pipet 10 ml of the acetylating reagent into a 125-ml glass-stoppered flask and add, accurately weighed, 4 to 5 meq of the acetylyzable sample. After a 5-minute reaction period, add 35 to 40 ml of 6:3:1, dimethylformamide-pyridine-water hydrolyzing solution. Allow the hydrolysis to proceed for 10 to 15 minutes, add 5 drops of 1% thymol blue indicator and titrate with alcoholic 0.55*N* potassium hydroxide solution to the blue end point. Run the appropriate blank.

Results obtained with this procedure appear in Table 4.

Dimethylformamide aids in solubilizing both the 1,2-dichloroethane and water. Its presence also sharpens the indicator end point.

Table 4. Results Obtained with Dichloroethane as the Acetylation Solvent

	Percentage of Theory[a]
Alcohols and phenols	
n-Butanol	99.6 ±0.7
2-Methoxyethanol	100.0 ±0.1
2-Propanol	99.6 ±0.4
Cyclohexanol	100.1 ±0.6[b]
2,4-Dimethyl-3-hexanol	100.4 ±0.5
Phenol	100.4 ±0.5
2-*tert*-Butylphenol	99.9 ±0.9
2,6-Diphenylphenol	99.7 ±0.4
Alkoxysilanes	
Diphenyldiethoxysilane	100.1 ±0.8
Diethyldiisopropoxysilane	99.6 ±0.4
7-Octenyltriethoxysilane	100.2 ±0.4
2-Methoxy-2-methyl-1-thio-2-silacyclopentane[c]	100.3 ±0.6
1,3-Di-*n*-propyl-1,1,3,3-tetraethoxydisiloxane	99.2 ±0.2
γ-Chloropropylmethyldiethoxysilane	100.2 ±0.5
Trivinyl-2-methoxyethoxysilane	99.4 ±0.4

[a] Average and average deviation of triplicate determinations.
[b] Acetylating reagent used within one hour of preparation.
[c] Bifunctional alkoxy- and mercaptosilane.

Phthalation

Method of P. J. Elving and B. Warshowsky

[*Reprinted in Part from Anal. Chem.* ***19,*** *1006* (*1947*)]

REAGENTS AND APPARATUS

The reagents used in the studies described were Mallinckrodt's analytical reagent grade phthalic anhydride, Barrett's refined grade 2A pyridine, 0.35*N* standard sodium hydroxide solution, and a 1% alcoholic solution of phenolphthalein indicator. The hydroxyl-containing compounds analyzed were distilled in most cases and a narrow-boiling fraction was used; the probable purity is indicated in Table 5.

Barrett's grade 2A pyridine as obtained commercially contains a significant amount of water and other substances that interfere with the accuracy of the determination. To eliminate these substances, the pyridine should be distilled

Table 5. Analysis of Hydroxyl Compounds by the Phthalization Method

Compound	Purity as Determined by Physical Constants (wt. %)	Purity by Phthalization Method (wt. %)		Recovery by Phthalization Method (%)
Methanol	99.3	100.0	100.2	100.8
Ethanol	100.0	100.7	100.7	100.5
	—	100.3	100.4	—
1-Propanol	98.4	98.1	97.9	99.5
2-Propanol	99.6	97.6	97.6	98.0
1-Butanol	100.0	100.6	100.6	100.6
2-Methyl-1-propanol (isobutyl alcohol)	—	101.8	101.2	—
Cyclohexanol	95–96	95.1	95.1	99–100
2-Ethylhexan-1-ol	99	99.1	99.1	100
2-Octanol	97–98	96.0	96.3	98–99
Ethylene glycol	99	98.7	98.7	99.5
Propylene glycol	100	99.4	99.6	99.5
Glycerol	95.5	87.9	87.9	92.0
	—	94.9[a]	94.3[a]	99.1
Benzyl alcohol	100	99.4	99.7	99.6

[a] Two hours at 100°C allowed for reaction.

from over barium oxide and only the portion distilling at 115°C used (5). The phthalization mixture is prepared by dissolving 20 grams of phthalic anhydride in 200 ml of purified pyridine; this solution is prepared fresh daily.

The apparatus required depends partly on the procedure selected. It was found satisfactory to carry out the reaction either under reflux or by the use of pressure bottles. Inasmuch as the pressure bottle technique is somewhat simpler and more rapid, this is described in detail below. For this technique, citrate of magnesia bottles of 1-pint capacity are used. An air bath set at 100 ±2°C is required; an ordinary laboratory drying oven will suffice.

It is preferable that the pyridine used be anhydrous and the reaction equipment, whether pressure bottles or flasks and condensers, be thoroughly dried. Unless these precautions are observed, low values may be obtained; water hydrolyzes the phthalic anhydride and may thereby reduce the concentration below the excess required.

PROCEDURE

For samples containing a high percentage of ethanol, carefully pipet a sample weighing 1.0 to 1.5 grams into a 50-ml volumetric flask containing

5. Note by S. Siggia—Reagent grade pyridine has been found to be usable as purchased.

30 to 40 ml of the purified anhydrous pyridine, which has been weighed, taking care to avoid wetting the neck of the flask; for monohydric alcohols of higher molecular weight and dilute solutions of ethanol, take samples of corresponding larger weight. After reweighing the solution, make it up to volume with pyridine and shake it to mix it thoroughly.

In certain cases where the sample is extremely volatile—that is, contains considerable material boiling below 40°C—weigh the sample directly in thin-walled ampoules. Then transfer the ampoule to the volumetric flask containing pyridine and crush the ampoule under the surface of the pyridine with a glass rod. Rinse the rod with pyridine on withdrawal, make the solution up to volume with pyridine, and mix the contents of the flask.

Pipet 25 ml of the phthalic anhydride solution into a clean, dry pressure bottle by means of an automatic Machlett pipet or buret or a Lowy pipet. Add to this 10 ml of the solution containing the sample. Place the sealed bottle containing the mixture in an air oven set at 100°C and heat it at that temperature for 1 hour.* At the end of this time, carefully release the pressure and add 50 ml of distilled water. After mixing the solution, cool it under the cold water tap and titrate it immediately with standard 0.35N sodium hydroxide, using phenolphthalein as an indicator.

Make a blank determination in the same manner on the reagents used.

CALCULATION

$$\% \text{ Hydroxyl} = \frac{V \times N \times 1.70}{W}$$

where W is the weight in grams of the sample in the aliquot taken; V is the volume in milliliters of standard sodium hydroxide solution used by sample, which equals the difference in volumes required by the blank and sample titrations; and N is the normality of the standard sodium hydroxide solution.

DISCUSSION OF PROCEDURE

SIZE OF SAMPLE. The weight of sample taken should be such as to have present a minimum of 100% molar excess of phthalic anhydride over the

* Use a covering on the bottle to protect the operator, should unexpected pressure develop in the bottle. This rarely happens, but the authors have seen some bottles rupture for unexplained reasons.

amount required. The presence of a sufficient excess of anhydride can be noted by the formation of a yellow color in the solution after heating for the prescribed length of time. Unless this color appears, the results may not be reliable. Failure of the color to appear indicates an insufficient excess of the reagent due to too large a sample or the presence of too much water.

ACIDIC SAMPLES. Separate portions of samples containing free acid or acidic groups in the hydroxyl-containing compounds should be titrated with the standard alkaline solution at room temperature, using phenolphthalein as indicator, and a suitable correction should be made in the volume of alkaline solution consumed in the phthalization procedure.

CONTACT TIME. Although 1 hour is specified for the phthalization reaction, in many cases this time can be reduced to 30 minutes or less. In dealing with mixtures containing substances that react with phthalic anhydride on prolonged heating, it would be worthwhile to determine the minimum reaction period necessary for the hydroxyl-containing compounds involved. The reaction mixture, after the addition of the water, should be cooled and titrated in a minimum amount of time to avoid the possibility of the phthalate esters formed hydrolyzing to any appreciable extent; ordinarily, the danger of such reaction is almost nil. There was no measurable consumption of phthalic anhydride due to polymerization, decomposition, or other reactions in blank samples heated up to 4 hours.

END POINT. Although the pink phenolphthalein end point is normally easily detected, it is masked to some extent in this determination by the yellow color of the final solution. Therefore, instead of a color change from colorless to pink, there is a gradual transition from yellow to brown to orange to pink. It is felt that a representation of the true end point is the first noticeable permanent color change of the solution—without having to continue the titration until the color is definitely pink. After several titrations this point can be detected without any difficulty. The use of the mixed indicator, thymol blue–cresol red, offered no appreciable advantage over the use of phenolphthalein. In the case of dark-colored solutions, the end point can be determined electrometrically.

Table 6 gives the results of analyzing one monohydroxyl and four polyhydroxyl compounds, which gave poor results. The purity of the samples of 2-methyl-2-butanol and of the 2-methyl-2,4-pentanediol exceeded 95%; the other three compounds were used without purification and were believed to be not less than 85% pure. Results for any given compound in Table 6 on the same horizontal line were obtained at the same time—that is, on the same batch of samples. The results for 2,3-butanediol are probably correct in view of the consistent values obtained for samples heated for 1 hour and for 4 hours, whereas the

Table 6. Analysis of Hydroxyl Compounds by the Phthalization Method

Compound	Apparent Purity, Weight Per-Cent for Different Reaction Times at 100°C		
	1 hour	2 hours	4 hours
1,3-Butanediol	84, 84	96, 89	
	79, 78	92, 91	93, 92
	75, 76		
2,3-Butanediol	89, 89	91, 90	89, 90
	89, 89		
2-Methyl-1,2-propanediol (isobutylene glycol)	54, 54	59, 59	
	53, 53	59, 60	64, 64
	54, 53		
2-Methyl-2-butanol (*tert*-amyl alcohol)	3, 3	5, 5	10, 11
2-Methyl-2,4-pentanediol (preparation I)	45, 46		
2-Methyl-2,4-pentanediol (preparation II)	43, 42	52, 51	56, 58
2-Methyl-2,4-pentanediol (preparation III)	51, 50	50, 57	60, 60

1,3-butanediol apparently required 2 hours of reaction for complete esterification. The results for the 2-methyl-1,2-propanediol are probably meaningless, since in the presence of even dilute acids, the compound dehydrates and rearranges readily to give isobutyraldehyde. The unsatisfactory results for 2-methyl-2-butanol and 2-methyl-2,4-pentanediol result from the ease with which tertiary alcohols are dehydrated in the presence of acidic catalysts. Investigation of the literature showed that 2-methyl-2,4-pentanediol can be dehydrated in the liquid phase in the presence of acidic catalysts to 2-methyl pentenols and 2-methyl pentadienes. The results obtained using stoichiometric molar ratios of 1:2, 1:1, and 2:1 of phthalic anhydride and 2-methyl-2,4-pentanediol in pyridine solution plus the data in Table 6, indicate that phthalic anhydride apparently causes the conversion of the methyl pentanediol to the methyl pentenol. The results obtained for 2-methyl-1,2-propanediol, which is also a tertiary alcohol, may be explicable in part on the basis of the same reaction; only about half the expected hydroxyl content is available for esterification. Furthermore, 2-methyl-2-butanol, which is very readily dehydrated, shows very little esterification. It is hoped that the reaction of polyhydroxyl compounds with phthalic anhydride will be investigated further.

To determine whether other substances possibly present in the liquid condensates obtained in catalytic organic reactions would affect the accuracy of the hydroxyl determination, synthetic mixtures were prepared

Table 7. Analyses of Synthetic Mixtures Containing Ethanol

Constituents of Mixture	Composition (wt. %)	Ethanol Found (%)
1. Ethanol	79.4	79.6, 79.7, 79.7
Water	20.6	
2. Ethanol	14.6	14.5, 14.7
Water	85.4	
3. Ethanol	75.2	75.3, 74.8, 75.0, 75.0
Water	19.6	
Acetaldehyde	5.2	
4. Ethanol	71.7	71.3, 71.1, 71.4
2,4-Hexadiene	4.7	
Water	18.7	
Acetaldehyde	4.9	
5. Ethanol	11.3	11.4, 11.3
Acetic acid	27.2	
Acetone	13.5	
Crotonaldehyde	24.0	
Ethyl acetate	17.6	
Phenol	6.3	

containing known amounts of water and of representative members of various types of organic functional groups, including carbonyl compounds, acids, esters, and unsaturated compounds. Ethanol was determined in the presence of these substances singly and in mixtures; results of this study, shown in Table 7, indicate that such substances, present in amounts likely to be encountered in reaction mixtures, do not interfere with the determination of the esterifiable hydroxyl group. Of particular interest is the accuracy attainable in mixtures containing as much as 85% water.

The authors have found that although the presence of large amounts of water do not appear to influence the esterification of ethanol in the phthalic anhydride method, water does adversely affect the esterification of alcohols in general. In fact, water reacts quite rapidly with phthalic anhydride in a pyridine medium, resulting in phthalic acid. The resulting free acid can apparently esterify ethanol but cannot quantitatively react with most hydroxyl groups.

The authors have also verified the fact that aldehydes and phenol do not interfere in this method. In addition to the aldehydes and phenol used by Elving and Warshowsky, we have used the method successfully in the presence of formaldehyde, propinaldehyde, and alkyl phenols.

PYROMELLITIC DIANHYDRIDE METHOD

Method of S. Siggia, J. G. Hanna, and R. Culmo, with Adaptations of R. Harper, S. Siggia, and J. G. Hanna

[*Anal. Chem.*, ***33***, *900* (*1961*); **37,** *600* (*1965*)]

Pyromellitic dianhydride (PMDA) has been found to combine the advantages of both the acetic and phthalic anhydrides. As in the case of phthalic anhydride, PMDA can be used in the presence of aldehydes; it is not volatile, it can be used to. determine alcohols in the presence of phenols, yet its rate of reaction is comparable to that of acetic anhydride. The time involved for analysis is approximately the same as that for the perchloric acid catalyzed acetic anhydride reaction, although the PMDA method does require a heating period.

REAGENTS

Pyromellitic dianhydride, 0.5*M*. Dissolve 109 grams of pyromellitic dianhydride in 525 ml of dimethylsulfoxide and then add 425 ml of pyridine.

Standard 1*N* sodium hydroxide solution.

Phenolphthalein indicator.

PROCEDURE

Pipet 50 ml of 0.5*M* pyromellitic dianhydride solution into a glass-stoppered 250-ml flask. Weigh a sample containing 0.010 to 0.015 equivalent of alcohol or amine and add to the reagent in the flask. Place the flask on a steam both, wet the stopper with pyridine, and seat loosely in the flask. Heat the contents for 15 to 20 minutes (30 minutes for polyglycols). Add 20 ml of water and continue the heating for 2 minutes. Cool the mixture to room temperature and titrate with 1*N* sodium hydroxide to the phenolphthalein end point. Treat a blank, from which only the sample is omitted, in the same manner.

$$\% \text{ Hydroxyl} = \frac{A \times N_{\text{NaOH}} \times 17.01 \times 100}{\text{Grams of sample} \times 1000}$$

where A is milliliters of sodium hydroxide solution used for the blank minus milliliters of sodium hydroxide solution used for the sample, corrected for any free acid or base in the sample.

When only small samples are available or when samples have low hydroxyl contents, 0.1M anhydride can be used. In these cases 0.2N sodium hydroxide is used as the titrant. When the more dilute system is used, it is good practice to make a 50% increase in the reaction time, to ensure complete reaction, although many of the common alcohols still react completely under the conditions described.

DISCUSSION AND RESULTS

Hydrolyzed pyromellitic dianhydride titrated with standard sodium hydroxide showed only one inflection in a potentiometric plot of volume versus pH. The midpoint of the maximum slope occurred at pH 9.1 to 9.2, indicating that phenolphthalein is a suitable indicator. Calculated on the basis of alkali consumed to this point, all four acid groups are neutralized.

Possible interference from aldehydes was checked by treatment of 2 to 3 grams each of formaldehyde, acetaldehyde, furfural, and acrolein according to the procedure. No anhydride was consumed in any case. Alcohols to which aldehydes were added in amounts equal to the alcohol were determined with no measurable interference from the aldehydes.

Tetrahydrofuran (THF) was the solvent for the pyromellitic dianhydride (PMDA) in the procedure as originally described. A solution of PMDA in THF was mixed with the sample and pyridine then added. PMDA shows a limited range of solubility in the usual solvents, including pyridine, and the addition of pyridine in the THF system caused some of the anhydride to precipitate. Although the reaction proceeded rapidly and quantitatively at the boiling point of the THF (65°C), the danger of bumping required constant attention during the heating period. One advantage of the THF solvent is that THF is partially evaporated during the heating, which causes a more rapid reaction in the concentrated mixture.

Table 8 gives results for the determination of alcohols and amines with THF as the solvent, Table 9 compares results obtained with THF and dimethylsulfoxide solvents, and Table 10 compares hydroxyl values determined using the two solvents with the phthalic anhydride results.

Further modifications of the PMDA method have been proposed (6). Imidazole is claimed to be superior to pyridine as a catalyst, being about

6. B. H. M. Kingston, J. J. Garey, and W. B. Hellwig, *Anal. Chem.*, **41,** 86 (1969).

Table 8. Determination of Alcohols and Amines with Tetrahydrofuran as Solvent

	% Purity	
	PMDA Method	Other Method
2-Propanol	97.7, 96.5, 96.7	97.2[a]
1-Butanol	100.4, 101.2	99.7[a]
1-Pentanol	94.3, 94.9	93.5[a]
3-Pentanol	101.9, 101.3	101.1[a]
1-Heptanol	100.7, 100.4	100.7[a]
1-Octanol	99.8, 98.5	99.4[a]
2-Octanol	98.9, 99.2	98.9[a]
Allyl alcohol	99.9, 99.5	99.4[a]
Cyclohexanol	99.8, 99.8	100.0[a]
1,2-Propanediol	98.4, 100.0, 100.0	101.3[a]
1,3-Butanediol	100.9	99.9[a]
Glycerol	96.6, 96.6[c]	96.2[a]
Aniline	98.5, 98.6	99.5[b]
2-Naphthylamine	100.2, 100.2	100.5[b]
1,2-Propanediamine	98.1, 98.2, 98.0	98.3[b]

[a] Acetylation procedure. See C. L. Ogg, W. L. Porter, and C. O. Willits, *Anal. Chem.*, **17,** 394–7 (1945); also see p. 12, this book.
[b] Acid titration.
[c] Sample contains 4.0% water by Fischer titration.

Table 9. Determination of Alcohols and Amines, Comparison of Results Obtained with Dimethylsulfoxide and Tetrahydrofuran as Solvents

	Purity, %	
Compound	Dimethyl-sulfoxide Solvent	Tetra-hydrofuran Solvent
Methanol	99.8, 99.8	99.8
2-Propanol	99.6, 99.7	99.7
1-Butanol	98.8, 98.8	99.3
3-Pentanol	100.0	100.0
1-Heptanol	99.5, 99.7	99.8
Triethylene glycol	99.2	100.0
1,2-Propanediol	99.6, 100.1	99.9
Isobutylamine	99.0, 100.0	99.0
Diisobutylamine	99.7, 100.1	100.1
2-Naphthylamine	98.3	98.2

Table 10. Hydroxyl Values of Polyglycols

Polyglycol	Hydroxyl Value, milligrams of potassium hydroxide per gram of sample		
	Dimethyl-Sulfoxide Solvent	Tetrahydro-furan Solvent	Phthalic Anhydride Method[a]
Poly G 3030 PG (triol mol-wt. range, 3000)	54.1, 54.6	54.5	54.3, 54.8
Poly G 4031 PG (triol mol-wt. range, 4000)	41.4, 41.7	41.5, 41.5	41.3, 41.4

[a] P. J. Elving and B. Warshowsky, see p. 22 of this book.

10 times more efficient. Imidazole does not catalyze the reaction of aldehydes and phenols; however it does cause partial reaction of tertiary alcohols and alkoxysilanes.

The PMDA method was adapted to the semimicro scale (7). A sample containing 0.4 to 0.6 meq. of hydroxy or amino compound is heated at 115°C for 30 to 40 minutes with 25 ml of 0.04*M* PMDA in dimethylsulfoxide. Ten ml of water is then added, and the mixture is heated for another 2 minutes. The acid is then titrated potentiometrically or to the phenolphthalein end point with 0.08*N* hydroxide solution. The method was used successfully for octadecanol, L-amphetamine, and polymeric glycols.

Acid Chloride Esterfication Method

There is one acid chloride that exhibits some utility over the anhydrides and with which the problem of stability of reagent is not too severe, although it is always present. This acid chloride is 3,5-dinitrobenzoyl chloride, a reagent that has been used for many years to prepare derivatives of alcohols for identification purposes.

For the general hydroxyl groups, the anhydride methods are preferred, mainly because of the stability and reproducibility of the reactivity of the reagent. The 3,5-dinitrobenzoyl chloride method, however, will quantitatively determine some stubborn hydroxyl groups more readily than the

7. W. Selig, *Microchem. J.*, **21,** 92 (1976).

anhydride methods. For example, sugars and tertiary hydroxyl compounds appear to esterify more readily by the acid chloride approach than by the anhydride approach. The only anhydride method that has had any success in this area is the perchloric acid catalyzed acetic anhydride method of Fritz and Schenk (p. 14). The Fritz and Schenk method, however, requires longer reaction times than the 3,5-dinitrobenzoyl chloride method.

Method of W. T. Robinson, R. H. Cundiff, and P. C. Markunas

[*Anal. Chem.*, ***33,*** *1030–4* (*1961*)]

The reaction of 3,5-dinitrobenzoyl chloride with an alcohol in pyridine solution is shown in eq. 1.

$$ROH + (NO_2)_2C_6H_3\text{—}\overset{O}{\overset{\|}{C}}\text{—}Cl + C_5H_5N \rightarrow (NO_2)_2C_6H_3\text{—}\overset{O}{\overset{\|}{C}}\text{—}OR + C_5H_5N{\cdot}HCl \quad (1)$$

The excess dinitrobenzoyl chloride is hydrolyzed by water, as indicated by eq. 2.

$$(NO_2)_2C_6H_3\text{—}\overset{O}{\overset{\|}{C}}\text{—}Cl + HOH + C_5H_5N \rightarrow (NO_2)_2C_6H_3\text{—}\overset{O}{\overset{\|}{C}}\text{—}OH + C_5H_5N{\cdot}HCl \quad (2)$$

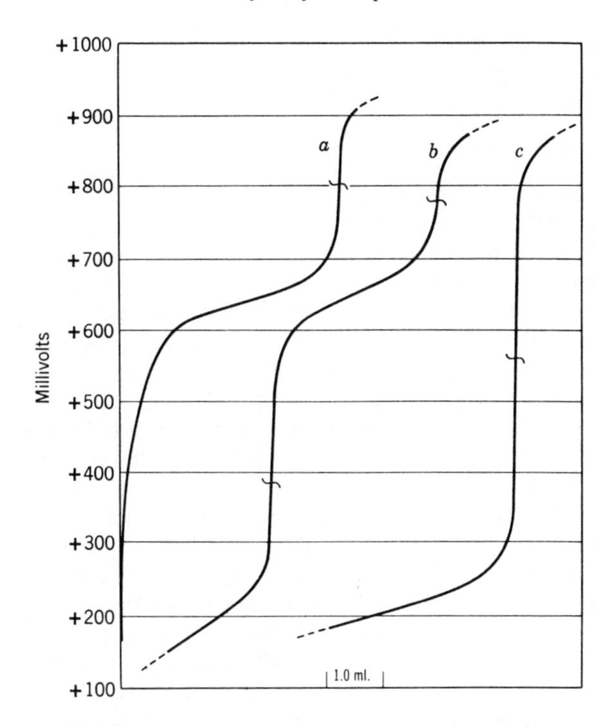

Fig. 1.1. Titration of components in determination of hydroxyl groups: *a*, ethyl=3,5-dinitrobenzoate; *b*, reaction mixture of 3,5-dinitrobenzoyl chloride and ethanol; *c*, 3,5-dinitrobenzoyl chloride.

As indicated in curve *b*, Fig. 1.1, the pyridinium hydrochloride and the dinitrobenzoic acid titrate simultaneously as strong acids, represented by the first inflection in the potentiometric curve, whereas the dinitrobenzoate titrates as a weak acid, represented by the second inflection in the curve. The amount of dinitrobenzoate formed is a measure of the organic hydroxyl content.

The reaction mixture changes from yellow to red at the first equivalence point, which provides the basis for a visual titration. That this is a valid end point was demonstrated by titration of separate pyridine solutions of 3,5-dinitrobenzoic acid and ethyl 3,5-dinitrobenzoate. In the first titration the color changed from yellow to brilliant red at the potentiometric end point. In the second titration addition of one drop of titrant produced the same red color.

Thus the hydroxyl content may be realized in three ways: by measurement of the dinitrobenzoyl chloride consumed as determined by a visual titration, by potentiometric titration through the first end point, or by a differentiating potentiometric titration in which the dinitrobenzoate formed is actually measured.

REAGENTS AND APPARATUS

TETRABUTYLAMMONIUM HYDROXIDE (8), 0.2*N* IN 7:1 BENZENE-METHANOL. Prepare and purify as described. Dissolve 160 grams of tetrabutylammonium iodide* in 300 ml of reagent grade absolute methanol. Place in an ice bath, add 80 grams of finely ground silver oxide, stopper the flask, and agitate intermittently for 1 hour. Filter through a sintered-glass funnel of fine porosity, rinse the flask, and precipitate with three 50-ml portions of cold benzene and add to the filtrate. Dilute the filtrate to 2 more liters with dry benzene. An equimolar portion of tetrabutylammonium bromide may be substituted for the tetrabutylammonium iodide. Use of the bromide permits the agitation time to be cut to 15 minutes.

To prepare an anion-exchange column, fill a 25×400 mm chromatographic tube about half full with Amberlite resin IRA-400 (OH^-), analytical grade. Pass 2*N* sodium hydroxide solution through the column until the effluent gives a negative halide test. Rinse with distilled water until the effluent is neutral to Alkacid test paper (Fisher Scientific Co., Pittsburgh). Pass 500 ml of absolute methanol through the column, followed by 500 ml of 10:1 benzene-methanol. Pass the tetrabutylammonium hydroxide solution through the column, collecting the effluent when basic to Alkacid paper. Allow the solution to pass through the column at the rate of 7 to 10 ml per minute. Collect the tetrabutylammonium hydroxide in a flask protected from carbon dioxide and moisture, and store in a reservoir protected from these. This solution is stable for at least 60 days; longer storage periods have not been tried.

PYRIDINE. Flash-distill technical grade pyridine from barium oxide, reflux the distillate over fresh barium oxide for 3 hours, and distill through a 50-cm upright, air-cooled column, protected from moisture. This material contains 0.02 to 0.04% water.

3,5-Dinitrobenzoyl chloride, 98 to 100% EK-2654, Distillation Products Industries. Finely grind in a mortar and store in a desiccator.

Precision Shell dual titrometer or equivalent pH meter.

ALCOHOL SAMPLES. Most liquid samples were distilled once; solid samples were analyzed as received. Estimated purity of all samples was 97 to 100%.

The remaining apparatus and reagents have been described.

PROCEDURE

For each series of analyses prepare a fresh 0.2*M* solution of 3,5-dinitrobenzoyl chloride by dissolving with gentle heating 1.15 grams in 25 ml of pyridine. Do not expose this solution to moist air unnecessarily.

8. Prepared as described by Cundiff and Markunas in *Anal. Chem.*, **30,** 1450 (1958), but modified slightly by the authors for this procedure.

* Obtained from Rymark Laboratories, Terre Haute, Ind.

To determine liquid samples, pipet approximately 4 meq of the hydroxyl compound into a tared 10-ml volumetric flask containing 3 ml of pyridine. To avoid error in weighing liquid samples, take care not to wet the neck of the flask with the sample. Reweigh and dilute to volume with pyridine. Pipet into a 125-ml $ Erlenmeyer flask, 4.0 ml of the dinitrobenzoyl chloride solution, then 1.0 ml of the sample dilution. Stopper tightly, swirl, and allow to stand 5 to 15 minutes at room temperature. Unstopper the flask and add 7 to 10 drops of water.

To determine solid samples, accurately weigh 0.4 meq of the hydroxyl compound directly into a 125-ml $ Erlenmeyer flask, add 4.0 ml of the dinitrobenzoyl chloride solution, stopper the flask, swirl gently to dissolve, and allow to stand 5 to 15 minutes at room temperature. Unstopper the flask and add 7 to 10 drops of water.

Prepare a blank solution by pipetting 4.0 ml of the dinitrobenzoyl chloride solution into a flask and immediately adding 7 to 10 drops of water.

VISUAL TITRATION. Add 40 ml of pyridine to the reaction mixture, heat nearly to boiling, cool, then titrate with 0.2*N* tetrabutylammonium hydroxide to the first definite and permanent red color. The titration is best performed with the titrant and solution protected from moisture and air and the tip of the buret immersed in the titrating solution. Titrate the blank in exactly the same manner. Use the difference in volumes between the blank and sample to calculate the percentage of hydroxyl.

POTENTIOMETRIC TITRATION. Add 25 ml of pyridine to the reaction mixture and heat nearly to boiling. Cool and transfer to a 250-ml beaker. Rinse the flask with two 10-ml portions of pyridine and add the washings to the beaker. Titrate potentiometrically under nitrogen. If a blank is determined, titrate only through the first inflection, using the difference in volumes between end points of the blank and samples to calculate the percentage of hydroxyl. If no blank is determined, titrate through both inflections and use the volume between the first and second end points to calculate percentage of hydroxyl. In the potentiometric titration, proceed slowly through the first end point, to avoid a false inflection.

ALTERNATIVE PROCEDURE

After addition of the 3,5-dinitrobenzoyl chloride and sample solution to the reaction flask, stopper the flask tightly and gently heat on a hot plate for 30 to 60 seconds (Caution!), remove, and allow to cool. Repeat three or four times if necessary, then follow the remainder of the procedure as described previously. The alternative procedure cannot be applied to most tertiary alcohols or the keto sugars.

EXPERIMENTAL

Figure 1.1 shows typical potentiometric titration curves of the various components involved in the determination of hydroxyl groups by the proposed procedure. All titrations were made in pyridine solution using tetrabutylammonium hydroxide titrant. Dinitrobenzoyl chloride titrates dibasically and dinitrobenzoic acid titrates monobasically with one inflection in their respective potentiometric curves.

Several concentrations of 3,5-dinitrobenzoyl chloride in pyridine were used to determine the minimum molar requirement for quantitative esterification of an alcohol. A minimum of 40% molar excess was required; however, because of possible interference from water, a 75 to 100% molar excess was used.

The data in Table 11 demonstrate that comparable results may be

Table 11. Analysis of Pure Samples by Visual and Potentiometric Titrations

	Purity Found, %		
		Potentiometric	
Compound	Visual	$V_B - V_1$[a]	$V_2 - V_1$[b]
Ethanol	99.33	99.33	99.59
2-Propanol	100.24	100.40	100.94
2-Methyl-2-propanol[c]	99.06	99.35	100.21
2-Methyl-2-butanol[d]	97.43	97.03	97.64
1-Octadecanol	99.50	99.39	99.50
Pentaerythritol[e]	99.04	100.19	100.32
Mannitol[f]	99.37	99.12	99.12
Triethylene glycol	100.19	100.01	100.25
Cholesterol	99.94	100.17	100.17
Dextrose	99.30	99.30	99.07
Sucrose	99.88	99.21	99.69
Cyclohexanone oxime	99.55	99.55	100.00
Thymol	100.39	100.39	
3,4-Dimethylphenol	100.31	100.54	
Isobutylamine	99.29	99.01	
Diphenylamine	99.85	100.29	

[a] Based on difference in volumes of blank and first endpoint of reaction mixture.

[b] Based on difference in volume from first to second endpoints of reaction mixture.

[c] Reaction time 24 hours at room temperature.

[d] Reaction time 48 hours at room temperature.

[e] Tetrabasic.

[f] Hexabasic.

obtained by using either the visual or potentiometric end points. The values listed in Table 11 as well as in all other tables are the average of at least two determinations by each means of titration.

Phenols and amines could not be determined by the difference in volumes between the two potentiometric end points of their reaction mixtures, because the dinitrobenzoate moieties of these compounds do not titrate quantitatively.

Table 12 lists the results of additional compounds determined by the visual titration method.

All hydroxyl groups in the glycols, with the exception of the one noted, were esterified by the reagent. Standard deviation for the procedure is 0.18 as determined by nine replicate analyses of octadecanol.

Table 13 lists results from determination of ethanol by the visual method in the presence of the types of compounds that most often interfere in hydroxyl determination. Reaction time was 5 minutes at room temperature unless otherwise noted.

Proportionately large amounts of benzaldehyde interfere. Since no interference was noted with 40% benzaldehyde, however, it can be assumed that the aldehyde interference is negligible. Interference can be expected from any compound that will react with the reagent. This accounts for the slightly high recovery noted in the presence of 2-methyl-2-propanol in the 5-minute reaction period. Extremely hindered tertiary alcohols, such as triphenylmethanol, will not interfere.

As indicated in Table 14, only 20 to 25% water can be tolerated in analysis of ethanol-water mixtures by the procedure described. Use of larger amounts of 3,5-dinitrobenzoyl chloride would appreciably increase this tolerance.

Of the compounds tested containing both active hydrogen and carbonyl groups, only the sugars could be quantitatively esterified by the proposed procedure. Interference was noted with acetamide, isovaleramide, succinimide, *p*-hydroxybenzoic acid, 3-hydroxy-2-naphthoic acid, benzoin, vanillin, α-benzoin oxime, and α-furil dioxime.

Although the sugars were quantitatively determined, only four of the five hydroxyl groups in the two keto sugars, fructose and sorbose, were esterified, whereas all hydroxyl groups in the aldo sugars and disaccharides were esterified.

DISCUSSION

The mechanism of the reaction that produces the red color at the first end point in the titration is not completely understood, although it is probably a type of quinoidal structure.

Table 12. Analysis of Other Compounds by Visual Titration Technique

Compound	Purity Found, %
Methanol	99.24
1-Propanol	98.91
1-Butanol	98.31
Isobutyl alcohol	99.18
Isopentyl alcohol	99.00
2-Propen-1-ol	99.05
Benzyl alcohol	97.79
Furfuryl alcohol	96.91
1-Tetradecanol	96.80
1-Hexadecanol (cetyl alcohol)	100.15
Solanesol	99.32
2-Butanol	97.99
Cyclohexanol	97.35
1-Menthol	99.92
Stigmasterol	99.03
Sclareol[a]	100.98
α-Terpineol[b]	99.64
Terpinol hydrate[b,c]	99.44
Glycerol	99.77
Propylene glycol	98.86
2-Hydroxy-2,5,5,8*a*-tetramethyl-1 (2-hydroxyethyl)-decahydro-naphthalene[d]	99.77
Tris(hydroxymethyl)aminomethane[b]	98.29
1(−)-Sorbose[e]	98.77
Fructose[e]	97.90
1-Naphthol	98.82
Benzylamine	98.32
Acetone oxime	100.30

[a] Reaction time, 96 hours at room temperature. Dibasic.

[b] Alternative procedure. Heating process repeated four times.

[c] Contains two tertiary hydroxyl groups. Only one esterified.

[d] Glycol of sclareolide. Contains one primary and one tertiary hydroxyl group. Only primary hydroxyl group esterified in 15 minutes at room temperature.

[e] Only 4 of 5 hydroxyl groups esterified. Must not be heated or allowed to react longer than 5 minutes at room temperature.

Table 13. Determination of Ethanol in Presence of Other Compounds

(0.4 mmole ethanol taken for each analysis)

Compound Present	Compound Present, mmole	Ethanol Recovered, %
2-Methyl-2-propanol	0.1	101.8
	0.3	101.8–102.7
Triphenylmethanol	0.3	99.8
Tribenzylamine	0.3	99.8
Acetone	0.1	99.5
	0.3	100.3
Cyclohexanone	0.4	99.3
Benzaldehyde	0.1	100.2
	0.3	97.6
	0.4	92.7
Benzaldehyde (23 minutes)	0.4	92.7

The red color that develops at the end point of a visual titration is easily and accurately determined; however it is not a sudden change from yellow to red. Rather, it is a transitional change from yellow to orange to red. The true end point, therefore, is taken as the first definite and permanent red color. At the end point, the intensity of the red color is directly proportional to the concentration of the titrant. For this reason, $0.2N$ tetrabutylammonium hydroxide proved more satisfactory than more dilute titrants.

The reaction time as stated for primary and secondary hydroxyl groups is dependent on the alcohol being determined as well as the reactivity of the dinitrobenzoyl chloride. For example, with one lot of dinitrobenzoyl chloride, pentaerythritol was completely esterified within 10 minutes, thymol in 15 minutes, and ethanol in 5 minutes. With another much more reactive lot of dinitrobenzoyl chloride the reaction time was cut to less

Table 14. Analysis of Ethanol-Water Mixtures

Added, mmole		Ethanol,	Ethanol Recovered,
Ethanol	Water	wt. %	%
0.42	0.11	90.17	99.57
0.41	0.25	80.80	99.37
0.43	0.38	74.42	85.42
0.47	0.53	69.17	63.91

than 1 minute with many primary and secondary alcohols. In general, it will be necessary for the analyst to test the reactivity of each lot of dinitrobenzoyl chloride to establish the necessary time of reaction. The 15-minute period should be the maximum time required. With less reactive dinitrobenzoyl chloride use of the alternative procedure should prove beneficial in analysis of many primary and secondary alcohols.

Benzoylation of all sugars was complete within 5 minutes at room temperature. The keto sugars could not be heated or allowed to react longer than 5 minutes before dilution, without sample degradation occurring. The reaction mixture of the keto sugars should be titrated by the visual method, because extremely weak inflections are realized in potentiometric titration of these compounds, making potentiometric equivalence point determinations unreliable. High molecular weight carbohydrate polymers, such as dextran or cellulose, cannot be accurately determined because of their limited solubility in pyridine.

Although not all tertiary alcohols can be stoichiometrically determined, enough such compounds can be determined to make application of this procedure worthwhile. The noted times required for quantitative reaction of those tertiary alcohols listed in Table 11 and 12 are maximum and were obtained using a lot of less reactive dinitrobenzoyl chloride. Use of a more reactive lot of dinitrobenzoyl chloride yielded quantitative results for 2-methyl-2-propanol after 3 hours, 2-methyl-2-butanol and 2-methyl-2-pentanol after 5 hours. The only two tertiary alcohols investigated that showed incomplete esterification were 3-ethyl-3-pentanol and triphenylmethanol. 3-Ethyl-3-pentanol showed only 60% reaction after 100 hours when the less reactive acid chloride was used, over 80% reaction after only 15 hours with the more reactive acid chloride. Triphenylmethanol showed practically no reaction after 100 hours with the less reactive reagent. With the exception of α-terpineol and terpinol hydrate, the alternative procedure cannot be applied to tertiary alcohols because, if heated, they will dehydrate in the presence of the dinitrobenzoyl chloride.

Because of the difference in reactivity or secondary alcohols and tertiary alcohols with 3,5-dinitrobenzoyl chloride, it should be possible to determine primary or secondary alcohols and tertiary alcohols simultaneously. In so doing, however, the reactivity of the dinitrobenzoyl chloride being used must be taken into consideration.

Determinations of Hydroxyl Compounds in the Presence of Primary and/or Secondary Amines

Primary and secondary amines will interfere with any of the esterification methods described previously for determining hydroxyl groups.

These amines react rapidly and quantitatively with anhydrides and in fact can be determined by any of the foregoing methods (see pp. 9–40).

Therefore in samples of hydroxyl compounds containing amine, the esterification methods will determine the total of hydroxyl plus primary and/or secondary amino groups. A direct titration of a separate sample with acid, in aqueous or nonaqueous media, however, will determine the amines. This amine value can then be subtracted from the esterification analysis to yield the hydroxyl value by difference. Since the acetylation values and the amine titration values are generally very accurate and precise, the hydroxyl analysis obtained by difference is generally quite accurate and precise. As the amine content rises and hydroxyl content falls, however, the difference between the two analyses decreases; thus the accuracy and precision also decrease.

$$\begin{array}{c} RNH_2 \\ \text{or} \\ (R)_2NH \end{array} + \begin{array}{c} RC(=O) \\ \diagdown \\ O \\ \diagup \\ RC(=O) \end{array} \rightarrow \begin{array}{c} RC(=O)NHR \\ \text{or} \\ RC(=O)N(R)_2 \end{array} + RC(=O)OH$$

There is a direct method for determining hydroxyl in the presence of amino groups (9) which consists of acetylating both groups forming the corresponding esters and amides. The excess anhydride and acetic acid are neutralized, and then advantage is taken of the fact that esters hydrolyze at a much faster rate than amides. A known excess of methanolic sodium hydroxide is added beyond the neutral point, and a controlled saponification is run. Many esters completely saponify before any noticeable amount of amide has reacted. This direct method is not any more rapid than the indirect method outlined earlier, however, also, on unknown systems the hydrolyzability of the amides obtained in the direct method is not known and can lead to questionable analysis. The indirect method is very generally applicable.

9. S. Siggia and I. R. Kervenski, *Anal. Chem.* **23,** 117 (1951).

Hydroxyl Groups on Adjacent Carbon Atoms (1,2-Glycols)

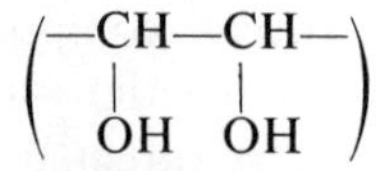

$$\left(\begin{array}{c}-\mathrm{CH}-\mathrm{CH}- \\ \mid \quad\quad \mid \\ \mathrm{OH} \quad \mathrm{OH}\end{array}\right)$$

Glycols can be determined by any of the esterification methods mentioned previously. Of these, the phthalic is the least desirable from the aspect of time, since some 2 hours of reaction time is generally needed. Also, the esterification methods do not differentiate these 1,2-dihydroxy compounds from other hydroxy compounds. There is, however, an oxidation approach that is very specific for compounds with hydroxyl groups on adjacent carbon atoms. This is the oxidation using periodic acid:

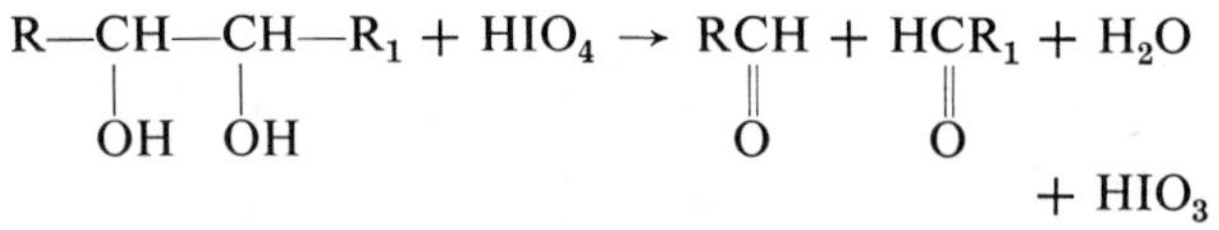

$$\underset{\mathrm{OH}\quad\mathrm{OH}}{\mathrm{R-CH-CH-R_1}} + \mathrm{HIO_4} \rightarrow \underset{\mathrm{O}}{\overset{}{\mathrm{RCH}}} + \underset{\mathrm{O}}{\mathrm{HCR_1}} + \mathrm{H_2O} + \mathrm{HIO_3}$$

This reaction is quite specific and clear-cut. Single hydroxyl groups or hydroxyl groups not on contiguous carbon atoms are generally not attacked.

Adapted from the Method of W. D. Pohle, V. C. Mehlenbacher, and J. H. Cook

[*Oil and Soap*, **22,** *115–9* (May *1945*)]

REAGENTS

Standard $0.1N$ sodium thiosulfate.

Oxidizing reagent. To a solution of 5 grams of periodic acid (HIO_4) in 200 ml of distilled water, add 800 ml of glacial acetic acid. The solution should be kept in a dark, well-stoppered bottle.

Potassium iodide solution. 200 grams per liter.

1% Starch indicator solution.

PROCEDURE

Weigh a sample containing approximately 0.0005 to 0.001 mole of dihydroxy compound into a glass-stoppered iodine flask. Add to the sample 100 ml of oxidizing reagent. Run a blank on the oxidizing agent alone. Allow the solution to stand for 30 minutes at room temperature.

(For most compounds 30 minutes is sufficient; however a few samples require 1 hour for complete reaction.) Then add 20 ml of potassium iodide solution, and titrate the liberated iodine with 0.1N sodium thiosulfate.

CALCULATIONS

$$\text{Milliliters for blank minus milliliters for sample} = A$$

$$\frac{A \times N\,\text{Thiosulfate} \times \text{mol. wt. of compound} \times 100}{\text{Grams of sample} \times 2000} = \%\ \text{dihydroxy compound}$$

The procedure as described was originated to determine various monoglycerides. It also works well, however, for other dihydroxyl compounds such as ethylene glycol, mannitol, glycerol (consumes 2 moles of periodic acid per mole of glycerol), dextrose, tartaric acid (needs to stand for 1 hour). Epoxide (—CH—CH—) compounds may also be determined

O

by this method, but a longer time ($1\frac{1}{2}$ hours) and an eightfold excess of periodic acid are necessary.

The titration for the sample should be more than 80% of the blank to make sure that enough reagent is present for complete oxidation because the iodate formed in the reaction also liberates iodine from potassium iodide. If all the periodic acid were reacted, the titration would be 75% of the blank. Caution should be taken where the reaction products are formaldehyde or formic acid, because these materials are subject to slow, although definite, oxidation at room temperature.

In general, periodic acid will not oxidize such compounds as olefins, alcohols, and aldehydes. Some compounds not containing adjacent hydroxyls are attacked. An example is 2-butyne-1,4-diol ($HOCH_2C{\equiv}C{\cdot}CH_2OH$). It oxidizes very slowly but significantly. Most amines are also attacked to some degree.

GLYCOLS IN THE PRESENCE OF GLYCERIN

Often samples must be analyzed that contain not only glycols but glycerin as well. Of course, the periodic acid method measures the total of the two quantitatively. It is possible, however, to determine glycerin separately and then obtain the glycol content by difference. When glycols react with periodic acid, only aldehydes are formed (see p. 00). When glycerin reacts, however, it yields formic acid as well as formaldehyde. The glycerin can be determined separate from the glycols by measurement of the formic acid.

$$\begin{array}{l} CH_2OH \\ | \\ CHOH \quad + 2HIO_4 \rightarrow 2CH_2O + H_2O \\ | \\ CH_2OH \qquad\qquad\qquad + HCOOH + 2HIO_3 \end{array}$$

Adapted from the Method of P. Bradford, W. D. Pohle, J. K. Gunther, and V. C. Mehlenbacher

[*Oil and Soap,* ***19,*** *189–93* (*1942*)]

REAGENTS

PERIODIC ACID SOLUTION. Dissolve 20 grams of periodic acid in one liter of water. If the solution is not clear, filter through sintered glass filter. Store the solution in a dark, glass-stoppered bottle. The oxidizing power of this solution decreases slowly with time. A blank must be run each day analyses are made.
SODIUM HYDROXIDE 0.1250*N*. Standardize with potassium acid phthalate using phenolphthalein indicator.
METHYL RED. Dissolve 0.1 gram in 100 ml of 95% alcohol.
Sodium hydroxide, approximately 0.05*N*.
Sulfuric acid, approximately 0.2*N*.

PROCEDURE

The amount of excess periodic acid in the glycerol-periodic acid reaction is critical. Therefore it is necessary to know the approximate concentration of the sample to be analyzed. Some times trial tests must be made to locate the approximate range. Refer to Table 15 for the size sample to be weighed.

All weighing must be accurately and rapidly made.

Weighings are conveniently made in a small beaker (or weighing bottle), pouring from this into a flask or beaker.

If the glycerin content is 30 to 100%, weigh into a 2-liter volumetric flask, make to volume with distilled water, mix well, pipet 50 ml into a 600-ml beaker, and cover with a watch glass.

If glycerin content is 10 to 30%, follow the same procedure, using a 500-ml volumetric flask instead of a 2-liter vessel.

If the glycerin content is below 10%, pour directly from the weighing beaker into 600-ml beaker, add about 50 ml of distilled water, and cover with a watch glass.

Table 15. Weight of Sample to be Taken for Analysis Based on the Glycerin Content

Per Cent Glycerol in Product to be Analyzed	Size of Sample to be Selected if Entire Sample is Used for Analysis	Size of Sample to be Selected if Same is to be Diluted to 2 liters and 50 ml. Withdrawn for Analysis	Size of Sample to be Selected if Same is to be Diluted to 500 ml. and 50 ml. Withdrawn for Analysis
100	0.1200 to 0.1500	4.8 to 6.0	
90	0.1330 to 0.1670	5.3 to 6.7	
80	0.1500 to 0.1880	6.0 to 7.5	
70	0.1720 to 0.2180	6.9 to 8.6	
60	0.2000 to 0.2500	8.0 to 10.0	
50	0.2400 to 0.3000	9.6 to 12.0	
40	0.3000 to 0.3750	12.0 to 15.0	
30	0.4000 to 0.5000	16.0 to 20.0	4.0 to 5.0
25	0.4800 to 0.6000		4.8 to 6.0
20	0.6000 to 0.7500		6.0 to 7.5
15	0.8000 to 1.000		8.0 to 10.0
10	1.200 to 1.5		12.0 to 15.0
8	1.5 to 1.88		15.0 to 18.8
7	1.72 to 2.18		
6	2.0 to 2.5		
5	2.4 to 3.0		
4	3.0 to 3.75		
3	4.0 to 5.0		
2	6.0 to 7.5		
1	12.0 to 15.0		
0.9	13.3 to 16.7		
0.8	15.0 to 18.8		
0.7	17.2 to 21.8		
0.6	20.0 to 25.0		
0.5	24.0 to 30.0		
0.4	30.0 to 37.5		
0.3	40.0 to 50.0		
0.2	60.0 to 75.0		

Add one drop of the methyl red indicator to the sample in the 600-ml beaker, acidify with the 0.2*N* sulfuric acid. Neutralize with the 0.05*N* sodium hydroxide to the yellow color that corresponds with a pH of about 6.2. If the color of this solution interferes with the detection of the color changes of the indicator, use the pH meter. Add with a pipet 50 ml of periodic acid, shake gently to effect thorough mixing, cover with a watch glass, and allow to stand for 1 hour at room temperature. At the same time prepare a blank containing only 50 ml of water, with no glycerol, and run along with the sample.

After allowing the sample to stand for 1 hour, dilute to 240 to 250 ml and titrate with the 0.125N sodium hydroxide, using a pH meter equipped with glass and calomel electrodes to determine the end point. The sample should be agitated with a stirrer during the titration.

A curve is plotted of pH versus milliliters of sodium hydroxide for the blank and sample titrations. The end points are read from these curves.

CALCULATION

$$\% \text{ Glycerol} = \frac{(X - Y) \times N \times 0.09206 \times 100}{W}$$

X = milliliters of sodium hydroxide to titrate sample
Y = milliliters of sodium hydroxide to titrate blank
N = normality of sodium hydroxide
W = weight of sample

NOTES ON METHOD

Cork should not be used to stopper any of the flasks used in this determination nor in any other way be allowed to come in contact with any of the materials used in or for the analysis, since periodic acid will react with cork.

The titration for the formic acid (titration of the sample minus the titration of the blank) must be not less than 30% and not more than 40% of the blank titration. This is to ensure the proper ratio of periodic acid to sample, that is, the correct excess of reagent.

Acidic Hydroxyl Groups (Enols, Phenols, Nitro-Alcohols)

Hydroxyl groups are known to be acidic, but most are too weakly acidic to be titrated with base. The hydroxyl group often exists, however, in a configuration that intensifies its acidity. For example, a hydroxyl group adjacent to a double bond ($\text{—}\underset{\underset{\text{OH}}{|}}{\text{C}}\text{=C—}$) such as occurs in enols and phenols is acidic enough for titration in nonaqueous media. Also, nitro groups are well known for their intensification of the acidity of hydroxyl groups; for example, in the nitro phenols the acidity of the hydroxyl group rises with the number of nitro groups on the molecule. The same is true in the aliphatic alcohols: that is, propanol is not titratable; however dinitropropanol is quite acidic and can be readily titrated.

Some of these acidic hydroxy compounds such as dinitrophenols, and trinitrophenols are strong acids and can be titrated readily in water. These are the exceptions, however; most of these acidic hydroxyl compounds are relatively weak acids and can best be titrated only in nonaqueous media. The best titration solvents for weak acids are generally the basic solvents such as pyridine, dimethylformamide, and ethylenediamine. However, the authors have been successful many times using acetone as the solvent with methanolic potassium hydroxide or potassium methylate. The basic solvents, however, have the widest application. The papers of Peal and Wyld (10) and Harlow, Noble, and Wyld (11) summarize the titration of weak acids in the various solvents.

The use of tetrabutylammonium hydroxide has recently extended the range of enol and phenol determination beyond that of Fritz (12), who used sodium methoxide dissolved in benzene-methanol and where dimethylformamide or ethylenediamine were tetration solvent media. The procedure of Fritz is a modification of the work of Moss, Elliot, and Hall (13), who used ethylenediamine as solvent and the sodium salt of ethanolamine as titrant. The quaternary hydroxide is an organic soluble very strong base. This base has been used successfully for very weak acids in many solvents as will be shown.

The procedure to be described is the most generally applicable system in the opinion of the authors of this book.

Method of R. H. Cundiff and P. C. Markunas

[*Reprinted in Part from Anal. Chem.*, **28,** *792–7* (*1956*)]

This method uses tetrabutylammonium hydroxide in benzene-methanol as titrant. Several anhydrous organic solvents are usable as titration media, namely: acetone, acetonitrile, pyridine, benzene-methanol, benzene-ethanol, benzene-isopropanol, *n*-butylamine, and dimethylformamide. Glass-calomel electrodes were satisfactory for potentiometric titrations, but a much inproved electrode system was obtained by a slight modification of the calomel electrode. Constant potentials were reached

10. V. Z. Deal and G. E. A. Wyld, *Anal. Chem.*, **27,** 47–55 (1955).
11. G. A. Harlow, C. M. Noble, and G. E. A. Wyld, *Anal. Chem.*, **28,** 787–91 (1956).
12. J. S. Fritz, *Anal. Chem.*, **24,** 674–5 (1952).
13. M. Moss, J. Elliot, and R. Hall, *Anal. Chem.*, **20,** 784 (1948).

rapidly, remained steady, and were reproducible. No precipitate formed in the titration of any acidic compound tested, and indicators could be used with many test compounds. In addition much better differentiation between various acidic groups within the same molecule or in mixtures was realized with this titrant than has been reported for other basic titrants.

In this work acids similar in strength to the mineral acids are designated "strong acids," those similar in strength to the unsubstituted monocarboxylic acids are designated "weak acids," and those similar in strength to phenol are designated "very weak acids."

PREPARATION OF TETRABUTYLAMMONIUM HYDROXIDE

The starting compound for the preparation of tetrabutylammonium hydroxide (Bu_4NOH) is tetrabutylammonium iodide (Bu_4NI). This compound, although available commercially, may be prepared by refluxing butyl iodide with *n*-tributylamine. Recrystallization from benzene yields a material of sufficiently high purity.

In some of the preliminary work tetrabutylammonium hydroxide was prepared by passing a benzene-methanol solution of tetrabutylammonium iodide through an Amberlite IRA-411(OH) anion-exchange resin. A satisfactory titrant was obtained in this manner, but it was not too suitable a preparation because the exchange capacity of the resin for iodide was poor, and it was exceedingly difficult to regenerate the hydroxide from the iodide form of the resin.

The classical preparation of the quaternary ammonium hydroxides is by reaction of an aqueous solution of a quaternary ammonium halide with an aqueous suspension of silver oxide. It was found that tetrabutylammonium hydroxide could be prepared in either methanol or ethanol by reaction of tetrabutylammonium iodide solution with an excess of silver oxide. The reaction proceeds readily at room temperature. The preparation of $0.1N$ tetrabutylammonium hydroxide is as follows.

PROCEDURE FOR TITRANT

Dissolve 40 grams of tetrabutylammonium iodide in 90 ml of absolute methanol. Add 20 grams of finely ground purified silver oxide, stopper the flask, and agitate vigorously for 1 hour. Centrifuge a few milliliters of the mixture and test the supernatant for iodide. If the test is positive, add 2 grams more of silver oxide and reagitate for 30 minutes; if negative, filter the mixture through a sintered-glass funnel of fine porosity. Rinse the reaction flask and funnel with three 50-ml portions of dry benzene, and add to the filtrate. Dilute the filtrate to 1 liter with dry benzene. Flush this solution for 5 minutes with prepurified nitrogen and store in a reservoir protected from carbon dioxide and moisture. The titrant remains stable on extended storage. Standardize against benzoic acid by the visual titration procedure.

Ethanol may be substituted for methanol in this procedure, but tetrabutylammonium iodide is more soluble in methanol, and there is little difference in the strength of the resulting titrants.

In certain situations 1 to 2% of a weak base impurity may cause difficulties by producing separate inflections in potentiometric titration curves, and such impurities have been found frequently in tetrabutylammonium hydroxide prepared by this procedure. Marple and Fritz (14) gave the following preparation method to eliminate these impurities.

Slurry 23 grams of purified silver oxide with 130 ml of a 75% methanol–25% water solution in an ice bath at 0°C. Add slowly, 32 grams of tetrabutylammonium bromide in 35 ml of pure methanol, and stir for 10 to 15 minutes. Filter the base solution through a coarse sintered-glass frit into a flask containing 0.5 gram of activated charcoal, mix well, and let settle for several hours. Filter the base solution through a fine sintered-glass frit into a flask containing a stirring bar. Remove the methanol by evaporation at a pressure of 25 mm Hg. Transfer the aqueous solution to a 250-ml graduate addition funnel and dilute to 200 ml with boiled distilled water. Add 30 to 40 ml of pure benzene, shake vigorously, and let the two phases separate completely. Pass the aqueous solution through a column containing 8 to 10 grams of strong base anion-exchange resin in the hydroxide form (Rohm and Haas Amberlite IRA 401) and wash with 20 ml of boiled distilled water. Collect the eluate in a 500-ml Erlenmeyer flask, protecting the eluate from carbon dioxide. Distill most of the water from the base at a pressure of 20 mm Hg until the crystalline hydrate is formed. When titrant is needed, distill water from the crystalline hydrate until the vapor pressure is between 7 to 10 mm Hg. The resulting solution will be approximately 2*M*. Dilute to 1 liter with a mixture of 20% isopropanol–80% benzene. The final solution is approximately 0.1*N*.

APPARATUS AND REAGENTS

Beckman general purpose glass electrode, No. 4990-80.

Beckman sleeve-type calomel electrode, No. 1170-71, modified by replacing

14. L. W. Marple and J. S. Fritz, *Anal. Chem.*, **34,** 796 (1962).

the saturated aqueous potassium chloride solution in the outer jacket with a saturated solution of potassium chloride in methanol; designated hereafter as methanol-modified calomel electrode.

Buret, 10 ml.

Tetrabutylammonium hydroxide, 0.1*N*, in 10 to 1 benzene-methanol, prepared as described previously.

Acetonitrile, technical grade.

Pyridine, technical grade.

Dimethylformamide, technical grade (Du Pont).

Benzene-isopropyl alcohol, 10:1. Mix 100 ml of isopropyl alcohol with 1 liter of benzene.

Thymol blue indicator solution. Dissolve 0.3 gram of thymol blue in 100 ml of isopropyl alcohol.

Azo violet indicator solution, a saturated solution of *p*-nitrobenzeneazoresorcinol in benzene.

Nitrogen, prepurified.

POTENTIOMETRIC TITRATIONS

PROCEDURE

Accurately weigh a sample that will require a titration of 2 to 10 ml into a 250-ml beaker. Add 50 ml of solvent and place in position under the glass and methanol-modified calomel electrodes. Fill the buret with 0.1*N* tetrabutylammonium hydroxide and cover the buret top with an Ascarite tube. Best results are obtained if the titrations are performed under a nitrogen blanket when basic solvents, such as dimethylformamide or pyridine, are used. Add the titrant in 0.1-ml increments until just before the equivalence potential is indicated, then in 0.05-ml increments. Continue the titration until the cell potential reaches a maximum and remains relatively constant on addition of titrant. Because most of the solvents contain acidic impurities, it is necessary to perform a blank titration.

Prepare a titration curve by plotting volume of titrant against millivoltage, and determine the equivalence point or points from this curve. When two or more inflections are present in the curve, use the difference between successive equivalence points to calculate the volume of titrant equivalent to the acids represented.

TITRATION OF KNOWN ACIDS

A number of acidic compounds were titrated following the potentiometric method. Pyridine was the best basic solvent for use with 0.1*N* tetrabutylammonium hydroxide titrant. The cell potentials were steadier and were reached more rapidly in this solvent than in others tested. Dimethylformamide was also used, but was not as satisfactory as pyridine. Acetonitrile was the best of the neutral solvents tested, although other neutral solvents including benzene-methanol, benzene-ethanol, benzene-isopropyl alcohol (all in 10:1 volume ratios), and acetone were usable.

The strongly basic solvents, *n*-butylamine and ethylenediamine, have not been studied exhaustively as yet because the potential range was so greatly compressed in these solvents. In dimethylformamide and pyridine the emf range is approximately 900 mV, whereas in ethylenediamine and *n*-butylamine this range is compressed to approximately 300 mV. In addition, the titrations in pyridine or dimethylformamide were much more definitive.

Good titration curves were obtained with glass-calomel electrodes; however, the sharpness of the inflections was markedly increased with the substitution of the methanol-modified calomel electrode. The fiber-type calomel electrode may be used with or without methanol modification, but occasionally it yields erratic readings because of precipitation of potassium chloride on the fiber. The antimony-calomel electrode system was also usable in pyridine and dimethylformamide solutions, as well as the antimony and methanol-modified calomel electrodes. These antimony-calomel electrode pairs were not nearly as satisfactory as the glass and methanol-modified calomel electrodes.

RESULTS

The titration curves of phenol in Fig. 1.2 indicate the comparative strength of tetrabutylammonium hydroxide when dissolved in water,

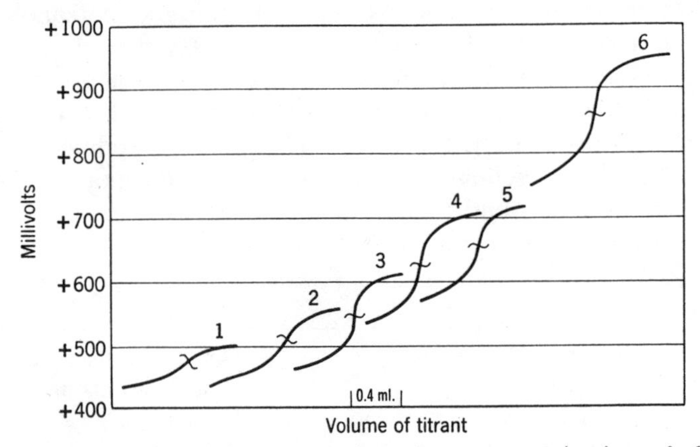

Fig. 1.2. Effect of solvent, titrant, and electrode system on titrations of phenol; 0.1*N* tetrabutylammonium hydroxide titrant.

isopropyl alcohol–water (9 : 1 by volume), or benzene-methanol (10 : 1 by volume). Curves 2, 4, and 6 illustrate the difference in inflection when phenol is titrated in benzene–isopropyl alcohol, acetonitrile, and pyridine with the same titrant. Comparison of curves 5 and 6 indicates the improvement of the inflection with the methanol-modified electrode. In addition to a sharper inflection, much steadier potentials were obtained when the modified electrode was used. The following tabulation gives the curve numbers in Fig. 1.2, along with the variables represented by each.

Curve	Solvent	Solvent for Titrant	Electrodes
1	Dimethylformamide	Water	Glass-calomel
2	Benzene–isopropyl alcohol	Benzene-methanol	Glass-modified calomel
3	Dimethylformamide	Water–isopropyl alcohol	Glass-calomel
4	Acetonitrile	Benzene-methanol	Glass-modified calomel
5	Pyridine	Benzene-methanol	Glass-calomel
6	Pyridine	Benzene-methanol	Glass-modified calomel

Figure 1.3 shows the titration curves for 2-mercaptobenzothiazole, succinimide, salicylaldoxime, acetyl acetone, and α-toluenethiol, all examples of very weak acids with the exception of 2-mercaptobenzothiazole. The samples were dissolved in pyridine and titrated with $0.1N$ tetrabutylammonium hydroxide in benzene-methanol using glass and methanol-modified calomel electrodes.

Figure 1.4 gives the titration curves for a series of di- and trihydroxybenzenes, in which the samples were titrated with $0.1N$ tetrabutylammonium hydroxide using the glass and methanol-modified calomel electrodes. A single inflection was obtained for each compound. Each of these compounds was found to be monobasic in this titration. Curves, solutes, and solvents are related as follows:

Curve	Solute	Solvent
1	Resorcinol	Dimethylformamide
2	Hydroquinone	Dimethylformamide
3	Dihydrodimethylresorcinol	Pyridine
4	Pyrogallic acid	Pyridine
5	Catechol	Pyridine

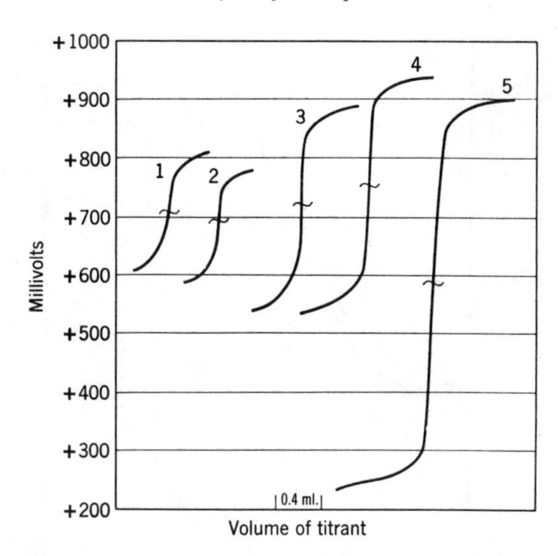

Fig. 1.3. Titration of very weak acids: 1, α-toluenethiol; 2, salicylaldoxime; 3, succinimide; 4, acetyl acetone; 5, 2-mercaptobenzothiazole.

Titration curves for *o*-, *m*-, and *p*-hydroxybenzoic acids are shown in Fig. 1.5. The solvent-titrant-electrode system was pyridine, $0.1N$ tetrabutylammonium hydroxide in benzene, methanol, and glass and methanol-modified calomel, respectively. Two inflections were obtained for *m*- and *p*-hydroxybenzoic acids, but only a single inflection representing the carboxyl group was obtained for *o*-hydroxybenzoic acid, indicating that the compound is monobasic in this titration.

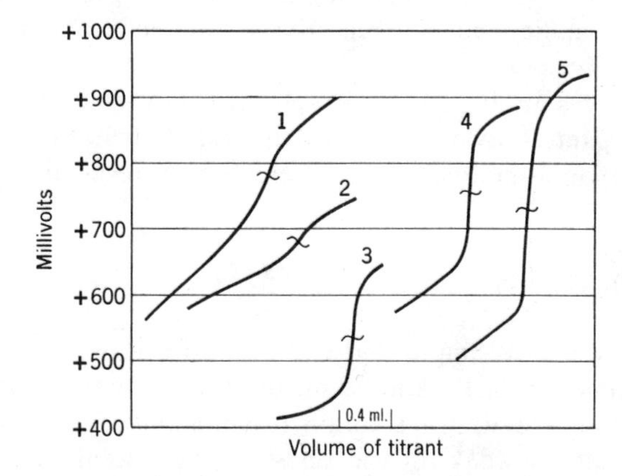

Fig. 1.4. Titration of di- and trihydroxybenzenes.

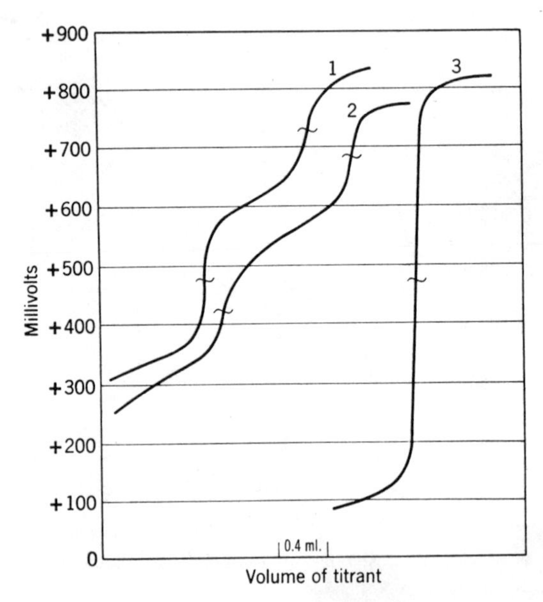

Fig. 1.5. Titration of hydroxybenzoic acids: 1, *m*-hydroxybenzoic acid; 2, *p*-hydroxybenzoic acid; 3, *o*-hydroxybenzoic acid.

Figure 1.6 shows the titration curves of acid mixtures with three inflections in each curve. Curves 2 and 3 represent the titration of a mixture of a strong acid, a weak acid, and a very weak acid. Curve 1 is for the titration of acetic and malic acids in pyridine solution. Malic acid has dissociation constants in water of the order of 10^{-4} and 10^{-6}, whereas acetic acid has a dissociation constant of the order of 10^{-5}. In this particular curve, the first and third inflections represent the two acid equivalents of malic acid, whereas the second inflection represents the acetic acid equivalent.

For the curves in Fig. 1.6 the solvent-titrant-electrode system was pyridine, $0.1N$ tetrabutylammonium hydroxide in benzene-methanol, and glass and methanol-modified calomel electrodes, respectively.

VISUAL TITRATIONS

Accurately weigh a sample that will require 6 to 8 ml of titrant into a 125-ml Erlenmeyer flask. Add 25 ml of solvent and 4 drops of thymol blue or azo violet indicator solution (thymol blue for weak monobasic acids, azo violet for weak dibasic acids and very weak acids). Titrate as rapidly as possible with $0.1N$ tetrabutylammonium hydroxide to the blue

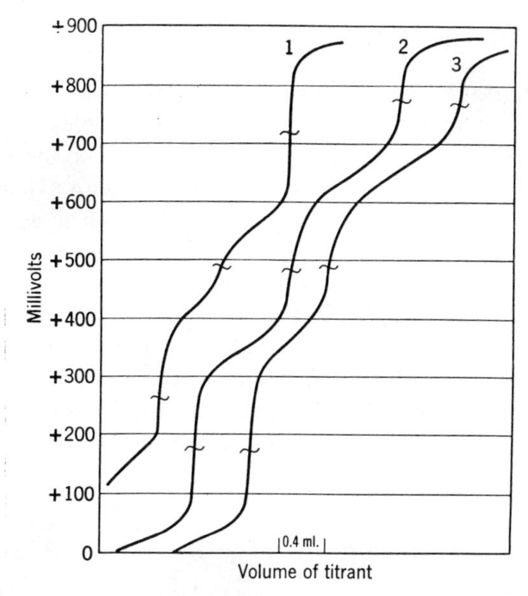

Fig. 1.6. Titration of acid mixtures: 1, acetic and malic acids; 2, *p*-toluenesulfonic acid, benzoic acid, phenol; 3, hydrochloric acid, acetic acid, *p*-cresol.

end point with thymol blue indicator, or to a violet or in some instances a blue end point with azo violet indicator. It is necessary either to make a blank titration for the solvent or to neutralize the solvent exactly prior to adding it to the sample.

RESULTS

A majority of the compounds titrated in pyridine could be determined by visual titration with either thymol blue or azo violet indicator. The visual change in most instances with thymol blue was from green to blue, and thymol blue is the indicator of choice when titrating weak acids. Azo violet was the preferred indicator when titrating very weak acids. The color at the end point for some compounds was violet; for others, blue. It is necessary to titrate potentiometrically first to ascertain whether a compound has an azo violet end point, and if the end point is blue or violet.

If a compound has a thymol blue end point in pyridine it will also have the corresponding visual end point in the neutral solvents such as acetonitrile, benzene-alcohol, or acetone. A great number of very weak acids will have an azo violet end point in acetonitrile.

PRECISION

The precision of the indicator titration procedure was tested by titrating 10 samples of 2-mercaptobenzothiazole and 10 samples of citric acid in acetonitrile solution to the thymol blue end point. The mean percentage purity of the 2-mercaptobenzothiazole was 98.82, with a standard deviation of 0.264. The mean percentage purity of the citric acid was 99.53, with a standard deviation of 0.213.

Similar tests were conducted on 1-naphthol and succinimide compounds that may be titrated to the azo violet end point. The mean percentage for 10 analyses of 1-naphthol in acetonitrile was 99.86, with a standard deviation of 0.166; the mean percentage for 10 analyses of 1-naphthol in pyridine was 99.65, with a standard deviation of 0.271. The mean percentage for 10 analyses of succinimide in pyridine was 99.13, with a standard deviation of 0.348.

COMPOUNDS TITRATED WITH 0.1*N* TETRABUTYLAMMONIUM HYDROXIDE

Listed below are some of the acid compounds that have been successfully titrated in pyridine with 0.1*N* tetrabutylammonium hydroxide in benzene-methanol. Glass and methanol-modified calomel electrodes were used throughout and thymol blue or azo violet indicators were used as indicated.

MONOBASIC COMPOUNDS HAVING THYMOL BLUE END POINT. Acetic acid, benzoic acid, nicotinic acid, salicylic acid, ammonium nitrate, ammonium acetate, *p*-nitrophenol, *o*-nitrophenol, 2,4-dinitrophenol, 2-mercaptobenzothiazole.

AMINO ACIDS HAVING THYMOL BLUE END POINT. Samples were dissolved in a minimum amount of water and pyridine was added.

Alanine, leucine, threonine, asparagine, methionine, hydroxyproline, glutamine, glycylglycine, glutamic acid, aspartic acid, arginine hydrochloride, histidine hydrochloride.

MONOBASIC COMPOUNDS HAVING AZO VIOLET END POINT. Phenol, *p*-benzylphenol, *o*-phenylphenol, 1-naphthol, 2-naphthol, 2,5-dimethylphenol, *p*-bromophenol, catechol, pyrogallic acid, dihydrodimethylresorcinol, succinimide, phthalimide, salicylaldoxime, acetyl acetone, *p*-chlorothiophenol, α-toluenthiol, 5-amino-2-benzimidazolethiol.

DIBASIC COMPOUNDS HAVING AZO VIOLET END POINT. *m*-Hydroxybenzoic acid, *p*-hydroxybenzoic acid, malic acid, oxalic acid, malonic acid, maleic acid, fumaric acid, succinic acid, *o*-phthalic acid, and sulfuric acid. Two inflections were observed in the titration curves of all these compounds.

Citric acid, a tribasic compound, has an azo violet end point coinciding with the potentiometric end point of its second equivalent, and thus would be considered dibasic in a visual titration.

MONOBASIC COMPOUNDS TITRATED POTENTIOMETRICALLY BUT WITH NO SUITABLE VISUAL END POINT. Thymol, hydroquinone, resorcinol, *p*-toluhydroquinone, *m*-cresol, *p*-cresol.

Determining Mixtures of Alcohols

A method exists for determining mixtures of alcohols utilizing the difference in their esterification rates. This method, described on pp. 826–832 in the chapter, "Utilizing Differential Reaction Rates to Resolve Mixtures of Compounds Containing the Same Functional Groups," will determine mixtures of primary and secondary alcohols, distinguish even primary and secondary hydroxyl groups on the same molecule, and also will distinguish alcohols of a homologous series—even members different by only one carbon atom. Other methods for differentiating primary and secondary alcohols are shown on pages 67–82.

PHENOLS ONLY

Bromination Method

Phenols can be determined by bromination according to the following reaction.

$$C_6H_5OH + 3Br_2 \rightarrow C_6H_2Br_3OH + 3HBR$$

The reaction works well, however, for only a rather narrow range of phenols. The acidimetric methods mentioned previously are the preferred methods.

Phenols will substitute bromine atoms only in unoccupied ortho and para positions. But bromine is notorious for substitution of hydrogens on organic molecules other than phenols. For example, hydrogens on any alkyl groups on the phenol can be substituted, although this substitution is much slower than the substitution of the ring hydrogen. Also, if phenols constitute only a portion of the sample, the nonphenolic portion may also substitute bromine, yielding high results. In addition, any material oxidized by bromine is an interference.

In spite of these shortcomings, however, the bromination method occasionally finds application in cases where the acidimetric method cannot be used; that is, a small amount of phenol in the presence of large quantities of other organic acids.

Method from Scott's Standard Methods of Chemical Analysis

(*Adapted with Permission from Wilfred Scott, Standard Methods of Chemical Analysis, 5th ed., Edited by N. Howell Furman, Vol. II, D. Van Nostrand Co., New York, 1939, p. 2253*).

REAGENTS

0.1*N* Bromate-bromide. 2.78 grams of anhydrous CP potassium bromate, and 10 grams of potassium bromide are dissolved in water and diluted to one liter.

PROCEDURE

Dissolve a sample containing 0.02 equivalent of compound in water and dilute to 1 liter. Dissolve an insoluble phenol in a minimum of aqueous sodium hydroxide before dilution. Take a 100-ml aliquot in a glass-stoppered iodine flask, and add 50 ml of bromate-bromide reagent. Shake the flask to mix the reactants and then add 5 ml of concentrated hydrochloric acid. Shake the solution again and allow to stand 15 minutes; then add 2 grams of potassium iodide, and titrate the liberated iodine with 0.1*N* thiosulfate.

Make a blank run on 50 ml of the bromate-bromide solution.

$$\text{Milliliters for blank minus milliliters for sample} = A$$

$$\frac{A \times N\ \text{thiosulfate} \times \text{mol. wt. of phenol} \times 100}{\text{Grams of sample} \times 2000 \times B} = \%\ \text{phenol}$$

where

B = number of bromine atoms that will substitute on the particular phenol being determined

The procedure as described in Scott's *Standard Methods of Chemical Analysis* was intended only for phenol. The procedure was found to work very well for other phenols, however. Materials that have been determined successfully are phenol, resorcinol, *p-tert*-amyl-, *p-tert*-butyl-, *p-n*-octyl-, and *o-n*-octylphenol.

Alkylated phenols should not be kept in contact with the bromine for more than 3 minutes, since bromine will tend to substitute on the alkyl groups and make the results slightly higher. Unsaturated compounds will of course interfere, since they add bromine. Aromatic amines will interfere also, since they rapidly substitute bromine; in fact, they can be determined by the foregoing method.

DIRECT TITRATION OF PHENOLS WITH BROMINE IN PROPYLENE CARBONATE

Adapted from the Method of R. D. Krause and B. Kratochvil

[*Anal. Chem.*, **45**, 844 (1973)]

Propylene carbonate (PC) has certain advantages as a solvent for the direct titration of phenols with bromine. Most phenols as well as their bromination products are soluble in PC. PC is resistant to chemical attack by halogens and its high dielectric constant, 65, increases reaction rates. It has a wide liquid range (−49 to 242°C), is colorless, odorless, and nontoxic, and is not appreciably hydroscopic. Its principal disadvantages are that it hydrolyzes fairly rapidly in the presence of strong acids or bases and that it has a moderately high viscosity.

REAGENTS

PROPYLENE CARBONATE Propylene carbonate from Jefferson Chemical Company was distilled at a pressure of 0.01 mm Hg on a 48 × 1 in. vacuum-jacketed column packed with nichrome helices (Podbielniak size C Heli Pak). The vacuum-jacketed still head contained a solenoid-operated glass valve set at a 10:1 reflux ratio. The purity of the distilled PC was monitored by ultraviolet spectroscopy; the fraction retained had an absorbance of less than 0.3 at 250 nm in a 1-cm quartz cell, measured against distilled water blank. In a typical distillation the first 800- and the last 200-ml charges were discarded.

PROCEDURE

Titrations were performed on a Metrohm Model E436 titrator equipped with a 5-ml buret. Bromine titrant was stored in a low-actinic flask fitted with a Teflon stopper. The flask was vented to the atmosphere only during withdrawal of titrant to minimize loss through volatilization. The rate of titrant delivery was about 2 ml per minute; this rate was automatically decreased in the end point region.

The titration cells were 4×10-cm cylindrical glass weighing bottles with machined Teflon convers. A 1-cm^2 platinum flag was used as the indicating electrode. The reference electrode, a silver wire in a $0.01N$ silver perchlorate in PC solution, was placed in a glass tube sealed to a short length of porous Vycor rod, and the rod was immersed in a bridge solution of $0.1M$ lithium perchlorate in PC. The bridge was separated from the solution being titrated by a 0.9 to 1,4-μm glass frit. Magnetic stirring was used. Samples typically ranged in size from 0.3 to 1 mM and were dissolved in approximately 25 ml of PC.

The exact concentration of a $0.05M$ solution of tetraethylammonium bromide in PC was determined by precipitation of silver bromide in a mixture of 25 ml of PC and 250 ml of $0.01M$ nitric acid in deionized water. The titrant, $0.5M$ bromine in PC, was standardized by titration of a 25-ml aliquot of this standard bromine solution. The titrations were made potentiometrically with the electrode system described. Titration of tetrabutylammonium bromide with bromine in PC yields a potential break at a 1:1 mole ratio corresponding to formation of tribromide ion. The relative standard deviation of replicate standardization titrations on the automatic titrator was 2 ppt or better. When stored in a closed system to minimize bromine volatilization and protected from light, the concentration of $0.5M$ bromine solutions in PC decreased less than 0.3% per day. In contrast, solutions exposed to normal laboratory illumination decreased on the order of 2% per day. Solutions prepared from commercial PC as received also decreased more rapidly than solutions prepared from distilled material.

DISCUSSION AND RESULTS

Addition of excess bromine to phenol in PC does not result in bromine substitution because PC is not a strong enough base to accept the protons released in the reaction. On addition of 3 or more equivalents of base such as pyridine to each equivalent of phenol in the solution, however, formation of the tribromophenol is rapid and complete. The stoichiometry is 6:1, due to the stability of tribromide ion in PC. The overall reaction in the presence of pyridine can be written

$$C_6H_5OH + 3Py + 3Br_2 \rightarrow C_6H_2Br_3O^- + 3PyH^+ + 3Br^- + H^+$$

where Py is pyridine and PyH^+ is pyridinium ion.

Stoichiometries in titrations of representative phenolic compounds are listed in Table 16, along with the effects of varying amounts of pyridine.

Table 16. Stoichiometries in Titration of Phenolic Compounds in Propylene Carbonate as a Function of the Amount of Pyridine Present[a]

Compound Titrated	Ratio, moles of pyridine to moles of phenol					
	2:1	3:1	4:1	6:1	8:1	10:1
Phenol	3.86	5.85	5.97	5.99	5.99	5.92
2-Naphthol	3.93	4.00	3.98	3.98	4.00	3.98
p-Nitrophenol	3.77	3.96	3.99	4.00	3.96	3.99
Salicylic acid	4.04	4.61	4.67	4.54	4.49	4.43
Methyl salicylate	3.73	4.13	4.05	—	4.02	—
p-Cresol	3.79	5.78	5.97	5.98	5.99	6.01
Thymol	3.84	5.74	5.97	5.98	5.99	5.98
Resorcinol	4.19	4.86	7.68	9.56	10.01	10.02

[a] Table values are moles of bromine per mole of phenolic compound at the end point.

Bromine uptake for brominations of phenol, 2-naphthol, *p*-nitrophenol, and methyl salicylate is as expected. In the presence of excess pyridine, phenol consumes 6 equivalents and the other compounds 4 equivalents of bromine. Thus all free positions ortho and para to the hydroxyl group are brominated stoichiometrically.

Salicylic acid consumes about 15% more bromine than the expected 4 equivalents. This may be due to partial substitution of the carboxylic group by bromine, a common side reaction in the bromination of hydroxybenzoic acids. Reduction of the temperature during titration to −20°C reduced the overconsumption to about 8%, but further reduction of the temperature gave poorly defined titration curves.

Titration stoichiometries for the bromination of *p*-cresol, thymol, and resorcinol were also higher than expected. With the assumption that the unoccupied ortho and para positions are the only ones brominated, the first two compounds should consume 4 moles and resorcinol 6 moles of bromine. Instead the amounts are 6 and 10. The additional uptake of bromine can be explained by assuming replacement of the hydroxyl-group protons of these compounds by bromine. For example, the reaction for resorcinol could be written

OH, OH (resorcinol) $+ 5Br_2 + 5Py \rightarrow$ (ring with OBr, Br, Br, H, OBr, Br) $+ 5Br^- + 5PyH^+$

These products would be analogous to the tribromophenol bromide ($C_6H_2Br_3OBr$) formed in aqueous solutions of phenol containing excess bromine (15). Thus compounds that have electron-releasing substituents on the aromatic nucleus in addition to the first hydroxyl group can undergo substitution of the hydroxyl proton. Indirect evidence of interaction of the hydroxyl groups is given by the disappearance of the O–H stretching band at $3380\ cm^{-1}$ when excess pyridine and bromine are added to a solution of resorcinol in PC.

Phenolic compounds that contain a proton-accepting site can be titrated in PC without addition of pyridine. For example, 8-hydroxyquinoline consumes exactly 2 equivalents of bromine. Although the stoichiometry of the reaction is 2:1 as in aqueous solution, the reaction products are tribromide ion and monosubstituted 8-hydroxyquinoline ion, not the dibromo derivative. To illustrate the reproducibility of the reaction, titrations of 0.6-meq. samples of 8-hydroxyquinoline with bromine in PC gave an average of 99.8% with a relative standard deviation of 1 ppt.

Water in small amounts does not affect any of the brominations significantly, but at levels of $0.1M$ or so it interferes. The effect of water on titration plots suggests that the stability of the tribromide complex is reduced greatly by water, presumably through bromide solvation by water.

In summary, an aprotic solvent such as PC can provide a useful medium for bromine substitution. Many reactions are rapid, and the scope appears to be broad. Unusual stoichiometries are observed with some phenols, and the formation in many cases of the tribromide doubles the sensitivity of the method.

Coupling Methods

Phenolic compounds will couple readily with diazonium compounds in alkaline solution. Compounds with active methylene groups will also couple, as will some amines.

$$\left[CH_3\text{-}C_6H_4\text{-}\overset{+}{N}\equiv N\right]Cl^- + C_6H_5OH \rightarrow$$

$$CH_3\text{-}C_6H_4\text{-}N{=}N\text{-}C_6H_4\text{-}OH + HCl \text{ removed by alkali}$$

15. W. M. Lauer, *J. Am. Chem. Soc.*, **48**, 442 (1926).

The coupling takes place in the para position on the phenol or amine unless this position is occupied, in which case the coupling may take place on the ortho position, although the reaction at the ortho position will usually be considerably slower. Some phenols and amines will couple first in the para position and then will commence to couple in the ortho position. The compounds containing active methylene groups will couple at these groups.

$$\left[CH_3\langle\bigcirc\rangle\overset{+}{N}{\equiv}N\right]Cl^- + \begin{array}{l} \quad O \\ \quad \| \\ \quad C{-\!\!-}N\langle\bigcirc\rangle \\ \;/ \qquad\;\; | \\ CH_2 \qquad\; | \\ \;\backslash \qquad\;\; | \\ \quad C{=\!=}N \\ \quad | \\ \quad CH_3 \end{array} \rightarrow$$

$$CH_3\langle\bigcirc\rangle N{=}NCH\begin{array}{l} \quad O \\ \quad \| \\ \quad C{-\!\!-}N\langle\bigcirc\rangle \\ \;/ \qquad\;\; | \\ \qquad\quad\;\; | \\ \;\backslash \qquad\;\; | \\ \quad C{=\!=}N \\ \quad | \\ \quad CH_3 \end{array} + HCl$$

The coupling procedures vary somewhat, depending on the reactivity of the compounds involved. Some compounds will couple in a solution at pH 5; others will couple only in very alkaline solution. The faster the coupling, the better the analysis, since the diazonium compound is unstable. The more alkaline the solution, the faster the coupling. However the more alkaline the solution, the faster the diazonium salt will decompose. In general, salts like sodium acetate and sodium bicarbonate will be sufficiently alkaline for most couplings. Sodium carbonate is sometimes used, whereas sodium hydroxide is seldom resorted to, since it decomposes the diazonium salt too readily.

The coupling procedures are, on the average, accurate to about $\pm 2\%$. These methods are not as accurate as we should like, but they are definitely useful in the determination of samples whose interfering action makes other methods of analysis impossible. The acetylation methods for amines and phenols described on pages 00–00 are more accurate; however if a sample contains some alcohol or other compounds that will acetylate, these methods are of no use. It is in cases like these that couplings can be used. Analytical coupling procedures are generally used in industrial analytical laboratories, especially those involved with dye intermediates.

The following procedure is a general coupling method that can be readily modified to deal with samples requiring different treatment.

REAGENTS

Two diazonium salts can be used as standards for coupling. These are the *p*-toluene- and *m*-nitrobenzene-diazonium chlorides.

0.1*N* *p*-TOLUENE DIAZONIUM CHLORIDE STANDARD SOLUTION. First, introduce 100 ml of 0.5*N* *p*-toluidine stock solution (53 grams of *p*-toluidine plus 131 ml of concentrated hydrochloric acid diluted to 1 liter with water) into a 500-ml volumetric flask, and reduce the temperature of the flask to 15 to 20°C. Next, add 50 ml of 1*N* sodium nitrite (temperature 15 to 20°C), and dilute the resultant solution to 500 ml. Shake the solution to assure thorough mixing. A slight excess of nitrous acid should be indicated on starch-iodide paper. Store the solution in the dark in an ice bath. After 30 minutes it is ready for use. Standardize the solution by using *m*-toluylene diamine (resublimed) or recrystallized 1-phenyl-3-methyl-pyrazolone-5. The solution of diazonium salt should not be used after 5 hours because of its instability.

0.1*N* *m*-NITROBENZENE DIAZONIUM CHLORIDE STANDARD SOLUTION. In a 500-ml volumetric flask cooled to 5°C, place 200 ml of a 0.25*N* *m*-nitroaniline stock solution (heat 34.5 grams of *m*-nitroaniline, 100 ml of concentrated hydrochloric acid, and 100 ml of water to bring about solution, and dilute to 1 liter with warm distilled water). Next, rapidly add 50 ml of 1*N* sodium nitrite that has been cooled to 5°C. Dilute the resultant solution to 500 ml with water that has been cooled to 5°C. It should give a positive test for nitrous acid when tested with starch-iodide paper and is ready for use after standing for 1 or 2 minutes. The solution should be stored in an ice bath in the dark. The solution should be practically colorless and not yellow, and it must not be more than slightly turbid. The solution is standardized in the same way as the *p*-toluene-diazonium chloride standard solution and should not be used after standing more than 5 hours.

PROCEDURE

STANDARDIZATION OF DIAZONIUM SOLUTIONS. Weigh about 0.6 gram of resublimed *m*-toluylenediamine or about 1 gram of recrystallized 1-phenyl-3-methyl-pyrazolone-5 accurately into a 600-ml beaker. Add to this 150 ml of water, 100 ml of 2*N* sodium acetate, and 10 ml of 1*N* acetic acid. Cool the solution to about 10°C.

Agitate the solution with a mechanical stirrer, and add the diazonium standard solution from a buret equipped with a water jacket through which water is circulating at about 10°C. The buret or the glass jacket should be of amber glass to minimize any decomposition of the diazonium salt by light.

Add the diazonium standard solution as rapidly as coupling will take place. No excess of diazonium salt should be allowed to accumulate, since there is danger of decomposition, which leads to high results. To test for excess diazonium salt in the solution, place a few drops of the solution on a piece of filter paper. About a centimeter away from the edge of the liquid mark, place a few drops of an indicator solution [1% resorcinol containing 0.5 gram of sodium carbonate; or 0.1% H acid (1-amino-8-naphthol-3-6-disulfonic acid) solution containing 5 ml of $2N$ sodium carbonate can be used to indicate excess diazonium salt]. Where the two liquid portions meet on the filter paper, a color will develop if excess diazonium salt is present; no color will be noted if the diazonium salt is absent.

To test for excess coupling agent (*m*-toluylenediamine or 1-phenyl-3-methyl-pyrazolone-5) in the reaction mixture use tetrazo dianisidine solution. First, tetrazotize 50 ml of $0.04N$ dianisidine hydrochloride exactly with $0.1N$ sodium nitrite (use starch-iodide paper to indicate excess nitrous acid). Put a few drops of this solution on a piece of filter paper about a centimeter away from the edge of a sample of reaction mixture, also placed on the filter paper. If a color is developed, it indicates that unreacted coupling agent is present in the reaction mixture.

Add the diazonium standard solution in portions, depending on the rate of coupling with the particular compound contained in the sample. Test the reaction mixture for excess coupling agent and diazonium salt after each addition. The additions of diazonium solution should be in increments of 0.25 ml when near the end point. Take the end point as the point at which a negative test for coupling agent and a negative test for diazonium salt are obtained. When this point is reached, read the buret, add a few drops more of diazonium solution, and test solution for diazonium salt. If a positive test is obtained, take the previous reading. If the dye is water soluble and the color interferes with the spot tests, add sodium chloride to salt out the dye and show a clear end point.

Analyze the samples in the same manner in which the diazonium solutions were standardized, but substitute the sample for the *m*-toluylenediamine or the pyrazolone. The sample should contain 0.003 to 0.005 mole of coupling material. Also, if the acetate–acetic acid buffer is not sufficiently alkaline for a satisfactory coupling rate, 100 ml of $2N$ sodium carbonate can be used. The diazonium salt solution should be standardized in the presence of the same base as is needed for the sample. Sodium hydroxide can be used for materials that couple very slowly or for samples soluble only in sodium hydroxide, but larger errors are more likely to be encountered when using this base, since the diazonium salt is readily decomposed by it.

$$\frac{\text{Milliliters of diazonium} \times N \text{ diazonium} \times \text{mol. wt. of coupler} \times 100}{\text{Grams of sample} \times 1000} = \%\ \text{coupler}$$

To determine diazonium salts add a 50-ml aliquot of $0.1N$ solution of *m*-toluylenediamine or 1-phenyl-3-methyl-pyrazolone-5 together with 100 ml of $2N$ sodium acetate and 10 ml of acetic acid. The pyrazolone is the better, since it couples in only one position whereas *m*-toluylenediamine may couple in two. The sample should contain 0.003 to 0.004 mole of diazonium compound. Then titrate the excess coupling agent with standard diazonium solution as shown earlier. The pyrazolone solution should be approximately $0.1N$, prepared with a weighed amount of recrystallized 1-phenyl-3-methyl-pyrazolone-5.

CALCULATIONS

Milliliters of diazonium solution used on 50-ml aliquot of pyrazolone solution minus milliliters of diazonium solution used on sample = A

$$\frac{A \times N \text{ Diazonium solution} \times \text{mol. wt. of diazonium salt} \times 100}{\text{Grams of sample} \times 1000} = \%\ \text{diazonium salt}$$

The procedure was tested with *m*-phenylenediamine, 1-naphthol-4-sulfonic acid; 7-amino-1-naphthol-4-sulfonic acid; 1- and 2-naphthol; phenol; resorcinol; 2-naphthol-3-6-disulfonic acid; phloroglucinol, aceto-acetanilide; 2,3-dihydroxynaphthalene; 2,3-dihydroxynaphthalene-6-sulfonic acid.

In this procedure, avoid adding the diazonium solution too fast, keep the temperature of the solutions below 10°C, and avoid excess alkalinity. Failure to observe each of these main precautions will lead to decomposition of diazonium salts and high results.

Determination of Trace Quantities of Hydroxy Compounds

PRIMARY AND SECONDARY ALCOHOLS

Method of D. P. Johnson and F. E. Critchfield

[*Reprinted in Part from Anal. Chem.*, **32,** *865–7* (*1960*) *with Modifications by D. P. Johnson via Private Communication to S. Siggia*]

The need for a specific method for determining trace quantities of alcohols in the presence of acid-hydrolyzable substances prompted an investigation to find a procedure requiring basic reaction conditions. Berezin (16) had reported the macrodetermination of alcohols by reacting them with 3,5-dinitrobenzoyl chloride (3,5-D) in pyridine and subsequently titrating the excess reagent with a standard base. Previous work has also shown that, in pyridine medium, 3,5-D reacts rapidly with active hydrogen atoms and that the products form highly colored quinoidal ions in certain basic nonaqueous media. This principle provided the basis for the method given below.

Experimental

REAGENTS

3,5-DINITROBENZOYL CHLORIDE SOLUTION (3,5-D). Dissolve 1 gram of reagent grade 3,5-dinitrobenzoyl chloride in 10 ml of redistilled pyridine, using a hot water bath to maintain solution. This solution should be prepared fresh immediately prior to use.

PROCEDURE

Prepare a pyridine (redistilled) solution of the sample to contain 2 to 50 γ of hydroxyl per milliliter. Transfer 2 ml of this solution to a 100-ml glass stoppered graduated cylinder and add 1 ml of the 3,5-D. Prepare a blank by adding 1 ml of 3,5-D to 2 ml of pyridine in another 100-ml graduated cylinder. Allow the sample and blank to stand for 15 minutes at room temperature and then add 25 ml of 2*N* hydrochloric acid. Add 20 ml of hexane by pipet to each of the cylinders, stopper the cylinders, and shake vigorously for 30 seconds. Allow the phases to separate

16. J. V. Berezin, *Dokl. Akad. Nauk SSSR*, **99,** 563–4 (1954).

completely and pipet 2 ml of the top layers into 25-ml graduated cylinders. Avoid getting any precipitate into the pipets. Add 10 ml of acetone and 0.3 ml of 2*N* aqueous sodium hydroxide to each. Shake well and let stand 3 to 5 minutes. Determine the absorbance of the sample with a suitable spectrophotometer at a wavelength of 575 nm based on a reading of zero for the blank. Note: if turbidity forms in the colored solution, add a pinch of sodium chloride crystals and mix before determining the absorbance. Determine the alcohol concentration from a previously prepared calibration curve.

CALIBRATION CURVE

The calibration curve was prepared with absolute ethanol and plotted as absorbance versus micrograms of hydroxyl. The hydroxyl value was used so that the same curve could be applied to a number of different alcohols being investigated. Beer's law is followed for concentrations up to 100 γ of hydroxyl.

DISCUSSION

The ability of the 3,5-dinitrobenzoyl functional group to form colored quinoidal ions provides a useful analytical tool for determining low concentrations of substances. This principle has been used for some time for determining trace quantities of amines (17) and alcohols. The mechanism of the color formation has been discussed by Porter (18). Depending on the nature of the reaction solvent and the base used, a variety of colors can be produced. In acetone the colors are usually purple or blue, whereas in pyridine and dimethylformamide they are predominantly red. The stability of the color is also dependent on the reaction conditions. In alcohols and acetone the colors are frequently unstable, whereas in dimethylformamide, they are usually more stable.

In the present method, pyridine has the twofold advantage of being an excellent solvent for the alcohol and of forcing the reaction to completion by absorbing the hydrogen chloride that is released. The latter function is of special significance, since the reaction is quantitative in 15 minutes at room temperature for all primary and secondary alcohols tested from C_1 to C_{20}.

Several bases, including potassium hydroxide, mono- and diamines, and tetrabutylammonium hydroxide, were investigated as color developers. Potassium hydroxide produced a very unstable color, presumably because

17. D. P. Johnson and J. B. Johnson, *Anal. Chem.*, **31,** 1373–4 (1959).
18. C. C. Porter, *Ibid.*, **27,** 805 (1955).

of saponification of the ester, whereas the colors produced by tetrabutylammonium hydroxide and ethylenediamine were only moderately stable. Diethylamine was the only monoamine tested that produced a color; however only an aged discolored sample of this base was effective. New material failed to produce any color. The mechanism of aging process is unexplained; upon standing, however, amines reportedly (19) extract silicates from glass containers. A 30-minute reaction period is required to obtain maximum sensitivity with diethylamine, and, unlike the other bases, it forms a purple color which exhibits maximum absorption at 415 nm.

PDA was found most suitable as a color developer. Although the color is less stable than that produced by diethylamine, it forms instantaneously and remains without significant change for 3 to 5 minutes, thus allowing ample time to determine the absorbance without loss of accuracy. Unlike diethylamine, a new sample of PDA is more efficient than an aged sample. As the PDA ages, there is a gradual decrease in the sensitivity of the color. The decrease is only slight, however, and if the calibration curve is checked periodically, the accuracy of the method is not significantly affected.

Table 17 lists a number of alcohols to which this method was successfully applied and gives factors for converting hydroxyl to the parent compound. *Tert*-Butanol was also tested but it failed to produce a color.

Table 17. Alcohols Determined by 3,5-D Method

Alcohol	Factor,[a] Hydroxyl to Compound	Lower Limit of Determination,[b] ppm
Methanol	1.88	2.8
Ethanol	2.71	4.1
2-Propanol	3.53	5.3
Butanol	4.36	6.5
2-Butanol	4.36	6.5
1-Pentanol	5.18	7.7
Hexanol	6.00	8.9
2-Ethylhexanol	7.65	11.4
5-Ethyl-2-heptanol	8.48	12.6
2-Eicosanol (mixed isomers)	17.54	26.1

[a] $\text{Factor} = \dfrac{\text{molecular weight of compound}}{17}.$

[b] Using a 1-gram sample and an absorbance of 0.01.

19. J. W. Cavett, Oral Presentation at 72nd Annual Meeting of Assoc. Offic. Arg. Chemists, Washington, D.C., October 1958.

Table 18. Recovery of Alcohols from Known Mixtures

Mixture	Alcohol, % by Weight		
	Added	Found	Recovered
Hexanol in methyl isobutyl ketone	0.189	0.177	98.6
Butanol in butyl ether	0.144	0.144	100.0
Ethanol in diethyl butyral	0.200	0.200	100.0
Ethanol in pentanedione	50.00	47.00	94.0
Ethanol in acetaldehyde	0.405	0.413	102.0
Ethanol in vinyl ethyl ether	0.033	0.034	103.0
Ethanol in ethyl acrylate	0.098	0.096	98.2
2-Propanol in isopropyl ether	0.824	0.800	97.0
Methanol in acetone	0.041	0.040	97.5
Ethanol in butylamine	48.00	49.90	104.0
Ethanol in water	0.600	0.600	100.0
		Av.	99.5
		Av. dev.	±2.2

The data in Table 18 demonstrate the applicability of the method to mixtures of alcohols with other materials. Included are acid-hydrolyzable substances such as vinyl ethers, acrylates, and acetals.

Several dihydroxy compounds, including ethylene glycol diethylene glycol, 2,2-dimethyl-1,3-butanediol, and 2,2-dimethyl-1,3-propanediol, were subjected to the method, but none responded quantitatively. Ethylene glycol and diethylene glycol failed to produce any color, whereas only about 50% reaction was obtained with the other materials. In contrast to ethylene glycol, its monomethyl ether, Methyl Cellosolve, reacts quantitatively with 3,5-D and under proper conditions can be determined by this method.

REACTION OF AROMATIC HYDROXYL COMPOUNDS

A brief investigation of the reaction of 3,5-D with aromatic hydroxyl compounds was conducted. Phenol and 1-naphthol responded quantitatively, but the colors were considerably less stable than those produced by aliphatic derivatives. There is evidence, however, that the diethylamine color may be sufficiently stable to measure. Hydroquinone failed to respond, but its monomethyl ether reacted quantitatively.

INTERFERENCES

Interfering compounds are generally restricted to those that consume the reagent. Water and primary and secondary amines interfere from this standpoint because 3,5-D reacts with each preferentially to alcohols. The products of these reactions, however, are not generally soluble in hexane, and their interference can usually be overcome by adding sufficient reagent. Water, for example, requires a 3,5-D-to-water weight ratio of 13:1, and this ratio must be exceeded before the alcohol will react. Consequently, under the conditions of this method, the water content in the reaction flask cannot exceed 7 mg. A similar tolerance can also be established for primary and secondary amines.

Adapted from the Method of M. W. Scoggins

[*Anal. Chem.*, ***36***, *1152* (*1964*)]

To avoid the difficulties sometimes observed with the unstable color formation of the Johnson and Critchfield method, the ultraviolet absorption spectra of the alkyl nitrobenzoate products were measured.

REAGENTS

Pyridine, freshly distilled Fisher Spectro Grade.
p-Nitrobenzoyl chloride, Eastman P 499. Solution prepared by dissolving 1 gram of reagent in 25 ml of pyridine. Prepare fresh daily.
Cyclohexane, Spectro grade.

PROCEDURE

Transfer 1 ml of liquid hydrocarbon sample to a 125-ml separatory funnel (Teflon stopcock) and add 2 ml of pyridine. (Scrub gaseous hydrocarbons through a known volume of pyridine in a small gas washing bottle at a rate of approximately 0.1 cubic foot per 15 minutes. Transfer a 2-ml aliquot of the pyridine solution to a separatory funnel.) Then add 1 ml of *p*-nitrobenzoyl chloride solution to the separatory funnel and mix by shaking. Allow the mixture to react at room temperature for 30 minutes. Add 25 ml of cyclohexane to the separatory funnel, mix well, and wash the mixture with 10 ml of 2*M* potassium hydroxide solution. Allow the phases to separate and discard the lower aqueous phase. Wash the cyclohexane with two 10-ml portions of $2M$ hydrochloric acid and follow this with two 10-ml alkali washes. Finally, wash the material with 10 ml of $2M$ hydrochloric acid and allow the phases to separate. Measure the absorbance of the cyclohexane in a 1-cm cell at 253 nm versus a blank treated in the same manner. Determine the alcohol concentration from a previously prepared calibration curve.

CALIBRATION

Prepare the calibration curve by treating 1-ml aliquots of cyclohexane-alcohol blends containing from 25 to 300 μg of alcohol per milliliter of solution in the same manner given previously. Plot the total absorbance values versus micrograms of alcohol. Beer's law is followed for concentrations up to 300 μg of alcohol per 25 ml of solution.

RESULTS

Table 19 presents the results of the analysis of synthetic blends of ethanol in cyclohexane. The average recovery is 100.4% in the concentration range of 25 to 200 ppm. Samples containing down to 5 ppm alcohol have been analyzed with good recovery. Table 20 gives the applicability of the method for higher alcohols. Results for these blends are based on the ethanol calibration. Therefore errors are magnified by the alcohol-ethanol molecular weight ratio.

Table 19. Analysis of Ethanol-Cyclohexane Blends with *p*-Nitrobenzoyl Chloride

Ethanol, ppm

Added	Average Recovery[a]	Standard Deviation
34.0	37.0 (6)	4.7
78.0	75.9 (7)	2.1
131.8	127.0 (6)	5.3
196.2	194.1 (6)	7.4
206.1	206.1 (6)	8.1
212.2	212.4 (6)	4.2
25.2	25.9 (3)	2.0

[a] Figures in parentheses indicate number of determinations.

Table 20. Analysis of Higher Alcohol Blends with *p*-Nitrobenzoyl Chloride

Alcohol	Alcohol, ppm Added	Recovered[a]	Standard Deviation	Recovery, %
1-Butanol	131.7	123.6	3.5	94
3-Methylbutan-1-ol	58.8	61.7	2.9	105
1-Hexanol	156.1	165.1	4.5	106
Cyclohexanol	114.5	115.5	7.1	101
2-Butanol	97.4	95.0	1.4	98
2-Propanol	66.2	68.0	3.2	103
1-Heptanol	155.4	160.8	11.4	103
1-Octanol	128.6	129.6	4.5	101
Dodecanol	155.5	164.4	3.1	106

[a] Average of four determinations.

DISCUSSION

p-Nitrobenzoyl chloride was selected as the esterification reagent because of its ready availability, high reactivity, and longer wavelength of maximum absorbance of its esters, 253 nm. Ultraviolet-absorbing species interfere at the wavelength of maximum absorption for ethyl 3,5-dinitrobenzoate, 220 nm.

Pyridine and *p*-nitrobenzoyl chloride have ultraviolet absorption bands at the same region as alkyl *p*-nitrobenzoates and must be removed from the cyclohexane phase prior to measurement.

SECONDARY ALCOHOLS IN THE PRESENCE OF PRIMARY ALCOHOLS

The basis of the method to be described is the oxidation of secondary alcohols to ketones. Primary alcohols are oxidized to the corresponding acid and therefore do not interfere. The analysis is made by colorimetric determination of the ketone formed using 2,4-dinitrophenylhydrazine forming the colored hydrazone.

$$R_2CHOH \xrightarrow{[O]} R_2C{=}O \xrightarrow{\text{2,4-}(NO_2)_2C_6H_3NHNH_2} R_2C{=}NNH\text{-}C_6H_3(NO_2)_2$$

Method of F. E. Critchfield and J. A. Hutchinson

[*Reprinted in Part from Anal. Chem.*, **32,** *862–5* (*1960*)]

Experimental

REAGENTS

ACID POTASSIUM DICHROMATE (APPROXIMATELY 0.3*N*). Dissolve 15 grams of reagent grade potassium dichromate in 500 ml of distilled water. Slowly add 360 ml of concentrated sulfuric acid and cool to prevent excessive heating. Dilute to 1 liter with distilled water and mix.

Hypophosphorous acid. Baker and Adamson technical grade, 50% solution.

CARBONYL-FREE METHANOL. Reflux 3 gallons of methanol containing 50 grams of 2,4-dinitrophenylhydrazine and 15 ml of concentrated hydrochloric acid for 4 hours, and collect the distillate until the head temperature reaches 64.8°C.

Pyridine stabilizer. Pyridine-water solution, 80 to 20 v/v.

2,4-DINITROPHENYLHYDRAZINE REAGENT. Suspend 0.05 gram of reagent grade 2,4-dinitrophenylhydrazine crystals in 25 ml of carbonyl-free methanol. Add 2 ml of concentrated hydrochloric acid and mix to effect solution. Dilute to 50 ml with carbonyl-free methanol.

OPTIMUM OXIDATION TIME FOR PURE ALCOHOLS

Pipet 15.0 ml of the potassium dichromate reagent into each of eight 250-ml Erlenmeyer flasks. Reserve four flasks for blank determinations. Into each of the other flasks introduce 2.0 to 3.0 meq. of the alcohol to be oxidized. If an aliquot of a dilution of the sample is used, introduce the same volume of solvent into the blank flasks. Allow one sample and one blank flask to stand for 30, 60, 90, and 120 minutes, respectively. At the end of the specified time, add 100 ml of distilled water and 10 ml of 15% potassium iodide to the sample and blank flasks. Titrate immediately with standard 0.1*N* sodium thiosulfate to a greenish yellow. Add a few milliliters of a 1% starch solution and continue to titrate to the disappearance of the starch-iodine color. The end point is a chromic blue. The optimum oxidation time should be selected as the time at which no further dichromate is consumed.

CALIBRATION

Into a 100-ml volumetric flask, add 50 ml of distilled water (or acetonitrile if specified in Table 21). Transfer 100 times the maximum sample specified in Table 21 to the volumetric flask and dilute to volume with the appropriate solvent and mix. Transfer 5.0, 10.0, 15.0, and 20.0-ml aliquots of the dilution into separate 100-ml volumetric flasks and dilute to volume with the solvent. Determine the absorbance of each of these standards by using 5-ml aliquots of the standards in place of the sample in the procedure described below.

Table 21. Reaction Conditions for Determination of Secondary Alcohols

Compound	Secondary Alcohol, Mg., Maximum	Oxidation Time, Minutes
2-Butanol	2.92	60 to 120[a]
3-Heptanol	6.16[b]	5 to 60
4-Heptanol	5.36[b]	5 to 75
2,5-Hexanediol	1.47	30 to 60
2-Hexanol	3.63	5 to 60
2-Propanol	1.96	5 to 60[a]
Isopropanolamine	2.17	120 to 210[a,c]
2-Octanol	4.06[b]	10 to 60
4-Octanol	5.47[b]	30 to 100
3-Pentanol	6.66	5 to 60

[a] At room temperature.
[b] Dilute in acetonitrile or use as cosolvent to effect solution in oxidation step.
[c] Allow 17 hours for quantitative reaction with 2,4-dinitrophenylhydrazine.

PROCEDURE

Pipet 15.0 ml of the potassium dichromate reagent into each of three 50-ml volumetric flasks, reserving one of the flasks for a blank determination. Into each of the other flasks, transfer sufficient sample so that the secondary alcohol content does not exceed the amount specified in Table 21. Prepare a dilution in water or redistilled acetonitrile if the optimum sample size is too small to be weighed accurately. If the sample is weighed directly and is insoluble in the reagent, add sufficient redistilled acetonitrile to effect solution. Add the same volume to the blank. Allow the samples and blanks to stand at 0°C, unless otherwise specified in Table 21 until the alcohols are oxidized quantitatively.

The optimum times for secondary alcohols are listed in Table 21, the optimum times for primary alcohols can be determined by the volumetric procedure. Immerse the flasks in an ice water bath and pipet 1.0 ml of the hypophosphorous acid into each flask, swirling the flasks during the addition. Remove the flasks from the bath and immerse them in a water bath at room temperature for 15 minutes. Dilute the contents of the flasks to volume with carbonyl-free methanol and mix. Transfer a 3.0-ml aliquot of each dilution to separate 50-ml volumetric flasks. Immerse the flasks in an ice water bath and pipet 3.0 ml of 4*N* potassium hydroxide into each flask. Remove the flasks from the bath and allow the contents to warm to room temperature. Pipet 3.0 ml of the 2,4-dinitrophenylhydrazine reagent into each flask, mix, and allow the flask to stand at room temperature for 30 minutes unless otherwise specified in Table 21. Pipet 15.0 ml of the pyridine stabilizer into each flask. Transfer 3.0 ml of freshly prepared 10% methanolic potassium hydroxide to each flask, stopper, and mix. Allow the flasks to stand at room temperature for 5 minutes and filter the solution through Whatman No. 40 filter paper. Collect the filtrate in separate 25-ml glass-stoppered graduated cylinders and allow the cylinders to stand for 10 minutes. Using a suitable spectrophotometer, measure the absorbance of the sample vs. the blank at 480 nm, using 1-cm cells. Read the concentration of secondary alcohol from the calibration curve.

DISCUSSION AND RESULTS

Reaction rate studies and a calibration curve were obtained for each secondary alcohol investigated. Table 21 lists the recommended reaction conditions for secondary alcohols to which this method has been applied. In all cases the maximum sample size is based on the amount of pure alcohol introduced in the oxidation step that corresponded to an absorbance of 0.6 under the conditions of the method. Table 21 shows that the oxidation time required for quantitative oxidation to the ketone varies from 5 to 120 minutes, depending on the alcohol oxidized. In most cases it was necessary to conduct the oxidation at 0°C to inhibit further oxidation of the ketone. The optimum oxidation time selected for the determination of a secondary alcohol in the presence of a primary alcohol will depend on the time necessary to oxidize both alcohols quantitatively—that is, secondary alcohols to ketones and primary alcohols to acids. The optimum time for oxidation of any primary alcohol–sample matrix can be conveniently determined by the volumetric procedure.

When acetonitrile is used as a cosolvent for samples insoluble in the potassium dichromate reagent, a redistilled grade should be employed, and the same volume should be incorporated in the blank.

With the exception of isopropanolamine, the ketones formed by the oxidation react quantitatively with 2,4-dinitrophenylhydrazine in less than 30 minutes. In the case of isopropanolamine, approximately 17 hours is required for quantitative reaction. Although this reaction time is long, the reaction can be conducted conveniently overnight.

In general, a separate calibration curve must be obtained for each secondary alcohol being determined. The data in Table 22 show the effect of the structure of the alcohol on the color reaction. In this table the sensitivity is expressed in terms of the absorbance obtained per microequivalent of secondary alcohol present as the ketone in the color reaction step. The differences in sensitivities obtained are due to the different sensitivities of the color reactions of the corresponding ketones. This was established by preparing calibration curves from the ketones. In all cases the same calibration curve was obtained from the ketone as from the alcohol. In general, ketones that contain a methyl group adjacent to the carbonyl give the most sensitive color reaction. The same effect was obtained with the corresponding alcohols. The most sensitive color reaction was obtained from 2,5-hexanediol. In this case the ketone would contain a methyl group adjacent to each carbonyl.

This method has been applied to the determination of 2-propanol in ethanol (Table 23). The lower limit of determination by this method is dependent on the sample matrix. For example, if the matrix is methanol, the sensitivity of the method is lower than for ethanol because methanol is oxidized to carbon dioxide by a three-step oxidation. Therefore, considerably more dichromate ion is consumed by methanol than in the two-step oxidation of ethanol to acetic acid. In general, the sample should not consume more than 85% of the dichromate ion. Because of this restriction on sample size, the lower limit of determination of 2-propanol in ethanol is approximately 0.02%.

An example of the determination of isopropanolamine in ethanolamine is also shown in Table 23. This analysis would be difficult to perform by other methods.

This method is not applicable to bifunctional secondary hydroxyl compounds in which the hydroxyls are separated by less than four carbon atoms. In general, the method is not applicable to the determination of

Table 22. Effect of Structure on Sensitivity

Compound	Structure	Absorbance per Equivalent
2-Butanol	$CH_3—CH(OH)—C_2H_5$	0.266
3-Heptanol	$C_2H_5—CH(OH)—C_4H_9$	0.188
4-Heptanol	$C_3H_7—CH(OH)—C_3H_7$	0.216
2,5-Hexanediol	$CH_3—CH(OH)—(CH_2)_2—CH(OH)—CH_3$	0.402
2-Hexanol	$CH_3—CH(OH)—C_4H_9$	0.281
2-Propanol	$CH_3—CH(OH)—CH_3$	0.308
Isopropanolamine	$CH_3—CH(OH)—CH_2—NH_2$	0.347
2-Octanol	$CH_3—CH(OH)—C_6H_{13}$	0.319
4-Octanol	$C_3H_7—CH(OH)—C_4H_9$	0.238
3-Pentanol	$C_2H_5—CH(OH)—C_2H_5$	0.132

Table 23. Analysis of Mixtures

Mixture	Composition, Weight, %		Recovery, %
	Added	Found	
2-Propanol in ethanol	0.04	0.02	50
	0.16	0.18	113
	1.25	1.23	98
Isopropanolamine in ethanolamine	0.43	0.47	109
	0.57	0.59	104
	1.02	0.97	95
Acetone in acetaldehyde	0.08	0.09	112
	0.16	0.17	106
	0.22	0.24	109

cyclic secondary alcohols or highly branched aliphatic alcohols. These compounds are not oxidized to ketones but to acids.

Any compound that is oxidized to a carbonyl and resists further oxidation will interfere in the method. Most ketones interfere quantitatively; however, suitable corrections can be made by utilizing the 2,4-dinitrophenylhydrazine method without the oxidation step.

This method is also applicable to the determination of low concentrations of ketones in the presence of aldehydes because the latter compounds oxidize to acids under the conditions of the method (Table 23). The determination of acetone in acetaldehyde is the only application that has been investigated to date; however, this type of determination should find considerable applicability.

TERTIARY ALCOHOLS IN THE PRESENCE OF PRIMARY AND SECONDARY ALCOHOLS

Adapted from the Method of M. W. Scoggins and J. W. Miller

[*Anal. Chem.*, ***38***, *612* (*1966*)]

Tertiary alcohols are determined by the action of hydriodic acid to form the corresponding tertiary alkyl halides.

$$R-\underset{R''}{\overset{R'}{\underset{|}{\overset{|}{C}}}}-OH + HI \rightleftharpoons R-\underset{R''}{\overset{R'}{\underset{|}{\overset{|}{C}}}}-I + H_2O$$

Extraction of the tertiary alkyl iodide into an organic solvent forces the reaction to completion. The reaction is almost as rapid and complete as neutralization. The tertiary alkyl iodides are measured by their characteristic ultraviolet absorption at 268 nm.

Tertiary alcohols through C_{10} were measured from the ppm level to the 50% level in the presence of a variety of oxygenated materials including primary and secondary alcohols. The method is applicable to both aqueous and hydrocarbon samples.

REAGENTS

Commercial hydriodic acid, 50 to 55%; or iodine-free 72% are satisfactory.

APPARATUS

Cary Recording Spectrophotometer, Model MS11, equipped with 1-cm quartz cells was used.

PROCEDURE

Transfer a quantity of hydrocarbon sample containing not more than 0.08 mM of tertiary alcohol to a 40-ml screw cap vial equipped with a polyethylene gasket in the cap and dilute to 25 ml with cyclohexane. For aqueous solutions, weigh 100 to 200 mg of sample into the vial and add 25 ml of cyclohexane. Add 3 ml of hydriodic acid and shake for 3 minutes. Transfer the contents of the vial to a separatory funnel, rinse the vial with 10 ml of water, and add the rinse water to the separatory funnel. Mix thoroughly, allow the phases to separate, and discard the aqueous phase. Add 10 ml of 1*M* sodium hydroxide solution and 3 drops of hydrogen peroxide and shake until the iodine color disappears. Transfer a portion of the hydrocarbon phase to a 1-cm cell and measure the absorbance versus a reagent blank by scanning from 300 to 240 nm. Convert the absorbance to tertiary alcohol concentration using a previously prepared calibration curve.

CALIBRATION

Prepare a calibration curve by diluting from 0 to 0.1 mmole of the appropriate test alcohol to 25 ml with cyclohexane and proceeding as above. Plot the absorbance versus millimoles of test alcohol.

RESULTS

Table 24 gives results of the analyses of synthetic blends of 2-methyl-2-propanol in cyclohexane. In the 50 to 2000 ppm range, the average recovery is 100.1%. Blends containing down to 10 ppm 2-methyl-2-propanol can be analyzed if the absorbance is measured in 1-cm cells. The method is somewhat less sensitive for higher molecular weight alcohols.

Table 24. Results for 2-Methyl-2-propanol

Blend	Solvent	Concentration, ppm Added	Found[a]	Recovery, %	Standard Deviation
A	Cyclohexane	45.9	45.5	99.1	1.4
B	Cyclohexane	91.8	90.4	98.5	1.7
C	Cyclohexane	183.6	183.5	99.9	0.5
D	Cyclohexane	459.0	454.4	99.0	2.5
E	Cyclohexane	712	725.5	101.9	6.6
F	Cyclohexane	2372	2399	101.1	17.6
		Average Recovery, wt. %		100.1	
G	Water	49.24	49.14	99.8	0.48

[a] Average of four determinations.

Table 24 also indicates the applicability of the method for aqueous blends of 2-methyl-2-propanol at the 50% level. Samples (100 mg) were analyzed by extracting with 100 ml of cyclohexane.

Table 25 shows the applicability of the method to higher molecular weight tertiary alcohols.

If the sample contains compounds such as propanones, with the $-\underset{\underset{O}{\|}}{C}-CH_3$ group, iodine-free hydriodic acid must be used to avoid high results by iodoform formation (Table 26). After 2 minutes of reaction, 2-methyl-2-propanol is completely converted, whereas 1-butanol and 2-butanol are only 2 to 3% converted. As a consequence, tertiary alcohols are determined in the presence of relatively high concentrations of primary and secondary alcohols (Table 26).

Table 25. Analysis of Tertiary Alcohol–Cyclohexane Blends[a]

Alcohol	Concentration ppm Added	Concentration ppm Found	Standard Deviation[b]	λ max,[c] nm	Molar Absorptivity,[d] 1/mole-cm
2-Methyl-2-propanol	489.2	457.1	25.1 (4)	269	591
2-Methyl-2-butanol[e]	424.6	380.9	23.8 (5)	267	602
2-Methyl-4-butyn-2-ol[f]	561.6	560.0	— (2)	—	—
2-Methyl-2-pentanol	571.2	560.7	19.7 (5)	268	613
3-Methyl-3-pentanol	195.5	212.2	1.5 (4)	266	689
2,4-Dimethyl-4-hexanol	459.1	460.3	4.6 (5)	267	636
4-Methyl-4-heptanol	671.2	653.7	7.8 (3)	269	618
4-Methyl-4-octanol	740.2	763.8	8.1 (3)	269	654
4-Methyl-4-nonanol	214.6	188.4	7.4 (5)	267	585

[a] 2,4-Dimethyl-4-hexanol calibration used for analysis.
[b] Figures in parentheses indicate number of determinations.
[c] Wavelength of maximum absorption of corresponding alkyl iodide.
[d] Molar absorptivity of corresponding alkyl iodide.
[e] Freshly distilled before standard solutions prepared.
[f] Hydrogenated prior to analysis.

1,2–DIOLS

The determination of micromolar quantities of 1,2-diols has been accomplished by the use of atomic absorption spectrophotometry as the end determination. The iodate formed in the oxidation of adjacent hydroxyl groups by periodic acid is separated by precipitation as silver iodate. The silver iodate is dissolved in ammonium hydroxide, and the resultant solution is analyzed for silver content by means of atomic absorption spectrophotometry.

Adapted from the Method of P. J. Oles and S. Siggia

[*Anal. Chem.*, ***46,*** *2197* (*1974*)]

REAGENTS

Prepare a solution of paraperiodic acid to contain 11 to 20 μM per milliliter. The exact concentration need not be determined.

2*M* Silver nitrate solution.

Nitric acid, 3 volumes of nitric acid and 1 volume of distilled deionized water.

Table 26. Effect of Diverse Organic Compounds on Recovery of Tertiary Alcohol

Compound	Ratio, diverse compound to tertiary alcohol[a]	Alcohol Recovery, % error
Acetone	250	33
Acetone[b]	250	1.5
Diethyl ketone	10	2.7
Diethyl ether	100	3.5
Ethyl acetate	100	3.8
n-Butyl bromide	100	4.0
1-Octene	1.5	1.5
1-Octene	5	4.6
1[c]-Octene	100	1.5
Caproic acid	100	5.7
n-Hexyl ether	100	3.3
Ethyl alcohol	300	3.0
Isobutyl alcohol	50	2.8
Cyclohexanol	50	7.4
n-Amyl alcohol	50	3.0
Dodecyl alcohol	50	6.4

[a] Tertiary alcohol concentration, 400 ppm of 2-methyl-2-propanol.

[b] Iodine-free hydriodic acid.

[c] Olefin hydrogenated prior to tertiary alcohol analysis.

APPARATUS

Measure absorbance at 328.1 nm. A Perkin–Elmer 403 Atomic Absorption Spectrophotometer was used, equipped with a single element hollow cathode lamp operated at 24 mA as the source of radiation.

PROCEDURE

Add a 1.00- to 2.00-ml sample of diol in water to a 6-in. test tube. The concentration of the diol should be between 0.10 and 4.00 μmole per milliliters. Add a volume of periodic acid reagent in water equal to the volume of sample added to the test tube. Mix thoroughly and place the test tube, protected from light, on a mechanical shaker for 10 to 30 minutes. Add a volume of nitric acid reagent that is exactly one-half the volume of the original sample (0.5–1.00 ml). Again mix thoroughly and add a volume of silver nitrate equal to the volume of nitric acid reagent. The volumes of reagents used are critical. Again place the test tube on the mechanical shaker for 10 to 30 minutes to allow the silver iodate to flocculate. Cool the test tube to −10 to −15°C in a mixture of dry ice–acetone or in a freezer, to suppress the solubility of silver iodate. Transfer the contents of the test tube to a fine frit (4–5.5 μm) Pyrex glass funnel and apply suction. Rinse the test tube three times with 4- to 5-ml portions of acetone-water (1:1, v/v) containing 0.2% (v/v) concentrated nitric acid maintained at approximately −15°C, each time allowing the rinsings to pass through the filter. Place the test tube containing the acetone-water mixture in crushed ice between rinsings. Add 3 ml of concentrated ammonium hydroxide to the test tube, place a clean 125-ml suction flask below the filter, and add the ammonium hydroxide to the filter. Apply suction and rinse the test tube, and filter with two portions of water. Transfer the contents of the flask to a 25.0-ml (0.1–1.0 μM 1,2-diol), 50-ml (1.0–2.0 μM 1,2-diol), or 100-ml (2.0–4.0 μM 1,2-diol) volumetric flask and dilute to the mark with rinsings from the suction flask. Then analyze the solution for silver content by atomic absorption spectrophotometry.

RESULTS AND DISCUSSION

The concentrations of nitric acid and silver nitrate are critical in this procedure. The addition of nitric acid is necessary to prevent the precipitation of silver periodate; however, an excess must be avoided because the solubility of silver iodate increases with increasing nitric acid concentration.

After addition of all reagents to a sample, final concentrations of nitric acid and silver ion of 2.0 and 0.34*M*, respectively were found to produce satisfactory results as indicated in Table 27. Samples that are very acidic or basic should be adjusted to pH 7 with dilute nitric acid or sodium hydroxide before addition of periodic acid. The optimum concentrations of reagents given may be varied by at least 10% without affecting recoveries, and the concentration of periodic acid may be varied from approximately 0.34 to 7.4 μM per milliliter. (Final concentration after addition of all reagents.)

The oxidation of some functional groups with periodic acid may result in the formation of products that are susceptible to further and, in some

Table 27. Determination of Adjacent Hydroxyl Groups by Atomic Absorption

Compound	Concentration, μM/ml	Hydroxyl Groups Taken, μM	Found, μM	Recovery,[a] %
Glycerol	3.53	3.53	3.46	98.0 ±1.1 (5)
1,2,6-Hexanetriol	1.93	1.93	1.90	98.4 ±4.4 (4)
7-(2,3-Dihydroxypropyl)-theophylline	1.66	1.66	1.64	98.8 ±2.4 (5)
1,2-Propanediol	3.39	3.39	3.28	96.8 ±0.9 (4)
trans-1,2-Cyclohexanediol	1.85	1.85	1.75	94.6 ±5.0 (5)
3-Chloro-1,2-propanediol	3.55	3.55	3.51	98.9 ±1.1 (5)
3-Piperidino-1,2-propanediol	1.42	1.42	1.39	97.9 ±2.5 (5)
1-Phenyl-1,2-ethanediol	1.61	1.61	1.58	98.1 ±1.1 (5)

[a] Figures in parentheses indicate number of determinations.

cases, slow oxidation by periodic acid. This phenomenon causes overconsumption of periodic acid and can result in recoveries significantly greater than 100% when the percentage of recovery is based only on the first oxidation step. An example of this phenomenon encountered in this work is the oxidation of tartaric acid with periodic acid.

$$\underset{\text{OH}\ \text{OH}}{\text{HOOCCHCHCOOH}} + HIO_4 \rightarrow 2\underset{\text{CHO}}{\text{COOH}} + H_2O + HIO_3$$

The glyoxylic acid produced in this reaction is subject to further oxidation by periodic acid, however at a much slower rate.

$$\underset{\text{CHO}}{\text{COOH}} + HIO_4 \rightarrow HCOOH + CO_2 + HIO_3$$

Two means are available to control this phenomenon so that results may become analytically useful. Conditions may be chosen for the reaction which are extremely mild—for example, low concentrations of reactant and periodic acid. If the rates of the two oxidation steps are significantly different, these mild conditions should allow the investigator to make differentiation so that essentially only iodate produced in the first and faster reaction will be determined—in this case, one mole of iodate produced per mole of tartaric acid. Alternately, conditions for the oxidation may be made relatively severe—by heating, for example, thereby forcing both reactions to completion; in this case 3 moles of iodate is produced per mole of tartaric acid present. Although the latter approach results in

Table 28. Determination of D-Tartaric Acid by Periodic Acid Oxidation

Concentration,[a] μM/ml					
Tartaric Acid	Periodic Acid	Temperature, °C	Reaction Time, min.	Iodate Found, μM	Recovery,[b,c] %
1.05	11.5	20	90	1.49	142 ±4.0 (5)
1.05	5.7	1	75	1.21	115 ±4.4 (5)
0.243[d]	1.0	1	60	0.554	114 ±4.6 (4)
0.243[d]	1.0	1	25	0.356	73.4 ±11 (4)
1.22	13.4	100	45	3.47	94.8 ±0.9 (8)

[a] Concentrations listed are those of tartaric acid and periodic acid before mixing.
[b] Figures in parentheses indicates number of determinations.
[c] Recoveries based on 1 mole of iodate produced per mole of tartaric acid present except for results obtained at 100°C.
[d] Two-ml sample taken.

threefold increases in sensitivity, the former procedure is preferred, since heating solutions of periodic acid results in loss of the selectivity of the reagent (20). Results obtained by both approaches for the determination of tartaric acid are illustrated in Table 28. The most precise and accurate results are those that result from heating. However, this procedure is not recommended in general for all cases when glyoxylic acid may be an oxidation product, since other products (or compounds in a sample) may undergo partial and irreproducible oxidation when heated with periodic acid.

Glycols have also been determined by argentometric determination of the silver iodate formed on addition of silver nitrate to the reaction mixture (21) and volumetric measurement of silver consumed.

Determinations of total 1,2-diol content in mixtures were made, and the results appear in Table 29. The specific behavior of periodic acid toward oxidation of 1,2-diols but not 1,3-diols may be noted from the results presented in this table. The procedure should be particularly useful for the determination of 1,2-diol impurities in compounds containing nonadjacent groups.

The results in Table 30 indicate the precision and accuracy that

20. G. Dryhurst, *Periodate Oxidation of Diols and Other Functional Groups*, Pergamon Press, New York, 1966, p. 72.
21. M. Pesez, *Bull Soc. Chim Fr.*, 148–9 (1956).

Table 29. Determination of Total 1,2-Diol Content in a Mixture

Mixture	1,2-Diol Concentration, μM/ml	Taken, μM	Found, μM	Recovery, %[a]
1,2-Propanediol, 0.572 μM+ 7-(2,3-dihydroxypropyl)theophylline, 0.336 μM+ 1,3-butanediol, 131.1 μM	0.908	0.908	0.869	95.7 ±1.9 (5)
1,2-Propanediol, 0.9698 μM+ 3-pyridino-1,2-propanediol, 0.6449 μM	1.61	1.61	1.56	96.9 ±2.1 (5)

[a] Figures in parentheses indicate number of determinations.

may be expected for determinations of 1,2-diols at the ppm level of concentration.

PHENOLS

The most sensitive and generally applicable method for trace quantities of phenol is the coupling with a diazotized aromatic amine yielding a colored azo dye.

$$C_6H_5OH + RN{\equiv}NCl \rightarrow RN{=}N{-}C_6H_4OH + HCl$$

Table 30. Determination of 1,2-Diols at the ppm Level

Compound	1,2-Diol Concentration, μM/ml	Taken, μM	Found, μM	Recovery,[a] %
1,2-Propanediol	0.339	0.339	0.349	103 ±5.0 (5)
3-Piperidino-1,2-propanediol	0.285	0.570	0.575	101 ±3.8 (7)

[a] Figures in parentheses indicate number of determinations.

Method Adapted from J. J. Fox and J. H. Gauge

[*J. Chem. Ind.*, **39**, *206T* (*1920–2*)]

REAGENTS

Sulfanilic acid (reagent grade or recrystallized). 7.6 grams per liter of distilled water.

Sodium nitrite. 3.4 grams per liter of distilled water.

Sulfuric acid. 1 part concentrated H_2SO_4 (sp. gr. 1.84) to 3 parts of distilled water.

8% Aqueous sodium hydroxide solution.

PROCEDURE

Diazotize the sulfanilic acid by adding 1 part of the sulfuric acid solution to 5 parts of the sulfanilic acid solution followed by 5 parts of the sodium nitrite solution. This solution of diazotized sulfanilic acid should be kept cool to minimize decomposition and should be prepared about 5 minutes before use.

The sample containing the phenol should be dissolved in water; or, if the sample is insoluble in water, the sample should be dissolved in as dilute sodium hydroxide solution as is possible to dissolve the phenol in question. If the sample is a complex mixture that is insoluble in water, it can be dissolved in a water-immiscible solvent such as benzene, petroleum ether, or carbon tetrachloride, and this solution can be extracted with aqueous caustic. This layer can then be used for the analysis.

To the sample solution or aqueous extract, add the 8% sodium hydroxide in the ratio of 5 ml of 8% sodium hydroxide per 100 ml of sample solution. The amount of sodium hydroxide is not critical except that the final solution should be alkaline and the standards and samples should be treated alike. To the alkaline sample solution, add enough diazotized sulfanilic acid to obtain optimum color development. These amounts of reactants and the optimum reaction time can only be determined experimentally for each sample. Carefully make sure that enough sodium hydroxide is present in the sample solution to neutralize the acid contained in the amount of sulfanilic acid reagent added. The final solution must be alkaline for the coupling reaction to proceed satisfactorily.

Large excesses of sulfanilic acid should be avoided, since the diazotized reagent will decompose in the alkaline media and can yield colored decomposition products.

The amount of sample and reagent used must be such that an adequate color is obtained for measurement. The color measurement can be made using Nessler tubes, optical color comparators, or spectrophotometers. Prepare color standards for the phenol in question by preparing known solutions and carrying them through the foregoing analysis. Then prepare calibration curves. It is well to emphasize that the amounts of reagents and conditions of analysis must be duplicated for both the standards and the unknowns. Otherwise, variations in color might occur; that is, some of the resulting colors have indicator properties and the colors vary with pH. Hence the pH of the final solution can be important.

This method will detect less than 1 ppm of cresols, phenol, and naphthols. Other phenols are detected in very low concentrations but no figures are available. It should be stated that the other components in the sample can affect the limit of detection quite significantly.

There is much latitude in the method as described, because of the wide range of phenols that can occur and the wide range of mixtures in which they might occur.

The phenols will tend to couple in the position para to the hydroxyl group. If this is occupied, they will generally couple in the ortho position but usually at a rate slower than para coupling. If the ortho and para positions are occupied, coupling may not occur.

Interferences consist of aromatic amines, many of which also couple with diazonium compounds to form dyes. Also, colors are sometimes formed between any excess of nitrous acid in the reagent that might react with amines or other nitrosatable compounds. Media that decompose diazonium compounds, that is, metallic salts, can also cause color to form, since the decomposition products are generally colored.

From the Study of L. R. Whitlock, S. Siggia, and J. E. Smola

[*Anal. Chem.*, **44**, *532* (*1972*)]

The work concerning diazo coupling in this report was carried out, in part, so that a complete and workable procedure for use with alkali fusion method for sulfonate analysis could be developed in detail (see p. 798). This includes a determination of the proper coupling reaction conditions for a wide variety of phenols and the ε and λ_{max} values for the azo dyes formed.

In addition, a critical survey of several diazo compounds, including some not previously used, was made to determine which give the best results for phenol measurement.

The problems commonly found in connection with the analytical utility of the coupling reaction are nearly all derived from the coupling reaction conditions that must be controlled. A more extensive use of the method for phenol measurement has been hindered for two reasons. First, there is a general lack of information defining the proper reaction conditions needed for the many possible phenol-diazo combinations. Second, there is a need for diazo reagents that are more stable and whose coupling reactions are less sensitive to reaction conditions, such as pH and time.

One amine chosen for study was sulfanilic acid, the reagent used by Fox and Gauge described previously. Sulfanilic acid gives a fairly active diazo (many times stronger than aniline), possesses high water solubility, and is easily diazotized; in addition, the reagent solution remains stable for several days if stored at 0°C in darkness.

Diazotized *p*-nitroaniline is considerably more electrophilic than diazosulfanilic acid and is known to couple with some less active phenols (22–24). The results obtained, however, indicated that this reagent is not useful for quantitative analysis for those phenols not already easily determined using diazosulfanilic acid. Difficulties included multiple coupling, much lower molar absorptivities, and the λ_{max} values shifted to considerably shorter wavelengths. Many of the dyes formed were water insoluble, requiring the use of organic solvents in the procedure. Also the diazonium salt has very limited stability. This reagent may be useful for qualitative detection, since partial coupling was observed with a larger number of phenols.

Diazotized 4-amino-1-naphthalenesulfonic acid was studied primarily to increase the molar absorptivity of the resulting azo dyes. This would increase the sensitivity of the method. The results obtained, however, indicated that this reagent, too, is not useful for quantitative phenol measurement.

Several distinct advantages were realized when diazotized *p*-phenylazoaniline was used for phenol analysis. This amine apparently has not been used before as a reagent for phenol analysis. It possesses a considerable amount of extended conjugation; once coupled to phenols, azo dyes with very high absorptivities are formed. Increases of 10,000 to 30,000 absorptivity units over dyes formed from diazosulfanilic acid were common. Besides the higher ε values obtained with this reagent, the pH dependence of the coupling reaction was eliminated. The optimum pH of coupling was 7.5 for all phenols studied. Similarly, the optimum time

22. R. H. DeMeio, *Science*, **108,** 391 (1948).
23. W. Lee and J. H. Trumbull, *Talanta*, **3,** 318 (1960).
24. J. A. Pearl and P. F. McCoy, *Anal. Chem.*, **32,** 1407 (1960).

between reaction and measurement was the same for each phenol. An additional advantage is that the diazonium salt is very stable. It was commonly used for a week or longer when stored on ice (0°C) and in darkness with no noticeable change in its ultraviolet spectrum.

REAGENTS

DIAZOTIZED SULFANILIC ACID, 5 m*M* SOLUTION. Dissolve 0.116 gram of sulfanilic acid sodium salt and 0.035 gram of sodium nitrite in approximately 50 ml of water. Cool the solution to 0°C on ice. Add 2 ml of 2*N* hydrochloric acid with vigorous stirring. The reaction is complete after a few minutes. Adjust the final volume to 100 ml with additional cooled (5°C) distilled water. If excess nitrous acid is present, as evidenced by the starch-iodide test paper, destroy it by adding a few drops of sulfanilic acid solution.

DIAZOTIZED *p*-PHENYLAZOANILINE, 5 *mM* SOLUTION. Dissolve 0.099 gram of *p*-phenylazoaniline in 10 ml of acetone and add 30 ml of water and 5 ml of 2*N* hydrochloric acid. Lower the temperature to 15°C. Add 0.035 gram of sodium nitrite in approximately 50 ml of water with stirring over a 20-minute period. A very slight excess of nitrous acid should be indicated by the starch-iodide paper. Adjust the final volume to 100 ml with cooled distilled water.

APPARATUS

All azo dye absorption measurements and spectra were recorded on a Perkin–Elmer 202, UV-visible spectrophotometer. Matched NIR silica cells (Beckman Instruments) with 1-cm path lengths were used to hold all solutions for measurement. Wavelength measurements were corrected to the 461-nm absorption band of holmium oxide glass.

PROCEDURE

Carry out the coupling reaction in a 50-ml volumetric flask by first adding a 5-ml sample solution aliquot to give a final concentration of 2 to 50 μM phenol. Next add 20 ml of water along with a predetermined amount of 0.1*M* sodium bicarbonate solution used to adjust the pH to the optimum value for coupling (Tables 31 and 32). When diazotized *p*-phenylazoaniline is used, add 25 ml of tetrahydrofuran to prevent precipitation of the azo dye formed and to accelerate the coupling reaction. Finally, add a 1.0-ml aliquot of the diazonium stock solution with thorough mixing, and adjust the solution volume to the mark.

Table 31. Reaction Conditions, λ_{max} and ε Values for Phenols Coupled with Diazotized Sulfanilic Acid

Compound	Reaction Time, min.	pH	Molar Absorptivity L/mole-cm	λ_{max}, nm
Phenol	2	8.5	21,000	450
o-Cresol	2	8.5	22,100	458
m-Cresol	2	8.5	18,200	434
Resorcinol	2	7.6	43,500	440
Phloroglucinol	2	7.1	51,000	440
m-Hydroxybenzoic acid	15	8.1	14,100	414
p-Hydroxybenzoic acid	15	7.8	13,800	430
m-Aminophenol	5	7.5	30,000	450
m-Chlorophenol	5	8.5	22,500	432
o-Iodophenol	2	7.5	25,200	448
o-Phenylphenol	2	7.5	19,300	460
α-Naphthol	15	7.1	25,000	520
β-Naphthol	15	7.1	21,800	492
2,7-Naphthalenediol	15	7.1	21,500	492

RESULTS AND DISCUSSION

When using diazotized *p*-phenylazoaniline, the reagent is light yellow above pH 7.0. An absorbance value between 0.05 and 0.10 was normally observed in the wavelength region used for the phenol measurement. This blank value was measured and subtracted from the dye absorbance value for each phenol determination.

Table 33 gives some analyses of phenol samples in water solution. Data using both diazotized sulfanilic acid and *p*-phenylazoaniline are included. The results were obtained by reacting known concentrations of the phenols and comparing their absorbances with calibration curves prepared for each of the phenols.

The advantage of this method of phenol measurement is that very small amounts can be determined. For example, assuming an ε of 5×10^4 and that 0.05 absorbance units can be measured accurately, the limit of detection is 1.0 micromole of phenol per liter of solution. For a cell volume of 2 ml, the minimum amount detectable is 0.3 μg.

The total phenolic content of a sample containing mixtures of phenols can be measured from a single coupling reaction. This is done by comparing the absorbance value of the unknown mixture with a calibration curve prepared using any chosen standard such as phenol. The

Table 32. Reaction Conditions λ_{max} and ε Values for Phenols Coupled with Diazotized *p*-Phenylazoaniline[a,b]

Compound	Molar Absorptivity L/mole-cm	λ_{max}, nm
Resorcinol	58,500	485
Phloroglucinol	85,000	495
m-Aminophenol	38,500	496
o-Cresol	27,500	535
m-Cresol	26,000	545
α-Naphthol	37,500	520
β-Naphthol	30,000	512
2,7-Naphthalenediol	30,000	510
1,5-Napthalenediol	38,000	650

[a] pH for reaction of each phenol was 7.5.
[b] Time between reaction and spectrophotometric measurement was 3 minutes for each phenol.

Table 33. Spectrophotometric Determination of Various Phenols by Formation of an Azo Dye

Compound	Diazonium	Moles ($\times 10^5$) Taken	Found
Phenol	*a*	0.90	0.91
m-Cresol	*a*	1.63	1.66
Resorcinol	*a*	1.31	1.28
p-Hydroxybenzoic acid	*a*	2.63	2.55
m-Hydroxybenzoic acid	*a*	1.15	1.11
o-Iodophenol	*a*	5.50	5.45
β-Naphthol	*a*	3.08	3.10
m-Aminophenol	*a*	4.21	4.26
o-Phenylphenol	*a*	1.05	1.04
Resorcinol	*b*	0.35	0.35
Phloroglucinol	*b*	0.24	0.25
o-Cresol	*b*	1.88	1.85
m-Aminophenol	*b*	2.00	2.04
α-Naphthol	*b*	1.65	1.66

[a] Diazosulfanilic acid.
[b] Diazo-*p*-phenylazoaniline.

resulting concentration is then expressed as percentage of the sample as phenol. For example, a solution containing $1.6 \times 10^{-5} M$ phenol, $1.1 \times 10^{-5} M$ *o*-cresol, and $2.5 \times 10^{-5} M$ *m*-cresol was determined to contain $4.9 \times 10^{-5} M$ as phenol. The error is about 5%.

The use of simultaneous equations permits the analysis for individual components present in a mixture. A unique application of this technique is shown for the analysis of the 1,5- and 2,7-naphthalenediol isomers using diazo-*p*-phenylazoaniline for coupling. The λ_{max} values for the 1,5- and 2,7-isomers are 650 and 510 nm, respectively, an unusually large separation. The simultaneous equations constructed on their azo dye absorption spectra with the appropriate substitution of molar absorptivities at each wavelength are as follows:

$$A_{510} = 8850C_{1,5} + 30{,}000C_{2,7} - B_{510}$$

$$A_{650} = 38{,}000C_{1,5} + 0C_{2,7} - B_{650}$$

where B_{510} and B_{650} are the reagent blank absorbance values for the diazonium solution alone, and $C_{1,5}$ and $C_{2,7}$ are the molar concentrations of each isomer. Table 34 presents results showing three different mixtures of these isomers.

Table 34. Spectrophotometric Analysis of Naphthalenediol Isomers by the Method of Simultaneous Equations

	1,5-Naphthalenediol, moles ($\times 10^5$)		2,7-Naphthalenediol, moles ($\times 10^5$)	
Mixture	Taken	Found	Taken	Found
A	0.75	0.72	1.15	1.20
B	1.50	1.44	1.73	1.78
C	2.25	2.16	2.88	2.89

2

Carbonyl Groups

There are two types of carbonyl compound: aldehydes, RCHO, and ketones, $\underset{\displaystyle O}{\overset{\|}{RCR_1}}$. These compounds occur commonly and in all sorts of situations. Hence there are many methods for determining carbonyl groups, each method attempting to overcome some of the limitations of the others. Presented below is a series of methods that make it possible to determine carbonyl groups in the majority of situations in which they might occur. These methods utilize the following reactions of the carbonyl group: oxime formation, bisulfite addition, oxidation (aldehydes only), Schiff base formation, and hydrazone formation.

Oxime Formation

The methods of analysis based on the reaction of carbonyl compounds with hydroxylamine or its salts have been the most widely studied and used for the determination of carbonyl compounds. These methods have the greatest range of application of any of the methods for determining this functional group. The reaction is as follows.

$$RC(=O)R_1 \xrightarrow{NH_2OH} RC(=NOH)R_1 + H_2O \quad (1)$$

$$RC(=O)R_1 \xrightarrow{NH_2OH \cdot HCl} RC(=NOH)R_1 + H_2O + HCl \quad (2)$$

In the foregoing reaction, R_1 in the case of aldehydes is a hydrogen atom. For methods using reaction 1, the hydroxylamine, which is a strong base, is titrated. The analysis is based on the amount of amine consumed. In reaction 2, the hydrochloric acid formed by the reaction is titrated.

Reaction 1 is quite rapid and complete, but it suffers from the disadvantage that hydroxylamine is not a stable reagent and is readily oxidized by atmospheric oxygen, making analysis difficult. Reaction 2 uses the hydroxylamine salt, which is a very stable reagent, but the reaction is a distinct equilibrium, which is a significant difficulty for some carbonyl compounds. Hence in the evolution of the oximation method of Fritz, Yamamura, and Bradford shown below, we see the attempts to harness the speed and completeness of reaction 1 and the stability of the reagent in reaction 2.

Hydroxylamine alone is never used for the reasons stated earlier, but hydroxylamine hydrochloride has been used for many years (1–5). The acid salt has been used successfully by the authors of this book for a wide range of aldehydes with no special systems at all. This procedure can be outlined as follows.

REAGENTS

0.5*N* or 0.1*N* Hydroxylamine hydrochloride in any of the following solvents: water, methanol, ethylene glycol, isopropanol, mixtures of the latter two, mixtures of glycol or isopropanol with benzene or petroleum ether.

0.5*N* or 0.1*N* Standard sodium hydroxide (aqueous or methanolic).

PROCEDURE

Place 50 ml of the hydroxylamine reagent in a reaction flask along with enough sample to yield a titration of approximately 20 ml with the titrant to be used. Allow the mixture to stand an appropriate length of time for the particular carbonyl compound being determined. Thirty minutes at room temperature is enough for some aldehydes (low aliphatic aldehydes). However, heating from $\frac{1}{2}$ to 2 hours is required for the more stubborn cases (generally ketones and hindered aldehydes). Reflux should be used in cases where heat is required with the solvent system such that the solvent or a portion thereof boils at a lower temperature than the carbonyl compound in the sample. This keeps the carbonyl material from lodging in the condenser; the solvent keeps it rinsed out of the condenser back into the reaction flask.

After the required length of time, rinse the solution from the reaction flask into an appropriate beaker using as minimum quantity of a suitable

1. A. H. Bennett, *Analyst*, **34,** 14–17 (1909).
2. C. T. Bennett and T. T. Cocking, *Ibid.*, **56,** 79–82 (1921).
3. H. Schutes, *Angew. Chem.*, **47,** 258–9 (1934).
4. W. M. D. Bryant and D. M. Smith, *J. Am. Chem. Soc.*, **57,** 57–61 (1935).
5. J. Maltby and G. R. Primavesi, *Analyst*, **74,** 498–502 (1949).

solvent (generally the same solvent as used for the reagent), and titrate the mixture potentiometrically using standard 0.1 or 0.5N sodium hydroxide solution, whichever best suits the sample.

The glass-calomel electrode system operates well in any of the solvent systems mentioned. A plot of apparent pH versus milliliters of sodium hydroxide solution yields the end point of the titration. Indicators cannot be used, since the system is too buffered to yield sharp pH charges at the end point. A blank run is recommended, since some solvents yield small blank values.

CALCULATION

$$\frac{\text{Milliliters of NaOH} \times N\ \text{NaOH} \times \text{mol. wt. of compound} \times 100}{\text{Grams of sample} \times 1000} = \%\ \text{carbonyl compound}$$

The foregoing method indicates the flexibility of the approach. The equilibrium of the oximation in the cases of some aldehydes and ketones, however, is so significant that this system cannot be applied. This is readily seen in the case of the phenyl ketones such as acetophenone. Also, for some ketones the system is too slow to be convenient. In addition, potentiometric titration is required, since the change in acidity at the end point of the titration is too gradual for the use of indicators.

To speed the reaction and eliminate the equilibrium problems, Bryant and Smith added pyridine to the system. This basic material ties up the hydrochloric acid produced in the reaction, increasing the rate and minimizing the equilibria. Even though indicators are used in this method, however, the indicator end points are not sharp, and potentiometric titration is still the preferred method for accurate and precise end point detection.

To further optimize the oximation approach, Higuchi and Barnstein (6) employed hydroxylamine acetate instead of the hydrochloride. Acetic acid was used as solvent and the hydroxylamine acetate was titrated as a base with perchloric acid used as a titrant. This system has a relatively rapid rate, but potentiometric titration is still required.

Fritz, Yamamura, and Bradford, in the method to be described, have been able to utilize an organic base to tie up the hydrochloric acid, to optimize the reaction completeness and to permit accurate and precise

6. T. Higuchi and C. H. Barnstein, *Anal. Chem.*, **28,** 1022 (1956).

titration with indicators. The method uses dimethylaminoethanol as the base and hydroxylamine hydrochloride in methanol-isopropanol as reagent. The base is consumed by the hydrochloric acid formed in the reaction, and the excess base is titrated with perchloric acid. The reactions are quite rapid, equilibria are generally overcome, and indicator end points are possible.

Method of J. S. Fritz, S. S. Yamamura, and E. C. Bradford

[*Reprinted in Part from Anal. Chem.*, ***31***, *260* (*1959*)]

REAGENTS

2-DIMETHYLAMINOETHANOL, 0.25*M*. Dissolve approximately 22.5 grams of freshly distilled 2-dimethylaminoethanol (Eastman Chemical Products, Inc., White Label or equivalent) in 2-propanol to make 1 liter of solution.
HYDROXYLAMMONIUM CHLORIDE, 0.4*M*. Dissolve 27.8 grams of the pure salt in 300 ml of absolute methanol and dilute to 1 liter with 2-propanol.
2-PROPANOL. Reagent grade, absolute.
MARTIUS YELLOW. Dissolve 0.0667 gram of Martius yellow (Harleco, Hartman-Leddon Co.) and 0.004 gram of methyl violet in ethanol and dilute to 50 ml with ethanol.
METHYL CELLOSOLVE. Merck & Co., Inc., reagent grade, or Union Carbide Chemicals Co.
PERCHLORIC ACID, 0.2*M*. Pipet 17.0 ml of 70% perchloric acid and dilute to 1 liter with Methyl Cellosolve. Standardize by titration of tris(hydroxymethyl)-aminomethane.
TRIS(HYDROXYMETHYL)AMINOMETHANE. Primary standard grade.
CARBONYL SAMPLES. The compounds analyzed were mostly Eastman White Label chemicals with an estimated purity of 98 to 100%. Some were purified by distillation or crystallization prior to analysis.

PROCEDURE

Weigh the sample containing 1.5 to 2.5 mM of reactive carbonyl into a 150-ml glass-stoppered flask. Add exactly 20 ml of 0.25*M* 2-dimethylaminoethanol, then add exactly 25 ml of 0.4*M* hydroxylammonium chloride. Stopper the flask, swirl gently to mix, and let stand the required length of time: 10 minutes at room temperature is sufficient for most aldehydes and simple aliphatic ketones. Check doubtful compounds, using a longer reaction time. Aryl ketones, hindered aliphatic compounds, and dicarbonyl compounds require an oximation period of 45 minutes or longer at 70°C. Add 5 drops of Martius yellow indicator and titrate with 0.2M perchloric acid. Take the change from yellow to colorless or blue-gray as the end point.

Determine the blank by titrating a similar mixture of 2-dimethylaminoethanol and hydroxylammonium chloride that has stood for the same period of time as the sample. Use the difference between the blank V_b and the sample titration V_s to calculate the percentage of the carbonyl compound in the sample.

$$\% \text{Carbonyl compound} = \frac{(V_b - V_s)(\text{molarity of } HClO_4)(\text{mol. wt.})}{10(\text{sample wt., grams})}$$

Quantitative results for the analysis of several aldehydes and ketones are given in Table 1. Most of the results obtained are in the range 98 to

Table 1. Determination of Aldehydes and Ketones

Compound	Reaction Time, min.	Reaction Temperature, °C	No. of Determinations	Found, %[b]
2-Acetonaphthone	120	Room	2	97.9 ± 0.2
	60	70	3	99.4 ± 0.1
	90	70	3	100.3 ± 0.3
Acetophenone	30	70	2	95.8 ± 0.0
	45	70	5	99.9 ± 0.1
	60	70	2	99.9 ± 0.1
Benzaldehyde	20[a]	Room	2	99.6 ± 0.1
Benzoin	120	70	2	98.6 ± 0.1
	180	70	3	99.3 ± 0.4
n-Butyraldehyde	20[a]	Room	2	98.5 ± 0.1
Cyclohexanone	30[a]	Room	2	98.5 ± 0.1
Cyclopentanone	30[a]	Room	2	98.4 ± 0.1
Dibenzylketone	30[a]	Room	3	95.8 ± 0.2[c]
Furfural	20[a]	Room	2	99.4 ± 0.2
p-Hydroxybenzaldehyde	5 to 15	Room	6	99.7 ± 0.2
Methyl isobutyl ketone	30[a]	Room	2	98.2 ± 0.1[c]
p-Nitrobenzaldehyde	20[a]	Room	2	99.4 ± 0.0
Salicylaldehyde	5 to 15	Room	4	99.6 ± 0.3
Tridecanone	5 to 15	Room	5	99.9 ± 0.2
Vanillin	5 to 15	Room	7	99.5 ± 0.2

[a] This reaction time is probably excessive.
[b] Precision index is average deviation from mean.
[c] Technical grade sample used.

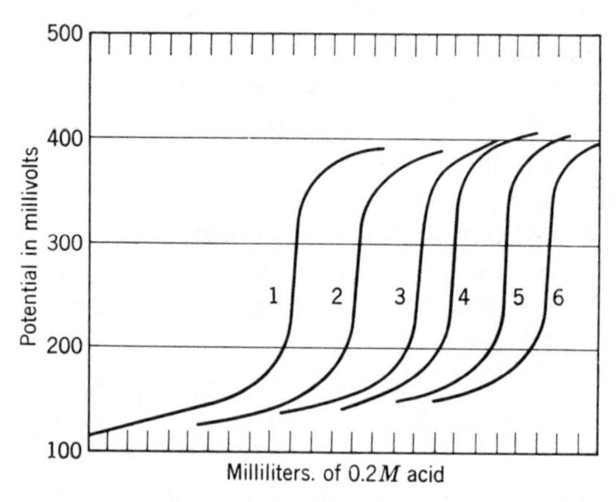

Fig. 2.1. Potentiometric curves for sample titrations: 1, cyclopentanone; 2, cyclohexanone; 3, methyl isobutyl ketone; 4, *n*-butyraldehyde; 5, *p*-nitrobenzaldehyde; 6, vanillin.

100%, which is the estimated purity of the samples taken. As a check, a vanillin sample was found to be 99.5 ±0.2% pure by oximation and 99.5 ±0.1% by acid-base titration. Although a 20- or 30-minute reaction time was used for many of these compounds, it was later found that many simple aldehydes and ketones react completely in 10 minutes or less.

Representative titration curves are plotted in Fig. 2.1. One striking feature of these curves is the sharpness of the break at the equivalence point. Also, the equivalence point potential is virtually the same for the various carbonyl compounds.

Unless the acid titrant is free of carbonyl impurities, there is a possibility of error, especially if the titration is performed slowly.

Better results are obtained if the solvents used are essentially free from water. Figure 2.2, for example, compares a blank titration with approximately 8% water present with the titration curve in anhydrous solution.

The reagent used shows little or no decomposition during oximation, even at 70°C.

Oximation does not distinguish aldehydes from ketones, since both react quantitatively. Also, in samples where the free carbonyl compound is in the presence of acetals, ketals, or vinyl ethers, oximation cannot be used because the hydroxylamine salts also react with these materials; in fact, oximation can also be used to determine acetals, ketals, and vinyl ethers (see pp. 510–11).

Bisulfite Addition

Bisulfite addition is another reaction of carbonyl groups that is used for determining these groups.

$$\mathrm{R\overset{\overset{\displaystyle O}{\|}}{C}R_1 + NaHSO_3 \rightleftharpoons R{-}\underset{\underset{\displaystyle OH}{|}}{\overset{\overset{\displaystyle SO_3Na}{|}}{C}}{-}R_1}$$

In this reaction R_1 can be a hydrogen atom in the case of aldehydes. As in the case of oxime formation, the bisulfite addition reaction is a distinct equilibrium. The equilibrium is more favorable for aldehydes than it is for ketones, hence aldehydes are more generally determinable with this reaction than are ketones; in fact, very few ketones can be measured with this approach.

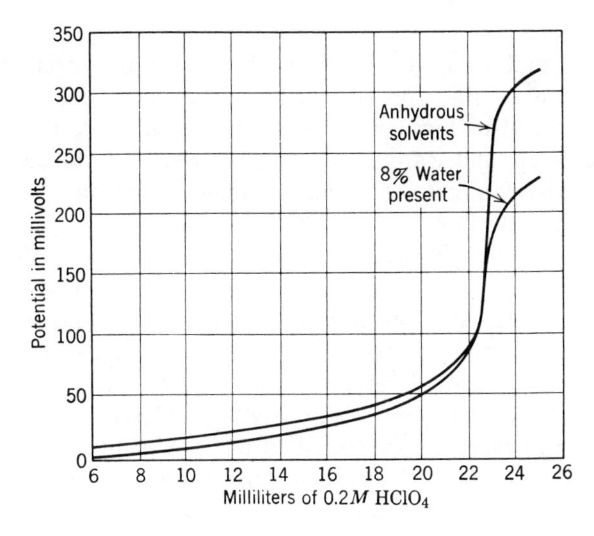

Fig. 2.2. Effect of water on titration curves for blank.

In the newer methods, sodium bisulfite is not used directly as the reagent for determining carbonyl groups because it is not sufficiently stable. The titer of its solution constantly drops with time. Instead, a mixture of sodium sulfite and sulfuric acid is used which generates sodium bisulfite. The sulfite and acid are mixed just prior to analysis to circumvent the instability of the generated bisulfite.

The end measurement is an acidimetric titration of the bisulfite system. An iodimetric titration was once used (7) but is now little used because the degradation of the bisulfite is due to oxidation; hence the instability of the reagent is more apparent with this end measurement. Also, iodine often significantly affects the equilibrium, since it reacts with the excess

7. A. Parkinson and E. Wagner, *Ind. Eng. Chem., Anal. Ed.*, **6,** 433–6 (1934).

bisulfite and causes the reaction to go more toward the free aldehyde and free bisulfite. The extent of the error depends on the equilibrium constant for the particular aldehydes, the concentrations of the various reaction-components, the amount of iodine added at any one time, and the time it is present in the system.

The evolution of the procedure of Siggia and Maxcy is described below. Seyewetz and Bardin (8) used sodium sulfite to determine aldehydes, but there were difficulties.

$$CH_3CHO + Na_2SO_3 + H_2O \rightleftharpoons CH_3C(H)(OH)(SO_3Na) + NaOH$$

There is an equilibrium involved in the foregoing reaction, and the reverse reaction is quite pronounced. In the determination of acetaldehyde, because of its low boiling point, much aldehyde is lost from the solution. Seyewetz and Bardin tried to overcome this difficulty by keeping the acetaldehyde content in their samples down to 7 to 8%. Also, they chilled the solutions to 4 to 5°C.

In attempts to use the Seyewetz and Bardin procedure, the disadvantage mentioned was very evident. The equilibrium caused trouble with the low-boiling aldehydes because of consistent loss of aldehyde and also caused trouble with the higher-boiling aldehydes because of the insolubility of the free aldehyde. The procedure can be used for determining formaldehyde (9) because of its relatively high solubility. A formaldehyde procedure of this type is also described by D'Alelio (10).

Romeo and D'Amico (11), realizing the equilibrium difficulties in the previous procedures, used sodium sulfite–potassium bisulfite mixtures to determine cinnamaldehyde and benzaldehyde, mentioning good results for cinnamaldehyde but poor results for benzaldehyde. They tried the method on some ketones, but results were generally poor.

The function of the bisulfite is to shift the equilibrium that occurs when sulfite alone is used. Because of the instability of bisulfite solutions, however, it was found advisable in our work to use sulfuric acid instead. An aliquot of standard sulfuric acid is added to a large excess of sodium sulfite solution to produce sodium bisulfite *in situ*. The acid is very stable and can stand for long periods of time without the titer changing

8. Seyewetz and Bardin, *Bull. Soc. Chim. Fr.* (3), **33,** 1000–2 (1905).
9. Seyewetz and Gibello, *Bull. Soc. Chim. Fr.* (3), **31,** 691 (1904).
10. G. F. D'Alelio, *Experimental Plastics and Synthetic Resins* Wiley, New York, 1946, p. 164.
11. G. Romeo and E. D'Amico, *Ann. Chim. Appl.*, **15,** 320–30 (1925).

significantly. The acid is not added to the sulfite until just before the aldehyde sample is introduced. The aldehyde reacts with the bisulfite, and the excess bisulfite is titrated with standard alkali. (The reaction might also be viewed as the aldehyde reacting with sulfite to liberate sodium hydroxide and the acid present consuming the sodium hydroxide, forcing the reaction to completion.) The large excess of sulfite is to keep the reaction essentially at completion as the excess bisulfite is titrated with alkali. In this system, the reaction is so near completion that no aldehyde can be detected above the solution, even in the case of low-boiling aldehydes such as acetaldehyde. This reagent also dissolves many insoluble, high-boiling aldehydes.

To detect the end point more accurately, it was found advisable to use a pH meter in the titration. A curve is plotted of pH versus milliliters of standard alkali added, and the end point is determined from the curve. As a shortcut, it can be noted that the end point in the case of each aldehyde comes at a rather definite pH (Table 2). (There is only slight

Table 2

Aldehyde	pH Range of Solution at Endpoint	Observed Value (moles)	Moles Sample Taken
Acetaldehyde	9.05–9.15	0.02455	0.02458
		0.02120	0.02118
		0.02858 (pH 9.1)[a]	0.02862
Propionaldehyde	9.30–9.50	0.02085	0.02090
		0.02037	0.02053
		0.01792 (pH 9.4)[a]	0.01811
Butyraldehyde	9.40–9.50	0.0249	0.0243
		0.0233	0.0236
		0.0275 (pH 9.45)[a]	0.0280
Cinnamaldehyde	9.50–9.60	0.0105	0.0106
(Takes on 2 moles bisulfite		0.0108	0.0110
per mole sample)		0.0120 (pH 9.55)[a]	0.0122
Crotonaldehyde	9.20–9.40	0.0250	0.0253
(Takes on 2 moles bisulfite		0.0189	0.0190
per mole sample)		0.0235 (pH 9.30)[a]	0.0238
Benzaldehyde	8.85–9.05	0.0163	0.0164
		0.0168	0.0173
		0.0147 (pH 8.95)[a]	0.0154

[a] Rapid method used.

Note. All samples were distilled in a Podbielniak fractionating column (manufacturer claims 100 plates) and analyzed less than 2 hours after distillation to minimize air and autoxidation of sample.

deviation with the size of sample.) Once this pH is known, each sample can be titrated to that pH, thus eliminating the necessity of plotting a curve for each sample.

The advantages of this procedure are that it is rather generally applicable to determining aldehydes and that it has overcome the difficulties of equilibrium, such as incomplete reaction, loss of sample, and poor titration. This procedure has also overcome the end point difficulties encountered in the previous procedures and has eliminated the necessity of using unstable standard solutions.

Adapted from the Method of Siggia and Maxcy

[*Ind. Eng. Chem., Anal. Ed.*, **19,** *1023 (1947)*]

REAGENTS

1*M* Sodium sulfite.
Standard 1*N* sodium hydroxide.
Standard 1*N* sulfuric acid.

PROCEDURE

To 250 ml of 1*M* sodium sulfite in a 500-ml glass-stoppered Erlenmeyer flask, add 50 ml of 1*N* sulfuric acid. As the acid is added, swirl the flask to prevent the loss of sulfur dioxide caused by localized overneutralization of the sodium sulfite. To this solution add a sample, sealed in a glass ampoule (see pp. 862–864), containing 0.02 to 0.04 mole aldehyde. Stopper the flask; for low-boiling aldehydes the stopper must be greased to prevent any loss. Then vigorously shake the flask to break the ampoule containing the sample. Some glass beads included in the flask with the ampoule will cause the ampoule to break more easily. Shake the flask for 2 to 3 minutes (5 minutes for the more insoluble aldehydes) to ensure complete reaction. Then transfer the contents quantitatively to a beaker. Insert electrodes from a pH meter in the solution, and stir the solution. Note the pH of the solution as standard 1*N* alkali is added to titrate the excess acid.

For accurate results, note and plot the pH reading versus milliliters of alkali added. The end point is determined from the plot. A more rapid method of determining the end point, though slightly less precise, is to add alkali until a pH is obtained corresponding to the pH at the end point for the particular aldehyde. This end point pH can either be taken from Table 2 or predetermined in cases of aldehydes not mentioned.

Sodium sulfite contains a small amount of free alkali as an impurity so that 250 ml of the solution consumes some acid. The blank is very small but not negligible; it amounts to about 0.4 to 0.5 ml of 1*N* acid per 250 ml of sulfite solution. On each carboy of sodium sulfite solution prepared, the free alkali should be accounted for or the aldehyde results will be slightly high. Rather than use a blank on the sulfite solution, it was found more satisfactory to add enough 1*M* sodium bisulfite to the sodium sulfite to neutralize the free alkali and bring the pH of the sulfite to 9.1. This procedure eliminates the need of a blank and may be done only once to each carboy of solution.

CALCULATIONS

A = Calculated amount of NaOH standard solution needed to titrate the 50 ml of standard acid used minus the milliliters of standard NaOH used to titrate the sample

$$\frac{A \times N\ \text{NaOH} \times \text{mol. wt. of compound} \times 100}{\text{Grams of sample} \times 1000} = \%\ \text{compound}$$

In the case of most of the aldehydes used earlier, the end point was sharp enough that the rapid method of just titrating to the pH of the end point could cause an error in end point determination of only ±0.2 to 0.3 ml. This error is not very significant and can be nullified by using a rather large sample of about 0.04 mole of aldehyde, which consumes 40 ml of 1*N* acid. It can readily be seen that a plot of pH versus milliliters of sodium hydroxide is not necessary once the pH at the end point of the particular aldehyde has been determined. The reproducibility of the procedure is ±0.2% if the entire curve is plotted and ±0.4% if the rapid method is used.

The end point in the determination of benzaldehyde is relatively poor (see Fig. 2.3), and the rapid method could cause an error of ±0.4 to 0.5 ml. Here a plot is necessary to determine the end point accurately.

Ketones cannot, on the whole, be determined by this method, even though they do form bisulfite addition products. Acetone, methyl ethyl ketone, quinone, naphthoquinone, and cyclohexanone were tried. Cyclohexanone is the only ketone tried that might be determined by this method. The titration curves for ketones indicate that the ketone does react with bisulfite, but there is no discernible end point (see Fig. 2.3). The pH climbs steadily on addition of the sodium hydroxide. In the

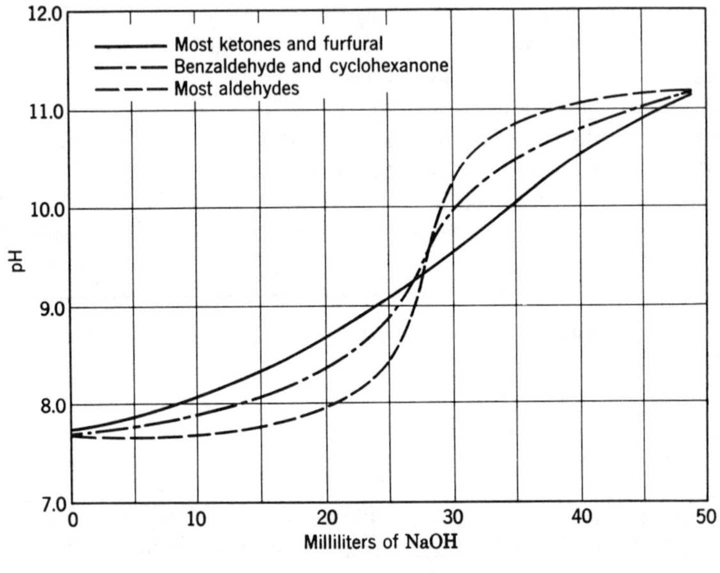

Fig. 2.3

determination of cyclohexanone, a poor end point is obtained, comparable to that of benzaldehyde. From the titration curves in the determination of ketones, it seems that the equilibrium between ketone and ketone-bisulfite addition product leans toward free ketone and sodium bisulfite to a great extent. As sodium hydroxide is added, the pH rises gradually, but no break occurs in the curves. When excess sodium bisulfite is consumed by the sodium hydroxide, the equilibrium is upset and some ketone-bisulfite addition product is consumed, together with the excess sodium bisulfite, and no end points are obtained. This explanation can be applied in the determination of furfural, which behaves like the ketones.

The poor end point in the determination of benzaldehyde can be attributed to the same cause, but in this instance the equilibrium is enough in favor of the addition product to make this aldehyde determinable by this method.

Ketones, in general, will not interfere in the determination of aldehydes if they are not present in excess of about 10 mole %. The aldehyde produces a break in the pH versus milliliters of sodium hydroxide curve, whereas the ketone does not. If ketones are present, the entire curve of pH versus milliliters of sodium hydroxide has to be plotted and the break in the pH curve determined to obtain the end point. As the Fig. 2.3 indicates, the presence of ketones can affect the pH at the end point, therefore can cause erroneous results if the rapid method of titrating to a definite pH value is used.

Acidic or basic impurities in the sample should be determined separately before the aldehyde procedure is applied, and the titration for aldehyde should be corrected for the presence of these impurities.

Acetals will not interfere in the procedure. These compounds hydrolyze in strong acid solution to yield acetaldehyde. However the pH of the sodium sulfite–sulfuric acid solution, in proportions described in the procedure, is about 6.8, and there is no noticeable hydrolysis of the acetals at this pH. Also, the aldehyde in the sample consumes the bisulfite so rapidly that the pH of the solution is raised to about 7.5 as soon as the sample comes in contact with the sodium sulfite–sulfuric acid solution, further lessening any possibility of hydrolysis.

Hydrazone Formation*

Hydrazines react quantitatively with carbonyl compounds to form the corresponding hydrazones,

$$RC(=O)R_1 + R_2NHNH_2 \rightarrow \begin{matrix} R \\ C{=}NNHR_2 \\ R_1 \end{matrix} + H_2O$$

where R and/or R_1 can be hydrogen atoms.

The use of hydrazines for the determination of aldehydes has been restricted by one or more factors, the most common of which are ease of oxidation of the hydrazines, use of aqueous reagents limiting analysis to water-soluble samples, and use of hydrazonium salts, which are acidic and will react with acetals as well as aldehydes. Employing dimethylhydrazine makes it possible to overcome these difficulties

Kleber (12) tried to determine some carbonyl compounds by addition of excess of phenylhydrazine and acidimetric determination of the excess. Phenylhydrazine is a poor reagent, owing to its ease of oxidation by atmospheric oxygen. Blanks can be run to account for the phenylhydrazine lost through oxidation during the reaction, but these blanks are too large for a satisfactory quantitative method.

12. C. Kleber, *Am. Perfumer Essent. Oil Rev.*, **6,** 284 (1912).

* Portions of this section have been reprinted with permission from the paper of S. Siggia and C. R. Stahl, Anal. Chem., **27,** 1975 (1955).

Ardagh and Williams (13) used phenylhydrazine for determining carbonyl compounds. Excess phenylhydrazine was added to a sample and the excess was determined iodometrically. Because of the instability of the reagents toward atmospheric and dissolved oxygen and also partial reaction of the hydrazones, the method is unwieldy. It requires oxygen-free conditions and an extraction to remove the hydrazone before the excess hydrazine can be determined. The work of Ardagh and Williams was an extension of the work of Von Meyer (14) and has been converted to a microprocedure (15).

Several procedures for carbonyl compounds use excess phenylhydrazine and decompose the excess hydrazine with Fehling's solution to liberate nitrogen, which is collected and measured (16–21). These procedures are used only for quantitative estimations and cannot be applied for precise and accurate measurement of most carbonyl compounds.

Hydrazine, as such, cannot conveniently be used, but hydrazine sulfate has been used to determine certain aldehydes (22–24). The acid liberated in the reaction with the carbonyl compound is measured. The reaction times are generally long and cannot be applied to acetal-containing samples, owing to the acidity of the reagent.

2,4-Dinitrophenylhydrazine has been used for volumetric methods. Clift and Cook (25) dissolved the hydrazone in excess standard caustic and back-titrated the excess. Espil and associates (26, 27) used titanous chloride to reduce the nitrohydrazones. The difficulties in these methods lie in the quantitative separation of the hydrazone and drawbacks in the use of titanous chloride. Schöeniger and Lieb (28, 29) determined the excess 2,4-dinitrophenylhydrazine with titanous chloride.

13. E. G. R. Ardagh and J. G. Williams, *J. Am. Chem. Soc.*, **47,** 2983–8 (1925).
14. E. Von Meyer, *J. Prakt. Chem.*, **36,** 115–19 (1887).
15. H. Lieb, W. Schöeniger, and E. Schivizhoffen, *Mikrochem. Mikrochim. Acta*, **35,** 407–11 (1950).
16. E. Fischer, *Ann.*, **190,** 101–8 (1878).
17. G. M. Ellis, *J. Chem. Soc.*, **130,** 848–51 (1927).
18. I. S. MacLean, *Biochem. J.*, **7,** 611–15 (1913).
19. L. Marks and R. S. Morrell, *Analyst*, **56,** 508–14 (1931).
20. H. Stracke, *Monatsh.*, **12,** 514 (1891).
21. *Ibid.*, **13,** 299 (1892).
22. L. Fuchs, *Sci. Pharm.*, **16,** 50–6 (1948).
23. L. Fuchs and O. Matzke, *Ibid.*, **17,** 1–11 (1949).
24. L. Monti and M. T. Masserizzi, *Ann. Chim. Appl.*, **37,** 101–5 (1947).
25. F. P. Clift and R. P. Cook, *Biochem. J.*, **26,** 1800–3 (1932).
26. L. Espil and I. Gévevois, *Bull. Soc. Chim. Fr.*, **5,** 17–18, 1532–5 (1938).
27. L. Espil and G. Mondillon, *Cot. R. Soc. Biol.*, **129,** 1187–8 (1938).
28. W. Schöeniger and H. Lieb, *Mikrochem. Mikrochim. Acta*, **38,** 165–7 (1951).
29. W. Schöeniger and H. Lieb, *Z. Anal. Chem.*, **134,** 188–91 (1951).

p-Nitrophenylhydrazine has been used (30) by isolating the derivative, reducing with stannous ion, and determining the excess stannous ion iodometrically. The procedure is long and suffers from the difficulty of complete isolation of the derivative.

2,4-Dinitrophenylhydrazine has been tried from a gravimetric standpoint—namely, removal of the hydrazone derivative and weighing it (31–37). The gravimetric methods give slightly low recovery and are limited generally to aqueous or partially aqueous media because of the lower solubility of the hydrazones in water.

2,4-Dinitrophenylhydrazine has been used also for colorimetric determination of carbonyl compounds (38–43).

Of the many hydrazine approaches used, the application of unsymmetrical dimethylhydrazine and 2,4-dinitrophenylhydrazine has proved the most successful. Neither reagent is significantly affected by oxidation.

UNSYMMETRICAL DIMETHYLHYDRAZINE METHOD

Unsymmetrical dimethylhydrazine gave a reagent that was stable for several weeks when stored in a dark bottle. The reaction with aldehydes is rapid and complete. This hydrazine also reacts with ketones, as is indicated by the evolution of heat and color development. Evidently, however, the basic strength of the hydrazones of the ketones is not too much different from the basic strength of the hydrazine itself. Therefore, on back-titration of the hydrazine, the excess hydrazine cannot be differentiated from the hydrazone formed, and a value for base is obtained that coincides with the amount of hydrazine initially added to the sample. This behavior makes possible the determination of aromatic (but not aliphatic) aldehydes in the presence of ketones (Table 3).

30. G. Petit, *Bull. Soc. Chim. Fr.*, **15,** 141–2 (1948).
31. H. Collatz and I. S. Neuberg, *Biochem. Z.*, **255,** 27–37 (1932).
32. B. J. Feinberg, *Am. Chem. J.*, **49,** 87–116 (1913).
33. O. Fernandez and L. Sócias, *Rev. Acad. Cienc. Exact. Fís. Nat. Madrid*, **28,** 330–3 (1932).
34. H. A. Iddles and C. E. Jackson, *Ind. Eng. Chem., Anal. Ed.*, **6,** 454–6 (1934).
35. H. A. Iddles, A. W. Low, B. B. Rosen, and R. T. Hart, *Ibid.*, **11,** 102–3 (1939).
36. E. M. Plein and C. F. Poe, *Ind. Eng. Chem., Anal. Ed.*, **10,** 78–80 (1938).
37. U.S. Pharmacopoeia XI, p. 353, 1936.
38. L. Barta, *Biochem. Z.*, **274,** 212–19 (1934).
39. H. K. Barrenscheen and M. Dreguss, *Ibid.*, **223,** 305–10 (1931).
40. G. R. Lappin and L. C. Clark, *Anal. Chem.*, **23,** 541–2 (1951).
41. W. C. Mathewson, *J. Am. Chem. Soc.*, **42,** 1277–9 (1920).
42. G. Matthiessen and H. Dahn, *Z. Ges. Exp. Med.*, **113,** 336–40 (1944).
43. M. F. Pool and A. A. Klose, *J. Am. Oil Chem. Soc.*, **28,** 215–18 (1951).

Table 3. Results Obtained in Analysis of Various Aldehydes

	% by Dimethyl-hydrazine	Reaction Time, hours	% by $NH_2OH \cdot HCl$[a]
Formaldehyde (aqueous solution)	36.6 ± 0.1 (4)[b]	0.25	36.6
Acetaldehyde	97.3 ± 0.8 (3)[b]	0.25	97.3
Propionaldehyde	96.6 ± 0.5 (3)[b]	0.25	96.5
Butyraldehyde	94.5 ± 0.5 (3)[b]	0.25	95.1
Cinnamaldehyde	96.6 ± 0.2 (3)[b]	0.50	97.1
Furfural	97.6 ± 0.2 (3)[b]	2.0	98.2
Crotonaldehyde	93.0 ± 0.2 (4)[b]	0.5	93.2[c]
Salicylaldehyde[d]	99.6 ± 0.3 (10)[e]	0.5	99.6 ± 0.4 (10)
Anisaldehyde	95.7 ± 0.4 (3)[e]	2.0	95.9
Benzaldehyde	92.7 ± 0.3 (3)[e]	2.0	94.7
m-Nitrobenzaldehyde	98.7 ± 0.1 (3)[e]	2.0	98.7
2,6-Dichlorobenzaldehyde	98.3 ± 0.5 (3)[f]	2.0	98.8
2,4-Dimethoxybenzaldehyde	99.1 ± 0.3 (3)[f]	2.0	97.7

[a] $NH_2OH \cdot HCl$ used was $0.5N$ in 1 to 1 methanol-water. Samples were refluxed 1 hour and then titrated potentiometrically with $0.5N$ NaOH.

[b] $0.2M$ DMH in ethylene glycol.

[c] Determined by bisulfite.

[d] Since salicylaldehyde is stable, it was possible to use it as a pure standard. Sample used was made from the bisulfite addition compound (Eastman Kodak), distilled once in laboratory before use.

[e] $1M$ DMH in ethylene glycol.

[f] $1M$ DMH in MeOH.

Figures in parentheses indicate number of determinations used. Precision indicated is as average deviation arrived at from number of determinations made.

Aromatic aldehydes yield hydrazones that exhibit no detectable basicity. Therefore on titration of the excess hydrazine, a sharp break in the titration curve is obtained when all the excess hydrazine is neutralized (Fig. 2.4). The aliphatic aldehyde hydrazones are noticeably basic, however; on back-titration of the excess hydrazine, a more gradual break in the titration curve is obtained (Fig. 2.4), and sometimes a break for the hydrazone is seen if the titration is carried further. When ketones are present, the hydrazone of the ketone buffers the solution very noticeably. The hydrazone of the aromatic aldehyde is so weakly basic that the break for the back-titration of the hydrazine is quite distinct. Thus although the break in the titration curve is subdued by the buffering action of the hydrazone of the ketone, the break for excess hydrazine is still visible. In

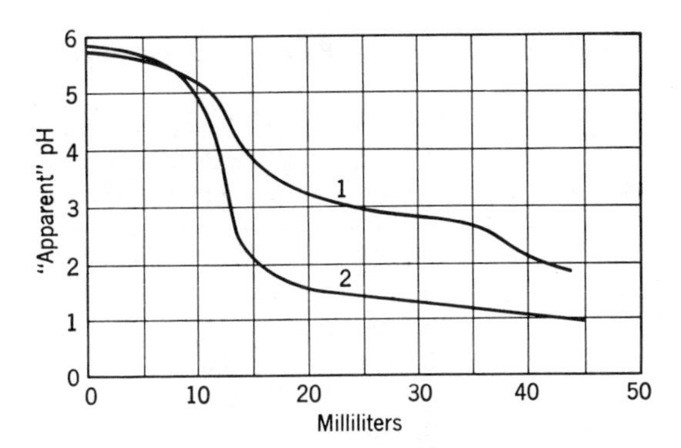

Fig. 2.4. Titration curves: 1, formaldehyde; 2, salicylaldehyde.

the case of the aliphatic aldehyde, however, the hydrazone of the ketone obliterates the already weak break.

Carboxylic acids, in general, do not interfere, since they are relatively weak acids compared with the mineral acid used in the titration (Table 4.) If a strong acid is present in a sample, a separate determination of this acid can be made, and a correction applied; or the sample can first be neutralized using the hydrazine reagent before the hydrazine for the aldehyde reaction is added. There is no evidence of hydrazide formation with carboxylic acids even after 4 hours of reaction time, which is far beyond the time intervals used in this study.

Acetals and ketals do not interfere in this procedure, since the alkalinity of the reagent prevents hydrolysis to the carbonyl compounds (Table 4). This is one advantage of this approach over hydroxylamine hydrochloride methods; another is ability to determine some aldehydes in the presence of ketones. The shape of the titration curves of all the aromatic aldehydes resembles those of salicylaldehyde and not those of the aliphatic aldehydes. The sharpness of the breaks makes possible the analysis with ketone present, since it can stand the buffering effect.

The reaction between the dimethylhydrazine and aldehydes was found to proceed the best in polar solvents. Water can be used, but a nonaqueous solvent was desired because many organic samples may not be water soluble and the titration breaks in the case of the aliphatic aldehydes were poor in water. Alcohols had the drawback of undergoing acetal formation with the aldehydes, giving low results. Ethylene glycol was found to operate satisfactorily both as solvent and as reaction medium and did not undergo any noticeable acetal formation. Alcohol was used in the determination of a few aldehydes (Table 3), since these aldehydes were not very soluble in the glycol and had carbonyl groups that were hindered

Table 4. Effect of Acids, Acetals, and Ketones on Aldehyde Determination

Mixture Analyzed	Aldehyde Added, grams	Acid Added, gram	Acetal Added, gram	Ketone Added, grams	Aldehyde Found Grams	Aldehyde Found %
Formaldehyde[a]	0.1660	—	—	—	0.0609	36.7
Formaldehyde[a]-formic acid	0.1684	0.0211	—	—	0.0611	36.3
	0.1758	0.0624	—	—	0.0643	36.6
Formaldehyde[a]-methylal	0.1830	—	0.0549	—	0.0670	36.6
	0.1631	—	0.1512	—	0.0595	36.5
Acetaldehyde	0.1062	—	—	—	0.1031	97.1
Acetaldehyde-acetic acid	0.1062	0.1283	—	—	0.1036	97.6
Acetaldehyde-dimethyl acetal	0.1062	—	0.1213	—	0.1036	97.6
	0.1062	—	0.0342	—	0.1040	98.0
Salicylaldehyde	1.3053	—	—	—	1.2962	99.3
Salicylaldehyde-salicylic acid	1.1999	0.1380	—	—	1.2008	100.1
Salicylaldehyde-cyclohexanone	1.2249	—	—	0.9580	1.2233	99.9
Salicylaldehyde-benzophenone	1.2046	—	—	1.5077	1.2008	99.7

[a] Baker's formaldehyde solution, 37%.

enough to undergo acetal formation slowly enough to prevent interference. Tetrahydrofuran and chlorobenzene were tried as solvents, but the reaction proceeded much too slowly. *tert*-Butyl alcohol was tried because this is still a polar solvent but does not undergo acetal formation very readily; the reaction was somewhat slow, and poor precision was obtained. Pyridine was also tried; the reaction proceeded rapidly, as indicated by the heat evolved and color formation, but the hydrazone and the excess hydrazine titrated together, making analysis impossible. Table 3 summarizes the results obtained with this method.

REAGENTS

Unsymmetrical dimethylhydrazine (Westvaco Chemical Division, Food Machinery and Chemical Corp., New York), 0.2*M* solution in ethylene glycol for aliphatic aldehydes, 1*M* solution in ethylene glycol for aromatic aldehydes, and 1*M* solution in methanol for disubstituted benzaldehydes. Standard 0.1 and 0.5*N* solutions of hydrochloric acid in methanol.

PROCEDURE

Pipet 25 ml of reagent into a glass-stoppered flask, and add the weighed sample (about 0.002 mole for aliphatic and 0.01 mole for aromatic aldehydes) to the reagent. The reaction is allowed to proceed at room temperature for 15 minutes or longer, depending on the aldehyde being determined (Table 3). When the reaction is complete, wash the solution into a 250-ml beaker with approximately 50 ml of methanol, and titrate the excess dimethylhydrazine potentiometrically with standard hydrochloric acid in methanol. Use hydrochloric acid, 0.1*N*, to titrate 0.2*M* hydrazine reagent, and 0.5*M* hydrochloric acid to titrate the 1*M* reagent. Determine a blank on 25 ml of reagent in the same manner. In all cases, use the glass-calomel electrode system.

Calculate percentage of aldehyde using the following equation:

$$\%\ \text{Aldehyde} = \frac{(\text{ml for blank}) - (\text{ml for sample}) \times N \text{ of HCl} \times \text{mol. wt.} \times 100}{\text{Weight of sample} \times 1000}$$

One molar reagent is necessary for determining aromatic aldehydes that react slowly with more dilute reagent, but it should not be used for aliphatic aldehydes, since a rise in temperature resulting from rapid reaction may cause loss of reagent. To keep the reaction time at a minimum, use 100% excess of reagent in all determinations.

The solubility of disubstituted benzaldehydes in ethylene glycol is not great enough to allow their determination in this solvent, but, since they do not form acetals under the conditions used, it is possible to determine them in methanol without loss of aldehyde. Most other aldehydes give low results in methanol because of acetal formation.

2,4-DINITROPHENYLHYDRAZINE METHOD

This procedure employs the standard identification reaction for carbonyl compounds. This reaction is essentially quantitative for many aldehydes and ketones, making the procedure quite generally applicable.

$$(NO_2)_2C_6H_3NHNH_2 + \begin{matrix} R \\ \diagdown \\ C{=}O \\ \diagup \\ R_1 \end{matrix} \rightarrow (NO_2)_2C_6H_3NHN{=}C\begin{matrix} R_1 \\ \\ R \end{matrix} + H_2O$$

where R and R_1 could be a hydrogen atom.

The reaction is also very specific; interferences consist mainly of materials that will oxidize the hydrazine to form tars, which are weighed with the hydrazones. This procedure is also applicable to the determination of acetals, ketals, and vinyl ethers. Since the hydrazine is in acid solution, it will hydrolyze the above-mentioned compounds to the corresponding aldehyde or ketone, which will then react with the reagent.

The procedure described below is designed for water-soluble samples only. Since such a small amount of carbonyl compound is necessary (40×10^{-5} mole), however, we can usually dissolve enough sample in water to yield enough carboxyl compound for analysis.

Methoḋ Adapted from the Procedure of H. A. Iddles and C. E. Jackson

[*Ind. Eng. Chem., Anal. Ed.*, **6**, *454–6 (1934)*]

REAGENTS

A saturated solution at 0°C of 2,4-dinitrophenylhydrazine in 2*N* aqueous hydrochloric acid solution. This solution contains about 4 mg hydrazine per milliliter.

2*N* Hydrochloric acid solution.

PROCEDURE

In a glass-stoppered flask, place 50 ml of reagent. To this add the sample which should contain approximately 40×10^{-5} mole of aldehyde. Allow the mixture to stand in an ice bath for 1 hour. In the case of the volatile carbonyl compounds such as acetaldehyde or acetone, it is advisable to shake the flask vigorously from time to time to ensure the reaction of any carbonyl material that may be in the atmosphere above the reagent. After the period of standing, filter the precipitate off into a tared Gooch crucible or into a tared sintered-glass funnel. Wash the precipitate with 2*N* hydrochloric acid, then with water, and dry it in a vacuum desiccator over sulfuric acid. The precipitates can usually also be dried in an oven at 100°C.

CALCULATIONS

$$\frac{\text{Weight of hydrazone}}{\text{Weight of sample}} \times \text{gravimetric factor} \times 100 = \%\ \text{carbonyl compound}$$

Used to test this procedure were acetaldehyde, acetone, methyl ethyl ketone, methyl *n*-propyl ketone, benzaldehyde, *p*-hydroxybenzaldehyde, salicylaldehyde, anisaldehyde, and vanillin. These aldehydes mentioned previously, normally considered water insoluble, are actually soluble enough in water (only 40×10^{-5} mole is required by the method) to be analyzable by this method.

The authors of this book have used propionaldehyde, formaldehyde, crotonaldehyde, furfural, methyl vinyl ether, dimethyl acetal, ethyl vinyl ether, and diethyl acetal in addition to some of the carbonyl compounds used by the originators of the procedure. The procedure was found to be reproducible to ±1% and accurate to ±1% in the case of the carbonyl compounds whose hydrazones are not very soluble at all. Slightly low results (−2 to −3%) were obtained for some aldehydes (acetaldehyde and propionaldehyde), because of the solubility of the hydrazone in water.

Oxidation Methods

OXIDATION WITH SILVER ION

The oxidation methods using silver ion are applicable to the determination of aldehydes only. Ketones do not undergo this reaction. Also, vinyl ethers and acetals generally do not hydrolyze under the condition of this approach, thus the aldehydes involved in these cases are not available to the reagents.

$$RCHO + Ag_2O \rightarrow RCOOH + 2Ag$$

There are two general approaches to the silver oxidation methods. One involves the use of Tollen's-type reagent (ammoniacal or amine silver oxide solution) and argentimetric determination of the excess unreacted silver ion. The other approach involves reaction with solid silver oxide and titration of the carboxylic acid formed in the reaction. The Tollen's method is the faster method, since all the reactants are in solution.

Using Tollen's Reagent (Ammoniacal)

Tollen's reagent has long been known as a qualitative reagent for aldehydes. This section indicates how it can be employed to measure aldehydes quantitatively.

This method uses silver oxide as a mild oxidizing agent for aldehydes.

$$R\text{—}CHO + Ag_2O \rightarrow R\text{—}COOH + 2Ag$$

The sample of aldehyde is reacted with a freshly prepared standard solution of a modified Tollen's reagent. After approximately 10 minutes at room temperature the excess ionized silver is titrated potentiometrically with potassium iodide using a silver and a calomel electrode with a potassium nitrate bridge. A sharp break in the titration curve like the one in Fig. 2.5 is obtained. Standard hydrochloric acid can be used if the solution is first neutralized with nitric acid. The breaks obtained with hydrochloric acid are not as sharp as those when potassium iodide is used. Also, in the neutralization a glass electrode must first be used; it is later exchanged for the silver electrode for the final titration.

Acetals, vinyl ethers, and ketones (with the exception of cyclohexanone, which reacts very slowly), do not interfere, nor do they have any effect on the titration curve. Acids in the aldehydes also do not interfere. Compounds that would interfere are those that have active halide groups.

This procedure has been tried on a number of aldehydes with results listed in Table 5. Since the aldehydes obtained were not of known purity, check analyses were run by known methods. With benzaldehyde, longer reaction time gives lower results. This effect may be due to a Cannizzaro-type reaction or to the formation of the aldehyde-ammonia. Salicylaldehyde reacts very slowly. After 1 hour the aldehyde is only slightly oxidized. It is possible that a longer reaction time and higher temperatures may bring higher results. However the reagents are unstable, and

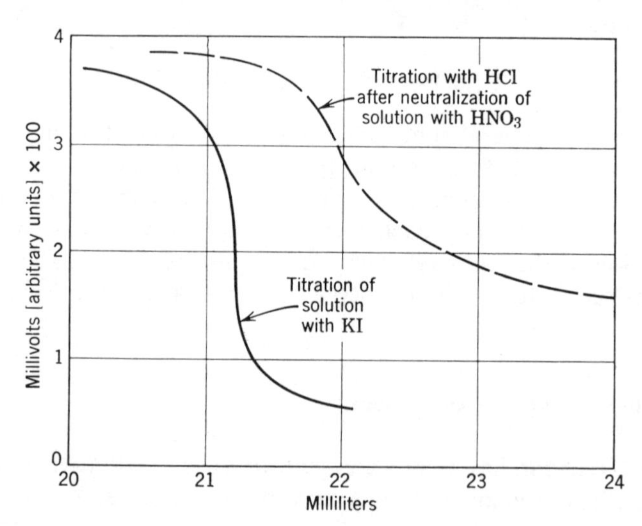

Fig. 2.5. Curves indicating the shape of the curve that does not vary in the systems but does vary if hydrochloric acid is used instead of potassium iodide. The use of hydrochloric acid and potassium iodide yields millivolt readings that are quite different. Also, different systems show some variance in millivolt readings.

Table 5

	Time min.	Found	Found—Other Method
Acetaldehyde	5	0.0749 g	0.0750 g[b]
		0.0740	
	15	0.0462	0.0451[b]
		0.0456	
		0.0463	
Butyraldehyde	20	91.4%[a]	96.5%[a,b]
	30	93.8	96.2
	67	95.5	97.5
	61	97.0	———
	72	96.5	96.7 av.
Formaldehyde	5	0.0501 g	0.0504 g[b]
	14	0.0499	
	35	0.0499	
Propionaldehyde	Approx. 10	0.0774 g	0.0771 g[c]
		0.0774	
Furfural	15	96.7%[a]	98.8%[a,d]
	30	98.2	
	45	98.5	
	60	98.2	
	60	97.9	
	35	98.2	
	45	97.0	
Crotonaldehyde	60	93.5%[a]	90.2%[a,d]
	60	91.7	
	60	93.2	
	120	93.0	

[a] These values are reported as percentages since the sample was weighed out undiluted. For the other aldehydes, a solution was made up, and aliquots were taken for each of the methods used.

[b] Bisulfite method of Siggia and Maxcy, *Ind. Eng. Chem., Anal. Ed.*, **19,** 1023 (1947).

[c] 2,4-Dinitrophenylhydrazine method of Iddles, Low, Rosen, and Harte, *ibid.*, **11,** 102 (1939).

[d] Hydroxylamine hydrochloride using the reagent in isopropyl alcohol and titrating liberated hydrochloric acid potentiometrically.

these trials should not be attempted without careful precautions (see Procedure).

Procedure of S. Siggia and E. Segal

[*Anal. Chem., 25, 640 (1953)*]

REAGENTS

6*N* Sodium hydroxide.
Standard 0.1*N* silver nitrate.
Concentrated ammonium hydroxide.
Standard 0.1*N* potassium iodide.

PROCEDURE

Make up the reagent for a single determination as follows. Pipet exactly 50 ml of 0.1*N* silver nitrate into a 250-ml glass-stoppered flask. Add 1 ml of 6*N* sodium hydroxide, and shake the mixture. Then add concentrated ammonium hydroxide dropwise to *just* dissolve the precipitate of silver oxide. Approximately 1 to 2 ml of concentrated ammonium hydroxide is necessary; shaking the mixture during the addition aids in dissolving the precipitate. Too large an excess of ammonium hydroxide makes the reagent less sensitive. Add a sample of aldehyde (0.0015–0.0020 mole) and shake the mixture for a few minutes. The formation of a silver mirror or brown turbidity is a sign of a positive reaction. Allow the solution to stand for the time listed in Table 5 for the particular aldehyde being determined; the reaction time should be predetermined if the aldehyde being determined is not included in this table.

The solution is then transferred to a 250-ml beaker, and the excess silver ion is titrated potentiometrically with 0.1*N* potassium iodide. A silver electrode and a calomel electrode with potassium nitrate bridge are used on the standard pH meter. (Model G Beckman pH meter was used in this work.)

The reagent is prepared fresh for each use. This is because the Tollen's reagent as prepared in this method has been known to deteriorate after standing some hours, forming a black solid that detonates violently on handling, even when wet. The reagent can be kept safely for at least 4 hours and has been kept 24 hours without trouble, but it is best to prepare it fresh for each use and to discard any excess. The reagent is safe to handle as long as it remains clear, but if a black turbidity forms, the flask should be wrapped in a towel, the operator keeping his face behind a suitable shield, and the solution poured down the drain.

The silver mirrors formed on the flask by the reaction should not be allowed to lie around. After the reaction mixture is transferred to the beaker, the mirror on the flask should be dissolved with concentrated nitric acid and the solution discarded.

CALCULATIONS

$$\frac{(A - \text{Titration}) \times N\,\text{KI} \times \text{mol. wt.}}{2 \times \text{Weight of sample} \times 1000} \times 100 = \%\ \text{aldehyde}$$

where A is the milliliters of standard potassium iodide necessary to react with 50 ml of standard $0.1N$ silver nitrate.

The foregoing procedure has the disadvantage of being applicable mainly to water-soluble samples. Some water-insoluble samples can be handled by shaking the reaction mixture vigorously while the reaction is proceeding. Dioxane and methanol were tried as solvents in this procedure, but even in the ratio of 1 : 1 with water, the reaction rate was very significantly decreased. Dioxane also has the disadvantage of containing enough aldehyde to necessitate a large blank correction.

Crotonaldehyde can be determined as is indicated in Table 5, but acrolein and cinnamaldehyde both showed difficulty, although their functional structure is essentially the same as that of crotonaldehyde. The cinnamaldehyde gave low results, which could be explained by its insolubility in water; the phenyl grouping could also be the cause for a slower reaction rate. Acrolein gave irreproducible results, but the sample was exceedingly bad. Because of its lachrymatory characteristics it was decided not to distill it. From the rate of mirror formation, however, acrolein was seen to react quite readily, and it may be determinable by this method. It may be, however, that some side reaction along with the oxidation may have been consuming aldehyde.

Using Silver-Amine Complex—Method of J. Mayes, E. Kuchar, and S. Siggia

[*Anal. Chem.*, ***36*****,** *934* (*1964*)]

The use of the silver-amine complex circumvents the difficulties of reagent hazards and sample solubility that exist with Tollen's reagent, which is generally applicable only to water-soluble aldehydes. The silver-amine reagent is stable even after half a month. Also, it can be used with organic solvents, and it provides the silver ion in a soluble form that is more advantageous as a reagent than the solid silver oxide.

Of the several amines tried, it was found that tertiary amines did not apparently complex the silver ion, since solubilization of silver oxide did not occur. Secondary amines did complex the silver, but the silver ion oxidized the amines to a degree that was undesirable. Primary amines were quite effective, especially in the C_1 to C_4 range. Of the several primary amines tried, *tert*-butylamine and isopropylamine performed best; *tert*-butylamine performed the better of the two.

Ethyl alcohol is used in this method as a solubilizing agent for water-insoluble aldehydes. The alcohol is very slowly oxidized by the reagent, but this does not become significant until the mixtures have stood for over 5 hours. The alcohol is not mixed with the reagent until the sample is introduced. Methyl alcohol was tried as a cosolvent, since it also was not too readily oxidized by the silver reagent.

The silver concentration of the reagent was found operable at $0.1N$ for many aldehydes, but $0.2N$ reagent was found to be a faster oxidant for the more stubborn compounds.

Table 6 summarizes the results obtained using this method with a range of aldehydes. It appears to be usable for determinining aldehydes in carboxylic acids (Table 7) and in ketones (Table 8). The method is being checked for determining aldehydes in acetals, which should work, since the reagent is alkaline and should not hydrolyze the acetals; however, the proof has not yet been obtained.

REAGENTS

Silver oxide, Fisher Scientific Co.

tert-Butylamine, $(CH_3)_3CNH_2$ Eastman White Label grade.

SILVER OXIDE REAGENT. To prepare 1 liter of approximately $0.2N$ reagent, add exactly 24.5 ±0.1 grams of silver oxide to a 1-liter reagent bottle. Insert a stirring bar and add 500 ml of deionized water. Add exactly 72.0 ±0.1 ml of *tert*-butylamine, stopper the bottle, and stir vigorously with a magnetic stirrer until all the silver oxide has dissolved (approximately 3 hours). The reagent prepared in this manner has an excess of silver oxide that can be removed by filtration. An excess of *tert*-butylamine should be avoided in the preparation of the reagent. After filtration, dilute the reagent with water to 1 liter for $0.2N$ or to 2 liters for $0.1N$ reagent. Use freshly prepared $0.2N$ reagent for aromatic aldehydes. The $0.1N$ reagent is adequate for aliphatic aldehydes and has been effective after having stood for 2 weeks.

Table 6. Determination of Aldehydes

Aldehyde	Time, min. Reagent 0.1*N*	Time, min. Reagent 0.2*N*	Found 0.1*N* Reagent Gram	Found 0.1*N* Reagent %	Found 0.2*N* Reagent Gram	Found 0.2*N* Reagent %	Found, Other Methods Gram	Found, Other Methods %
Formaldehyde	Filtered immediately	—	0.1228	36.1	—	—	1.7145	36.3
		—			—	—	1.7117	36.3 [a]
	5	—	0.1228	36.1	—	—	1.3895	36.3
	10	—	0.1228	36.1	—	—		36.2
	10	—	0.0994	36.0	—	—		36.2 [b]
	10	—	0.1301	35.9	—	—		36.3
	10	—	0.1406	36.2	—	—		36.2
	30	—	0.1228	36.1	—	—		
n-Propionaldehyde	30	—	0.1186	98.4	—	—	1.0550	98.5 [a]
	30	—	0.0997	95.4	—	—	0.9238	98.2
	30	—	0.1020	98.5	—	—		89.3
								87.6 [b]
								87.4
n-Butyraldehyde	10	—	0.1071	98.0	—	—	1.0642	97.5
	10	—	0.1101	96.7	—	—	1.1029	98.0 [a]
	10	—	0.1065	98.1	—	—	1.1680	97.9
	30	—	0.1071	98.0	—	—		94.7 [b]
	60	—	0.1071	98.1	—	—		92.1
n-Valeraldehyde	30	—	0.1389	96.0	—	—	1.2989	97.2
	45	—	0.1389	96.2	—	—	1.4002	97.4 [a]
	60	—	0.1453	95.6	—	—	1.4817	97.4
	135	—	0.1406	96.4	—	—		
Benzaldehyde	20	20	0.1476	89.4	0.1476	97.9	1.7076	99.3
	60	45	0.1476	95.6	0.1476	98.7	1.5472	99.2 [a]
	135	90	0.1476	95.6	0.1476	99.2	1.5959	99.4
	180	90	0.1476	97.1	0.1619	96.1		
	—	90	—	—	0.1314	99.0		
	60	130	0.0874	91.4	0.1048	93.4	1.2791	94.9 [a]
	70	—	0.0963	88.4	—	—	1.0788	94.5
	120	—	0.0874	95.3	—	—		
Crotonaldehyde	65	30	0.0907	88.0	0.0907	93.2	0.9991	98.6
	—	60	—	—	0.0907	97.9	1.0153	98.6 [a]
	—	90	—	—	0.0907	99.4	0.9659	98.7
	—	90	—	—	0.0976	100.0		
	—	120	—	—	0.1024	98.2		
p-Chlorobenzaldehyde	—	30	—	—	0.1561	97.5	2.1186	96.8
	—	60	—	—	0.1561	97.2	2.0159	97.4 [a]
	—	120	—	—	0.1561	97.3	3.5704	97.8

[a] Hydroxylamine hydrochloride, using the reagent in 2B alcohol and titrating the liberated hydrochloric acid.
[b] Bisulfite method of Siggia and Maxcy pp. 82–5.

Table 7. The Determination of *n*-Butyraldehyde in the Presence of Glacial Acetic Acid, Using 0.2*N* Reagent

n-Butyraldehyde, wt. % Added	Reaction Time	*n*-Butyraldehyde, wt. % Found
5.26	30 min.	5.23
23.7	45 min.	23.0
46.2	45 min.	46.0
69.0	45 min.	68.8
91.4	45 min.	91.8

Table 8. Determination of *n*-Butyraldehyde in the Presence of Methyl Ethyl Ketone, Using 0.1*N* Reagent

Methyl Ethyl Ketone, wt. %	*n*-Butyraldehyde, wt. %		
	Reaction Time	Found	Added
94.2	30 min.	5.67[a]	5.70
	128 min.	5.89	
71.8	30 min.	27.2	27.5
	1 hour	27.3	
50.1	30 min.	48.6	48.7
	110 min.	49.1	
29.7	30 min.	69.0	68.8
	95 min.	69.1	
6.0	30 min.	91.6	91.8
	87 min.	92.3	

[a] Sample was not diluted.

PROCEDURE

Add a sample containing 0.015 to 0.020 mole of aldehyde to a 100-ml volumetric and dilute to the mark with 2B ethanol. Add a 10-ml aliquot of this sample to a 250-ml glass-stoppered, Erlenmeyer flask containing 50 ml of the silver-amine reagent. Shake the mixture for a few minutes. The formation of a brown turbidity or a silver mirror is a sign of a positive reaction. Allow the solution to stand for the desired time for the particular aldehyde (see Table 6) with occasional shaking. Then filter the reaction mixture through a sintered-glass crucible or funnel with adequate washing of flask and funnel. Acidify the filtrate with concentrated nitric acid, 5 ml when the 0.1*N* silver reagent is used and 10 ml when the 0.2*N* reagent is used. To the mixture, add 2 ml of the ferric indicator, and titrate the excess silver ions with the standard potassium thiocyanate to the orange-brown end point. Run a blank on the reagent.

If the samples are expected to contain carboxylic acids in amounts greater than 10%, add 6*N* sodium hydroxide solution to the 0.2*N* silver-amine reagent (10-ml 6*N* sodium hydroxide per liter silver reagent). This should not be done routinely, since the sodium hydroxide retards the reaction; hence the reason for adding the sodium hydroxide only to the 0.2*N* reagent.

The silver mirrors can be removed with concentrated nitric acid.

CALCULATION

$$\frac{(A-B)\times N_{KCNS}\times \text{mol. wt.}\times 100}{\text{Weight of sample}\times 2000}=\%\ \text{aldehyde}$$

where

A = milliliters for blank titration

B = milliliters for sample titration

RESULTS AND DISCUSSION

The saturated aliphatic amines from C_1 to C_4 were found to form soluble complexes with silver oxide in aqueous solution. It was established that the availability of the silver oxide as an oxidant depended on the amine that formed the complex. The most effective primary amines were isopropylamine and *tert*-butylamine. The isopropylamine complex does not appear to be as strong an oxidant as the *tert*-butylamine complex; therefore it may find application in systems where the stronger oxidant might not be workable.

The 0.1N reagent is applicable to the aliphatic aldehydes; however 0.2N must be used to analyze the aromatic aldehydes because the oxidation proceeds rather slowly with the 0.1N reagent (Table 6).

The procedure was applied to a mixture of *n*-butyraldehyde in glacial acetic acid (Table 7) and in methyl ethyl ketone (Table 8). In samples containing large quantities of the acid, it is advisable either first to neutralize the acid in the sample or to add 10 ml of 6N sodium hydroxide solution per liter of 0.2N silver oxide reagent. The former approach is preferred.

It was necessary to use the 0.1N reagent and a sample size such that 50% of the reagent was consumed for *n*-butyraldehyde in the presence of methyl ethyl ketone. The data indicated that a more concentrated reagent oxidized the ketone slowly.

Addition of a 10-ml aliquot of an aldehyde-free acetal to a 50-ml aliquot of the 0.2N reagent shows the absence of measurable interference from the acetal.

Alcohols do not interfere. Interference can be expected from some halogenated organic compounds.

Aromatic aldehydes can be determined, as shown in Table 6. However some difficulty was experienced with anisaldehyde and cinnamaldehyde; low results were obtained.

Using Silver Oxide

Silver oxide can be used in the solid state to oxidize aldehydes to the corresponding acids. Since the reaction is a two-phase one, the analysis takes longer than the Tollen's-type methods shown previously.

Method of H. Siegel and F. T. Weiss

[*Reprinted in Part from Anal. Chem.*, **26,** *917–9* (*1954*)]

The procedure of Mitchell and Smith (44), involving the use of silver oxide and employing acidimetric determination, gives low recovery with formaldehyde and suffers from interference if acids or esters are present. The use of a column packed with silver oxide for the determination of aldehydes, as employed by Bailey and Knox (45) does not overcome the interference resulting from the hydrolysis of esters. In the experience of Siegel and Weiss, after suitable modification the argentimetric method published a number of years ago by Ponndorf (46) has provided a safe, rapid, and reliable method for the determination of aldehydes.

The Ponndorf method involves the reaction of aldehyde with silver oxide, formed *in situ* by addition of sodium hydroxide to the aqueous or alcoholic reaction mixture containing dilute silver nitrate. No side reactions—such as the Cannizzaro reaction or aldol condensation—are expected, since the reaction mixture is kept dilute with respect to the sample and is made strongly basic only in the last stages of oxidation. The unreduced silver ion is determined in the filtered reaction mixture, after acidification to redissolve silver oxide. A number of modifications have been introduced in Ponndorf's method to make it more rapid and convenient, including final titration with thiocyanate reagent according to Volhard.

APPARATUS AND MATERIALS

A shaking machine, of suitable construction to accommodate two or more 100 or 250-ml volumetric flasks.

A water bath, maintained at 60 ±2°C.

Ethyl alcohol, absolute. If appreciable carbonyl compounds or other reactive impurities are present, purify the solvent by distilling over excess solid silver oxide.

44. J. Mitchell Jr., and D. M. Smith, *Anal. Chem.*, **22,** 746 (1950).
45. H. C. Bailey and J. H. Knox, *J. Chem. Soc.*, **1951,** 2741.
46. W. Ponndorf, *Ber.*, **64,** 1913 (1931).

PROCEDURE

Pipet 25.0 ml of 0.1N silver nitrate solution into a 100-ml volumetric flask. Add a quantity of sample containing approximately 0.5 mM of aldehyde. If the sample is volatile or the carbonyl content is high, weigh the required amount in a glass ampoule. If the sample is not volatile from water or alcohol, the sample containing 5 mM of aldehyde may be dissolved in 100 ml of water or alcohol and a 10-ml aliquot of the solution may be taken for analysis. Add 5 ml of 0.5N sodium hydroxide solution and shake the mixture on a shaking machine for 15 minutes. At the end of this time, add 2 ml of 0.5N sodium hydroxide solution and continue the shaking for 10 minutes. Add 10 ml of 6N sodium hydroxide solution and repeat the shaking for the same period of time. Acidify the reaction mixture with 5 ml of 18N sulfuric acid solution. After allowing the mixture to cool to room temperature, dilute to the mark with distilled water. Filter the mixture through a dry Whatman No. 41 filter paper into a 400-ml beaker. Pipet 50.0 ml of the filtrate into a 500-ml glass-stoppered Erlenmeyer flask and add 4 ml of ferric alum indicator. Titrate with 0.05N thiocyanate solution until the end point is approached, as indicated by a more slowly fading red color. Stopper the flask, shake vigorously for 20 to 30 seconds, and continue the titration until 1 drop produces a reddish coloration that does not fade upon swirling or vigorous shaking. Carry out a blank determination by following the procedure as described but omitting the sample.

APPLICABILITY

The argentimetric method has been found to give an accuracy of ±2% with a number of pure aldehydes (see Table 9). High values were obtained with acrolein, presumably because of the partial oxidation of the olefinic bond. All of the ketones tested, with the exceptions of cyclohexanone and cyclopentanone, showed little reactivity under the conditions of the method (Table 10). Cyclohexanone reacts to the extent of 2% and cyclopentanone to the extent of 13%. Aldehydes in the presence of ketones have been determined in the range of 5 to 90 wt.% to within ±3%, but poor accuracy was obtained for the determination of a few tenths of 1% of aldehyde.

In examining the method for possible interfering substances, esters, carboxylic acids, and monohydroxyl alcohols were found to be without

47. American Society for Testing and Materials, D 1154–51T.
48. J. Mitchell Jr., D. M. Smith, and W. M. D. Bryant, *J. Am. Chem. Soc.*, **63,** 573 (1941).

Table 9. Determination of Aldehydes

Aldehyde Tested	Aldehyde Found, wt. % Hydroxylamine Method	Argentimetric Method	Recovery[a] %
Formaldehyde	27.4[b]	27.0	98.5
		27.0	98.4
Acetaldehyde	95.5[b]	94.2	98.5
		94.2	98.5
		94.2	98.5
Propionaldehyde	94.4[b]	94.2	99.8
		94.2	99.8
		93.6	99.1
n-Butyraldehyde	93.5[c]	95.2	101.8
		94.9	101.5
Acrolein	95.1[b]	99.9	104.9
		99.4	104.4
Benzaldehyde	99.3[c]	99.2	99.9
		97.7	98.4
Valeraldehyde	87.9[c]	86.2	98.9
		88.0	100.1

[a] Ratio of results by argentimetric method to hydroxylamine method.
[b] Aqueous hydroxylamine hydrochloride method (47).
[c] Fischer reagent-carbonyl method (48).

effect (Table 11). Polyhydroxyl compounds interfere, but this interference can generally be overcome by separating the aldehyde from the glycol with steam distillation, followed by determining the aldehyde content of the distillate. Acetals interfere to the extent of 0.04 to 2.2%, calculated as acetaldehyde. α-Epoxides were briefly tested, using propylene oxide as a representative compound, and no interference was observed.

MERCURIMETRIC OXIDATION

The oxidation of aldehydes using mercuric ion does not differ markedly from that involving silver ion. The advantages and limitations are about the

Table 10. Determination of Aldehydes in Presence of Ketones by Argentimetric Method

Ketone Present	Aldehyde Present	Ketone Added, wt. %	Aldehyde, wt. % Added	Aldehyde, wt. % Found
Acetone	—	100	0.0	<0.1[a]
	Acetaldehyde	99.8	0.2	0.4
		99.7	0.3	0.4
		94.7	5.3	5.3, 5.2
		73.3	26.7	25.8, 25.6
		40.7	59.3	56.3, 54.7
	Propionaldehyde	90.9	9.1	9.2, 9.2
		60.1	39.9	39.8, 40.2
Methyl ethyl ketone	—	100	0.0	<0.2[a]
	Acetaldehyde	99.7	0.3	0.8, 0.7, 0.6, 0.7
	Propionaldehyde	87.0	13.0	13.4, 13.7
		65.5	34.5	34.5, 34.7
		9.3	91.7	91.2, 90.3
Diethyl ketone	—	100	0.0	<0.2[a]
Methyl isobutyl ketone	—	100	0.0	<0.2[a]
Methyl isopropyl ketone	—	100	0.0	<0.2[a]
Methyl *n*-amyl ketone	—	100	0.0	<0.3[a]
Diisobutyl ketone	—	100	0.0	<0.3[a]
Ethyl *n*-butyl ketone	—	100	0.0	<0.3[a]
Cyclohexanone	—	100	0.0	2.[a]
Cyclopentanone	—	100	0.0	13.[a]

[a] Calculated as aldehyde isomeric to ketone tested.

same. The mercuric oxidative reagent takes the form of Nessler's reagent or a modified form thereof.

$$RCHO + K_2HgI_4 + 3KOH \rightarrow RCOOK + Hg^{\circ} + 4KI + 2H_2O$$

The method involves iodimetric determination of the metallic mercury found in the reaction.

Study of J. E. Ruch and J. B. Johnson

[*Reprinted in Part from Ruch and Johnson, Anal. Chem.*, **28,** *69–71* (*1956*)]

Previous investigators who have attempted the determination of aldehydes by some mercurimetric procedure have recommended the

Table 11. Determination of Aldehyde in Presence of Various Substances by Argentimetric Method

Substance Tested	Substance Added, wt. %	Propionaldehyde, wt. % Added	Propionaldehyde, wt. % Found
Methyl alcohol	99.7	0.28	0.27
Ethyl alcohol	99.7	0.28	0.27
Isopropyl alcohol	99.7	0.28	0.25
	99.5	0.56	0.55
Ethylene glycol	100.0	0.0	0.2
	97.7	2.3	4.2
	99.2	0.77	>3
Triethylene glycol	99.8	0.23	0.26
Mannitol	100.0	0.0	>22
	100.0	0.0	<0.1[a]
	89.8	10.2	9.4[a]
Formic acid	33.0	67.1	66.2
Acetic acid	97.9	2.1	1.9
Propionic acid	97.8	2.2	1.8
Lactic acid	97.5	2.5	2.5
n-Caprylic acid	97.6	2.4	2.3
Ethyl acetate	97.6	2.4	2.3
Diallyl phthalate	97.8	2.2	2.1
Benzyl benzoate	98.1	1.9	1.8
Diallyl maleate	97.8	2.2	3.0
Methyl benzoate	98.0	2.0	1.9
Methylal	100	0.0	0.04[b]
n-Propylal	100	0.0	0.2[b]
	64.3	35.7[c]	36.2
Dimethyl acetal	100	0.0	1.1[b]
	68.2	31.8[c]	32.3
Diethyl acetal	100	0.0	2.0[b]
	70.4	29.6[c]	30.2
Di-*n*-butyl acetal	100	0.0	0.4[b]
	72.3	27.7[c]	28.6
Propylene oxide	88.4	11.6[d]	11.9
	88.8	11.2[d]	11.4

[a] Results obtained by steam distillation of aldehyde.
[b] Results calculated as acetaldehyde, wt. %.
[c] Acetaldehyde substituted for propionaldehyde.
[d] Acrolein substituted for propionaldehyde.

method primarily for the estimation of formaldehyde (49–51). In addition, Bougault and Gros (52) have reported the determination of furfural, benzaldehyde, and piperonal, and Goswami, Das-Gupta, and Ray (53), Goswami and Das-Purkaystha (54), and Goswami and Shaha (55) have estimated sugars with various degrees of success using empirical factors.

These investigators all employed an alkaline solution of potassium mercuric iodide, K_2HgI_4, as an oxidizing agent. In the reaction, aldehyde is oxidized to the corresponding acid whereas mercuric ion is reduced to free mercury. Both isolation and nonisolation methods have been proposed for the determination of the free mercury. In the opinion of Ruch and Johnson it is best to acidify the reaction mixture and react the free mercury with a measured excess of iodine. The amount of iodine consumed is a stoichiometric function of the free mercury which, in turn, is a measure of the aldehyde originally present. Agar is employed as a protective colloid to maintain the free mercury in a finely divided state, thus promoting its reaction with iodine.

The name "mercural reagent" has been coined to differentiate the reagent from other potassium mercuric iodide preparations such as Nessler's reagent. "Mercural" signifies a mercuric oxidation of aldehydes.

REAGENTS

MERCURAL REAGENT. To 1830 ml of distilled water contained in a 1-gallon jug, add 150 grams of reagent grade potassium chloride, 240 grams of U.S. Pharmacopeia grade mercuric chloride (mercury bichloride), 642 grams of reagent grade potassium iodide, and 1000 ml of an aqueous 40% by weight potassium hydroxide solution. Shake the contents after each addition to ensure complete solution. This reagent is stable and does not deteriorate on standing. The slight amount of yellow or brown precipitate that may form is assumed to be due to ammonium ion in the reagents; however it is not detrimental to the effectiveness of the reagent.

AGAR SOLUTION, 0.1%. Add 3.0 grams of Difco Bacto-Agar to 300 ml of boiling distilled water. Continue heating with occasional swirling until the solid has dissolved and the resulting solution is essentially clear. Cool and dilute to 3

49. E. R. Alexander and E. J. Underhill, *J. Am. Chem. Soc.*, **71,** 4014–19 (1949).
50. J. Bolle, J. Jean, and T. Jullig, *Mém. Services Chim. État (Paris)* **34,** 317–20 (1948).
51. W. Stüve, *Arch. Pharm.* **244,** 540 (1906).
52. J. Bougault and R. Gros, *J. Pharm. Chim.* **26,** 5–11 (1922).
53. M. Goswami, H. N. Das-Gupta, and K. L. Ray, *J. Indian Chem. Soc.*, **12,** 714–18 (1935).
54. M. Goswami and B. C. Das-Purkaystha, *Ibid.*, **13,** 315–22 (1936).
55. M. Goswami and A. Shaha, *Ibid.*, **14,** 208–13 (1937).

liters with additional distilled water. Add 0.1 gram of mercuric iodide as a preservative, and shake vigorously for a few seconds.

Acetic acid, analytical grade.

Iodine, approximately 0.1*N*.

Starch indicator, 1.0% solution.

Standard 0.1*N* sodium thiosulfate.

Methanol, commercial grade. Carbide and Carbon Chemicals Co.

SAMPLING

Unless direct sample addition is specified, introduce the sample into a tared 50-ml volumetric flask containing 30 ml of the required solvent (methanol that has been neutralized to bromothymol blue indicator, or distilled water) using a hypodermic syringe fitted with a 3-in. needle and chilled if necessary to facilitate transfer. Stopper the flask and swirl to effect solution. An acetaldehyde dilution must be allowed to stand for approximately 15 minutes, with occasional venting to the atmosphere to reach equilibrium before recording the gross weight. The gross weight of dilutions of other aldehydes may be determined immediately. Dilute to the mark with additional solvent and mix thoroughly. A 5-ml aliquot of this dilution should contain not more than 3.0 meq. of aldehyde. Fill the pipet by pressure to avoid loss of aldehyde.

If the sample is weighed directly into the reagent, care must be exercised to shake the flask vigorously at once, to intimately mix the contents and prevent localized side reactions.

PROCEDURE

The determination is best performed in 500-ml, Erlenmeyer glass-stoppered flasks that are fitted with 24/40 ground-glass joints. Prepare sample and blank flasks by adding 50 ml of mercural reagent to each. Consult Table 12 for the proper reaction temperature and, if necessary, cool each of the flasks in a wet ice bath for 10 minutes. With constant swirling during the addition, introduce an amount of sample containing not more than 3.0 meq. of aldehyde using the procedure specified in Table 12. If a dilution is used, add a similar amount of solvent to the blank. Allow the flasks to stand together at the temperature and for the length of time specified in Table 12. Add 50 ml of agar solution to each flask and swirl vigorously for approximately 1 minute to disperse the mercury precipitate, then add 25 ml of glacial acetic acid with constant agitation during the addition. If the sample contains acetaldehyde, allow

Table 12. Sampling Procedure and Reaction Conditions for Determination of Aldehydes by Mercural Procedure

Compound	Maximum Sample Size for Pure Material, grams[a]	Reaction Time, min.[b]
Acetaldehyde	0.66	5 to 60
Acetaldol	1.3	5 to 60[e]
Acrolein	0.84[c]	180 to 240[d]
Benzaldehyde	1.6[c]	15 to 60[d]
Butyraldehyde	1.1	30 to 60
2-Ethylbutyraldehyde	0.15[e]	15 to 60[f]
Formaldehyde	0.45	1 to 60
Glutaraldehyde	0.75	15 to 60
Hexaldehyde	0.15[e]	30 to 60[f]
Isobutyraldehyde	1.1[c]	5 to 60[d]
Methacrolein	0.90[c]	15 to 60[d]
Propionaldehyde	0.87	15 to 60

[a] Use distilled water as a solvent in the sample dilution unless otherwise specified.

[b] Minutes at room temperature unless otherwise specified.

[c] Use methanol, which has been neutralized to bromothymol blue indicator, as the dilution solvent.

[d] Minutes in a wet-ice bath (0 to 3°C).

[e] Add the sample directly to the sample flask, stopper, and immediately shake the contents vigorously by hand for 1 minute prior to the mechanical shaking.

[f] Minutes on a mechanical shaker.

the flasks to stand at room temperature for approximately 15 minutes before proceeding. The standing period is not required for samples of other aldehydes. Pipet exactly 50 ml of approximately $0.1N$ iodine into each flask, using pressure to fill the pipet. Stopper each flask and shake vigorously until all the gray mercury precipitate goes into solution. If necessary, place on a mechanical shaker for 5 minutes. Carefully remove each stopper, rinse any adhering liquid into the flask, and rinse down the inside walls of the flask with distilled water. Titrate with standard $0.1N$ sodium thiosulfate until the brown iodine color begins to fade. Add a few milliliters of starch indicator solution and continue the titration just to the disappearance of the blue color, approaching the end point dropwise while swirling constantly. From the difference between blank and sample titrations the percentage of aldehyde present in the sample can be calculated; one aldehyde group consumes two equivalents of iodine:

$$—CHO \equiv Hg^{\circ} \equiv I_2 \equiv 2S_2O_3^{--}$$

Hence for monoaldehydes the equivalent weight is one-half the molecular weight.

DISCUSSION

The reagent originally investigated was of the composition usually specified as Nessler's reagent, although, generally speaking, no two authors use the same formulation. However, it was found at this point that Nessler's reagent would not quantitatively oxidize most aldehydes.

A study of the reagent was therefore initiated to determine its optimum composition. Experiments were conducted to determine the effect of the following variables: concentration of potassium mercuric iodide complex, concentration of potassium hydroxide, and ratio of potassium iodide to mercuric chloride. In each case a sample of acetaldehyde was reacted for 1 hour at room temperature with approximately 50 ml (70 grams) of reagent. Results showed 10 to 20% by weight of the potassium mercuric iodide complex in solution gave quantitative results. Likewise, a potassium hydroxide content of 10 to 20% by weight afforded a quantitative oxidation of acetaldehyde. Higher percentages of either component caused solubility difficulties, whereas lesser amounts resulted in incomplete reaction. Variation of the potassium iodide–mercuric chloride ratio indicates that best results were obtained when the ratio of iodide to mercuric ions was slightly higher than the 4:1 of the potassium mercuric iodide complex. Any ratio less than 4:1 tended to produce an undesirable precipitate of mercuric iodide, whereas a ratio significantly higher than 5:1 not only gave low results, but also impaired the effectiveness of the agar used as a protective colloid, yielding a mercury precipitate that was less reactive with iodine.

On the basis of these experiments a reagent was formulated to contain 16% by weight potassium mercuric iodide, 13% by weight potassium hydroxide, and approximately 1 gram of excess potassium iodide per 50 ml of reagent.

Using this reagent, experiments were undertaken to establish the necessary reaction conditions for pure aldehydes. Water-soluble aldehydes were sampled in the form of aqueous dilutions and oxidized at room temperature. As a mutual solvent for higher molecular weight aldehydes, methanol has proved satisfactory. Its exact use depends on the particular aldehyde, but the usual procedure is to employ neutralized methanol as a dilution solvent and conduct the reaction at the temperature of a wet ice bath (0–3°C) to prevent any oxidation of the methanol. In some instances direct addition of sample to reagent, accompanied by shaking, is the best procedure. Table 12 gives the most suitable methods of sampling, reaction conditions, and sample size for a number of aldehydes for which this procedure has been found satisfactory.

RESULTS

Comparable data on the purity of a number of aldehydes were obtained by the mercural procedure and a hydroxylamine hydrochloride–triethanolamine method (56). The average result, the precision attained, and the number of determinations for each sample are given in Table 13.

The standard deviation for the determination of acetaldehyde using

Table 13. Purity Determinations on Aldehydes by Mercural and Hydroxylamine Procedures

Compound	Purity by Mercural Procedure[a], %	Purity by Hydroxylamine Procedure[b], %
Acetaldehyde	98.9 ± 0.3 (5)	98.9 ± 0.3 (4)
Acetaldol	101.5 ± 0.2 (2)	101.6 ± 0.3 (3)
Acrolein	98.8 ± 0.3 (5)	99.0 ± 0.0 (2)
Benzaldehyde	95.3 ± 0.2 (8)	95.3 ± 0.2 (5)
Butyraldehyde	98.0 ± 0.5 (11)	97.7 ± 0.5 (7)
2-Ethylbutyraldehyde	96.5 ± 0.3 (3)	96.9 ± 0.1 (2)
Formaldehyde	35.9 ± 0.1 (6)	35.7 ± 0.1 (2)
Glutaraldehyde	26.3 ± 0.05 (4)	—
Hexaldehyde	95.4 ± 0.2 (4)	94.7 ± 0.3 (2)
Isobutyraldehyde	97.7 ± 0.3 (9)	97.4 ± 0.1 (2)
Methacrolein	90.7 ± 0.1 (3)	90.6 ± 0.1 (2)
Propionaldehyde	97.1 ± 0.0 (4)	96.8 ± 0.4 (5)

[a] Figures in parentheses represent number of determinations.
[b] Hydroxylamine hydrochloride-triethanolamine (54).

56. Carbide and Carbon Chemicals Co., South Charleston, W. Va., unpublished method.

aqueous dilutions was 0.39% for 13 degrees of freedom on a sample whose average purity was 97.5%. The sampling error was not significant.

The procedure has been modified and found suitable for the determination of trace amounts of aldehydes in organic compounds. For example, determinations of acetaldehyde in ethylene oxide and propionaldehyde in propylene oxide have been performed successfully.

INTERFERENCE STUDIES

Many organic compounds do not interfere with this procedure, permitting the determination of aldehyde in the presence of most acids, ketones, esters, acetals, ethers, alcohols, epoxides, and organic chlorides.

Oxidation studies were conducted on methanol, ethyl alcohol, isopropyl alcohol, and butyl alcohol, both at room temperature and at the wet ice bath temperature. Methanol is slowly attacked by the reagent at room temperature, but is completely resistant to oxidation at 0 to 3°C and is, therefore, a preferred nonaqueous solvent for some aldehydes, as indicated in Table 12. Isopropyl alcohol is the worst offender, not only because it is oxidized even at 0 to 3°C, but also because its oxidation product, acetone, complexes the mercuric ion. Ethyl and butyl alcohols are only slightly oxidized at 0 to 3°C. Studies indicate that the oxidation of alcohols by mercural reagent follows the mass action law, enabling one to compensate for this deleterious reaction by using a reagent diluted 50/50 with distilled water, adding a similar amount of alcohol to the blank, and performing the oxidation in a wet ice bath, allowing a suitably longer reaction time to offset the dilution of reagent and reduction in temperature. Errors introduced by this procedure are not serious when alcoholic samples containing only a few per cent aldehyde are involved. Samples containing esters require the same conditions, because they are saponified to alcohols by the potassium hydroxide in the reagent.

Some vinyl compounds are known to interfere with this procedure by adding iodine, thus yielding a high result. This method has been found applicable to the determination of acrolein (acrylaldehyde) and methacrolein (methacrylaldehyde) (see Table 12), whereas crotonaldehyde has been analyzed with an accuracy of within ±2%. However, no satisfactory results have been obtained on unsaturated aldehydes containing more than four carbon atoms—for example, 2,4-hexadienal (sorbaldehyde), 2-ethylcrotonaldehyde, and 2-ethyl-3-propylacrolein. Therefore the determination of unsaturated aldehydes or of aldehyde in any mixture containing an unsaturated compound must be checked for interference.

Acetone reacts with mercuric ion in the following manner (57):

$$Hg^{++} + 2CH_3{-}\underset{\underset{O}{\|}}{C}{-}CH_3 \rightleftharpoons Hg(CH_3{-}\underset{\underset{\underset{|}{O}}{|}}{C}{=}CH_2)_2 + 2H^+$$

In the presence of the alkaline reagent and excess mercuric ions, this equilibrium reaction is displaced to the right, depositing the mercuric ion–acetone complex as a yellow solid. On acidification the reaction is reversed, proceeding to the left. This reversal must be complete, as indicated by the absence of the yellow precipitate, or else iodine is consumed.

Lower temperatures induce precipitation or even resinification of the mercuric ion–acetone complex, hence greater solubility difficulties are experienced at 0 to 3°C than at room temperature.

To illustrate the effect of the presence of acetone, a series of blank determinations was made as specified in the method, using reaction conditions of 30 minutes at 0 to 3°C. From 0 to 3.0 grams of acetone was added to each flask. With the addition of up to 0.3 gram of acetone, a yellow precipitate was formed that easily dissolved on acidification. More then 0.3 gram of acetone caused deposition of a resin requiring additional potassium iodide to effect solution. Hence the acetone tolerance of this method is approximately 0.3 gram for determinations performed in a wet ice bath.

Because a portion of the mercuric ion contained in 50 ml of reagent is complexed by 0.3 gram of acetone, it was then necessary to prove that a sufficient amount of reagent was still available for the quantitative determination of aldehyde. Results on the determination of propionaldehyde in the presence of 0.3 gram of acetone show that quantitative oxidation is attained even when the maximum sample size of propionaldehyde is taken.

Methyl ethyl ketone complexes mercuric ion to a much smaller degree than acetone, whereas methyl isopropyl ketone and ethyl butyl ketone are practically inert.

Hydroxy ketones constitute a positive interference, as do other easily oxidized substances or anything that consumes iodine. Conversely, oxidizing agents such as peroxides are likely to produce low results, either by competing with mercuric ion in the oxidation of aldehyde or by oxidizing iodide to iodine.

As a rule, the amount of acid or ester that can be tolerated must not be so great as to neutralize more than one-third of the potassium hydroxide

57. J. B. Fernandez, L. T. Snider, and E. G. Rietz, *Anal. Chem.* **23,** 899–900 (1951).

in the reagent, whereas no more than one-half of the mercuric ion content should be reduced and/or complexed.

HYPOIODITE OXIDATION

Hypoiodite has been used to determine methyl carbonyl compounds ($RCCH_3$ with C=O, i.e. $R\overset{\overset{\large O}{\|}}{C}CH_3$; where R can be H) (58–61). The method has a very narrow range of applicability, however, and is included in this text mainly for completeness. The main problem is one of interferences; hypoiodite is an oxidizing agent strong enough to oxidize many noncarbonyl compounds.

Romijn Procedure for Formaldehyde (61)

Measure the sample, containing up to 0.16 gram of formaldehyde, into an iodine flask. Add sufficient distilled water to make the volume about 100 ml. Then introduce 30 ml of 3*N* sodium or potassium hydroxide, followed by 75 ml of 0.2*N* iodine solution. Stopper the flask and allow to stand at room temperature for 30 minutes. Add about a 5-ml excess (ca. 95 ml) of 1*N* sulfuric acid and titrate the liberated iodine immediately with 0.1*N* sodium thiosulfate, using starch as an indicator. Run at least one blank with each set of samples.

CALCULATION

$$\frac{\text{Net milliliters} \times N(Na_2S_2O_3) \times 1.50}{\text{Weight of sample}} = \%\ \text{HCHO}$$

Messinger Procedure for Acetone (61)

Measure the sample, containing up to 2 mM of acetone, into an iodine flask. Add sufficient water to make the volume about 400 ml. Then pipet

58. G. Romijn, *Z. Anal. Chem.* **36,** 18–24 (1897).
59. J. Messinger, *Ber.*, **21,** 3366–7s (1888).
60. C. O. Haughton, *Ind. Eng. Chem., Anal. Ed.*, **9,** 167–8 (1937).
61. J. Mitchell et al. *Organic Analysis* Vol. I, pp. 267–9 Wiley-Interscience Publishers, New York, 1953, reprinted with permission.

50 ml of 1N sodium hydroxide into the solution and allow the homogenized mixture to stand at room temperature for 5 minutes. At the end of this time, slowly add 50 ml of 0.2N iodine solution. During the addition, shake the flask constantly with a swirling motion. Allow the mixture to stand for 10 minutes at room temperature. Then add 51 ml of 1N sulfuric acid, that is, required amount plus 1 ml, and titrate the excess iodine immediately with 0.1N sodium thiosulfate to the starch end point. Run a blank with each set of samples.

CALCULATION

$$\frac{\text{Net milliliters} \times N(\text{Na}_2\text{S}_2\text{O}_3) \times 0.967}{\text{Weight of sample}} = \%\ \text{acetone}$$

HYPOBROMITE OXIDATION

Adapted from the Method of M. H. Hashmi and A. A. Ayaz

[*Anal. Chem.*, ***36,*** *384* (*1964*)]

Methyl ketones and acetaldehyde can be titrated directly with a standard hypobromite solution to a Bordeaux indicator end point. Because the titration procedure prevents excess hypobromite, complications resulting from the formation of halogenated organic acids and tetrahalo compounds, which are often noted with other haloform procedures, are eliminated. The method is especially suitable for low concentrations.

REAGENTS

SODIUM HYPOBROMITE REAGENT (62). Dissolve 16 grams of sodium hydroxide in 40 ml of water containing 4 ml of bromine and dilute to 1 liter. To standardize the solution (63) relative to combined bromite and hypobromite, add an excess of standard 0.1N arsenious oxide solution to 5 ml of the reagent solution. After 5 minutes add 4 to 5 grams of sodium bicarbonate, followed by dilute acetic acid to neutralize the sodium hydroxide. Neutralization is indicated by the free effervescence produced by each drop of acetic acid. Titrate the solution with standard iodine, using starch as the indicator. The reagent is reasonably stable for 2 weeks if stored in an amber bottle, but if kept beyond that time it should be restandardized.

Bordeaux indicator, 0.2% aqueous solution.

62. M. H. Hashmi, E. Ali and M. Umar, *Anal. Chem.*, **34,** 988 (1962).
63. M. H. Hashmi and A. A. Ayaz, *Anal. Chem.*, **35,** 908 (1963).

PROCEDURE

Add 3 ml of $3N$ sodium hydroxide followed by 2 ml of water to a known volume (2 ml) of ketone or acetaldehyde solution. Titrate with standard hypobromite solution with 3 drops of Bordeaux solution as an internal indicator. The end point indication is a change of the light pink color to colorless or to a faint yellow.

RESULTS AND DISCUSSION

Only hypobromite or bromite acts on ketones and acetaldehyde; bromate does not.

$$CH_3COR + 3NaOBr \rightarrow CHBr_3 + RCOONa + 2NaOH$$

$$2NaOBr \rightarrow NaBrO_2 + NaBr$$

$$2CH_3COR + 3NaBrO_2 + 3NaBr \rightarrow 2CHBr_3 + 2RCOONa + 4NaOH$$

$$CH_3COR + NaBrO_3 \rightarrow \text{no reaction}$$

For all calculations, then, the combined strength of hypobromite and bromite ions in the reagent solution is used, and calculations are made on the assumption that 3 moles of bromine is used per mole of the carbonyl compound. Table 14 presents typical results.

Large quantities of alcohol interfere, but when the alcohol-carbonyl compound ratio is 2:1, the interference is practically negligible. Organic compounds that normally respond to the haloform reaction will interfere. Low results were obtained when the method was tried for propionaldehyde and butyraldehyde.

Schiff Base Formation

Carbonyl compounds react with primary amines to form the corresponding imino compound or Schiff base.

$$R\overset{\overset{O}{\|}}{C}R_1 + R_2NH_2 \rightarrow \underset{R_1}{\overset{R}{C}}{=}NR_2 + H_2O$$

(R and, or R_1 can be H)

The formation of the Schiff base as a basis for determining aldehydes has not been used to any great degree. This reaction has undoubtedly been considered by many people to be usable because it is rapid, and the

Table 14. Determination of Ketones and Acetaldehyde by Hypobromite Titration

Carbonyl Compound	Ketone or Acetaldehyde Present, mole/liter	Ketone or Acetaldehyde Found mole/liter	Difference, %
Ethyl methyl ketone	0.10738	0.10692	−0.43
	0.05370	0.05320	−0.93
	0.02685	0.02714	+1.08
Methyl *n*-propyl ketone	0.10000	0.09910	−0.90
	0.08000	0.07990	−0.12
	0.04000	0.04038	+0.95
Methyl isopropyl ketone	0.05440	0.05428	−0.22
	0.03260	0.03250	−0.30
	0.02176	0.02168	−0.36
Methyl *n*-butyl ketone	0.01732	0.01748	+0.92
	0.01385	0.01375	−0.72
	0.01154	0.01144	−0.86
Methyl isobutyl ketone	0.06950	0.06900	−0.72
	0.05210	0.05130	−1.50
	0.04170	0.04150	−0.48
Methyl *tert*-butyl ketone	0.01578	0.01568	−0.70
	0.01052	0.01032	−1.90
	0.00789	0.00794	+0.63
Acetaldehyde	0.04300	0.04270	−0.70
	0.03640	0.03616	−0.66
	0.02600	0.02562	−1.40

excess of amine used should be easily determined. It has been found, however, that several problems arise when this reaction is tried to determine aldehydes, and these problems could account for the little that has been published. These obstacles have been overcome to some degree, and it has been found that the reaction can be applied to determine some aldehydes with a good degree of accuracy and precision. The method is also fast and is not involved. The method has some other attributes: it is run in a nonaqueous medium, which permits solution of the more insoluble samples; it can be used in the presence of some ketones; and it employs a reaction in an alkaline medium, which is desirable in the handling of samples such as acetals that liberate aldehydes in acid media.

It was found that there is a definite equilibrium present in the above-named reaction. This equilibrium is probably the reason this reaction was not used successfully earlier. A nonaqueous system had to be used to make the reaction proceed far enough toward completion to be usable. The only water present is that formed in the reaction. It was also noted that the Schiff bases formed deteriorate with strong acids, and one would titrate the original amount of amine put in the system if mineral acids were used for the titration of the excess amine. It is for this reason that salicylic acid is used in the procedure of Siggia and Segal described below to titrate the excess amine. Salicyclic acid has a dissociation constant of 1×10^{-3} (in water) and is strong enough to effect a good titration of the excess amine without causing the Schiff base to hydrolyze back to the free amine and aldehyde. (This behavior could be looked on as connected with the equilibrium present in the reaction. The strong acids tie up the free amine very effectively, causing the equilibrium to be shifted to the left, whereas the salicyclic acid does not bind the amine strongly enough to cause the reaction to reverse itself.) The Schiff base itself is a much weaker base than the original amine, so that it causes no trouble in the titration.

Only aliphatic amines are usable with this method, since we are titrating with a relatively weaker acid than that normally used. Aromatic amines are too weak to be titrated effectively with salicylic acid. Lauryl amine, the amine used in the procedure described below, was finally chosen because of its high boiling point and its availability. Butyl amine was the first amine tried, and it worked satisfactorily except that, owing to loss of butyl amine from the reagent through evaporation, repeated standardization was necessary.

From the titration curves obtained (see Fig. 2.6), it is evident that

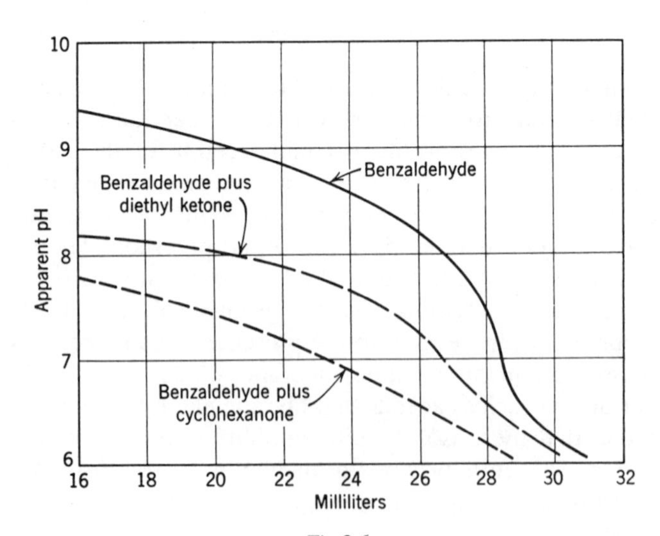

Fig. 2.6

ketones undergo a reaction with the aliphatic amines to the corresponding Schiff base, but the equilibrium is in favor of the free amine and ketone. The titration curves obtained in these cases are flattened, indicating free amine being liberated as the excess is titrated. In some ketones, the break in the curve is completely obscured. Methyl ketones (acetone, methyl ethyl ketone, etc.) and cyclohexanone are in the latter category. With diethyl ketone, however, a distinct break is visible in the curve for the total amount of amine introduced. With ketones present in the aldehydes, the same holds true; methyl ketones and cyclohexanone will obliterate the end point through their weak tieing up of amine, but the higher ketones do not affect the final results for the aldehyde, although they do diminish the intensity of the break. If a small amount of aldehyde is present in a large amount of ketone, it would be presumed that the break might be diminished to the extent of being completely obscured.

This method does not work on aliphatic aldehydes, except for the formaldehyde where it works well. From the titration curves obtained, it is concluded that the Schiff bases of the aliphatic aldehydes and the aliphatic amines are not stable enough or, better, that the equilibrium in these cases is too far to the left for the reaction to be used for analytical purposes. No titration curves could be obtained for acetaldehyde, propionaldehyde, or butyraldehyde. Cinnamaldehyde is peculiar in this respect in that it is essentially an aliphatic aldehyde with the phenyl group on the chain, yet it will react completely enough to be used. Crotonaldehyde, which is similar to the cinnamaldehyde except that the phenyl group is replaced with a methyl group, does not show any titration break at all.

Interferences in this method consist of acids that are approximately as strong as, or stronger than, the salicylic acid used in the titration. A weak acid will not interfere in that the stronger salicylic acid will titrate the amine away from it. (We must speak of "weaker" and "stronger" acids in this case, since we are working in a nonaqueous medium in which the dissociation constants of the various acids are not known.) Acid anhydrides and acid halides interfere in that they consume amine.

Procedure of S. Siggia and E. Segal

[*Anal. Chem.*, ***25***, *830* (*1953*)]

REAGENTS

Solution of lauryl amine in ethylene glycol-isopropanol mixture containing 2 moles lauryl amine per liter of solvent. (The lauryl amine is purchased as Armeen 12D from Chemical Division of Armour & Co.)

Standard 1*N* salicylic acid in ethylene glycol–isopropanol, standardized against alcoholic sodium hydroxide.

PROCEDURE

Weigh a 0.02-mole sample of the aldehyde into a 100 to 150-ml glass-stoppered flask. Add exactly 20 ml of the lauryl amine solution, replace the cover, and shake the mixture for a few minutes. Allow the mixture to stand for 1 hour. Then transfer the solution with ethylene glycol–isopropanol to a 250-ml beaker, and titrate the excess lauryl amine potentiometrically with 1*N* salicylic acid.

Also run a blank.

CALCULATION

$$\frac{(\text{Blank} - \text{titration}) \times N \text{ salicylic acid} \times \text{mol. wt.}}{\text{Grams of sample} \times 1000} \times 100 = \%\ \text{aldehyde}$$

Table 15. Primary Amines

	Time	Sample	% Found	% Found Other Method
Formaldehyde	$1\frac{1}{2}$ hr.	Portion of 35% solution	36.5	36.3[a]
	Over week end	weighed in	36.1	
Benzaldehyde	1 hr.	0.02 mole	98.7	96.3[b]
		weighed in	99.2	
Salicylaldehyde	1 hr.	0.02 mole	99.6	99.0[b]
		weighed out	99.8	
			99.2	
			99.4	
	1 hr.	Aliquot portion used containing 0.02 mole	99.0	
Cinnamaldehyde	1 hr.	0.02 mole	99.3	98.6[a]
		weighed out	99.9	
Furfural	1 hr.	0.02 mole	99.5	99.3[b]
		weighed out	98.0	98.8

[a] Bisulfite method of Siggia and Maxcy, *Ind. Eng. Chem., Anal. Ed.*, **19,** 1023 (1947).

[b] 2,4-Dinitrophenyl hydrazone method of Iddles, Low, Rosen, and Harte, *ibid.*, **11,** 102 (1939).

Miscellaneous Methods

Several miscellaneous methods have been used but have never achieved prominence because of one or more shortcomings relative to the approaches described in detail previously.

METHONE METHOD (ALSO KNOWN AS DIMEDON METHOD)

Methone (5,5 dimethyl dihydroresorcinol) has been used as a qualitative reagent for the identification of aldehydes, but it has achieved only limited use for the quantitative determination of these compounds.

The main application of the methone is for determining formaldehyde, but acetaldehyde has also been determined. Yet this method has no advantages over the procedures mentioned earlier for determining these aldehydes except for determining formaldehyde in the presence of other aldehydes. This method is described below. Ketones and sugars do not react with dimedon under ordinary conditions.

The reaction can be written as follows:

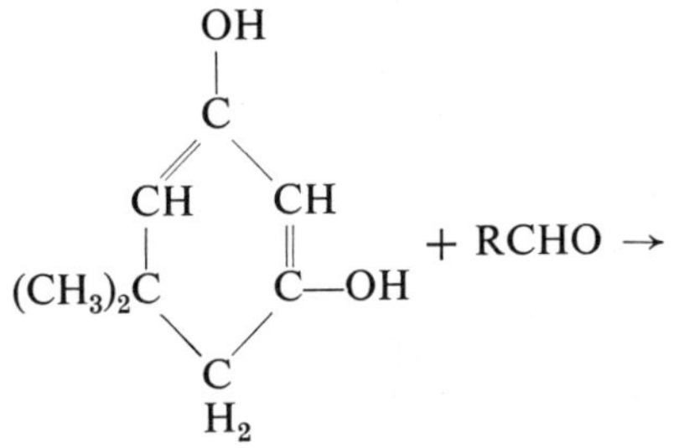

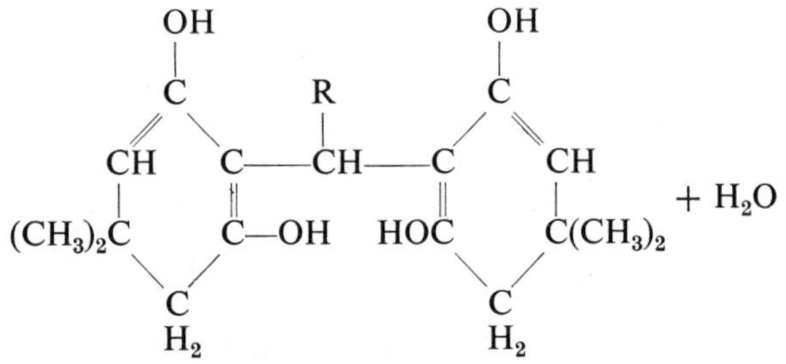

```
           OH
           |
           C
         //  \
       CH     CH
       |      ||        + RCHO →
(CH3)2C       C—OH
        \    /
          C
          H2
```

```
                    OH                     OH
                    |                      |
                    C        R             C
                  //  \      |           /   \\
                CH     C ——— CH ——— C          CH
                |      ||            ||        |           + H2O
         (CH3)2C       C—OH      HOC           C(CH3)2
                 \    /              \        /
                   C                    C
                   H2                   H2
```

Formaldehyde (64) and methone react quantitatively in neutral, alkaline, or mildly acidic aqueous or alcoholic solutions to form methylene bismethone, as indicated in the foregoing equation. The product of this

64. J. F. Walker, *Formaldehyde*, 2nd ed., Reinhold, New York, 1953, pp. 391–3, reprinted in part with permission.

equation is almost completely insoluble in neutral or mildly acidic aqueous solutions; 100 ml of water dissolves only 0.5 to 1.0 mg at 15 to 20°C. It is a crystalline material melting at 189°C. It is soluble in alkali, behaving as a monobasic acid, and it can be titrated with standard alkali in alcohol solution.

Other aliphatic aldehydes such as acetaldehyde also form condensation products of low solubility by reaction with methone. These products differ from the formaldehyde derivative in that they are readily converted to cyclic hydroxanthene derivatives by treatment with glacial acetic or dilute sulfuric acid. These xanthene derivatives are not soluble in alkali and can thus be separated from the formaldehyde product. Acetaldehyde reacts with methone to give ethylidenedimethone (m.p. 139°C) which is 0.0079% soluble in water at 19°C. The conversion of this product to the alkali-insoluble xanthene derivative is indicated below:

```
          H                      H
          O                      O
          |                      |
          C          CH3         C
        //  \         |        /  \\
    HC        C-------C-------C      CH
    |         ||      |       |      |                Acid
(CH3)2C       C-OH    H  HO---C      C(CH3)2         ------>
        \   /                  \   /
          C                      C
          H2                     H2

                          H          H
                          O  H    CH3O
                          |    \  /   |
                          C     C     C
                        //  \  / \  /  \\
                    HC       C     C     CH
                    |        ||    ||    |              + H2O
               (CH3)2C       C     C     C(CH3)2
                       \   /   \  /  \  /
                         C      O     C
                         H2           H2
```

The acetaldehyde derivative, ethylidenedimethone, differs from the formaldehyde product in that it behaves as a dibasic acid when titrated with alkali at 70 to 75°C.

In determining formaldehyde with methone, the reagent is best added as a saturated aqueous solution or as a 5 to 10% solution in alcohol. In the latter case, care must be taken to avoid addition of large quantities of reagent because of the limited solubility of methone in water and because high concentrations of alcohol will interfere with the precipitation of the formaldehyde derivative. The formaldehyde solution to be analyzed should be neutral or mildly acid. Weinberger (65) claims that addition of salt increases the sensitivity of the test and states that agitation speeds up the precipitation of methylene derivative. Vörlander (66) allows a reaction period of 12 to 16 hours at room temperature for complete precipitation. By Weinberger's accelerated method (65) it is claimed that formaldehyde bismethone is precipitated in 15 minutes when present at a concentration of 4 ppm. In the absence of other aldehydes the precipitate may be filtered off, washed with cold water, and dried to constant weight at 90 to 95°C for gravimetric measurement. Each gram of precipitate is equivalent to 0.1027 gram of formaldehyde. Yoe and Reid (67) report improved accuracy if the precipitation is carried at pH 4.6 in a sodium acetate–hydrogen chloride solution.

When acetaldehyde or other aliphatic aldehydes are present, their methone derivative must be separated from the precipitate first obtained. This may be accomplished by shaking solution and precipitate with $\frac{1}{15}$ volume of cold 50% sulfuric acid for 16 to 18 hours; or the filtered precipitate (moist or vacuum-dried) can be heated with 4 to 5 times its volume of glacial acetic acid on a boiling water bath for 6 to 7 hours, after which it is treated with an excess of ice water to precipitate the products. The formaldehyde bismethone can then be removed from the acid-treated precipitates with dilute alkali and reprecipitated by acidification (66). When acetaldehyde is the only other aldehyde present, it may be determined by weighing the dried alkali-insoluble precipitate, 1 gram of which is equivalent to approximately 0.1180 gram of acetaldehyde. Since ethylidene bismethone is more soluble than the methylene derivative, the method is not as accurate for acetaldehyde as formaldehyde.

The volumetric titration technique developed by Vörlander (66) is readily applied to the measurement of moist precipitated methylene bismethone by dissolving in alcohol and titrating with caustic at room temperature. Temperatures of 70°C do not affect the results of this titration. One ml of normal alkali is equivalent to 0.030 gram of formaldehyde. Another variant involves titration of a given volume of methone solution followed by titration of an equal volume of the same solution, which has been reacted with the formaldehyde sample. The difference in these titers is equivalent to the formaldehyde titer, since 2 moles of methone is equivalent to 2 liters of normal caustic, whereas 1 mole of methylene bismethone (equivalent to 2 moles of methone) is equivalent to

65. V. Weinberger, *Ind. Eng. Chem., Anal. Ed.*, **3,** 365–6 (1931).
66. D. Vörlander, *Z. Anal. Chem.*, **77,** 32–7 (1929).
67. J. J. Yoe and L. C. Reid, *Ind. Eng. Chem., Anal. Ed.*, **13,** 238–40 (1941).

only 1 liter of normal caustic. If acetaldehyde is present, the final titration must be carried out at 70°C, since 1 mole of ethylidene bismethone titrates as a dibasic acid at this temperature and this does not interfere with the analysis, since it behaves the same as 2 moles of unreacted methone.

CANNIZZARO REACTION

A method is reported (68) utilizing the Cannizzaro reaction which is applicable to mainly aromatic aldehydes, but a few alphatic aldehydes (isobutyraldehyde, isovaleraldehyde, heptaldehyde) are determinable with long reaction times of several hours.

$$2RCHO + KOH \rightarrow RCH_2OH + RCOOK + H_2O$$

NESSLER REAGENT METHODS (SEE MERCURIMETRIC OXIDATION METHODS, PP. 126–36)

GRIGNARD METHODS (SEE ACTIVE HYDROGEN METHODS, PP. 478–88)

LITHIUM ALUMINUM HYDRIDE METHODS

Lithium aluminum hydride will quantitatively reduce aldehydes and ketones and can be used to determine these compounds. Higuchi et al. (69, 70) applied this approach by adding an excess of hydride in a tetrahydrofuran solution to the sample and titrating the excess electrometrically with ethanol or propanol in dry benzene.

The main difficulty with this approach is the nonspecificity of lithium aluminum hydride. It will react with any active hydrogen-containing compound, such as alcohols, primary and secondary amines and amides, water, mercaptans, and acids; and will reduce many other variety of compounds including amides, esters, nitrites.

HYDROGEN PEROXIDE OXIDATION

Hydrogen peroxide oxidation is mainly applicable to the determination of only formaldehyde. The other aldehydes react, but generally too slowly for quantitatively determination.

68. L. Palfray, S. Sabetay, and D. Sontag, *Chem. & Ind.* **29,** 1037–8 (1933).
69. T. Higuchi, C. J. Lintner, and R. H. Schleif, *Science*, **III,** 63–4 (1950).
70. C. J. Lintner, D. A. Zuck, and T. Higuchi, *J. Am. Pharm. Assoc.*, **39,** 418–20 (1950).

Blank and Finkenbeiner (71) determined formaldehyde by heating with peroxide for 5 minutes at 60°C. Other aldehydes interfered but could not be determined.

MacCormac and Townsend (72) were able to determine a few of the lower aldehydes by carrying out the oxidation in a pressure bottle on a steam bath.

AMMONIA REACTION

Aldehydes undergo reaction with ammonia much the same as in the case of amines (see p. 138). The equilibrium in this reaction, however, is much in favor of the reactants such that the reaction is of little value except for determining formaldehyde (73–76).

POTASSIUM CYANIDE REACTION

Potassium cyanide will undergo a sort of cyanhydrin reaction with aldehydes, but quantitative reaction is achieved only with formaldehyde; the other aldehydes generally interfere.

$$HCHO + KCN \rightarrow \underset{\displaystyle OK}{\overset{}{CH_2CN}}$$

The procedure (77) consists of adding excess cyanide reagent to the sample and then adding a known quantity of silver nitrate in excess of the remaining cyanide present. The excess silver ion is then titrated with thiocynate.

SODIUM BOROHYDRIDE REACTION

Sodium borohydride will reduce aldehydes to the corresponding alcohol, and this reaction can be used analytically. The general approach is

71. O. Blank and H. Finkenbeiner, *Ber.*, **31,** 2979–81 (1898); **32,** 2141 (1899).
72. M. MacCormac and D. T. A. Townsend, *J. Chem. Soc.*, **1940,** 151–6.
73. L. Legler, *Ber.*, **16,** 1333–7 (1883).
74. A. G. Craig, *J. Am. Chem. Soc.*, **23,** 638–43 (1901).
75. A. Foschini and M. Talenti, *Z. Anal. Chem.*, **117,** 94–9 (1939); **118,** 94–7 (1940).
76. J. Buchi, *Pharm. Acta Helv.*, **6,** 1–54 (1931).
77. G. Romijn, *Z. Anal. Chem.*, **36,** 18–24 (1897).

to add excess of the borohydride and then to determine the excess iodimetrically (78) or by evolving the hydrogen (79). A newer approach (80) is to titrate the aldehyde directly with the borohydride using a photometric end point. All these methods, although workable, suffer from the shortcoming of the nonspecificity of the borohydride. Other classes of compounds are reduced by the reagent and, hence, interfere. Ketones and acids are the most common interferences. In light of the other methods available to determine aldehydes, the borohydride method will probably not achieve much popularity.

PERMANGANATE AND DICHROMATE OXIDATION

The oxidants permanganate and dichromate have also been reported for determining aldehydes (81–84). They are of little value, however, since both will oxidize many organic materials other than aldehydes, to cause interference.

Methods for Trace Quantities of Carbonyl Compounds

Traces of carbonyl compounds can often be determined by ultraviolet absorption techniques, especially the compounds where the carbonyl double bond is in conjugation with a carbon-to-carbon double bond, that is, acrolein, crotonaldehyde. Also, polarography can be used for determining trace quantities of aldehydes. But trace carbonyl materials as a class of compounds are very amenable to colorimetric determination, since they undergo quantitative reactions with colored products, and these reactions are generally quite specific for carbonyl compounds. The color reactions the most widely used for this purpose are the 2,4-dinitrophenylhydrazone formation and the Schiff reaction.

COLORIMETRIC 2,4-DINITROPHENYLHYDRAZONE METHOD

The reaction between carbonyl compounds and 2,4-dinitrophenylhydrazine is shown on p. 113, where the gravimetric determination of

78. D. A. Lyttle, E. H. Jensen, and W. A. Struck, *Anal. Chem.*, **24,** 1843 (1952).
79. M. Sobotka and H. Trutnovsky, *Microchem. J.*, **3,** 211 (1959).
80. E. Cochran and C. A. Reynolds, *Anal. Chem.*, **33,** 1893 (1961).
81. J. Hetper, *Z. Anal. Chem.*, **50,** 343–70 (1911); **51,** 409–29 (1912).
82. H. M. Smith, *Analyst*, **21,** 148–51 (1896).
83. T. Harrington, T. H. Boyd, and G. W. Cherry, *Analyst*, **71,** 97–107 (1946).
84. M. Nicloux, *Bull. Soc. Chim. Fr.* (3), **17,** 839–40 (1897).

carbonyl materials is discussed. The procedure described here utilizes the same reaction, except that the colored hydrazone product is measured and related back to the concentration of the original carbonyl compound.

Method of G. R. Lappin and L. C. Clark

[*Reprinted in Part from Anal. Chem.*, **23,** *541–2* (*1951*)]

The addition of a solution of sodium or potassium hydroxide to an alcoholic solution of a 2,4-dinitrophenylhydrazone produces a very intense wine-red color, presumably because of the formation of the resonating quinoidal ion I. A similar quinoidal ion has been suggested for

$$R{-}CH{=}N{-}NH{-}C_6H_3(NO_2){-}NO_2 \xrightarrow{\text{base}} R{-}CH{=}N{-}N{=}C_6H_3(NO_2){=}N(O^-)O$$

I

the colored solution formed when base is added to the phenylhydrazone of a nitroaromatic aldehyde (85). This color reaction has been made the basis of a very sensitive method for the estimation of ketosteroids in biological extracts (86). Herein is reported the extension of the method to the quantitative determination of traces of aldehydes or ketones in water, organic solvents, or organic reaction products. The method is most useful in the range of carbonyl concentration from 10^{-4} to 10^{-6} M, wherein few if any other methods give reliable results or are of general application.

Absorption spectra were run on alkaline alcoholic solutions of a number of 2,4-dinitrophenylhydrazones. It was found that the position of the maximum as well as the value of E_{max} were nearly independent of the structure of the carbonyl compound (with exceptions noted below) and were independent of the concentration of base as long as a sufficient excess was present. The colors formed were relatively stable, although slow fading over a period of several days was noted. Beer's law was obeyed in the concentration range studied. The value of E_{max} determined

85. F. Chattaway, S. Ireland, and A. Walker, *J. Chem. Soc.*, **1925,** 1851.
86. L. Clark and H. Thompson, unpublished research.

Table 16. Position and Values of E_{max} for Various Compounds

Compound	Maximum, mμ	$E_{max.} \times 10^{-4}$
Acetaldehyde	478	2.72
Acetone	476	2.66
Acetophenone	480	2.71
Anisaldehyde	480	2.70
Acetylacetone	480	5.42
Acetthienone	480	2.71
Benzaldehyde	481	2.72
Butyraldehyde	480	2.73
Cinnamaldehyde	480	2.70
Cyclohexanone	480	2.69
Cyclopentanone	480	2.68
3,5-Dichlorobenzaldehyde	480	2.70
Furfural	479	2.72
9-Heptadecanone	480	2.68
p-Hydroxyacetophenone	480	2.70
Methyl cyclopropyl ketone	476	2.69
Methyl ethyl ketone	480	2.75
Methyl phenyl diketone	480	5.46

for a large number of compounds averaged 2.72×10^4 at 480 nm. Table 16 gives more exact values for a number of compounds.

For actual analysis it was found unnecessary to isolate the phenylhydrazone. If it was prepared in solution, using an excess of 2,4-dinitrophenylhydrazine, the addition of base converted the excess reagent to a very light yellow substance, the absorption of which was corrected for by using a blank determination.

PREPARATION OF REAGENTS

CARBONYL-FREE METHANOL. To 500 ml of CP methanol were added about 5 grams of 2,4-dinitrophenylhydrazine and a few drops of concentrated hydrochloric acid. After refluxing for 2 hours, the methanol was distilled through a short Vigreux column. If kept tightly stoppered, the methanol remains suitable for use for several months.

2,4-DINITROPHENYLHYDRAZINE SOLUTION. A saturated solution in carbonyl-free methanol was prepared, using 2,4-dinitrophenylhydrazine that had been twice recrystallized from this solvent. This solution should not be used more than a week or two after preparation.

POTASSIUM HYDROXIDE SOLUTION. Ten grams of potassium hydroxide was dissolved in 20 ml of distilled water, and the solution was made up to 100 ml with carbonyl-free methanol. This solution will keep indefinitely.

PROCEDURE

The unknown or its solution should not be more than $10^{-3}M$ in carbonyl. In such dilute solutions the phenylhydrazone will not precipitate at room temperature. The solution must be neutral or very weakly acidic to prevent precipitation of potassium salts when the base solution is added.

To 1.0 ml of the unknown or its solution in carbonyl-free methanol, add 1.0 ml of the 2,4-dinitrophenylhydrazine reagent and 1 drop of concentrated hydrochloric acid. Stopper the tube loosely and heat in a water bath at 50° for 30 minutes or at 100°C for 5 minutes. After cooling, add 5.0 ml of the potassium hydroxide solution. The almost black solution that results rapidly clears to the characteristic wine-red color. Simultaneously make a blank determination, using 1.0 ml of the carbonyl-free methanol in place of the sample.

Determine the optical density of the solution using a Beckman Model DU spectrophotometer. Adjust the instrument for 100% transmittance for the solution from the blank determination, no further correction for the blank being necessary. Take the measurement at 480 nm and make the calculations using the average value of E_{max}. In later work the instrument was standardized using acetophenone and a graph was constructed to allow direct reading of carbonyl concentration from the observed optical density.

DISCUSSION

The method has been found to be applicable to a large number of aldehydes and ketones, both aliphatic and aromatic, as well as to some diketones. The only interfering structures so far encountered are nitroaromatic groups and conjugation of the chalcone-type ketones. Compounds containing such groups can still be determined by using the same compound for standardization. Accuracy of the order of 2 parts per hundred was obtained in the range of 5×10^{-6} to $10^{-4}M$ carbonyl. Carbonyl concentrations as low as $5 \times 10^{-7}M$ can be detected qualitatively.

Lappin and Clark have successfully used the method in the following applications: in the determination of carbonyl compounds in water solution, followed by chromatographic adsorption to concentrate the hydrazone. This concentrated hydrazone is then eluted from the column and compounds formed in certain rearrangement reactions (87, 88); in the

87. G. Lappin, *J. Am. Chem. Soc.*, **71,** 3966 (1949).
88. G. Lappin, unpublished research.

qualitative identification of aldehydes and ketones in an organic qualitative analysis course (for this purpose the intense color due to larger concentrations of carbonyl compounds makes it easy to distinguish visually between trace impurities and a major component); and in the determination of the number of carbonyl groups in a compound of known molecular weight (88).

The method of Lappin and Clark can be extended to limits lower than those specified earlier by the use of large samples for hydrazone formation, followed by chromatographic adsorption to concentrate the hydrazone. This concentrated hydrazone is then eluted from the column and the color intensity is measured. The foregoing technique has been used successfully in the laboratory of S. Siggia (89) to determine hydrazones down to the 0.1 ppm level on 100-gram samples. The chromatographic procedure used was essentially that of Gordon et al. (90). Keep in mind, however, that chromatographic conditions and adsorbents vary with the hydrazone being adsorbed and with the medium in which the hydrazone is contained. Thus no general chormatographic method can be described.

SCHIFF REACTION

The Schiff reaction is one commonly used to determine trace quantities of aldehydes. The reaction is as follows:

The Schiff reaction is not optimal for analytical purposes. It is subject to fading of the pink or red color of the product. This fading is attributed to the reaction of the colored quinoid product with the sulfurous acid to form the aldehyde bisulfite addition compound, although this is not known exactly. Also, this reagent is prone to give false positive values.

The Schiff reagent is usable, however, if it is adaptable to the system being analyzed. No specific method can be cited, since each system to be analyzed requires reaction conditions of its own.

Hoffpouir et al. (91) used the Schiff base reagent in determination of formaldehyde. Tobie (92) used it for determining aldehyde groups or sugars. The authors of this text have been successful in using essentially Tobie's method to determine traces of acetaldehyde and hydroxybutyraldehyde in aqueous systems.

The Tollen's reagent proved to be suitable for the determination of

89. P. M. Thomas, unpublished work.
90. B. E. Gordon, F. Wopat, Jr., H. D. Burnham, and L. C. Jones, *Anal. Chem.*, **23,** 1754–8 (1951).
91. C. L. Hoffpouir, G. W. Buckaloo, and J. D. Guthrie, *Anal. Chem.*, **15,** 605–6 (1943).
92. W. C. Tobie, *Anal. Chem.*, **14,** 405–6 (1942).

micromolar quantities of aldehyde if atomic absorption spectrophotometry is used to analyze the reduced silver content.

Adapted from the Method of P. J. Oles and S. Siggia

[*Anal. Chem.*, **46**, *911* (*1974*)]

APPARATUS

Absorbances were measured at 328.1 nm with a Perkin–Elmer 403 Atomic Absorption Spectrophotometer.

A fine frit (4–5.5 μm) Pyrex glass funnel was used for all filtrations. The use of a medium frit (10–15 μm) funnel results in significant losses of the silver precipitate; therefore the larger size was not used.

REAGENTS

TOLLEN'S REAGENT. Pipet 5.00 ml of 0.50*M* silver nitrate solution into a 50-ml beaker; add exactly 1.00 ml of 3*M* sodium hydroxide solution and agitate the contents. Then add dropwise 2.00 ml of 1 : 1 ammonia (sp. gr. 0.90) and water to dissolve all the silver oxide present. This reagent must be prepared fresh and should be used immediately after its preparation. A highly explosive black precipitate forms when the reagent stands for a time; therefore all unused reagent should be discarded within 4 hours of its preparation.

PROCEDURE

Add a 0.100- to 1.00-ml sample of the aldehyde in water to a 6-in. test tube. Depending on the reactivity of the aldehyde, the concentration should be between 0.25 and 4.00 μm per milliliter. Under minimum lighting conditions, add a volume of Tollen's reagent equal to the volume of the sample. Mix the contents of the test tube thoroughly, put the test tube in a light-tight container, and place it on a mechanical agitator for the specified time (Table 17). After this time has elapsed, transfer the contents of the test tube to a fine fritted glass funnel and apply suction. Rinse the test tube with two 5-ml portions of 1 : 1 ammonia-water, to dissolve any silver oxide present, and then with two 5-ml portions of water; in both cases allow the rinses to pass through the filter. Place a clean 125-ml flask below the filter and add 6 ml of 1 : 1 concentrated nitric acid–water to the test tube to dissolve the adhering silver mirror.

Table 17. Analysis of Aldehydes by Oxidation with Tollen's Reagent

Compound	Aldehydes Concentration, μM/ml	Taken, μM	Found, μM	Recovery, %	Relative Standard Deviation[a]	Minimum Reaction Time, min.
Formaldehyde	3.28	1.64	1.65	101	±2.5 (5)	30[b]
Propionaldehyde	3.47	1.74	1.72	98.8	±3.9 (5)	135
Butyraldehyde	1.13	1.13	1.13	100	±4.4 (4)	120
p-Nitrobenzaldehyde	0.990	0.990	1.00	101	±4.5 (5)	25
m-Nitrobenzaldehyde	1.029	1.029	1.03	100	±5.8 (5)	30
m-Cycanobenzaldehyde	1.004	1.004	0.97	96.6	±5.2 (5)	45
p-Chlorobenzaldehyde	1.007	1.007	1.02	101	±5.8 (5)	90
m-Methoxybenzaldehyde	0.985	0.985	1.02	103	±5.5 (5)	120
Benzaldehyde	3.93	1.96	1.98	101	±1.2 (3)	60[c]
3,5-Dimethoxybenzal-dehyde	1.001	1.001	1.01	101	±3.0 (5)	100
p-Nitrobenzaldehyde	0.990	0.099	0.094	95	±4.2 (3)	25
p-Nitrobenzaldehyde[d]	1.07	1.07	1.04	97.3	±5.7 (4)	25
p-Nitrobenzaldehyde[e]	1.00	1.00	1.00	100	±3.2 (4)	120

[a] Figures in parentheses indicate number of determinations.
[b] $0.1M$ $AgNO_3$ used in preparing reagent.
[c] $0.25M$ $AgNO_3$ used in preparing reagent.
[d] Analysis carried out in presence of 750 ppm potassium chloride.
[e] Sample in ethanol.

Transfer this solution to the filter and upon dissolution of the silver precipitate, apply suction. Rinse the test tube with two 5-ml portions of water and pass through the filter and collect the rinsings. Transfer the contents of the suction flask to a 50-ml volumetric flask and dilute to volume with water. When less than 0.50 μm of aldehyde is present, adjust the acid and water volumes so that the final volume of the sample is 10.0 ml. Analyze the solution by atomic absorption spectrophotometry for the silver content. Prepare a calibration curve by taking aliquots of the stock silver nitrate solution and diluting to the same final volumes as the sample.

RESULTS AND DISCUSSION

CHOICE OF SILVER OXIDANT. The silver–*tert*-butylamine complex (p. 119) that has been applied successfully on the millimolar scale was found to be unsuitable on the micromolar scale. Low and irreproducible results were obtained with the silver–*tert*-butylamine complex. It is postulated that in the presence of *tert*-butylamine, the following reaction may occur:

$$HCHO + H_2NC(CH_3)_3 \rightleftharpoons H_2C{=}NC(CH_3)_3 + H_2O$$

The imine formed in the reaction is not oxidizable by the silver reagent;

therefore the consumption of aldehyde by *tert*-butylamine results in low recoveries. The use of Tollen's reagent does not result in low recoveries, since the reaction of ammonia and formaldehyde (to form hexamethylenetetramine) would be unfavorable in low formaldehyde concentrations; therefore the oxidation reaction proceeded to completion.

It was found that in an open vessel, also under vacuum during the filtering operation, the following reactions occur:

$$Ag(NH_3)_2^+ \rightleftharpoons Ag^+ + 2NH_3$$
$$NH_3 + H_2O \rightleftharpoons NH_4^+ + OH^-$$
$$2Ag^+ + 2OH^- \rightleftharpoons Ag_2O + H_2O$$

The precipitation of Ag_2O was found to be a source of a high blank value at the start of the study. Washing the precipitate with ammonia-water dissolves any Ag_2O present and was found to reduce the blank value to the order of 0.06 to 0.08 μM aldehyde.

The results for aliphatic and aromatic aldehydes appear in Table 17. The determination of 0.099 μM of *p*-nitrobenzaldehyde was carried out with a final sample volume of 10.0 ml. In all other cases, samples were diluted to 50.0 ml so that the silver concentration would lie in the linear region of the calibration curve for silver. The reactivity of aliphatic aldehydes, other than for formaldehyde, was observed to be much less than the substituted benzaldehydes at the concentrations studied. For this reason, their determination is best accomplished by using smaller amounts of more concentrated reagent. The results in Table 18 indicate that the time required for the quantitative recovery of the aldehyde is drastically reduced by adding less reagent, thereby reducing the dilution

Table 18. Effect of Dilution of Aldehyde by Adding Various Amounts of Reagent: Analysis of 1.00 μM per milliliter of *p*-Chlorobenzaldehyde

Time, min.	Recovery, % 1.00 ml Reagent Added[a]	Recovery, % 0.50 ml Reagent Added[b]
15	70	76
30	79	100
60	88	100
75	100	100

[a] Final approximate concentrations are: $[Ag^+] = 0.3F$, $[RCHO] = 0.5$ μmole/ml, $[OH^-] = 0.37F$, $[NH_4OH] = 1.9F$.

[b] Final approximate concentrations are: $[Ag^+] = 0.2F$, $[RCHO] = 0.67$ μmole/ml, $[OH^-] = 0.25F$, $[NH_4OH] = 1.3F$.

Table 19. Determination of Total Aldehyde Concentration in Mixtures of Aldehydes

	Aldehyde				
Mixture[a]	Concentration, μM ArCHO/ml	Taken, μM	Found, μM	Recovery, %	Relative Standard Deviation[b]
0.535 μM *m*-nitro-benzaldehyde + 0.428 μM *p*-nitro-benzaldehyde	0.963	0.963	0.97	101	±4.6 (5)
0.299 μM *m*-cyano-benzaldehyde + 0.529 μM 3,5-di-methoxybenzaldehyde	1.655	0.828	0.82	100	±5.5 (4)

[a] Analysis accomplished with reagent prepared from 1*M* $AgNO_3$.
[b] Figures in parentheses indicate number of determinations.

of the aldehyde to a minimum. The Tollen's reagent used in this study was prepared by pipetting 5.00 ml of 1.0*M* silver nitrate into a beaker and adding 1.00 ml of 6*M* sodium hydroxide solution and 2.00 ml of concentrated ammonia. The volume of this reagent found to produce the optimum time required for the quantitative recovery of *p*-chloro-benzaldehyde was 0.50 ml.

The results of the analysis of some mixtures of aldehydes appear in Table 19. It was learned that each aldehyde in a mixture reacts independently, and the reaction time required for the quantitative recovery of the total aldehyde content of the sample is approximately the same time required for the quantitative recovery of the slower reacting component in a mixture.

THE HAMMETT RELATIONSHIP. The existence of a Hammett relationship (Fig. 2.7) among substituted benzaldehydes was considered as a means of systematizing their reactivity. The Hammett relationship is defined as follows:

$$\log k/k' = \rho\sigma$$

where k is the rate constant of the substituted benzaldehyde, k' is the rate constant of benzaldehyde, ρ is a measure of the sensitivity of the reaction rate to changes in the σ value of the substituent, and σ is the Hammett substituent constant. Since k is inversely proportional to the half-life of

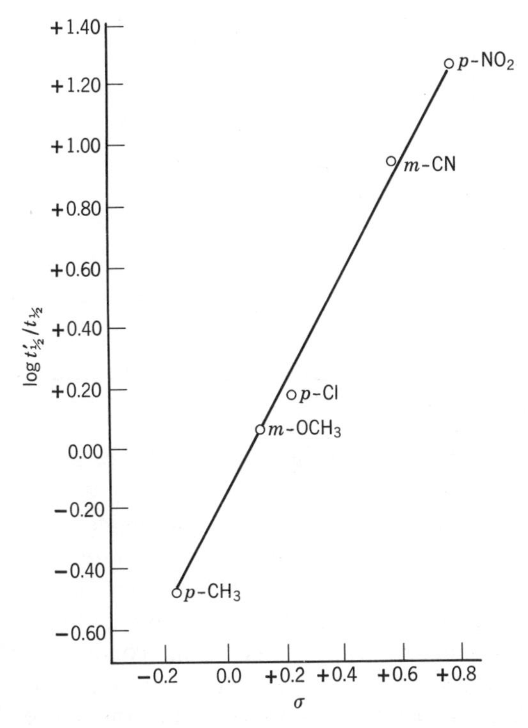

Fig. 2.7. The Hammett relationship for the oxidation of substituted benzaldehydes with Tollen's reagent.

the reaction $t_{1/2}$, the equation may be rewritten

$$\log t'_{1/2}/t_{1/2} = \rho\sigma$$

Under the conditions used in this study, $\rho = +1.8$, indicating that substituents will greatly affect the reactivity of the substituted benzaldehyde. The practical significance of the large absolute value of the slope in Fig. 2.7 is illustrated by a comparison of the rate curves for *p*-nitrobenzaldehyde ($\sigma = +0.778$) and *p*-tolualdehyde ($\sigma = -0.170$) in Figs. 2.8 and 2.9. A comparison of rate curves for 3,5-dimethoxybenzaldehyde and *p*-chlorobenzaldehyde (Figs. 2.10 and 2.11) indicate that, in some cases, multiple substituents will have an additive effect on the reactivity, since $\sigma = +0.227$ for *p*-chlorobenzaldehyde and $\sigma = 0.115 \times 2 = +0.230$ for 3,5-dimethoxy. *p*-Methoxybenzaldehyde ($\sigma = -0.268$) does not undergo a quantitative reaction with Tollen's reagent either at the micromolar or the millimolar level. Since the oxidation is slow, the side reactions of ammonia and the aldehyde to form the substituted hydroxybenzamide

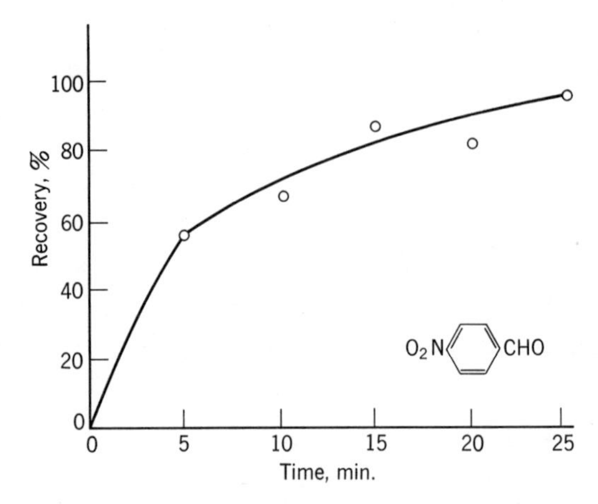

Fig. 2.8. Reaction versus time curve for *p*-nitrobenzaldehyde.

may occur to a significant extent (93). *p*-Acetamidobenzaldehyde ($\sigma =$ 0.00) should react at approximately the same rate as benzaldehyde. However quantitative recoveries were not obtained, presumably because *p*-acetamidobenzaldehyde hydrolyzes under the experimental conditions to *p*-aminobenzaldehyde, which may then condense with another molecule of *p*-aminobenzaldehyde to form the resultant imine. Further *p*-aminobenzaldehyde ($\sigma = 0.66$) would be quite unreactive. Vanillin (4-hydroxy-3-methoxybenzaldehyde) did not react under the experimental

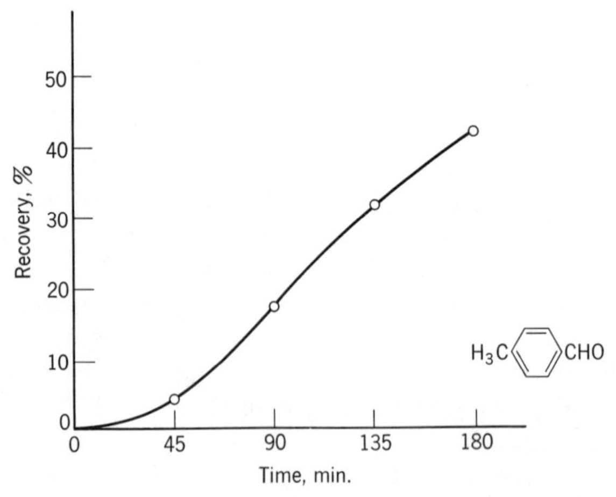

Fig. 2.9. Reaction versus time curve for *p*-tolualdehyde.

93. Y. Ogate, A. Kawasaki, and N. Okumura, *J. Org. Chem.*, **29,** 1986 (1964).

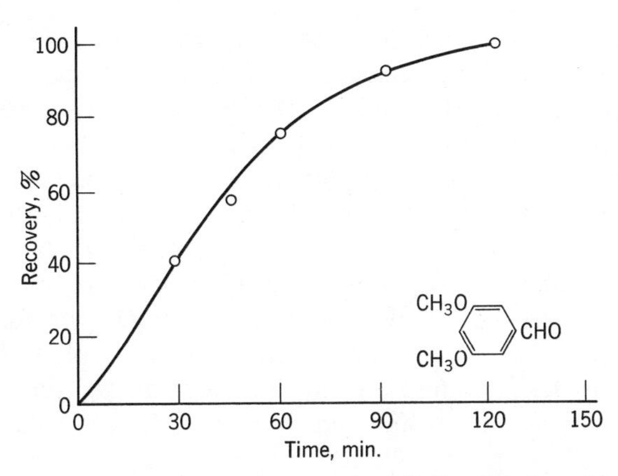

Fig. 2.10. Reaction versus time curve for 3,5-dimethoxybenzaldehyde.

conditions after a period of 3 hours. The additive effect of *p*-O^- ($\sigma = -1.00$) and *m*-OCH_3 ($\sigma = 0.115$) may be used to predict that overall vanillin should be quite unreactive at the concentration in this work (1.00 μM/ml).

The results of the analysis of *p*-nitrobenzaldehyde (1.07 μM/ml) containing 750 ppm potassium chloride appear in Table 17. It was observed that sufficient ammonia is present in the reagent to just dissolve any silver chloride that precipitated. Larger amounts of chloride present in a sample would probably require additions of ammonia, which would decrease the rate of reaction.

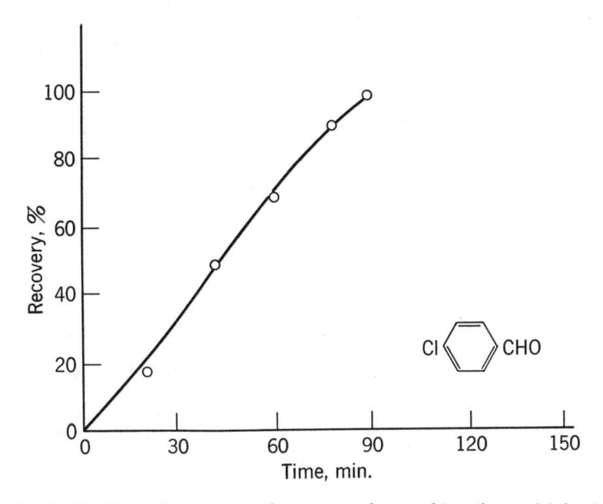

Fig. 2.11. Reaction versus time curve for *p*-chlorobenzaldehyde.

A solution of *p*-nitrobenzaldehyde (1.00 μM/ml) in pure ethanol was prepared and subsequently analyzed for aldehyde content to determine the effect of the solvent on reactivity. Blank values were somewhat higher in this solvent (equivalent to about 0.4 μM RCHO); however they were reproducible and were not a function of time. When 1.00 ml of the Tollen's reagent is added to a 1.00-ml sample of ethanol, a precipitate of silver oxide is immediately observed. The addition of approximately 0.2 ml of 1:1 ammonia-water is required to dissolve the silver oxide precipitate. The results of this determination appear in Table 17. Two hours is required to obtain quantitative recovery of 1.00 μM of *p*-nitrobenzaldehyde. Therefore the use of ethanol as a solvent for aldehyde in this procedure is not recommended at the aldehyde concentrations of the order of 1.00 μM per milliliters.

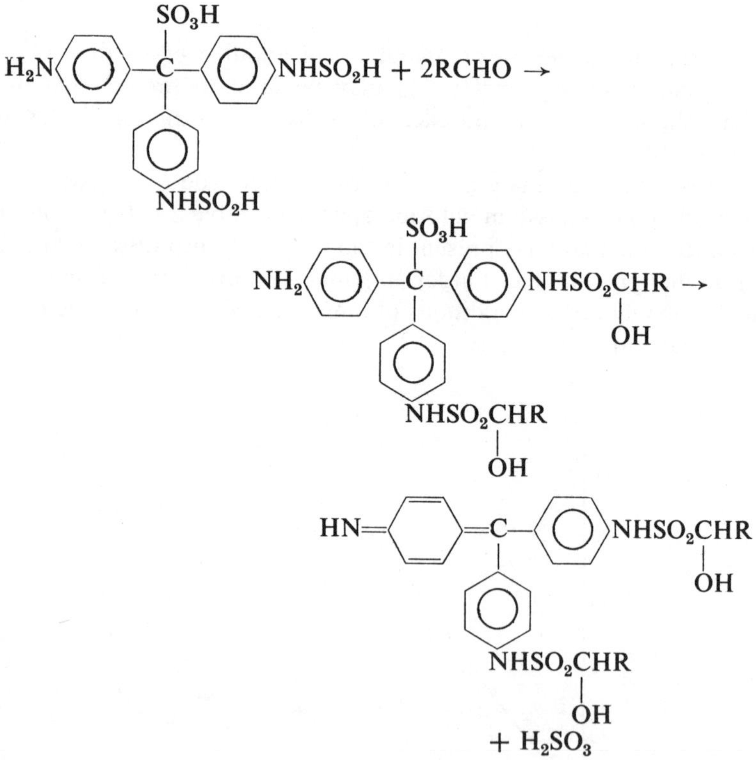

SO_3H

H_2N—C$_6$H$_4$—C—C$_6$H$_4$—$NHSO_2H$ + 2RCHO →

$NHSO_2H$

SO_3H

NH_2—C$_6$H$_4$—C—C$_6$H$_4$—$NHSO_2CHR$ →

OH

$NHSO_2CHR$

OH

HN=C$_6$H$_4$=C—C$_6$H$_4$—$NHSO_2CHR$

OH

$NHSO_2CHR$

OH

+ H_2SO_3

3

Carboxylic Acids, Salts, Esters, Amides, Imides, Chlorides, and Anhydrides

Carboxylic Acids

Carboxylic acids are generally moderately strong acids. The bulk of the existing carboxylic acids have dissociation constants of 10^{-5}, of the order of acetic acid or benzoic acid. The structure of the acid has a minor influence on the acid strength of the carboxyl group, but substituents on the molecule can have marked affects on the acidity. For example, all the aliphatic acids, branched and straight chained, have quite similar dissociation constants (see Table 1) (1). Substituents such as halogens, nitro groups, hydroxyl groups and other carboxyl groups, however, noticeably intensify the acidity of the carboxyl groups. Table 2 shows the effect of substituents in acetic and benzoic acids.

The effect of these substituents on acid strength is a function of proximity of these substituents to the carboxyl group. Hence on aliphatic acids the alpha substituents show the strongest effect (Table 3) and on aromatic acids ortho substituents have a more enhanced effect than meta or para (Table 4).

It is quite evident from the dissociation values in Table 3 that the majority of carboxylic acids are readily titratable with sodium hydroxide using no special conditions, reagents, or indicating systems.

The pK_a values also indicate that various carboxylic acids can be differentiated by titration with base. For example, *o*-phthalic acid can be differentiated from benzoic acid by a potentiometric titration. The phthalic acid yields two inflection points in the titration, one for each acid group. Benzoic acid yields only one inflection point, which coincides with the second inflection point for phthalic acid. Hence the phthalic acid content in a mixture can be computed from the first inflection point and the total acid from the second. The benzoic acid content is obtained by difference.

An aqueous system is usually adequate if the sample dissolves in water. If the sample is insoluble in water, it may dissolve in excess aqueous caustic, and the excess caustic is titrated. A potentiometric titration is

1. E. A. Brande and F. C. Nachod, *Determination of Organic Structures by Physical Methods*, Academic Press, New York, 1955, pp. 573–688.

Table 1

Acid	pKa
Acetic	4.76
Propionic	4.88
n-Butyric	4.82
iso-Butyric	4.86
n-Valeric	4.86
iso-Valeric	4.78
Hexanoic	4.88
Heptanoic	4.89
Octanoic	4.90
Nonanoic	4.95

advisable when one is working with acids of unknown strength; an appropriate indicator can be chosen when the pH at the titration break has been determined.

For water-insoluble samples or samples that give poor titration curves in an aqueous system, a nonaqueous system can be used. Much sharper breaks are obtained for weak acids in acetone, dimethylformamide, and methanol-benzene, than are obtained in water. Out of these three, at least one solvent can be found that will dissolve stubborn samples.

In the acetone solvent, the sample can be titrated with 0.1*N* alcoholic

Table 2[a]

Substituent	pKa	
	Acetic Acid	*o*-Benzoic Acid
	4.76	4.20
—Cl	2.86	2.94
—Br	2.86	2.85
—OH	3.83	2.98
$—NO_2$	1.68	2.17
—COOH	3.40 (malic acid)	2.98 (*o*-phthalic acid)
—C≡N	2.43	—
—SH	3.67	—
$—NH_2$	—	4.98
$CH_3NH—$	—	5.33
$(CH_3)_2N—$	—	8.42

[a] From Ref. 1.

Table 3[a]

Substituent	Propionic pKa	Butyric pKa
None	4.88	4.82
α-Cl	2.80	2.84
β-Cl	4.08	4.06
γ-Cl	—	4.52
α-Br	2.98	2.99
β-Br	4.02	—
γ-Br	—	4.58
α-OH	3.86	4.22
β-OH	4.51	4.52
γ-OH	—	4.72
α-SH	3.70	—
β-SH	4.34	—

[a] From Ref. 1.

Table 4

Acid	Substituent Group	pKa		
		Ortho	Meta	Para
Benzoic (pKa = 4.2)	None	—	—	—
	—Cl	2.94	3.83	3.99
	—Br	2.85	3.81	4.00
	—OH	2.98	4.08	4.58
	—COOH	2.98	3.46	3.51
	$—NO_2$	2.17	3.45	3.44
	$CH_3NH—$	5.33	5.10	5.04
	$(CH_3)_2N—$	8.42	5.10	5.03

(methanol) caustic. The ordinary glass and calomel electrodes can be used in potentiometric titrations in this solvent.

In dimethylformamide (2), the sample can be titrated with 0.1 to 0.2*N* sodium methylate in benzene-methanol. (About 5 grams of sodium is cleaned with methanol and then dissoved in 100 ml of absolute methanol. Cooling in ice water may be necessary at times to slow down the reaction. When all the sodium has reacted, 150 ml of methanol and 1500 ml of benzene are added.) Thymol blue in a 0.3% solution in methanol can

2. J. S. Fritz, *Acid-Base Titrations in Nonaqueous Solvents*, G. Frederick Smith Co.. Columbus, Ohio, 1952, pp. 28–9.

often be employed as indicator. The glass and calomel electrodes can be used in this solvent, as can the antimony-calomel electrodes.

Benzene-methanol (3) is a good solvent for titrating acids. The titrant is the same as that used in dimethylformamide. The authors claim that an antimony versus a calomel electrode should be used, and that a little lithium chloride should be added to decrease the resistance in the solutions. However, S. Siggia has been able to use the glass and calomel electrodes satisfactorily without electrolyte. There may be a difference in behavior of the electrodes depending on the materials being determined, so it is best to keep both electrode systems in mind, in case one fails to operate in the analysis being made.

If the particular carboxylic acids being analyzed happen to be unusually weak, that is, dissociation constants smaller than 10^{-6}, the nonaqueous titration systems used for enols and phenols (pp. 46–57) can be applied. These cases, however, do not arise often.

TITRATION OF DILUTE AQUEOUS SOLUTIONS OF AMINO ACIDS

The carboxyl group of amino acids is not strong enough in aqueous solution to permit direct titration with standard base because of the influence of the amino group present in the molecule. However, the amino group may be effectively masked by reaction with an aldehyde to form a Schiff base. By this reaction, as in the elimination of the basicitly of primary amines in mixtures of amines (p. 581), the titration of the acid group is possible.

$$\underset{\displaystyle RCHCOOH}{\overset{\displaystyle NH_2}{|}} + HCHO \rightarrow \underset{\displaystyle RCHCOOH}{\overset{\displaystyle N{=}CH_2}{|}} + H_2O$$

A procedure based on this reaction was first proposed by Sorensen (4) and was subsequently adopted and widely used by many workers, as noted by Taylor (5). The following method was developed for amino acids in lemon juice but is generally adaptable.

Method of C. E. Vandercook, A. Rolle, and R. M. Ikeda

[*Adapted from J. Assoc. Off. Agr. Chem.*, ***46****, 353* (*1963*)]

Titrate a quantity of 37% formaldehyde, which can be used within an hour, to a pH of 8.4 with 0.04*N* sodium hydroxide solution. Use a pH meter to make the measurements. Neutralize a 25-ml aqueous aliquot of the sample containing 1 to 3 meq of amino acid to pH 8.4. Make the final

3. J. S. Fritz and N. M. Lisicki, *Anal. Chem.*, **23,** 589–91 (1951).
4. S. P. L. Sorensen, *Biochem. Z.*, **7,** 45 (1907).
5. W. H. Taylor, *Analyst,* **82,** 488 (1957).

adjustment with 0.04*N* acid or base. Add 10 ml of the formaldehyde solution and titrate the resulting acidity to pH 8.4 with standard 0.04*N* sodium hydroxide solution.

Carboxylic Acid Salts

Salts of carboxylic acids are generally weak bases; however the ionic strength of the salt can be relatively evaluated from the strength of the corresponding free acid. Thus the salts of the same cation with the acids listed in Table 1 will all have essentially the same character. As the acid strength of the free acid increases, the basic strength of the salts decreases.

The most applicable method for determining carboxylic acid salts is direct titration with standard acid. The common carboxyl salts, the sodium and potassium salts of unsubstituted acids such as those in Table 1, however, are too weak to be titrated accurately and precisely in aqueous media. Indicator end points are vague, and potentiometric end points are correspondingly poor. Titration of these salts is best carried out in nonaqueous media. The only carboxylic acid salts that can be adequately titrated in water solution are the salts of amino acids. The amino group enhances the basicity of the salt.

The solvents that are most useful for titrating carboxylic acid salts are glacial acetic acid and mixtures of glycols with other solvents. The acetic acid solvent has the advantage of generally making possible the sharpest end points. The glycol-solvent mixtures do not usually yield as sharp end points as acetic acid, but they are generally better solvents, and the mixtures can be varied to dissolve a wide range of materials. Furthermore, the glycol-solvent mixtures usually permit better differentiation of multiple bases present in the same mixture; acetic acid generally yields only one end point, including all bases in the sample.

In addition to the titration methods, there is a combustion method for determining carboxylic acid salts. This involves burning the sample, in which case the carboxyl salt is converted to the corresponding carbonate. The carbonate is then determined. This combustion approach is not as general or as simple to apply as is the titration approach; however it has application to insoluble carboxyl salts and also to mixtures not otherwise amenable to titration.

TITRATION METHOD USING GLYCOL–NONGLYCOL SOLVENT MIXTURES

The solvent mixtures used in this analysis cover a wide range of compositions. They consist of either ethylene or propylene glycol mixed with almost any other solvent, that is, alcohols, hydrocarbons (aromatic and aliphatic), chlorinated compounds, and ethers. In general, the composition of these mixtures can vary widely, but when nonpolar cosolvents are used, the glycol content of the solvent mixtures should be greater than 25% to obtain potentiometric measurements. Furthermore, with nonpolar cosolvents such as benzene, chloroform, and petroleum ethers, propylene glycol is the glycol of choice, since it will mix with these solvents. Ethylene glycol is not miscible with the nonpolar solvents.

The most generally applicable solvent mixture of this type is ethylene glycol–isopropanol, which has a wide range as a solvent and yields good end points for carboxylic acid salts.

Method of S. Palit

[*Ind. Eng. Chem., Anal. Ed.,* ***18,*** *246–251* (*1946*)]

This method involves the use of special solvent mixtures in which the end point during standard acid addition is sharper than in water. These solvents are mixtures of a glycol-type solvent and a solvent for hydrocarbons, such as hydrocarbons themselves, alcohol, and chlorinated hydrocarbons. The solution can be titrated directly with hydrochloric acid or perchloric acid dissolved in the same solvent mixture. Indicators or a pH meter can be used to indicate the end point.

Glycols are good solvents for salts of carboxylic acids; this solubility is attributed to hydrogen bonding between the solvent and solute. The hydrocarbons in the solvent mixture contribute to the solution of salts of long-chain fatty acids and also contribute to the sharpening of the end point. The best solvent mixture for most general applications is ethylene glycol or propylene glycol and isopropyl alcohol. In the determination of salts of long-chain fatty acids, such as sodium stearate, the isopropyl alcohol does not have sufficient dissolving power. More powerful solvents such as butyl or amyl alcohol, chloroform, or dioxane have to be used.

The proportions of the solvents that can be used successfully extend over a wide range; 15 to 70% isopropyl alcohol by volume is satisfactory. However, a 1:1 mixture by volume of ethylene or propylene glycol with isopropanol is used as a standard solvent medium; although in general, 20% by volume of isopropanol is slightly better as to sensitivity and solubility.

Any free strong base contained in the carboxylic acid salt can be determined, since on addition of the acid the free alkali will yield a break in the pH versus milliliters of standard acid curve at a higher pH than the salt. A pH meter or separate indicators for the free base and the carboxylic acid salt can be used to indicate the two end points.

Both hydrochloric and perchloric acids give satisfactory results. The latter is more advantageous in titrating concentrated solutions, since the perchlorate formed is more soluble in the special solvent than the chloride and consequently, does not produce a turbidity during the titration.

PROCEDURES

POTENTIOMETRIC METHOD. Weigh a sample of about 0.01 to 0.04 mole of salt into a 150-ml beaker together with 40 ml of a 1:1 ethylene glycol-isopropyl alcohol or other appropriate solvent mixture, depending on the material being determined. (See foregoing discussion for details.) Use the standard Beckman pH meter electrodes (glass and calomel). Adjust the electrodes with an aqueous buffer at pH 7. Rinse the electrodes with water, gently wipe with cleansing tissue, and immediately immerse in the solution. Steady readings are usually obtained in 2 minutes. Add 0.5 to 1.0*N* acid (perchloric or hydrochloric) in the same solvent mixture as used for the sample from a 50-ml buret. The readings are taken in the usual manner. The pH readings have no absolute significance in these solvents, but the sharp break in the pH versus milliliters of acid curve is the change sought.

INDICATOR METHOD. Weigh a sample of about 0.004 mole of salt into a 125-ml Erlenmeyer flask and dissolve in a minimum of 1:1 ethylene glycol–isopropanol or other convenient solvent. Quick solution can be brought about by adding 10 ml of the glycol to soften the sample, warming the sample if necessary. After the sample is well swollen, add an equal volume of isopropanol or other solvent (usually chororform is used as alternate) and dissolve the sample. Add 3 to 5 drops of a 0.05% alcoholic solution of methyl red (or methyl orange), and titrate the solution with standard 0.2*N* perchloric acid in 1:1 glycol–isopropanol solution to the pink color. (Chloroform sometimes requires a blank because of the free hydrochloric acid it may contain.)

Sodium salts of acetic, propionic, butyric, oleic, stearic, cinnamic, and benzoic acids give satisfactory results by these methods. Salts of acids stronger than formic acid ($K = 2.1 \times 10^{-4}$) cannot be titrated without losing some accuracy.

Acidic or basic impurities in the sample will interfere. Weak bases such as aromatic amines can be determined by this method (see pp. 533–35, 546).

TITRATION IN GLACIAL ACETIC ACID

Glacial acetic acid has been used very successfully for titrating carboxylic acid salts. The details for this procedure can be found on page 545, where the titration of amines is discussed. The procedure was tested with sodium acetate, sodium propionate, sodium benzoate, sodium stearate, sodium citrate, potassium formate, potassium oxalate, and calcium gluconate, and was found to be very neat and easily applied. Very good indicator end points are obtained, but potentiometric titration can also be used with the ordinary pH meter and glass and calomel electrodes. Precisions of ±0.3% are easily attained. This solvent system usually gives sharper titration curves than the glycol-isopropanol system, but the glycol-isopropanol system has the advantage of the flexibility of solvent for dissolving many different types of samples.

COMBUSTION METHOD

Adapted from the Combustion Method of Siggia and Maisch

[*Ind. Eng. Chem., Anal. Ed.*, ***20,*** *235–6* (*1948*)]

This method is rather simple in application and principle. The salt of a carboxylic acid is oxidized to the corresponding carbonate, and the carbonate is determined acidimetrically.

REAGENTS

0.5*N* Aqueous sulfuric acid.
0.5*N* Aqueous sodium hydroxide.
Phenolphthalein indicator—methyl red indicator.

PROCEDURE

Weigh a sample containing about 0.02 equivalent of carboxylic acid salt into a platinum crucible. Platinum is used because the carbonate will react with porcelain. Ignite the contents until all traces of carbon have been burned off. Cool the crucible and drop it into a 250-ml beaker containing 50 ml of standard 0.05*N* sulfuric acid. Use a watch glass to cover the beaker and prevent loss of solution by spraying during the evolution of carbon dioxide. Boil the solution for 20 to 30 minutes to drive off all the dissolved carbon dioxide. Cool the contents and titrate the excess acid with standard $0.5N$ sodium hydroxide, phenophthalein indicator being used. With barium and calcium salts, methyl red indicator must be substituted and hydrochloric acid should be used instead of sulfuric acid, since it is more efficient in effecting solution of the carbonates. Barium and calcium sulfates, being insoluble, make solution of the carbonate in sulfric acid difficult. The ignition residues dissolve readily in hydrochloric acid. To avoid loss of hydrochloric acid during boiling, however, it is best to titrate the excess acid with standard alkali as soon as the ignition residue is dissolved. Then make the solution barely acid with hydrochloric acid. The solution can then be boiled to eliminate the carbon dioxide without appreciable loss of hydrochloric acid. Titrate the solution to the end point.

CALCULATIONS

A = milliliters of alkali needed to titrate all standard acid used in analysis
B = milliliters of alkali used in titration of sample
$A - B$ = milliliters of alkali equivalent to carboxylic acid salt $= C$

$$\frac{C \times N\ \text{NaOH} \times \text{mol. wt. salt} \times 100}{\text{Grams of sample} \times 1000 \times B} = \%\ \text{carboxylic acid salt}$$

B = valence of cation on salt × number of cation atoms per molecule of salt

To test this procedure the following were used: sodium acetate, sodium benzoate, sodium citrate, sodium succinate, sodium caprylate, sodium palmitate, sodium laurate, sodium caprate, sodium potassium tartrate, potassium succinate, potassium acid phthalate, calcium acetate, calcium gluconate, calcium citrate, calcium stearate, barium acetate, and barium tartrate.

Esters of Carboxylic Acids

SAPONIFICATION METHODS

Esters may be very simply determined using the saponification reaction

$$\text{R}\overset{\text{O}}{\overset{\|}{\text{C}}}\text{OR}_1 + \text{NaOH} \rightarrow \text{RC}(=\text{O})\text{—ONa} + \text{R}_1\text{OH}$$

A known excess amount of sodium hydroxide is added to the sample and the excess is determined by titration with acid. Since esters vary widely in reactivity, the saponification conditions used must be varied accordingly.

Cyclic esters (lactones—$(CH_2)_n$ ring structure with C(=O)—O—C)

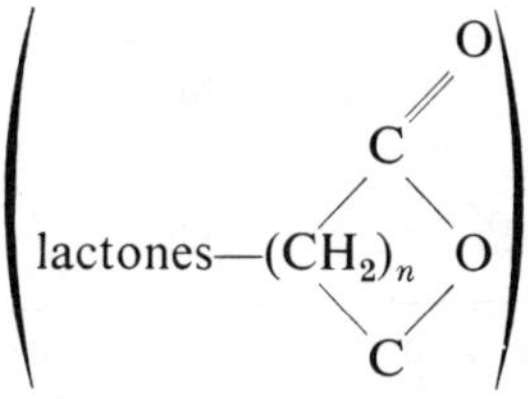

react very rapidly with alkali, in fact, so rapidly that they can often be titrated directly with standard alkali as we would titrate a free acid. At the other extreme of reactivity, polymeric esters, especially those where the ester groups are on a polymeric carbon chain

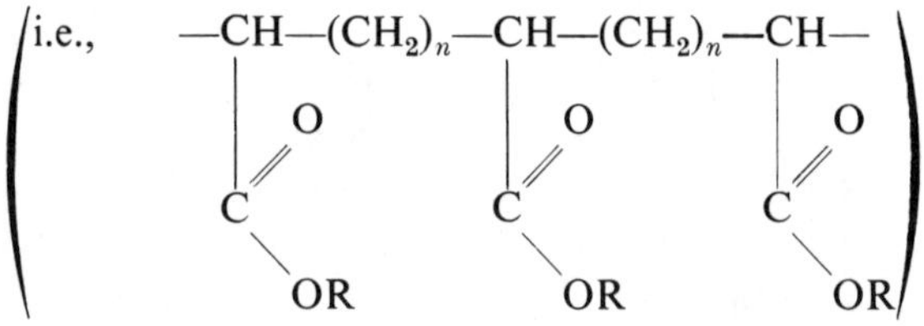

$$\left(\text{i.e.,}\quad -\underset{\underset{OR}{|}}{\overset{}{CH}}\!\!-\!(CH_2)_n\!-\!CH\!-\!(CH_2)_n\!-\!CH- \text{ with each CH bearing } -C(=O)OR\right)$$

are very difficult to saponify, probably because of steric hindrance problems. Polyesters, namely, condensation polymers of polyfunctional acids and polyfunctional alcohols,

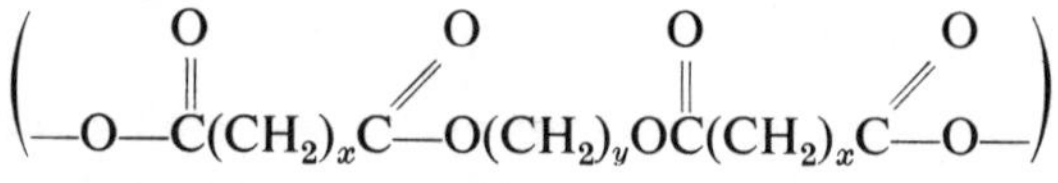

$$\left(-O-\overset{O}{\overset{\|}{C}}(CH_2)_xC(=O)-O(CH_2)_yO\overset{O}{\overset{\|}{C}}(CH_2)_xC(=O)-O-\right)$$

are not as stubborn to saponification as the aforementioned polymeric esters. In the latter case, the chains are broken by saponification and, as the chain gets smaller, the reactivity increases.

PROCEDURE

Weigh a sample containing about 0.01 mole of ester into a 250-ml glass-stoppered Erlenmeyer flask with a condenser to fit the ground joint for use with volatile or water-insoluble samples. To the sample, add 50 ml of 0.5*N* sodium hydroxide, aqueous if the sample is soluble in water or alcoholic if the sample is insoluble in water. Heat the solution for 2 hours on a steam bath (longer if the particular ester can be saponified only with difficulty). The heating should be under reflux for volatile samples or if alcoholic sodium hydroxide is used. After the specified length of time, titrate the excess alkali with standard $0.5N$ acid, using phenolphthalein indicator.

For esters that saponify with difficulty, amyl alcohol can be used as a solvent instead of methanol. Amyl alcohol has a higher boiling point, and this will accelerate the reaction. Potassium hydroxide should be used in amyl alcohol because it has a greater solubility than sodium hydroxide in this solvent. Also, concentrations of potassium hydroxide up to $5N$ can be used for saponifying very stubborn esters. When this is done, however, the potassium hydroxide must be standardized each day and a blank should be heated for the same length of time as the sample, since some hydroxide is lost under these extreme conditions either by reaction with the glass or with some impurity in the amyl alcohol.

For samples insoluble in the alcoholic solvents alone, benzene can be included. It is a good idea to dissolve the sample in enough benzene that when the alcoholic base is added, the sample will remain in solution. For titrating the excess base in this solvent system as well as in the amyl alcohol, a standard solution of hydrochloric acid in 1:1 ethylene glycol–isopropanol can be used. Use phenolphthalein indicator for most cases, but potentiometric titration with the glass and calomel electrodes can also be employed. These electrodes operate well in all these solvents except in the benzene-alcohol systems when the amount of benzene exceeds the amount of alcohol.

To determine small quantities of esters, the normality of the alkali can be reduced to as low as $0.01N$. However in these cases care must be taken to account for reaction of the alkali with the glass apparatus. At these low concentrations, the small amount of reaction becomes significant. Alkali-resistant glass must be used, and blank analyses must be run to provide corrections for any reaction with the glass. It is well in these cases to avoid long reaction times.

CALCULATIONS

$$\text{Milliliters of acid to titrate 50 ml of the alkali minus milliliters of acid for sample} = A$$

$$\frac{A \times N \text{ acid} \times \text{mol. wt. of ester} \times 100}{\text{Grams of sample} \times 1000} = \%\ \text{ester}$$

The value of A should be corrected for any free base or acid in the sample.

The procedure was tested with ethyl acetate, diethyl succinate, dibutyl phthalate, glycerol monoacetate, methyl salicylate, methyl acrylate, and ethyl benzoate.

Amides constitute an interference, since they will also hydrolyze with alkali to yield carboxylic acid salts. The foregoing procedure can be used to determine many amides. A longer time and stronger alkali for the hydrolysis, however, are generally required.

$$RCONH_2 + NaOH \rightarrow RCOONa + NH_3$$

The liberated ammonia can be dissolved in an excess of standard acid and the excess acid titrated, or the ammonia can be boiled off and the excess alkali used in the hydrolysis can be titrated.

TRACE QUANTITIES OF CARBOXYLIC ESTERS

As seen earlier, small quantities of esters can be determined by saponification, but this approach is not practical below the 100-ppm level of ester content. In this low area, a colorimetric approach is preferred.

Colorimetric Method

[*Adopted from R. F. Goddu, N. F. LeBlanc, and C. M. Wright, Anal. Chem.,* ***27,*** *1251–5 (1955)*]

An organic ester, acid chloride, or anhydride may form a hydroxamic acid by the reactions shown in eqs. 1 to 3. Most hydroxamic acids combine with ferric ion to form characteristic red to purple chelate complexes (eq. 4) that can be measured spectrophotometrically. Work is currently under way to define more clearly the value of n in the ferric-hydroxamate complex of eq. 4 (6).

$$RCOOR' + NH_2OH \xrightarrow{OH^-} R{-}\overset{\overset{\displaystyle O}{\|}}{C}{-}NHOH + R'OH \qquad (1)$$

$$RCOCl + NH_2OH \longrightarrow R\overset{\overset{\displaystyle O}{\|}}{C}NHOH + HCl \qquad (2)$$

$$(RCO)_2O + NH_2OH \longrightarrow R\overset{\overset{\displaystyle O}{\|}}{C}{-}NHOH + RCOOH \qquad (3)$$

$$1/nFe^{+++} + R\overset{\overset{\displaystyle O}{\|}}{C}{-}NHOH \longrightarrow R{-}\underset{\underset{\displaystyle O}{\|}}{C}{-}\underset{\underset{\displaystyle O}{|}}{N}{-}H + H^+ \qquad (4)$$

(the two O atoms coordinated to Fe/n)

6. W. W. Brandt, personal communication.

Hydroxamic acids were first reported in 1869 by Lossen (7), who observed that they could be rearranged to isocyanates (the well-known Lossen rearrangement). It was not until 1934, however, that the analytical possibilities of the hydroxamic acids were widely exploited, when Feigl and his co-workers (8) reported a spot test for esters and anhydrides based on the color reaction described. Since then, the reaction has been used to detect and determine many types of esters (9–14), amides (15–17), anhydrides (8–18), and nitriles (17). Very little of this previous work has been concerned with a systematic study of the reaction variables. To ascertain optimum conditions for analytical determinations, test runs should be made with the compounds under investigation.

APPARATUS

A Beckman Model B spectrophotometer with 1-cm cells was used for this work. For routine work a colorimeter with a 525-nm filter is satisfactory.

Erlenmeyer flasks with $\mathfrak{T}$ 19×22 glass joints, 25-ml capacity.

Small condensers with $\mathfrak{T}$ 19×22 glass joints.

REAGENTS

Hydroxylamine hydrochloride, 12.5%, in methanol.

Sodium hydroxide, 12.5%, reagent grade, in methanol.

Both hydroxylamine hydrochloride and sodium hydroxide are prepared by refluxing 12.5 grams of the solid material with 100 ml of methanol for a few minutes. The sodium hydroxide solutions are usually cloudy because of precipitated sodium carbonate. The hydroxylamine hydrochloride solution is about 1.8*M* and the sodium hydroxide is 3.1*M*.

FERRIC PERCHLORATE, STOCK SOLUTION. Dissolve 5.0 grams of ferric perchlorate (nonyellow, G. F. Smith Chemical Co.) in 10 ml of 70% perchloric acid and 10 ml of water. Dilute to 100 ml with anhydrous 2B alcohol, cooling under a tap as the alcohol is added.

Alternatively weigh out 0.8 gram of iron wire into a 50-ml beaker. Add 10 ml of 70% perchloric acid and heat on a hot plate at low heat until the iron dissolves. Be careful, for the iron dissolves very rapidly when the acid is hot. Cool the

7. H. Lossen, *Ann.*, **150,** 314 (1869).
8. F. Feigl, V. Anger, and O. Frehden, *Mikrochemie*, **15,** 12 (1934).
9. F. E. Bauer and E. F. Hirsh, *Arch. Biochem.*, **20,** 242–50 (1949).
10. R. E. Buckles and C. J. Thelen, *Anal. Chem.*, **22,** 676 (1950).
11. U. T. Hill, *Ind. Eng. Chem., Anal. Ed.*, **18,** 317–19 (1946).
12. U. T. Hill, *Ibid.*, **19,** 932–3 (1947).
13. A. G. Keenan, *Can. Chem. Process Ind.*, **29,** 857–8 (1945).
14. A. R. Thompson, *Aust. J. Sci. Res.*, **3A,** 128–35 (1950).
15. F. Bergman, *Anal. Chem.*, **24,** 1367 (1952).
16. J. G. Polya and P. L. Tardew, *Anal. Chem.*, **23,** 1036 (1951).
17. S. Soloway and A. Lipschitz, *Anal. Chem.*, **24,** 898 (1952).
18. W. M. Diggle and J. C. Gage, *Analyst*, **78,** 473 (1953).

beaker and transfer the contents to a 100-ml volumetric flask wich 10 ml of water and dilute to volume with anhydrous 2B alcohol, cooling under a tap as the alcohol is added.

REAGENT SOLUTION

Prepare the reagent solution by adding 40 ml of stock solution to a 1-liter volumetric flask. Add 12 ml of 70% perchloric acid and dilute to volume with anhydrous 2B alcohol. The dilution should be carried out by adding the alcohol in 50- to 100-ml increments and cooling between each addition until the perchloric acid has been diluted to about 10% of its orginal concentration. The ferric ion concentration of this solution is 5.7 mM and the acid concentration is 0.16*M*.

All esters were used without extensive purification and were of a purity roughly equivalent to Eastman grade of Distillation Products Industries.

RECOMMENDED PROCEDURES

PROCEDURE FOR DETERMINATION OF ESTERS. Prepare the alkaline hydroxylamine reagent by mixing equal volumes of 12.5% hydroxylamine hydrochloride and 12.5% methanolic sodium hydroxide and filtering off the precipitated sodium chloride on Whatman No. 40 paper. The clear filtered reagent solution is usable for 4 hours.

The sample should be dissolved in anhydrous 2B ethanol or one of the other solvents mentioned below. The concentration of substance to be determined should be between 0.01 and 0.001*M*. Pipet 5 ml of the sample solution into a 25-ml flask with a ground-glass joint. (If a more concentrated solution is to be analyzed, or if a calibration curve is to be drawn up, a smaller volume of sample should be used and enough solvent added to make the total volume of 5 ml in the flask—e.g., 2 ml of sample and 3 ml of solvent.) Add 3 ml of the filtered alkaline reagent solution to each sample flask and to a blank that contains 5 ml of solvent. Add a boiling chip to each flask, then place the flasks on a hot plate on low heat, and attach reflux condensers. Reflux the samples for 5 minutes. Then remove the flasks from the hot plate (the condensers are not washed down), cool to room temperature, and wash the contents into a 50-ml volumetric flask with the ferric perchlorate reagent. Make the samples up to volume with the reagent. Shake the flasks to ensure complete solution of the initially precipitated ferric hydroxide. After several minutes read the absorbance of the samples against the reagent blank on a suitable spectrophotometer or colorimeter. The wavelength of maximum absorbance varies according to the type of ester. Draw up a calibration curve using the same ester as that to be determined or, less preferably, an ester of the same acid as the acid portion of the ester to be determined. Use the same solvent in all cases.

If a mixture of anhydride and ester is to be analyzed for both constituents, a calibration curve for the anhydride must also be run under the conditions for the ester determination. The concentration of anhydride is determined under neutral hydrolysis conditions. Then the absorbance due to that concentration of anhydride under basic hydrolysis conditions is subtracted from the observed absorbance prior to calculating the ester content.

PROCEDURE FOR ANHYDRIDES AND LACTONES IN PRESENCE OF ESTERS. Prepare the hydroxylamine reagent by neutralizing a portion of the methanolic hydroxylamine hydrochloride to a phenolphthalein end point by the addition of the 12.5% methanolic sodium hydroxide. Filter the precipitated sodium chloride on Whatman No. 40 paper. The clear filtered reagent is usable for at least 4 hours.

The sample, if anhydride, should be dissolved at a concentration of 0.01 to 0.001*M* in benzene that has been suitably dried—for example, dried over anhydrous calcium sulfate for 24 hours. Lactones and esters may be dissolved in any of the ethers, alcohols, or hydrocarbons referred to below.

The procedure for carrying out the analysis is exactly the same as that used for the ester determination, except that a 10-minute reflux time should be used. Calibration curves should be made with known concentrations of anhydride or lactone by plotting the observed absorbance against concentration. The calibration curves are not always straight lines under these neutral conditions.

INVESTIGATION OF HYDROXYLAMINOLYSIS CONDITIONS. The determination of esters involves the reaction of an ester with hydroxylamine in alkaline solution to form a hydroxamic acid. An acid ferric perchlorate solution is then added to form the colored chelate complex.

Previous investigators had found that many esters would react at room temperature; thus the first variable considered was reaction time at 25°C, rather arbitrarily using 12.5% solutions of hydroxylamine hydrochloride and sodium hydroxide in methanol, concentrations used by Thompson (14). It was found that for acetate esters dissolved in anhydrous 2B alcohol maximum color development was obtained in a minimum of 15 minutes (Fig. 3.1).

With other esters, such as fatty acid esters and esters of aromatic acids, however, maximum color development was not obtained after 30-minute reaction time in alkaline solution at 25°C. Accordingly, the effects of temperature and time were studied over a wider range (Table 5). At

Table 5. Effect of Time and Temperature on Hydroxamic Acid Formation

	Methyl Oleate A_{520}			Methyl *p*-Toluate A_{550}		
Time, min.	At 26°C	At 55°C	At Reflux	At 26°C	At 55°C	At Reflux
5	0.670	0.858	0.963	0.290	0.630	1.058
10	0.808	0.880		0.480		
15	0.860	0.753		0.625		
30	0.940			0.845		

Color development conditions were hydrogen ion concentration 0.3*M* and ferric ion concentration 2.4 mM. A_λ = absorbance at wave length λ.

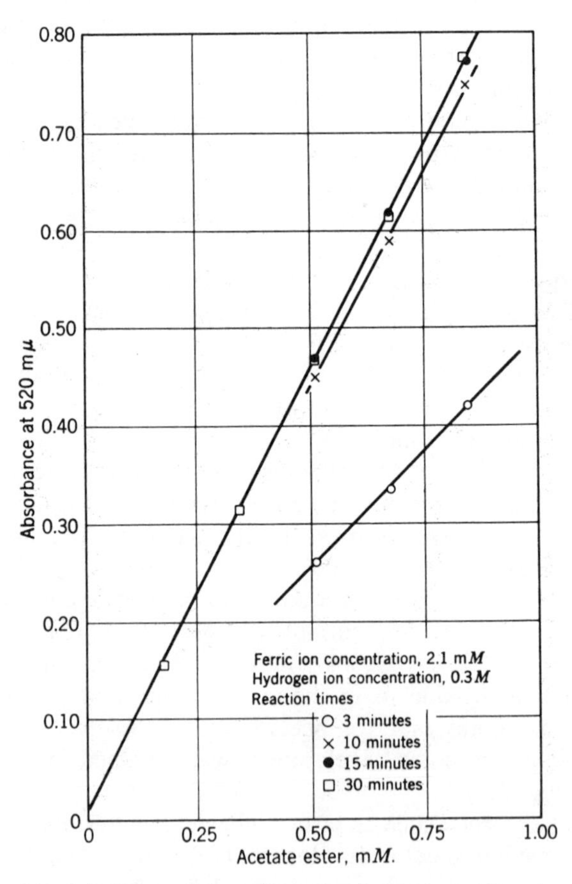

Fig. 3.1. Effect of time of reaction between acetate ester and hydroxylamine at 25°C.

elevated temperatures the reaction to form the hydroxamic acid proceeds more swiftly, but prolonged exposure to high temperature may cause decomposition of the hydroxamic acid. For general use, a 5-minute reflux (temperature approximately 72°C) has been found most satisfactory and applicable to all cases in which it is possible to form a colored ferric hydroxamate complex. Room-temperature reactions may be used for limited number of esters.

INVESTIGATION OF COLOR DEVELOPMENT CONDITIONS. There are two independent variables in the color development conditions—the concentration of ferric ion used to complex hydroxamic acid and the concentration of hydrogen ion present in the solution.

An investigation of the effect of the concentration of ferric ion necessary for maximum color development (Fig. 3.2) showed that maximum

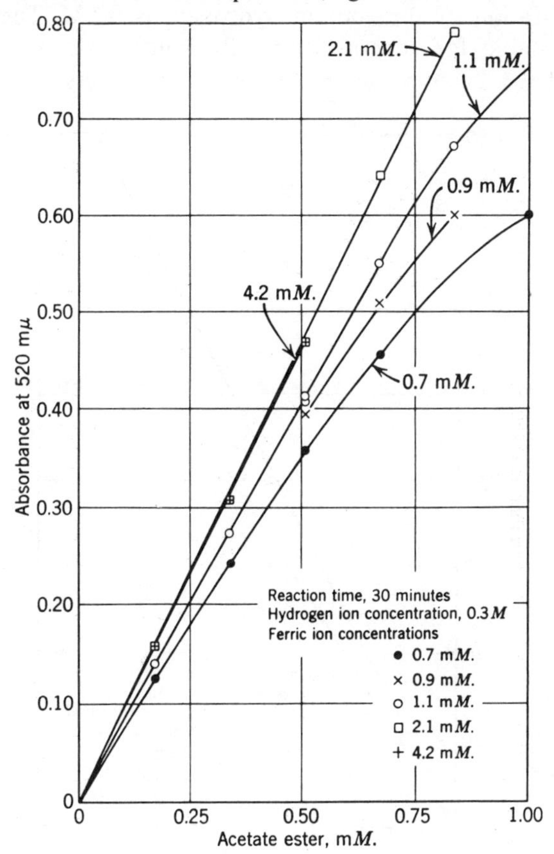

Fig. 3.2. Effect of ferric ion concentration on sensitivity of acetate ester determination.

color is obtained with ferric ion concentrations equal to or greater than 2 mM. In the final procedure the ferric concentration is 4.8 mM. Since the usual concentrations of esters in the color development solution are from 0.1 to 1.0 mM, there is at least a 4.8-fold molar excess of ferric ion present at all times.

The second variable in the color development was the amount of excess acid desirable over that necessary to neutralize the sodium hydroxide required for the saponification. The curves in Fig. 3.3 indicate that too high an acidity will hinder color development, but at acidities less than 0.6*M* satisfactory colors are obtained. A closer investigation of the effect of acidity (Table 6) indicates that at low acidities the color from a given sample is marginally more intense and that the color is much more stable. As a result, a procedure was adopted in which the acid concentration after neutralization of the sodium hydroxide is 0.1*M*. The colors thus formed are stable for several hours. Previous ferric hydroxamic acid procedures have been plagued with color instability unless hydrogen

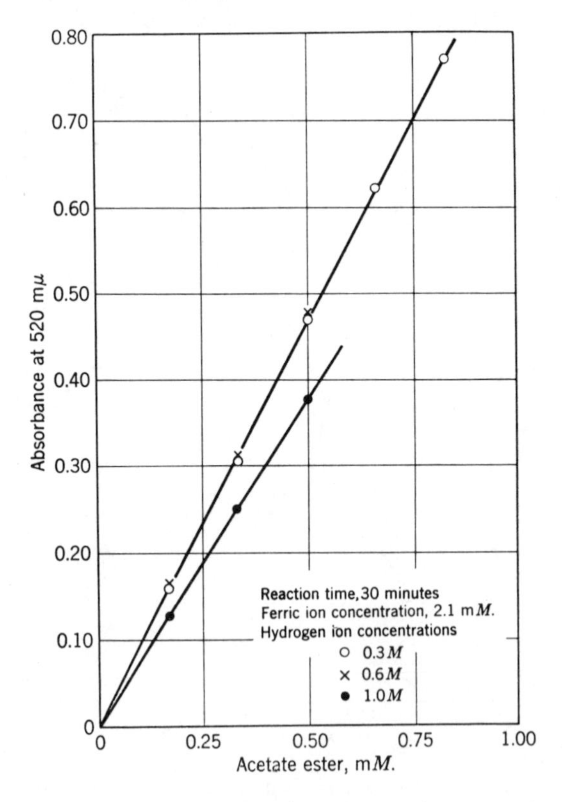

Fig. 3.3. Effect of hydrogen ion concentration on sensitivity of acetate ester determination.

Table 6. Effect of Hydrogen Ion Concentration on Color Intensity and Stability of Ferric–Acethydroxamic Acid Complex

(4.8 mM. ferric ion; 0.76 mM. butyl acetate)

Molarity of Perchloric Acid	Absorbance at 530 mμ	Fading Rate, % per hour
0.036	0.805	Negligible
0.09	0.810	0.05 to 0.10
0.22	0.780	—
0.46	0.760	—
0.71	0.755	0.8
0.95	0.748	—
1.20	0.750	6.0

peroxide was added to eliminate the excess hydroxylamine, as suggested by Hill (12).

SOLVENTS. Since solvents other than ethanol are often necessary or desirable, several have been tried (Fig. 3.4). A solution of the ferric reagent and ester sample in isopropyl alcohol is in every way comparable to ethanol. A benzene solution of the sample may be used with the

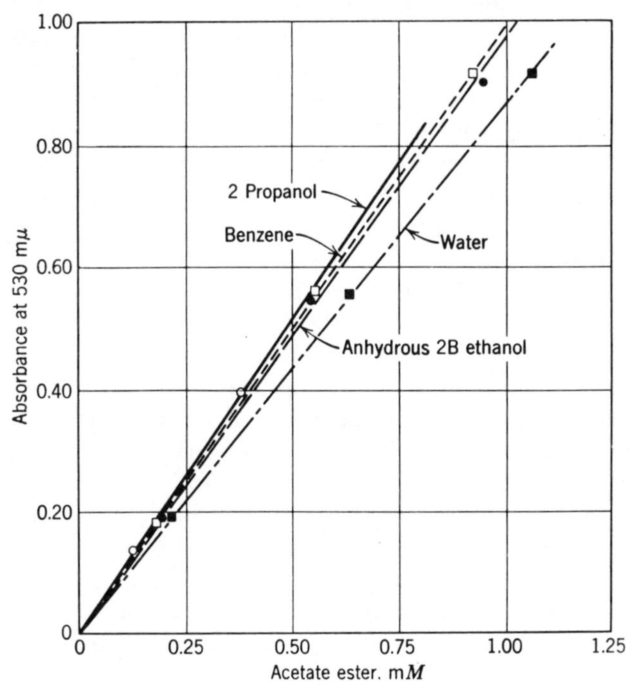

Fig. 3.4. Effect of solvent on ferric hydroxamate complex.

remainder of reagents as in the procedure written above. Dioxane, if properly purified, will probably be a suitable solvent. Methylene chloride–ethanol mixtures and petroleum ether have also been used successfully Diethyl ether has been used by Thompson (14) after extensive purification. Presumably other ethers and higher alcohols would be useful as solvents for the esters if purified. Water solutions of esters may be analyzed but at a slight loss in sensitivity, perhaps because of a competition between water and the hydroxamic acid for the ferric ion. The same solvent should always be used for the calibration curve as for the determination.

SPECTROPHOTOMETRIC DATA. The absorption maxima and molar absorptivities obtained with a wide variety of esters are shown in Table 7. The ferric complexes of most aliphatic hydroxamic acids have their absorption maxima at 530 nm. The ferric hydroxamate complexes of aromatic esters have broad absorption maxima at 550 to 560 nm. Ferric hydroxamate complexes of esters of acids containing conjugated double bonds or more than one carboxyl group may have maxima at slightly different wavelengths. The molar absorptivities are essentially the same for esters of the same acid and are additive for esters of polyhydroxyl alcohols. Esters of dicarboxylic acids have approximately twice the absorptivity found for similar monocarboxylic acids. Esters of resin acids form no color at all, perhaps because hydrolysis conditions are too weak, allowing other esters to be determined when present in admixture.

DETERMINATION OF ANHYDRIDES IN PRESENCE OF ESTERS. Anhydrides of carboxylic acids also react in a manner similar to esters under the alkaline conditions discussed, forming 1 mole of hydroxamic acid per mole of anhydride that reacts. A method for the selective determination of anhydrides was developed based on the use of neutral hydroxylamine to form the hydroxamic acid. This reaction is similar in principle to one recently described for determining anhydrides in the presence of esters using a standard solution of morpholine as a base, and measuring the amount of morpholine remaining after the amide has been formed (19). The weak base morpholine, or in this case hydroxylamine, is strong enough to react with the anhydride, yet no reaction occurs with the ester. About 65% of the color obtained under alkaline conditions is developed using a 10-minute reflux time with the neutral reagent. The effect of appreciably increasing the reflux time was not investigated. It is assumed but not proved that the incomplete reaction of the anhydride is due to a competing reaction with the methanol that is present as the solvent for the hydroxylamine. Of the esters tested, only phenolic esters, peroxyesters, lactones, and formates react. It has been reported that esters of

19. J. N. Hogsett, H. W. Kacy, and J. B. Johnson, *Anal. Chem.*, **25**, 1207 (1953).

Table 7. Molar Absorptivities of Ferric Hydroxamates Formed from Different Esters

Esters	Wavelength, mμ	Molar Absorptivity, × 10^3
Ethyl formate	520	1.06
Ethyl acetate	530	1.10
n-Butyl acetate	530	1.06
n-Amyl acetate	530	1.05
Phenyl acetate (impure)	530	0.99
Triacetin	530	3.33
Ethyl propionate	530	1.02
γ-Butyrolactone	530	1.11
Methyl *n*-butyrate	530	1.06
n-Butyl *n*-butyrate	530	1.05
n-Amyl *n*-butyrate	530	0.91
Dimethyl malonate	520	1.72
Dimethyl maleate	520	1.53
Dimethyl adipate	530	2.04
Pentaerythritol tetracaproate	530	3.89
Methyl oleate	530	1.00
Methyl benzoate	550	1.13
n-Butyl benzoate	550	1.08
Benzyl benzoate	550	1.13
Methyl *p*-toluate	550	0.94
Dimethyl *o*-phthalate	540	1.48
Dimethyl isophthalate	540	2.42
Dimethyl terephthalate	540	2.37

electronegatively substituted acids such as the chlorinated acetic acids also react with hydroxylamine under neutral conditions (20). By applying the neutral reagent to mixtures of anhydrides and esters, or lactones and esters, a quantitative determination of the anhydride or lactone content is possible. Separate calibration curves must be used for the anhydride or lactone under basic and neutral conditions (Fig. 3.5). Several representative analyses of mixtures are presented in Table 8. Both the precision and accuracy are reasonably good when it is considered that most such applications will be made to minor constituents.

INTERFERENCES. Acids, most amides, and nitriles do not interfere with these procedures. The hydroxylaminolysis conditions are not severe enough to cause reaction of the latter two classes of compounds. Acid chlorides, of course, react in both procedures. High concentrations of

20. S. Soloway, personal communication.

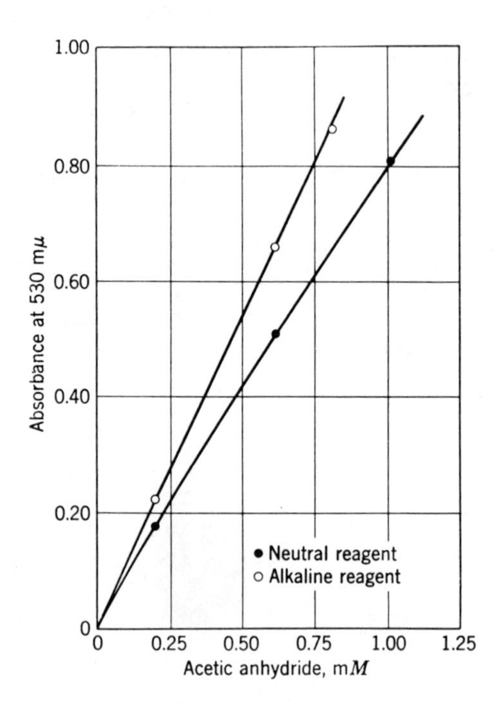

Fig. 3.5. Calibration curves for acetic anhydride in dry benzene.

Table 8. Mixture Analyses

No.	Anhydride Added, mg.	Anhydride Found, mg.	Anhydride Recovery, %	Ester Added, mg.	Ester Found, mg.	Ester Recovery, %
			Acetic Anhydride-*n*-Butyl Acetate			
1	3.10	3.17	102	2.80	2.85	102
	3.10	3.20	103	2.80	2.80	102
			Toluic Anhydride-Methyl *p*-Toluate			
1	2.00	1.85	93	8.10	—	—
2	6.00	5.90	98	4.05	—	—
3	2.37	2.25	95	5.87	5.82	99
	2.37	2.13	90	5.87	5.94	101
4	3.01	2.95	98	5.06	5.30	105
	3.01	3.00	100	5.06	5.25	104

carbonyls react with the hydroxylamine, probably necessitating the use of a higher hydroxylamine concentration. Transition elements such as copper, nickel, and vanadium react with hydroxamic acids to form colored chelate compounds which would interfere. It is possible that vanadium might be substituted advantageously for iron in the color development portion of the procedure above (6). Ions that complex ferric iron, such as chloride, tartrate, acetic acid, and water, may appreciably affect the intensity of the color in both the ester and anhydride methods.

EXTENSIONS OF METHOD. Presumably this ester method could be used for the determination of amides and nitriles by using a higher-boiling solvent such as propylene glycol as suggested by Soloway and Lipschitz (17). By using an excess of acetic anhydride, substituted hydroxylamines may also be determined spectrophotometrically. The sensitivity of the method possibly may be increased by using the ultraviolet absorption spectra of the ferric–hydroxamic acid complexes.

Other compounds that may be converted to hydroxamic acids and possibly determined by their ferric hydroxamate colors are sulfonic acids, aldehydes, nitro compounds, and isocyanates (21, 22).

Carboxylic Acid Amides

Carboxylic acid amides can be considered as esters of carboxylic acids with amines or ammonia. The reactions of the esters and the amides are quite similar except that amides have the amine or ammonia added where the ester has the alcohol. It can be generally stated that the amides tend to be somewhat less reactive than the corresponding ester.

Amides, as shown previously, can be determined by saponification (p. 172). The range of application of the saponification reaction is not as wide for amides as it is for esters, however. Primary amides $\left(\mathrm{RC}\begin{matrix}\mathrm{O}\\ \mathrm{NH_2}\end{matrix}\right)$ are the most amenable to analysis by saponification, but even then, there are a fair number of primary amides too resistant to quantitative saponification. Secondary amides $\left(\mathrm{RC}\begin{matrix}\mathrm{O}\\ \mathrm{NHR_1}\end{matrix}\right)$ and ertiary amides

21. F. Mathis, *Bull. Soc. Chim. Fr.*, **1953,** D-9.
22. H. L. Yale, *Chem. Rev.*, **33,** 209 (1943).

$$\left(\begin{array}{c} O \\ \| \\ RC-N \begin{array}{l} R_1 \\ R_2 \end{array} \end{array} \right)$$ hydrolyze, but with difficulty.

The nitrogen atom in amides gives the analyst a handle that is not present in esters and can be used to determine these materials. The nitrogen atom contributes a basicity that can be titrated directly in special solvent media. However amides are significantly less basic than amines. Amides can be reduced to amines, which are easily titrated; esters reduce to the corresponding alcohols or ethers, and these products are not basic.

TITRATION METHODS FOR AMIDES

Potentiometric Titration Method of D. C. Wimer Using Acetic Anhydride Solvent

[*Reprinted in Part from Anal. Chem.*, ***30,*** *77–80* (*1958*)]

This is one method for direct titration of amides as bases. It involves the use of acetic anhydride as a solvent and perchloric acid as a titrant.

APPARATUS

A Precision-Dow Recordomatic Titrometer, Model K-3-247, equipped with 50-ml feed pumps, was used in all titrations. A glass electrode (Beckman No. 4990-80) and a sleeve-type calomel electrode (Beckman No. 1170–71) were equilibrated by soaking in acetic anhydride for 12 hours prior to use. To minimize liquid junction potentials and to promote reproducibility, the aqueous bridge in the calomel cell was replaced with a $0.1M$ solution of anhydrous lithium perchlorate in acetic anhydride. Lithium chloride proved to be too insoluble in acetic anhydride for use as a supporting electrolyte in the bridge solution.

REAGENTS AND SOLUTIONS

Lithium perchlorate, anhydrous salt, is available from the G. Frederick Smith Chemical Co., Columbus, Ohio.

Acetic anhydride, ACS reagent grade.

Glacial acetic acid, ACS reagent grade.

Dioxane, purified by distillation from lithium aluminum hydride or from sodium wire.

Perchloric acid, 70% vacuum distilled, available from the G. Frederick Smith Chemical Co., Columbus, Ohio.

PERCHLORIC ACID. A 0.1N solution in acetic acid is prepared by dissolving approximately 9 ml of 70% perchloric acid in acetic acid, adding 25 ml of acetic anhydride, and diluting to 1 liter with acetic acid. The solution is allowed to stand 24 hours prior to use. The titrant is standardized either visually (23) or potentiometrically against primary standard potassium acid phthalate dissolved in acetic acid.

Perchloric acid in dioxane is prepared and standardized by the procedure of Fritz (24) and may be used as an alternative titrant (see p. 550).

All samples were analyzed as received without further purification. The compounds shown in Table 9 are research samples and the purest grade of commercially available amides.

Table 9. Titration of Amides in Acetic Anhydride

Compound	Purity, %	Max. $\Delta E/\Delta V$, Mv./Ml. (Approx.)
Formamide[a]	99.4, 99.6	160
N-Methylformamide	100.2, 99.9	120
N,N-Dimethylformamide	98.6, 98.6	190
N,N-Diallylformamide	97.4, 97.3	100
Acetamide	97.7, 97.3	120
N-Methylacetamide	97.4, 97.3	120
N,N-Dimethylacetamide	96.6, 96.6	300
Acetyl *n*-butylamine (N-*n*-butylacetamide)	98.7, 98.3	180
Acetyl di-*n*-butylamine (N,N-di-*n*-butylacetamide)	98.8, 98.7	190
N,N-Diethylacetamide	99.9, 99.4	280
Thioacetamide	99.0, 98.5	470
N,N-Diethylacetoacetamide	98.2, 98.7	220
N-(*p*-Nitrophenylethyl) acetamide	98.1, 98.0	120
N-β-(3-ethoxy-4-methoxyphenylethyl) acetamide	97.0, 96.5	260
N-β-(3-benzyloxy-4-methylphenyl) acetamide	90.1, 90.1	200
β-Hydroxyethylacetamide	97.9, 98.4	120
N,N′-Dimethylurea[b]	98.7, 98.4	300
N,N,N′,N′-Tetramethylurea	99.6, 99.4	560
Propionamide	95.4, 96.0, 95.8	140
Isobutyramide	93.8, 93.6	120
α-Mercaptoisobutyramide[c]	85.0, 83.5	100
Hexanamide	97.0, 97.2	100
n-Valeramide	97.5, 96.3	140
α-Methylnonamide	93.8, 93.8	100
n-Dodecylamide	98.7, 98.4	200
β-Phenylvaleramide	96.3, 96.3	120
Cyclohexylbutyramide	95.0, 94.5	140
N,N-Dimethylcyanamide	99.5, 99.6	130
Malonamide	98.5, 98.5	120
N,N,N′,N′-Tetraethylphthalamide	99.0, 99.1	260
Acrylamide[d]	86.5, 86.1, 86.7	120
Crotonamide[d]	90.7, 91.5, 91.5	170
2-Furamide[d]	51.0, 51.2, 51.0	100
N-Formylpyrrolidine	99.8, 99.3	300
N-Formylpiperidine	97.3, 97.8	280
N-Acetylpiperidine	94.8, 95.3	470
N-Acetylmorpholine	96.8, 96.4	160
N-Formylmorpholine	98.3, 98.9	80
N-Acetyldimethylphenylethylamine	98.5, 98.9	400
Hexamethylphosphoramide	98.2, 98.2	500
Tripyrrolidylphosphoramide	99.9, 100.3	500

[a] Titrated with $HClO_4$ in dioxane.
[b] Titrated at 5°C to prevent acetylation.
[c] Impure sample.
[d] Probable reaction with acetic anhydride.

PROCEDURE AND RESULTS

Dilute a 0.006- to 0.009-mole sample to 100 ml with acetic anhydride in a volumetric flask. Transfer a 10-ml aliquot to a tall-form beaker, add 100 ml of acetic anhydride, and carry out the titration with 0.1N perchloric acid in acetic acid. The end point of the titration may be determined by inspection or by calculating maximum change in potential for small increments of perchloric acid added. It is convenient to calculate

23. W. Seaman and E. Allen, *Anal. Chem.*, **23,** 592–4 (1951).
24. J. S. Fritz, *Anal. Chem.*, **22,** 578–9 (1950).

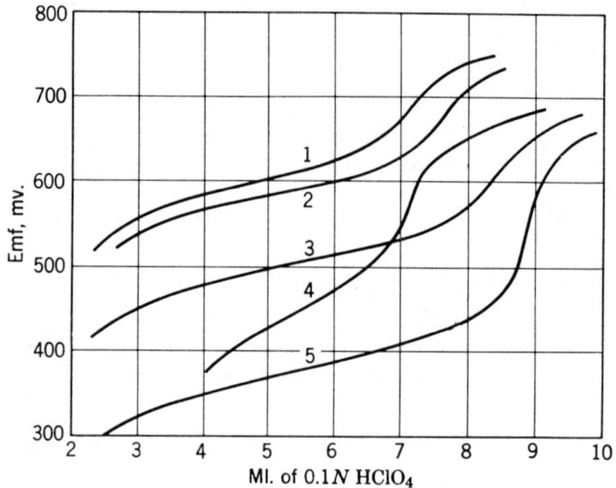

Fig. 3.6. Titration of various types of amides in acetic anhydride: 1, β-phenylvaleramide; 2, *N,N*-dimethylcyanamide; 3, malonamide; 4, *N,N,N′,N′*-tetraethylphthalamide; 5, tripyrrolidylphosphoramide.

maximum ΔE for constant values of ΔV equal to 0.05 ml since this increment can be easily estimated from the recorder chart paper.

For all results reported in Table 9, perchloric acid in acetic acid was used as the titrant. Perchloric acid in dioxane, however, appears to be the titrant of choice for formamide. The reasons for the much sharper end point observed are not yet apparent. Significant end point improvement

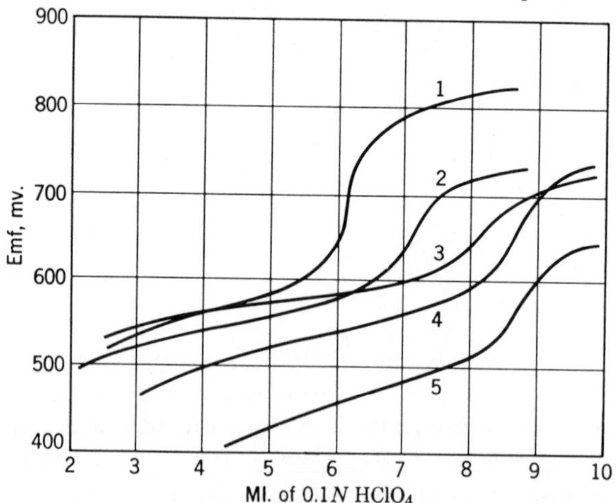

Fig. 3.7. Titrations of acetylated and formylated amines in acetic anhydride: 1, *N*-acetylmorpholine; 2, *N*-acetyldimethylphenylethylamine; 3, *N*-formylpiperidine; 4, *N*-formylpyrrolidine; 5, *N*-acetyl-di-*n*-butylamine.

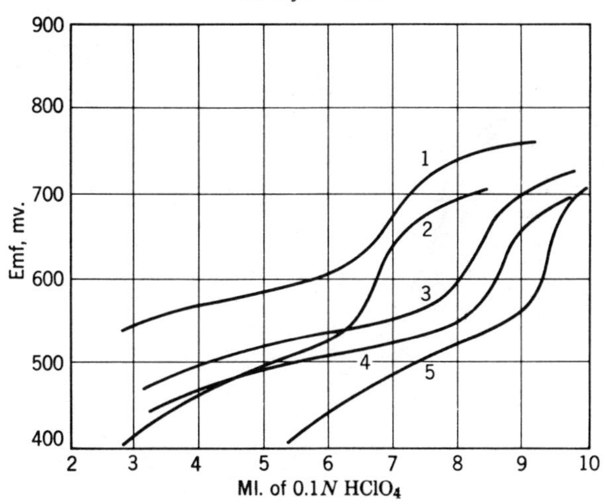

Fig. 3.8. Titration of aliphatic amides in acetic anhydride: 1, *N,N,N′, N′*-tetramethylurea; 2, *N,N*-dimethylformamide; 3, acetamide; 4, *N,N*-diethylacetoacetamide; 5, *n*-dodecylamide.

was not observed for most amides when titrated with perchloric acid in dioxane. Approximate values of maximum $\Delta E/\Delta V$ were calculated and are presented in Table 9 for comparison purposes. Figures 3.6 to 3.8 show representative curves for some amides, acetylated amines and formylated amines, and aliphatic anhydrides varying from "weak" to "strong" in relative base strength. This is merely an arbitrary classification based on the magnitude of the first derivative.

A 0.1*N* solution of perchloric acid in acetic anhydride have maximum changes in inflection at the end point. (The 0% acetic acid intercepts in Fig. 3.9 were determined with this reagent.) The solution is not hazardous to prepare or use, but it darkens on standing and is not a suitable titrant. Perchloric acid in acetic acid proved to be a satisfactory titrant. Maximum sensitivity is achieved by titrating in a volume of anhydride such that the amount of acetic acid introduced from the titrant is low compared to the total solution volume. Ten percent acetic acid by volume may be tolerated without seriously reducing end point inflections. The titration might be regarded as the reaction of a Lewis acid, "acetyl perchlorate," with an amide to form a salt.

$$\mathrm{R\overset{\overset{\displaystyle O}{\|}}{C}NH_2 + CH_3CO^+CLO_4^- \rightleftharpoons R\overset{\overset{\displaystyle O}{\|}}{C}NH_2CH_3CO^+CLO_4^-} \qquad (5)$$

The magnitude of the maximum $\Delta E/\Delta V$ obtained for a particular amide may depend largely on the extent of salt formation, as proposed in

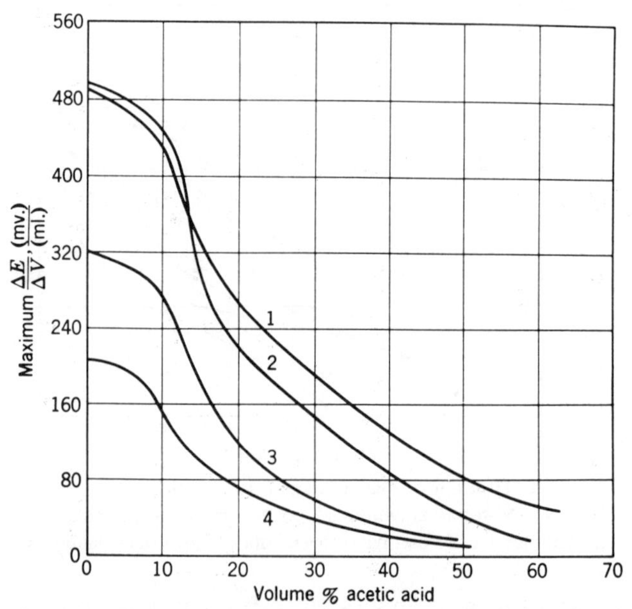

Fig. 3.9. Effect of acetic acid on end point potentials in titrations of amides in acetic anhydride: 1, thioacetamide; 2, *N*-acetylpiperidine; 3, *N*-formylpyrrolidine; 4, *N,N*-dimethylformamide.

eq. 5. Other factors, of course, will determine the ultimate success or failure of the determination.

Following completion of this work, 0.1*N* perchloric acid in dioxane was found to be a suitable alternative titrant. No similar equilibrium effect is observed here, as it was for acetic acid–acetic anhydride mixtures of perchloric acid. Increasing dilution of the acetic anhydride with large amounts of dioxane merely reduces sensitivity by lowering the dielectric constant; this procedure renders the electrode system less reproducible.

Certain amides cannot be determined by direct titration in acetic anhydride. Diamides of dibasic acids are practically insoluble in acetic anhydride. Malonamide and tetrasubstituted phthalamides are apparent exceptions. *N*-Phenyl or α-phenyl substitution results in electron withdrawal and nearly complete reduction in basic properties. β-Phenyl-substituted amides exhibit sharp inflections. Unsaturated amides, where the double bond is conjugated with the carbonyl group, appear to react with acetic anhydride. Sharp breaks were observed, but quantitative results could not be obtained. Trifluoromethylformamide, cyanamide, and tertiary amides of the configuration $—N\begin{matrix} R \\ R \end{matrix}$, where R is $CH_3\overset{O}{\overset{\|}{C}}—$, exhibited no measurable basic properties toward perchloric acid–anhydride mixtures. Except as noted in Table 9, no evidence of reaction of amides with acetic anhydride was observed. Reproducible titration curves were obtained on samples that had been in contact with the solvent 1 to 2 hours. Hydroxy-substituted amides showed no evidence of *O*-acetylation at room temperature.

Potentiometric Titration Method of C. Streuli Using Nitromethane as Solvent

(*See pp. 551–58 under Amines*)

Photometric Titration Method

[*Adapted from T. Higuchi, C. H. Barnstein, H. Ghassemi, and W. E. Perez, Anal. Chem.*, **34,** *400–3 (1962), reprinted*]

This method also involves direct titration of amides, but in glacial acetic acid instead of acetic anhydride, with Wimer's method. As in most methods, there are advantages and disadvantages. The glacial acetic acid is a less reactive solvent than the anhydride, hence may be more generally applicable to amides or hydrazides that tend to react with anhydride. But since acetic acid does not intensify the basicity of the amides to the same extent as the anhydride does, simple potentiometric titration is not possible; photometric titration must be used.

Despite the fact that titration of various weak bases in acetic acid has been studied extensively for a number of years, its application to systems essentially nonbasic in water has been relatively limited. Although the applicability of the method to both qualitative and quantitative determination of amides has been particularly stressed, a number of other functional groups have also been investigated as a part of this study.

The basic relationships governing interactions between acids and bases in acetic acid and the use of photometric titration plots have been pointed out (25–27). Conventional Type II photometric titration plots are based on the following relationship (28)

$$\frac{1}{X}=\left(\frac{K_{ex}}{S}\right)\left(\frac{I_b}{I_a}\right)+\frac{1}{S} \tag{6}$$

where X is the amount of acid added at any point, S is the amount of acid equivalent to the total base in solution, K_{ex} the indicator exchange constant (28), and I_b/I_a the ratio of indicator base to indicator acid. Upon consideration of the normality of standard perchloric acid titrant and the

25. T. Higuchi and K. A. Connors, *J. Phys. Chem.*, **64,** 179 (1960).
26. T. Higuchi, J. A. Feldman, and C. R. Rehm, *Anal. Chem.*, **28,** 1120 (1956).
27. I. M. Kolthoff and S. Bruckenstein, *J. Am. Chem. Soc.*, **78,** 1 (1956).
28. K. A. Connors and T. Higuchi, *Anal. Chem.*, **32,** 93 (1960).

total solution volume, the amounts X and S are readily determined from the volumes of standard acid added.

Unmodified Type II plots do not yield satisfactory straight lines for very weak bases because of the high degree of solvolysis present even in the presence of excess perchloric acid. This problem was solved in one manner by Connors and Higuchi (28) by plotting photometric differences between results obtained for high initial base concentration and those for low base concentration. This approach also obviated to some extent difficulties presented by the presence of variable quantities of water. For the present study it was felt that a simpler technique based on complete elimination of water by addition of a known small excess of acetic anhydride in the presence of free perchloric acid permitted more straightforward interpretation.

Thus in systems where a significant extent of solvolysis exists, X', the added concentration of perchloric acid solution, would be

$$X' = C_{BHClO_4} + C_{HClO4} \tag{7}$$

And S', the total initial base concentration would, on addition of the mineral acid, be $S' = C_{BHClO_4} + C_B$. It is apparent then that eq. 7 must be modified for these systems to

$$\frac{1}{X' - C_{HClO_4}} = \left(\frac{K_{ex}}{S'}\right)\left(\frac{I_b}{I_a}\right) + \frac{1}{S'} \tag{8}$$

Whereas K_{ex} may be obtained directly from the slope of the linear plot of eq. 7, it is necessary to know C_{HClO_4} to plot the data according to eq. 8.

The concentration of the free acid C_{HClO_4} may be determined from the indicator color and indicator constant obtained from titration of a blank. Since the indicator is the only base titrated in a blank solution, the C_{HClO_4} value corresponding to a given indicator ratio can be determined from the indicator perchlorate formation equilibrium

$$I + HClO_4 \rightleftharpoons IHClO_4 \tag{9}$$

$$K^{IHClO_4} = \frac{C_{IHClO_4}}{C_I C_{HClO_4}} \tag{10}$$

$$C_{IHClO_4} = \left(\frac{1}{K_f^{IHClO_4}}\right)\left(\frac{I_a}{I_b}\right) \tag{11}$$

When (I_a/I_b) was plotted versus C_{HClO_4} in the blank solution, a straight line passing through the origin (29) as predicted by eq. 11 was obtained. It is evident, therefore, that the amount of perchloric acid consumed in

29. C. R. Rehm and T. Higuchi, *Anal. Chem.*, **29**, 367 (1957).

indicator perchlorate formation is negligible; thus values of C_{HClO_4} corresponding to various indicator ratio values are obtainable from blank titration plots for insertion into eq. 8. In the absence of water linear plots of eq. 28 yield reproducible exchange constants for very weak bases with Sudan III indicator in acetic acid.

Because water behaves like a base in acetic acid, its presence in the solvent system is undesirable, since it interferes with the interaction between a weak base and the reference acid. The solvent that was employed in all titrations performed in this study was a solution of 0.25M redistilled acetic anhydride in acetic acid, which has been purified by the method of Eichelberger and LaMer (30). Since many substances known to possess basic properties in acetic acid react with the anhydride, only compounds having nonacylable basic functions are reported here.

EXPERIMENTAL (31)

The titration vessel is a small Erlenmeyer flask equipped with an outlet tube sealed tangentially at the lowest part of the side and an inlet tube which enters the bottom of the flask at its center. These two tubes are bent to be parallel (Fig. 3.10).

The absorption cell varies with the photometer used. A modified 1-cm cylindrical cell (Pyrocell Manufacturing Co., 207-11 East 84th St., New York N.Y. 10028) used with the Cary recording spectrophotometer, having inlet and outlet tubes instead of the usual single opening, is shown

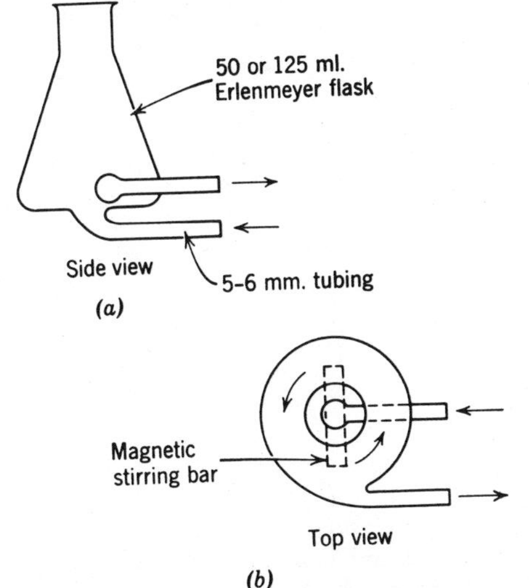

Fig. 3.10. Titration flask. Arrows show direction of liquid flow.

30. W. C. Eichelberger and V. K. LaMer, *J. Am. Chem. Soc.*, **55**, 3633 (1933).

in Fig. 3.11. This cell is also suitable with the Beckman DU and B instruments, if the cell holder is modified to hold the cylindrical cuvette. The test tube cells supplied with the Bausch & Lomb Spectronic 20 colorimeter may be adapted for use in this inexpensive instrument. A rubber stopper receives the inlet and outlet tubes, which terminate above the level of the light path (Fig. 3.11). The surface of the solution remains at the level of the outlet tube.

The titration flask and absorption cell are connected with short lengths of flexible tubing: rubber, Tygon, or Vinyl (available through medical supply houses as blood transfusion tubing). For titrations in very acidic solvents, such as acetic acid, the last is preferred. The slightly basic properties of rubber and Tygon prevent their use in the titration of very weak bases.

The titration stirring bar is placed in the flask, which is supported over a magnetic stirrer. The solution (25 to 35 ml in a 50-ml titration flask) is circulated through the tubing to the cell and back to the flask by the action of the stirring bar (Fig. 3.10). Complete mixing is achieved in 20 to 25 seconds with both the cylindrical and the tube-type cell. Figure 3.12 shows the rate of mixing with the cylindrical cell.

Modification of the spectrophotometer is usually unnecessary or slight. A small hole drilled in the cell compartment cover of the Cary, for example, allows the flexible tubing to pass into the compartment. A black cloth thrown over the tubing excludes stray light, as with other instruments.

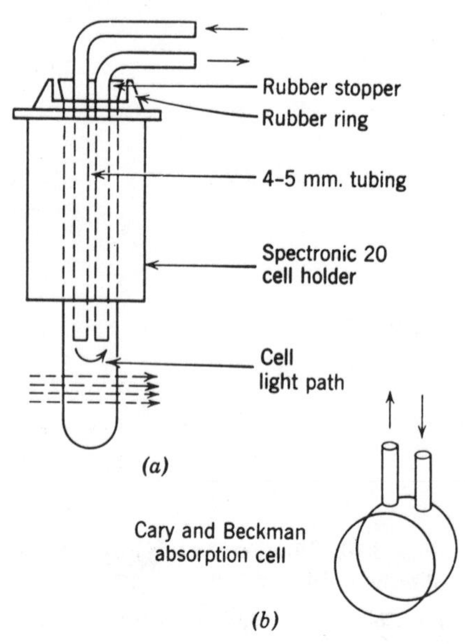

Fig. 3.11. (*a*) Bausch & Lomb Spectronic 20 absorption tube modified for photometric titrations. (*b*) Cylindrical cell for titrations with Cary and Beckman spectrophotometers.

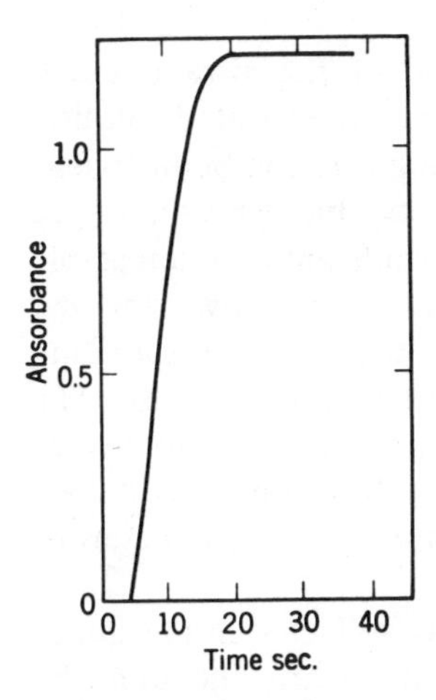

Fig. 3.12. Time required to attain homogeneity of titration solution using cylindrical cell.

The titration apparatus is particularly convenient if a closed system is necessary. Atmospheric moisture and carbon dioxide are excluded by passing the buret tip through a stopper. A positive pressure may be maintained by feeding dry air or nitrogen into the flask.

The assembly constitutes a very modest investment if the Spectronic 20 is used as the measuring instrument; the total cost is comparable with that of common potentiometric titration equipment.

In Fig. 3.13 the ratios of the base form of Sudan III to its acid form during titration of *N,N*-dimethylcapramide with approximately 0.5*N* perchloric acid are plotted according to eqs. 7 and 8. In the figure the *X*

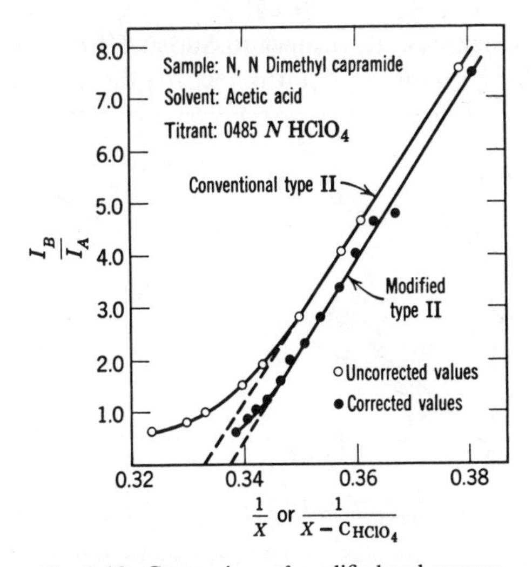

Fig. 3.13. Comparison of modified and conventional Type II plots.

values are expressed in terms of actual volume of the titrant added ranging from 2.630 to 3.090 ml. The final volume of the titrated solution was approximately 30 ml, the volume change during titration being essentially negligible. It is evident from the curvature exhibited by the upper curve that even for this relatively strong base, a significant and detectable degree of solvolysis exists. The extrapolated end point shown for the simpler relationship is slightly less than 0.04 ml too large, corresponding to a little more than a per cent error (high) for the unmodified case. The residual end curvature for the corrected plot is probably due to a very small tendency of these amides to take up a second proton. This departure from linearity, however, does not prevent extremely precise determination of the stoichiometric end point.

The relative basicities of these stronger bases are reflected in the slopes of the straight line portions of these plots and are nearly the same by either plot. This is not the case, however, for extremely weakly basic compounds. In Fig. 3.14 the photometric data obtained for chloracetamide are again plotted in both ways. In this instance the conventional plot is in considerable error, and correct values can be obtained only from the modified plots.

Diethyl ether was the weakest of the compounds that gave evidence by this method of possessing basic properties. The modified Type II plot of data for the photometric titration of diethyl ether appears in Fig. 3.15. Although the plotted points exhibit a considerable degree of scatter, their distribution describes a sufficiently straight line to allow for calculation of the exchange constant, $K_{ex} = 1400 \pm 200$.

The relative basicities with respect to Sudan III of various amides and other organic compounds as determined by these plots are shown in

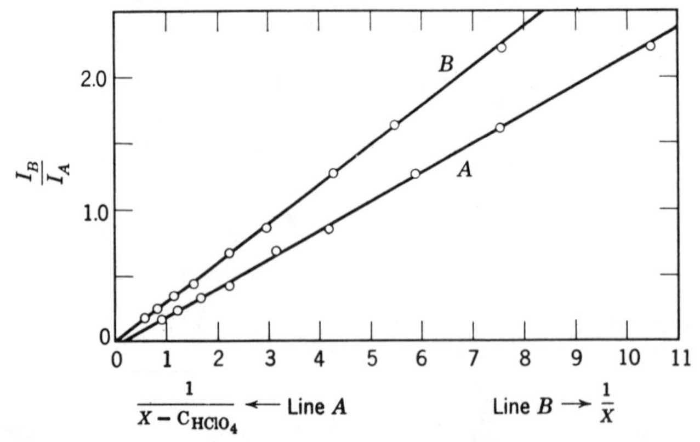

Fig. 3.14. Plots for photometric titrations of chloroacetamide with 0.4926*N* perchloric acid. Indicator is Sudan III; line *A*, modified Type II plot.

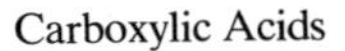

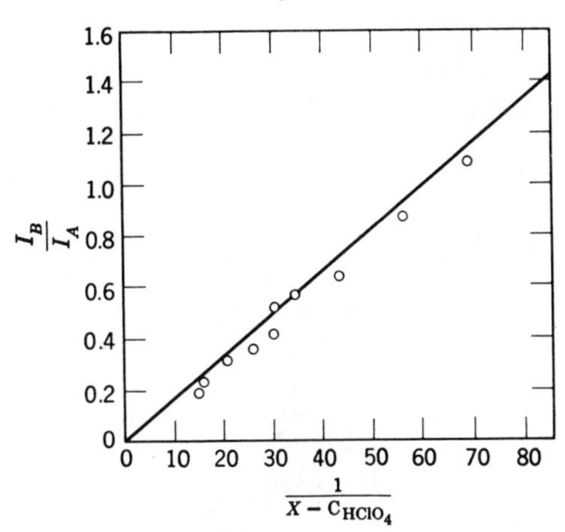

Fig. 3.15. Modified Type II photometric plot for titration of diethyl ether with 0.4703*N* perchloric acid. Indicator is Sudan III.

Table 10. Since the perchlorate formation constant for Sudan III is of the order of 700 (26), the perchlorate constants for all the compounds listed can be estimated readily from the table.

DISCUSSION

NONBASIC COMPOUNDS. Six of the compounds in Table 10 are evidently nonbasic, with exchange constants with respect to Sudan III estimated to be in excess of 500 or of 2000, depending on the concentrations of the solutions analyzed.

Loss of basicity of the amide function of nicotinamide probably follows as a consequence of protonation of the relatively strongly basic amine nitrogen adjacent to it. The apparent nonbasicity of phenobarbital in acetic acid is in agreement with previous findings of Higuchi and Connors (25). Dichloroacetamide and trichloroacetamide show no evidence of possessing basic properties, presumably because of the inductive effects of the adjacent chloro substituents. The same polar effect is probably responsible in part for the lack of basicity of acetylated chloramphenicol, which is a mono *N*-substituted dichloroacetamide.

ACETANILIDE SERIES. Substitution of the negative groups, —Cl and —COOH, in the meta and para positions of acetanilide produced a marked decrease in the basicity of the parent compound as expected. Hammett substituent constants σ, obtained from the literature (32), were observed

31. C. Rehm, J. I. Bodin, K. A. Connors, and T. Higuchi, *Anal. Chem.*, **31,** 483 (1959), reprinted in part.
32. L. P. Hammett, *Physical Organic Chemistry*, McGraw-Hill, New York, 1940.

Table 10. Basicities of Amides and Other Organic Compounds in Acetic Acid Relative to Sudan III

Compound	K_{ex}
N,N-Dimethylacetamide	0.011
1-Acetylpiperidine	0.014
N,N-Dimethylcaprylamide	0.019
N,N-Dimethylcapramide	0.019
N,N-Dimethyllauramide	0.020
N,N-Dimethylmyristamide	0.020
N-Methylacetamide	0.023
N,N-Dimethylpalmitamide	0.024
Acetamide	0.10
Lauramide	0.11
N,N-Dimethyl-2-naphthamide	0.12
N,N-Dimethylbenzamide	0.13
Pelargonamide	0.13
N-Methylformamide	0.13
N,N-Dimethylformamide	0.14
2-Pyrrolidone	0.14
N,N-Dimethyl-1-naphthamide	0.20
2,6-Dimethylacetanilide	0.31
N-Methylbenzamide	0.37
Formamide	0.53
Acetanilide	0.61
Acetyl-α-naphthylamine	0.64
Acetyl-β-naphthylamine	0.64
Benzamide	0.84
β-Naphthamide	0.86
α-Naphthamide	1.6
p-Chloroacetanilide	1.8
m-Acetamidobenzoic acid	2.6
m-Chloroacetamide	2.7
p-Acetamidobenzoic acid	4.9
2,4,6-Tribromoacetanilide	16
Benzanilide	18
Chloroacetamide	18
Diethyl ether	1400
Nicotinamide (amide N)	>500
Phenobarbital	>500
Acetonitrile	>2000
Dichloroacetamide	>2000
Trichloroacetamide	>2000
Chloroamphenicol (acetylated)	>2000

Table 11. Comparison of Effect of N-Alkylation on Basicities of Several Amides and Ammonia

Compound	$CH_3C(=O)—$	$HC(=O)—$	$CH_3(CH_2)_{10}C(=O)—$	$C_6H_5C(=O)—$ (benzoyl)	1-naphthoyl ($C_{10}H_7C(=O)—$)	2-naphthoyl ($C_{10}H_7C(=O)—$)	R = H—
	R						K_b Values
R—NH_2	0.10	0.53	0.11	0.84	1.6	0.86	1.8×10^{-5}
R—$NHCH_3$	0.023	0.13	—	0.37	—	—	4.4×10^{-4}
R—$N(CH_3)_2$	0.011	0.14	0.02	0.13	0.2	0.12	5.0×10^{-4}

to vary linearly with the logarithms of the exchange constants, yielding a specific reaction constant ρ of 1.4 ±0.1.

2,6-Dimethylacetanilide ($K_{ex} = 0.31$) is a stronger base than acetanilide ($k_{ex} = 0.61$), whereas 2,4,6-tribromoacetanilide ($K_{ex} = 18$) is, relatively, a very weak base. These differences suggest that substituents on the ring influence the basicity of the anilides significantly but relatively weakly. The increased ability of the ortho-dimethylated compound to attract perchloric acid may be partially ascribable to steric interference toward coplanarity of the ring with the resonating amide function.

EFFECT OF *N*-ALKYLATION. The influence on amide basicity of methyl substitution on the amide nitrogen is also evident from Table 11. In the series of formamides and acetamides it is evident that methyl substitution of one hydrogen of the amide resulted in markedly enhanced basicity, whereas similar substitution of the second hydrogen produced little further change in basicity. This effect more or less parallels the changes in the basicity of the parent amine as shown in the last column of Table 11.

EFFECT OF AROMATIC SUBSTITUTION. Substitution of an aromatic group reduced the basicity of the amide function by virtue of its tendency to attract the unshared electrons of the amide nitrogen and oxygen. Acetanilide, acetyl α-naphthylamine, and acetyl β-naphthylamine appear to possess unusual base strength relative to acetamide; however if comparison is made with basicities of the amines that are obtained on hydrolysis, ammonia is at least 50,000 times more basic than aniline or the naphthylamines, but acetamide is only 6 times more basic than acetanilide or the *N*-acetyl naphthylamines. This may be due at least in part to the lack of coplanarity between the ring system and the resonating amide plane.

DETERMINATION BY MEANS OF REDUCTION TO CORRESPONDING AMINES

Adapted from the Method of S. Siggia, and C. R. Stahl

[*Anal. Chem.*, **27,** *550* (*1955*)]

This method can be applied in instances where, for one reason or another, direct titration cannot. This approach makes use of the reduction of amides to amines using lithium aluminum hydride (33).

$$2R\overset{\overset{\displaystyle O}{\|}}{C}NR'_2 + LiAlH_4 \rightarrow 2RCH_2NR'_2 + LiAlH_2$$

The amine formed is steam-distilled from the reaction mixture and

33. R. F. Nystrom and W. G. Brown, *J. Am. Chem. Soc.*, **70,** 3738 (1948).

titrated. It was first thought that the amine formed could be titrated in the reaction mixture. It was supposed that one break would be obtained for the strong bases (metal hydroxides) and a second break for the weaker base (amine). The results obtained by this approach were very high, however, probably because aluminum hydroxide is not as strong a base as was first supposed. In view of these results, the steam-distillation step was added to the analysis.

In the determination of fatty acid amides, the corresponding amines formed on reduction steam-distill very slowly; therefore ethylene glycol was used instead of water in the distillation.

The glycol distillation is used for the determination of the sodium *N*-lauroylsarcosinate ($C_{11}H_{23}C(=O){-}N(CH_3)CH_2COONa$). The amine formed in the reduction of this compound is distillable from a caustic solution (the Kjeldahl distillation). The sarcosinate contains a carboxyl group that in the resultant amine should be in the form of the salt in the alkaline solution. This amino acid salt should be difficult to distill. It is presumed that in the reduction step, however, the carboxyl group is reduced to the alcohol, so that the resultant amine would be $C_{11}H_{23}CH_2N(CH_3)CH_2CH_2OH$, which is distillable. Amides of the type $RC(=O){-}N(R_1)CH_2CH_2SO_3Na$ cannot be determined by this method. Evidently the sulfonic acid group is not satisfactorily reduced, and the salt of the amine will not distill. Very low results are obtained on these compounds.

This method is applicable to a large range of amides, as seen in Table 12. Primary, secondary, and tertiary amides of low molecular weight acids and fatty acids, as well as difunctional amides, were successfully determined. A cyclic amide (methyl pyrrolidone) was also determined. Unsatisfactory reductions were noted for acrylamide, *N-tert*-butyl acrylamide, and urea. In the case of *N,N*-diphenyl acetamide, the resulting amine (diphenylethylamine) is too weakly basic to titrate, even in solvents designed for titrating weak bases.

Many functional groups react with lithium aluminum hydride, but very few result in the formation of a volatile base. Nitriles, imides, and aliphatic nitro compounds are the only others known to form an amine on reduction with the hydride. The aliphatic nitro compounds are rarely

Table 12. Results

Compound	%	% by N	Procedure
Acetyl diethylamine (N,N-diethylacetamide)	99.2	97.7	A
	98.1		
	98.2		
	98.4		
	98.4		
N-butyramide	95.4	97.1	A
	97.2		
	96.2		
N-methyl acetamide	93.5	95.2	A
	92.8		
	93.8		
	94.2		
N-methyl pyrrolidone	98.2	97.8	A
	98.6		
	98.1		
Acetamide	100.6	99.8	A
	100.8		
	100.3		
Propionamide	101.1	100.4	A
	100.8		
	100.3		
	100.0		
	100.0		
γ-Hydroxybutramide	96.3	96.1	A
	96.0		
	95.8		
Methyl glycolamide	95.6	96.3	A
	95.2		
	95.4		
N,N′-dimethyl oxamide	99.9	100.8	A
	99.8		
	99.5		
	99.8		
Benzamide	98.5	100.8	A
	100.0		
	99.7		
Formamide	99.0	100.6	A
	98.9		
	98.6		

Compound	%	% by N	Procedure
Dimethyl formamide	90.7	93.4	A
	89.5		
	92.2		
Octanamide	96.4	96.6	A
	98.1		
	97.5		
Hexadecanamide	95.2	95.1	B
	94.6		
	94.6		
	94.9		
Methyl stearamide	97.4	98.3	B
	98.6		
	100.0		
	98.0		
Stearamide	92.4	94.5	B
	92.7		
	93.7		
Dodecanamide	96.0	96.6	B
	95.8		
	94.2		
Methyl dodecanamide	94.4	97.6	B
	92.3		
	93.1		
Succinimide	99.3	98.8	A
	99.9		
N-lauroyl sarcosine	100.5	99.1	B[a]
	100.2		
	100.3		
	99.8		
Sodium N-lauroyl sarcosinate	30.27	—	B[a]
	30.11		
	30.04		
Sodium N-lauroyl sarcosinate[c]	94.7	95.8	B[a]
	93.7		
	93.7		
	94.6		
	94.4		

[a] Reflux time for reduction step was increased to 1.5 hours for these compounds.
[b] A 30% aqueous solution of sodium N-lauroyl sarcosinate. Each sample was dried at 100°C before reduction.
[c] Sample contained 5.7% water as determined by Karl Fischer titration.

found in amide samples; nitriles are sometimes found in samples of primary amides.

Lithium aluminum hydride will reduce nitriles to the corresponding amine, and this will result in high values for the amide. The reduction of nitrile does not proceed to completion for all nitriles; therefore this would not be a good general method for nitriles. Benzonitrile, butyronitrile, capronitrile, and chlorobenzonitrile were determined satisfactorily by this approach (see Table 13). Nitriles that were tried and could not be reduced completely to amine in a convenient length of time were acetonitrile, acrylonitrile, succinonitrile, adiponitrile, phenylacetonitrile, 3-butenenitrile, γ-phenoxybutyronitrile, lactonitrile, *m*-nitrobenzonitrile, and 1-naphthonitrile.

It is advisable, therefore, if nitriles are suspected in a sample of a primary amide, to use the procedure of Mitchell and Ashby (p. 204) for

Table 13. Nitrile Results

Benzonitrile	98.90	98.67	A
	98.69		
	98.81		
Butyronitrile	96.15	98.86	A
	96.40		
	96.72		
N-capronitrile	98.45	99.65	A
	99.07		
	97.76		
p-Chlorobenzonitrile	99.95	102.06	A
	98.93		
	99.76		

determining the primary amide. Because nitriles seldom occur in secondary and tertiary amides, possible interference is minimized.

Compounds that contain active hydrogen atoms as well as alkyl halides, esters, epoxides, and azoxy compounds will consume hydride. If such compounds are present in quantity, enough hydride must be present in the reaction mixture to convert all the amide to amine. It is impossible to run dilute aqueous solutions of amides by this method, because too large an excess of reagent would be needed. Samples containing 10% water were successfully run by this method, however. Samples containing 50% water showed results which were 10% low in amide, but using more hydride reagent may make it possible to run 50% aqueous solutions. It may be possible to run samples containing 25% water, but this was not tried.

The precision of the analysis to be described is usually within ±2% and very often within ±1% for the various amides tried. The accuracy is about the same as the precision; in many of the amides tried, however, the precision of the nitrogen analysis (Dumas) used to assay the amides is poorer than the precision of the reduction method being tested. For the amides of nitrogen content below 10%, the accuracy values are controlled not by the reduction method but by the nitrogen method (the Dumas nitrogen analyses are reproducible to ±0.2% of nitrogen).

The analysis is very simple to carry out and the apparatus is, for the most part, standard laboratory equipment. Precautions are required in handling lithium aluminum hydride; these are described on every package of the material. It is well not to use a hydride sample that is too old or one that has been stored in a loosely stoppered container; it may be too hydrolyzed to be effective.

REAGENTS

LITHIUM ALUMINUM HYDRIDE. Ten grams of lithium aluminum hydride is refluxed with 500 ml of anhydrous diethyl ether for several hours. If the hydride is finely divided, it will dissolve in a relatively short time. Insoluble products, formed by the reaction of impurities in the ether with the lithium aluminum hydride, settle on cooling, and the clear solution can be pipetted off as needed. The solution should be protected from atmospheric moisture. The usable life of the solution is about one month.

Standard 0.02*N* sulfuric acid.

Standard 0.02*N* sodium hydroxide.

6*N* Sodium hydroxide.

Methyl purple indicator (Fleisher methyl purple, Burrell Corp., Pittsburgh).

Ethylene glycol.

Isopropyl alcohol.

DISTILLATION APPARATUS

The distilling apparatus used in Procedure A is the standard Kjeldahl steam distillation equipment.

The distilling apparatus in Procedure B consists of a 200-ml round-bottomed flask connected to a Kjeldahl bulb, which is attached to a water condenser by a 75° connector. A stopcock and funnel are sealed on the connector at the bend so that ethylene glycol can be dropped into the flask (see Fig. 3.16).

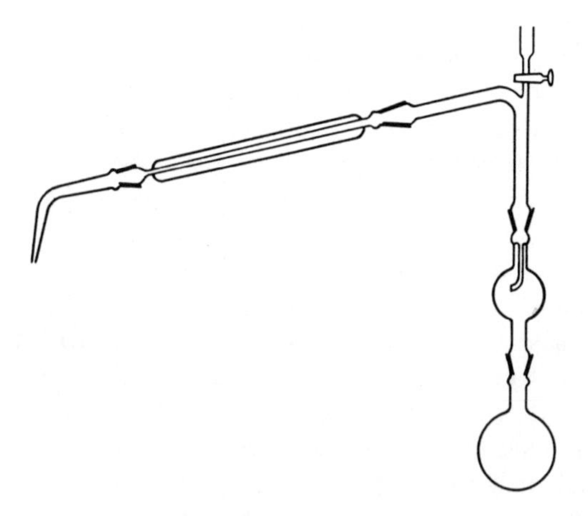

Fig. 3.16. Distillation apparatus.

PROCEDURE A

Weigh exactly a sample containing approximately 0.0006 mole of amide, place it in a 100-ml Kjeldahl flask, and add 5 ml of lithium aluminum hydride reagent. Allow the solution to stand for 15 minutes at room temperature to ensure complete reduction of the amide; then attach the flask to the Kjeldahl distillation apparatus. Place a 200-ml Erlenmeyer flask containing exactly 50 ml of 0.02*N* sulfuric acid on the apparatus so that the end of the condenser is below the surface of the acid. Add water dropwise to the reaction flask until the excess lithium aluminum hydride is decomposed. Add 10 ml of 6*N* sodium hydroxide and carry out steam distillation as in a Kjeldahl determination. Collect about 50 ml of distillate in the 0.02*N* sulfuric acid and titrate the excess acid with standard 0.02*N* sodium hydroxide to the green end point of methyl purple indicator. Calculate percentage of amide as follows:

$$\frac{\text{(Titration for 50 ml of acid minus titration sample)} \times N \text{ of NaOH} \times \text{mol. wt. amide} \times 100}{\text{Weight of sample} \times 1000} = \%\ \text{amide}$$

PROCEDURE B

Place a weighed sample containing approximately 0.0006 mole of amide in a 200-ml round-bottomed flask, and add 10 ml of lithium aluminum hydride reagent. reflux the mixture on a steam bath for 30 minutes. Cool the flask to room temperature and decompose the excess reagent by adding water dropwise. After the reagent is completely decomposed, wash the sides of the flask with about 10 ml of water, and add 5 ml of 6*N* sodium hydroxide. Add a few boiling chips and 25 ml of ethylene glycol before attaching the flask to the distilling apparatus. Distill the solution at a rapid rate nearly to dryness and add 25 ml of ethylene glycol through the stopcock on the connector at such a rate that boiling does not stop. Continue the addition and distillation of 25-ml portions of ethylene glycol until 100 ml has been distilled. Wash the condenser with approximately 50 ml of hot isopropyl alcohol, and titrate the amine contained in the distillate and washings potentiometrically with 0.02*N* sulfuric acid. Calculate the percentage of amide in the following manner:

$$\%\ \text{Amide} = \frac{\text{Milliliters of } H_2SO_4 \times N \text{ of } H_2SO_4 \times \text{mol. wt.} \times 100}{\text{Weight of sample} \times 1000}$$

DETERMINATION OF PRIMARY AMIDES BY REACTION WITH 3,5-DINITROBENZOYL CHLORIDE

Adapted from J. Mitchell and C. E. Ashby

[*J. Am. Chem. Soc.*, ***67,*** *161–4* (*1945*)]

Primary amides react with acid chlorides to yield the corresponding acyl amide. Acid halides are generally unstable, hence are not satisfactory reagents; the 3,5-dinitrobenzoyl chloride is sufficiently stable to be used (see Hydroxyl Group, pp. 67–71). The procedure to be described is of particular value where the primary amide content is desired in a mixture containing secondary and tertiary amides. The foregoing titrimetric and reduction methods would not be able to differentiate the two types.

$$RC(=O)NH_2 + (NO_2)_2C_6H_3C(=O)Cl \rightarrow RCN + (NO_2)_2C_6H_3COOH + HCl$$

REAGENTS

2*M* 3,5-DINITROBENZOYL CHLORIDE. The reagent solution is prepared by dissolving 461 grams of 3,5-dinitrobenzoyl chloride (Eastman Kodak) in enough purified, anhydrous 1,4-dioxane to yield 1 liter of solution. The reagent solution is treated with activated carbon and rapidly filtered, care being taken to protect it from exposure to moisture. The final solution should be no darker than light yellow or else there will be difficulty in determining the end point color change.

CP Dry pyridine.

CP Dry methanol.

0.5*N* SODIUM METHOXIDE IN METHANOL. This is prepared directly from sodium in anhydrous methanol. This solution should be standardized each day by titrating an aliquot with 0.5*N* standard acid, phenolphthalein indicator being used, or ethyl bis-2,4-dinitrophenyl acetate indicator (a saturated solution in 1:1 acetone–ethyl alcohol).

PROCEDURE

Weigh a sample containing about 10 meq. of amide into a 250-ml glass-stoppered Erlenmeyer flask containing 15 ml of the 3,5-dinitrobenzoyl chloride reagent and 5 ml of pyridine. Place the flask together with a blank in a water bath at 60°C for 30 minutes (70°C for 1 hour when amides of dibasic acids are determined). At the end of this time remove the flasks and cool in ice. Decompose the excess aroyl chloride with dry methanol, which is added in two portions, first 2 ml and, after 5 minutes, an additional 25 ml. Titrate the solution and blank with 0.5*N* standard sodium methoxide, using ethyl bis-2,4-dinitrophenyl acetate indicator. Phenolphthalein can be used, but the end point is not as sharp as with the dinitrophenyl acetate because of the yellow-orange color of the solution.

The net increase in acidity of the sample over the blank, after correction for free acid and water present in the original sample, is equivalent to the quantity of primary amide present.

CALCULATIONS

$$\text{Milliliters of sample minus milliliters of blank} = A$$

$$\frac{A \times N\ CH_3ONa \times \text{mol. wt. of amide} \times 100}{\text{Grams of sample} \times 1000} = \%\ \text{primary acid amide}$$

Interfering substances in the foregoing procedure consist mainly of free acid and water. These materials should be determined in the original samples and the corrections applied to the results of the amide analysis. The water hydrolyzes the acid chloride to liberate 3,5-dinitrobenzoic acid and hydrochloric acid. Secondary and tertiary amides generally do not interfere in the analysis. Amines and alcohols do not affect the results beyond the fact that they may inactivate some of the reagent.

To test this procedure, the following primary amides were used: formamide, acetamide, propionamide, butyramide, isobutyramide, *n*-valeramide, heptamide, succinamide, glutaramide, adipamide, benzamide, salicylamide, *p*-nitrobenzamide, phthalamide, furoamide.

The procedure is generally accurate and precise to ±0.5 to 1.0%.

HYPOBROMITE TITRATION OF ALIPHATIC AMIDES

From the Method of W. R. Post and C. A. Reynolds

[*Anal. Chem.*, ***36,*** *781* (*1964*)]

The first step in the Hofmann reaction, the production of *N*-bromoamide by the action of hypobromite on amide, is used as the basis for a quantitative spectrophotometric titration procedure for primary aliphatic amides. A sample of amide is dissolved in an aqueous solution that is $0.1M$ in potassium bromide and is strongly buffered at pH 10 with borax. The solution is titrated with a standard solution of calcium hypochlorite, and the hypobromite ion produced *in situ* and the amide immediately react to form the *N*-bromoamide. The titration is followed by observation of the absorbance of hypobromite ion at 350 nm.

REAGENTS AND APPARATUS

Prepare an approximately $0.1N$ solution of calcium hypochlorite by dissolving reagent grade calcium hypochlorite in water. Filter to remove solid calcium carbonate. Standardize the solution iodometrically. The strength decreases approximately 0.2% per week.

Prepare a borate buffer by adding a concentrated carbonate-free sodium hydroxide solution to a saturated solution of sodium tetraborate until a pH of 10.0 is reached.

The photometric titration apparatus (34) consisted of a Beckman Model DU monochromator and an ultraviolet light source, a cubical glass cell of 125-ml capacity, and an American Instrument Company photoelectric microphotometer unit.

A calibrated 10-ml microburet was used, and both mechanical and magnetic stirring was employed.

PROCEDURE

Pipet a 10-ml aliquot containing 0.15 mM of amide into a quartz titration cell and follow with 5 ml of $1M$ potassium bromide solution and 30 ml of the borax buffer. Titrate the mixture with the standard hypochlorite solution. Take absorbance readings at 350 nm when equilibrium has been reached after the addition of each increment of titrant. Perform blank titrations in the same manner, but add 10 ml of water in place of the amide aliquot.

RESULTS AND DISCUSSION

Table 14 gives the results for the titration of a number of representative aliphatic amides. Although most of the results reported were obtained with 0.15-mM samples, variation in the sample size from 0.08 to 0.34 mM did not influence either the average result or the precision.

34. R. W. McKinney and C. A. Reynolds, *Talanta*, **1,** 46 (1958).

Table 14. Photometric Titration of Primary Aliphatic Amides

Amide[a]	Number of Trials	Average Purity, %	Standard Deviation[b]
Formamide	4	98.9	0.4
Acetamide	16	99.6	−0.4%
Propionamide	4	100.6	0.6%
n-Butyramide	4	99.4	1.0
n-Valeramide	4	99.7	0.9
α-Chloroacetamide	4	100.7	0.4
Trichloroacetamide	3	103.0	0.0
Adipamide	4	96.5	0.3
Acrylamide	3	98.6	0.6
Furamide	6	100.5	1.3
Nicotinamide	3	101.0	0.1
Succinamide	3	98.7	0.6

[a] Sample 0.15 mM, except for adipamide and furamide, for which 0.08 mM was taken.

[b] In the case of acetamide and propionamide, relative error is reported.

The reaction can be represented as follows:

$$RC(=O)NH_2 + OBr^- \rightarrow RC(=O)NHBr + OH^-$$

At 350 nm hypobromite is the major absorbing species, whereas *N*-bromoamide absorbs only slightly.

If too long a time is taken for the titration in the case of propionamide and the higher primary aliphatic amides, the action of hypobromite on the bromoamide and on the subsequent reaction products results in a decrease in absorbance of the hypobromite and causes errors in the final results. To prevent error, the hypobromite is added quickly from a buret until enough has been added to overrun the end point; absorbance readings due to the excess hypobromite are then taken at three or four increments of titrant beyond this point. The best straight line drawn through these three or four points constitutes the second slope of the titration curve. The first slope of the titration curve in the case of these amines was constructed by measuring the absorbance of a freshly prepared solution of *N*-bromopropionamide at 350 nm, calculating a molar absorptivity for this compound, and then calculating the absorbance of the titration solution, assuming a stoichiometric yield of *N*-bromopropionamide. This standard first slope of the titration curve was used in all results reported for propionamide, *n*-butyramide, *n*-valeramide, and adipamide. Because of the very small slope of this initial portion of the titration curve, changes in concentration of the amide being titrated caused almost no change in the final percentage purity calculated from the results of the titration.

In the cases of all the other amides, except acrylamide, the first slope of the titration curve was experimentally measured each time because the subsequent attack of the *N*-bromoamide occurred so slowly that no appreciable oxidation of the products by hypobromite took place within at least 20 minutes.

Acrylamide and hypobromite continue to react beyond the theoretical formation of *N*-bromoacrylamide, rendering direct titration of this amide impossible. A number of other amides could not be determined by this procedure for a variety of reasons. Nonamide and oxamide are not soluble enough in water to produce a solution capable of titration. The formation of the *N*-bromo derivative of isobutyramide proceeds so slowly compared with the rearrangement of *N*-bromoisobutyramide that no titration was feasible. In the case of malonamide, consistently high results were always obtained, indicating that partial bromination of the methylene carbon atom was also taking place.

Esters, alcohols, nitriles, carboxylic acids, and ethers do not interfere; but amines, aldehydes, and methyl ketones were oxidized under the conditions of the titration and high results were obtained. However preoxidation of the sample solution with bromine in neutral solution removes the interference of all three of these functional groups. Most aromatic amides and *N*-methylformamide interfere in this procedure but the presence of higher *N*-alkyl amides and the di-*N*-alkylamides do not cause any interference.

Method of R. D'Alonzo and S. Siggia

[*Reprinted in Part from Anal. Chem.*, ***49,*** *262* (*1977*)]

This procedure is based on a modified Hofmann reaction.

$$RCONH_2 + 2Ba(OH)_2 + Br_2 \rightarrow RNH_2 + BaCO_3\downarrow + BaBr_2 + 2H_2O$$

Barium hypobromite and primary amides react quantitatively to produce the amine plus insoluble barium carbonate. Similarly, barium hypobromite and imides react to produce barium carbonate and the barium salts of the resulting amino acids. The separated insoluble barium carbonate is dissolved in nitric acid, and its barium content is determined by flame emission spectrometry. The method applies to primary amides both aliphatic and aromatic in the presence of secondary and tertiary amides.

The method is highly selective because few other functional groups produce insoluble barium salts.

REAGENTS

BARIUM HYPOBROMITE SOLUTION. Add 200 ml of distilled deionized water to 0.03 mole of barium hydroxide ($Ba(OH)_2 \cdot 9H_2O$). Stir for 1 hour and filter through a fine frit (4–5.5 μm) Pyrex glass funnel. Add 0.01 mole of reagent grade liquid bromine. The final concentration of barium hypobromite should be between 0.025 and 0.075*M*. Higher concentrations result in serious side reactions with aromatic amides. The hypobromite solution is not stable and should be discarded after 24 hours if not used.

NITRIC ACID REAGENT. Mix 1 volume of concentrated nitric acid and 1 volume of distilled deionized water.

BARIUM STOCK SOLUTION. Transfer a weighed sample of 0.10 to 0.15 gram of dry barium carbonate to a 1-liter volumetric flask, dissolve with the least of 1:4 nitric acid–water and dilute to volume with distilled deionized water.

POTASSIUM CHLORIDE STOCK SOLUTION. Dissolve 38 grams of potassium chloride (Fisher-Certified ACS-0.001% Ba) with sufficient distilled deionized water to make 1 liter of solution.

STANDARD BARIUM SOLUTION FOR CALIBRATION. Prepare with appropriate aliquots of barium stock solution, add 10 ml of potassium chloride stock solution, and dilute to volume with distilled deionized water in 100-ml volumetric flasks.

APPARATUS

A Perkin-Elmer 403 Atomic Absorption Spectrometer operated in the flame-emission mode was used to measure intensities in the ultraviolet region at a wavelength of 276.8 nm. A nitrous oxide–acetylene flame was used.

PROCEDURE

Pipet a 2-ml sample of amide solution into a 15×150 mm test tube. The concentration of amide should be between 1.00 and 4.00 μM per milliliter. Add an equal volume of barium hypobromite reagent and seal the top of the test tube with Parafilm. Mix the contents of the test tube and make a small pin hole in the Parafilm. Place the test tube in a water bath maintained between 70 and 75°C for 10 to 15 minutes. After this heating period, remove the test tube from the bath and place it in an ice water bath for 5 minutes. Remove the test tube from the ice water bath and allow the contents to come to room temperature before filtering.

Filtration of cold solutions results in longer filtration duration, therefore higher blank values, because the reagent is in contact with atmospheric carbon dioxide for a longer period. Transfer the contents of the test tube to a fine frit (4–5.5 μm) Pyrex glass funnel along with two 5-ml washings from the test tube. Allow each washing to pass completely through the filter before the next portion is added; otherwise, high blank values will result because of incomplete washing of excess barium hydroxide. When filtration is complete, transfer a clean suction flask to the vacuum filtering apparatus and add 10 ml of 1:1 concentrated nitric acid–water to the filter funnel from the test tube to dissolve the barium carbonate. After 1 to 2 minutes, add an additional 5 ml of distilled deionized water to the funnel and apply suction. Then use two 15-ml portions of water from the test tube to wash the filter of final traces of barium and nitric acid. Transfer the contents of the suction flask to a 100-ml volumetric flask containing 10 ml of potassium chloride solution and dilute to volume with rinsings from the suction flask. Run a blank for each glass filter funnel used in the determinations. Prepare the blanks in the same manner as the samples with distilled deionized water in place of the amide solution. Use water from the same batch that was used to prepare the amide solution to prevent errors due to differences in dissolved carbon dioxide content. Analyze both the sample and blank solutions for barium content by flame-emission spectrometry.

RESULTS AND DISCUSSION

OPTIMUM CONCENTRATION OF REAGENTS. The concentration of bromine in the reagent is critical in this procedure. A high concentration of bromine will result in high recoveries for aromatic amides because of the serious side reaction of bromination, which yields an insoluble organic precipitate. The resulting mixed precipitate of a brominated aromatic and barium carbonate becomes difficult to wash free of excess barium hydroxide in the reagent. A final concentration of $0.025M$ barium hypobromite after the addition of reagent to sample was found to give satisfactory results for selected amides and imides, as indicated in Table 15. The recommended low concentration of bromine and high alkaline concentration were found to eliminate the problem of bromination of all aromatic amides tested, with the exception of salicylamide. The phenolic moiety interferes because of its ability to readily form an insoluble brominated phenol that includes barium from the reagent resulting in high recoveries. If the sample does not contain any aromatics, the concentration of bromine may be as high as $0.15M$ without any serious effects.

CALIBRATION CURVE. The optimum working range used for barium was 2

Table 15. Determination of Primary Amides by Flame-Emission Spectrometry

Compound	Concentration, μM/ml	Taken, μM	Found, μM	Recovery,[a] %
		Amides		
Acetamide	1.705	3.41	3.41	100.0 ± 5.8 (5)
Propionamide	2.79	5.58	5.31	95.2 ± 5.1 (9)
Butyramide	1.425	2.85	2.80	98.2 ± 4.8 (4)
Isobutyramide	2.685	5.37	5.17	98.3 ± 5.5 (5)
Valeramide	1.095	2.19	2.19	100.0 ± 8.6 (6)
Isovaleramide	1.75	3.50	3.44	98.3 ± 6.5 (5)
Adipamide	1.04	2.08	2.07	99.5 ± 5.5 (5)
2-Chloroacetamide	2.82	5.64	3.19	56.6 ±13.0 (6)
Acrylamide	3.835	7.67	8.05	105.0 ± 1.8 (5)
Methacrylamide	2.545	5.09	5.03	98.8 ± 6.0 (4)
Benzamide	2.35	4.70	4.64	98.7 ± 3.2 (5)
p-Nitrobenzamide	2.75	5.50	5.61	102.0 ± 3.0 (6)
p-Toluamide	2.155	4.31	4.44	103.0 ± 5.1 (5)
o-Toluamide	2.05	4.10	4.14	101.0 ± 4.1 (6)
Salicylamide	2.265	4.53	6.80	150.0 ± 2.6 (6)
Nicotinamide	1.805	3.61	3.58	99.2 ± 6.5 (6)
Pyrazinamide	1.615	3.23	3.29	102.0 ± 3.6 (6)
Succinimide	1.38	2.76	2.76	100.0 ± 6.5 (5)
Phthalimide	0.805	1.61	1.63	101.0 ± 2.7 (5)

[a] Figures in parentheses indicate number of determinations.

to 1 ppm, which corresponds to approximately 1.5 to 7.0 μM of amide. Because barium is partially ionized in the nitrous oxide–acetylene flame, potassium chloride solution is added to all standards and samples to suppress ionization. The optimum concentration of potassium was determined experimentally as 2000 mg per liter for a 2-ppm solution of barium.

SOLUBILITY OF BARIUM CARBONATE. The solubility of barium carbonate in pure water at 20°C is reported to be 0.002 gram per 100 cm^3 (35). This corresponds to 0.10 μM of barium carbonate per milliliter. Because of the high concentration of barium in the reagent, however, the actual solubility of barium carbonate is much less than that reported for pure water. Water used to wash the precipitate was maintained at 0°C to minimize losses due to solubility.

SOURCE OF THE BLANK. Two main factors were found to contribute to the

35. R. C. Weast, Ed., *Handbook of Chemistry and Physics*, Chemical Rubber Company, Cleveland, 1970–1971, p. B-70.

unexpected high blank: barium carbonate production resulting from contact of the reagent with atmospheric carbon dioxide, and adsorption of barium from solution on the glass frit filter. Adsorption of barium on the glass frit was found to be directly related to the flow rate of the filter used (Fig. 3.17). The large difference between the slowest and fastest filters indicated that adsorption was the chief contributor to the blank. A second study was done to determine the amount of blank attributable to adsorption. The extrapolated value of the y-axis of a blank versus time curve is the amount of barium in the blank when the reagent is exposed to atmospheric carbon dioxide at zero time or simply the blank due to adsorption. The data for the curve were obtained with the same filter and were found to be reproducible. The blank value due to adsorption divided by the corresponding value for the same filter in Fig. 3.17 is the percentage of the total blank due to adsorption. This value was found to be 83%; that is, 17% of the blank resulted from atmospheric carbon dioxide interference. Thus 5% of the total barium carbonate produced from a 5-μM sample of amide is due to carbonate production from atmospheric carbon dioxide. Because the blank due to adsorption is reproducible and the blank due to atmospheric carbon dioxide may vary, the actual "working blank" is due only to atmospheric carbon dioxide. This value, as stated, is on the order of an acceptable 5%, which is reflected in the good accuracy of the results. No efforts were made to reduce this value by working in a carbon dioxide-free atmosphere because of the undesirable increase in total analysis time that would have resulted.

UTILITY OF THE HOFMANN REACTION. The Hofmann reaction in the past has not been regarded as a useful analytical reaction for two reasons: (1) it has always been considered to be a synthetic method, and (2) the resulting amine is not always produced quantitatively. Table 16 gives

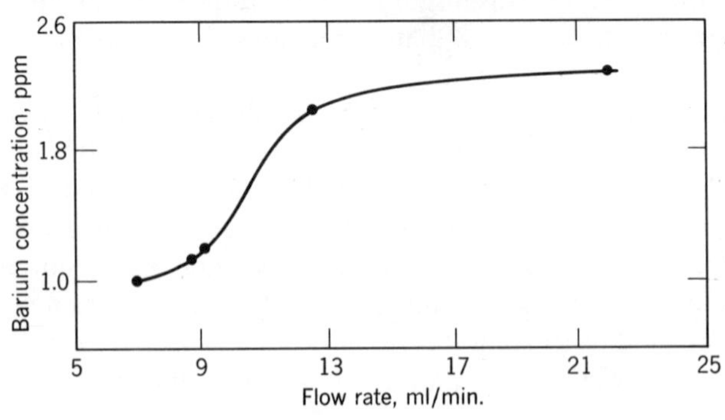

Fig. 3.17. Effect of filter flow rate on concentration of barium blank.

Table 16. Product Recovery Comparison for the Hofmann Reaction

	Recovery, %	
Compound	Barium Carbonate Flame Emission[a]	Amine Literature[b]
Acetamide	100	70–80
Propionamide	95	85
n-Butyramide	98	90
Isovaleramide	98	90
Benzamide	99	Good
Salicylamide	150	70
Acrylamide	105	Poor[c]
Methacrylamide	99	Poor[c]
2-Chloroacetamide	57	Poor[c]

[a] Values reported are from Table 15.
[b] Values reported are for isolated amine, Ref. 36.
[c] Yield obtained in aqueous media.

literature values for isolated amine produced (36) compared with values for barium carbonate produced for the same amides. It can be seen that for many aliphthatic and aromatic amides, both the amine and the carbonate are produced in excellent yields. In many cases, however, particularly where other functionalities are present in the molecule, the production of carbonate is quantitative and the production of amine is poor. This is particularly true when the other functionality is subject to the reactive bromine in the reagent such as halohydrin formation or even oxidation of double bonds by hypobromite (37). 2-Chloroacetamide was the only compound tested that gave a low recovery of barium carbonate. This is not surprising because α-haloamides have been reported to give low yields of the expected amine with many side reactions producing aldehydes, ketones, and *gem*-dihalides as the common products (38).

OTHER INTERFERENCES. No interferences were observed from secondary and tertiary amides and nitriles as long as the reaction temperature did not exceed 75°C. At higher temperatures (95–100°C) these compounds have been observed to result in high barium carbonate recoveries. Isocyanates will interfere by hydrolysis to produce barium carbonate, but these compounds are seldom found in combination with primary amides.

AMIDE MIXTURES. Successful determinations for total amide content in

36. R. Adams, Ed., *Organic Reactions*, Vol. III, Wiley, New York, 1946, p. 267.
37. A. Weerman, *Ann.*, **401,** 1 (1913).
38. J. March, *Advanced Organic Chemistry*: *Reactions, Mechanisms, and Structure,* McGraw-Hill, New York, 1968, p. 817.

Table 17. Determination of Total Primary Amide in Mixtures

Mixture	Amides			
	Concentration, μM/ml	Taken, μM	Found, μM	Recovery,[a] %
Acrylamide, 3.83 μM, +merthacrylamide, 2.55 μM	6.38	6.38	6.34	99.4 ±2.4 (6)
Butyramide, 2.41 μM, +isobutyramide, 2.61 μM	5.02	5.02	4.77	95.0 ±4.8 (7)
Propionamide, 1.50 μM, +butyramide, 1.41 μM	2.91	2.91	2.86	98.3 ±6.6 (7)

[a] Figures in parentheses indicate number of determinations.

three-amide mixtures have been accomplished. The results are reported in Table 17. This method should be of particular value for the determination of total primary amide content at trace levels in a mixture containing esters and anhydrides.

REACTION TIME. Completeness of reaction as a function of time for benzamide was studied. The results indicate that the reaction is complete after 12 minutes of heating at 70°C. Benzamide was chosen because alkylamides and benzamides substituted with electron-donating groups, such as methyl, react more readily. A longer reaction time of 20 minutes was required for the determination of *p*-nitrobenzamide because the electron-withdrawing nitro group retards the rearrangement.

HYPOCHLORITE TITRATION OF PRIMARY AMIDES

Method of W. R. Post and C. A. Reynolds

[*Reprinted in Part from Anal. Chem.*, **37,** *1171* (*1965*)]

This procedure can be used for both aliphatic and aromatic amides. The sample of amide is dissolved in a dioxane–water mixture that is 1*M* in hydrochloric acid and titrated with a standard solution of calcium hypochlorite to an amperometric end point.

REAGENTS AND APPARATUS

CALCIUM HYPOCHLORITE SOLUTION. Prepare an approximately 0.5*N* aqueous solution from reagent grade calcium hypochlorite. Filter to remove solid calcium carbonate. Standardize the solution iodometrically.

The amperometric titration apparatus consisted of a dry cell potential source, a divider, an RCA Model WV-848 microammeter, a Leeds and Northrup pH meter, a rotating platinum electrode, and a saturated calomel electrode.

PROCEDURE

Using aqueous hydrochloric acid, dilute a 25-ml aliquot of sample solution containing 0.5 mM of amide in dioxane to exactly 150 ml, so that the final solution is approximately $1M$ in hydrochloric acid and 20% dioxane. Set the applied potential to +0.4 volt versus a standard calomel electrode, and add the calcium hypochlorite titrant in 0.5-ml increments. Take the resulting current readings after the addition of each increment of titrant, corrected for dilution, and plot against the volume of titrant. Draw the best straight line through the experimental points and take the intersection of these lines as the end point.

RESULTS AND DISCUSSION

Table 18 presents the results of the titration of a number of primary amides. Although these results were obtained with 0.5-mM samples of amide, as small a sample as 0.05 mM was titrated with $0.05N$ calcium hypochlorite with only slight decrease in accuracy and precision. A titration can be completed in about 10 minutes.

The excess chlorinating agent slowly decomposes after the end point of the titration has been passed. The rate of decomposition is apparently 1% per minute. The following technique was developed to eliminate errors due to the slight instability of the chlorinating agent. A direct titration was carried out, but the first experimental point after the end point was taken with only a slight excess (usually about 25%) of hypochlorite. The remaining point, which determined the second segment of the titration curve, was then assumed to be constant if the current decreased less than 1 μA per minute.

Table 18. Amperometric Titration of Primary Amides

Amide	Number of Trials	Average Apparent Purity, %	Standard Deviation, %
Benzamide	8	99.6	0.6
o-Toluamide	9	100.6	0.9
p-Toluamide	5	97.0	0.6
Acetamide	6	99.6	1.3
Propionamide	3	100.0	0.7
n-Butyramide	3	100.4	0.5

The procedure can be used successfully for primary aliphatic amides and for aromatic amides that are not substituted or are substituted with only alkyl groups. The presence of other functional groups substituted on the aromatic ring in general causes interference in the determination of the amide group. Electron-withdrawing groups such as nitro and carboxyl reduce the rate of *N*-chlorination and also affect the equilibrium. Electron-donating groups activate the ring so that chlorination of the ring takes place readily, and the exact stoichiometry of the titration reaction becomes uncertain. In addition, substituted groups such as amino, carbonyl, and hydroxyl are easily oxidized by hypochlorous acid in the titration medium.

CYCLOPROPYLAMIDES DETERMINED BY A MODIFIED MERCURIC ACETATE PROCEDURE

The titration methods for weakly basic amides are not generally applicable to cyclopropylamides because of their nearly neutral and, in some cases, weakly acidic character. The mercuric acetate procedure of Johnson and Fletcher (p. 519) with slight modifications can be used for these compounds.

Method of J. G. Theivagt

[*Reprinted in Part from Anal. Chem.*, **33**, *1391* (*1961*)]

PROCEDURE

Use the procedure as described by Johnson and Fletcher, except allow the reaction mixtures to stand at room temperature instead of at −10°C used for vinyl ethers.

DISCUSSION AND RESULTS

The mechanism for the methoxymercuration of cyclopropylamides was presumed to be similar to that suggested for cyclopropane by Levina and Kostin (39), who showed that the ring opening occurred between the substituted and unsubstituted carbon atoms, with the methoxy group adding to the substituted carbon.

39. R. Y. Levina and V. N. Kostin, *J. Gen. Chem., USSR*, **23,** 1054 (1953).

$$Hg(OCOCH_3)_2 + RCONHCH\begin{matrix} CH_2 \\ | \\ CH_2 \end{matrix} \longrightarrow$$

$$RCONH\overset{\displaystyle OCH_3}{\underset{\displaystyle CH_2CH_2HgOCOCH_3}{C}} + CH_3COOH$$

Table 19 lists several compounds of a related nature which were tested under the conditions of the general method. Only the *N*-allyl amide analogs would cause any appreciable interference if present as impurities.

Table 20 gives results for various cyclopropylamides. Secondary cyclopropylamides reacted completely in all cases within 30 minutes at room temperature except for *N*-cyclopropyl-2,3,4,5,6-pentachlorbenzamide and *N*-cyclopropylisonicotinamide, which required refluxing. On the other hand, tertiary cyclopropylamides were much less reactive. Table 20 includes two tertiary cyclopropylamides that gave no reactions.

DETERMINATION OF TRACES OF AMIDES

Trace quantities of amides can be determined colorimetrically in two ways—by the hydroxamic acid–ferric complex, much like the esters, and

Table 19. Compounds Unreactive in Mercuric Acetate Procedure

Compound
4-Amino-3,5-dichlorobenzamide
4-Amino-3,5-dichlorobenzoic acid[a]
4-Amino-3,5-dichloro-*N*-isopropylbenzamide
4-Amino-*N*-cyclopentyl-3,5-dichlorobenzamide
4-Amino-*N*-cyclobutyl-3,5-dichlorobenzamide
4-Amino-*N*-(cyclopropylmethyl)-3,5-dichlorobenzamide
4-Amino-*N*-cyclopropyl-3,5-dichloro-*N*-(hydroxyethyl)benzamide
4-Amino-*N*-benzyl-*N*-cyclopropyl-3,5-dichlorobenzamide
N-Allyl-4-amino-3,5-dichlorobenzamide[b]
Cyclopropyl ketone
Cyclopropyl ketoxime

[a] Acid function corrected by titration.

[b] One-tenth of theoretical addition occurred.

Table 20. Determination of Cyclopropylamides by Mercuric Acetate Titration

Compound	Percentage
N-Cyclopropylbenzamide	98.3, 98.7
4-Amino-*N*-cyclopropyl-3,5-dichlorobenzamide	99.0, 99.0
N-Cyclopropyl-3,4,5-trimethoxybenzamide	99.0, 99.0
4-Bromo-*N*-cyclopropylbenzamide	97.3, 96.9
4-Chloro-*N*-cyclopropylbenzamide	96.8, 96.5
N-Cyclopropyl-2,3,4,5,6-pentachlorobenzamide[a]	99.4, 99.8
N-Cyclopropyl-4-(trifluoromethyl)benzamide	98.3, 98.7
N-Cyclopropylcyclopropanecarboxamide	98.3, 98.5
N-Cyclopropylisonicotinamide[a]	96.0, 95.9
N-Cyclopropyl-3-methylenecyclobutanecarboxamide[b]	92.4, 92.1
N-Cyclopropyl-2-furamide	99.0, 98.6
N,N'-Biscyclopropylphthalamide[c]	98.1, 98.2
N-Cyclopropyl-3,4,5-trimethoxycinnamide[d]	99.0, 100.0
N-Cyclopropyl-3,5-dichloro-4-ethoxybenzamide	102.0, 101.8
N-Cyclopropylphenylcarbamate	97.6, 97.7
N,N'-Biscyclopropyl-2-methyl-2-*n*-propyl-1,3-propanedioldicarbonate[c]	100.1, 100.0
N-Benzyl-*N*-cyclopropylurethane	102.5, 102.5
1-Cyclopropyl-3-*o*-tolylurea	96.9, 96.9
1,3-Biscyclopropylurea[c]	94.6, 94.8
1-Cyclopropyl-3-(*p*-toluenesulfonyl)urea[e]	97.5, 97.5

[a] Refluxed in reagent for 30 minutes.
[b] Both methylene and cyclopropyl groups reacted.
[c] Both cyclopropylamide groups reacted.
[d] Reacted for 30 minutes at −10°C.
[e] Corrected for acid function by titration.

also by hydrolysis and colorimetric determination of the ammonia or amine produced.

Ferric-Hydroxamate Method

[*Adapted from F. Bergmann, Anal. Chem.*, **24**, *1367-9* (*1952*)]

REAGENTS

The reagents used differed but slightly from those described by Hestrin (24).

Solution 1. Hydroxylamine sulfate, 2*N*
Solution 2. Sodium hydroxide, 3.5*N*
Solution 3. Hydrochloric acid, 3.5*N*
Solution 4. Ferric chloride, 0.74*M* in 0.1*N* hydrochloric acid solution

PROCEDURE

All amides used in this investigation were dissolved in water at concentrations of 5 or $10 \times 10^{-3} M$. The alkaline hydroxylamine reagent (2 ml), prepared by mixing equal volumes of solutions 1 and 2, and graded volumes of the amide solution, with the addition of enough water to give a total volume of 3 ml, was kept at various temperatures for different periods of time. The reaction mixture was then rapidly cooled to room temperature, 1 ml each of solutions 3 and 4 were added, and the extinction was determined in a Klett-Summerson photoelectric colorimeter, using filter No. 54 (spectral range 500–570 nm). A No. 50 filter (spectral range 470–530 nm) was used for fluoroacetamide, because fluoroacethydroxamic acid has its maximum of absorption near 500 nm.* Readings were carried out within 5 minutes. The color of the complexes fades slowly, showing a decrease of 15 to 20 Klett units per micromole per hour. In the case of fluoroacethydroxamic acid, fading progressed much faster, about 50 Klett units per micromole per hour. Each experiment comprised five different concentrations of amide in order to test the linear relationship between extinction and concentration. This relationship was found to hold in all cases.

In preliminary experiments the reaction with hydroxylamine was carried out at various pH values. For both acetamide and nicotinamide the rate of reaction increased with increasing pH, but above pH 13 the hydrolysis by hydroxyl ion became faster than the interaction with hydroxylamine. Therefore, the standard mixture described previously with a pH of about 13.8 was used for all determinations.

RESULTS

The reaction of hydroxylamine with amides was studied at three different temperatures, namely 26, 60, and 100°C. It was found that the conditions for the maximum colorimetric value were different for each compound. Figures 3.18 and 3.19 show representative curves. Pure acetohydroxamic acid gave a value of 105 under standard conditions. Acetamide reached this value after interaction for 8 hours at 26°C, but at 60°C the maximum reached after 2 hours was only 90, and at 100°C, 65 (reached after 10 minutes). Furthermore, the curves showed that at room temperature the maximum persisted over a period of several hours, at 60°C only for 30 minutes, and at 100°C less than 5 minutes. Since other amides behaved in a similar fashion, the following rules can be established.

* Any standard colorimeter may be used for this analysis.

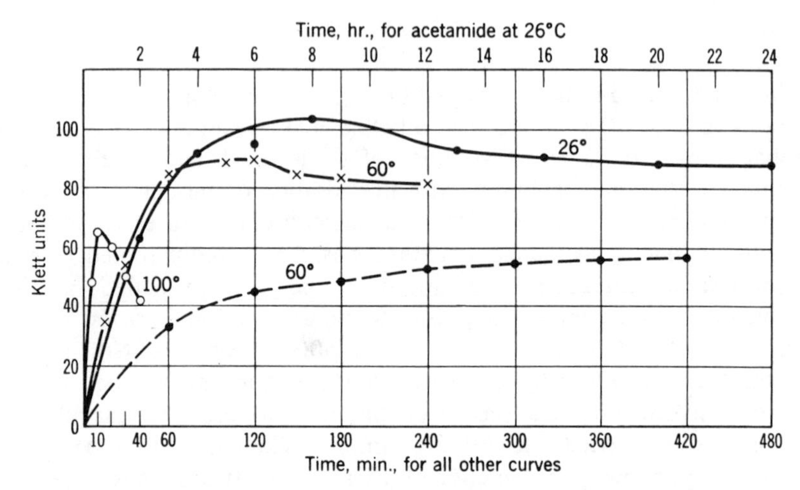

Fig. 3.18. Rate of reaction of acetamide (full curves) and *N*-methylacetamide (dashed curve) with hydroxylamide at various temperatures. Note that for this reason the initial slopes for the acetamide curves at 26 and 60°C are not identical.

The rate of reaction between amides and hydroxylamine increases with temperature, but a competitive reaction—namely, the hydrolysis of the amide—becomes more and more preponderant, thus depressing the yield of hydroxamic acid.

If the reaction time is extended beyond the optimal period required for maximum colorimetric yield, a gradual decomposition of the hydroxamic acid already formed is observed. Therefore, at elevated temperatures, the optimal period is considerably shortened, because the energy of the carbon–nitrogen bond in amides and hydroxamic acids is nearly the same and because conditions that lead to faster hydrolysis of the former also produce more rapid decomposition of the latter.

The results obtained with about 20 amides are represented in Table 21. We can now relate, in many cases, the rate of reaction to the specific structure of an amide. For example, formamide reached its maximum at 26·C in less than 1 hour, acetamide after 8 hours. Substitution of amide hydrogen reduced the speed of reaction considerably. *N*-Methylacetamide (Fig. 3.18) reached its maximum after 7 hours (60°C) and 24 hours (26°C), respectively, as compared to 2 and 8 hours for acetamide itself. The corresponding figures for formamide were 10 and 40 minutes, for dimethyl-formamide, 40 and 300 minutes, respectively. In accordance with this observation, acetylglycine and peptides gave a slow reaction and low colorimetric values. A similar relationship was found among the derivatives of nicotinamide. Nicotinamide itself reached its maximum value of 52 per micromole per milliliter after reaction for 8

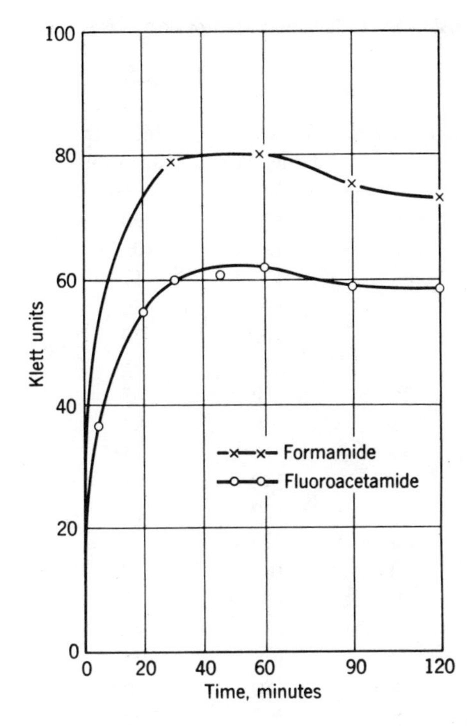

Fig. 3.19. Reaction with hydroxylamine at 26°C.

hours at 26°C, whereas its *N,N*-diethyl derivative (coramine) gave a maximum value of 6 Klett units per micromole after 8 hours at 60°C. On the other hand, the quaternary pyridinium salt (see below) behaved like the parent compound.

$CONH_2$

N^+

CH_3 $(CH_3SO_4)^-$

Substitution in the acyl residue also exerted a pronounced influence. This can be demonstrated in two ways:

Fluoroacetamide showed the fastest reaction of all acetamides, reaching the speed of reaction of formamide (Fig. 3.19).

Acetanilide (Fig. 3.20), which was used as representative of *N*-arylamides has been reported to react sluggishly in alcoholic solution. In aqueous solution there was rapid reaction with hydroxylamine, which reached its maximum at 60°C after 3 hours. The reaction of N^4-acetylsulfanilamide was similar, but slower.

Table 21. Optimal Conditions for Conversion of Amides into Hydroxamic Acids

Name of Compound	Temperature, °C	Reaction Time, Min.	Klett Reading, Units per Micromole
Acetamide	60	120	90
	26	480	103
N-Methylacetamide	60	420	57
Acetanilide	60	180	70
N^1-Acetylsulfanilamide	60	240	70
Acetylglycine	60	240	35
Fluoroacetamide	26	60	62
Formamide	26	60	80
	60	10	75
Dimethylformamide	26	240	45
Succinimide	60	120	85
Caprolactam	60	420	41
Asparagine	60	180	38
Glutamine	60	180	35
Glutathione	60	120	48
Glycyclglycine	60	120	25
Nicotinamide	26	480	45
N^1-Methylnicotinamide methosulfate (I)	26	360	45
Nicotinic acid methylamide	60	240	30
Coramine (nicotinic acid diethylamide)	60	480	6
Pantothenic acid, calcium salt	26	300	89
Barbitone	100	45	1.7
Pentobarbitone	60	300	1.5
Phenobarbitone	100	120	7.5
Evipan, sodium	100	30	9

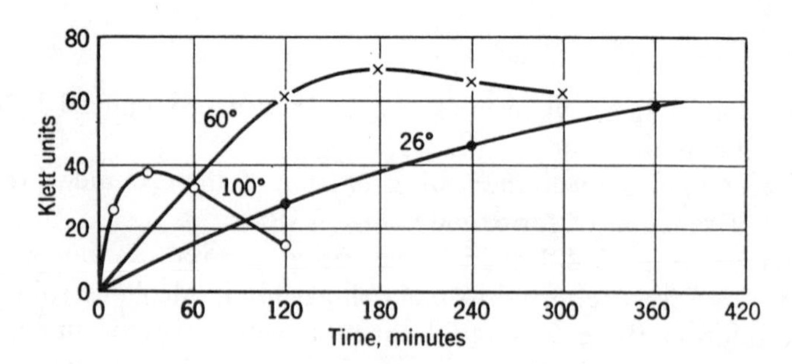

Fig. 3.20. Reaction of acetanilide (in aqueous solution) with hydroxylamine.

Succinimide, in agreement with theoretical predictions, gave only one equivalent of hydroxamic acid—that is, the maximum value of 85 was comparable to the value obtained for ethyl hydrogen succinate (90). This the author ascribes to the fact that the alkaline reagent converts the amide into the sodium salt of the monoamido acid, which then reacts with hydroxylamine.

Carboxylic Imides

Imides can sometimes be determined by hydrolysis, but this is quite rare. They can be titrated as acids quite readily, however, by the procedure described for enols on pages 46–57. Table 22 shows imides that have been determined successfully. The table was obtained from the same source as the procedure. Imides have also been determined using the modified Hofmann reaction developed by D'Alonzo and Siggia (see p. 208). This procedure, however, is sensitive to the presence of amides as well.

Table 22

Compound	Solvent	Indicator	% Theoretical
Dithiobiuret	DMF[a]	Azo violet	99.8
			99.7
Hydantoin	DMF	Azo violet	99.9
			99.6
Phthalimide	DMF	Azo violet	99.0
			99.2
Succinimide	DMF	Azo violet	99.7
			99.9
Thiobarbituric acid	DMF	Thymol blue	99.8
			99.7

[a] Dimethylformamide.

Carboxylic Acid Chlorides (40)

In most existing methods for the determination of acid chlorides no attention is given to the determination of the hydrogen chloride and the free carboxylic acid they contain. In fact, the hydrogen chloride present is

40. Reprinted in part with permission from C. R. Stahl and S. Siggia, *Anal. Chem.*, **28,** 1971–3 (1956).

often an interference which is determined as acid chloride. Thus, in the sodium acetate titration of Usanovich and Yatsimirshiĭ (41), the sodium methylate titration of Fritz (42), and the silver nitrate titration of Drahowzal and Klamann (43), any hydrogen chloride present titrates, giving high values for acid chloride. Mitchell and Smith (44) were able to determine acid chlorides without interference from hydrogen chloride by making the chloride react with pyridine, hydrolyzing with water, and titrating the excess water. Ackley and Tesoro (45) determined the free carboxylic acid in acid chloride by making the chloride react with ammonia, neutralizing with hydrochloric acid, and extracting and titrating the carboxylic acid. This method, however, is useful only for chlorides of long-chain fatty acids. Pesez and Willemart (46) determined both hydrogen chloride and free carboxylic acid by indirect means. The subtraction of two large figures to obtain the small values for hydrogen chloride and free carboxylic acid in this method makes its use undesirable. Klamann (47) determined the hydrogen chloride content of acid chlorides by a direct titration with silver nitrate in acetone; however, the carboxylic acid determination is also indirect.

Since the amounts of hydrogen chloride and free carboxylic acid present in acid chlorides are usually small, it is desirable to determine these compounds directly. In the present method this is accomplished by titrating the hydrogen chloride in chlorobenzene-ether with *N*-tripropylamine, and the free carboxylic acid with sodium hydroxide after the acid chloride has reacted with *m*-chloroaniline. In the latter titration, two potential breaks are obtained (Fig. 3.21). One corresponds to the neutralization of the amine hydrochloride formed from the acid chloride and the hydrogen chloride, and the other represents the neutralization of free carboxylic acid. Thus both hydrogen chloride and free carboxylic acid are determined by direct titration, and the only value obtained by difference is that for acid chloride, which is calculated by subtracting the free hydrogen chloride value from the amine hydrochloride titration. This subtraction introduces no serious error, because a small value is subtracted from a large value.

Although amines have been used in several previous methods—Klamann (47), Ackley and Tesoro (45), and Pesez and Willemart (46)—the amines employed in those methods were too strongly basic to allow

41. M. Usanovich and K. Yatsimirshii, *J. Gen. Chem.* (*USSR*), **11,** 957–8 (1941).
42. J. S. Fritz, *Anal. Chem.*, **23,** 589–91 (1951).
43. F. Drahowzal and D. Klamann, *Monatsh.*, **82,** 470–2 (1951).
44. J. Mitchell and D. M. Smith, *Chemical Analysis,* Vol. V, *Aquametry,* Wiley-Interscience, New York, 1948, pp. 369–71.
45. R. R. Ackley and G. C. Tesoro, *Ind. Eng. Chem., Anal. Ed.*, **18,** 444–5 (1946).
46. M. Pesez and R. Willemart, *Bull. Soc. Chim. Fr.*, **1948,** 479–80.
47. D. Klamann, *Monatsh.*, **83,** 719–23 (1952).

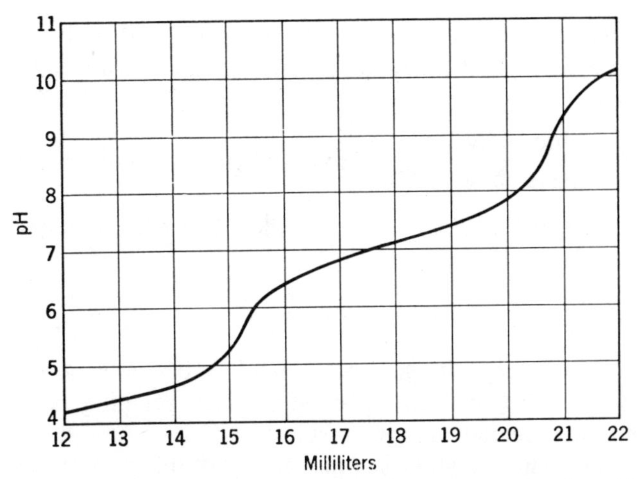

Fig. 3.21. Titration of amine hydrochloride plus carboxylic acid with sodium hydroxide.

differential titration of the amine hydrochloride and free carboxylic acid. After several aromatic amines had been tried, *m*-chloroaniline was selected for use in the present method. The nitroanilines were found to be too weakly basic to prevent loss of hydrogen chloride during the reaction, and *p*-chloroaniline caused poor breaks on titration. The naphthylamines caused poor breaks and also gave trouble by precipitating during the titration. *m*-Chloroaniline was found to be a strong enough base to prevent loss of hydrogen chloride but not too strong to interfere with the titration.

REAGENTS

CP Acetone, anhydrous diethyl ether, chlorobenzene, standard $0.5N$ and $0.1N$ sodium hydroxide, standard $0.1N$ *N*-tripropylamine in chlorobenzene, and freshly distilled *m*-chloraniline are used.

The tripropylamine (TPA) may be standardized against maleic acid dissolved in acetone or against a solution of dry hydrogen chloride in ether-chlorobenzene, which has been standardized by dissolving an aliquot in acetone-water and titrating with standard $0.1N$ sodium hydroxide. If the maleic acid standardization is used, it should be remembered that only one carboxy group on the maleic acid can be titrated with tripropylamine.

PROCEDURE A

Dissolve an exactly weighed sample of acid chloride containing not more than 0.001 mole of hydrogen chloride in a 1:1 mixture of ether-chlorobenzene and titrate potentiometrically with standard $0.1N$ tripropylamine in chlorobenzene, using a 10-ml buret. Only the hydrogen chloride titrates, and the break occurs between 350 and 150 mV (Fig. 3.22). The percentage of hydrogen chloride is calculated using the following equations:

$$\frac{\text{Milliliters} \times N\ \text{TPA}}{\text{Weight of sample} \times 1000} = \text{moles of HCl per gram of sample}$$

$$\text{Moles of HCl per gram of sample} \times 100 \times 36.5 = \%\ \text{HCl}$$

PROCEDURE B

Weigh a sample containing approximately 0.01 mole of acid chloride plus free carboxylic acid in a glass-stoppered weighing bottle or in a sealed ampoule. Sealed ampoules should be used for all volatile acid chlorides. Place the weighed sample in a 250-ml glass-stoppered flask containing 5 ml of *m*-chloroaniline and 25 ml of acetone. If the sample is weighed in a weighing bottle, remove the stopper just before placing the sample in the flask and slide the weighing bottle down the side of the flask, so that the sample and reagent do not mix. Seal the flask with stopcock grease and shake to mix the sample and reagent. Cool the contents of the flask slightly below room temperature in running tap water or in ice, and allow the flask to stand 5 minutes to ensure complete reaction.

After the reaction is complete, remove the stopper and add 5 ml of distilled water. Replace the stopper and shake the flask to dissolve the amine hydrochloride formed in the reaction. Then wash the contents of the flask into a beaker with acetone. Using a pH meter equipped with a glass-calomel electrode system, add standard $0.5N$ sodium hydroxide from a buret until the pH of the solution is approximately 4. Read the volume of $0.5N$ sodium hydroxide, then titrate the solution potentiometrically with standard $0.1N$ sodium hydroxide.

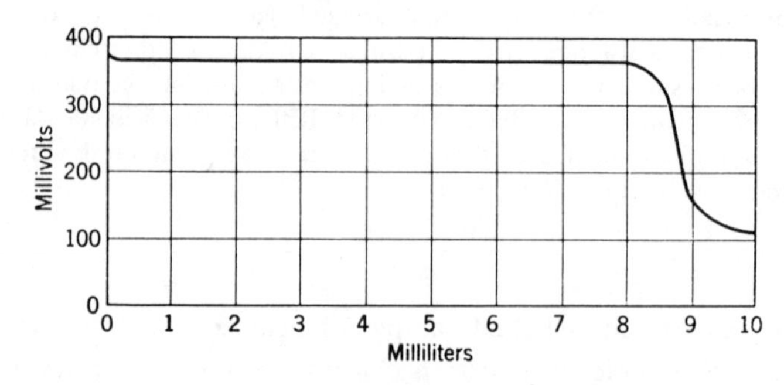

Fig. 3.22. Titration of hydrochloric acid with tripropylamine.

The first break, which occurs between pH 5 and 6, represents the neutralization of the amine hydrochloride formed by the reaction of the *m*-chloroaniline with the acid chloride and free hydrogen chloride. The equivalents of $0.1N$ sodium hydroxide plus the equivalents of $0.5N$ sodium hydroxide added, equal the equivalents of acid chloride plus hydrogen chloride in the sample. The second break occurs between pH 8 and 9.5 and represents the titration of the free carboxylic acid (Fig. 3.21).

Two concentrations of standard sodium hydroxide are used. A rather large sample is required to obtain a sufficiently large titration for the free carboxylic acid, which is usually present in relatively small quantities. The total titration would be large if $0.1N$ hydroxide were used, and the titration for free carboxylic acid would be small if $0.5N$ hydroxide were used in the total titration. In this procedure, the bulk of the acidity is neutralized with $0.5N$ reagent, and the $0.1N$ hydroxide is used to carry the titration through the two neutral points. The percentage of acid chloride is calculated as follows from the first break:

$$\frac{(\text{Milliliters of } 0.5N\ \text{NaOH} \times N) + (\text{milliliters of } 0.1N\ \text{NaOH} \times N)}{\text{Weight of sample} \times 1000} = \text{moles of acid chloride} + \text{HCl per gram of sample}$$

$$(\text{Moles of acid chloride} + \text{HCl per gram}) - (\text{moles of HCl per gram}) = \text{moles of acid chloride per gram}$$

$$\text{Moles of acid chloride per gram} \times 100 \times \text{mol. wt. of acid chloride} = \%\ \text{acid chloride}$$

The percentage of free carboxylic acid is calculated using the following equation:

$$\frac{(\text{Milliliters to second break minus milliliters to first break}) \times N \text{ of NaOH} \times \text{mol. wt. of acid} \times 100}{\text{Weight of sample} \times 1000} = \%\ \text{free carboxylic acid}$$

DISCUSSION AND RESULTS

All reagents were used as obtained, except the *m*-chloroaniline, which was distilled. After distillation the *m*-chloroaniline can be used for 2 months or more without redistillation. The acetone employed contained about 0.2% water, but the acid chlorides react rapidly enough with the

Table 23. Analysis of Synthetic Mixtures of Acid Chlorides and Carboxylic Acid

Synthetic Mixture	Added, %		Found, %	
	Acid Chloride	Carboxylic Acid	Acid Chloride	Carboxylic Acid
Benzoyl chloride	99.2	0.8	98.6	0.8
Benzoic acid	97.5	2.5	97.5	2.7
	96.6	3.4	96.4	3.5
	96.5	3.5	96.2	3.7
	93.7	6.3	93.3	6.4
	92.5	7.5	92.2	7.5
	92.0	8.0	91.4	7.8
	90.6	9.4	90.6	9.5
Lauroyl chloride	97.3	2.7	97.5	2.6
Lauric acid	94.4	5.6	94.2	5.2
	93.0	7.0	93.0	6.9
	90.1	9.9	90.0	9.7
Acetyl chloride	98.7	1.3	98.3	1.4
Acetic acid	97.0	3.0	96.0	3.2
	95.0	5.0	94.7	5.2
	90.6	9.4	90.5	9.4
Palmitoyl chloride	98.7	1.3	98.1	1.5
Palmitic acid	96.7	3.3	96.7	3.2
	93.5	6.5	93.0	6.8
	91.8	8.2	91.9	8.4
Fatty acid chloride	96.9	3.1	97.0	3.1
(av. mol. wt. = 293.5)	95.1	4.9	95.2	4.7
Stearic acid	92.8	7.2	92.9	6.9
Propionyl chloride	97.1	2.9	97.2	3.2
Propionic acid	95.1	4.9	94.5	5.0
	91.9	8.1	91.8	8.2
Benzoyl bromide	98.1	1.9	98.0	2.0
Benzoic acid	97.0	3.0	96.8	3.2
	94.6	5.4	94.4	5.2
	89.6	10.4	89.3	10.5

m-chloroaniline to prevent the occurrence of any hydrolysis. Acetone dried over Drierite gave the same results as undried acetone.

Synthetic mixtures of acid chlorides and the corresponding acids were prepared and analyzed by Procedure B (Table 23) and synthetic mixtures of acid chlorides and hydrogen chloride were analyzed by Procedure A (Table 24) to test the method. The acid chlorides used in these mixtures were distilled from the best available samples to obtain acid chlorides free

Table 24. Determination of Hydrogen Chloride in Presence of Acid Chlorides

Acid Chloride	Hydrogen Chloride, %	
	Added	Found
Acetyl chloride	2.02	2.07
	3.85	3.93
	6.14	6.30
	16.50	17.08
Propionyl chloride	1.95	1.94
	3.49	3.65
	6.72	6.19
	14.72	14.47
Lauroyl chloride	0.72	0.70
	1.67	1.65
	3.08	3.24
	4.00	4.21
Palmitoyl chloride	1.24	1.37
	2.35	2.47
	3.45	3.60
	7.88	8.49
Benzoyl chloride	1.42	1.57
	2.64	2.75
	5.10	5.17
	10.53	10.64
Naphthoyl chloride	1.14	1.21
	2.10	2.14
Fatty acid chloride	0.11	0.13
(av. mol. wt. = 293.5)	0.81	0.80
	1.87	1.83

of hydrogen chloride and carboxylic acid. Because of the difficulty of adding known amounts of dry hydrogen chloride directly to acid chlorides to obtain synthetics of known composition, the hydrogen chloride–acid chloride synthetics were prepared by adding aliquots of an ether solution of hydrogen chloride of known composition to weighed samples of acid chlorides.

Results obtained for succinyl chloride and naphthoyl chloride are given in Table 25. No synthetic mixtures were prepared for these compounds because for succinyl chloride, the second chloride group reacts incompletely and the free carboxylic acid values obtained cannot be used. The unreacted acid chloride group forms carboxylic acid when water is added, and high results are obtained for free carboxylic acid. For naphthoyl chloride, Stahl and Siggia were unable to obtain a pure sample from which to prepare synthetic mixtures.

The analysis of three acid bromide samples was attempted, but good results were obtained only for benzoyl bromide. In the analysis of benzoyl bromide 25 ml of ether was used in Procedure B instead of 25 ml of acetone. Low results for acid bromide were obtained when acetone was used. Results obtained for acetyl and valeryl bromides varied greatly.

Carboxylic Acid Anhydrides

Carboxylic anhydrides are very reactive materials, but their reactions are also common to free carboxylic acids. Since many anhydrides contain free carboxylic acids, a reaction must be used that is very rapid for the

Table 25. Analysis of Succinyl and Naphthoyl Chlorides

Acid Chloride	Hydrogen Chloride, %	Acid Chloride, %	Carboxylic Acid, %
Succinyl chloride	0.0	99.4	—
		99.4	—
		99.1	—
		99.1	—
		99.2	—
		99.2	—
		98.9	—
Naphthoyl chloride	0.0	93.6	6.3
		93.7	6.3
		93.6	6.3

anhydrides and very slow for the acid. Such a reaction is amide formation using morpholine or aniline as the amine reagents.

Morpholine is preferred to aniline as the amine reagent, since it is a stronger base and will yield sharp indicator end points, whereas the aniline method shown below can only be used with a potentiometric determination of the excess aniline. The aniline method, however, will determine the anhydrides of strong acids such as maleic, whereas the morpholine method will not.

Morpholine Method

[*Adapted from J. B. Johnson and G. L. Funk, Anal. Chem.,* **27,** *1464–5 (1955)*]

Earlier methods for the determination of carboxylic acid anhydrides were based on the simultaneous measurement of the acid and the anhydride. Either the aniline reaction of Radcliffe and Medofski (48) or the sodium methylate titration of Smith and Bryant (49) was used in conjunction with a total hydrolysis with sodium hydroxide to arrive at the anhydride and acid content of a sample.

Two direct methods for the determination of anhydrides, which do not involve the measurement of the acid originally present in the sample, have been reported. Smith, Bryant, and Mitchell developed a procedure employing the Karl Fischer reagent (50, 51). The method described here also measures the anhydride independent of the acid content of the sample. It possesses the speed and convenience of an indicator titration and excellent precision for both high and low concentrations of anhydride.

$$(RC(=O))_2O + HN(CH_2CH_2)_2O \rightarrow RC(=O)N(CH_2CH_2)_2O + RC(=O)OH$$

Morpholine reacts with carboxylic acid anhydrides to produce equimolar quantities of amide and acid. In a methanolic medium using mixed methyl yellow–methylene blue indicator, all the components of the reaction are neutral except morpholine. If a measured excess of morpholine

48. L. G. Radcliffe and S. Medofski, *J. Soc. Chem. Ind.* (*London*), **36,** 628 (1917).
49. D. M. Smith and W. M. D. Bryant, *J. Am. Chem. Soc.*, **58,** 2452 (1936).
50. D. M. Smith, W. M. D. Bryant, and J. Mitchell, Jr., *Ibid.*, **62,** 608 (1940).
51. D. M. Smith, W. M. D. Bryant, and J. Mitchell, Jr., *Ibid.*, **63,** 1700 (1941).

reacts with a sample containing anhydride, the anhydride reacts preferentially with morpholine, and the excess can be titrated with standard methanolic hydrochloric acid. The morpholine consumed, represented by the difference in titration between a blank and sample, is a measure of anhydride. The free acid in a sample can be obtained by determination of the total acidity using soldium hydroxide and then subtracting the anhydride value.

REAGENTS

HYDROCHLORIC ACID, 0.5N METHANOLIC SOLUTION. Transfer 84 ml of 6N hydrochloric acid to a 1000-ml volumetric flask and dilute to volume with methanol. Standardize daily against standard 0.5N sodium hydroxide using phenolphthalein indicator. The reagent is best handled in an automatic buret assembly.
MORPHOLINE SOLUTION, 0.5N METHANOLIC SOLUTION. Transfer 44 ml of redistilled morpholine to a 1-liter reagent bottle and dilute to 1 liter with methanol. Fit the bottle with a two-hole rubber stopper and through one hole insert a 50-ml pipet so that the tip extends below the surface of the liquid; through the other hole insert a short piece of glass tubing to which is attached a rubber atomizer bulb.
METHYL YELLOW–METHYLENE BLUE MIXED INDICATOR. Dissolve 1.0 gram of methyl yellow (*p*-dimethylaminoazobenzene) and 0.1 gram of methylene blue in 125 ml of methanol.

PROCEDURE

Carefully pipet 50 ml of the morpholine solution into each of two 250-ml glass-stoppered flasks. Fill the pipet by exerting pressure in the reagent bottle with the atomizer bulb.

Reserve one of the flasks as a blank and introduce not more than 20 meq. of anhydride into the second flask. Swirl the flask to effect solution. Allow both the sample and blank to stand at room temperature for the time indicated for each specific anhydride (Table 26).

Add 4 or 5 drops of indicator to each flask and titrate with the 0.5N acid to the disappearance of the green color. At this point the color is best described as amber.

The difference in titration between blank and sample is a measure of anhydride.

RESULTS

The data obtained on eight anhydrides are shown in Table 26. All samples used were commercial grade materials containing some of the

Table 26. Purity of Anhydrides by Morpholine Method

Anhydride	Minimum Reaction Time at Room Temperature, Minutes	Purity, wt. %[b]	Acid Content, wt. %[c]	Total, wt. %
Acetic	5	99.7 ± 0.1 (5)	0.2 (2)	99.9
Butyric	5	98.0 ± 0.2 (4)	1.9 (2)	99.9
Chrysanthemum	5	98.6 ± 0.1 (2)	0.5 (2)	99.1
2-Ethylhexanoic	30	98.9 ± 0.2 (6)	0.1 (2)	99.0
Glutaric	5	97.1 ± 0.1 (3)	3.4 (2)	100.5
Phthalic	5	99.6 ± 0.1 (3)	0.3 (2)	99.9
Propionic	5	99.5 ± 0.1 (3)	0.5 (2)	100.0
Succinic	5	96.9 ± 0.1 (5)	2.4 (2)	99.3

[a] 0.5*N* reagents.

[b] Figures in parentheses indicate number of analyses.

[c] Calculated from difference between Radcliffe-Medofski (48) and morpholine reactions.

corresponding acid and, in some cases, small but significant amounts of other impurities. To show the recoveries to be expected from the morpholine procedure, the amount of acid in each sample is also shown. This was determined by measuring the acid after reaction with aniline according to the procedure of Radcliffe and Medofski (48) and correcting for the anhydride present in the sample.

Table 27 shows the accuracy and precision obtained in the determination of low concentrations of acetic anhydride in acetic acid.

Table 27. Determination of Acetic Anhydride in Glacial Acetic Acid by the Morpholine Method

	wt. %[a]
Sample 1	
Anhydride added	0.065
Anhydride found	0.066 ± 0.002 (5)
Sample 2	
Anhydride added	0.011
Anhydride found	0.016 ± 0.007 (4)

[a] Figures in parentheses indicate number of analyses.

DISCUSSION

REACTION RATES. Rate studies of the reaction of acetic anhydride with 0.02, 0.1, and $0.5N$ morpholine indicate that the reaction is practically instantaneous, being quantitative in less than 30 seconds. With the exception of 2-ethylhexanoic anhydride, which requires a 30-minute reaction, all the anhydrides in this work react quantitatively within 5 minutes at room temperature, and in most cases the reaction is complete in less than 1 minute.

POTENTIOMETRIC TITRATION. Potentiometric titration studies showed that the indicator end point occurs on the steep break of the curve. The true equivalence point does not exactly coincide with the indicator end point described, but the error is not significant, particularly in view of the fact that the blank titration is in error to nearly the same degree. A number of indicators have been tried, but none has been found superior to a methyl yellow–methylene blue mixture.

The method is not applicable to anhydrides if the acids have ionization constants in water greater than 2×10^{-2}, since such acids are somewhat acidic to the indicator in a methanol medium. Thus maleic and citraconic anhydrides cannot be substituted satisfactorily in these cases.

EFFECT OF SOLVENTS. Siggia and Hanna (52) (see pp. 236–9) were unable to obtain satisfactory indicator end points in acetic acid medium when aniline was used as the reactant. As a result they resorted to potentiometric titrations in an ethylene glycol-2-propanol, medium. The poor end points in the acetic acid medium were undoubtedly due to the enhanced basicity of the resulting amide. Amides are appreciably more basic in acetic acid than in methanol solution.

REACTION WITH OTHER AMINES. Morpholine has been selected for use in the method because of its high reactivity with anhydrides and because it is a secondary amine that should be subject to less interferences than compounds having primary amino groups. Obviously, in certain cases other amines, primary or secondary, could be substituted for morpholine as the reagent.

LOW CONCENTRATIONS. For the determination of low concentrations of anhydride, it is convenient to use $0.1N$ reagents. For the determination of acetic anhydride in glacial acetic acid, a 10-ml sample was made to react with 50 ml of $0.02N$ morpholine. Excess morpholine was titrated with $0.1N$ methanolic hydrochloric acid. Samples in excess of 10 ml are not recommended, inasmuch as the resultant amides tend to buffer the end points if more than this amount of acid is present.

INTERFERENCES. The only known interferences are compounds that can

52. S. Siggia and J. G. Hanna, *Anal. Chem.*, **23,** 1717 (1951).

react with morpholine to destroy its basic behavior. These compounds include ketene and diketene, which react quantitatively under certain conditions, and acid chlorides, which can react quantitatively only in a dilute solution, because they react with both morpholine and methanol. If a dilute solution of an acid chloride is added slowly to the reagent while the latter is being swirled, the morpholine reacts preferentially with the acid chloride. Under these conditions the equivalent weight of the acid chloride is one-half the molecular weight as a mole of amide and a mole of mineral acid are formed. Mineral acids, of course, interfere. In certain cases the interference is quantitative and appropriate corrections may be applied.

New Indicator for the Determination of Organic Acid Anhydrides by the Morpholine Method

Adapted from the Study of J. E. Ruch [*Anal. Chem.*, **47**, *2057* (*1975*)]

Dimethyl yellow (methyl yellow, 4-dimethylaminoazobenzene) has been termed a carcinogen by the federal Occupational Safety and Health Administration and is so described in the United States Government Occupational Safety and Health Standards (53).

As a replacement, 4,4′-bis(4-amino-1-naphthylazo)-2,2′-stilbenedisulfonic acid was found to be superior to dimethyl yellow in both visual testing and in matching the equivalence point of both blank and sample. In methanol medium the color transition is both vivid and sharp.

PREPARATION OF INDICATOR

The indicator solution was prepared by weighing 0.050 gram of 4,4′-bis(4-amino-1-naphthylazo)-2,2′-stilbenedisulfonic acid, (bis-4-azo), Eastman No. 7089, and 0.010 gram of brilliant yellow, Eastman No. 837, into a 2-ounce vial. Next, 1.5 ml of 0.1*N* sodium hydroxide was added to the vial by pipet, and the mixture was stirred well with a small stirring rod. Distilled water, 3.5 ml, was added and the contents stirred again. The object was to break up all small particles to facilitate solution in the dilute sodium hydroxide solution. The mixture was transferred to a storage bottle, and the vial was rinsed with 45 ml of methanol. The rinsings were added to the bottle, the bottle was capped and shaken to effect thorough mixing.

53. *Fed. Reg.*, Jan. 29, 3756–97 (1974).

PROCEDURE

Use 7 to 8 drops of the bis-4-azo–brilliant yellow indicator in place of dimethyl yellow–methylene blue in the morpholine method for anhydrides. The sharpest indicator color transition is from reddish-orange to purple. Very slight overtitration (0.01–0.02 ml) produces a pure blue.

RESULTS

Table 28 compares and summarizes data obtained from visual titrations of four anhydrides using bis-4-azo and dimethyl yellow. Excellent agreements (0.03-ml deviation) were obtained between the color transitions and the potentiometric end points.

Table 28. Comparison of the Two Indicator Results

	Bis-4-Azo–Brilliant Yellow,			Dimethyl Yellow–Methylene Blue,		
Anhydride	Anhydride	Acid	Total	Anhydride	Acid	Total
Acetic	98.27	1.54	99.71	98.53	1.31	99.84
	98.17	1.54	99.61	98.28	1.31	99.59
Phthalic	95.44	4.45	99.99	95.21	4.65	99.86
	95.47	4.45	100.02	95.34	4.65	99.99
Sucninic	90.77	9.42	100.19	90.35	9.66	100.01
	90.44	9.42	99.86	90.44	9.66	100.10
Itaconic	97.31	2.34	99.65	97.85	1.70	99.55
	97.17	2.34	99.51	97.80	1.70	99.50
	97.20	2.34	99.54	97.99	1.70	99.69
	97.21	2.34	99.55	97.46	1.70	99.16

Aniline Method

[*Adapted from S. Siggia and J. G. Hanna, Anal. Chem.*, **23,** *1717* (*1951*)]

Malm and Nadeau (54) employed aniline successfully to determine acetic anhydride in acetylating mixtures for cellulose. This reaction can also be applied to determining other carboxylic anhydrides in the presence of their acids. This approach also circumvents the difficulties exhibited by the hydrolytic and titration procedures for determining anhydrides (55–57).

54. C. J. Malm and G. F. Nadeau, U.S. Patent 2,063,324 (Dec. 8, 1936).
55. D. M. Smith and W. M. D. Bryant, *J. Am. Chem. Soc.*, **58,** 2452–4 (1936).
56. D. M. Smith, W. M. D. Bryant, and J. Mitchell, Jr., *J. Am. Chem. Soc.*, **62,** 608–9 (1940).
57. D. M. Smith, W. M. D. Bryant, and J. Mitchell, Jr., *J. Am. Chem. Soc.*, **63,** 1700–1 (1941).

Aniline reacts with carboxylic acid anhydrides according to the following equation:

$$\begin{array}{l} RC(=O) \\ \quad \backslash \\ \quad\; O \\ \quad / \\ RC(=O) \end{array} + \phi NH_2 \rightarrow RC(=O)NH\phi + RC(=O)OH$$

The carboxylic acids produced in the reaction or the free carboxylic acids present in the original sample are weak acids; their bond with the aniline is so weak that the aniline can be titrated away from the carboxylic acids by the strong acid used to titrate the excess aniline. In this way the aniline consumed in the reaction is a measure of the anhydride, and the free carboxylic acid present in the samples does not offer any interference.

Succinic acid was found to be an exception and does interfere in determining succinic anhydride. This interference, however, was not a result of the neutralization of the aniline by the acid, but in this case the reaction conditions to react the anhydride completely had to be so vigorous that the succinic acid liberated by that reaction as well as the free succinic acid present also reacted with the aniline to form the amide. The reaction between the acid and the aniline is not very fast, but the error introduced is significant. When running pure succinic anhydride, the results were about 5% high; also, when pure succinic acid was treated with aniline under the same conditions as the anhydride, there was a very significant consumption of aniline.

Along the same lines, Table 29 indicates that maleic and phthalic acids would have been interferences, had it not been that the anhydrides react completely under relatively mild conditions.

Ethylene glycol–isopropanol (58) was used as solvent medium to accentuate the end point in the titration of the excess aniline. Titration in glacial acetic acid was tried, but difficulty with end points was noticed in analyzing some anhydride-acid systems. The glycol-isopropanol could be applied to all the systems studied.

REAGENTS

Ethylene glycol–isopropyl alcohol mixture, 1:1.

Standard 0.2*N* hydrochloric acid in ethylene glycol–isopropyl alcohol mixture,

58. S. Palit, *Ind. Eng. Chem.* (*Anal. Ed.*), **18,** 246–51 (1946).

Table 29

Anhydride	Reaction Time, Minutes	Reaction Temperature	% Found	Acid-Anhydride Mix: % Anhydride Added	Acid-Anhydride Mix: % Anhydride Found	Aniline Recovered after Reaction with Acid Alone
Acetic	5	Room	100.2	77.8	77.5	99.2
	5	Room	99.9			
	5	Room	99.8			
	5	Room	100.1			
	5	Room	99.7			
Propionic	5	Room	100.1	60.9	60.6	99.3
	5	Room	99.5			
	5	Room	99.6			
Maleic	15	Room	99.4	91.0	90.7	99.7
	15	Room	99.4			
	15	Room	99.4			
	15	Room	99.6			
	15	100°	110.4			
	90	Room	100.3			
Phthalic	15	100°	100.5	50.8	50.3	100.0
	15	100°	99.8			
	15	100°	99.5			
	30	100°	102.0			
	30	100°	101.2			
Camphoric	5	100°	21.6			
	45	100°	100.0	87.5	88.0	99.3
	45	100°	100.1			
	45	100°	100.3			
Butyric	5	Room	99.8	87.2	86.5	98.8
	5	Room	100.1			

Note. All samples were distilled or recrystallized until their carbon and hydrogen analyses were within ±0.2% of the theoretical.

19 ml of concentrated hydrochloric acid diluted to 1 liter with 1:1 ethylene glycol–isopropyl alcohol.

CP Aniline.

PROCEDURE

Weigh accurately a sample containing approximately 0.004 mole of acid anhydride in a 20 × 150 mm test tube. If the sample contains an acid anhydride that requires heat for complete reaction, weigh it in a 50 ml condenser flask. Add aniline to the sample, drop by drop, until 0.9 gram has been added. Accurately weigh the amount of aniline used. Allow the sample to stand in the test tube 5 minutes. Attach the condenser flask to a condenser in a reflux position and immerse it in a beaker of boiling water for the required length of time. Transfer the reaction mixture quantitatively from the test tube or the condenser flask to a 150-ml beaker with 1:1 ethylene glycol–isopropyl alcohol mixture. Add ethylene glycol–isopropyl alcohol mix until the volume is approximately 50 ml. Use a pH meter to indicate the apparent pH after each addition of acid as the sample is titrated with 0.2*N* hydrochloric acid prepared in the ethylene glycol–isopropyl alcohol mixture. Determine the neutralization point by plotting the apparent pH against milliliters of acid. Run a blank on the aniline by titrating an accurately weighted amount, approximately 0.4 gram potentiometrically, with 0.2*N* hydrochloric acid in the ethylene glycol–isopropyl alcohol mixture.

CALCULATIONS

$$\frac{(x-a)\times N\text{HCl}\times \text{mol. wt. of acid anhydride}}{1000\times \text{Weight of sample}}\times 100 = \%\ \text{acid anhydride}$$

where

a = milliliters of acid used to titrate excess aniline

x = milliliters of acid needed to titrate total amount of aniline used

The aniline used should not be assumed to be 100% but should be assayed so that this value is a correct one. The aniline can be assayed by titration as described in the foregoing procedure.

Titration of Anhydrides with Tetrabutylammonium Hydroxide

*Adapted from the Method of C. A. Lucchesi, L. W. Kao, G. A. Young, and H. M. Chang [Anal. Chem., **46**, 1331 (1974)]*

This method involves a room-temperature indicator titration of a sample in pyridine with tetrabutylammonium hydroxide (TBAH) in benzene-methanol as the titrant. Under these conditions the anhydrides tested behave as monobasic acids.

REAGENTS

The pyridine (Eastman) and benzene (Fisher Spectranalyzed grade) were dried over 15% by weight of Linde molecular sieves 4A before use.

The 0.1*N* TBAH titrant was prepared as described on page 48.

Thymol blue indicator was used as a 0.3% (w/v) solution in absolute methanol.

Azo violet indicator was a saturated solution of *p*-nitrobenzeneazoresorcinol (Fisher) in dry benzene.

APPARATUS

The titrations were carried out with a 10-ml microburet fitted with Drierite on top and a one-hole rubber stopper attached to the tip. The titration vessel was either a 125- or 250-ml iodine flask. Indicator solutions were added with a 1-ml tuberculin syringe.

PROCEDURE

STANDARDIZATIONS. The TBAH solution is standardized against NBS benzoic acid. Weigh from 0.05 to 0.1 gram of benzoic acid into a dry iodine flask. Add about 20 ml of pyridine, displace the air with dry nitrogen, and cover the flask with a glass stopper. When the acid has completely dissolved, add about 0.2 ml of thymol blue indicator with the syringe. Place the flask under the rubber stopper at the buret tip and immediately titrate the solution with the 0.1*N* TBAH to the blue end point color of the indicator. Record the volume of titrant and correct for the volume of titrant used for a blank. A typical blank titration for the pyridine solvent was 0.04 ml.

TITRATION. Perform the TBAH titrations on an amount of sample that would require from 0.5 to 0.9 meq. of titrant. Do the determinations in the same way as the standardization above, but use 2 drops of azo violet indicator for succinic acid and for succinic acid–anhydride samples. All the acids tested except succinic were monoprotic in the procedure; succinic was diprotic. Each anhydride has an equivalent weight equal to its molecular weight. The anhydride content of the acid-anhydride mixtures may be calculated with the following equation:

$$\%\ \text{Anhydride} = \left(\frac{B}{W} - \frac{1}{E}\right)\left(\frac{E \times M}{E - M}\right) \times 100$$

where B = net milliequivalents of TBAH

W = sample weight in milligrams

E = equivalent weight of the acid

M = molecular weight of the anhydride

RESULTS

Table 30 gives results obtained with the TBAH titrations for a number of anhydride samples along with the tripropylamine titration results for the acid contents of the same samples. The precision of the method for a

Table 30. Comparison of TBAH and Tripropylamine Titrations of Anhydride Samples

Anhydride Sample	Found by TBAH, %	Acid Found by Tripropylamine Titration, %
Acetic	98.0	2.1
	97.7	2.0
Benzoic I	82.1	18.1
	82.6	17.9
Benzoic II	94.0	5.5
	94.2	5.6
Maleic	98.5	1.0[a]
	98.1	1.1[a]
Phthalic	95.7	5.2 5.0[a]
	96.0	5.0 5.1[a]
Succinic	96.9	4.8
	96.4	4.9

[a] Method of Siggia and Floramo (p. 244). The other acid results were obtained by the method of Greenhow and Jones (p. 249).

single determination, as measured by the standard deviation of the six sets of duplicates, is 0.27%.

Table 31 presents the recovery data for the five acid-anhydride systems studied. Over the concentration range, the average anhydride recovery was 99.7%.

Table 31. Recovery Data for TBAH Titrations of Acid-Anhydride Mixtures

Acid-Anhydride Mixture		Anhydride, %		
		Present	Found	Recovery
Acetic	A	94.5	94.9	100.4
	B	76.1	75.9	99.7
	C	51.4	53.4	103.9
	D	24.8	25.9	104.4
	E	13.3	13.7	103.0
Benzoic	A	84.7	85.3	100.7
	B	55.0	55.1	100.2
	C	9.2	8.6	93.5
Maleic	A	89.4	88.9	99.4
	B	61.7	60.8	98.5
	C	14.7	14.9	101.4
	D	94.7	94.3	99.6
	E	88.8	88.2	99.3
Phthalic	A	84.8	84.0	99.1
	B	58.3	58.8	100.9
	C	8.7	8.2	94.3
	D	94.9	94.5	99.6
	E	88.9	88.2	99.2
Succinic	A	86.5	87.4	101.0
	B	60.5	61.7	102.0
	C	9.1	8.5	93.4

DISCUSSION

Titration of each of the anhydrides studied showed that 1 mole of quaternary base is consumed per mole of anhydride in reaching the end point in the titration. The so-called quaternary hydroxide is known to consist of equimolar amounts of methoxide and hydroxide (59), and the following reaction is suggested:

$$(RCO)_2O + (Bu_4N)(OCH_3) \rightarrow RCOOCH_3 + (Bu_4N^+)(RCOO^-)$$

59. M. L. Cluett, *Anal. Chem.*, **31,** 610 (1959).

If this is the reaction, then as the methoxide initially present is consumed, more is produced by the reaction of hydroxide and methanol in the titrant. Another possible mechanism that gives the same 1:1 stoichiometry involves the reaction of methanol in the titrants to give the methyl ester and the carboxylic acid followed by neutralization of the carboxylic acid with the quaternary base.

Traces of Carboxylic Acid Anhydrides

The method of Goddu, Leblanc, and Wright, given in the section on carboxylic esters (pp. 172–183) was also devised for determining anhydrides and anhydrides in the presence of esters. The details appear on the stated pages.

Determination of Free Acids in Some Anhydrides

DIRECT TITRATION METHOD

The free carboxylic acid in some anhydrides such as maleic or phthalic (or any acid-anhydride system where the dissociation constant of the free acid is 10^{-3} or greater) can be determined by direct titration with a tertiary amine.

The procedure to be described yields a method of titrating the free acid in maleic and phthalic anhydrides directly without interference from the anhydride. Two tertiary amines have been used successfully as bases to titrate the acids. These amines are tri-*n*-propyl amine and *N*-ethyl piperidine. Being tertiary amines, these do not react with the anhydride. Also, being fairly strong bases (dissociation constant approximately 10^{-4} in water), they achieve a neutralization of the acid that can be followed potentiometrically in the solvents used. Dimethylformamide can also be used, but the breaks obtained in the titration curves are not as strong as those obtained in the ketone solvents. This is probably because the tertiary bases are not as strong bases in the dimethylformamide as they are in the ketones.

The tri-*n*-propyl amine and *N*-ethyl piperidine are the strongest tertiary bases that could be found, and the ketone solvents were the best solvents that could be found as far as intensifying the acidity and basicity of the materials encountered in the system is concerned. In spite of these facts, the only acids that can be titrated under these conditions are maleic and phthalic. (Dissociation constants for the first hydrogens are 10^{-2} and 10^{-3}, respectively, as measured in water. The second hydrogens are too weak to titrate; therefore only one break will be seen in the curves.) Acids with dissociation constants of 10^{-4} to 10^{-5} (in water) were tried, but none of these were titratable in these solvents with the tertiary bases used. The procedure is then limited to acids of a strength of 10^{-3} or greater, and the only acids that might interfere in the analysis are acids of these strengths. Acids tried were acetic (dissociation constant 10^{-5}), benzoic (6×10^{-6}), succinic (7×10^{-5}), and camphoric. Malonic acid (10^{-3} for first hydrogen) is titratable, but the anhydride, carbon suboxide, is so rarely used that there is little need for this procedure for malonic acid.

Adapted from Procedure of Siggia and Floramo

[*Anal. Chem.*, ***25,*** *797* (*1953*)]

REAGENTS

Dry acetone or methyl ethyl ketone (CP acetone was found to be dry enough for use in this procedure).

0.1*N* Tri-*n*-propyl amine (Sharples) in acetone; or

0.1*N* *N*-Ethyl piperidine (for penicillin G determination, sold by Eastman Kodak) in acetone.

Standardize both solutions against pure maleic acid, by an aqueous sodium hydroxide titration. Remember in the calculation of the normality of these solutions that only one carboxyl group is being titrated by the amines. The second carboxyl group of both the maleic and phthalic acid is too weak to show a break in the titration curve.

PROCEDURE

Dissolve enough sample in CP acetone to contain approximately 0.002 mole of acid if possible. This will yield approximately a 20-ml titration. For samples where the free acid content is very low, it would be impractical to take samples large enough to yield 0.002 mole of acid. In these cases, take smaller samples and use smaller capacity buret, for the titration. For the samples containing 0.1% maleic acid, shown in Table 32, take a 10-gram sample, which gives approximately a 1-ml titration with the 0.1*N* reagents. It is well to indicate for the calculation that the reagents are titrating only one carboxyl group of the acids.

Table 32

System	% Acid (Theoretical*) (Remainder of Sample is Anhydride)	% Acid Found	
Maleic acid-maleic anhydride	100.0	99.8	(TPA)
	100.0	99.5	(TPA)
	100.0	99.4	(EP)
	100.0	100.0	(EP)
	71.43	71.36	(TPA)
	54.72	54.74	(TPA)
	37.67	37.53	(TPA)
	10.78	10.81	(TPA)
	1.19	1.12	(EP)
	0.12	0.11	(EP)
Phthalic acid-phthalic anhydride	100.0	99.3	(TPA)
	100.0	99.9	(TPA)
	100.0	99.7	(TPA)
	100.0	99.3	(EP)
	75.00	74.75	(TPA)
	10.08	10.21	(EP)
	1.11	1.02	(TPA)
	0.22	0.21	(EP)

EP indicates that ethyl piperidine reagent was used.

TPA indicates that tripropyl amine reagent was used.

* Acids were assayed by titration with sodium hydroxide and tested for any anhydride by method of Siggia and Hanna [*Anal. Chem.*, **23,** 1717 (1951)]. The anhydrides were recrystallized from benzene in which the acids are only very sparingly soluble, whereas the anhydrides are very soluble. The crystals were dried in a vacuum desiccator, and a blank titration was run by the preceding method. Both anhydrides assayed about 0.03% free acid, which was accounted for in the theoretical values quoted.

CALCULATION

$$\frac{\text{Milliliters of standard amine soln.} \times N \text{ amine} \times \text{mol. wt. of acid}}{\text{Weight of sample} \times 1000} \times 100 = \%\ \text{free acid}$$

Salt (lithium chloride) enhancement of acidity in nonaqueous solvents has been used to extend the foregoing method of Siggia and Floramo to acids with pk_a values as high as 5.5. It is claimed that the reaction,

$$HA + LiCl \xrightarrow{CH_3CN} LiA + HCl$$

is displaced to the right, forming weakly dissociated lithium carboxylate and hydrochloric acid. This stronger acid can be titrated with the tertiary amine.

Adapted from the Method of H. W. Wharton

[*Anal. Chem.*, **37,** *730* (*1965*)]

TITRANT

Tri-*n*-propylamine (Distillation Products Industries) was prepared as 0.05*M* in reagent grade acetone and standardized against reagent grade succinic acid (Matheson Coleman and Bell) or primary standard benzoic acid. Standardizations were made by potentiometric titration in acetonitrile made 0.035*M* (saturated) in lithium chloride.

SOLVENT

Acetonitrile (Matheson, Coleman, and Bell) was stored over anhydrous calcium sulfate to thoroughly dry it; then it was passed through a column of fresh anhydrous calcium sulfate (1 liter of acetonitrile per 100 grams of calcium sulfate) saturated with dried reagent grade lithium chloride (Mallinekrodt Chemical Works). Final lithium chloride concentration at saturation was 0.034 to 0.035*M*.

SAMPLES

All samples (0.1–0.3 meq.) were dissolved in 30 ml of lithium chloride–acetonitrile solvent or prepared as approximately 0.1*N* in acetonitrile or a 60/40 chloroform-acetone solvent for pipetting into 30 ml of lithium chloride–acetonitrile solvent.

ELECTRODES

A glass-sleeve type saturated calomel electrode system was used. The aqueous potassium chloride internal electrolyte in the calomel electrode was replaced by a methanol solution of sodium chloride (saturated).

RESULTS AND DISCUSSION

Figure 3.23 plots titration curves for three typical monocarboxylic acids titrated with tri-*n*-propylamine in the presence and absence of lithium chloride.

Table 33 gives typical recoveries of acids alone and in intentional mixtures with acid anhydrides, along with the appropriate anhydride analyses as determined independently by the method of Johnson and Funk (p. 232). The standard deviations are ±0.30% in the 0 to 100% acid range and 0.41% anhydride in the 60 to 100% anhydride range. The precision of the free acid analysis tends to increase with increasing chain length of the monocarboxylic acid, since $\Delta E_{e.p.}$ also increases.

Table 34 summarizes the potential breaks at the end points for the titration of a variety of acids. In the absence of lithium chloride, only acidic groups having pK_a (H_2O) values less than 3.13 (citric acid) could be titrated, supporting the observations of Siggia and Floramo (p. 244). With lithium chloride present, monocarboxylic acids with pK_a values up to 5.5 (picolinic acid) could be titrated. In addition, all other acidic groups in a molecule having one acidic group with a pK_a value less than 5.5 were also titratable up to pK_a of 11.9 (HPO_4^{-2}).

Studies of other salts and solvent systems were made to enhance

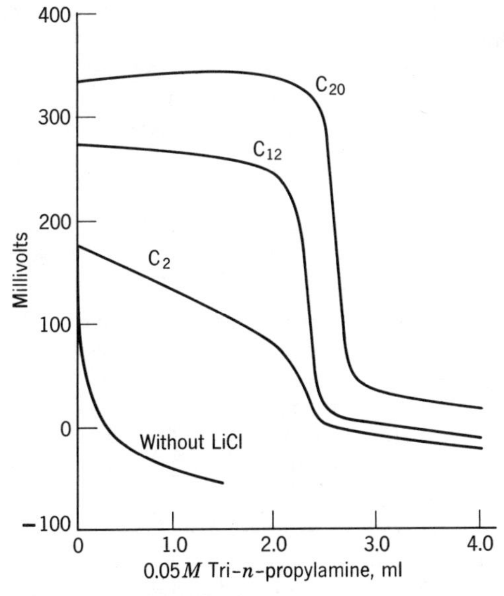

Fig. 3.23. Titration of typical monocarboxylic acids in acetonitrile with and without lithium chloride; tri-*n*-propylamine as titrant.

Table 33. Typical Results for Analyses of Carboxylic Acids, Anhydrides, and Mixtures of Acids and Anhydrides

Acid Samples	Acid Content		
	Taken, mg	Found, mg	Recovery, %
Acetic	31.5	32.4	103.8
	120.0	124.8	104.0
Propionic	37.3	37.7	100.9
	148.2	149.1	100.8
Stearic	157.5	158.4	100.5
	569.0	563.0	99.0
Maleic	57.7	57.9	100.3
	220.0	217.0	98.6
Succinic	59.3	60.0	101.2
	336.2	339.2	100.8

Acid Anhydride Samples	Acid Content, %	Anhydride Content, %	Total, %
Acetic anhydride	3.5	97.6	101.1
Propionic anhydride	32.3	67.3	99.6
Stearic anhydride			
A	0.37	102.3	102.7
B	1.10	99.6	100.7
Maleic	0.61	98.5	99.1
Succinic	7.75	91.7	99.5

Known Mixtures of Acids and Anhydrides + Anhydride	Acid Content			Anhydride Content		
	Taken,[a] meq.	Found, meq.	Recovery, %	Taken,[b] meq.	Found, meq.	Recovery, %
Acetic	0.338	0.340	100.6	0.297	0.294	99.0
Propionic	0.409	0.408	98.1	0.229	0.227	99.2
Maleic	0.302	0.306	101.3	0.294	0.291	99.0
Succinic	0.323	0.346	105.0	0.287	0.285	99.4

[a] Corrected for anhydride assay value as reported above.

[b] Corrected for acid assay value and acid derived from anhydride as reported above.

Table 34. Typical End Point Breaks ($\Delta E_{e.p.}$) for Titration with Tri-*n*-Propylamine in Acetonitrile Containing Lithium Chloride

Acid	pK_a (H_2O)	$\Delta E_{e.p.}$, mV	Number of Equivalents Titrated
Oxalic	1.27, 1.47	437[a]	2[b]
Malonic	2.86, 5.70	255[a]	2
Adipic	4.41, 5.28	280[a]	2
Sebacic	4.55, 5.52	415[a]	2
Citric	3.13, 4.76, 6.40	310[a]	3[b]
Lactic	3.86	270[a]	1
Phenylphosphoric	1.83, 7.07	475[a]	2[b]
Nicotinic	4.78	60[c]	1
Picolinic	5.50	185[c]	1
Benzoic	4.17	95[a]	1
Phosphoric	2.1, 7.2, 11.9	385[c]	3[b]
Acetic	4.76	75[a] (55[c])	1
Hexanoic	4.85	120[a] (110[c])	1
Dodecanoic		230[a] (190[c])	1
Octadecanoic		290[a] (250[c])	1

[a] Glass-calomel (saturated sodium chloride in methanol) electrode pair.
[b] 1 equivalent titrated in absence of lithium chloride.
[c] Glass-calomel (0.1*M* lithium chloride in methanol) electrode pair.

acidities of low molecular weight fatty acids and to increase the precision and accuracy of the titration. The potential break at the equivalence point for acetic acid was only 75 mV per milliliter of 0.05*M* titrant (Table 34), the lower limit for practical use. In the foregoing study Wharton suggested the possibility of using calcium perchlorate, but this salt is difficult to dehydrate completely. Yet if the salt is not completely anhydrous, the anhydride may be hydrolyzed. Greenhow and Jones (see below) found barium perchlorate, which can be dried by heat at 140°C, to be a satisfactory replacement for lithium chloride for short-chain carboxylic acids.

Adapted from the Method of E. J. Greenhow and R. L. P. Jones

[*Analyst* **97,** *346* (*1972*)]

REAGENTS

ACETONE. Analytical reagent grade acetone was dried over molecular sieve 4A before use.

ACETONITRILE. Reagent grade acetonitrile was dried over molecular sieve 4A before use.

ACRYLONITRILE, 99%. This was used as received.

ANHYDROUS BARIUM PERCHLORATE, 99%. This was dried at 140°C before use.

TRI-*n*-PROPYLAMINE, 0.25*M* SOLUTION IN ACETONE. This was standardized against reagent grade succinic acid (25–30 mg) or benzoic acid (50–60 mg) in acetonitrile (30 ml) containing anhydrous barium perchlorate (0.2 gram) by potentiometric titration as described under Procedure.

APPARATUS

A closed, magnetically stirred, 100-ml titration cell fitted with an inlet and an outlet for inert gas was used.

The calomel electrode in the glass-calomel system contained methanol saturated with sodium chloride and had a porous ceramic membrane.

PROCEDURE

Weigh the sample expected to contain 1 to 4 meq. of acid into a dry 100-ml calibrated flask, dissolve it in the titration solvent, and make the volume to the mark with the same solvent. Transfer 25 ml of the solution with a pipet into the titration cell, displace the air with dry nitrogen, add about 0.2 gram of anhydrous barium perchlorate, stir the mixture for 2 minutes to allow the exchange reaction to proceed, and titrate the solution potentiometrically with standard tri-*n*-propylamine solution. Make the blank titrations on 50 ml portions of solvent.

RESULTS AND DISCUSSION

Table 35 compares the enhancement of the acidity of acetic acid by lithium chloride and barium perchlorate in acetonitrile and acrylonitrile.

Table 35. Comparison of Enhancement of Potential Response by Salts in Acetonitrile and Acrylonitrile Using Glass Modified Calomel Electrode System

Solvent	Enhancing Salt	Potential Range ±0.5 ml of Equivalence, mV	$[\Delta mV/\Delta ml]_{MAX}$
CN_3CN	LiCl	100	170
CH_3CN	$Ba(ClO_4)_2$	240	650–1100
C_2H_3CN	$Ba(ClO_4)_2$	250	700–1400

Acrylonitrile is marginally superior to acetonitrile in terms of sharpness of the end point and is generally preferable to the latter because it is more readily available in a pure, dry, acid-free form.

Table 36 gives results for the titrations of acetic acid–acetic anhydride mixtures in acetonitrile with barium perchlorate as enhancing salt. Accuracies are better than 1% for acetic acid alone when 25-mg samples are used, of the order of 2% for 5% of the acid in the anhydride, and 10 to

Table 36. Potentiometric Titration of Acetic Acid with Tripropylamine in the Presence of Acetic Anhydride and Barium Perchlorate

Sample			Acetic Acid		
Acetic Acid					
Added, mg	In Anhydride, mg	Acetic Anhydride, grams	Found, mg	Found, %	Taken, Columns 1+2,%
32.0	—	0	31.7	99.1	100.0
32.0	—	0	32.1	100.3	100.0
53.4	40.1	5.3400	99.6	1.85	1.73
53.4	40.1	5.3400	104.6	1.94	1.73
62.3	45.0	6.0017	105.6	1.74	1.77[a]
62.3	45.0	6.0017	114.0	1.88	1.77[a]
42.3	15.8	2.1120	61.9	2.87	2.70
42.3	15.8	2.1120	61.9	2.87	2.70
129.9	45.0	6.0016	159.1	2.59	2.85[a]
129.9	45.0	6.0016	178.8	2.92	2.85[a]
58.0	8.1	1.0740	66.3	5.86	5.84
58.0	8.1	1.0740	65.3	5.77	5.84
0	—	6.0000[a]	43.6	0.73	—
0	—	6.0000[a]	45.2	0.75	—
0	—	6.0000[a]	45.7	0.76	—
0	—	6.0000[a]	44.2	0.74	—
0	—	6.0130[a]	107.2	1.78	—
0	—	6.0130[a]	105.6	1.76	—

[a] Acrylonitrile as solvent; all other determinations had acetonitrile as solvent.

20% for 1% of the acid in anhydride. Higher fatty acids were determined alone or in the presence of their anhydrides with the enhancement technique. In general, the end points were as sharp or sharper than those obtained with acetic acid; typical results appear in Table 37.

Table 37. Potentiometric Titration of Samples of Acids and of Acids in Anhydrides in the Presence of Barium Perchlorate–Acrylonitrile Solvent

Sample	Acids			Potential Range ±0.5 ml of Equivalence, mV	$[\Delta mV/\Delta mL]_{max}$ ±10%
	Taken, grams	Found, mg	Found, %		
Propionic	5.6426	36.6	0.65	270	750
anhydride	5.6430	37.2	0.66		
Succinic	0.6762	14.4	2.14	350	2500
anhydride	0.7709	15.7	2.04		
Phthalic	0.4424	6.8	1.54	390	5000
anhydride	1.8443	28.5	1.55		
Butyric	0.0302	30.4	100.7	220	780
acid	0.0331	32.8	99.1		
Benzoic	0.0549	Standardization		280	1180
acid	0.0566	of titrant			

DETERMINATION OF CARBOXYLIC ACIDS AND ANHYDRIDES

Morpholine–Carbon Disulfide Method

[*Adapted from F. E. Critchfield and J. B. Johnson, Anal. Chem.*, **28,** *430–6 (1956)*]

A satisfactory review of methods available for the analysis of carboxylic acid–anhydride mixtures was given in a paper by Johnson and Funk (60); see pages 231–35. The method presented by these authors is based on the reaction of morpholine with an anhydride to give one mole each of acid and amide. In this method a known excess of morpholine is added and the excess is determined by a nonaqueous titration. For the determination of free acid, a total value must be obtained by an independent procedure.

The method presented here was made possible by the discovery that primary and secondary amines can be titrated as the corresponding dithiocarbamic acids. In the method finally adopted, the sample containing anhydride is added to a measured excess of morpholine solution. The free acid present in the sample and the acid formed by reaction of anhydride are titrated with standard sodium hydroxide. Under the conditions of the titration the excess morpholine does not interfere. The

60. J. B. Johnson and G. L. Funk, *Anal. Chem.* **27,** 1464 (1955).

addition of carbon disulfide at the equivalence point of this titration converts the excess morpholine to the corresponding dithiocarbamic acid. This acid is then titrated with standard sodium hydroxide. The difference between a blank and a sample for the second titration is a measure of the anhydride originally present. The difference between the two determinations, in milliequivalents per gram, is a measure of the free acid present in the sample. A potentiometric titration curve for the determination of acetic anhydride by this method is given in Fig. 3.24.

REAGENTS

Carbon disulfide, reagent grade.

Isopropyl alcohol, commercial grade, Carbide and Carbon Chemicals Co.

Morpholine, 0.2*N* in acetonitrile. Transfer 17 ml of morpholine to a 1000-ml volumetric flask and dilute to volume with acetonitrile, Carbide and Carbon Chemicals Co. commercial grade.

Thymolphthalein indicator, 1.0% pyridine solution.

Sodium hydroxide, 0.1*N*.

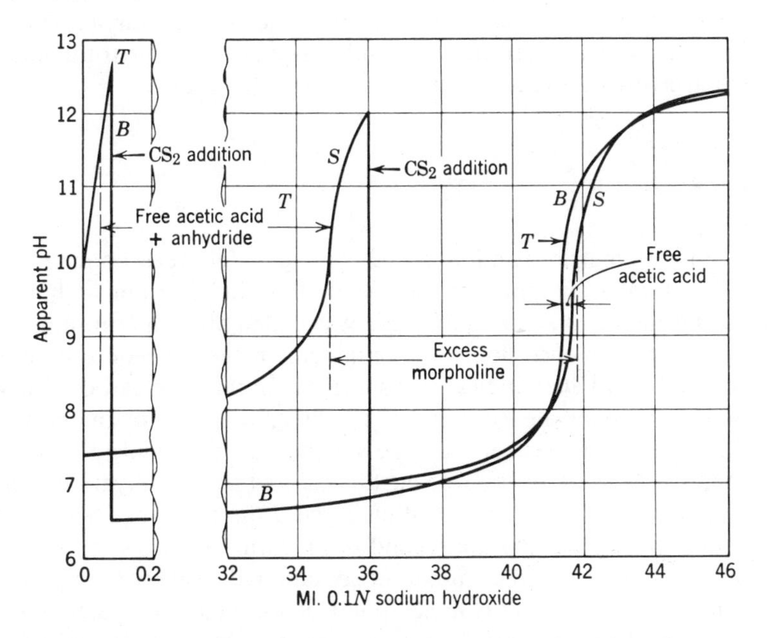

Fig. 3.24. Potentiometric titration curves for determination of acetic anhydride by morpholine–carbon disulfide method: *B*, blank; *S*, sample; *T*, thymolphthalein end point.

PROCEDURE

Pipet 25 ml of the 0.2*N* morpholine solution into each of two 250-ml glass-stoppered Erlenmeyer flasks. Reserve one of the flasks as a blank. Into the other flask introduce an amount of sample that contains between 2.0 and 3.5 meq. of acid anhydride and acid. Allow both the sample and blank to stand at room temperature for 15 minutes. To each flask add 75 ml of isopropyl alcohol, followed by 5 or 6 drops of thymolphthalein indicator. Titrate with 0.1*N* sodium hydroxide to the first blue color stable for at least 15 seconds. Record these titrations and zero the buret for both the sample and the blank. Do not overtitrate the end point. Add 20 ml of water to the blank. Pipet 5 ml of carbon disulfide into each flask and swirl to mix the contents thoroughly. Titrate the contents of each flask with 0.1*N* sodium hydroxide. Swirl the flasks during the titration to prevent a local excess of titrant in the titration medium. The end point selected should be the first blue or blue-green color stable for at least 1 minute. The difference between the blank and sample for the second titration is a direct measure of the anhydride originally present. The difference between the first determination, and the second, in milliequivalents per gram, is a measure of the free acid in the sample. The free acid can also be calculated from the difference between the total volumes of sodium hydroxide used for the sample and the blank.

DISCUSSION

The potentiometric titration curves in Fig. 3.24 were obtained by a procedure similar to that just described. For the titration of both the sample and the blank, carbon disulfide was added after the titration curve for the first equivalence point was obtained. In the procedure described earlier, thymolphthalein indicator is specified. In this case carbon disulfide is added at the equivalence point of the first titration. For the curves in Fig. 3.24 the difference between the sample and the blank, for the first titration, is a measure of the free acetic acid and acetic anhydride present in the sample. For the sample titration curve *S* the difference between the first equivalence point and the second is a measure of excess morpholine. The difference between the amount of sodium hydroxide consumed in the titration of the second equivalence point of the blank, curve *B*, and the excess morpholine for the sample, is a measure of acetic anhydride. Using the indicator technique, the calculations are simplified by leveling the buret at zero after the titration to the first end point. By this procedure the first titration is obtained by subtracting the blank from the sample; the second titration calculation is the blank minus the sample. Point *T* in Fig. 3.24 represents the thymolphthalein indicator end point. Although the indicator end point is slightly higher than the potentiometric end point, the error introduced is nearly the same for both the sample and the blank. The use of phenolphthalein is not recommended because the difference in the potentiometric and indicator end points for the first titration is too large.

In the method presented here, 0.2*N* morpholine in acetonitrile and 0.1*N* sodium hydroxide titrant are specified. The titration is carried out using 75 ml of isopropyl alcohol as a cosolvent. To prevent precipitation of the dithiocarbamic acid on the addition of carbon disulfide, a small amount of water must be incorporated in the medium. Sufficient water is introduced by the titrant for the titration of the sample, but for the blank, 20 ml of water must be added before the addition of carbon disulfide. Stronger reagents such as 1*N* morpholine and 0.5*N* sodium hydroxide cannot be used if a procedure similar to the preceding is employed. When water in the titrant is added to a titration medium consisting of 25 ml of 1*N* morpholine in acetonitrile and 75 ml of pyridine, morpholine becomes basic to thymolphthalein indicator. Also, the dithiocarbamic acid formed on the addition of carbon disulfide cannot be solubilized by water. Satisfactory results have been obtained with these stronger reagents when the procedure is modified, by preparing the morpholine reagent using pyridine as the solvent and using a titration medium consisting of 50 ml of pyridine, 25 ml of water, and 50 ml of isopropyl alcohol.

Table 38 lists several anhydrides for which the purities were determined using the method just described. Data are also shown for the determination of the free acid present in each of these samples. In most

Table 38. Purity of Anhydrides by Morpholine–Carbon Disulfide Method

Anhydride	Purity, % by Wt.[a]			Total, % by Wt.	
	Morpholine-Carbon Disulfide	Morpholine[a]	Acid Content, % by Wt.[a]	Morpholine-Carbon Disulfide	Morpholine[b]
Acetic	98.8 ± 0.1 (4)	98.9	0.4 ± 0.1 (4)	99.2	99.3
Butyric	97.3 ± 0.2 (2)	96.9	2.7 ± 0.1 (2)	100.0	99.6
Chrysanthemum	97.4 ± 0.2 (2)	97.8	1.6 ± 0.0 (2)	99.0	99.4
Endomethylenetetrahydrophthalic (carbic)	96.0 ± 0.0 (2)	—	1.5 ± 0.0 (2)	97.5	—
2-Ethylhexanoic	90.9 ± 0.2 (2)	91.1	8.9 ± 0.0 (2)	99.8	100.0
Maleic	98.2 ± 0.1 (4)	—	0.9 ± 0.03 (4)	99.1	—
3-Methylglutaric	98.5 ± 0.05 (2)	99.5	1.5 ± 0.05 (2)	100.0	101.0
Phthalic	99.5 ± 0.2 (2)	99.4	0.6 ± 0.05 (2)	100.1	100.0
Propionic	96.5 ± 0.1 (2)	96.9	3.3 ± 0.0 (2)	99.8	100.2
Succinic	95.7 ± 0.1 (2)	95.2	4.4 ± 0.05 (2)	100.1	99.6

[a] Figures in parentheses indicate number of determinations.
[b] Procedure of Johnson and Funk (60).

Table 39. Analysis of Mixtures of Anhydrides and Their Corresponding Acids by Morpholine–Carbon Disulfide Method

Acid-Anhydride Mixture	Added, % by Wt.			Found, % by Wt.		
	Acid	Anhydride	Total	Acid	Anhydride	Total
Maleic	25.4	74.0	99.4	25.7	74.3	100.0
	53.6	46.0	99.6	53.7	46.0	99.7
Acetic	53.8	45.5	99.3	53.9	45.4	99.3
	21.7	77.6	99.3	21.9	77.7	99.6

cases the purities are compared with results obtained using the method of Johnson and Funk (60). The precision is in the order of ±0.1% for the determination of purity. Table 39 gives the analyses of several known mixtures of carboxylic acids and their corresponding anhydrides. For these data the accuracy of the method is within ±0.2% for both the acid and the anhydride determinations.

The method is applicable to the determination of a wide variety of anhydrides. Maleic anhydride can be determined by the method if 0.2*N* morpholine solution is used. With 1.0*N* morpholine, addition across the double bond occurs. The method of Johnson and Funk (60) cannot be used for this anhydride because maleic acid is acidic to the indicator used in this method. Of the compounds investigated, only acrylic and chloroacetic anhydride could not be determined by this method. Halogen-substituted anhydrides react with morpholine to liberate a halogen acid; acrylic compounds add morpholine across the unsaturation. The method also has merit in that both the acid and anhydride determinations are obtained using a single sample and titrant.

NITRILES

From the Method of D. H. Whitehurst and J. B. Johnson

[*Anal. Chem.*, ***30,*** *1332* (*1958*)]

In this method the reaction of nitrile and alkaline peroxide to form the amide was used as the basis of an analytical procedure for simple aliphatic nitriles. In the initial reaction of nitrile, excess hydrogen peroxide and potassium hydroxide, some of the amide is simultaneously converted to the corresponding acid salt. Concentration of the alkaline reagent converts the remaining amide completely to the acid salt. The excess potassium hydroxide is then titrated with standard sulfuric acid to the phenolphthalein end point.

DETERMINATION OF PURITY OF NITRILES

Pipet 50 ml of 1.0*N* potassium hydroxide solution into each of two 300-ml glass-stoppered, alkali-resistant flasks. Pipet 100 ml of 3% hydrogen peroxide into each flask. Reserve one of the flasks as a blank. Introduce an amount of sample that contains 6 to 10 meq. of nitrile into the other flask. Allow the sample and the blank to remain at room temperature for 5 minutes with occasional swirling. Add a few glass beads to each flask (boiling stones cause erratic results), and attach each to a 40-cm glass column (10-mm diameter, 24/40 glass joint). Use silicone stopcock grease on the joints. Apply heat and evaporate the sample and blank to a volume of approximately 10 ml each. Do not evaporate to dryness. Cool and wash each glass column with 100 ml of water and collect the washings in the respective flasks. Drain the columns, remove the flasks, and pipet exactly 50 ml of 0.5*N* sulfuric acid into each. Add 6 to 8 drops of phenolphthalein indicator and titrate with standard 0.5*N* sulfuric acid.

DETERMINATION OF LOW CONCENTRATIONS OF NITRILES IN WATER

Pipet exactly 25 ml of 0.2*N* potassium hydroxide solution into each of two 300-ml glass-stoppered, alkali-resistant flasks. Pipet 20 ml of the 30% hydrogen peroxide into each flask. Add 200 ml of the water sample from a graduate to one of the flasks. Add 200 ml of distilled water to the other flask and reserve it for a blank determination. Allow the flasks to stand at room temperature for 5 minutes with occasional swirling. Add a few glass beads to each flask and attach to a 40-cm glass column. Grease each joint with silicone stopcock grease. Apply heat and evaporate to 2 ml or less. Do not evaporate to dryness. Cool and wash each condenser with 50 ml of distilled water and collect the washings in the respective flasks. Drain the condensers and remove the flasks. Add phenolphthalein and titrate with standard 0.1*N* sulfuric acid.

DISCUSSION

Complete conversion of nitrile to amide and acid salt occurs in 5 minutes at room temperature for each pure nitrile listed in Table 40. If the peroxide volume is reduced to 50 ml the reaction of acetonitrile, propionitrile, and succinontirile is complete, but it is approximately 3% low for butyronitrile.

Table 40. Purity of Nitriles

Compound	Average Purity, wt. %	Number of Determinations	Standard Deviation
Acetonitrile	100.5	34	0.5
Propionitrile	99.7	13	0.3
Butyronitrile	100.0	28	0.7
Succinonitrile	100.7	17	0.5

It was necessary to add 30% peroxide to a 200-ml water sample to produce the 3% solution of hydrogen peroxide.

Because only part of the amide formed from the nitrile is converted to the acid salt in the 5-minute reaction, it is necessary to saponify the remaining amide by a second reaction. In the case of acetonitrile, approximately 70% of the amide is converted to potassium acetate in the first reaction. Concentration of the potassium hydroxide until its concentration is at least $2N$ produces complete saponification of the amide formed from acetonitrile, butyronitrile, propionitrile, and succinonitrile. In the case of acetonitrile, evaporation to a volume of 50 ml (a concentration of $1N$) resulted in an indicated purity of 97%. However evaporation continued to a final volume of 10 to 25 ml produced an indicated average purity of 100.5% for the same material. Evaporation to approximately 2 ml is necessary to obtain $2N$ potassium hydroxide solution for low concentrations of nitriles in water.

All compounds that are oxidized to an acid under the conditions of the reaction will interfere. Certain compounds (e.g., acetaldehyde and formaldehyde) will oxidize quantitatively, and a correction can be made if they are present. Methanol, ethanol, and 2-propanol interfere only slightly and can be tolerated in small quantities. Most esters and amides give quantitative reactions and can be determined independently. Amines such as ethanolamine and 2-ethylhexylamine do not interfere if they form azeotropes or steam distill with water during the evaporation step.

Results as high as 120% were obtained for purities of benzonitrile, acrylonitrile, ethylene cyanohydrin, and 3-methoxypropionitrile. Approximately 2 moles of potassium hydroxide was consumed for each mole of lactonitrile; however quantitative purity values could not be obtained under the conditions of the reaction.

Nitriles added to water to obtain concentrations in the 5- to 10-ppm range showed the recoveries listed in Table 41. A standard deviation of 0.15 ppm was obtained for acetonitrile in the 5-ppm range.

Table 41. Low Concentrations of Nitriles

Compound	Concentration, ppm	
	Added	Found
Acetonitrile	1001	988
	502	500
	50	58
	25	24
	5.6	5.0[a]
Propionitrile	6.2	8.8
Butyronitrile	5.6	4.1
Succinonitrile	5.3	5.7

[a] Average of 10 or more determinations, standard deviation of 0.15 ppm.

Method of D. C. White

Reprinted in part from [Analyst ***96****, 728 (1971)]*

This work was undertaken to find conditions such that cyanide, if present, will be included quantitatively with nitriles as total hydrolyzable nitrogen. Complete hydrolysis of cyanide was not obtained with the described Whitehurst and Johnson procedure but was realized when a larger amount of hydrogen peroxide was used. Distillation and titration of ammonia produced served as the end determination. A separate determination of cyanide permits calculation of nitrile by difference.

The reaction was formulated as follows:

$$RCN + H_2O_2 \xrightarrow{OH^-} R\overset{\overset{NH}{\|}}{C}OOH + H_2O_2 \longrightarrow$$

$$RCONH_2 + O_2 + H_2O$$

REAGENTS

Hydrogen peroxide, 30%.
Sodium hydroxide solution, 5*N*.
Boric acid solution, about 4%.
Hydrochloric acid solution, 0.05*N*.

MIXED INDICATOR. (*a*) Dissolve 0.166 gram of methylene blue in ethanol; (*b*) dissolve 0.250 gram of methyl red in ethanol and make to 100 ml in each case. Mix equal volumes of solutions *a* and *b*. The exact proportions are important.

APPARATUS

The apparatus is illustrated in Fig. 3.25. The quartz wool plug in the side arm from the flask is essential to prevent the carryover of alkaline spray formed as a result of the decomposition of the hydrogen peroxide during the initial heating of the reaction mixture.

PROCEDURE

Measure 30 ml of hydrogen peroxide solution into the reaction flask and add sufficient sodium hydroxide solution to neutralize any acidity in the sample plus 1 ml in excess. If the acidity of the sample is unknown, titrate a separate aliquot with 1*N* sodium hydroxide with phenolphthalein as indicator. Pipet a volume of sample containing about 0.7 meq. of nitrogen, but not more than 10 ml, into the flask. Then add water to make the volume to 10 ml if necessary. Attach the flask to the rest of the apparatus and allow to stand 10 minutes. Place a 250-ml flask containing 10 ml of boric acid solution and 3 drops of indicator below the condenser, the end of the condenser dipping below the surface of the solution. Add 40 ml of sodium hydroxide solution and 10 ml of water through the top funnel, mix by swirling, then heat very gently until the effervescence ceases. Remove the source of heat if at any time the effervescence becomes vigorous. Then increase the heat and distill off the ammonia until about 30 ml of liquid remains in the reaction flask. Titrate the ammonia with 0.05*N* hydrochloric acid to the neutral gray end point. It is

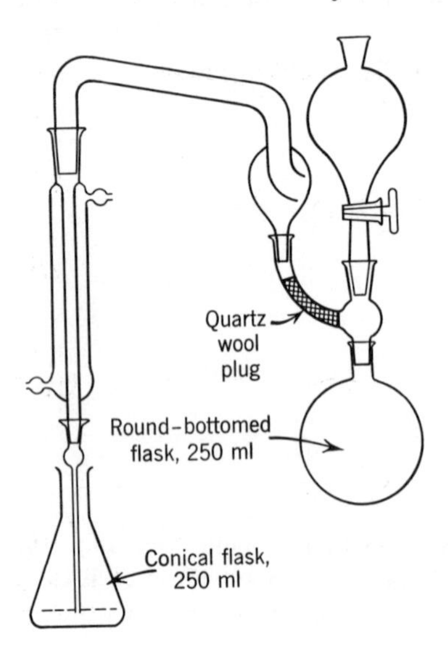

Fig. 3.25. Apparatus for the determination of total hydrolyzable nitrogen.

advisable at this juncture to add a further 20 ml of water to the distillation flask, replace the titration flask below the condenser, and distill a further 10 ml. A small additional titration is usually obtained. Carry out a blank on all the reagents used.

CALCULATION

$$\text{Nitrogen (grams per 100 ml)} = \frac{(T-B) \times F \times 14.008 \times 1.03}{2 \times 100 \times \text{Sample volume}}$$

where T is the sample titration in milliliters, B is the blank titration in milliliters, F is the factor for the $0.05N$ hydrochloric acid solution, and 1.03 is a factor assuming a 97% recovery of ammonia from all the components.

RESULTS

Some results appear in Table 42. The mean recovery was 97% with a standard deviation of 1.0%. It was therefore decided to assume an

Table 42. Analytical Results for Individual Components

Solution	Nitrogen in Solution, grams per 100 ml	Nitrogen Taken, mg	Nitrogen Found, mg	Recovery, %
Acetonitrile in $0.2N$ H_2SO_4	0.1521	7.61	7.36	96.7
		15.21	14.89	97.9
Acrylonitrile in $0.2N$ H_2SO_4	0.1058	5.29	5.09	96.2
		5.29	5.07	95.8
		10.58	11.09	95.4
	0.2398	11.99	11.63	97.0
			11.53	96.2
	0.1930	19.20	18.63	97.0
			18.45	96.1
	0.1953	19.43	19.05	98.0
KCN in $0.2N$ H_2SO_4	0.2076	10.38	10.05	96.8
			10.14	97.7
KCN in neutral solution	0.2145	10.73	10.33	96.3
	0.2103	10.52	10.12	96.2
HCN in $0.2N$ H_2SO_4	0.1450	7.25	7.17	98.9
			7.12	98.2
		14.50	14.09	97.2
			14.17	97.7
			14.16	97.7

Table 43. Analysis of Mixtures

Solution	Nitrogen in Solution, grams per 100 ml	Nitrogen		
		Taken, mg	Found, ×1.03 mg	Recovery, %
Acrylonitrile + HCN in 0.2N H_2SO_4	0.770	7.70	7.75	100.6
			7.66	99.5
	0.1165		11.65	100.0
			11.59	99.5
Acetonitrile, acrylonitrile, HCN in 0.2N H_2SO_4	0.1241	6.20	6.21	100.2
			6.18	99.7
		12.41	12.40	99.9
			12.47	100.5

average recovery of 97% for the components in the analysis of unknown mixtures. Table 43 gives recoveries of mixtures.

4

Alkoxyl and Oxyalkylene Groups

$$RO\text{—} \quad \text{and} \quad \text{—O}\overset{\overset{\displaystyle R}{|}}{C}H(CH_2)_nO\text{—}$$

The determination of the stated groups is accomplished through a common reaction system namely, hydrolytic splitting using hydriodic acid. The alkoxyl groups yield the corresponding alkyl iodide, which can be determined acidimetrically or iodimetrically. The oxyalkylene groups yield the corresponding alkylene diiodide and alkyl iodide, which can be determined as such or can be allowed to react further to yield one mole of free iodine per oxyalkylene group (see p. 279 for equations). This liberated iodine can be titrated and thus used as a measure of oxyalkylene groups.

Alkoxyl Groups

$$ROR' \xrightarrow{HI} RI + R'I + H_2O$$

$$RC(=O)OR' \xrightarrow{HI} \left[RC(=O)I + H_2O \right] + R'I$$

$$\longrightarrow RC(=O)OH + HI$$

ACIDIMETRIC APPROACH

Method of R. H. Cundiff and P. C. Markunas

[*Reprinted in part from Anal. Chem.*, **33**, *1028–30* (*1961*)]

The alkyl iodide formed on reaction of the alkoxyl compound with hydriodic acid reacts with pyridine according to the following equation:

$$R\text{—}I + C_5H_5N \rightarrow [C_5H_5N\text{—}R]^+I^-$$

The alkyl pyridinium iodide thus formed and the hydriodic acid carried along are resolved by a differentiating titration with tetrabutylammonium hydroxide.

A typical potentiometric curve, obtained from titration of a hydriodic acid–*n*-butyl iodide mixture, is shown in Fig. 4.1. The first inflection represents neutralization of the hydriodic acid; the second represents neutralization of the iodide. The volume difference between the two end points is a measure of the alkoxyl content. The alkyl iodides do not behave as acids in neutral solvents such as acetone, methyl isobutyl ketone, or acetonitrile.

The first equivalence point in the titration corresponded to the visual change of azo violet indicator from orange to red, and the second

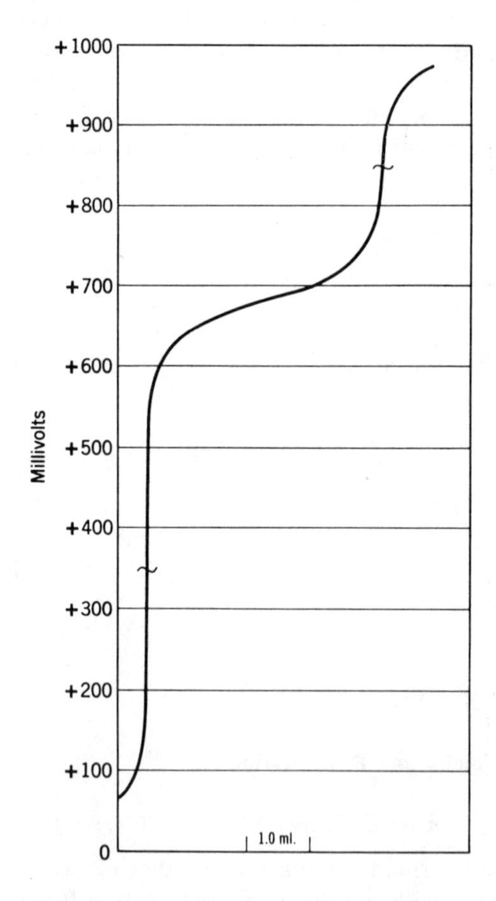

Fig. 4.1. Titration of hydriodic acid–*n*-butyl iodide mixture.

equivalence point corresponded to the visual change from red to violet. Thus the titration may be performed visually using a single indicator.

APPARATUS AND REAGENTS

ALKOXYL APPARATUS. A diagram of this apparatus appears in Fig. 4.2. Any conventional alkoxyl apparatus may also be used, although the scrubber is superfluous.

TETRABUTYLAMMONIUM HYDROXIDE, 0.02*N*. Prepare 0.1*N* tetrabutylammonium hydroxide as described on page 00. Add 20 ml of methanol to 200 ml of this solution and dilute to 1 liter with benzene.

PYRIDINE. Flash-distill technical grade pyridine from barium oxide using an upright condenser, discarding the first and last 10% of the distillate.

HYDRIODIC ACID. Merck & Co., Inc., reagent grade, 55 to 58% HI, specific gravity 1.7. No additional purification is necessary.

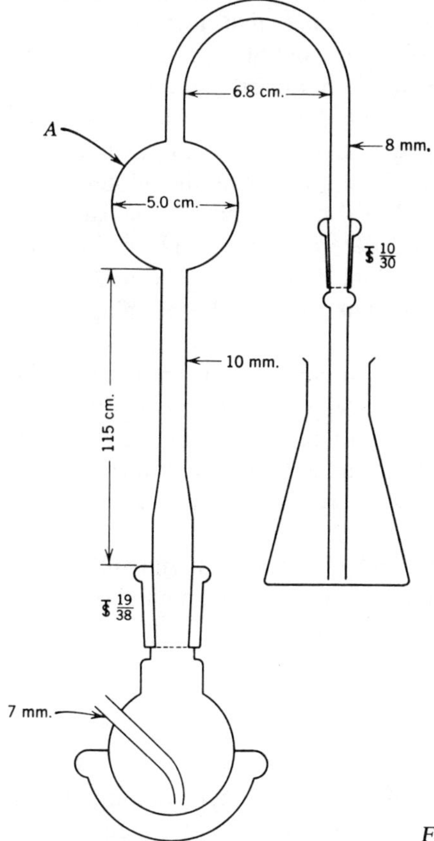

Fig. 4.2. Alkoxyl apparatus.

AZO VIOLET INDICATOR SOLUTION. Dissolve 0.5 gram of *p*-nitrobenzeneazoresorcinol in 100 ml of pyridine.
XYLENE. Analytical reagent grade.

PROCEDURE

Accurately weigh 10 to 15 mg of the solid alkoxyl compound and transfer to the reaction flask. Weigh volatile samples in gelatin capsules. Add 0.5 ml of xylene to the flask and dissolve the solid samples, heating if necessary. Add 5.0 ml of hydriodic acid and a few boiling stones. Lightly grease the standard-taper joints and connect the flask to the apparatus. Place 50 ml of pyridine in an Erlenmeyer flask and allow the delivery tip to extend below the surface of the solvent. Pass nitrogen through the system, adjusting the initial rate to one bubble per second. Apply heat with a heating mantle, and adjust the heat so that the top of the condensate is just above bulb *A*. Continue this nitrogen rate for 20 minutes, then increase the rate to 2 to 3 bubbles per second for the remainder of the reaction period. Allow a minimum amount of condensate to pass into the receiver, and if fuming is observed in the receiver, lower the nitrogen rate and adjust the heat so that fuming is minimized. Continue the reaction an additional 25 minutes for methoxyl determination, an additional 40 minutes for ethoxyl determination, an additional 100 minutes for propoxyl and butoxyl determination, and an additional 160 minutes for *S*-methyl determination. Disconnect the Erlenmeyer flask and rinse the exit tube with pyridine, adding the washings to the receiving flask.

Gently boil the pyridine solution for 2 minutes, cool, then titrate. The titration may be performed potentiometrically or as follows. Add 2 drops of azo violet indicator solution to the pyridine solution and titrate under nitrogen to a red end point, record the volume, then titrate to a violet end point. The volume difference between the red and violet end points is a measure of the alkoxyl content.

Perform blank analyses with each change of reagents, and subtract the volume difference between the red and violet end points from that obtained in the sample analysis.

EXPERIMENTAL

ALKOXYL COMPOUNDS ANALYZED. A series of alkoxyl compounds was analyzed by the described procedure. If the compound tested was less than 98% pure, it was recrystallized or chromatographed until the purity was greater than 98%. Solid samples were weighed directly into the reaction flask; liquid samples were weighed in gelatin capsules. Results of these analyses (Table 1) are the average of a minimum of two determinations.

ATTEMPTED RESOLUTION ON METHOXYL–ETHOXYL MIXTURES. It is necessary to boil pyridine solutions of ethyl, propyl, and butyl iodide to obtain stoichiometric conversion to the corresponding alkyl pyridinium iodide. Methyl iodide reacts quantitatively at room temperature. This indicated that it might be possible to resolve methoxyl and ethoxyl mixtures.

In this investigation, the receiving flasks were chilled during the reaction. One pyridine solution was titrated while still at a low temperature; a second solution was boiled, cooled, then titrated. The first titration indicated the methoxyl content; the second, the total alkoxyl content, and the ethoxyl content was obtained by difference. Imprecise results were noted in all variations of the technique tried. The temperature of the receiving solution is the critical factor. If the temperature is held too low, reaction of the methyl iodide is incomplete; if held too high, small amounts of ethyl iodide react. It seems entirely possible that sufficiently sensitive conditions could be established that precise quantitative results could be obtained. Even so, qualitative distinction between methoxyl and higher alkoxyl groups can easily be realized by this means.

DISCUSSION

That iodine and sulfides do not interfere was established by adding iodine crystals to one reaction mixture, and iodine and ferrous sulfide to a second reaction mixture. The blank value from each of these reactions was the same as that obtained with these additives absent. This was tested further by adding ferrous sulfide and iodine to a reaction mixture in the analysis of methyl cellulose. The result for methoxyl content was identical to that obtained with the interferents absent and corresponded to the theoretical value.

Phenol and acetic anhydride or propionic anhydride were tried in conjunction with hydriodic acid in the hydrolysis of the alkoxyl group. Phenol could not be used, since any phenol entering the receiver titrated simultaneously with the alkyl pyridinium iodide. The anhydrides could be used without the phenol, since the acid formed and carried into the receiver titrated separately from both the hydriodic acid and the iodide. No particular increase in efficiency was noted, however, when the anhydrides were used.

Table 1. Determination of Alkoxyl Content in Organic Compounds

Compound	% Alkoxyl Theory	% Alkoxyl Found
CH_3O—		
3-Methoxy-4-hydroxybenzaldehyde	20.40	20.47
p,p′-Dimethoxybenzophenone	25.64	25.58
m-Methoxyphenol	25.00	25.24
p-Methoxyphenol	25.00	25.37
4-Methoxy-2-nitroaniline	18.46	18.40
4′-Methoxy-2-(*p*-methoxyphenyl) acetophenone	24.22	24.39
1-(*p*-Methoxyphenyl)-1-propyl palmitate	7.67	7.75
Methyl *p*-aminobenzoate	20.53	20.74
1-Naphthyl methyl ether	19.62	20.00
2-Naphthyl methyl ether	19.62	20.03
p-Methylanisole	25.40	25.98
m-Methoxyanisole	44.92	44.90
2,4-Dimethylanisole	22.79	22.80
3,4-Dimethylanisole	22.79	23.11
3,5-Dimethylanisole	22.79	22.95
1,3-Dimethoxy-5-methylbenzene	40.78	40.41
1,2-Dimethoxy-4-allylbenzene	34.63	35.07
Methocel, Dow	30.0[a]	30.06
C_2H_5O—		
3-Ethoxy-4-hydroxybenzaldehyde	27.12	27.07
p-Diethoxybenzene	54.22	54.61
p-Ethoxyacetanilide	25.14	25.11
2-Ethoxynaphthalene	26.16	25.97
p-Ethoxybenzoic acid	27.12	27.63
Ethyl *p*-aminobenzoate	27.28	27.47
Ethocel, Dow	46.3[a]	46.04
C_3H_7O—		
n-Propylcellulose, Dow	51.0[a]	50.77
C_4H_9O—		
p-Dibutoxybenzene	65.77	66.10
Butyl *p*-aminobenzoate	37.83	37.99
CH_3S—		
Methionine	31.57	31.49

[a] Standard cellulose ether samples supplied by The Dow Chemical Co. and analyzed by Samsel and McHard method (1).

1. E. P. Samsel and J. A. McHard, *Ind. Eng. Chem., Anal. Ed.* **14,** 750 (1942).

Xylene not only dissolves the test compound, but also boils at a temperature sufficiently high to aid in carrying the alkyl iodide to the receiver. The presence of xylene in the mixture also minimizes the fuming of hydriodic acid.

It was not necessary to purify further the hydriodic acid as obtained commercially or to use special storage precautions. The more dilute commercial hydriodic acid solutions, with and without preservative, can also be used, although the reaction time must be increased.

The procedure was not evaluated with respect to the highly volatile compounds such as diethyl ether, although it is believed that if they are weighed in gelatin capsules and xylene is used in the reaction mixture, additional precautionary steps are unnecessary. If this should prove unsuitable, substitution of a similar apparatus with a water condenser, for use during the hydrolysis, should ensure quantitative results with this type of compound.

One distinct advantage of this procedure is that only 2 hours are required for determination of the propoxyl and butoxyl groups. The efficient procedure of Shaw (see p. 273), for example, requires a minimum of 3 hours for quantitative conversion of the butoxyl group, whereas other procedures require longer reaction periods and even more extensive modification of the apparatus.

IODIMETRIC APPROACH (FOR METHOXY AND ETHOXYL ONLY)

Combined Method of Elek, and Samsel and McHard

[*Ind. Eng. Chem., Anal. Ed.*, **11, 174,** *(1939); ibid.*, **14,** *754 (1942).* (*Adopted with Permission from the Description in Niederl and Niederl's Micromethods of Quantitative Organic Analysis, 2nd ed., Wiley, New York, 1942, pp. 239–44*)]

$$(a)\ RI + Br_2 \rightarrow RBr + IBr$$
$$(b)\ IBr + 3H_2O + 2Br_2 \rightarrow HIO_3 + 5HBr$$
$$(c)\ HIO_3 + 5HI \rightarrow 3I_2 + 3H_2O$$

REAGENTS

POTASSIUM ACETATE IN ACETIC ACID. Dissolve 100 grams of CP anhydrous potassium acetate in 1 liter of solution containing 900 ml of glacial acetic acid and 100 ml of acetic anhydride.

BROMINE SOLUTION. Dissolve 1 ml of bromine in 29 ml of the foregoing potassium acetate reagent. This solution should be prepared fresh daily.
SODIUM ACETATE. Dissolve 250 grams of CP anhydrous sodium acetate in 1 liter of distilled water.
AQUEOUS RED PHOSPHORUS SUSPENSION. Suspend 30 grams of CP red phosphorus in 50 ml of 5% cadmium sulfate.

Potassium iodide, CP.

Formic acid, reagent grade, 90%, sp. gr. 1.20.

Hydriodic acid, CP constant boiling mixture, b.p. 126 to 127°C (57% hydriodic acid).

10% Sulfuric acid. Add 60 ml of concentrated sulfuric acid to 940 ml of distilled water.

0.1*N* standard sodium thiosulfate.

Starch indicator.

CARBON DIOXIDE. A commercial cylinder with a reducing valve is a good source.

APPARATUS

Complete apparatus (Fig. 4.3) can be purchased from Scientific Glass Apparatus Co., Bloomfield, New Jersey.

PROCEDURE

Fill trap (*A*) two-thirds with the red phosphorus suspension. The phosphorus trap is to keep any free iodine from getting into the receiver. The cadmium sulfate is to remove any hydrogen sulfide formed.

$$2P + 3I_2 \rightarrow 2PI_3$$

$$PI_3 + 3H_2O \rightarrow 3HI + H_3PO_3 \text{ (2)}$$

Fill receiver (*B*) with bromine solution to a level representing one-third of the bulb portion (*C*). Place the sample in the reaction vessel together with a boiling chip and a crystal of phenol. Liquid samples should be weighed in gelatin capsules. To the sample, add 6 ml of constant boiling hydriodic acid, immediately attach the reaction flask to the condenser, and moisten the joint with hydriodic acid. Circulate water through the condenser and heat the reaction flask. Bubble a slow stream of carbon dioxide through the solution at a rate of about 2 bubbles per second. An oil bath at 130 to 150°C or a hot-spotter can be used as the heating

2. E. P. Samsel and J. A. McHard, *Ind. Eng. Chem., Anal. Ed.*, **14,** 754 (1942).

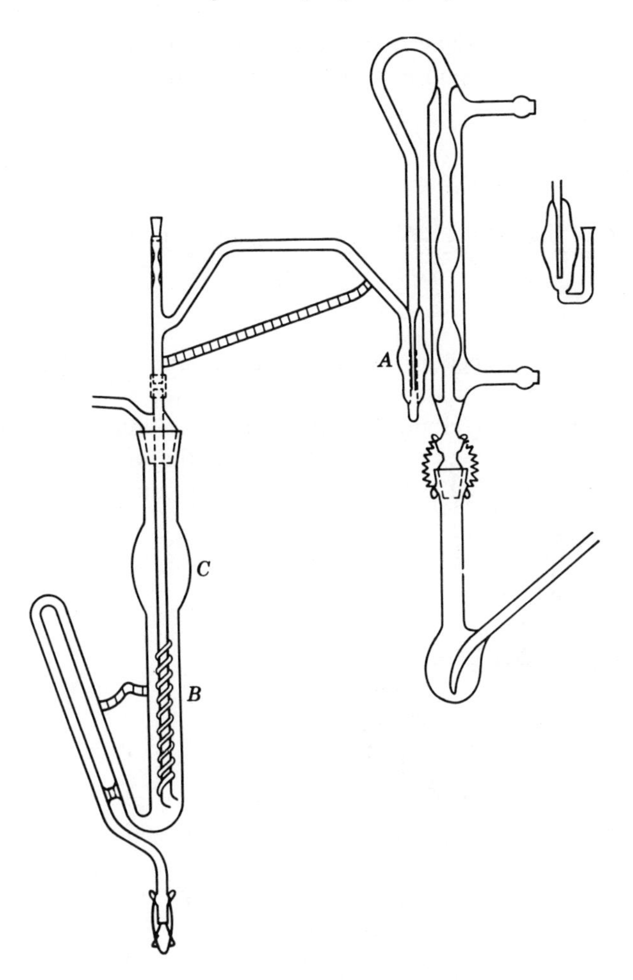

Fig. 4.3. Methoxyl-ethoxyl apparatus.

element. After 45 minutes (1 hour for ethoxyl determination) quantitatively transfer the contents of the receiver to a 500-ml flask containing 10 ml of sodium acetate solution. Dilute the solution to about 125 ml with water, and add formic acid dropwise until the bromine color is discharged; then add 3 additional drops. After the solution has stood for 3 minutes, add 3 grams of potassium iodide and 15 ml of 10% sulfuric acid. Titrate the liberated iodine with 0.1N sodium thiosulfate, using starch indicator. A blank should be run on the phenol alone. In most cases, however, the blank is so small as to be unnecessary.

In a few cases the sample is not attacked completely by the hydriodic acid. This is probably because of insolubility in the acid. In these cases, a 1:1 phenol–hydriodic acid mixture should be tried.

CALCULATIONS

$$\frac{\text{Milliliters of thiosulfate} \times N \text{ thiosulfate} \times \text{mol. wt. of alkoxyl group} \times 100}{\text{Grams of sample} \times 6000} = \%\ \text{alkoxyl}$$

The foregoing procedure is reproducible and accurate to ±0.2%. Methyl and ethyl alkimides interfere, since they are determined in a similar manner. Propoxyl and butoxyl groups also interfere because their iodides, which are formed, are sufficiently volatile to be partially distilled with the methyl or ethyl iodide, causing high results.

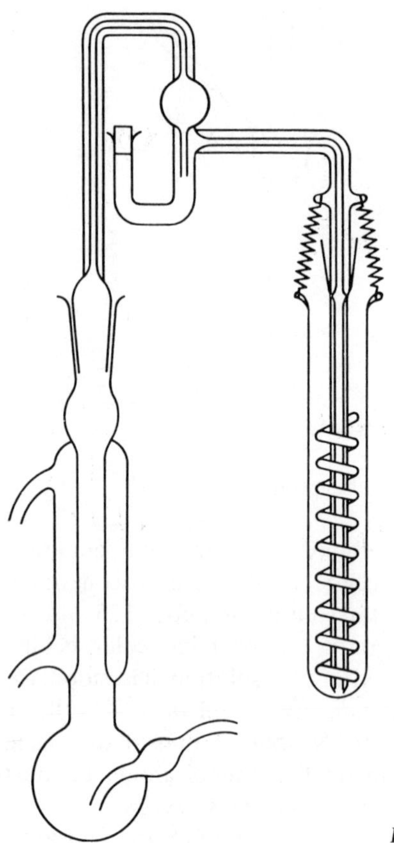

Fig. 4.4. Propoxyl-butoxyl apparatus.

The procedure was tested with ethoxy- and methoxybenzene, dimethylformal, dimethylacetal, 1,1,3-trimethoxybutane, sodium ethylate, and 2,4-dimethoxy-2-methylpentane.

IODIMETRIC APPROACH FOR PROPOXYL–BUTOXYL GROUPS

The method described for the determination of methoxyl and ethoxyl groups can be applied in this case if the apparatus is modified as shown in Fig. 4.4. This modification (3) makes it possible to distill the propyl and butyl iodides quantitatively into the receiver by eliminating a great deal of the tubing connecting the reactor and the receiver.

IODIMETRIC APPROACH FOR HIGHER ALKOXYL GROUPS

Adopted from the Method of S. Ehrlich-Rogozinski and A. Patchornik

[*Anal. Chem.*, ***36,*** *840* (*1964*)]

This method eliminates the distillation step and enables the quantitative determination of alkoxyl groups in compounds having four to 26 carbons. After the alkoxyl compound and hydriodic acid have reacted, the alkyl iodide formed is extracted with benzene and treated with aniline. The anilinium iodide obtained is titrated with standard sodium methoxide solution.

APPARATUS

Centrifuge tubes, 30-ml capacity for semimicro determination and 10-ml for the micro determination, fitted with No. 14 ground glass joints.

Apparatus for treatment of alkyl halide (Fig. 4.5) includes a boiling flask *A* with a capacity of 15 ml. Part *B* is 9 cm long and has a side vessel *E* with a capacity of about 2 ml for the microdetermination and about 5 ml for the semimicro procedure. During the reaction the apparatus is at an angle (Fig. 4.5). At the end of the reaction the apparatus is tilted to permit benzene in *E* to run back into flask *A*.

REAGENTS

Hydriodic acid, 70%, sp. gr. 1.96.

Sodium methoxide in benzene-methanol prepared as follows according to the

3. B. M. Shaw, *J. Soc. Chem. Ind.*, **66,** 147–9 (1947).

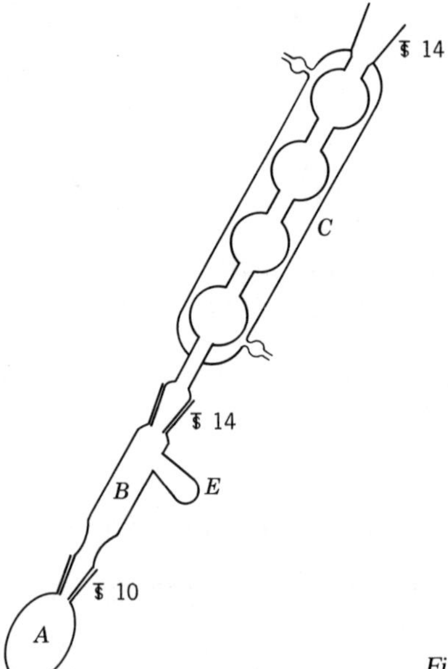

Fig. 4.5. Apparatus for reaction of alkyl halides.

directions of Fritz and Lisicki (4). Wash 6 grams of sodium by dipping in methanol for about 15 seconds and dissolve immediately in 100 ml of methanol. Protect the solution from atmospheric carbon dioxide while the sodium is dissolving; if necessary, cool the solution in cold water to prevent the reaction from becoming too violent. When all the sodium has reacted, add 150 ml of methanol and 1500 ml of benzene and store the reagent in borosilicate glassware protected from carbon dioxide. Determine the concentration of this stock solution by titration of benzoic acid dissolved in a mixture of 3 volumes of benzene to 1 volume of methanol. Dilute the solution with 3:1 benzene-methanol to obtain 0.05*N* solution for the semimicro and 0.01*N* for the micro procedure.

PROCEDURE

FOR ALCOHOLS, SEMIMICRO. Weigh 0.05 to 0.3 mM of the alkoxyl compound into a centrifuge tube of 30-ml capacity. Weigh liquids in capillaries. Add 1 ml of 70% hydriodic acid and 20 mg of red phosphorus. Immediately close the tube with a greased ground-glass stopper and clamp it tight. Heat the tube in boiling water for 1 hour. Cool the tube in an ice bath, remove the stopper, and add 7.00 ml of benzene. Shake well, and add 2 to 3 drops of an aqueous suspension of thymol blue. Cautiously with cooling, neutralize excess hydriodic acid with 6*N* aqueous sodium

4. J. S. Fritz and N. M. Lisicki, *Anal. Chem.*, **23,** 589 (1951).

hydroxide (normally about 2 ml). Ensure complete extraction by vigorous shaking. After centrifugation, transfer a 5.00-ml aliquot of the benzene layer to the boiling flask. Neutralize to the blue end point with 0.01*N* sodium methoxide. Add 5 ml of aniline, neutralized similarly. Attach the boiling flask to the apparatus and reflux the contents for 30 minutes. After cooling, return the benzene in the side tube *E* to the reaction flask *A*. Wash the apparatus with 2 ml of neutralized ethanol. Perform the titration in flask *A* with 0.05*N* sodium methoxide to the blue end point of thymol blue.

FOR ALCOHOLS, MICRO. Weigh the sample containing 0.01 to 0.05 mM of the alkoxyl compound and use 0.5 ml of 70% hydriodic acid. Use 3 ml of benzene for extraction. Treat a 2.00-ml aliquot with 2 ml of aniline in the apparatus with a side arm trap of 2-ml capacity. Perform the titration with 0.01*N* sodium methoxide.

FOR ESTERS. Dissolve 0.05 to 0.3 mM of the ester in 0.5 ml of propionic acid or anhydride for the determination on the micro scale. Add 0.5 ml of 70% hydriodic acid. Proceed as for the semimicro procedure for alcohols.

For the micro analysis, use 0.01 to 0.05 mM of ester, 0.3 ml of propionic acid or anhydride and 0.3 ml of 70% hydriodic acid and proceed as for alcohols.

Determine blanks in all cases by carrying out the appropriate procedure in the absence of the alkoxyl compound.

RESULTS AND DISCUSSION

To determine the range of applicability of this method, a series of *n*-alkyliodides from C_4 up to C_{16} was refluxed with aniline and the resulting anilinium iodides were titrated with sodium methoxide. In all cases tested, quantitative results were obtained with a maximum relative error of 1%.

The same series of *n*-alkliodides was treated also by the semimicro procedure described for alcohols. Recovery was quantitative within a maximum relative error of 2%. Methyl-, ethyl-, propyl-, and butyliodides gave recoveries between 85 and 95%. This may be explained by the relatively higher water solubilities of these alkyliodides as well as their greater susceptibility to basic hydrolysis during neutralization of excess hydriodic acid. Such a variety of excellent methods exist for the determination of C_1 to C_4 alkoxyl groups that no effort was made to apply this procedure to those four groups.

Table 2 presents the results for a series of commercially obtained, analytically pure alcohols, analyzed as described.

Table 2. Determination of Higher Alcohols, from the Data of Ehrlich-Rogozinski and Patchornik[a]

Alcohol	Average Recovery,[b] %
1-Pentanol	97.5 (2)
1-Hexanol	98.9 (2)
1-Heptanol	98.5 (3)
1-Octanol	100.2 (4)
1-Nonanol	98.7 (2)
1-Decanol	99.5 (2)
1-Undecanol	99.1 (2)
1-Dodecanol	99.9 (2)
1-Tetradecanol	99.3 (2)
1-Pentadecanol	101.1 (2)
1-Hexadecanol	100.6 (4)
1-Heptadecanol	101.4 (2)
1-Octadecanol	100.0 (5)
1-Docosanol	101.5 (2)
1-Hexacosanol	99.3 (3)

[a] Patchornik and A. Ehrlich-Rogozinski, *Anal. Chem.*, **33,** 803 (1961).

[b] Figures in parentheses indicate number of determinations.

Table 3. Determination of Esters, from the Data of Ehrlich-Rogozinski and Patchornik[a]

Ester	Average Recovery,[b] %
n-Hexyl-*p*-hydroxybenzoate	96.8 (3)
n-Heptyl-*p*-hydroxybenzoate	99.8 (6)
n-Octyl-*p*-hydroxybenzoate	99.2 (4)
n-Nonyl-*p*-hydroxybenzoate	99.6 (5)
n-Dodecyl-*p*-hydroxybenzoate	99.1 (5)
Dioctylphthalate	98.6 (3)
n-Hexylacetate	101.1 (5)
n-Heptylacetate	100.5 (3)
n-Octylacetate	99.3 (3)
n-Nonylacetate	99.6 (6)
n-Decylacetate	98.2 (4)

[a] A. Patchornik and A. Ehrlich-Rogozinski, *Anal. Chem.*, **33,** 803 (1961).

[b] Figures in parentheses indicate number of determinations.

Table 4. Accuracy of the Method

Compound	Taken, mg	Found, mg	Recovery, %
n-Hexylacetate	28.96	29.32	101.2
	28.63	28.70	100.2
	37.29	36.74	98.5
	42.36	42.79	101.0
	47.20	47.59	100.8
	49.49	49.57	100.2
	49.03	48.85	99.6
	41.14	41.33	100.5
	42.75[a]	42.92	100.4
	42.58[a]	42.64	100.2
	42.87[a]	43.54	101.6
	57.63	57.04	99.0
	56.02	56.85	101.5
	63.12	63.74	101.0
	53.65	54.91	102.3
	12.33	12.33	100.0
	12.18	12.58	103.4
	8.95	9.16	102.3
		Average	100.8
1-Hexadecanol	62.55	61.61	98.5
	54.75	54.05	98.7
	58.37	58.10	99.5
	46.38	45.50	98.1
	52.50	52.22	99.5
	53.86	53.37	99.1
	65.05	65.64	100.9
		Average	99.2
1-Octadecanol	49.45	49.41	99.9
	38.14	37.93	99.5
	54.27	53.59	98.7
	39.63	39.48	99.6
	54.84	53.42	97.4
	68.16	67.25	98.7
		Average	99.0

[a] With 70% HI only.

Esters are easily converted to alkyliodides, but in some cases their solubilities in aqueous hydriodic acid are low. Phenol is usually recommended as the solvent; but with this procedure it interfered with the titration because a small amount is extracted by benzene and titrated as acid. Both propionic acid and propionic anhydride are suitable solvents which do not interfere.

The results for a series of esters appear in Table 3.

The accuracy of the method is shown in Table 4. The results indicate a relative error of about 1%.

Oxyalkylene Groups

The oxyalkylene groups can be represented as $—O\underset{|}{\overset{R}{C}}H(CH_2)_nO—$ where R can be a hydrogen atom and n can be 1 as in the case of the oxyethylene group. The formula is written as it is to cover all members. The discussion to follow is divided into two parts. One is the determination of the oxyalkylene groups on a molecule as in $R(OCH_2CH_2)_xOH$, where there are x oxyethylene groups and this quantity is determined. The second portion of the discussion deals with polyoxyalkylene compounds as total entities, where the analytical reagent is acting on the oxyalkylene group making analysis possible, but the reaction proceeds in a rather empirical stoichiometry. In the latter cases, the percentage of compound can be obtained, but the percentage of oxyalkylene group cannot.

DETERMINATION OF OXYALKYLENE GROUPS

Method of S. Siggia, A. C. Starke, J. J. Garis, and C. R. Stahl

[*Reprinted in part from Anal. Chem.*, **30**, *115–6 (1958)*]

Very few methods exist for the determination of oxyalkylene groups. Dichromate oxidation (5, 6) has been used, but this approach lacks the

5. H. B. Elkins, E. D. Storlazzi, and J. W. Hammond, *J. Ind. Hyg. Toxicol.*, **24,** 229 (1942).
6. H. W. Werner and J. L. Mitchell, *Ibid.*, **15,** 375–6 (1943).

desired specificity. Morgan (7) who used hydriodic acid to split the oxyethylene groups of polyethylene glycol ethers and esters, found ethylene to be liberated and also an alkyl iodide presumed to be ethyl iodide. The alkyl iodide was caught in silver nitrate solution and determined via a Volhard-type titration. The ethylene was caught in a separate receiver and determined by using bromine. The total of ethylene and alkyl iodide represents the total oxyethylene groups in the sample.

Morgan postulated the reaction to proceed as follows:

$$\text{—}(CH_2CH_2O)_x\text{—} + 2xHI \rightarrow xICH_2CH_2I + xH_2O$$

$$ICH_2CH_2I \xrightarrow{\text{decomp.}} CH_2{=}CH_2 + I_2$$

Some of the diiodoethane is thought to react with hydriodic acid:

$$ICH_2CH_2I + HI \rightarrow CH_3CH_2I + I_2$$

Early work in the authors' laboratory had indicated that a stoichiometric amount of iodine was formed during treatment of polyoxyalkylene-containing compounds with hydriodic acid. A method was devised around the measurement of the liberated iodine. This work substantiates the postulations of Morgan. This approach permits determination of oxyalkylene groups in a shorter time, with less manipulation than with the earlier method, and with much simpler equipment. The method was extended beyond the oxyethylene groups to oxypropylene groups as well.

REAGENTS

Hydriodic acid (55–58%, sp. gr. 1.7) as used for methoxyl determinations is required. It is preferable to use hydriodie acid with as little free iodine as possible, in order to obtain low blanks and results with optimum precision and accuracy. Freshly opened bottles of hydriodic acid have a free iodine content equivalent to 2 to 4 ml of $0.1N$ thiosulfate per 5 ml of hydriodic acid. The free iodine increases rapidly once the bottle is opened, and it is not advisable to use acid with a free iodine content equivalent to over 10 ml of $0.1N$ thiosulfate per 5 ml of acid. This impure acid will cleave the ether and ester linkages and can be used, but the high blanks prevent optimum results. Hydriodic acid can be distilled to lower its free iodine content; however, it was found more expedient to purchase the acid in quarter-pound bottles. Each bottle lasts for a few determinations, and not enough time elapses for the blank to become excessive. Hydriodic acid containing hypophosphorous acid as a stabilizer must not be used.

Aqueous potassium iodide solution, 20%.

Standard sodium thiosulfate, $0.1N$.

7. P. W. Morgan, *Ind. Eng. Chem., Anal. Ed.*, **18,** 500 (1946).

Carbon dioxide. Cylinder gas or dry ice in a Dewar flask can be used. Unopened cylinders should be rapidly vented to the atmosphere until frost forms on the nozzle of the valve. This reduces the oxygen content of the remaining gas and results in lower blanks.

PROCEDURE

Into a 50-ml round-bottomed flask, pipet 5 ml of hydriodic acid. The flask contains a ground-glass joint to accommodate a vertical condenser, and is equipped with a side arm through which carbon dioxide can be passed to blanket the solution. Add a weighed sample containing 0.001 to 0.002 mole of oxyalkylene group to the hydriodic acid. The sample is best weighed in a tared glass thimble (1-ml beaker works well) and then drop it into the acid, thimble and all. Connect the vertical condenser with a thin grease seal at the outermost edge to cause a good seal. Avoid too much grease, since iodine tends to dissolve in the excess grease.

Commence the flow of carbon dioxide, and keep it at a rate of a few (1–5) bubbles per second. Use a bubbler in the carbon dioxide line to avoid excessive amounts of gas, to prevent iodine from being swept out of the system, causing low results. The system has to be kept under an atmosphere of carbon dioxide to avoid air oxidation of the iodide ion to free iodine, which would yield high, irreproducible blanks. After allowing a few minutes for the system to be covered with a blanket of carbon dioxide, commence heating. Gently boil the sample solution for 90 minutes; vigorous boiling causes loss of iodine through the condenser. Ninety-minute boiling was sufficient for the most stubborn compounds encountered in this study; 45 minutes was satisfactory for ethylene glycol, but slightly low values were obtained for the Cellosolve and Carbitol samples listed in Table 5.

Concurrent with the sample, run a blank in the same manner and in duplicate equipment. Include a glass bead in the blank to avoid bumping. Feed carbon dioxide from the same source into the system containing the blank. Heat the blank for the same length of time, because the blank is sizable and variation must be kept at a minimum for optimum results. It was found advantageous from a time standpoint to run several samples at one time, along with one blank. These are all purged at one time, by using a manifold of glass tubing to deliver the carbon dioxide from one cylinder. The use of one cylinder is emphasized, as carbon dioxide from cylinders contains oxygen which affects the blank. Different cylinders would contain different amounts of oxygen.

After the 90-minute boiling period, wash down the walls of the condenser with 15 ml of 20% potassium iodide solution. Dissolve any

Table 5. Determination of Oxyalkylene Groups

	Assay, %
Butyl Carbitol (diethylene glycol monobutyl ether)	99.9
	97.5
	98.9
Phenyl Cellosolve (ethylene glycol monophenyl ether)	98.5
	97.8
Methyl Carbitol (diethylene glycol monomethyl ether)	98.4
	99.0
	98.6
Carbowax 400 (polyethylene glycol)	94.5[a]
	93.5[a]
Dioxane	97.3
	99.5
	97.0
Polypropylene glycol	97.1[b]
	94.4[b]
	96.1[b]
Ethylene glycol	98.2
	98.1
	99.2
Propylene glycol	87.2[c]
	90.2[c]
	88.3[c]
Diethylene glycol dimethyl ether	102.4
	105.2
	105.2
Hydroxyethyl acetate	99.1
	99.5
	99.2
Stearic acid ester of polyethylene glycol, 5.24 moles of ethylene oxide per mole of acid	100.9[d]
	103.2[d]
	102.6[d]
8.00 moles of ethylene oxide per mole of acid	100.2[d]
	98.1[d]
	99.1[d]
Stearic acid ester of polypropylene glycol, 8.4 moles of propylene oxide per mole of acid	94.9[d]
	97.1[d]

[a] % as (OCH_2CH_2). Theoretical value for oxyethylene content is 96.4 as determined from hydroxyl group determination and corrected for terminal groupings.

[b] % as ($—OCH_2\underset{CH_3}{\underset{|}{CH}}—$). Theoretical value for oxypropylene content is 98.4 determined from hydroxyl group determination and corrected for terminal groups.

[c] Assay by periodic acid method for 1,2-glycol came to 91.4%.

[d] Based on theoretical value calculated from synthesis ratios used.

crystals of iodine that may have formed in the condenser with the potassium iodide. Then wash the condenser with two 10-ml portions of water and disconnect from the flask. Rinse the tip and add this washing to the flask. Wash the contents of the flask into an Erlenmeyer flask and titrate with 0.1*N* thiosulfate to the disappearance of the iodine color. Some samples that contain large organic nuclei leave a tarry residue as a button. This residue, which is visible either in the titration flask or in the reaction flask, usually contains a measurable amount of iodine dissolved in it. The button should be dissolved in methanol, and any iodine present should be titrated with thiosulfate. Add this increment to the original titration.

DISCUSSION

Table 5 shows results on various oxyethylene and oxypropylene ethers and esters. In the case of dioxane, the sample boiled below the boiling point of the hydriodic acid. To avoid low results with this sample, it was necessary to circulate ice water through the condenser, and apply a loose-fitting cork to the top of the condenser. Carbon dioxide was allowed to pass into the flask only at 5-minute intervals, to avoid flushing out significant amounts of sample. After 45 minutes of gentle boiling, another 5 ml of hydriodic acid was added through the condenser to rinse any condensed dioxane back into the reaction flask.

The Butyl Carbitol, Phenol Cellosolve, and Methyl Carbitol were purified by distillation. Distillation cuts were used whose carbon and hydrogen analyses checked with the theoretical values. The other samples were used as received.

Although 1,2-dihydroxy compounds operated satisfactorily with this method, compounds with more than two adjacent hydroxyl groups could not be determined. Thus glycerol, 1,2,4-trihydroxybutane, and dextrose gave results to which no stoichiometry could be attached. Dihydroxy compounds in which the hydroxyl groups are not adjacent to one another do not behave like 1,2-dihydroxy compounds. When 1,4-butanediol was used, only 4% of the theoretical amount of iodine was liberated. Evidently the diiodobutane is very stable. Compounds in which the oxyalkylene group is connected to a nitrogen [$R_2N(CH_2CH_2O)_xH$, where R can be hydrogen and x can be 1 or greater] cannot be entirely decomposed to liberate iodine. Ethanolamine and diethanolamine liberated no significant amounts of iodine. For polyglycol amines (reaction products of amines and ethylene oxide), the results corresponded to the total oxyethylene groups on the molecule minus 1. This signifies that the hydriodic acid attacks the ether linkages but not the carbon–nitrogen links. Erratic results were obtained with epoxides.

DETERMINATION OF POLYOXYALKYLENE COMPOUNDS IN COMPLEX SYSTEMS

Since polyoxyalkylene compounds are used as food additives and detergents they are often found in very complex mixtures. Methods with high selectivity are required in these cases. The methods described here have a greater selectivity than the method of Siggia, Starke, Stahl, and Garis presented earlier, but their general accuracy and precision is somewhat lower.

The methods given here have only been used for polyoxyethylene-type compounds; however they may well apply to other polyoxyalkylene compounds (i.e., polypropylene glycol ether).

Oliver and Preston (8) used a precipitation technique employing phosphomolybdic acid. Shaffer and Critchfield (see methods below) describe a gravimetric and colorimetric method for polyethylene glycols using silicotungstic acid and phosphomolybdic acid. Schönfeldt (see method below) describes a volumetric method using potassium ferrocyanide to precipitate the polyglycol from the mixture, the residual ferrocyanide is then titrated.

Phosphomolybdic and Silicotungstic Acid—Methods of C. B. Shaffer and F. H. Critchfield

[*Adapted from Anal. Chem.*, ***19***, *32–4* (*1947*), *Reprinted in Part*]

In the course of a considerable amount of investigation, the only reaction of any practical analytical significance that the polyethylene glycols have been observed to undergo is the formation in acid solution of highly insoluble complexes with the heteropoly inorganic acids, such as phosphomolybdic and silicontungstic, in the presence of a heavy metal cation such as barium. No explanation is offered of the mechanism of this reaction, but it has proved useful as a basis for the quantitative determination of the soluble polyethylene glycols. Two methods have been devised.

In the gravimetric method, the polyglycol is precipitated in hydrochloric acid solution with silicotungstic acid and barium chloride, and the

8. J. Oliver and C. Preston, *Nature*, **172**, 820 (1953).

precipitate is filtered, washed, dried, and ignited at 700°C in a muffle furnace. The residue, consisting of the mixed oxides of barium, silicon, and tungsten, is weighed. The amount of polyglycol originally present in the sample is calculated from the weight of residue by means of an empirical factor determined from known quantities by this method. This procedure is suitable for quantities of polyglycol of the order of 5 to 100 mg, where an ordinary macroanalytical balance is used.

In the colorimetric modification the polyglycol is precipitated from the sample in a small centrifuge tube by the addition of barium chloride and phosphomolybdic acid. The precipitate is isolated and washed by repeated centrifugation, following which it is digested in concentrated sulfuric acid. The digest is diluted, neutralized, and made up to a definite volume, in an aliquot of which molybdenum is determined. The useful range of this method is of the order of 0.05 to 1.0 mg of polyglycol at a minimum concentration of 0.01 mg per milliliter. Phosphomolybdic is substituted for silicotungstic acid in this case because molybdenum can be determined somewhat more satisfactorily than tungsten.

Owing to the variation in useful range of the two methods, the colorimetric modification has been applied principally to whole blood and plasma, whereas the gravimetric technique is used almost exclusively for urine.

REAGENTS

SILICOTUNGSTIC ACID, 10%. Dissolve 10 grams of silicotungstic acid

$$(4H_2O \cdot SiO_2 \cdot 12WO_3 \cdot 22H_2O)$$

in a small quantity of water and neutralize with 10% sodium hydroxide to a methyl red end point. Dilute to 100 ml.

HYDROCHLORIC ACID, 1–4. Dilute 1 volume of concentrated hydrochloric acid to 4 volumes with water.

Barium chloride, 10%.

Phosphomolybdic acid, 10%.

PHENYLHYDRAZINE SULFATE (FOR MOLYBDENUM DETERMINATION). Dilute 3 ml of concentrated sulfuric acid to 60 ml with water and add 3 ml of freshly distilled phenylhydrazine. Shake well to dissolve the precipitate and dilute to 100 ml. Store in a brown glass bottle and refrigerate when not in use.

METHODS

PRETREATMENTS OF SAMPLES. Filtrates of whole blood or plasma that are completely free from traces of protein are suitable for analysis, provided they are not prepared with tungstic or molybdic acids. In the present work, preference has been given to Somogyi's (9) zinc sulfate–barium hydroxide precipitation of plasma proteins. Sulfate obviously will interfere in the gravimetric method, and although theoretically it should not affect the colorimetric method, it is considered desirable to remove it in this case as well. Despite careful balancing of the zinc sulfate and barium hydroxide solutions, it has been the experience of Schaffer and Critchfield that filtrates of plasma prepared in this manner invariably contain excess sulfate. The latter may be removed by precipitation with barium immediately prior to precipitation of the polyglycol.

In the analysis of urine by the gravimetric method, Fiske's ferric acetate treatment has been found satisfactory for the removal of interfering substances, with the exception of sulfates, from average, normal urines. The procedure is carried out as described by Hoffman (10), who states that the basic ferric precipitate removes not only phosphates but also protein, lipoids, and any debris present, along with a large part of the urinary pigment. The filtrate from the ferric acetate precipitation must undergo further treatment for the removal of sulfate. For this purpose, dilute an aliquot of the filtrate to 50 ml, add 4 ml of concentrated hydrochloric acid, and hydrolyze and precipitate sulfates according to Folin's directions (11) for the determination of total sulfate in urine. After removal of the barium sulfate by filtration, the filtrate, including washings, is ready for precipitation of the polyglycol contained therein according to the procedure given below.

The experience of handling a large number of control blood and urine samples has demonstrated that these methods of pretreatment are effective in removing substances ordinarily present in the body fluids that might form insoluble products with silicotungstic or phosphomolbdic acids. Specimens from subjects receiving certain types of medication—for example, cinchona alkaloids—would present individual problems. In some instances, as in the foregoing example, the interference could be removed by ether extraction of the sample. In connection with the question of interfering materials, it seemed desirable to test the reaction of certain metabolites and related substances with silicotungstic acid. It was found that glycine, tyrosine, methionine, cystine, cysteine, choline, creatinine, uric acid, allantoin, phenol, catechol, and hydroquinone in 50-mg amounts form no precipitate with silicotungstic acid under the conditions of this determination.

9. M. Somogyi, *J. Biol. Chem.*, **160,** 69–73 (1945).
10. W. S. Hoffman, *J. Biol. Chem.*, **93,** 787–96 (1931).
11. J. P. Peters and D. D. Van Slyke, *Quantitative Clinical Chemistry*, Vol. II, *Methods*, Williams and Wilkins, Baltimore 1932, pp. 771, 893.

GRAVIMETRIC PROCEDURE. Place the sample, freed of interfering materials, in a 600-ml beaker and add 10 ml of 1:4 hydrochloric acid and 10 ml of 10% barium chloride solution. For ease of subsequent filtration, it is desirable that the quantity of polyglycol present be less than 0.1 gram. In the case of urine treated according to the foregoing scheme, sufficient hydrochloric acid and barium chloride are already present at the conclusion of the procedure, that further quantities of these reagents need not be added. Dilute the contents of the beaker to about 250 to 300 ml and bring to a boil on a burner or hot plate. At this point precipitate the polyglycol with 10 ml of silicotungstic acid added slowly from a pipet. Boil the solution a few seconds to flocculate the precipitate, and then cover the beaker and set it aside for 8 to 12 hours. At the end of this time, filter the precipitate with suction in a tarred Gooch crucible previously ignited at 700°C for 30 minutes and cooled in room air. Wash the precipitate thoroughly with a minimum of 200 ml of distilled water, dry it in an oven at 110°C for several hours, and transfer it to a muffle furnace at 700°C for consecutive 30-minute periods to constant weight. The temperature of ignition is important, since it is the optimal temperature that minimizes incomplete dehydration of the silicon oxide on the one hand and volatilization of tungsten oxide on the other (12).

It is not necessary to boil the solution upon adding the silicotungstic acid, since the precipitate appears to form almost equally well in the cold. The gravimetric factor will be very slightly different in the latter case. In filtering the precipitate, it is important to prevent the crucible from running dry at any time during the process. If the precipitate is sucked dry, it cakes and breaks into fragments, through the crevices of which the wash water channels.

Table 6 lists the results of a series of known amounts of Carbowax compounds 1000 and 6000 added to urine and recovery by the gravimetric method.

COLORIMETRIC PROCEDURE. Place 10 ml of a protein-free filtrate of plasma from which sulfate has also been removed in an ordinary 15-ml graduated centrifuge tube, to which are added, in the order given, 1 ml of 1:4 hydrochloric acid, 1 ml of 10% barium chloride, and 1 ml of 10% phosphomolybdic acid. The solution must be stirred after each addition with a thin glass rod.

When addition of the reagent has been completed, allow the tube to stand for at least 1 hour, during which time a characteristic flocculent greenish precipitate forms by the interaction of the phosphomolybdic acid and any polyglycol present. Then centrifuge the tube at about 2500 rpm

12. J. R. Spies, *Ind. Eng. Chem., Anal. Ed.*, **9,** 46–7.

Table 6. Recoveries by Gravimetric Method of Carbowax Compounds Added to Human Urine

	Carbowax Compound 1000		Carbowax Compound 6000	
Carbowax Compound Added mg.	Average Recovery in Duplicate Determinations %	Graduated Values[a] mg.	Average Recovery in Duplicate Determinations %	Graduated Values[a] mg.
20	100.8	20.1	96.4	19.5
30	99.8	30.2	98.7	29.7
40	101.7	40.3	101.3	39.9
50	101.0	50.4	99.6	50.0
60	100.6	60.4	101.7	60.2
70	99.7	70.5	98.9	70.4
80	100.9	80.6	101.3	80.6

[a] Graduated values obtained by substitution of "Carbowax compound added" in straight-line equations calculated from recovery data by method of least squares.

for 10 minutes, and slowly and carefully draw the supernatant solution off without disturbing the precipitate. For siphoning off this fluid, a capillary device such as the one described for washing calcium precipitates (11) is very useful. Wash the precipitate and tube with two portions of 0.1*N* hydrochloric acid. In each washing allow 3 ml of the acid to flow slowly from a pipet down the sides of the tube, so that the walls are washed about their entire circumference. Then fragment the precipitate with the glass rod and suspend in the acid, rinse down the rod and the walls of the tube with about 7 ml of distilled water, and recentrifuge the tube and repeat the washing process.

After the final centrifugation, draw off the liquid and transfer the precipitate quantitatively to a 100-ml Kjeldahl flask with a minimum amount of water. Next add 3 ml of concentrated sulfuric acid to the contents of the flask, and digest the precipitate with nitric and perchloric acids in the usual fashion. At the completion of this process, cool the sulfuric acid residue, dilute with about 20 ml of water, and neutralize with 40% sodium hydroxide to a phenolphthalein end point. At this point, add 1 or 2 drops of dilute sulfuric acid to bring the mixture just to the acid side of the indicator. Then make up the solution to 100 ml in a volumetric flask.

Molybdenum is determined in the neutralized, diluted digest as follows (13, 14).

Place a 10-ml aliquot followed by 5 ml of the phenylhydrazine solution in an Evelyn colorimeter tube and mix the solutions by swirling. Close the tube with a clean rubber stopper pierced by a fine capillary and place it in a hot-water bath at 81 ±2°C. Allow the tube to remain in the bath exactly 15 minutes, then remove and allow it to cool to room temperature. Read the transmission of the contents in the colorimeter using filter 490, and setting the galvanometer to 100 with a blank carried through all the steps of the procedure. Then read the polyglycol content of the original sample from a standard curve prepared from a series of known quantities carried through the identical procedure. A plot of concentration against extinction is linear within the range of the instrument. Any standard colorimeter or spectrophotometer can be substituted for the Evelyn type.

Table 7 lists the results of a series of known amounts of Carbowax compounds 1000 and 6000 added to plasma and recovery by the colorimetric method.

Table 7. Recoveries by Colorimetric Method of Carbowax Compounds Added to Rabbit Plasma

Carbowax Compound Added mg. %	No. of Determinations	Carbowax Compound 1000			Carbowax Compound 6000		
		Mean Recovery %	Range %	Graduated Values[a] mg. %	Mean Recovery %	Range %	Graduated Values[a] mg. %
10	5	98	92–100	9.5	103	100–108	9.9
20	5	100	100	19.6	100	100	20.1
40	5	98	98	39.8	99	95–102	40.6
60	5	101	100–102	60.0	103	98–105	61.0
80	5	99	96–100	80.2	102	99–104	81.4
100	5	101	100–103	100.4	102	101–104	101.9

[a] Graduated values obtained by substitution of "Carbowax compound added" in straight-line equations calculated from recovery data by method of least squares.

DISCUSSION

Phosphotungstic, phosphomolybdic, or silicotungstic acid will form insoluble complexes with all the Carbowax compounds in acid aqueous

13. G. von Hevesy and R. Hobbie, *Z. Anorg. Allgem. Chem.*, **212,** 134–44 (1933).
14. E. Montignie, *Bull. Soc. Chim. Fr.* (4), **47,** 128 (1930).

solution. In the case of silicotungstic acid, polyglycols 200, 300, and 400 do not react with the formation of insoluble products in dilute acid aqueous solution where hydrogen is the only cation present. In this connection it became of interest to ascertain at what point in the polyglycol series the capacity to form precipitates with barium and silicotungstic acid commences. Therefore, approximately 200-mg amounts of di-, tri-, tetra-, penta-, and hexaethylene glycols were submitted to the procedure above. It was found that pentaethylene glycol is the lowest member of the series to give this precipitate. This finding, however, does not exclude the possibility that some sort of reaction may take place with the first three glycols that does not involve formation of an insoluble product.

REACTION OF COMPOUND 4000 WITH SILICOTUNGSTIC ACID. Examined microscopically, the turbidity produced by silicotungstic acid with Carbowax compound 4000 in pure, dilute hydrochloric acid solution may be seen to consist of finely divided oily droplets, which, in the course of several hours' standing, coalesce to form a colorless, greasy film on the bottom of the container. If, however, sodium or ammonium ions are introduced into the solution, either by the addition of their chlorides or by previous partial neutralization of the silicotungstic acid with the corresponding hydroxides, a white, flocculent, solid precipitate forms which may be filtered. Although no attempt has been made to determine sodium or nitrogen quantitatively in these precipitates, they are presumed to be the sodium and ammonium salts of the polyethylene glycol—silicotungstic acid complex, since they do not form where these cations, or others except hydrogen, are absent. They, as well as the barium derivative, are referred to hereinafter as "salts" for want of better knowledge of their composition.

Samples of the isolated sodium salt melted at 234 to 236°C (Fisher-Johns, uncorrected). Samples of the ammonium salt melted at 179 to 181°C, and analyzed 18.45% carbon, and 62.5% residue after ignition at 700°C. It is assumed that this residue is $SiO_2 \cdot 12WO_3$. No melting point is reported for the barium salt because it was consistently observed to undergo slight charring at 282 to 283°C.

Small quantities of the barium salt of the Carbowax compound 4000–silicotungstic acid complex were prepared for elementary analyses by precipitating the polyglycol under the conditions of the analytical procedure set forth earlier. The precipitates were washed with distilled water by repeated centrifugation and dried in a desiccator at room temperature. Drying was completed in a vacuum pistol at a temperature of 100°C and a pressure of 0.005 mm Hg. The average of replicate analyses showed hydrogen, 2.160%; carbon, 11.685%; and residue, 77.225%. Chlorine was absent. The residue itself, consisting of the mixed oxides of barium, silicon, and tungsten, analyzed an average of 9.265% barium in duplicate determinations, and was carbonate free. This figure approximates a theoretical value of 8.72% barium for a residue of the proportions $2BaO \cdot SiO_2 \cdot 12WO_3$.

Accurate interpretation of the composition of the unignited barium derivative from the carbon and hydrogen values just reported is obviously impossible without a knowledge of the exact mean molecular weight of Carbowax compound 4000. This molecular weight has variously been estimated at 3000 (Menzies Wright) and 3590 (acetyl value). If we accept the former figure as the correct one, we may calculate an almost exact ratio of 1 mole of polyglycol to 4 moles of mixed oxides—that is, 1 glycol to 4 $(2BaO \cdot SiO_2 \cdot 12WO_3)$.

In the present state of our knowledge it is impossible to say with any certainty whether any water of composition or crystallization remains in the compound under the conditions of drying employed. However it would seem from the slight discrepancy between the residue value obtained in the elementary analyses (77.225%) and that which may be calculated from the gravimetric factor (78.09%) that some water was present in the samples submitted for analysis. This inference is further borne out by the fact that the calculated value for percentage of polyglycol in the compound, based on percentage of hydrogen found, is higher than that based on either percentage of carbon or percentage of residue. If water of composition or crystallization were present, the result based on the hydrogen value would be expected to be high.

DETERMINATION OF GRAVIMETRIC FACTORS FOR DIFFERENT CARBOWAX COMPOUNDS. Samples of all four Carbowax compounds, 1000, 1540, 4000, and 6000, were dehydrated in benzene and stored in a desiccator over phosphorus pentoxide. From these samples standard solutions of 2 grams per liter in distilled water were prepared and aliquots of each were analyzed in quadruplicate by the gravimetric procedure. From the results of the analyses, a gravimetric factor was calculated that defined the amount of each Carbowax compound represented per amount of residue weighed: Carbowax compound 6000, 0.2712; Carbowax compound 4000, 0.2808; Carbowax compound 1540, 0.2487; Carbowax compound 1000, 0.2544.

VARIATION OF GRAVIMETRIC FACTOR AMONG DIFFERENT BATCHES OF CARBOWAX COMPOUND 4000. The gravimetric factors listed obviously can apply strictly only to the particular samples of polyethylene glycols under observation. Since the Carbowax compounds used industrially are blends of production batches, the proportions of the components of which are varied to maintain constant physical properties, it seemed necessary to indicate the probable limits within which the factor for a given compound may vary, in view of the possible use of this method in estimating these compounds in miscellaneous commerical preparations.

Accordingly, samples were obtained of 10 consecutive production batches of Carbowax compound 4000. Eight of these specimens were in the form of fine flakes; the other two resembled hard waxes. Portions of each were crushed and desiccated *in vacuo* overnight over phosphorus pentoxide, and a stock solution containing approximately 2 grams per liter was prepared for each sample. Aliquots of each stock solution were then submitted to the analytical procedure, and the ratio of ash to Carbowax compound was calculated. On ignition, these 10 examples produced residues ranging from 3.56 to 3.78 (mean 3.661 ±0.004) milligrams per milligram of desiccated polyglycol.

Ferrocyanide Method of N. Schönfeldt

[*Adapted from J. Am. Oil Chemists Soc.*, **32,** *77–9 (1955), Reprinted in Part*]

The starting point of this investigation was the observation of von Baeyer and Villiger (15) that ferrocyanic acid $H_4[Fe(CN)_6]$ gives addition products with diethyl ether. Several modifications led to the following method.

REAGENTS

0.25*M* potassium ferrocyanide, reagent grade, containing 0.5 gram of anhydrous sodium carbonate per liter.

Ammonium sulfate-solution containing 400 grams of recrystallized ammonium sulfate per liter.

Sodium chloride, reagent grade.

Hydrochloric acid, reagent grade, sp. gr. 1.18.

1% diphenylamine (1 gram + 99 grams sulfuric acid, sp. gr. 1.84).

2% potassium ferricyanide (2 grams + 98 ml distilled water).

0.075*M* zinc sulfate, reagent grade.

For washing of the precipitate, the following solution is used: 840 ml of distilled water, 240 grams of sodium chloride, and 80 ml of hydrochloric acid, sp. gr. 1.18.

PROCEDURE

Place a 100-ml solution containing essentially not more than 0.3 gram of the ethylene oxide adduct in a 300-ml Erlenmeyer flask, and add 10 ml

15. A. von Baeyer and V. B. Villiger, *Ber.*, **34,** 2679 (1901).

of hydrochloric acid (sp. gr. 1.18) and 15 grams of sodium chloride. Shake the mixture until all the salt is dissolved. Then add 5.0 ml of potassium ferrocyanide. Shake the Erlenmeyer flask again, and, after letting it stand for a few minutes, filter the precipitate and wash it with 25 ml of washing solution.

After washing, add 5 ml of ammonium sulfate solution, 5 drops of 2% potassium ferricyanide, and 5 drops of 1% diphenylamine to the filtrate, and titrate it without delay with 0.075*M* zinc sulfate (15). The solution becomes greenish, and at the end point it changes to blue-violet.

The zinc sulfate solution should be standardized against 100 ml of blank solution. An empirical factor f is calculated for each ethylene oxide adduct by carrying a series of known quantities of the ethylene oxide adduct in question through the procedure. A standard curve is thereby constructed. The empirical factor f can be calculated from the equation of $f = x/c - b$ where x is the amount of ethylene oxide adduct, c the initially added amount of ferrocyanic acid (determined by titration with zinc sulfate solution), and b the amount of ferrocyanic acid left in the filtrate and estimated as above. Thus $c - b$ is the amount required for the precipitation of the ethylene oxide adduct. By knowing f, the ethylene oxide compound content of a sample can be calculated from the standard curve.

NOTES

1. The solutions of potassium ferrocyanide, diphenylamine, and potassium ferricyanide should be kept in dark bottles.
2. The potassium ferrocyanide solution is discarded when more than 4 days old.
3. Some adducts may be difficult to determine because of their poor dispersibility. In some instances this can be counteracted by adding the potassium ferrocyanide solution immediately after the addition of hydrochloric acid and before the sodium chloride.
4. Titrate first with moderate haste and dropwise near the end point.
5. When washing a highly dispersible precipitate, the washing solution should be added cautiously in small portions.
6. It is easier to observe the end point of the titration with daylight or a daylight lamp.
7. Do not allow the samples to stand for a long time before titration.
8. Generally the filtration can be carried out with filter paper No. 3. However in some cases No. 00 may be preferable for obtaining a clear filtrate.

Table 8 illustrates the results obtained with the method. The first column shows the amount of ethylene oxide adducts. The other columns

Table 8. Amount of Ferrocyanic Acid (grams) Required for Precipitation

Adduct gram	un. ethyl. ox. per mole *p*-octyl phen.					un. ethyl. ox. per mole oleyl alcohol			un. ethyl. ox. per mole oleyl amine		
	6.3	9.2	12.4	15.6	17.8	6.2	9.4	12.4	6.5	20.1	30.3
0.30	0.097	0.124	0.157	0.164	0.177	0.086	0.132	0.149	0.115	0.171	0.192
	0.096	0.124	0.156	0.163	0.176	0.085	0.131	0.149	0.111	0.169	0.191
0.24	0.082	0.102	0.128	0.135	0.144	0.079	0.108	0.120	0.101	0.142	0.156
	0.080	0.102	0.126	0.136	0.144	0.081	0.107	0.120	0.102	0.143	0.157
0.12	0.039	0.054	0.068	0.070	0.073	0.039	0.055	0.059	0.054	0.074	0.077
	0.040	0.054	0.067	0.069	0.072	0.040	0.054	0.059	0.050	0.072	0.075
0.06	0.019	0.027	0.036	0.037	0.038	0.018	0.026	0.031	0.022	0.037	0.041
	0.020	0.029	0.034	0.037	0.035	0.019	0.026	0.031	0.024	0.037	0.043
0.03	0.012	0.015	0.016	0.017	0.017	0.008	0.013	0.016	0.014	0.019	0.021
	0.011	0.015	0.018	0.017	0.017	0.008	0.013	0.016	0.011	0.018	0.022
1.00 m (calc.)	0.304	0.451	0.552	0.573	0.594	0.306	0.441	0.507	0.407	0.599	0.666
m.d.	7.9%	3.3%	4.7%	2.6%	2.4%	7.9%	2.3%	2.5%	7.6%	2.5%	4.5%

contain the amounts of ferrocyanic acid necessary for the precipitation of adducts with different chain length. These amounts have been calculated in the following way. Assume that the filtrate after precipitation and washing of the ethylene oxide adduct consumes m ml of $0.075M$ zinc sulfate. Furthermore, let us assume that the solution containing 5.0 ml of $0.25M$ ferrocyanic acid and the other additions on standardizing requires 23.0 ml of $0.075M$ zinc sulfate. The theoretical value for the above mentioned amount of ferrocyanic acid lies at 25.0 ml of $0.075M$ zinc sulfate. This gives a correction factor at 25.0/23.0. On multiplying m with this factor and subtracting from 25.0, the number of milliliters of zinc sulfate is obtained. This number multiplied by 0.0108 gives the corresponding amount of ferrocyanic acid in grams. It is advisable to standardize the zinc sulfate solution daily against the blank.

The adducts in Table 8 were prepared by condensing various amounts of ethylene oxide with commerical grade *p*-octyl phenol, oleyl alcohol, and oleyl amine. On examining this table, it can be noted that the reproducibility was good. The accuracy in the range 0.03 to 0.3 gram per 100 ml is sufficient; at concentrations below 0.01 gram per 100 ml it is less good. From the data in Table 8 the mean (m) and the mean deviation (m.d.) for 1.00 gram of adduct have been calculated.

From Table 8 the mole ratio of adduct to ferrocyanic acid was calculated for *p*-octyl phenol + 6, 9, 12, 15, and 18 units ethylene oxide as 1 to 0.8, 1.3, 1.9, 2.3, and 2.8, respectively; for oleyl alcohol + 6, 9, and 12 units as 1 to 0.8, 1.3, and 1.8, respectively; and for oleyl amine + 6, 9, and 12 units as 1 to 0.9, 1.4, and 1.9 respectively. This ratio is the same at a constant number of ethylene oxide units.

On knowing the molecular weight of the hydrophobic part (M) and the number of ethylene oxide units in the adduct (c), it is possible to calculate from Table 8 the amount of ferrocyanic acid (a) required for precipitation of one gram mole of ethylene oxide:

$$a = \frac{b}{c(M+44c)}$$

where b signifies the amount in grams of ferrocyanic acid required for the precipitation of 1 gram of adduct.

The following values in grams were obtained: for p-octyl phenol + 6.3, 9.2, 12.4, 15.6, and 17.8 units, 26.4, 30.4, 33.4, 33.4, and 33.4, respectively; for oleyl alcohol + 6.2, 9.4, and 12.4 units, 26.8, 32.2, and 33.2, respectively; and for oleyl amine + 6.5, 20.1, and 30.3 units, 34.4, 34.4, and 35.2, respectively. Since the molecular weight for ferrocyanic acid is 215.96, the mean of the values approaches one sixth of it.

With two exceptions (some are the lowest numbers in their series), the values for a show comparatively good agreement.

CONCLUSIONS

The reaction between the ethylene oxide adducts and the ferrocyanic acid takes place on the ethylene oxide chain and is independent of the hydrophobic portion of the compound. The mole ratios as well as the a values show for three different substances, with different hydrophobic parts, comparatively good agreement, making plausible an analogy to von Baeyer and Villiger's observation on diethyl ether that the reaction takes place between the oxygen atoms of the ethylene oxide chain and ferrocyanic acid.

The a values make it furthermore plausible that one mole of ferrocyanic acid is required for the precipitation of an adduct containing six units of ethylene oxide. This indicates that complete precipitation of a longer chain is easier than of a shorter one. This probably leads to an explanation as to why in Table 8 just the lowest numbers in the series show the greatest m.d. and why the above-mentioned two deviations for the a values were found at the lowest numbers.

DETERMINATION OF TRACES OF POLYOXYALKYLENE COMPOUNDS

These methods must not only be selective (since the matrix of the sample usually is a complex system such as a biological fluid, a cleaning bath, or oil well drilling mud), but they must also be sensitive in very low concentration ranges. One approach is colorimetric, based on complexation of the cobaltothiocyanate ion by polyoxyalkylene compounds. The other method is nephelometric, based on the precipitation of polyoxyalkylene compounds with potassium mercuric iodide reagent. The authors of this book have found the nephelometric method to be the more widely applicable for trace quantities. The nephelometric method is used extensively to monitor cleaning formulations and also oil well drilling muds.

Nephelometric Method of B. T. Kho and H. J. Stolten

[*Published by General Aniline and Film, 435 Hudson St., New York, N.Y. (0014)*]

This method permits the determination of the concentration of water-soluble nonionic surfactants of known chemical structure. Although the sensitivity of this test varies with the nonionic material being tested, for most nonionics the lower limit of determination is 1 to 10 ppm.

The procedure presented here calls for visual comparison of an unknown with standards. However the method should be adapted to instrumentation wherever possible. In fact, instrumental comparison of turbidities is recommended in the analysis of samples containing relatively high concentrations of nonionic.

SCOPE

In general, this analysis is applicable to all water-soluble ethylene oxide-containing nonionic surfactants and to polyglycols. Such products include polyethylene glycols and the ethylene oxide adducts listed below.

Water-Soluble Ethylene Oxide Adducts Derived from
Alkylphenols
Aliphatic alcohols
Aliphatic acids
Fatty amines
Alkyl mercaptans
Polyalkylene glycols
Fatty amides

This analysis is suitable for quality control of formulated products and for accurately controlling the concentration of nonionic surfactants in processing operations. One of the chief assets of this method is its specificity for nonionics. The only substances known to form precipitates or turbidity with Nonionic Reagent, thus interfere with this method, are alkaloids, germicidal quaternaries, and certain proteinaceous materials.

BASIS OF THE METHOD

Nonionic Reagent (K_2HgI_4) forms a white turbidity with nonionic surfactants in very dilute solutions. The simplified equation for this reaction may be written as follows:

$$K_2HgI_4 + \text{nonionic} \rightarrow K_2HgI_4 \cdot \text{nonionic}$$

Turbidity formation is essentially completed in 15 minutes; however several hours may be required for development of maximum turbidity as measured by precise instrumental methods. The turbidity remains fine and well dispersed for some time.

Turbidity increases with increasing concentration of nonionic as is shown in Fig. 4.6, which plots concentration of Igepal CO-630* against the turbidity formed by the reaction Igepal CO-630 + Nonionic Reagent.

The procedure presented here was designed for determination of a water-soluble nonylphenol ethylene oxide adduct. It should be modified to suit the requirements of each type of analysis. See section entitled "Suggestions for Using the Method."

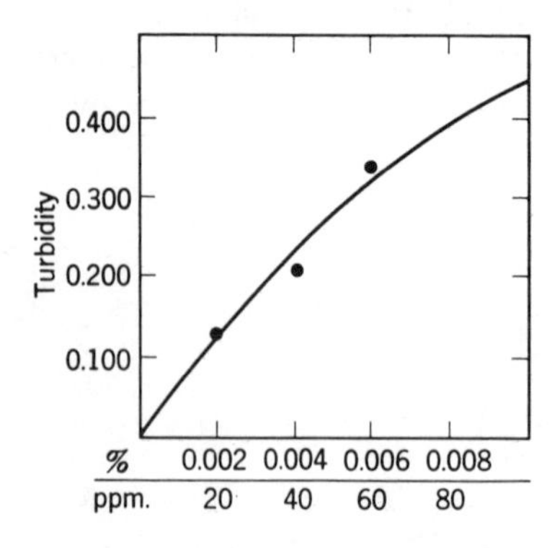

Fig. 4.6. Concentration of Igepal CO-630. Optical density measured on Beckman DU spectrophotometer at 460 nm, 15 minutes after start of reaction.

* Igepal CO-630 is nonylphenol to which has been added 9 moles of ethylene oxide per mole of phenol. The name Igepal is a registered trademark of General Aniline and Film Corp.

APPARATUS

Four 25-ml graduate cylinders.
Two 1-ml volumetric pipets.
One 2-ml volumetric pipet.
One 3-ml volumetric pipet.
Four 32-ml flat-bottom, clear glass vials.

REAGENT

NONIONIC REAGENT.* Contains 39.25 grams per liter of potassium iodomercurate (K_2HgI_4 or $2KI \cdot HgI_2$); it is prepared by dissolving 13.55 grams of mercuric chloride ($HgCl_2$) and 50.00 grams of potassium iodide (KI) in 1 liter of water.

STANDARD

An aqueous solution containing 0.05% of the nonionic surfactant under test and similar in all other respects to the sample to be analyzed.

PROCEDURE

Quantitatively dilute or dissolve the sample, to obtain an aqueous test solution in which the concentration of active nonionic falls in the range 0.05 to 0.15%. Add 5 ml of Nonionic Reagent to each of four 25-ml graduates (I, II, III, and IV, respectively).† Add with a volumetric pipet 1, 2, and 3 ml of standard to graduates I, II, and III, respectively. Add with a volumetric pipet 1 ml of test solution to graduate IV. With clean water dilute the reactants in each of the four graduates to 25 ml; subsequently invert each graduate once or twice to mix the contents. Allow mixtures to stand for at least 10 minutes. Pour the contents of each graduate into a clear 32-ml vial. Observe the turbidities against a dark background by looking down at the bottom of the vial through the surface of the liquid.

Compare turbidity IV with turbidities I, II, and III and determine which one it most nearly matches. This establishes the number of milliliters of standard that is equivalent to 1 ml of test solution. When turbidity IV lies between two of the others, it is possible to estimate to 0.5 ml of standard.

* Mayer's Reagent.

† Because of its toxicity, Nonionic Reagent should *not* be pipetted by mouth.

CALCULATION

% Nonionic in sample = $T \times F \times (0.05)$

where T = milliliters of standard that produced turbidity most nearly like that of test solution

F = dilution factor relating test solution to sample

SUGGESTIONS FOR USING THE METHOD

AGES OF TURBIDITIES SHOULD BE ESSENTIALLY THE SAME. It is necessary to minimize differences in the ages of the turbidities to be compared. Therefore, it is important to minimize delay between each addition of nonionic (standard and test solution) to reagent; the order of the steps in this procedure was established with this in view.

The turbidity formed by the reaction of Nonionic Reagent with nonionic surfactants increases with time, as shown in Fig. 4.7, which plots time against the turbidity formed by the reaction of Igepal CO-630 with Nonionic Reagent. The time for development of maximum turbidity varies with the particular nonionic being determined, concentration of nonionic, temperature, and other materials present. In most cases, however, turbidity adequate for visual analysis is formed after 10 minutes; it is merely necessary that the ages of the turbidities of standard and test solution be essentially the same.

When comparison of turbidities is made by instrument, age of turbidity is a variable that must be specified in the standard calibration curve.

NONIONIC REAGENT IS TOXIC AND SHOULD NOT BE PIPETTED BY MOUTH. It is suggested that Nonionic Reagent be measured by pouring it into the graduates to the desired (5-ml) mark.

THE STANDARD WILL HAVE TO BE ADJUSTED TO SUIT THE REQUIREMENTS OF EACH ANALYSIS. Where greater sensitivity in visual comparison of turbidities is required, the standard turbidities should be prepared to cover a more narrow range of concentrations of nonionic.

There is some evidence that in the determination of certain nonionics, greater precision can be obtained by using three standards rather than

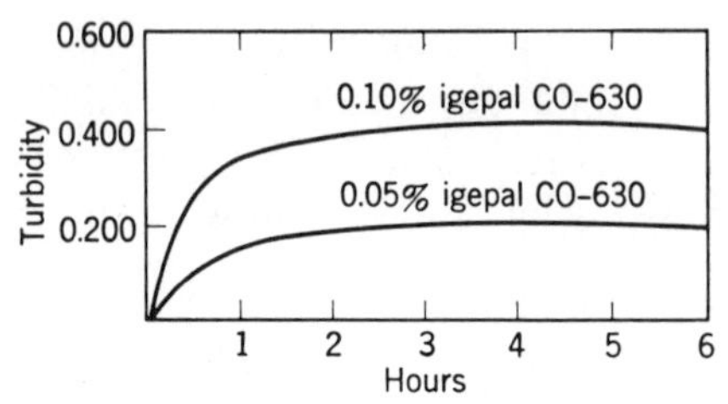

Fig. 4.7. Optical density measured on Beckman Du spectrophotometer at 460 nm.

three aliquots of one standard. Using three standards permits the initial formation of standard and test solution turbidities to take place in equal volumes of solution.

The standard in the foregoing procedure was chosen arbitrarily from the range of concentrations suitable for this test when applied to nonylphenol ethylene oxide adducts. Depending on the system in which the nonionic is used, however, a standard may be made equivalent to some convenient measure of nonionic in sample, such as pounds per gallon or pounds per barrel.

TURBIDITY OBTAINED MAY BE TOO DENSE OR TOO SPARSE. These directions are suitable for the determination of many nonionics. When the formation of turbidity by a nonionic is relatively great or small, however, the concentrations of standard and (conversely) the degree of dilution of sample must be adjusted upward or downward.

When the concentration of nonionic is very small, the amount of sample reacted with reagent can be increased (to 20 ml if necessary); the calculation should be adjusted accordingly.

COLORED SAMPLES MAY BE ANALYZED. The extent of dilution of sample is, in most cases, sufficiently great to nullify the interference with the test by color. If the presence of color in the sample prevents accurate comparison of turbidities, however the following steps can be taken:

1. Prepare standard and test solution controls (CS and CTS, respectively; see Fig. 4.8); that is, pipet 1 ml, of standard and 1 ml of test solution into 25-ml graduates, respectively, and (refraining from adding reagent) dilute the contents of each to 25 ml.

2. Pour contents of each graduate into a separate comparator tube. View test solution turbidity (TST) with standard control held beneath it; view standard turbidities (ST) with test solution control held beneath them.

When color interferes with instrumental comparison of turbiditics, use

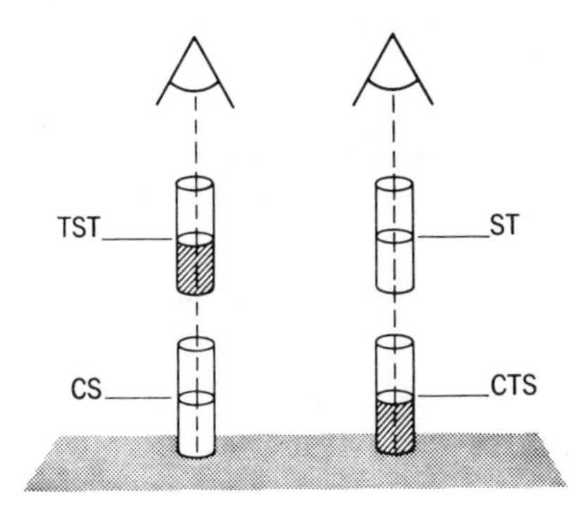

Fig. 4.8

controls as blanks or, preferably, choose a wavelength at which the color in question does not interfere.

THE NEPHELOMETER IS RECOMMENDED FOR INSTRUMENTAL MEASUREMENT OF TURBIDITIES. This instrument measures scattered light that is directly proportional to turbidity. However photometers and colorimeters of the filter or dispersion types can also be used.

The use of an instrument permits the establishment of a standard calibration curve relating turbidity to concentration of nonionic. Subsequent measurements need be made on test solution alone; the concentration of nonionic test solution is simply read from the standard curve. Moreover, instrumental comparison of turbidities is recommended for analysis of samples that contain a high content of nonionic (e.g., formulated products). Since such samples must be greatly diluted to bring the concentration of nonionic into the range of measurable turbidities, a large dilution factor must be entered into the calculation. The net result is a significant magnification of the absolute error (in measurement of turbidity). Instrumental measurement of turbidities greatly diminishes the absolute error.

NONIONIC SURFACTANTS MAY BE ANALYZED IN THE PRESENCE OF ANIONICS. The presence of anionic surfactants tends to diminish the turbidity produced by nonionics. Analysis by this method is still possible, however, provided the standard contains the same ratio of anionic to nonionic as does the sample.

EFFECTS OF ENVIRONMENT. *Temperature.* Turbidity decreases with increasing temperature. However this effect is negligible at room temperature.

pH. Analysis of nonionics with Nonionic Reagent may be conducted over a wide pH range; however the pH values of standard and test solution should be approximately the same.

Concentration of Electrolytes. Electrolytes (including polyphosphate builders) and water hardness do not prevent the use of this method. But it is advisable to make up standards having the same type and concentration of electrolytes known to be in the sample.

Colorimetric Cobaltohiocyanate Method—Adapted from D. J. Morgan

[*Reprinted in Part from Analyst,* ***87,*** *233–4 (1962)*]

Gnamm (16) and van der Hoeve (17) showed that poly(ethylene oxide) condensates also form blue precipitates with ammonium cobaltothio-

16. H. Gnamm, *Die Lösungs und Weichmachungsmittel,* 6th ed., Wissenschaftlische Verlagsgesellschaft m.b.H., Stuggart, 1950, p. 336.
17. J. A. Van der Hoeve, *Rec. Trav. Chim. Pays-Bas,* **67,** 649 (1948).

cyanate. Brown and Hayes (18) developed a quantitative method based on this reaction and involving use of a modified version (200 grams of ammonium thiocyanate *plus* 30 grams of hydrated cobalt nitrate) of van der Hoeve's reagent. They extracted the complex into chloroform and measured the optical density of the solution at 620 or 318 nm, use of the latter wavelength giving increased sensitivity.

It was found that the complexes formed by several poly(ethylene oxide) surfactants were insoluble in chloroform, but could readily be extracted into benzene. After evaporation of the benzene, the complex was decomposed with water, and the cobalt in the aqueous solution was determined as its complex with nitroso-R salt, optical-density measurements being made at 500 nm (19). The sensitivity achieved by this procedure was about as high as that attained by making direct measurements of the blue complex at 318 nm; thus the proposed method should be valuable to laboratories equipped with a visible spectrophotometer. Furthermore, if Brown and Haye's modified reagent (18) is used and cobalt is determined in this way, it should be possible to increase the sensitivity still further.

REAGENTS

AMMONIUM COBALT THIOCYANATE SOLUTION. Dissolve 174 grams of ammonium thiocyanate and 2.8 grams of cobalt nitrate hexahydrate in water, and dilute to 1 liter.

HYDROCHLORIC ACID–NITRIC ACID MIXTURE. Dilute a mixture of 25 ml of concentrated hydrochloric acid and 5 ml of concentrated nitric acid to 100 ml with water.

Nitroso-R salt solution, 0.05% w/v, aqueous.

Sodium acetate solution, 50% w/v, aqueous.

PROCEDURE

Measure approximately 5 ml of the thiocyanate solution into a stoppered 25-ml cylinder. If the surfactant is soluble in benzene, add an aliquot of benzene solution containing about 0.5 mg; if the surfactant is water soluble, add not more than 0.5 ml of its aqueous solution. Dilute to about 10 ml with benzene, shake the cylinder vigorously for 2 minutes, and allow the two layers to separate completely. Blow the benzene layer into a 100-ml beaker, and add a further 5 to 10 ml of benzene to the aqueous layer. Invert the cylinder a few times, allow the layers to separate, and again blow the benzene layer into the beaker.

18. E. G. Brown and T. J. Hayes, *Analyst*, **80**, 755 (1955).

19. E. B. Sandell, *Colorimetric Determination of Traces of Metals*, 3rd ed., Wiley-Interscience, New York and London, 1959, p. 419.

Table 9. Sensitivity and Reproducibility of Results

Optical-density measurements were made in 4-cm cells at 500 mμ; the final volume of solution was 10 ml

Surfactant	Amount of Surfactant Present, mg.	Optical Density Found	Optical Density Per Unit Concentration
A	1	0.278, 0.280, 0.272 (mean 0.277)	2.77
	2	0.535, 0.540, 0.536 (mean 0.536)	2.68
	3	0.805, 0.825, 0.800 (mean 0.810)	2.70
	4	1.13, 1.10, 1.10 (mean 1.11)	2.78
	5	1.46, 1.39, 1.44 (mean 1.43)	2.81
B	0.1	0.158, 0.147, 0.159 (mean 0.155)	15.5
	0.2	0.314, 0.320, 0.322 (mean 0.318)	15.8
	0.3	0.462, 0.470, 0.460 (mean 0.464)	15.5
	0.4	0.642, 0.634, 0.630 (mean 0.635)	15.9
	0.5	0.795, 0.800, 0.804 (mean 0.800)	16.0

Cover the beaker with a watch glass, cautiously evaporate its contents to dryness, and add 5 ml of water and 0.5 ml of the acid mixture. Add exactly 1 ml of nitroso-R salt solution and then 2 ml of sodium acetate solution, cover the beaker with the watch glass, and boil for 1 minute. Add 1 ml of concentrated nitric acid, replace the watch glass, and boil for another minute. Allow the beaker to cool in the dark, transfer its contents quantitatively to a 10-ml calibrated flask, and dilute to the mark with water. Measure the optical density of this solution at 500 nm against a reagent blank solution prepared from 5 ml of water, 0.5 ml of the acid mixture, and so on. Calculate the concentration of nonionic surfactant present from a calibration graph plotted from results for the particular surfactant being determined.

RESULTS

Table 9 shows that, for two typical poly(ethylene oxide) surfactants, the optical density per unit concentration of surfactant was independent of concentration, but that the sensitivity of the reaction (i.e. the weight of cobalt complexed by unit weight of detergent) was about 6 times greater for one than for the other. For this reason, it is necessary to identify the particular nonionic surfactant before plotting the calibration graph.

Measurements of the reagent blank solution against water indicated that the blank value was about 16 times greater at the wavelength of maximum absorption of the cobalt complex, namely, 425 nm, than it was at 500 nm (0.550 vs. 0.035 unit of optical density). For this reason, measurements were made at 500 nm, even although the extinction coefficient of the cobalt complex at this wavelength is only about two-thirds of its value at 425 nm.

5

Epoxide Groups (Oxirane Oxygen)

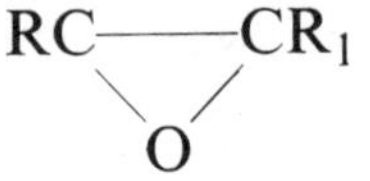

$$\underset{\diagdown\ \ O\ \ \diagup}{RC\text{———}CR_1}$$

Hydrochlorination Methods

The principle used for the bulk of the determinations of epoxide groups is hydrochlorination forming the

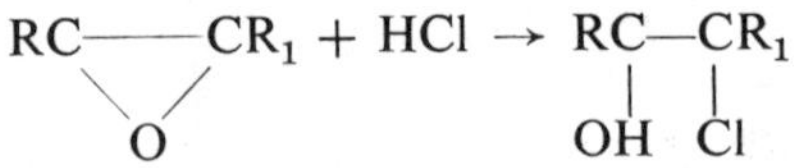

$$\underset{\diagdown\ O\ \diagup}{RC\text{———}CR_1} + HCl \rightarrow \underset{OH\ \ Cl}{RC\text{—}CR_1}$$

corresponding chlorohydrin. The reaction can be run in various solvents—water, alcohol, Cellosolve, ethyl ether, dioxane, pyridine, and pyridine-chloroform. Jungnickel, Peters, Polgar, and Weiss in the study shown made an evaluation of the various systems. The references to the original methods are contained in Table 1.

Study of J. L. Jungnickel, E. D. Peters, A. Polgar, and F. T. Weiss

[*Organic Analysis, Vol. 1, Wiley-Interscience, New York, pp. 133–49**]

AQUEOUS MAGNESIUM CHLORIDE HYDROCHLORINATION

The hydrochlorination reagent is a saturated magnesium chloride solution, which is $0.1N$ in hydrochloric acid. It is prepared by shaking 1000 grams of CP magnesium chloride hexahydrate with 300 ml of distilled water and adding 8.0 ml of concentrated hydrochloric acid. The mixture is shaken at room temperature until saturated and allowed to settle for at least 2 hours. At the end of this time, the supernatant liquid is decanted through glass wool and stored in a glass-stoppered bottle. Then 50 ml of the aqueous hydrochlorination reagent is pipetted into a 250-ml glass-stoppered flask. Because of the viscosity of the reagent, a consistent drainage period should be maintained. A weighed amount of

* Reprinted in part with permission. Abridged by the authors of this text.

Table 1. Summary of Essential Features of α-Epoxy Methods Involving Hydrochlorination

Method	Aqueous Magnesium Chloride-Hydrochlorination	Alcoholic Magnesium Chloride-Hydrochlorination	Hydrochloric Acid-Cellosolve	Hydrochloric Acid-Ethyl Ether	Hydrochloric Acid-Dioxane	Pyridinium Chloride-Pyridine	Pyridinium Chloride-Chloroform
References	1, 2, 3	3	Du Pont[a]	4	3, 5	3, 6, 7	3
Reagents	Saturated aqueous magnesium chloride containing 0.1*N* HCl	0.5*N* HCl-magnesium chloride in ethanol	0.2*N* HCl in Cellosolve	0.2*N* dry HCl in ethyl ether	0.2*N* HCl in dioxane	0.2 and 1*N* HCl in pyridine	0.2 and 1*N* dry HCl in pyridine and chloroform
Apparatus required	Capped bottle	Glass-stoppered flask	Capped bottle and oven	Glass-stoppered flask	Glass-stoppered flask	Flask and reflux apparatus	Flask and reflux apparatus
Reaction temperature	Room temperature	Room temperature	65°C	Room temperature	Room temperature	Reflux, *ca.* 115°C	Reflux, *ca.* 60°C
Reaction time	15–30 min.	30 min.	4 hrs.	3 hrs.	15 min.	20 min.	30 min. (2 hrs. for resins)
Titrant	Aqueous 0.1*N* NaOH	Aqueous 0.5*N* NaOH	Aqueous 0.1*N* NaOH	Aqueous 0.1*N* NaOH	Methanolic 0.1*N* NaOH	Aqueous 0.5*N* or methanolic 0.1*N* NaOH	Methanolic 0.1 or 0.5*N* NaOH
Indicator	Methyl orange	Bromocresol green	Bromothymol blue	Phenolphthalein	Cresol red	Phenolphthalein	Phenolphthalein

[a] Private communication.

1. O. F. Lubatti, *J. Soc. Chem. Ind.*, **51,** 361T (1932).
2. O. F. Lubatti, *Ibid.*, **54,** 424T (1935).
3. Unpublished data from the laboratories of Jungnickel, Polgar, Peters, and Weiss.
4. D. Swern, T. W. Findley, G. N. Billen, and J. T. Scanlan, *Anal. Chem.*, **19,** 414 (1947).
5. G. King, *Nature*, **164,** 706 (1949).
6. T. F. Bradley (to Shell Development Company), U. S. Patent 2,500,600 (1950).
7. S. O. Greenlee (to Devoe and Raynolds Company, Inc.), U.S. Patent 2,502,145 (1950).

sample containing from 0.001 to 0.002 equivalent of α-epoxide is added, and the flask is shaken and allowed to stand 15 to 30 minutes. The stopper and neck of the flask are rinsed with not more than 20 ml of distilled water. Several drops of a 0.1% methyl orange indicator solution are added, and the mixture is titrated with standard $0.1N$ aqueous sodium hydroxide solution.

ALCOHOLIC MAGNESIUM CHLORIDE HYDROCHLORINATION

The hydrochlorination reagent is a $0.5N$ hydrochloric acid solution in ethanol that is saturated with magnesium chloride. The reagent is prepared by mixing 45 ml of concentrated hydrochloric acid with 1 liter of 95% ethanol; 50 ml of this solution is pipetted into a 500-ml glass-stoppered flask containing 40 grams of magnesium chloride hexahydrate, and the mixture is shaken to obtain a solution saturated with salt. A sample containing from 0.010 to 0.015 equivalent of α-epoxide is weighed into the flask, which is stoppered, swirled to mix the contents, and allowed to stand at room temperature for 30 minutes. After the addition of 100 ml of distilled water and a few drops of bromocresol green indicator solution, the contents of the flask are titrated with standard $0.5N$ aqueous sodium hydroxide solution to the blue-green end point.

HYDROCHLORIC ACID–CELLOSOLVE

The hydrochlorination reagent $0.2N$ hydrochloric acid in Cellosolve, is prepared by adding 1.6 ml of concentrated hydrochloric acid per 100 ml of commerical Cellosolve (ethylene glycol monoethyl ether). A sample containing 0.001 to 0.002 equivalent of α-epoxide is dissolved in 25 ml of Cellosolve in a pressure bottle. Next 25 ml of the hydrochlorination reagent is pipetted into the bottle, which is then stoppered and heated (with occasional swirling) in an oven for 4 hours at 65°C. After cooling the bottle to room temperature, 50 ml of methylisobutyl ketone and 100 ml of distilled water are added, and the bottle is restoppered and shaken vigorously. The contents are then titrated with standard $0.1N$ aqueous sodium hydroxide solution to the bromothymol blue end point.

HYDROCHLORIC ACID–ETHYL ETHER

The hydrochlorination reagent is prepared by passing anhydrous hydrogen chloride into absolute ethyl ether until an approximately $0.2N$

solution is obtained. A sample containing not more than 0.002 equivalent of α-epoxide is weighed into a glass-stoppered flask. The sides of the flask are washed down with 5 ml of absolute ethyl ether and 25 ml of the hydrochlorination reagent is pipetted into the flask, which is stoppered, mixed, and allowed to stand for 3 hours at room temperature. A few drops of phenolphthalein indicator solution and 50 ml of 95% ethanol are added, and the contents of the flask titrated with standard 0.1*N* aqueous sodium hydroxide solution.

HYDROCHLORIC ACID–DIOXANE

The hydrochlorination reagent is a solution of hydrochloric acid in dioxane, having a strength of approximately 0.2*N*. It is prepared by pipetting exactly 1.6 ml of concentrated hydrochloric acid into 100 ml of purified dioxane contained in a dark bottle equipped with a screw cap lined with Teflon. The reagent is mixed thoroughly and inspected for homogeneity, since an excessive water content in the dioxane makes solution incomplete. The hydrochlorination reagent is prepared only for immediate use because it deteriorates on standing.

The dioxane, used for preparation of the hydrochlorination reagent, is freed of active impurities by refluxing the technical material with 3% of its weight of potassium hydroxide pellets for 3 hours while a slow stream of nitrogen is bubbled through the liquid. At the end of this time, the mixture is flash-distilled without a fractionating column. The portion of the distillate boiling below 98°C is discarded and the remainder, up to 75% of the charge, is collected in a can or dark bottle under nitrogen. Addition of an inhibitor, such as Ionol (a trisubstituted phenol available from Shell Chemical Corp.) in concentration of 0.1%, helps stabilize the solvent. The nitrogen atmosphere is replenished each time the container is opened. Dioxane prepared and stored in this manner usable for at least 3 months.

Neutralized ethanol, used in the titration mixture, is prepared by adding 1 ml of cresol red indicator solution (0.1 gram of sodium salt in 100 ml of 50% ethanol) to 100 ml of denatured, anhydrous 3A alcohol and neutralizing to the first violet color of the end point with 0.1*N* methanolic sodium hydroxide solution.

A weighed quantity of sample containing from 0.002 to 0.004 equivalent of α-epoxide is added to a 250-ml flask into which 25 ml of hydrochlorination reagent has been pipetted. The flask is swirled to effect dissolution, and the mixture is allowed to stand at room temperature for 15 minutes. At the end of this time, 25 ml of neutralized ethanol, containing cresol red, is added from a graduated cylinder, and the excess acid is titrated with standard methanolic $0.1N$ sodium hydroxide solution to the first violet color of the end point.

The cresol red indicator, in the alcoholic-dioxane solution, changes from pink to yellow slightly short of the change from yellow to violet, which is chosen as the end point.

With materials such as some resins, which are difficult to dissolve, a maximum of 5 grams of sample is dissolved in 25 ml of purified dioxane by heating to 40°C. The solution is cooled to room temperature prior to addition of hydrochlorination reagent.

PYRIDINIUM CHLORIDE–PYRIDINE

The hydrochlorination reagent, a $0.2N$ solution of hydrochloric acid in pyridine, is prepared by cautiously pipetting 17 ml of CP concentrated hydrochloric acid into 1 liter of CP pyridine and mixing thoroughly. Then 25 ml of the pyridine hydrochlorination reagent is pipetted into a 250-ml flask, equipped with a standard taper joint. A weighed amount of sample, containing from 0.002 to 0.003 equivalent of α-epoxide, is added and dissolved by heating the mixture at about 40°C. After dissolution is complete, the mixture is refluxed, under a condenser, on a hot plate for 20 minutes. The flask and contents are cooled, 6 ml of distilled water is added, together with 0.2 ml of phenolphthalein indicator solution, and the titration is made with standard $0.1N$ methanolic sodium hydroxide solution to a definite pink color.

PYRIDINIUM CHLORIDE–CHLOROFORM

The hydrochlorination reagent is a $1N$ solution of pyridinium chloride in chloroform and is preapred, in 1-liter quantities, by adding exactly 75 grams of CP anhydrous pyridine and approximately 400 ml of chloroform to a 2-liter cylinder. The cylinder is weighed, cooled in an ice water bath, and dry hydrogen chloride bubbled into it slowly. At intervals of several minutes, the hydrogen chloride flow is stopped, the cylinder removed from the bath, wiped dry, and weighed to establish the rate of flow of the hydrogen chloride. When approximately 35 grams has been introduced, the flow is stopped, the mixture is warmed to room temperature, and the vapors are expelled with a stream of dry air. A 10-ml aliquot, added to a small volume of water, is titrated with standard $0.5N$ methanolic sodium hydroxide solution to the phenolphthalein end point. From this titration the weight of pyridine necessary to produce a 5% excess is calculated. This amount of pyridine is added, and the solution is diluted with chloroform to a strength of approximately 1*N*.

To verify that the reagent has been properly prepared, two 25-ml portions of the reagent are titrated with 0.5*N* methanolic sodium hydroxide, one portion having been boiled gently in a hood for 15 minutes. The two titrations should agree within 0.1 ml, indicating no loss of acid on boiling.

A weighed quantity of sample containing from 0.010 to 0.015 equivalent of α-epoxide is added to a 250-ml flask containing a pipetted 25-ml volume of 1*N* hydrochlorination reagent. The mixture is boiled on a hot plate, under a reflux condenser, for 30 minutes. The flask is cooled to room temperature and diluted with 10 ml of distilled water. Phenolphthalein indicator solution is added and the mixture is titrated with standard 0.5*N* methanolic sodium hydroxide solution to a definite pink color.

For analysis of Epon resins and other α-epoxy compounds of relatively low epoxide content, the use of more dilute hydrochlorination reagent is advantageous. In this determination the reagent is diluted to 0.2*N* with chloroform; to 25 ml of the dilute reagent, a sample containing from 0.002 to 0.003 equivalent of α-epoxide is added, and the mixture is refluxed for 2 hours. The titration is made in the same manner as previously, but using standard 0.1*N* methanolic sodium hydroxide solution.

Comparison of Hydrochlorination Methods

The hydrochlorination methods described in the previous sections were evaluated in laboratories of Jungnickel et al. (3). A number of typical α-epoxides of best obtainable purity were analyzed by the several procedures. The characteristics considered to be most important were: (1) applicability to a variety of typical α-epoxides, (2) precision and accuracy, (3) degree of interference from various substances, and (4) operating expediency. These aspects of the methods tested are discussed.

APPLICABILITY TO VARIOUS α-EPOXIDES

The applicability of a method depends on several factors. First, the reagents should provide good solvent properties for the sample, particularly if the sample is a solid. Thus the aqueous magnesium chloride–hydrochlorination method has a limited scope because only such α-epoxides as ethylene oxide, propylene oxide, and glycidol are readily soluble in the aqueous salt solution, although less soluble α-epoxides may dissolve slowly and consequently may react, but at a somewhat slower rate. The alcoholic magnesium chloride–hydrochlorination method is more general, in that many higher α-epoxides, such as glycidyl ethers and epichlorohydrin, are soluble in the reagent. High molecular weight polyepoxides such as the more highly polymerized (solid) Epon resins, however, are not soluble in the alcoholic medium; consequently the method is not practical for such materials. The hydrochloric acid–ethyl ether reagent also is a good solvent for many α-epoxides but is not satisfactory for Epon resins. The hydrochloric acid–Cellosolve reagent is a fairly good solvent for Epon resins, but most samples go into solution rather slowly, requiring heating and mixing over a considerable length of time. Pyridine, chloroform, and dioxane are all good solvents for α-epoxides, including Epon resins, and the reagents prepared using these solvents have been found to dissolve nearly all types of samples tested.

Another factor in the applicability of a method depends on the extent to which undesirable side reactions occur. In these hydrochlorination methods, any reaction of the α-epoxide is undesirable if it does not result in the net consumption of one mole of acid per equivalent of α-epoxy group. For example, under certain conditions some α-epoxides tend to isomerize to the corresponding carbonyl compounds, and some α-epoxides or their chlorohydrins are susceptible to hydrolysis or alcoholysis. These reactions offer competition for the α-epoxide against the desired hydrochlorination reaction and give low recoveries.

Examination of comparative results obtained on a number of typical α-epoxides, given in Tables 2 and 3, showed the following:

1. Ethylene and propylene oxides reacted essentially quantitatively with several reagents. Recoveries by the alcoholic magnesium chloride–hydrochlorination method, however, were approximately 2% low, possibly because of hydrolysis or alcoholysis. The slightly low results obtained on ethylene oxide by the hydrochloric acid–ethyl ether method were probably due, at least in part, to loss from evaporation during the long reaction period.

2. Epichlorohydrin reacted essentially quantitatively in all the methods except the hydrochloric acid–ethyl ether and the aqueous magnesium chloride–hydrochlorination methods. Low recovery appeared to be the result of incomplete reaction by the former method, and the result of limited solubility by the latter method. There was some evidence that the hydrochloric acid–dioxane method yielded values for epichlorohydrin that were slightly closer to theoretical than the two pyridinium chloride methods, and it was evident that the alcohol magnesium chloride–hydrochlorination method gave slightly low recoveries.

Table 2. Determination of Purified α-Epoxides by Various Methods (each reported value is the mean of two or more determinations)

Compound Determined	Apparent Purity, Wt. %, for Following Methods					
	Aqueous Magnesium Chloride-Hydro-chlorination	Alcoholic Magnesium Chloride-Hydro-chlorination	Hydro-chloric Acid-Ethyl Ether	Hydrochloric Acid-Dioxane	Pyridinium Chloride-Pyridine	Pyridinium Chloride-Chloroform
Reagent strength	0.1*N*	0.5*N*	0.2*N*	0.2*N*	1*N*	1*N*
Reaction conditions	15 min., 25°C	30 min., 25°C	3 hrs., 25°C	15 min., 25°C	20 min., reflux	30 min., reflux
Ethylene oxide	98.9	96.9	97.8	99.4	98.9	99.2[c]
Propylene oxide	98.7	96.4	—	99.2	99.2	99.8[c]
Isobutylene oxide[a]	ca. 30[b]	73	98.0	92.9	91.3	98.2[c]
Butadiene monoxide	83	86	94.5	95.1	90	97.6[c]
Styrene oxide	—	70	88	87	93.1	96.9
Epichlorohydrin	97.8	98.3	94.3	99.3	98.5	98.9
Glycidyl allyl ether	—	97.4	100.5	99.5	99.5	99.7
Glycidyl isopropyl ether	—	97.8	100.4	99.5	99.7	99.9
Glycidyl *tert*-butyl ether	—	—	—	—	—	98.7
Glycidyl diethyl-carbinyl ether	—	—	—	—	—	97.7
Glycidyl phenyl ether	—	97.8	99.4	99.3	99.5	99.4
Glycidyl *o*-cresyl ether	—	—	—	—	—	99.8
Glycidyl *p*-chloro-phenyl ether	—	—	—	—	—	99.6
Glycidyl dichloro-phenyl ether	—	—	—	—	—	97.0
Glycidyl benzoate	—	—	—	—	—	99.1

[a] All values corrected for 2.22 weight per cent water found in this sample by means of Karl Fischer reagent.
[b] Rapidly fading endpoint.
[c] Reaction conditions were: 30 minutes at 60°C in pressure bottles.

Table 3. Determination of α-Epoxide in Epon Resins by Various Methods (each reported value is the mean of two determinations)

Resin	α-Epoxy Value Found, eq./100 g., for Following Methods:			
	Hydrochloric Acid-Cellosolve	Hydrochloric Acid-Dioxane	Pyridinium Chloride-Pyridine	Pyridinium Chloride-Chloroform
Reagent strength	0.2*N*	0.2*N*	0.2*N*	0.2*N*
Reaction conditions	4 hrs., 65°C	15 min., 25°C	20 min., reflux	2 hrs., reflux
Epon 562[a]	0.630	0.646	0.647	0.649
Epon 828	0.476	0.499	0.501	0.502
Epon 834	0.345	0.363	0.364	0.364
Epon 864	0.258	0.270	0.269	0.269
Epon 1001	0.178	0.185	0.186	0.186
Epon 1004	0.100	0.105	0.106	0.106
Epon 1007	0.0525	0.0561	0.0562	0.0559
Epon 1009	0.0336	0.0354	0.0370[b]	0.0346

[a] Condensate of glycerol and epichlorohydrin; other listed products are bis-phenolepichlorohydrin condensates.

[b] Heavy precipitation of sample during the titration partially obscured the endpoint.

3. Glycidyl ethers reacted to essentially the same extent with all reagents tested except the alcoholic magnesium chloride–hydrochlorination reagent, which gave recoveries that were 1 to 2% low. With glycidyl ethers of known high purity, all other methods tested gave quantitative recovery.

4. Isobutylene oxide and butadiene monoxide gave markedly low recoveries in aqueous and aqueous-alcoholic media (the two magnesium chloride methods). Results were considerably higher, although still low, in reaction mixtures containing only small amounts of water (the hydrochloric acid–dioxane and pyridinium chloride–pyridine methods). When essentially anhydrous reagents were employed (the hydrochloric acid–ethyl ether and pyridinium chloride–chloroform methods), the recoveries were nearly quantitative for isobutylene oxide by both methods and for butadiene monoxide by the chloroform method. The water content of the reaction mixture, therefore, appeared to affect the recovery of these two substances. On the basis of this evidence, it seemed likely that these α-epoxides are readily hydrated or that their chlorohydrins are readily hydrolyzed. It is known that α-epoxides containing a tertiary carbon atom are hydrated much more rapidly than other α-epoxides (8).

5. With styrene oxide, there appeared to be an effect somewhat different from that observed with isobutylene oxide and butadiene monoxide. Recoveries were quite low when essentially "free" hydrochloric acid (the reactive agent in the alcoholic magnesium chloride–hydrochlorination, hydrochloric acid–ethyl ether, and hydrochloric acid–dioxane methods) was present in the reagent, particularly when a considerable amount of water was also present. When less severe reagents containing the hydrochloric acid as the pyridinium salt were employed, however, the recoveries were much higher. The pyridinium chloride–chloroform method gave values that were nearly theoretical for the pure material. These results seemed to indicate that styrene oxide is an example of an α-epoxide which undergoes isomerization to a carbonyl compound quite readily in the presence of an acid catalyst. This conclusion is in agreement with observations reported by other authors (9, 10) showing that styrene oxide rearranges in acid solution to give an aldehyde that forms the usual bisulfide addition product.

6. Dieldrin, an insecticide produced by J. Hyman and Company, Division of Shell Chemical Corp., is a condensed-ring α-epoxide with the structure

8. A. Eltekow, *Ber.*, **16,** 395 (1883).
9. A. Klages, *Ber.*, **38,** 1969 (1905).
10. S. Winstein and R. B. Henderson, "Ethylene and Trimethylene Oxides," in R. C. Elderfield, *Heterocyclic Compounds*, Vol. I, Wiley, New York, 1950, pp. 1–60.

```
            CH      CCl
          / | \    / | \
        CH  |  CH    |  CCl
       /|   |   |    |   ||
     O  |  CH2  |   CCl2 ||
       \|   |   |    |   ||
        CH  |  CH    |  CCl
          \ | /    \ | /
            CH      CCl
```

This material gave no reaction when treated according to the hydrochloric acid–dioxane method or the pyridinium chloride–chloroform method. However this stable epoxide does react quantitatively with anhydrous hydrogen bromide in dioxane.

7. Epon resins (see Table 3) were found to give very nearly the same α-epoxy values when analyzed by the hydrochloric acid–dioxane method and the two pyridinium chloride methods. The hydrochloric acid–Cellosolve method, however, yielded values that were approximately 3 to 8% low, as compared to the other three methods, apparently because of partial reaction of the α-epoxy group with the alcoholic function of the Cellosolve solvent. Because of this fundamental deficiency of the hydrochloric acid–Cellosolve method, no further work was done with the method. The magnesium chloride–hydrochlorination methods and the hydrochloric acid–ethyl ether method are not applicable to these resin samples because of the limited solvent properties of these reagents.

8. Aqueous solutions of sparingly soluble α-epoxides (epichlorohydrin and glycidyl ethers), when reacted according to several of the methods, gave approximately the same results regardless of the reagent employed. The pyridinium chloride–chloroform method, however, could not be used for aqueous solutions because of the formation of two liquid phases, and consequent incomplete reaction. The results obtained by the hydrochloric acid–dioxane method and the two magnesium chloride–hydrochlorination methods were slightly lower than by the pyridinium chloride–pyridine method, probably indicating that large amounts of water lower the recoveries to a small extent.

In the course of studying the application of methods for aqueous samples, it was observed that aqueous solutions of epichlorohydrin and glycidyl ethers are unstable. The α-epoxy values for several such aqueous solutions were found to decrease markedly on standing for a few days at room temperature. Presumably, hydration of the α-epoxy group occurs even in neutral solution. For this reason, samples containing substantial amounts of water should be analyzed for α-epoxide as soon as possible after sampling.

PRECISION AND ACCURACY

The only available means for determining the accuracy of the methods tested was to examine the values obtained for the purest obtainable samples of various α-epoxides. The recoveries found for the purified compounds tested, discussed in the previous section, are summarized in Table 2.

The precision of the methods employing the described procedures was estimated from the results of replicate determinations on samples of ethylene oxide, propylene oxide, epichlorohydrin, glycidyl ethers, and Epon resins. Estimates based on these results appear in Table 4.

Table 4. Estimated Precision of Various Hydrochlorination Methods for α-Epoxide Determination

Estimated Precision, % of Mean, for Samples of:	Aqueous Magnesium Chloride-Hydrochlorination	Alcoholic Magnesium Chloride-Hydrochlorination	Hydrochloric Acid-Ethyl Ether	Hydrochloric Acid-Dioxane	Pyridinium Chloride-Pyridine	Pyridinium Chloride-Chloroform
Ethylene and propylene oxides	±0.8	±0.8	±1	±0.5	±0.4	±0.4
Epichlorohydrin	(*a*)	±0.8	—	±0.3	±0.4	±0.3
Glycidyl ethers	(*a*)	±0.8	±0.6	±0.2	±0.4	±0.3
Epon resins	(*b*)	(*b*)	(*b*)	±0.8	±0.8	±0.8

[a] Method applicable to dilute aqueous solutions of these compounds only, for which precision is approximately 3%.
[b] Method not applicable.

INTERFERING SUBSTANCES

The effects of various materials on the determination of α-epoxides have been studied rather extensively, with particular emphasis on the hydrochloric acid–dioxane and pyridinium chloride–chloroform methods. A general summary of the results of tests for interfering materials for all the methods evaluated is given in Table 5 and is discussed in the following sections.

WATER

The presence of sufficient water (about 0.2 gram) in the pyridinium chloride–chloroform method to produce a two-phase system gave low

Table 5. Summary of Results of Tests for Interfering Materials in the x-Epoxy Hydrochlorination Methods[a]
(weights of substances tested, 1–50 grams)

Compound	Aqueous Magnesium Chloride-Hydrochlorination	Alcoholic Magnesium Chloride-Hydrochlorination	Hydrochloric Acid-Dioxane	Hydrochloric Acid-Ethyl Ether	Pyridinium Chloride-Pyridine	Pyridinium Chloride-Chloroform
Water	N	N	Nq	—	N	Sq
Acetals	S	S	S	S	S	S
Acids	N	N	N	N	N	N
Alcohols						
C_1 to C_4	N	N	S	—	S	S
Above C_4	N	N	N	N	N	N
Aldehydes						
Saturated	S	S	S	S	S	S
α,β-Unsaturated	L	Lq	L	L	L	L
Amines	S	S	Sq	Sq	Sq	Sq
Chlorohydrins	N	N	N	N	N	N
β-Epoxides	—	—	L	—	—	—
Esters	Nq	Nq	Nq	Nq	Nq	Nq
Ethers	N	N	N	N	N	N
Hydrocarbons	—	—	N	N	N	N
Ketones						
Saturated	N	N	N	N	N	N
α,β-Unsaturated	L	Lq	L	L	L	L
Nitriles	N	N	N	N	N	N
Peroxides						
Hydrogen	N	N	N	N	N	N
Alkyl	N	N	N	N	N	N
Benzoyl	L	L	Lq	N	—	—
Phenols	—	—	—	—	—	S

[a] N = no interference. S = slight interference. L = large interference. q = qualified (see text). — = no data available.

α-epoxy values because the hydrochloric acid was estracted into the aqueous phase almost quantitatively, whereas the α-epoxide remained in the chloroform phase. Similar results might be expected in the hydrochloric acid–ethyl ether method, but this has not been investigated.

Water, when present in rather large amounts, was found to have a small effect on recovery of α-epoxides by the hydrochloric acid–dioxane method. In the preparation of the hydrochloric acid–dioxane reagent, care must be taken to avoid the presence of too much water or too much aqueous hydrochloric acid because of an unusual salting-out effect of hydrohalogen acids (11), resulting in a phase separation in the reagent. When more than 0.5 gram of water was added to the reagent-sample mixture (in addition to the small amount of water normally present in the reagent), the recoveries of α-epoxides tended to be somewhat low. The extent of lowering of recovery, however, was generally rather minor (usually 1–2%), unless very large amounts of water (more than 2 grams) were added. It is probable, however, that the recovery of substances that characteristically give low values in partially aqueous reagents (e.g., isobutylene oxide and butadiene monoxide) would be lowered further when large amounts of water are present. Nevertheless, the presence of more than 0.5 gram of water in a sample taken for analysis by the

11. W. T. Grubb and R. C. Osthoff, *J. Am. Chem. Soc.*, **74,** 2108 (1952).

hydrochloric acid–dioxane method usually is a relatively uncommon occurrence, except for aqueous mixtures of ethylene oxide or propylene oxide or for dilute aqueous solutions of higher α-epoxides, all of which can be analyzed by the aqueous magnesium chloride–hydrochlorination method or by other aqueous chemical methods.

It was considered advisable to test briefly an essentially anhydrous hydrochloric acid–dioxane reagent, particularly because the low results obtained for isobutylene oxide, butadiene monoxide, and styrene oxide might be explained by hydrolysis of the α-epoxy group or of the chlorohydrin. However it was found that the extent of reaction at 15 minutes became lower as more anhydrous conditions were maintained (3). These results are not surprising in view of the fact that Swern et al. (4), using a reagent consisting of anhydrous hydrogen chloride in ethyl ether, found it necessary to allow reaction periods of 2 to 3 hours for various α-epoxides, whereas King (5) and the results obtained in the laboratories of Jungnickel et al., with aqueous hydrochloric acid in dioxane, showed complete reaction in 5 or 10 minutes. The explanation for the requirement of a certain amount of water in the reaction mixture may possibly lie in the necessity for dissociation of the ether–hydrochloric acid complex before hydrochlorination of the α-epoxy group can occur.

The presence of water has no appreciable effect on the determination of α-epoxides by the pyridinium chloride–pyridine method. Also, it has very little effect on the two magnesium chloride–hydrochlorination methods, except perhaps when present in very large amounts.

ORGANIC COMPOUNDS

A larger number of classes of organic compounds were found to cause no interference in the α-epoxy methods: acids, higher alcohols, amines, higher esters, ethers, hydrocarbons, saturated ketones, and nitriles were generally found to be without effect on the methods. Water-insoluble materials can be expected to lower the results by the aqueous magnesium chloride–hydrochlorination method for epoxy derivatives that are not easily extracted into water. The presence of more than 4 or 5 grams of C_1 to C_4 alcohols tended to lower the results by several of the methods, presumably because of partial alcoholation of the α-epoxy ring. This effect was small with higher alcohols. Saturated aldehydes and the corresponding acetals offered slight and somewhat erratic interference in the methods. α,β-Unsaturated aldehydes and ketones, such as acrolein, methacrolein, crotonaldehyde, and mesityl oxide, appeared to interfere in all methods tested by consuming acid. But a modification of the alcoholic magnesium chloride–hydrochlorination method was found to give good results for several epoxides in the presence of α,β-unsaturated aldehydes. The procedure was modified by reducing the reaction temperature to 0°C and limiting the reaction time to specified times of 1, 2, 3, or 5 minutes; a plot of apparent α-epoxide values against reaction time, when extrapolated to zero time, was found to yield an accurate value. It is likely that the aqueous magnesium chloride–hydrochlorination method can be similarly modified by reducing the temperature and reaction time and using the extrapolation technique described earlier. It was found that the extent of interference by unsaturated carbonyl compounds in the hydrochloric acid–dioxane method could be reduced, but not eliminated, by such chilling and shortening of reaction time.

Large amounts of strong amines, such as the alkyl amines, interfere in all methods except those using magnesium chloride reagent, because the indicators used in the other methods do not allow a quantitative correction to be made. It is possible that indicators changing in the pH range of 3 to 5, or potentiometric titration, might allow the use of any of the methods with samples containing large amounts of stronger amines. With smaller amounts of strong amines, the hydrochloric acid–dioxane method was found to give results of approximate accuracy, provided 50 ml of water in place of the 25 ml of ethyl alcohol was added to the reaction mixture prior to titration. Aliphatic, aromatic, and cyclic ethers (except 1,3-oxides) did not interfere in any of the nonaqueous methods. 1,3-Oxides (β-epoxides) are partially hydrochlorinated by the hydrochloric acid–dioxane method; a sample of 2-allyloxy-1,3-propylene oxide gave approximately 56% reaction in 15 minutes at room temperature. No tests for interference by β-epoxides have been made for other hydrochlorination methods. Because β-epoxides are seldom encountered (10), this source of interference usually can be disregarded.

Hydrogen and alkyl peroxides did not interfere in any of the several methods tested. Benzoyl peroxide interfered in the hydrochloric acid–dioxane method and the magnesium chloride–hydrochlorination methods, however, presumably through hydrolysis to produce benzoic and perbenzoic acids. The effect of benzoyl peroxide in the pyridinium methods was not determined. It is significant, however, that Swern et al. (4) found that benzoyl peroxide did not interfere at all in the determination of α-epoxides by means of anhydrous hydrogen chloride in ethyl ether (3 hours at room temperature) even though the titration of excess hydrogen chloride was performed using aqueous sodium hydroxide solution. Thus the hydrolysis of benzoyl peroxide possibly can be avoided by reducing the temperature of the reaction mixture and/or maintaining more nearly anhydrous conditions during the reaction period.

Phenol interfered in the pyridinium chloride–chloroform method by causing poor end points and low recoveries of epichlorohydrin; but xylenol did not show these effects.

Organic acids did not interfere in any of the several methods tested. It is necessary, however, to make a separate determination of the acid content and to apply the appropriate correction in the determination of α-epoxide. When the acid content is high, relative to the α-epoxide content, the precision and accuracy of the α-epoxy determination are, of course, lowered.

Formate and acetate esters are partially hydrolyzed to produce organic acids when treated with reagents containing water, as in the magnesium chloride methods, the pyridinium chloride–pyridine method, and the hydrochloric acid–dioxane method. The extent of hydrolysis is greatly decreased by use of nearly anhydrous reagents such as hydrogen chloride in ethyl ether and pyridinium chloride in chloroform, but is largely dependent on the quantity of water present during the reaction period. The extent of interference by reactive esters probably can be decreased in the hydrochloric acid–dioxane method and the two magnesium chloride–hydrochlorination methods by reducing the temperature and/or the reaction time. Another possible means for avoiding the interference of reactive esters is suggested: instead of determining the unconsumed hydrochloric acid by acidimetric titration, the chloride ion remaining might be determined by the Volhard method or some other argentimetric procedure. Less readily hydrolyzed esters, such as ethyl isovalerate, dimethyl malonate, butyl stearate, ethyl oleate, and benzyl benzoate, were found to give no interference in the acidimetric hydrochloric acid–dioxane and pyridinium chloride–chloroform methods. The applicability of one of the preferred methods (hydrochloric acid–dioxane) to a complex mixture is demonstrated by data in Table 6.

OPERATING EXPEDIENCY

The hydrochlorination methods are all relatively simple in operation, but are not equally rapid. One of the most time-consuming is the hydrochloric acid–ethyl ether method, which requires a 3-hour reaction period, the remainder of the methods tested employ reaction periods ranging from 15 to 30 minutes, except the pyridinium chloride–chloroform method, which requires 2 hours for resins. The need for reflux in the pyridinium methods adds to the time requirement for equipment operation and maintenance for those methods. The preparation of reagent is cumbersome in the hydrochloric acid–ethyl ether and the pyridinium chloride–chloroform methods, since cylinder hydrogen

Table 6. Analysis of Partially Epoxidized Olive Oil for α-Epoxide and Other Functional Groups to Obtain Oxygen Balance

Analysis	Method (Ref.)	Value Found, eq./100 g.	Oxygen,[a] wt. %
Acidity	Phenolphthalein	0.0088	0.28
Ester value	Saponification (12)	0.363	11.62
Hydroxyl value	$LiAlH_4$ (13)	0.096	1.53
Carbonyl value	NH_2OH-Fischer (14)	0.0470	0.75
Peroxide value	NaI (15)	0.0043	0.07
Water	Fischer reagent (16)	0.0053	0.09
α-Epoxy value	HCl-dioxane	0.1372	2.20
Oxygen	Sum of the above	—	16.54
Oxygen	Direct determination (17)	—	16.68
Recovery of total oxygen content			99.2%

[a] Calculated from analyses, assuming 16 grams of oxygen per equivalent of hydroxyl, carbonyl, peroxide, water, and α-epoxide, and 32 grams of oxygen per equivalent of acid and ester.

chloride is required. Especially time-consuming is the need for the close equivalence of hydrogen chloride and pyridine in the latter method. The methods most rapidly and easily applied are the two magnesium chloride methods, the hydrochloric acid–dioxane method, and the pyridinium chloride–pyridine method.

DIRECT TITRATION OF EPOXY COMPOUNDS INVOLVING OTHER HALIDE REAGENTS

Durbetaki (18) described a direct titration of epoxy compounds with hydrogen bromide in acetic acid to a crystal violet end point. This method was adopted as standard by both ASTM (D-1652-67) and ADCS (Cd 9-57). It gives good results if particular attention is given to protecting the reagent in a closed system during storage and use and if the reagent is standardized often.

12. D. T. Englis and J. E. Reinschreiber, *Anal. Chem.*, **21,** 602 (1949).
13. F. A. Hochstein, *J. Am. Chem. Soc.*, **71,** 305 (1949).
14. J. Mitchell, Jr., and D. M. Smith, *Aquametry*, Wiley-Interscience, New York, 1948, p. 328.
15. C. D. Wagner, R. H. Smith, and E. D. Peters, *Anal. Chem.*, **19,** 976 (1947).
16. J. Mitchell, Jr., and D. M. Smith, *Aquametry*, Wiley-Interscience, New York, 1948, p. 71.
17. V. A. Campanile, J. H. Badley, E. D. Peters, E. J. Agazzi, and F. R. Brooks, *Anal. Chem.*, **23,** 1421 (1951).
18. A. J. Durbetaki, *Anal. Chem.*, **28,** 2000 (1956).

Two subsequent procedures (19, 20) are based on the generation of hydrogen bromide or iodide *in situ* from the corresponding quaternary halides. The sample mixtures are titrated with standard perchloric acid solution to a crystal violet end point. The quaternary halide used by Dijkstra and Dahmen (19) for glycidyl ethers and esters was cetyltrimethylammonium bromide. Jay (20) used tetraethylammonium bromide and iodide for epoxy compounds and aziridines. Both methods have the advantage over the method of Durbetaki of providing for the use of a stable and readily available titrant. The two procedures are quite similar, and the details of the Jay procedure follow.

Adapted from the Method of R. Jay

[*Anal. Chem.*, ***36***, *667* (*1964*)]

REAGENTS

STANDARD 0.1*N* PERCHLORIC ACID IN GLACIAL ACETIC ACID. Mix 8.5 ml of 75% perchloric acid ($HClO_4$) with 300 ml of glacial acetic acid and add 20 ml of acetic anhydride (AcOH). Dilute to 1 liter with glacial acetic acid and allow to stand overnight. Standardize against potassium acid phthalate.

TETRAETHYLAMMONIUM BROMIDE REAGENT. Dissolve 100 grams of tetraethylammonium bromide (NEt_4Br) in 400 ml of glacial acetic acid. Add a few drops of crystal violet indicator. Compensate for any slight indicator blank by titrating dropwise with the standard perchloric acid solution to the end point color change.

TETRABUTYLAMMONIUM IODIDE REAGENT, 10% IN CHLOROFORM. Dissolve 50 grams of tetrabutylammonium iodide (NBu_4I) in 500 ml of chloroform. Store in the dark. This reagent is stable providing it is not preneutralized with perchloric acid reagent or exposed to light.

PROCEDURE

Weigh a sample estimated to contain 0.6 to 0.9 meq. of oxirane or aziridine into a 50-ml Erlenmeyer flask. Dissolve in about 10 ml of chloroform. Acetone, benzene, or chlorobenzene may be used as solubility considerations dictate. Add 10 ml of the quaternary halide reagent and 2 or 3 drops of crystal violet indicator. Titrate to a definite color change with standard 0.1*N* perchloric acid solution from a 10-ml microburet. In some cases where sharp visual end points are not seen, a potentiometric titration may be desirable. The reagent blank is usually negligible but should be checked occasionally.

19. R. Dijkstra and E. A. M. Dahmen, *Anal. Chim. Acta*, **31**, 38 (1964).
20. R. R. Jay, *Anal. Chem.*, **36**, 667 (1964).

DISCUSSION AND RESULTS

Either halide reagent can be used for epoxides, but the quaternary bromide is satisfactory for virtually all the materials usually encountered and is recommended over the iodide because of economy and better stability. The iodide is preferred for aziridines because it gives more rapid reactions and sharper end points.

Table 7. Determination of Epoxy Compounds

	Epoxy Equivalent Weight			
Compound	$HClO_4$-NEt_4Br	$HClO_4$-NBu_4I	HBr-AcOH[a]	HCl-Dioxane[b]
Epon 820 (Shell)[c]	190	—	194	193
Epon 828 (Shell)[c]	188, 186, 186, 186	186, 187, 186	—	—
Araldite TSWR-375 Ciba)	180, 181, 182, 180, 179	179, 179, 179, 180	—	181, 181, 181
Epoxy 201[d] (Union Carbide)	152, 151, 151, 151	—	—	152
ERLB-0500[e] (Bakelite)	113[f]	—	—	114
KP-90 Butyl epoxy stearate (Food Machinery Corp.)	418, 416, 417, 415, 418, 418, 417, 419, 418, 419	—	414, 415, 414, 413, 413, 415	420
Butyl glycidyl ether	133.5, 133.6, 133.7, 133.6	134.8, 134.0, 133.7, 134.4	—	—
Epon 1031 (Shell)[g]	257.2, 257.3, 257.4	—	Failed[h]	—
ERRA 0153[g] (Bakelite)	260.0, 259.3	—	Failed[h]	—

[a] Direct titration with hydrogen bromide in acetic anhydride, Ref. 18.
[b] Excess hydrochloric acid–dioxane, back-titration.
[c] Commercial epoxy resins, epichlorohydrin-bisphenol-A type.
[d] 3,4-Epoxy-6-methylcyclohexylmethyl-3,4-epoxy-6-methylcyclohexanecarboxylate.
[e] *p*-(2,3-Epoxypropoxy)-*N*,*N*-di(2,3-epoxypropyl)aniline.
[f] Corrected for tertiary amine obtained by separate titration.
[g] A solid epoxy resin containing structures of the type 1,1,2,2-tetrakis(2,3-epoxypropoxyphenyl)ethane.
[h] Satisfactory analysis impossible because of poor or fading end points.

The larger excess of bromide and the higher acid strength of perchloric acid afford somewhat more rapid oxirane titrations and sharper end points than those obtained with the hydrogen bromide–acetic acid titrant.

Some typical data are presented in Table 7. In particular it should be noted that epoxy resins Epon 1031 and ERRA 0153 were readily analyzed by the quaternary bromide technique, whereas the hydrogen bromide–acetic acid method failed to give discernible end points with these essentially tetrafunctional resins.

Acids of strengths comparable to acetic acid or weaker do not interfere. Bases will interfere unless corrections are made.

Other Approaches to Epoxide Determination

The other approaches to epoxide determination are sparse and do not have the advantages of the hydrochlorination methods. They are mentioned here for the sake of completeness.

Epoxides can be hydrolyzed to glycols, and the glycols can be determined using periodic acid (see pp. 42–43). Eastham and Latremouille (21) used this approach for determining ethylene oxide.

Hydrogen sulfide (22), alkali hydrogen sulfides and thiols (22), sodium sulfite (22, 23), and sodium thiosulfate (22, 24) can also be used.

$$R\text{—}\underset{\diagdown}{\overset{H}{C}}\text{——}\underset{\diagup}{\overset{H}{C}}\text{—}R_1 \;(\text{ring closed by O}) + NaX + H_2O \rightarrow R\text{—}\underset{OH}{\overset{H}{\underset{|}{C}}}\text{—}\underset{X}{\overset{H}{\underset{|}{C}}}\text{—}R_1 + NaOH$$

where X = the sulfur containing groups.

The final measurement is accomplished via acid-base titration.

TRACES OF EPOXY COMPOUNDS

Adapted from the Method of H. E. Mishmash and C. E. Meloan

[*Anal. Chem.*, **44,** *835* (*1972*)]

The reactions used in this method involve first the cleavage of the α-epoxide in the presence of a mineral acid to form the glycol. The glycol

21. A. M. Eastham and G. A. Latremouille, *Can. J. Res. B*, **28,** 64 (1950).
22. A. Tchitchibabine and M. Bestougeff, *Chem. Zentr.*, **106,** I, 3619 (1935).
23. D. Swan, *Anal. Chem.*, **26,** 878–80 (1954).
24. W. C. J. Ross, *J. Chem. Soc.*, **1950,** 2257.

is then cleaved with an excess of periodate. An iodometric determination of the excess periodate was used with a spectrophotometric measurement of the deep blue, starch-triiodide complex.

REAGENT

CADMIUM IODIDE–STARCH REAGENT. Dissolve 11 grams of cadmium iodide (CdI_2) in 400 ml of distilled, deionized water, and boil gently for 15 minutes to expel any iodine. Dilute to 800 ml and add 15.0 grams of Superlose HAA-11-HV. Continue the boiling and stirring for 15 minutes. Filter and dilute the reagent to 1 liter with distilled, deionized water.

PROCEDURE

Use a range of concentration in 1:1 (v/v) glyme-water varying from 0 to 2.25 μM of oxirane compound to make a calibration curve. Place the samples in 100-ml volumetric flasks. Add 50 ml of standard periodate solution (100 ppm). Seal the flasks with "poly" stoppers and place in a water bath at 45°C for 30 to 40 minutes. Withdraw the flasks and allow to cool. Add 2 ml of 1*N* sodium hydroxide solution and 10 ml of pH 4.5 buffer. Fill the flasks to about 90 ml with distilled, deionized water. Add 1 ml of cadmium iodide–starch solution and dilute to 100 ml with deionized water. Allow 20 minutes for color development.

RESULTS AND DISCUSSION

The procedure was used successfully for propylene oxide, 1,2-butylene oxide, (epoxyethyl)benzene(styrene oxide), 1,2-epoxy-5,6-*trans*-9,10-*cis*-cyclododecadiene, 16,17-epoxydesoxycorticosterone acetate, and epoxybutyl stearate. The wavelength of maximum absorbance for all but epoxybutyl stearate was 590 nm. Because of the greater insolubility of the epoxybutyl stearate, more glyme was needed, and the wavelength absorbance shifted to 560 nm. The method failed for Dieldrin, and this was attributed to dechlorination of the highly chlorinated compound by acid at elevated temperature, with the produced chlorine reducing periodic acid and iodic acids to free iodine.

Interferences include α-dicarbonyl, α-hydroxy carbonyl, α-amino alcohols, and any other of the interferences normally involved in iodometric methods.

The problem of peroxides forming in the glyme solvent, causing an error, was eliminated by passing the glyme through a column of activated alumina. The purified glyme was then stored over alumina.

There are a few tests for specific epoxy compounds. Deckert (25) determined traces of ethylene oxide in air using hydrochlorination and the color changes in acid-base indicators. Gunther et al. (26) determined traces of ethylene oxide as derived from a specific insecticide, with lepidine as the color-producing agent.

25. W. Deckert, *Angew. Chem.*, **45,** 559, 758 (1932).
26. F. A. Gunther et al., *Anal. Chem.*, **23,** 1835 (1951).

6

Organic Peroxides

Organic peroxides are rather easily determined because of their oxidative properties. It should be noted, however, that the reducibility of organic peroxides varies widely; some peroxides reduce easily, whereas others are quite stable and can only be reduced with vigorous reagents and conditions (e.g., diethyl peroxide). Therefore, since it cannot be predicted which reducing conditions are required for any peroxide, a known sample of the peroxide must be tried with the diverse methods. A severe complication in peroxide analysis is that standard peroxide samples are not often obtainable, hence the analyst cannot readily check the applicability of the available methods to the peroxide in question.

Of the number of reducing agents that can be applied to the determination of peroxides, iodide ion is the most commonly used. The iodine liberated in the reaction can be titrated or it can be measured colorimetrically. The iodometric titration methods are the most popular and cover a myriad of conditions and solvents in an attempt to cover the range of peroxide solubilities and reactivities. Table 25 (p. 370) summarizes the analytical methods.

Other reducing agents that have been used are ferrous, stannous, arsenious, and titanous ions and organic reducing agents such as hydroquinone, leuco methylene blue, and certain diamines. These reductants are measured either volumetrically or colorimetrically as indicated in Table 25.

The methods shown are the most widely applicable in the experience of the authors. Because of the singularity of many peroxides, however, the reader is cautioned that there is no "general method." Table 25 should be consulted to evaluate the various methods for each particular problem.

Iodometric Methods

Study of C. D. Wagner, R. H. Smith, and E. D. Peters

[*Reprinted with Permission from Anal. Chem.*, ***19,*** *976–9* (*1947*)]

Iodometric methods as well as ferrous ion methods have been widely used for the determination of peroxides in oxidized organic materials.

Most of the proposed iodometric methods involve reduction of peroxides by iodide ion in acetic acid solution with or without addition of a strong mineral acid. The well-known procedures of Wheeler (1), Lea (2), Marks and Morrell (3), Liebhafsky and Sharkey (4), Taffel and Revis (5), and Stansby (6) are examples of such methods, some of which use chloroform or carbon tetrachloride to aid in dissolving the sample. Kokatnur and Jelling (7) proposed a procedure employing isopropyl alcohol as solvent, in conjunction with a small amount of acetic acid and potassium iodide. Nozaki (8) has reported a procedure in which acetic anhydride is used as the solvent.

There are many differences of opinion concerning the reliability of iodometric methods. It has been well established that many peroxycarboxylic acids, diacyl peroxides, hydroperoxides, and other peroxide compounds can be determined quantitatively by iodometric methods, although a few reports exist doubting the accuracy of certain iodometric methods on such simple peroxides as cyclohexene and Tetralin peroxides. The most common criticism has been that iodine liberated in the reaction disappears by addition to olefinic double bonds; this assertion has been based on dependence of results on sample size with certain materials and on test experiments involving reaction of free iodine rather than triiodide ion. Wheeler (1) found that when iodide ion was present, iodine did not add to certain unsaturated peroxide-free oils, but proof was lacking that triiodide did not react with olefins in the presence of peroxides. Panyutin and Gindin (9) formulated a procedure by which they measured iodide added, iodine found, and iodide unoxidized, thus hoping to compensate for iodine addition; but the procedure itself was so different from the usual procedures that information presented in that report did not prove or disprove iodine addition under normal conditions of iodometric methods.

Another criticism that might be directed at iodometric methods, particularly those employing strong acid, is the possibility of reduction of organic compounds other than peroxides, including possible reduction of any alkylene diiodides formed by addition of iodine to olefin. Atmospheric oxygen is known to cause high results with most iodometric methods, particularly in the presence of strong acids, and methods

1. D. H. Wheeler, *Oil & Soap*, **9,** 89 (1932).
2. C. H. Lea, *Proc. Roy. Soc.* (*London*), **108B,** 175 (1931).
3. S. Marks and R. S. Morrell, *Analyst*, **54,** 503–8 (1929).
4. H. A. Liebhafsky and W. H. Sharkey, *J. Am. Chem. Soc.*, **62,** 190–2 (1940).
5. A. Taffel and C. Revis, *J. Soc. Chem. Ind.*, **50,** 87T (1931).
6. M. E. Stansby, *Anal. Chem.*, **13,** 627 (1941).
7. V. R. Kokatnur and M. Jelling, *J. Am. Chem. Soc.*, **63,** 1432–3 (1941).
8. K. Nosaki, *Ind. Eng. Chem., Anal. Ed.*, **18,** 583 (1946).
9. P. S. Panyutin and L. G. Gindin, *Bull. Acad. Sci. USSR.*, **1938,** No. 4, 841.

employing sodium bicarbonate or solid carbon dioxide (4, 10) were designed to overcome this defect. The Kokatnur-Jelling method (7) was reported not to be influenced by atmospheric oxygen.

For general use the method proposed by Kokatnur and Jelling (7) has been found the most suitable of the iodometric methods tested. [Since this work was completed, Nozaki (8) reported that use of acetic anhydride in place of acetic acid overcomes the disadvantages inherent in the use of the latter as solvent.] Modifications found advantageous are (1) replacement of potassium iodide by sodium iodide, which is much more soluble in the reaction mixture and makes possible a higher concentration of iodide, tending to increase the reaction rate and to decrease possible addition of iodine to unsaturated materials; and (2) elimination of water from the reaction mixture to avoid the low results obtained on autoxidized diolefins in the presence of water. [Lips, Chapman, and McFarlane (11) noted that presence of water markedly decreased results obtained on fats by an iodometric method, and Liebhafsky and Sharkey (4) observed that water retards peroxide reduction by iodide.] To avoid possible interference by oxygen in the experimental work, the reaction mixture was kept under a blanket of carbon dioxide, and sample and reaction mixture were deaerated prior to analysis; however such precautions have been found unnecessary for general use.

Experiments using the modified iodide method show that results on known hydroperoxides are accurate, since wide variation in reaction conditions produces little change in results, which are already close to theoretical. The fact that peroxide was not volatilized by passage of carbon dioxide gas was proved by tests using *tert*-butyl hydroperoxide. It was found that appreciable amounts of water must be present in the mixture at the titration end point in order to avoid high results due to the slowness of reaction between iodine and thiosulfate and consequent overtitration, especially when the titer is small. This effect had been noted by Liebhafsky and Sharkey (4) while studying a method utilizing acetic acid as solvent.

Variation in reaction conditions produced little change in results on materials containing no autoxidized conjugated diolefins, indicating that the method is accurate when applied to such materials; this is consistent with observations that mono-olefin peroxides and initially formed ether peroxides are, in fact, hydroperoxides. Results on diolefins and ascaridole, however, showed wide variation with changes in reaction conditions, indicating that such bridge-type peroxides are not determined accurately by this method.

Experiments have demonstrated that under the conditions of the

10. P. D. Bartlett and R. Atschul, *J. Am. Chem. Soc.*, **67,** 816 (1945).

11. A. Lips, R. A. Chapman, and W. D. McFarlane, *Oil & Soap*, **20,** 240 (1943).

method, iodine does not add to mono-olefins and that, in the absence of peroxides, it does not add to diolefins. One experiment indicated that iodine might by absorbed by diolefins in the presence of diolefin peroxides. The sodium iodide–isopropyl alcohol method was found to be as useful in all cases as sodium iodide–acetic acid methods and to have the advantages of somewhat more general applicability and comparative freedom from interference by atmospheric oxygen.

EXPERIMENTAL ANALYTICAL PROCEDURE

Into a 250-ml Erlenmeyer flask equipped with a gas inlet tube, introduce 40 ml of dry isopropyl alcohol, 2 ml of glacial acetic acid, and up to 10 ml (usually 5 ml) of the sample, containing up to 2 meq. of peroxides. Connect the flask to a reflux condenser and pass carbon dioxide gas through the mixture for 3 minutes. Stop the carbon dioxide flow, heat the solution to reflux, add through the condenser 10 ml of isopropyl alcohol saturated with sodium iodide, and heat the mixture at gentle reflux for 15 minutes ±30 seconds. Resume the carbon dioxide flow, disconnect the flask from the condenser, and titrate the contents immediately with $0.1N$ sodium thiosulfate to the disappearance of the yellow color (use $0.01N$ thiosulfate for very low peroxide concentrations).

EXPERIMENTAL

MATERIALS USED. Tetrahydronaphthyl hydroperoxide (Tetralin peroxide) was prepared by air oxidation of Tetralin at 75°C, crystallization, and recrystallization. α,α-Dimethylbenzyl hydroperoxide (cumene peroxide) was prepared from cumene by a method similar to that of Hock and Lang (11a). Hydrogen peroxide was a sample of Baker's CP 30% (in water), assaying 29.9%. *tert*-Butyl hydroperoxide was prepared by careful distillation of Union Bay State material and anlayzed close to 96.5% by the procedure given previously. Benzoyl peroxide was obtained from Lucidol Corporation. Ascaridole, obtained from Eastman Kodak Company, was said to be better than 99% pure, and analyzed 72.3% carbon, 9.8% hydrogen (theory 71.4% carbon, 9.6% hydrogen).

The following materials were autoxidized by allowing them to stand under oxygen in diffused daylight until analysis indicated a reasonably high peroxide content (air was used in the case of 2-pentene): diisobutylene, consisting of about 75% 2,2,4-trimethyl-1-pentene and

11a. H. Hock and S. Lang, *Ber.*, **77B,** 257 (1944).

25% other isomeric octenes, was not redistilled before autoxidation. 2-Pentene, prepared by dehydration of *sec*-amyl alcohol, was carefully distilled. The product probably contained a small amount of 1-pentene. Cyclohexene, Eastman Kodak Company material, was dried over potassium carbonate and potassium hydroxide and redistilled, the fraction distilling between 83.0 and 83.3°C being retained. Tetralin, also obtained from Eastman Kodak Company (practical grade), was redistilled, the fraction boiling between 204.3 and 207.3°C being retained. Methylpentadiene, prepared by dehydration of 2-methyl-2,4-dihydroxypentane, was carefully distilled before allowing it to autoxidize. It consisted of about 70% 2-methyl-1,3-pentadiene and 30% 4-methyl-1,3-pentadiene. Isoprene, redistilled, was 97.1% pure (determined by freezing point). Diethyl ether was anhydrous Merck reagent.

EFFECT OF WATER ON THE TITRATION. It was found that the reaction between sodium thiosulfate and triiodide ion is rather slow unless 5 to 10% of water is present in the reaction mixture. When the titration volume is reasonably large, sufficient water is furnished by the thiosulfate solution, but with low titers water must be added before titration. This effect is shown in Table 1, which summarizes experiments in which known

Table 1. Effects of Absence of Water on Accuracy of Iodometric Titration

Volume of Standard Iodine Solution ml.	Calculated Titer, 0.1N $Na_2S_2O_3$ ml.	Titer Found, 0.1N $Na_2S_2O_3$	
		Anhydrous ml.	5 ml. of H_2O Added ml.
50	11.53	11.57	11.56
25	5.77	5.75	5.78
10	2.31	2.81	2.37
2	0.46	1.02	0.51

amounts of a standard isopropyl alcohol solution of iodine contained in the reaction mixture were titrated hot, using a constant total amount of reaction mixture [x ml of standard solution iodine in isopropyl alcohol, $(50-x)$ ml of isopropyl alcohol, 2 ml of glacial acetic acid, and 2 grams of sodium iodide].

ACCURACY OF METHOD AND EFFECT OF MANIPULATIVE VARIABLES ON RESULTS. The known peroxides were tested to study the effect of variations in heating time, reaction time, sample size, amount of iodine, and amount of water (Table 2). These tests indicate that (1) the amount of iodide is not critical, provided a large excess is present, (2) reaction

Table 2. Effects of Manipulative Variables in the Sodium Iodide–Isopropyl Alcohol Method on Results for Known Peroxides[a]

Variable	Tetralin Peroxide, %[b]		*tert*-Butyl Hydroperoxide, %[b]		Hydrogen Peroxide, %[b]	Benzoyl Peroxide, %[b]		Ascaridole, %[b]
	Anhydrous	5 ml. of H_2O before Titration	Anhydrous	5 ml. of H_2O before Titration	5 ml. of H_2O before Titration	Anhydrous	5 ml. of H_2O before Titration	Anhydrous
Reaction time								
2 min.	103.7	97.3	96.7	96.5	29.7	100.2	98.8	7.3
5 min.	105.1	97.1	96.5	96.5	29.6	100.2	98.0	11.2
15 min.	102.4	97.3	95.8	96.0	29.4	100.7	98.5	25.7
Sample size								
2 ml.	105.7	96.2	103.7	95.1	29.9	104.0	98.3	55.6
5 ml.	102.4	97.3	95.8	96.0	29.4	100.7	98.1	25.7
10 ml.	96.7	97.2	96.2	96.2	—	—	—	—
Amount of iodide[c]								
2 grams	102.2	96.9	—	96.3	29.4	99.4	98.8	25.1
7 grams	99.4	98.1	—	97.5	29.7	102.3	100.2	28.6
Effect of 5 ml. of H_2O								
Added at start	97.2		96.0		—	100.7		3.7
Added after reflux	97.0		96.0		—	98.5		18.4
No H_2O	102.4		95.8		—	98.5		25 7
Effect of heat								
30-min. reflux before I^- added	97.4	93.5	—	94.2	29.0	84.3	86.2	23.5
0-min. reflux before I^- added	102.4	97.3	95.8	96.0	29.4	100.7	98.5	25.7

[a] Dissolved in benzene (C.P. thiophene-free), about 0.1*N* except hydrogen peroxide, which was diluted with water.
[b] Per cent recovery based on 100% purity. "Anhydrous" refers to anhydrous conditions of experimental method. Other columns of data obtained by adding 5 ml. of water to reaction mixture just before titration.
[c] Method modified: sodium iodide dissolved by refluxing in isopropyl alcohol-acetic acid, followed by heating to reflux, and addition of sample through condenser. Carbon dioxide used in usual way before and after refluxing.

time and sample size are not critical for the peroxides tested (except for ascaridole, which reduces slowly), (3) only the ascaridole analysis was markedly affected by the presence of water during the reaction, and (4) except for benzoyl peroxide, the peroxides are stable toward heat. Consistent results close to theoretical with the hydroperoxides indicate that the method is accurate for their determination. Additional support for this conclusion is given by consistent results of 94% obtained on a sample of α,α-dimethylbenzyl hydroperoxide prepared by autoxidation.

The experimental procedure was applied with variations to the autoxidized materials,with results shown in Table 3. With materials containing no conjugated diolefins, results were affected little by changes in the experimental conditions, whereas results with autoxidized isoprene and methylpentadiene were critically dependent on the conditions employed. The diolefin peroxides react slowly and incompletely, like ascaridole, and results showing dependence on sample size and amount of iodide ion demonstrate that the rate of reduction depends on the concentration of iodide ion in the mixture rather than on the concentration of the peroxides. Water present during the reflux period causes lower values to

Table 3. Effects of Manipulative Variables in the Sodium Iodide–Isopropyl Alcohol Method on Peroxide Numbers of Autoxidized Materials

Variable	Peroxide Number, Milliequivalents per Liter						
	Diiso-butylene	2-Pentene	Cyclo-hexene	Tetralin	Diethyl Ether	Methyl-pentadiene	Isoprene
Reaction time							
2 min.	11.0	40.0	83.6	114	12.0	22	31
5 min.	11.8	40.2	86.7	115	12.1	33	52
15 min.	11.6	40.4	85.0	113	12.0	61	92
Sample size[a]							
2 ml.	—	—	—	112	11.7	99	88
4 ml.	11.6	42.6	85.0	—	—	—	—
5 ml.	—	—	—	109	12.0	55	48
10 ml.	9.8	38.3	83.3	—	12.0	—	—
Amount of iodide[b]							
2 grams	11.2	44.8	84.9	117	12.1	67	94
7 grams	11.8	42.4	84.9	118	12.6	227	128
Effect of 5 ml. of H_2O							
Added at start	8.6	38.8	82.3	113	11.9	19	15
Added after reflux	9.2	39.6	82.6	113	12.1	57	78
No H_2O	11.6	40.4	85.0	113	11.7	61	92
Effect of heat							
30-min. reflux before I^- added	12.6	39.8	81.6	109	11.8	59	82
No reflux before I^- added	11.6	40.4	85.0	113	12.1	61	92

[a] Except in case of conjugated dienes, indicated variation of results with sample size can be ascribed chiefly to the fact that no water was added before titration.
[b] Method modified, see *c*, Table 2.

be obtained in some cases, and to prevent this effect, the exclusion of water during the reflux period is recommended.

The results demonstrate conclusively that peroxides in autoxidized conjugated diolefins differ radically from those in other materials; the latter are now generally accepted to be hydroperoxides, whereas Bodendorff (12) has demonstrated that conjugated diolefins form peroxides largely by 1,4-addition to produce intramolecular peroxides of the ascaridole type, or polymeric peroxides by intermolecular peroxy-bridging to the 1- or 4-positions of other molecules:

Intramolecular peroxide from conjugated diolefin

Intermolecular polymeric peroxide from conjugated diolefin

12. K. Bodendorff, *Arch. Pharm.*, **271,** 1 (1933).

Such peroxides would be expected to react differently from the simple alkenyl hydroperoxides, which are formed with mono-olefins by oxidative attack at the carbon atom alpha to a double bond. The fact that ascaridole reacts in a way very similar to the autoxidized diolefins is to be expected in view of Bodendorff's observations that the monomeric product formed from the autoxidation of the conjugated diolefin, 1,3-menthadiene, is chemically and structurally similar to ascaridole. In addition to these observations and the fact that the hydroperoxides react very rapidly with iodide ion, another point to be noted is the stability of all the peroxides in autoxidized materials. In no case did a significantly decreased final value result from refluxing the sample in the reaction mixture for 30 minutes before addition of iodide ion. Any slight decreases obtained may possibly be due to reduction of the peroxide by the alcohol.

INTERFERENCES

IODINE ABSORPTION BY UNSATURATED MATERIALS. To test the magnitude of this potential source of error, samples of isoprene and cyclohexene were fractionally distilled into a receiver through which oxygen-free nitrogen was slowly passed. Ten-milliliter portions of each of these peroxide-free materials were subjected to two modifications of the prescribed procedure. In one, 10 ml of isopropyl alcohol containing about 0.2 meq. of iodine was added following the addition of the iodide, and in the other the iodine solution was substituted for the iodide solution. In another set of experiments 5 ml of water was added to the mixture at the start. Results (Table 4) demonstrate that in the absence of iodide ion a considerable fraction of the iodine is lost, particularly in the presence of water. This is in agreement with observations by Margosches, Hinner, and Friedmann (13) on the effect of water on rate of iodine addition to unsaturated materials. When iodine is combined with iodide in the form of triiodide ion, no evidence of loss is observed (with peroxide-free material).

Inasmuch as the amount of iodine lost is approximately proportional to the amount originally present, the failure of loss to occur in the presence of iodide is undoubtedly due to the formation of unreactive triiodide ion and resultant small concentration of free iodine.

To determine whether a peroxide such as benzoyl peroxide could catalyze addition of iodine as triiodide to olefins, a weighed quantity of sodium iodide was dissolved by refluxing in 40 ml of isopropyl alcohol and 2 ml of acetic acid in the presence of carbon dioxide; 5 ml of water

13. B. M. Margosches, W. Hinner, and L. Friedmann, *Ber.*, **57,** 996 (1924).

Table 4. Addition of Iodine to Unsaturated Materials Under Conditions of Analysis

Material	I_2 or I_3^- Added, Equivalent × 10^3	I_2 or I_3^- Found after Reflux, Equivalent × 10^3	
		In Anhydrous Medium	In Presence of 5 ml. of H_2O
Isoprene	0.1890 I_2[a]	0.1600	0.1400
	0.0756 I_2[a]	0.0601	0.0432
	0.0378 I_2[a]	0.0249	0.0161
Cyclohexene	0.1760 I_2[a]	0.1519	0.0971
Isoprene	0.2278 I_3^-	0.2274	0.2260
	0.0911 I_3^-	0.0906	0.0907
	0.0456 I_3^-	0.0459	0.0457
Cyclohexene	0.2050 I_3^-	0.2049	0.2050

[a] Determined by refluxing concentrated solutions 15 minutes and titrating. The two dilute solution concentrations were calculated from dilution factors. About 0.02 milliequivalent of iodine was lost from concentrated iodine solutions themselves on refluxing.

was added in some cases. Then 5 ml of purified isoprene and 10 ml of a standard benzoyl peroxide solution in benzene were added, and the mixture was refluxed 15 minutes and titrated, maintaining a blanket of carbon dioxide at all times. Results are tabulated in Table 5. No peroxide-catalyzed addition of iodine is observed.

Nevertheless, definite evidence was still lacking as to whether iodine as triiodide is absorbed by diolefins in the presence of diolefin peroxides. To test this point, a 1-ml sample of autoxidized methylpentadiene was subjected to analysis in the usual fashion in the presence of 4 ml of freshly

Table 5. Addition of Triiodide Ions to Isoprene in the Presence of Benzoyl Peroxide

NaI Present, grams	Benzoyl Peroxide Added, eq. × 10^3	I_3^- Found after Reflux, Equivalent × 10^3	
		In Anhydrous Medium	In Presence of Water
2	0.871	0.863	0.862
7	0.871	0.877	0.867

distilled peroxide-free methylpentadiene and 4 ml of benzene. Results of 0.309 and 0.341 meq., respectively, were obtained, which may indicate some absorption of iodine in the presence of the diolefin peroxide (lower results with greater concentration of diolefin).

POSSIBILITY OF REDUCTION OF OLEFINS OR ALKYLENE DIIODIDES DURING ANALYSIS. If hydrogen iodide should add to olefins to produce alkyl iodides, or if iodine should add to produce alkylene diiodides, reduction of such organic iodides by hydrogen iodide would produce iodine and lead to high results. To test whether this is at all likely, the method was applied to 1 ml of autoxidized methylpentadiene both with and without the addition of 2 ml of pure ethyl iodide. Titers were 16.3 ±0.2 ml of 0.01*N* thiosulfate, indicating that in the presence of diolefin peroxides simple alkyl iodides are not reduced by the iodide reagent.

INTERFERENCE BY AIR. A sample of methylpentadiene, analyzed in the prescribed way, gave 2.3 meq. of peroxide per liter. When an air stream was substituted for carbon dioxide, a value of 3.3 meq. per liter was obtained, and instability of the end point was troublesome. Approximately the same absolute difference in results was noted on diolefin samples of higher peroxide content, whereas no significant difference was noted with cyclohexene and other simple olefins. It is concluded that exclusion of air is essential only if the method is used for analyzing diolefin samples of low peroxide content.

PROCEDURE RECOMMENDED FOR GENERAL USE

The following method, which is closely similar to the Kokatnur-Jelling method, is recommended for general use on materials containing no conjugated diolefins.

Into a 250-ml Erlenmeyer flask introduce 40 ml of dry isopropyl alcohol, 2 ml of glacial acetic acid, and the sample (up to 10 ml, ordinarily 5 ml). Heat to reflux, add 10 ml of isopropyl alcohol saturated at room temperature with sodium iodide (prepared by refluxing 25 grams of sodium iodide with 100 ml of isopropyl alcohol), reflux 5 minutes, add 5 ml of water, and titrate with 0.1 or 0.01*N* sodium thiosulfate. Blank determinations on the reagents will be nil unless oxidizing impurities are present in the isopropyl alcohol: therefore a blank determination on each new batch of alcohol is sufficient.

Although the method is not recommended for use on conjugated diolefins, it is sometimes useful for obtaining precise empirical results, provided an atmosphere of carbon dioxide is maintained in the flask during the analysis.

COMPARISON WITH METHODS EMPLOYING ACETIC ACID AS SOLVENT

The following two methods, employing acetic acid as solvent, were compared with the recommended method described previously.

ACETIC ACID, REFLUX METHODS. Into a 250-ml Erlenmeyer flask equipped with a gas inlet tube, introduce 50 ml of glacial acetic acid and 5 ml of sample. Connect to a reflux condenser, pass carbon dioxide through the solution slowly for 2 minutes, stop the flow, and heat the mixture to reflux. Add 2 ml of a saturated aqueous solution of sodium iodide through the condenser, reflux for 15 minutes, start the carbon dioxide flow, add 100 ml of distilled water, cool the contents to room temperature, and titrate with thiosulfate to the disappearance of the yellow color, maintaining gentle carbon dioxide flow during the titration. Make a blank determination on the reagents in a similar manner.

ACETIC ACID, ROOM-TEMPERATURE METHOD. This method is the same as the reflux method, except that after carbon dioxide is passed through the solution for 2 minutes; then add the iodide, stop the carbon dioxide flow, stopper the flask, and allow it to stand in the dark fòr 15 minutes at room temperature before resuming the carbon dioxide flow, diluting with water, and titrating.

Comparative results are shown in Table 6. It is evident that except for ascaridole, results are generally lower when acetic acid is used, and results by the acetic acid reflux method are ordinarily lower than are those by the acetic acid room-temperature method. For the diolefins, the dark-colored polymer that forms, particularly on heating, obscures the end point. The lower results by the method employing heat are due to destruction of peroxides during the period of heating to reflux, as shown by the following experiment. A sample of cyclohexene gave a value of 95.9 meq. per liter by the room-temperature acetic acid method and 85.7 meq. per liter by the acetic acid reflux method. When the reaction mixture was refluxed without iodide for 15 minutes, cooled rapidly, then treated according to the room-temperature acetic acid method, a value of only 72.2 meq. per liter was obtained. It seems likely that the peroxides are less stable in acetic acid than in isopropyl alcohol.

It is concluded that acetic acid as solvent offers no advantages over isopropyl alcohol, other than elimination of the reflux operation. Disadvantages are greater interference by air, occasional slight destruction of peroxides, need for great dilution with water before titration, and increased tendency for diolefinic materials to polymerize to dark-colored products. Use of a strong acid in addition to acetic acid would probably accentuate these difficulties.

Table 6. Comparison of Results by Sodium Iodide–Isopropyl Alcohol and Sodium Iodide–Acetic Acid Methods

Material Analyzed[a]	Solvent and Condition		
	Isopropyl Alcohol, Reflux	Acetic Acid, Room Temperature	Acetic Acid, Reflux
	Apparent Peroxide Content, %		
Tetralin peroxide	97.3	96.3	91.1
tert-Butyl hydroperoxide	96.0	96.8	95.7
Hydrogen peroxide (29.9%)	29.4	29.6	29.4
Benzoyl peroxide	99.4	99.4	85.2
Ascaridole	25.7	34.1	71.4
	Peroxide No., Milliequivalents per Liter		
Diisobutylene	36.9[b]	27.5	34.7
2-Pentene	45.9	41.6	40.4
Cyclohexene	95.1[c]	91.8	80.6
Tetralin	126.1	123.8	109.3
Diethyl ether	12.2	21.4	11.3
Methylpentadiene	53.4	(20)[d]	(100)[d]
Isoprene	68.3[c]	(20)[d]	(100)[d]

[a] With known peroxides samples were about 0.1*N* solutions in benzene, except that hydrogen peroxide was diluted with water. Autoxidized materials were used directly.

[b] Method modified by adding 5 ml. of water before titration.

[c] Application of isopropyl alcohol method at room temperature gave: cyclohexene, 83.4 me./l., isoprene, 5.3 me./l.

[d] Reaction mixture was very dark in color, with much polymer; results are very much in doubt.

Study of R. D. Mair and A. J. Graupner

[*Anal. Chem.*, ***36***, *194* (*1964*)]

The suitability of the foregoing procedure of Wagner, Smith, and Peters for all easily reduced peroxides was confirmed. Some minor changes, including the combining of acetic acid and isopropyl alcohol as a single solvent and dropping the step of boiling the sample and solvent before the addition of the iodine reagent, were recommended. A convenient, three-group classification of peroxides according to their reactivities was then based on this method, designated Method I. As shown in Table 7, this scheme defines easily reduced peroxides, all those which are quantitatively reduced by Method I in 5 minutes as Class I; peroxides that

Table 7. Classification of Peroxides According to Reactivity by Method I

Class	Description	Definition
I	Easily reduced	Iodine liberated stoichiometrically in 5 minutes or less[a]
II	Moderately stable	Iodine liberated, but stoichiometric reduction requires more than 5 minutes[b]
III	Difficult to reduce	No iodine liberation, even with prolonged reflux[c]

[a] Includes peracids, diacyl peroxides, all hydroperoxides, and certain other compounds.

[b] Includes most peresters, aldehydes, and ketone peroxides.

[c] Includes *trans*-annular peroxides, diaralkyl peroxides, di-*tert*-alkyl peroxides, and mixed aralkyl-*tert*-alkyl peroxides.

are reduced only with difficulty, all those which are inert to Method I, are Class III. Some of the Class II (moderately stable) peroxides can be determined by Method I if extended reaction periods are used; for others, the reductions will be impractically slow. The latter and all Class III peroxides require more rigorous methods.

REAGENTS

Granular soldium iodide, reagent grade.
Isopropyl alcohol, 99% pure.

APPARATUS

Special reaction flasks (Fig. 6.1) were used to maintain an inert atmosphere. A six-place refluxing setup was used with cool-joint Liebig condensers. The setup was usually placed in a laboratory hood to remove acetic acid fumes and to protect the reactants from bright sunlight. A gas flow that just dimpled the surface of the solvent was used.

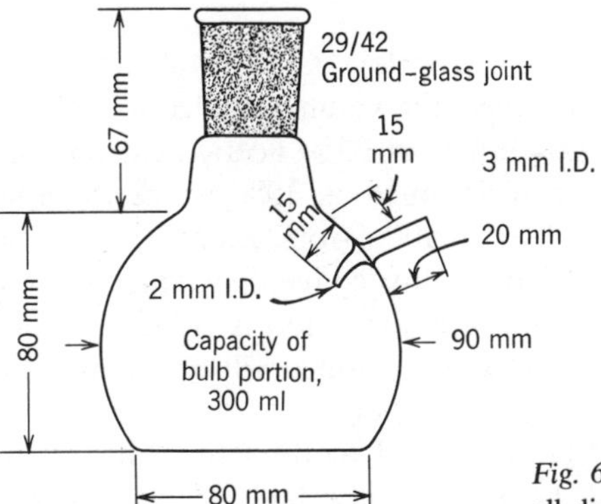

Fig. 6.1. Reaction flask for Methods II and III; all dimensions are approximate.

PROCEDURES

The procedures are outlined in Table 8.

Table 8. Outline of Procedures for Methods II and III

Procedure	Method	
	II	III
1. Reflux 50 ml of acetic acid briefly[a]	yes	yes
2. Cool, add, and partially dissolve 6 grams of sodium iodide[b]	yes	yes
3. Add 3.0 ml of water[c]	yes	no
4. Add sample aliquot[d] in acetic acid or xylene[e]	yes	yes
5. Add 2.0 ml of 37% HCl; put on to boil immediately[f]	no	yes
6. Reflux[g] at least	20 min.	50 min.
7. Add 100 ml of water and titrate[h]	yes	yes
8. Run blank determinations[i]	yes	yes

[a] Inert gas control, cut off whenever flask contents are put on to boil, maintain slow flow at all other times.

[b] Mallinckrodt analytical reagent granular.

[c] If sample is expected to contain but little peroxide, add replicate 1- to 2-meq. portions of I_3^- to both sample and blank reaction flasks at this point.

[d] Up to 2.5 meq.

[e] Purified by being passed through active alumina. Lower-boiling solvents are avoided, to prevent lowering of the reaction temperature.

[f] And cut off inert gas to avoid undue loss of HI.

[g] Temperature about 120°C.

[h] With standard 0.1*N* sodium thiosulfate, starch may be used as an indicator, but it is not necessary.

[i] Blanks for Methods II and III should be, respectively, 0.01 to 0.02 and 0.05 to 0.07 meq. of iodine.

DISCUSSION AND RESULTS

Method I (sodium iodide–acetic acid–6% water) is remarkably tolerant of variations in the procedure. For instance, the volume of alcohol solvent may range up to 100 ml or more. As much as 25% acetic acid may be permitted in the reaction mixture, and as much as 25% water. Sample size may vary by a factor of at least 2500, from more than 5 to 0.002 meq. The sample may be added at any convenient stage of the procedure. And, finally, although quantitative reaction is frequently obtained after only a 1- or 2-minute reflux, the results will be identical if refluxing is continued for an hour or even longer.

Method II was not intended to be a general method for difficult-to-reduce peroxides; rather, it had been carefully tailored to provide interference-free determinations of a specific group of compounds, the diaralkyl peroxides. However it turns out to be generally useful for the Class II peroxides of Table 7 and for all Class III except the di-*tert*-alkyls in that table.

There is one circumstance in which compounds like 7-cumyl alcohol and α-methylstyrene can interfere in Method II: when the sample under analysis contains no peroxidic components and should therefore give a result of nil. In such a situation, the cited compound will be partially reduced and will give a false indication that peroxides are present. A typical example is the 10 to 14% iodine liberated by the *p*-diisopropylbenzene alcohol, compound 4 (Table 9).

When the sample contains more than traces of peroxides, the attack is effectively suppressed by the liberated iodine. The mechanism has not been determined. At any rate, protection from such interference at low peroxide levels in Method II can be gained by the addition of known amounts of I_3^- (Table 8 footnote [c]).

Following the work on Method II, development of a more drastic method was undertaken in which the liberation of iodine by 7- and 8-cumyl alcohols and by 9-methylstyrene would be quantitative rather than partial. With such a method, aralkyl hydroperoxides like cumene hydroperoxide could be expected to liberate 2 moles of iodine rather than the 1 mole, which they produce in Method I or II; and diaralkyl peroxides such as dicumyl peroxide should liberate 3 moles rather than 1. In addition, it would be anticipated that the di-*tert*-alkyl peroxides would respond quantitatively to liberate 1 mole of iodine.

A method with such capabilities was obtained by elimination of the added water from the Method I reagents and replacement with concentrated hydrochloric acid. One other change was essential, to cut off the flow of inert gas during the reflux reaction period to prevent HI from being swept from the reaction flask.

Several comments are in order regarding the recommended procedure for Method III outlined in Table 8. If a sample contains aralkyl hydroperoxides, it is best to delay addition of the hydrochloric acid for a few minutes until the hydroperoxide has been reduced, to avoid partial destruction of the sample through acid cleavage. With the di-*tert*-butyl peroxide, on the contrary, hydrochloric acid should be added promptly to avoid losses through volatility of the sample.

Not more than 2.0 ml of 37% hydrochloric acid should be used. Such additional acid would give a weaker rather than a stronger reducing action, since it would increase the water content of the system.

Methods I, II, and III have been applied to many pure materials, and the results are given in Tables 9 to 11.

For Method I the relative standard deviation $\hat{\sigma}$, calculated with 21

Table 9. Comparison of Methods I, II, and III for Aralkyl Peroxides and Related Compounds

Compound	Method	Recovery, %
α,α-Dimethyl benzyl alcohol (7-cumyl alcohol)	III	99.0, 98.8
α-Methylstyrene	III	97.0, 96.4
Cumene hydroperoxide	I	100.5, 100.6
	III	100.6, 100.6
p-Diisopropylbenzene alcohol	III	95.0, 94.9
p-Diisopropylbenzene hydroperoxide	I	98.5, 98.7
	II	99.6, 100.1
	III	100.5, 99.9
p-Diisopropylbenzene alcohol hydroperoxide	I	99.9, 99.9
	II	99.9, 99.9
	III	99.7, 100.0
p-Diisopropylbenzene dihydroperoxide	I	99.0
	II	99.1, 99.1
	III	98.7, 98.7
m-Diisopropylbenzene alcohol hydroperoxide	I	98.4, 98.4
	II	99.2, 99.3
	III	98.6, 98.7
m-Diisopropylbenzene dihydroperoxide	I	99.1
	II	99.4
	III	97.8
1,3,5-Triisopropylbenzene dihydroperoxide	I	98.0, 98.1
	II	98.6, 98.5
	III	97.8
1,3,5-Triisopropylbenzene trihydroperoxide	I	95.0, 94.8
	II	98.0, 96.8
	III	97.6, 96.3
Dicumyl peroxide	II	92.3, 92.7
	III	96.4, 96.3
p-Diisopropylbenzene peroxide	II	89.4, 89.4
	III	89.3, 89.2

degrees of freedom (d.f.) was 0.23%; for Method II with 26 d.f., $\hat{\sigma}$ was 0.30%; and for Method III with 54 d.f., $\hat{\sigma}$ was 0.32%.

From the data in Table 9 it is clear that aralkyl hydroperoxides liberate exactly 1 mole of iodine per mole in Methods I and II, and exactly 2 moles per mole in Method III (the additional mole through reduction of the intermediate alcohol or substituted styrene formed from the hydroperoxide). It is also clear that such alcohols and styrenes have an equally consistent behavior, whether initially present in a sample or formed as intermediates; they are inert in Methods I and II and liberate exactly 1 mole of iodine per mole in Method III.

Table 10. Iodine Liberation in Method III by Terpenes and Related Compounds

Compound	Reaction Time, min.	Apparent Percentage Reacted
Monofunctional monocyclic terpenoids		
Cavomenthene	60	2.4
3-Menthene	65	1.0
Menthol	120	1.9
cis-Dihydro-α-terpineol	60	2.1
	120	2.5
Difunctional monocyclic terpenoids		
Dipentene	15	101.0
	75	101.0
Terpinolene	15	99.0
	75	99.9
α-Terpineol	40	99.9
	80	99.4
1,8-Terpin hydrate	15	99.0
	60	98.5
	120	99.5
Bicyclic terpenoids, reactive		
1,8-Cineole	15	99.8, 100.5
	70	100.1
	100	99.6
Δ^3-Carene	15	98.3, 98.4
	100	98.5, 97.4
Bicyclic terpenoids, unreactive		
Camphene	15	4.0, 4.0
	60	9.4
	120	12.9
	180	17.1
Borneol	120	6.6
Fenchyl alcohol	120	3.7
Bicyclic terpenoids, isomerization		
α-Pinene	15	21.2
	110	26.8
β-Pinene	15	20.6
	110	26.4
cis-Methylnopinol	15	25.4
	120	25.0
trans-Methylnopinol	15	27.4
	120	25.4

Table 11. Iodine Liberation by Miscellaneous Compounds in Method III

Compound	Reaction Time, min.	Apparent Percentage Reacted
p-Diisopropenylbenzene	30	99.5 } of 2 moles
	60	98.5 } of 2 moles
1-Phenylcyclohexene	45	7.8
	90	15.5
α,α-Dimethylbenzyl alcohol (7-cumyl alcohol)	50	99.0, 98.8
1-Phenylcyclohexanol	60	4.0
1-Methylcyclohexanol	60	2.9
Styrene	45	63.4
	120	77.3
Styrene oxide	60	161, 167, 167
	120	179
trans-Stilbene	60	5.1
Dimethoxystilbene	30	23.7
	60	38.2
Anethole	30	19.9
	60	37.3
Anisole	60	2.7

At the end of Table 9 it can be seen that *p*-diisopropylbenzene peroxide liberated precisely 3 times as much iodine in Method III as in Method II, but with less than 90% of exact stoichiometry. Dicumyl peroxide exceeded this performance in Method III, however, liberating almost 3 moles of iodine per mole. It is uncertain which peroxide is the more typical.

Many nonperoxidic compounds are reduced under the conditions of Method III either partially or quantitatively (Table 10 and 11).

Ferrous Thiocyanate Colorimetric Method

Study of C. D. Wagner, H. L. Clever, and E. D. Peters

[*Reprinted with Permission from Anal. Chem.*, ***19,*** *980–2* (*1947*)]

Ferrous ion has been used extensively as a reducing agent for the determination of organic peroxides. One of the common methods involving ferrous ion, originally proposed by Young, Vogt, and Nieuwland (14), comprises reduction of the sample in an acidified solution of ferrous thiocyanate in methanol, followed by measurement of the depth of color of the ferric thiocyanate produced. Bolland, Sundralingam, Sutton, and Tristram (15) used this method with slight variations in rubber research, and Farmer et al. (16) based theoretical conclusions on results using this method. Lips, Chapman, and McFarlane (11) used a similar method with acetone as solvent in place of methanol.

Young, Vogt, and Nieuwland (14) based their claim for accuracy of the method on determinations of pure succinyl peroxide. Bolland et al. (15) obtained analyses close to theoretical on succinyl peroxide, dihydroxyheptyl peroxide, and cyclohexene peroxide. However Lea (2), using a method similar to that of Lips, Chapman, and McFarlane (11), discovered that when air was carefully removed from both sample and reagent, the results were only 10 to 30% of those obtained without such treatment; he concluded that atmospheric oxygen and dissolved oxygen were oxidizing ferrous ion in the presence of peroxide (although little atmospheric oxidation takes place in the absence of peroxide) and, therefore, that the results included very large positive errors.

In view of the conflicting reports of accurate values on known peroxides on the one hand, and oxidation by air to produce high results, on the other hand, experiments were undertaken to resolve these differences. By using the method of Bolland, Sundralingam, Sutton, and Tristram, essentially accurate results were obtained on pure *tert*-butyl hydroperoxide, α,α-dimethyl-benzyl hydroperoxide (cumene peroxide), and tetrahydronaphthyl hydroperoxide (tetralin peroxide). With autoxidized monoolefins good agreement was obtained between results by the sodium iodide–isopropyl alcohol and colorimetric methods. In all colorimetric analyses wherein oxygen was carefully excluded, however, the observations of Lea (2) were substantiated. Furthermore, it was found that no additional color developed when air was passed through the mixture of oxygen-free sample and reagent, which indicates complete destruction of peroxides largely by decomposition rather than reduction. It therefore appears that dissolved oxygen is necessary in this method, either to speed up the reduction reaction or to inhibit the ferrous ion-catalyzed decomposition.

14. C. A. Young, R. R. Vogt, and J. A. Nieuwland, *Ind. Eng. Chem., Anal. Ed.*, **8,** 198–9 (1936).
15. J. L. Bolland, A. Sundralingam, D. A. Sutton, and G. R. Tristram, *Trans. Inst. Rubber Ind.*, **17,** 29 (1941).
16. E. H. Farmer, G. F. Bloomfield, A. Sundralingam, and D. A. Sutton, *Trans. Faraday Soc.*, **38,** 348 (1942).

As a result of these studies, the colorimetric method is recommended as the best for frequent analyses of materials containing only small amounts of peroxides. For materials (except diolefins) containing large amounts of peroxides, the iodometric method is more precise and accurate. The colorimetric method is not well suited for materials containing polymer, particularly diolefins, because of turbidity caused by insoluble high polymers.

ANALYTICAL METHOD

APPARATUS

A Spekker photoelectric absorptiometer was used equipped with blue No. 6 and green No. 5 filters and absorption cells allowing the light to pass through a layer of solution 1 cm thick.*

REAGENT

FERROUS THIOCYANATE SOLUTION. Dissolve 1 gram of CP ammonium thiocyanate and 1 ml of 25% by weight of sulfuric acid in 200 ml of CP deaerated methanol, and shake the resulting solution with 0.2 gram of finely pulverized ferrous ammonium sulfate. Place the decanted reagent (prepared fresh daily) in a brown glass-stoppered bottle.

PROCEDURE

Introduce 1 ml of sample preferably containing between 0.0001 and 0.0007 meq. of reactive peroxide into a 25-ml volumetric flask. (If necessary, use 1 ml of a suitable methanol dilution of the sample.) Fill to the mark with ferrous thiocyanate reagent, mix well, and fill the colorimeter cell with the mixture. At 10 minutes from the time of mixing read the optical density and convert the value, corrected by a blank determination, to terms of concentration by comparing it with the optical density calibration curve obtained by use of corresponding known amounts of standard ferric chloride solutions.

* Note by Siggia. Any color comparator or spectrophotometer can be adapted, usually with little or no change in conditions.

EXPERIMENTAL

MATERIALS USED

Samples of tetralin peroxide, cumene peroxide, *tert*-butyl hydroperoxide, and 30% hydrogen peroxide and autoxidized samples of diisobutylene, 2-pentene, cyclohexene, methylpentadiene, and diethyl ether were identical with those used in experiments reported in a previous paper (17). (See pp. 325–36).

APPLICABILITY AND ACCURACY

The listed materials were analyzed by the colorimetric method and also by the sodium iodide–isopropyl alcohol method (17). Results given in Table 12 show that the two methods agree reasonably well in most cases.

Table 12. Accuracy of Colorimetric Peroxide Method Compared to Sodium Iodide–Isopropyl Alcohol Method

Material	Sodium Iodide-Isopropyl Alcohol Method	Colorimetric Method
Apparent Peroxide Content, %		
tert-Butyl hydroperoxide	99.8	105
Cumene peroxide	94.2	112
Tetralin peroxide	95.4	95
Hydrogen peroxide, 30%	29.2	24
Peroxide No., Milliequivalents per Liter		
Cyclohexene	121	120
2-Pentene	51	48
Diisobutylene	19.7	19
Diethyl ether	12.2	7.4
Methylpentadiene	61	266

Poor agreement was obtained on ether and methylpentadiene (a conjugated diolefin). In the latter case, the reduction reaction was slow and incomplete with both methods, indicating that both disagreeing results were of only empirical value. Benzoyl peroxide also reacted only slowly with the ferrous thiocyanate reagent. Since results close to theoretical

17. C. D. Wagner, R. H. Smith, and E. D. Peters, *Anal. Chem.*, **19,** 976–9 (1947).

(considering the low precision of the colorimetric method) were obtained on the pure hydroperoxides, there is reason to believe that the colorimetric method is fundamentally accurate when applied to autoxidized materials containing only isolated double bonds.

EFFECTS OF SAMPLE SIZE AND REACTION TIME

The hydroperoxides react quickly; the optical density increases rapidly during the first 2 minutes and thereafter remains constant. However the peroxides in methylpentadiene, which are presumably of the bridge type described by Bodendorff (12), react very slowly (Table 13). Sample size does not seem to influence results appreciably on any of the materials tested (Tables 13 and 14). The absence of an effect of sample size with methylpentadiene seems odd in view of the pronounced effect of this variable on results for conjugated diolefins by the sodium iodide method.

Table 13. Effects of Sample Size and Reaction Time on Results for Autoxidized Methylpentadiene by Colorimetric Peroxide Method

Sample Dilution in Methanol	Peroxide No., Milliequivalents per Liter[a]				
	2.5 min.	5 min.	10 min.	20 min.	40 min.
1:500	170	230	300	365	430
1:1000	160	230	300	360	410

[a] Referred to original sample before dilution.

EFFECTS OF OXYGEN CONCENTRATION

To remove dissolved oxygen from the sample and reagent before mixing, an apparatus similar to that described by Lea (2) was used (Fig. 6.2).

Nitrogen, from which oxygen had been removed by passage over copper at 700°C, was passed for 15 minutes through the 1 ml of sample in the cylinder and then through the reagent in the funnel. At the end of this period the reagent was allowed to mix with the sample, and a portion of the mixture was removed (with nitrogen being bubbled through the solution) via a nitrogen-filled pipet to fill a colorimeter cell, through which nitrogen was passed by means of a special cover (Fig. 6.3). After a glass stopper was placed in the ground joint and the stopcock closed, the

Table 14. Effects of Sample Size on Results by Colorimetric Peroxide Method

Sample	Dilution	Peroxide No., Milliequivalents per Liter
tert-Butyl hydroperoxide	40.2 mg./l.	0.96[a]
in methanol	20.1 mg./l.	0.48[a]
Cyclohexene	1:200	98 to 104
	1:400	108 to 112
Diisobutylene	1:25	16 to 17
	1:50	16 to 17
2-Pentene	1:200	62 to 70
	7:4000	46 to 57
Diethyl ether	1:5	2.35
	1:10	2.20

[a] Results quoted on dilute solution of pure peroxide; all others refer to original sample before dilution.

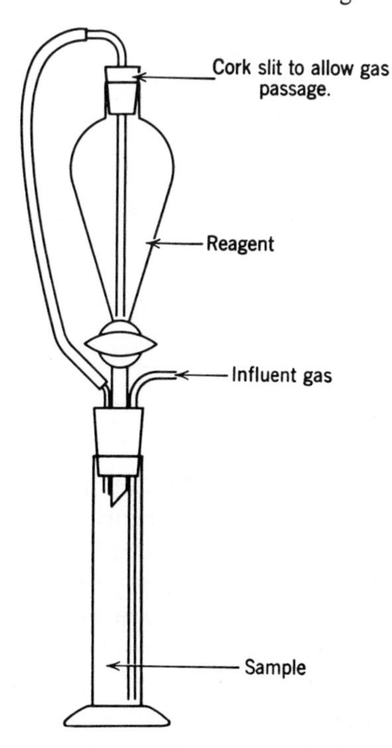

Fig. 6.2. Apparatus for changing dissolved gas in reagent and sample before mixing.

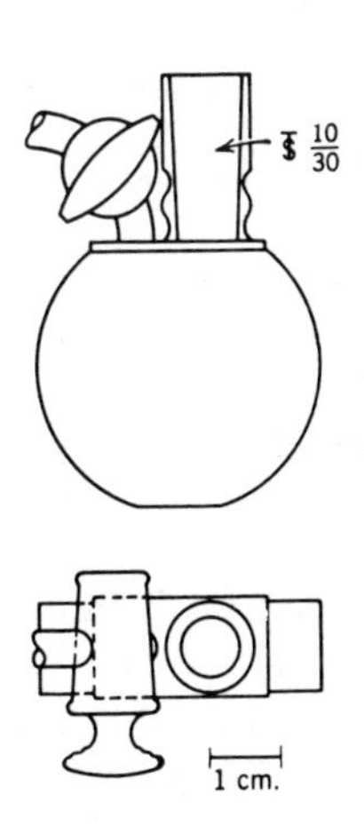

Fig. 6.3. Calorimeter cell with cover for air exclusion.

tubing conveying the nitrogen was removed and the optical density of the solution determined.

The third column of Table 15 substantiates the observation of Lea (2) that much lower results are obtained in the absence of oxygen. As a check on the effect of the passage of gas through the materials, air was passed through the sample and reagent before mixing in exactly the same way as

Table 15. Effects of Oxygen Concentration on Results by Colorimetric Peroxide Method

Sample	Prescribed Procedure	10 Min. Air, Then Mixed	10 Min. N_2, Then Mixed	10 Min. N_2, Mixed, Then 5 Min. Air	5 Min. O_2, Then Mixed
	Apparent Peroxide Content, %				
tert-Butyl hydroperoxide in methanol	94	96	63	56	—
Tetralin peroxide in methanol	98	68	4	6	91
29.4% hydrogen peroxide in methanol	23.5	—	15.4	15.4	—
	Peroxide No., Milliequivalents per Liter				
Cyclohexene	120	120	20	44	108
2-Pentene	48	53	11	11	—
Diisobutylene	19	21	4	4	—
Methylpentadiene	266	246	193	180	—
Diethyl ether	2.3	—	0.8	0.8	—

for nitrogen. The second column shows that good agreement is obtained with the results by the prescribed method except for pure tetralin peroxide, which is somewhat anomalous. When air was added after mixing the deaerated sample and reagent, no deepening of color occurred, as shown in the fourth column. This demonstrates that the peroxide (nondiolefin) is completely and rapidly destroyed in the presence of ferrous ion by both reduction and decomposition. In the presence of molecular oxygen, the reduction reaction takes place much more rapidly than the decomposition reaction; in the absence of molecular oxygen the reverse is true, and only a small quantity of ferric thiocyanate is formed. Use of oxygen in place of nitrogen gave approximately normal values, verifying the conclusion that a certain amount of oxygen is necessary to obtain quantitative results and that the effect is not one of oxidation of ferrous ion by molecular oxygen.

In the absence of oxygen, lower values are also obtained on methylpentadiene, although the degree of lowering is less marked. The presence or absence of oxygen seems to have no influence on the rate of reduction of the peroxides remaining after mixing, since the optical density after mixing increases slowly and at about the same rate, whether oxygen is present or not. This suggests that the hydroperoxides are destroyed rapidly upon mixing the reagent and sample, leaving only the bridge-type peroxides intact.

Ferrous-Titanous Reduction Method

Study of C. D. Wagner, R. H. Smith, and E. D. Peters

[*Reprinted with Permission from Anal. Chem.*, ***19,*** *982–4* (*1947*)]

In 1931 Yule and Wilson (18) reported that peroxides in gasolines could be quickly and conveniently determined by reducing them with ferrous thiocyanate in aqueous acetone and titrating the resulting ferric thiocyanate with standard titanous solution. They used results obtained in this way chiefly as a measure of the gum-forming tendencies of the gasoline and stated that other methods were less suitable, although they recognized that the results by the ferrous-titanous method did not represent actual peroxide content. The latter fact was evident in view of dependence of results on sample size and given the observation that certain materials treated with ferrous sulfate contained peroxides capable of oxidizing iodide ion. Their choice of the ferrous-titanous method over the iodometric methods was based on the greater sensitivity of the ferrous-titanous method compared with that of the Marks and Morrell method (3), with which high and variable blanks are obtained [the blanks by the modified Kokatnur and Jelling method (7, 17) are negligible].

Evidence has been obtained (17, 19) that the sodium iodide–isopropyl alcohol method adapted from that of Kokatnur and Jelling (7) and the ferrous thiocyanate colorimetric method of Young, Vogt, and Nieuwland (14) are both basically accurate for the determination of peroxides in autoxidized materials which contain no conjugated diolefins. It was found that both methods give only empirical results on conjugated diolefin samples because of the slow rate at which such peroxides are reduced and possibly because of side reaction. In continuation of this study and in view

18. J. A. C. Yule and C. P. Wilson, *Ind. Eng. Chem.*, **23,** 1254 (1931).
19. C. D. Wagner, H. L. Clever, and E. D. Peters, *Anal. Chem.*, **19,** 980–2 (1947) (See pp. 342–49).

of the wide use of the Yule and Wilson method, particularly in the petroleum industry, studies were made to determine its fundamental accuracy in the determination of peroxides in autoxidized materials and to compare it with other methods. Since the correction factors and dilutions prescribed by Yule and Wilson (18) were admittedly arbitrary, they have not been applied in the work described by Wagner, Smith, and Peters.

Specific peroxides of known purity, such as tetrahydronaphthyl hydroperoxide (tetralin peroxide), α,α-dimethylbenzyl hydroperoxide (cumene peroxide), *tert*-butyl hydroperoxide, benzoyl peroxide, and hydrogen peroxide gave results ranging from 12 to 80% of theoretical, with rather precise values for any given material. Results on autoxidized materials (nondienes) were generally from 20 to 60% of those by the sodium iodide method. Therefore, the evidence was strong that results by the ferrous-titanous method are generally very low. Yule and Wilson ascribed this tendency to decomposition of peroxides in the presence of ferrous ion to form nonoxidizing products (such reactions with tetralin peroxide and other peroxides are well known), and stability of some peroxides toward reduction by ferrous ion.

Yule and Wilson did not report the effect of molecular oxygen concentration on results by their method, but Lea (2), employing a similar method, proved that values with air excluded were only a fraction of those obtained by the usual procedure performed in the presence of air. He concluded that atmospheric and dissolved oxygen was oxidizing ferrous ion in the presence of peroxide and that in the absence of peroxides such an oxidation was insignificant. The effect of lack of oxygen has previously been recognized by Wagner, Smith, and Peters, who had drawn similar conclusions. However in no cases were the values obtained known to be higher than the theoretical ones.

In the light of later studies with the colorimetric method (15) indicating that values obtained on autoxidized rubber samples were in no case higher than warranted by the amount of oxygen absorbed by the rubber, it seemed more likely that oxygen either catalyzes the reduction reaction or inhibits the ferrous ion-catalyzed decomposition reaction. Failure to obtain quantitative reduction of peroxides of known purity by the Yule-Wilson method, even in the presence of oxygen, is ascribed to too low a ratio of dissolved oxygen to peroxide, this ratio being of the order of only one-hundredth that obtaining with the ferrous thiocyanate colorimetric method. The dependence of results on sample size (peroxide concentration) is thought to be closely related to the oxygen effect.

With the assumption that the relative results obtained by this method can be correlated with other properties of the material, hence can be useful, the extent of dependence of precision on the manipulative variables was studied. On conjugated diolefins, results were found to be critically dependent on sample size, time of reaction, and temperature; on the rapidly reacting nondienes, only sample size had any effect.

ANALYTICAL METHOD

REAGENTS

Ferric chloride solution, standard 0.1*N*.

FERROUS THIOCYANATE SOLUTION. Dissolve 5 grams of ferrous sulfate and 5 grams of ammonium thiocyanate in 500 ml of distilled water and 500 ml of acetone. Add about 1 gram of pure iron wire and 5 ml of concentrated sulfuric acid, and expel air by passing hydrogen or carbon dioxide through the solution. Store the solution in an all-glass system under hydrogen; use when the red color has disappeared.

TITANOUS SULFATE SOLUTION, STANDARD 0.01*N*. Heat a mixture of 20 ml of CP concentrated sulfuric acid and 80 ml of water to 70°C and add in small portions 0.6 gram of titanium hydride powder (obtainable from Metal Hydrides, Inc., Beverly, Mass., under the trade name Altertan). When the reaction subsides, boil the solution on a hot plate for 2 minutes, then pour it into about 900 ml of distilled water, freshly deaerated with carbon dioxide. After the undissolved material has settled, siphon the supernatant liquid into a dark storage bottle previously filled with carbon dioxide. Store the solution under hydrogen, dispense from an all-glass system, and standardize daily against the standard ferric chloride solution in the presence of thiocyanate ions.

PROCEDURE

Measure 50 ml of ferrous thiocyanate solution into a 250-ml glass-stoppered Erlenmeyer flask and discharge any red color with the minimum amount of titanous solution. Bring the solution to 25 ± 2°C, add the sample containing up to 5 meq. of peroxide, stopper the flask, shake vigorously for 5 minutes ±5 seconds, and titrate with standard titanous solution to the disappearance of the red color.

EXPERIMENTAL

MATERIALS USED

Samples of tetralin peroxide, cumene peroxide, *tert*-butyl hydroperoxide, 30% hydrogen peroxide, benzoyl peroxide, and ascaridole, and autoxidized samples of diisobutylene, 2-pentene, cyclohexene, tetralin, methylpentadiene, isoprene, and diethyl ether; these were identical with those used in experiments reported in Ref. 17.

ACCURACY

The method was applied to several of the known peroxides to check its accuracy. As shown in Table 16, the recoveries with hydroperoxides are invariably far below theoretical or below that for the sodium iodide–isopropyl alcohol method.

Table 16. Comparison of Results on Known Peroxides by Ferrous-Titanous and Sodium Iodide–Isopropyl Alcohol Methods

Peroxide	Peroxide Concentration G./l.[a]	Peroxide Purity Found	
		Sodium Iodide Method	Ferrous-Titanous Method
		Per cent by weight	
Tetralin peroxide	7.92	98.3, 98.4	25.5, 27.0
Cumene peroxide	13.86	94.2	73.5
tert-Butyl hydroperoxide	6.59	98.3, 98.5	78.7, 79.1
Hydrogen peroxide (29.9%)	11.59	29.4	12.2
Benzoyl peroxide	10.10	98.7, 99.0	13.6, 13.9
Ascaridole	8.40	25.7	31.7

[a] Benzene (C.P. thiophene-free) used as solvent except in case of hydrogen peroxide, which was dissolved in water. 5 ml. of solution used in all cases.

Of the peroxides tested, only tetralin peroxide and cumene peroxide were formed by natural autoxidation, but the results obtained on those demonstrated that the ferrous-titanous method is inaccurate on even the simplest peroxides, whereas the iodometric method gives satisfactory accuracy (17). The fact that the ferrous-titanous results are dependent on sample size indicates, however, that it may be possible to obtain theoretical results on samples containing only very small amounts of peroxides. Comparative data obtained on typical samples of autoxidized materials are shown in Table 17.

Comparison of the analyses of the nondiolefinic samples shows considerable similarity with results on pure hydroperoxides. This result is expected, since it has been shown that the initial products of autoxidation

Table 17. Comparison of Results on Autoxidized Materials by Ferrous-Titanous and Sodium Iodide–Isopropyl Alcohol Methods

Material[a]	Peroxide No., Milliequivalents per Liter	
	Sodium Iodide Method	Ferrous-Titanous Method
Diisobutylene	35.7, 36.9	25.2
2-Pentene	40.8, 41.6	16.0, 15.9
Cyclohexene	85.0, 84.5	20.6, 20.6
Tetralin	91.4, 90.8	18.6, 18.0
Diethyl ether	12.2, 12.4	7.2, 7.4
Methylpentadiene	65.9, 68.5	88.0, 89.2
Isoprene	58.8, 57.1	67.7, 66.1

[a] 5 ml. of sample used in all cases.

of olefins are, in fact, hydroperoxides. On the other hand, Bodendorff (12) has shown that conjugated diolefins form bridge-type peroxides by 1,4-addition of molecular oxygen intramolecularly or intermolecularly. These substances react slowly with ferrous ion as well as with iodide ion, the results with both methods being purely empirical. Ascaridole, a pure monomeric 1,4-bridge-type peroxide, behaves in the same manner, reacting slowly and incompletely with both iodide ion and ferrous ion.

OXYGEN EFFECT

To test the oxygen effect, noted by Lea (2), experiments were performed in which carbon dioxide or oxygen was passed through the reagent for 2 minutes before adding the sample and shaking the mixture, and also during the titration. Results, compared with those obtained with an air atmosphere, are shown in Table 18. Blank corrections, which were always very small, were applied in each case.

The dependence on oxygen concentration is very evident. Moreover, as was the case with the colorimetric method (19), no further coloration appears upon passage of air or oxygen into the mixture after a sample of an autoxidized nondiolefin has been reduced under carbon dioxide and titrated. This indicates that the peroxides have been entirely destroyed, although they have been only partially reduced by ferrous ion. This is evidence for the conclusion that oxygen either catalyzes the reduction reaction or inhibits the decomposition reaction.

Table 18. Effects of Oxygen Concentration on Results by Ferrous-Titanous Method

	Purging Gas Used		
	CO_2	Air	O_2
Apparent Peroxide Content, %[a]			
Tetralin peroxide	6.4	24.0	56.9
Cumene peroxide	58.3	73.2	—
tert-Butyl hydroperoxide	57.1	61.8	64.2
Hydrogen peroxide	11.7	12.2	13.8
Benzoyl peroxide	15.4	18.8	18.0
Ascaridole	9.2	31.7	39.2
Peroxide Number, Milliequivalents per Liter			
Diisobutylene	1.3	2.2	2.4
2-Pentene	5.2	17.6	30.7
Cyclohexene	6.7	20.6	46.8
Tetralin	2.2	16.9	41.1
Methylpentadiene	2.4	23.8	23.8
Isoprene	9.0	60.0	65.3
Diethyl ether	3.6	7.4	—

[a] Hydrogen peroxide diluted with water; all other known peroxides dissolved in C.P. benzene. 5-ml. samples used.

In contrast to the colorimetric method, results even in the presence of pure oxygen are considerably below theoretical. These results are understandable if we assume that the concentration of oxygen necessary for quantitative reduction of the peroxides is dependent on the peroxide concentration; the ratio dissolved oxygen to peroxide is roughly one-hundredth as great with the ferrous-titranous method as with the colorimetric method (19).

DEPENDENCE ON MANIPULATIVE VARIABLES

In the light of the data, the ferrous-titanous method appears to be empirical and useful only if values obtained by it are reproducible and correlate well with some property of the test material significant to application, such as explosion hazard or gum-forming tendency. Such correlations can be tested only by experiments on the material in question. To test the dependence of precision on the experimental conditions of the method, various materials were analyzed, using variations in time of shaking, sample size, and reaction temperature. Results (Tables 19 and

Table 19. Effects of Time of Shaking and Sample Size in Determination of Known Peroxides by Ferrous-Titanous Method

	Apparent Peroxide Content, %W of Original Sample					
	Time of Shaking			Sample Size		
	1 min.	5 min.	15 min.	10 ml.	5 ml.	2 ml.
Tetralin peroxide	23.1	22.1	24.5	20.6	22.1	39.1
tert-Butyl hydroperoxide	80.8	80.6	80.6	68.8	80.6	90.0
Hydrogen peroxide[a]	14.8	14.1	14.0	13.5	14.1	14.5
Benzoyl peroxide	6.5	15.0	31.9	—	15.0	27.1
Ascaridole	31.7	31.7	30.4	29.2	31.7	38.7

[a] Solutions in water; all others made up in C.P. benzene, about 10 grams per liter. Hydrogen peroxide original solution actually about 30%. 5 ml. samples used unless otherwise specified.

20) show that the behavior of peroxides in conjugated diolefins is in sharp contrast with the behavior of those obtained from compounds with isolated double bonds or ether linkages. Peroxides in conjugated dienes are slow reacting, and results are affected by sample size, temperature,

Table 20. Effects of Manipulative Variables on Results for Autoxidized Materials by Ferrous-Titanous Method

	Peroxide No., Milliequivalents per Liter							
	Time of Shaking			Sample Size			Reaction Temperature[a], °C	
Material	1 min.	5 min.	15 min.	10 ml.	5 ml.	2 ml.	15	30
Diisobutylene	2.0	2.2	2.2	2.0	2.2	—	2.2	2.2
2-Pentene	17.7	17.6	16.5	—	15.7	21.5	17.4	15.6
Cyclohexene	29.0	29.4	27.0	—	29.4	45.8	23.5	21.6
Tetralin	21.8	20.8	22.0	—	20.9	40.0	18.2	16.8
Diethyl ether	8.3	8.2	7.3	7.5	8.2	8.6	—	—
Isoprene	25	60	103	36	64	99	37.8	70.1
Methylpentadiene	43	90	140	—	90	144	13.7	29.3

[a] Obtained in special water-jacketed flask, with reagent and flask brought to temperature before addition of sample.

and oxygen concentration. Peroxides in the other materials react almost instantaneously, and results depend only on oxygen concentration and, to a lesser extent, on sample size. These results are consistent with the fact that peroxides in mono-olefins, and probably those in ethers, are the rapidly reacting alkyl hydroperoxides, whereas those in conjugated diolefins are the slow-reacting bridge-type peroxides. It is evident that with the latter, precise results are obtainable only if variables in the analysis are controlled rather closely. Benzoyl peroxide is a special case, reacting fairly rapidly with iodide ion but only slowly with ferrous ion. Ascaridole, although it is a typical diolefin peroxide monomer, seems to react rapidly in contrast to autozidized diolefins. The slow-reacting diolefin peroxides may be the polymeric peroxide types mentioned by Bodendorff (12) rather than the monomeric bridge-type forms.

Leuco Methylene Blue Method

Method of M. I. Eiss and P. Giesecke

[*Reprinted in Part from Anal. Chem.*, ***31,*** *1558–60* (*1959*)]

A method using leuco methylene blue as a reagent for peroxides was proposed by Ueberreiter and Sorge in 1956 (20). The leuco base was not available in a usable form, however, because of its extreme instability. It was difficult to synthesize and troublesome to store.

Benzoyl leuco methylene blue was found to be a new sensitive colorimetric reagent for organic peroxides. In a benzene–trichloroacetic acid solution, it reacted with peroxides and hydroperoxides to form the characteristic methylene blue color (Fig. 6.4).

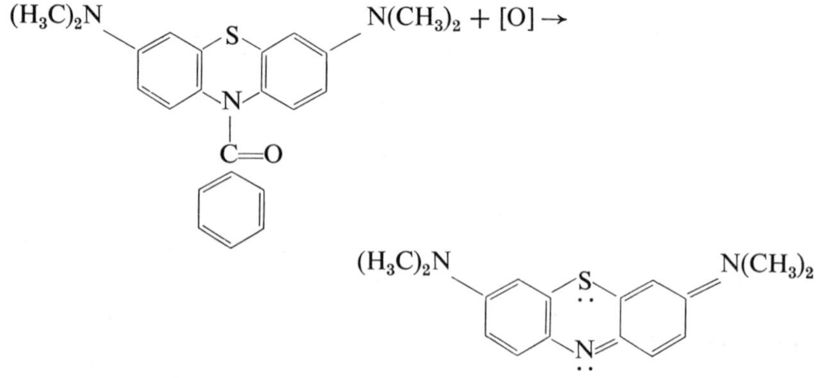

$(H_3C)_2N$ ... $N(CH_3)_2 + [O] \rightarrow$ ... $(H_3C)_2N$ ... $N(CH_3)_2$

Benzoyl leuco methylene blue

Methylene blue cation

20. G. Sorge and K. Ueberreiter, *Angew. Chem.*, **68,** 486–91 (1956).

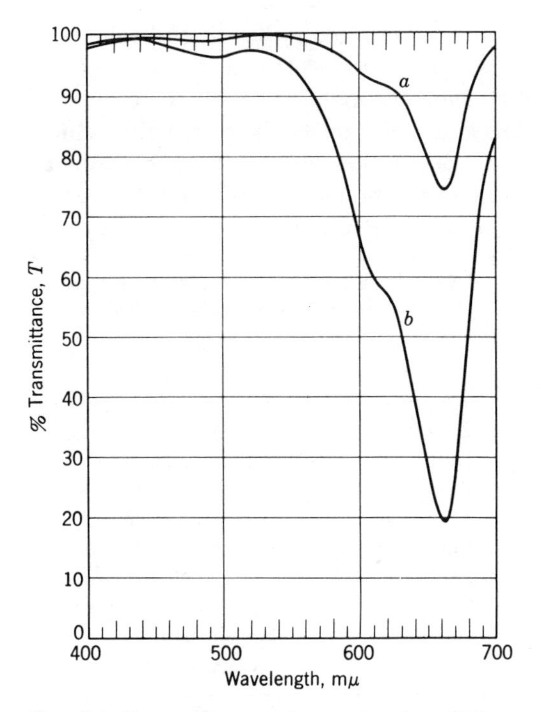

Fig. 6.4. Spectrophotometric curves of methylene blue system: *a*, no active oxygen; *b*, 3γ active oxygen.

The reagent was stable in its crystalline form and storable in a refrigerator under normal conditions. It was not greatly affected by air, and in benzene solution it could be stored in a brown bottle at room temperature. Benzoyl leuco methylene blue was affected by ultraviolet light (it turned blue rapidly, therefore had to be kept out of sunlight) and, to a lesser degree, by heat and artificial light. However the benzoyl leuco dye–peroxide reaction was fairly slow. A hydroperoxide of the *tert*-butyl type required 36 hours to react at room temperature, and benzoyl peroxide took over 120 hours.

An attempt was made to find a method of accelerating the color development by speeding up the decomposition of the peroxide and, at the same time, maintaining a good level of accuracy. Several heating variations were attempted, but even though the reaction was accelerated, the results were not quantitative. Light was also found to increase the reaction rate, but it caused a marked decrease in reproducibility.

It was thought that amines and cobalt naphthenate would increase the rate of color development by quickly decomposing the peroxides. Several amines and amides were tried, but they did not have the required speed.

Cobalt naphthenate was reported to be a useful compound for decomposing peroxides. It did not increase the rate of this reaction, however. An investigation of all available metallic naphthenates was undertaken. Table 21 lists the metals and their effects on the reaction. Zirconium was found to give the fastest reaction time and still have a reasonable reagent blank. Zinc and lead produced some increase of color development but were not as good as the zirconium. Cerium and manganese could possibly have

Table 21. Effects of Metallic Naphthenates on Benzoyl–Methylene Blue System

Metals[a]	Effects
Zirconium	Excellent acceleration
Lead	Some acceleration
Zinc	Some acceleration
Manganese	Excessive reagent blank
Cerium	Excessive reagent blank
Iron	Highly colored, green
Copper	Highly colored, green
Cobalt	No acceleration
Calcium	No acceleration

[a] One drop commercial naphthenate in 52 ml. of benzene-trichloroacetic acid solution.

been used in lower concentrations but, under these particular conditions, have an excessive reagent blank.

In general, the metallic naphthenates are thought to react with hydroperoxides and peroxides in a manner similar to the following lead reaction:

$$2R{\cdot}OOH \xrightarrow{Pb^{+2}} RO\cdot + ROO\cdot + H_2O$$

The RO· and ROO· free radicals would then oxidize benzoyl leuco methylene blue to the methylene blue cation. However it is not yet known whether zirconium reacts with peroxides in the same way as lead because of the reluctance of the former to exist in several valence states.

REAGENTS AND APPARATUS

Benzene, reagent grade.

Zirconium naphthenate, 0.24%. Dilute 1 ml of commerical (6%) zirconium naphthenate to 25 ml with benzene.

Trichloroacetic acid, reagent grade, 0.5% in benzene.

Benzoyl leuco methylene blue (obtained from the National Cash Register Co., Dayton, Ohio). Dissolve 0.05 gram in 100 ml of benzene. Store in dark bottle.

All spectrophotometric measurements were made with a General Electric recording spectrophotometer with 1-cm cells. A colorimeter can also be used with Corning filters 2403 and 3962.

PROCEDURE

Prepare standard peroxide and hydroperoxide solutions by dissolving weighed amounts of the pure materials in benzene. To obtain a calibration pipet aliquot portions of peroxide solution into a 25-ml volumetric flask that contains 15 to 20 ml of 0.5% trichloroacetic acid–benzene solution, and add 0.3 ml of 0.24% zirconium naphthenate and 1 ml of leuco dye. Fill the flask to the mark with benzene, mix thoroughly, and protect it from light immediately. Allow the flasks to stand in the dark for a designated time (Table 22) at 24 ±1°C. Measure the transmittance against water at 662 nm on a spectrophotometer using 1-cm cells. Run a reagent blank with each set of standards.

The same procedure is used for sample analysis. If the peroxide species is unknown, the time for the peroxide decomposition to reach completion must be checked experimentally by measuring the absorbance until it no longer increases. Results are then calculated as percentage of active oxygen.

RESULTS AND DISCUSSION

Table 22 indicates the various peroxides and hydroperoxides tested and the time required for each reaction to reach completion. It is clearly seen that zirconium aids in the decomposition of the peroxides, thereby accelerating the color reaction. The color development time of *tert*-butyl hydroperoxide was reduced from 36 hours to 30 minutes and the benzoyl peroxide from 120 to 30 hours. Benzoyl peroxide decomposed much more slowly that the other peroxides tested, even with the addition of zirconium. For the faster reacting hydroperoxides (cumene and *tert*-butyl), most of the color was developed within the first 10 minutes (Fig. 6.5).

The reaction product of benzoyl leuco methylene blue and benzoyl peroxide, lauroyl peroxide, cumene hydroperoxide, and *p*-menthane

Table 22. Reaction Times and Absorbances for Peroxides Tested

Name	Formula	% Active O	Time for Complete Color Development at 25°C	Absorbance, 1 mg./100 ml., 1-cm. Cell
tert-Butyl hydroperoxide	$H_3C{-}C(CH_3)_2{-}OOH$	17.8	30 min. (36 hr.[a])	Approx. 16
Cumene hydroperoxide	$CH_3{-}C(CH_3)(C_6H_5){-}OOH$	10.5	40 min. (38 hr.[a])	9.7 ± 0.25[b]
p-Menthane hydroperoxide	$CH_3{-}C(OOH){-}C_6H_4{-}CH(CH_3){-}CH_3$	9.3	2 hours	(8.8 ± 0.15)[b]
Lauroyl peroxide	$H_{23}C_{11}{-}C({=}O){-}O{-}O{-}C({=}O){-}C_{11}H_{23}$	4.02	5 hours	(4.6 ± 0.10)[b]
Benzoyl peroxide	$C_6H_5{-}C({=}O){-}O{-}O{-}C({=}O){-}C_6H_5$	6.6	30 hr. (120 hr.[a])	(6.2 ± 0.12)[b]

[a] Time required for complete color development without use of zirconium naphthenate.
[b] 95% confidence limit.

hydroperoxide followed Beer's law up to 1 ppm. However, *tert*-butyl hydroperoxide deviated somewhat, and a calibration curve of concentration versus absorbance was necessary for this compound.

The validity of this method should be verified experimentally before applying it to peroxides other than those tested. For example, dialkyl peroxides, such as di-*tert*-butyl peroxide (20), do not react with benzoyl leuco methylene blue.

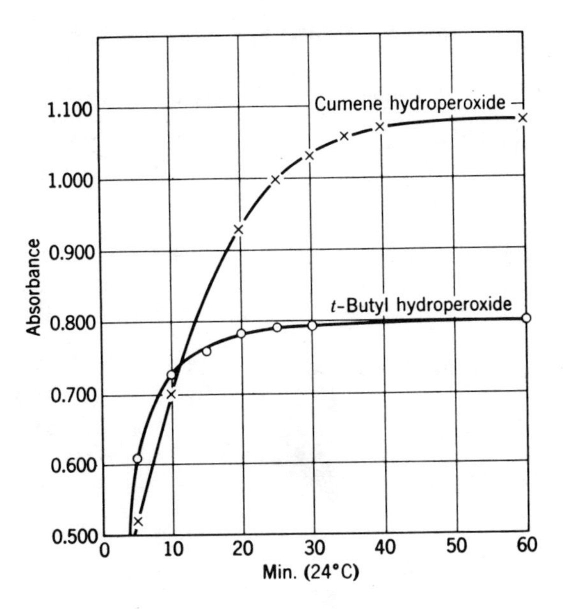

Fig. 6.5. Rate of color development.

The reaction was heat sensitive; therefore, all work was done at 24 ± 1°C. At 30°C, the results became very erratic. Because the color was also light sensitive, the solutions were stored in the dark while the color was developing. Artificial light caused irregular results, and sunlight ruined the determination completely by causing very excessive reagent blanks.

The reagents were all stable in benzene solution and were usable for 3 to 4 weeks. The leuco dye was kept in a brown bottle away from direct light. The only reagent concentration that appeared to be critical was the zirconium naphthenate. If too much was added, the reagent blank became excessive.

Although there was a waiting period for color development, it was simple to run six to ten determinations at the same time. The reaction was sensitive, and active oxygen was determined down to less than 0.5 mg. Table 23 lists analyses of the five standard peroxides. All commerical peroxides used as standards were analyzed iodometrically. The precision of the method was obtained by running 7 to 16 replicate samples for each peroxide. For the four compounds that follow Beer's law, the 95% confidence limits range from ±2.6 to ±1.7%.

Table 23. Analyses of Peroxide Standards

Name	Present, γ	Found, γ
Lauroyl peroxide	11.0	10.9
	17.0	16.8
	22.0	21.6
Benzoyl peroxide	9.9	10.2
	19.8	20.0
	36.0	35.8
p-Menthane hydroperoxide	13.1	13.1
	26.3	25.9
	32.8	33.5
Cumene hydroperoxide	3.8	3.6
	7.6	7.3
	15.2	14.9
tert-Butyl hydroperoxide	5.9	5.7
	11.8	11.9
	23.6	24.3

Determination of Peroxides Using Arsenious Oxide

Need often arises for the determination of peroxides in media in which methods involving the liberation of iodine from potassium iodide cannot be used. In the discussion of the foregoing procedure it was shown that the only common organic solvents in which the iodometric methods will work efficiently are 2-propanol and ethyl alcohol. For peroxides in unsaturated organic compounds, such as may occur in polymerization work, the iodometric methods cannot be used because they present the difficulties of oxidation and substitution.

In this method, arsenious oxide is used to reduce the peroxide. This method works in many common organic solvents and also can be used to determine peroxides in monomers and polymers. Reichert and coworkers (21) used arsenious oxide to determine various inorganic peroxides.

21. J. S. Reichert, S. A. McNeight, and H. W. Rudel, *Ind. Eng. Chem., Anal. Ed.*, **11,** 194–7 (1939).

Procedure of S. Siggia

[*Ind. Eng. Chem., Anal. Ed.*, **19,** 872 (1947)]

REAGENTS

0.1*N* Arsenious oxide standard solution containing 25 grams of sodium bicarbonate per liter of solution.
1*N* Sulfuric acid.
Sodium bicarbonate.
0.1*N* Standard iodine solution.

PROCEDURE

Introduce 25 ml of a standard 0.1*N* arsenious oxide solution containing 25 grams of sodium bicarbonate per liter, into a 125-ml Erlenmeyer flask. To this add a sufficient amount of peroxide-containing solution to yield about 0.005 to 0.010 gram of active oxygen. Where the sample solutions are immiscible with water, add ethanol until a homogeneous solution is obtained. When the sample is a polymer, dissolve it in benzene, and add the ethanol as before. Precipitation of the polymer does not affect results noticeably. If the peroxide content in the polymer is greater than 5% of benzoyl peroxide, however, it is best to pour off the supernatant liquid from the precipitated polymer, redissolve the polymer, and reprecipitate it with ethanol. Then combine the extracts.

Add boiling chips and concentrate the solution to about 25 ml with a stream of air flowing over the liquid to increase the rate of evaporation. Add water until the volume is about 40 ml. Continue this procedure until practically all the monomer, alcohol, or other organic solvents are replaced with water. The solution is made barely acid with 1*N* sulfuric acid, and then add 0.5 gram of sodium bicarbonate. Chill the solution and titrate the excess arsenious oxide with standard 0.05 to 0.1*N* iodine to the appearance of the yellow iodine color. Large amounts of organic solvent in the water cause the final end point to develop very slowly. The end point is very sharp if most of the organic solvents are removed.

This method was sufficiently sensitive so that 0.008 gram of active oxygen could be accurately determined to within ±0.5%. Hydrogen peroxide was also used together with the benzoyl peroxide to test the procedure.

No interference was noted when iodine was added to the excess arsenious oxide in samples that contained unsaturated compounds because the arsenious oxide reacts instantaneously with iodine. In addition, repeated boiling with water removed most of the organic material that might have interfered by reaction with iodine. Both these factors contributed to the lack of interference.

For peroxides in polymers, it was found that not much peroxide was trapped in the polymer upon precipitation with ethanol. Benzoyl peroxide in polymers, in amounts less than 3%, could be determined with no noticeable loss of peroxide because of occlusion during the precipitation. In polymer samples containing more than 5% benzoyl peroxide, it was found best, after precipitation of the polymer with the ethanol, to pour off the supernatant liquid, redissolve the polymer in benzene, and reprecipitate it with ethanol, and then to combine the two liquid portions before analysis. This achieves a more complete extraction of the peroxide.

Table 24 shows recovery data for benzoyl peroxide in various mixtures.

Table 24

Results	Grams of Active Oxygen	
	Observed	Calculated
Benzoyl peroxide in CH_3OH	0.00387 0.00396	0.00393
Benzoyl peroxide in acetone	0.01304 0.01315 0.01303	0.01320
Benzoyl peroxide in benzene	0.00400 0.00397	0.00394
Benzoyl peroxide in styrene	0.00582 0.00581 0.00584	0.00581
Benzoyl peroxide in isobutyl vinyl ether	0.01322 0.01308	0.01323
Benzoyl peroxide in methyl methacrylate	0.01332 0.01335	0.01323
Benzoyl peroxide in polystyrene	0.00579 0.00581	0.00581

The benzoyl peroxide samples used were recrystallized from acetone.

Colorimetric Method for Trace Peroxide Using *N,N*-Dimethyl-*p*-phenylenediamine

Procedure of P. Dugan

[*Reprinted in Part from Anal. Chem.* **33,** *696–8* (*1961*); *Ibid., 1630–1* (*1961*)]

REAGENTS AND APPARATUS

A stock solution of *N,N*-dimethyl-*p*-phenylenediamine sulfate (Eastman) was prepared by dissolving 0.3 gram in 10 ml of redistilled water in a 100-ml volumetric flask; it was brought to the mark with CP absolute methanol.

Standard solutions of lauroyl peroxide (Lucidol) and benzoyl peroxide (Lucidol) were prepared by dissolving in benzene to give final concentrations in the range of 5 to 100 μg per milliliter of benzene. The stock peroxides were assayed by iodometric titration with sodium thiosulfate according to a method supplied by Lucidol and based on a molecular weight of 404 for lauroyl peroxide, and 242 for benzoyl peroxide.

All colorimetric determinations were carried out on a Beckman Model DU or DK 1 spectrophotometer employing matched 1-cm silica absorption cells.

PROCEDURE

Mix 2 ml of the *N,N*-dimethyl-*p*-phenylenediamine sulfate reagent with 2 ml of benzene containing either lauroyl or benzoyl peroxide in a concentration of 5 to 100 μg per milliliter.

Allow the mixture to react at 25°C for 30 minutes. Then determine absorbance (or %*T*) of the solution at 560 nm using a suitable reagent blank without peroxide as the reference.

RESULTS

Figure 6.6 shows spectral curves obtained by holding benzoyl peroxide constant at 40 μg per milliliter and varying amounts of *N,N*-dimethyl-*p*-phenylenediamine sulfate reagent. There was little difference between per cent transmittance at 525 and 560 nm using 0.1 and 0.3% reagents. Transmittance was greater when 0.5% reagent was used. A more highly colored reagent blank is present in the reference beam as the reagent concentration is increased.

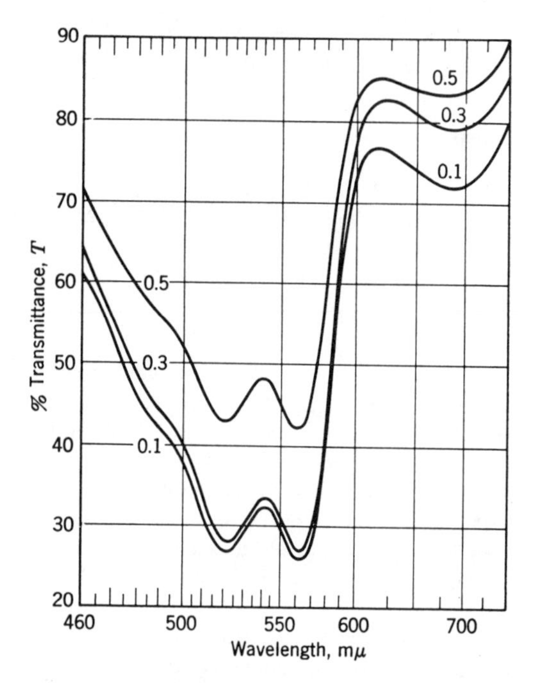

Fig. 6.6. Absorption spectra of reaction product employing three concentrations of reagent.

Figure 6.7 shows the spectrum of the reagent solution color prior to reaction with peroxide, when compared to a solvent blank in the reference beam.

Figure 6.8 shows values obtained with varied concentrations of benzoyl peroxide against a reagent blank. The reaction follows Beer's law between 5 and 30 μg per milliliter of benzoyl peroxide but deviates somewhat at 40 μg per milliliter.

Figure 6.9 shows a color change 1 hour after the spectrum in Fig. 6.7 was recorded. Greater absorption is indicated in the 400- to 525-nm range, whereas less absorption is indicated at 560 nm, although the color in the reaction vessel appears darker to the eye.

Figure 6.10 shows percentage transmission (% T) values obtained by varying the concentration of lauroyl peroxide versus a reagent blank in the reference cell. The reaction follows Beer's law in the concentration range of 10 to 100 μg of lauroyl peroxide per milliliter when measured at 560 nm.

The following observations have also been made:

Reagent in HOH + lauroyl peroxide in benzene + CH_3OH → rapid color development.

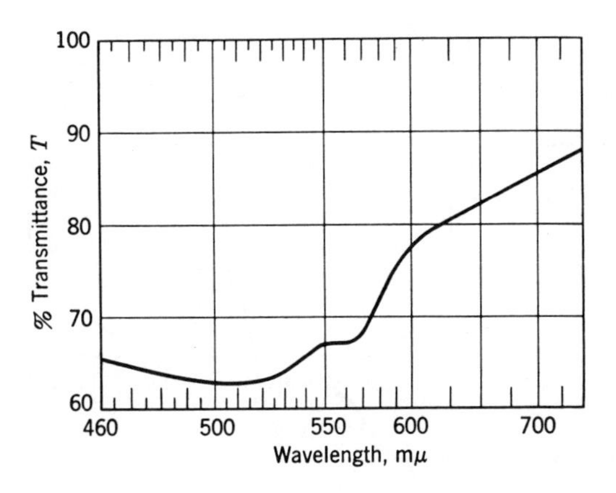

Fig. 6.7. Spectrum of reagent solution versus solvent blank.

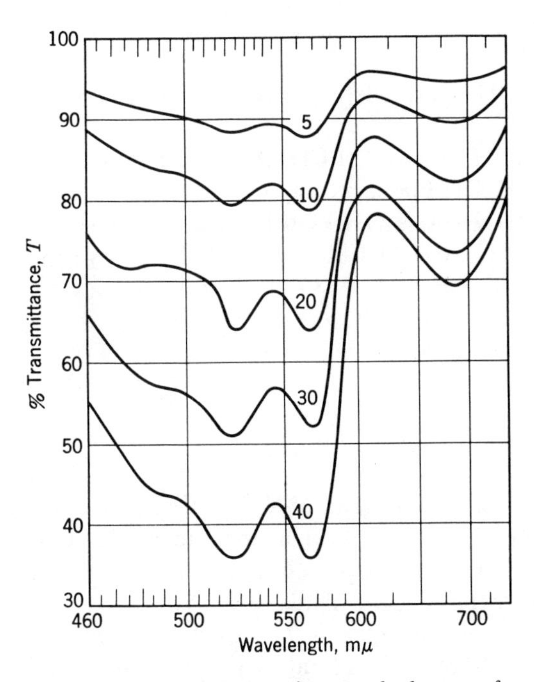

Fig. 6.8. Spectra representing standard curve for benzoyl peroxide. Values represent micrograms of peroxide per milliliters of solvent.

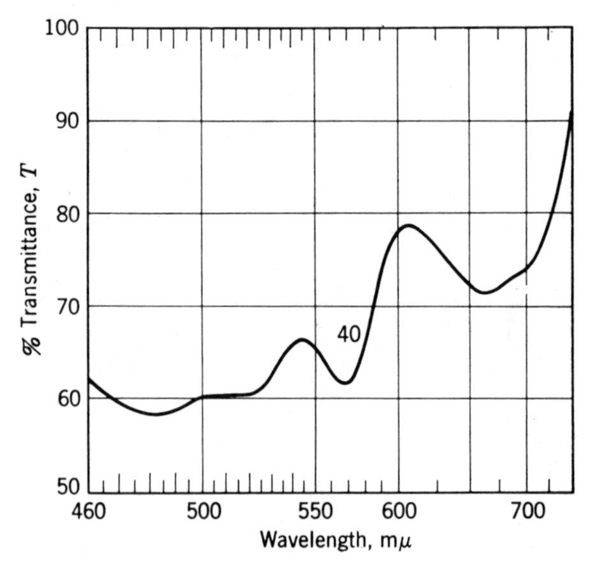

Fig. 6.9. Spectrum showing change in absorption 1 hour after recording in Fig. 6.8. Values represent micrograms of peroxide per milliliter of solvent.

Reagent in HOH + peroxide in benzene (no CH_3OH) → immiscible, no color reaction.

Reagent in HOH + benzene + CH_3OH → very slow color reaction.

Reagent in CH_3OH + benzene → very slow color reaction.

Reagent in HOH + CH_3OH + lauric acid → very slow color reaction.

Reagent in HOH + CH_3OH + lauryl alcohol → very slow color reaction.

Reagent in 90% $CH_3OH + H_2O_2$ → very slow color.

Color complex + 0.1*N* H_2SO_4 → no effect.

Color complex + 0.1*N* NaOH → colorless.

DISCUSSION

Concentration of the reagent *N*,*N*-dimethyl-*p*-phenylenediamine sulfate did not appear to be critical in the range of 0.1 to 0.3%. Higher concentrations than necessary probably decrease the sensitivity by causing excessive color in the reagent blank.

The use of 2 ml of reagent plus 2 ml of the benzene-peroxide solution was convenient, although other ratios proved satisfactory, provided the solutions were miscible.

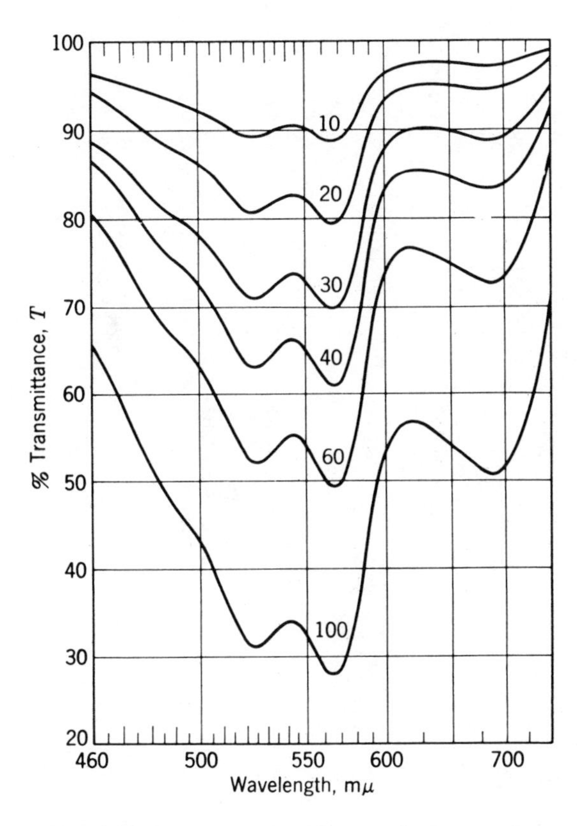

Fig. 6.10. Spectra representing standard curve for lauroyl peroxide. Values represent micrograms of peroxide per milliliter of solvent.

The ternary mixture benzene-methanol-water was examined by altering proportions of each solvent until a cloud point was reached. Completely miscible proportions were reacted with benzoyl peroxide, and light transmittance of the reaction was determined. Greater color developed with an increased amount of water in solution.

Benzene is a convenient solvent for organic peroxides. However the reaction will proceed in other solvents such as mineral oil.

Time of reaction is critical, which implies that other factors that influence rate of reaction will influence color development. Temperature was therefore held constant at 25°C during the investigation, although its influence was not specifically studied. The effect of light on the reaction

Table 25. Peroxides (22)

Type of Analysis	Reduction Reagent	Reduction Time (min.)	Reduction Temperature	Solvent	Final Colorimetric or Volumetric Reagent (if any)	Reference
Volumetric	KI (including CdI_2)	60	RT[a]	Aqueous ethyl alcohol	$Na_2S_2O_3$	23, 24, 25, 26
Volumetric	KI	Few	RT[a]	Acetic acid	$Na_2S_2O_3$	3, 4, 27
Volumetric	NaI	5–20	RT[a]	Acetic anhydride	$Na_2S_2O_3$	8
Volumetric	KI	2–5	Incipient boiling	2-propanol	$Na_2S_2O_3$	7
Volumetric	NaI	15	Reflux	2-propanol	$Na_2S_2O_3$	17
Volumetric	KI	1	—	$CHCl_3$ or $ClCH_2CH_2Cl$	$Na_2S_2O_3$	28, 29
Volumetric	KI (After reaction with $MnCl_2$)	25–300	—	—	$Na_2S_2O_3$	30
Volumetric	KI	15–60	RT[a]	tert-butyl alc.-CCl_4	$Na_2S_2O_3$	31
Volumetric	$Fe(NH_4)_2(SO_4)_2$	—	—	Acetic acid	$K_2Cr_2O_7$	32
Volumetric	$Fe(NH_4)_2(SO_4)_2$	15	—	$H_2O—C_2H_5OH$	$HgNO_3$	33
Volumetric	$FeSO_4$	—	—	1:1 Acetone-H_2O	$TiCl_3$	18, 27, 34, 35
Volumetric	Hydroquinone	—	—	—	Excess I_2 added and excess titrated with $Na_2S_2O_3$	36
Volumetric	As_2O_3	—	Boil	Ethanol water	I_2	37
Volumetric	$SnCl_2$	—	95°	Sample used as solvent	$FeCl_3$	34, 38
Volumetric	H_2(PtO_2 or Pd on charcoal used as catalyst)	—	—	—	—	39
Colorimetric	Fe^{++}	0–5	RT[a] to incipient boiling	Absolute CH_3OH	CNS^-	14, 19
Colorimetric	Fe^{++}	15	RT[a]	Benzene-CH_3OH	*o*-Phenanthroline	40
Colorimetric	Fe^{++}	5–75	RT[a]	$C_2H_5OH—CHCl_3$	CNS^-	41
Colorimetric	Fe^{++}	—	—	Benzene-CH_3OH or acetone	CNS^-	42, 43, 44
Colorimetric	$Ti(SO_4)_2$	—	—	—	None needed	45
Colorimetric	Luminol	5	—	Aqueous	Na_2CO_3	46
Colorimetric	I^-	—	—	—	I_2 color measure	47
Colorimetric	HI	—	—	Aqueous	Thiofluorescein	48
Colorimetric	3,5 Dichloro-4,4′ dihydroxy phenylenediamine	10	100°	—	None needed	49
Colorimetric	Leuco-methylene blue	—	—	Aqueous	None needed	20
Colorimetric	Leuco-methylene blue	30 min.–30 hr.	RT[a]	Benzene-trichloroacetic acid	None needed	50
Colorimetric	N,N-dimethyl-*p*-phenylenediamine	5	RT[a]	THF-methanol	None needed	51

[a] Room temperature.

can be seen by comparing the following relative values, which were obtained by holding identical lauroyl peroxide-containing reaction mixtures: (1) in the dark (64% *T*), (2) under fluorescent lamp (54% *T*), and (3) in direct window light (30% *T*). Light also affected the *N,N*-dimethyl-*p*-phenylenediamine reagent solution in the same manner. Reagent solutions were therefore prepared fresh daily and held in the dark. Reactions were carried out in subdued light, not total darkness.

The color failed to develop in the presence of dicumyl peroxide even when iron, cobalt, and rare earth naphthenates were present as catalysts. The effects of metallic naphthenates were not tested with lauroyl or benzoyl peroxide. Steric hindrance is thought to be responsible for failure of dicumyl peroxide to catalyze the color reaction.

The reaction appears to be: reagent + aqueous methanol → blue complex. This reaction is greatly accelerated by the presence of lauroyl or benzoyl peroxide. Although the reaction mechanism is not known, it does

22. S. Siggia, *Handbook of Chemical Analysis*, L. Meites. Ed., McGraw-Hill, New York, 1963, p. 112–113.
23. G. Dindgren and R. Vesterberg, *Svensk Form. Tids.*, **47,** 17–25 (1943).
24. A. W. Rowe and E. P. Phelps, *J. Am. Chem. Soc.*, **46,** 2078–85 (1924).
25. R. I. Schoetzow, *J. Am. Pharm. Assoc.*, **22,** 412–3 (1933).
26. R. Van Winkle and W. G. Christiansen, *J. Am. Pharm. Assoc.*, **18,** 1247–50 (1929).
27. J. Risbey and H. B. Nisbet, *Analyst*, **70,** 50–1 (1945).
28. N. Dresdov and L. Starikova, *Myasnaya Ind. SSSR*, **22,** No. 3, 52–5 (1951).
29. L. Starikova, *Myasnaya Ind. SSSR*, **24,** No. 2, 72–3 (1953).
30. J. Mattner and R. Mattner, *Z. Anal. Chem.*, **134,** 1–8 (1951).
31. L. Hartmann and M. D. L. White, *Anal. Chem.*, **24,** 527–9 (1952).
32. E. M. Tanner and T. F. Brown, *J. Inst. Petrol.* **32,** 341–50 (1946).
33. B. N. Chalishasar and C. E. Spooner, *Fuel*, **36,** 127–8 (1957).
34. H. Koch and H. Pohl, *Brennstoff-Chem.*, **19,** 201–4 (1938).
35. C. D. Wagner, R. Y. Smith, and E. D. Peters, *Anal. Chem.*, **19,** 982–4 (1947).
36. S. R. Rofikiv and N. Y. Sibiryakova, *Izvest. Akad. Nauk Kaz. SSR, Ser. Kim.*, **1956,** No. 9, 13–22.
37. S. Siggia, *Ind. Eng. Chem., Anal. Ed.*, **19,** 827 (1947).
38. H. Hock and L. Schrader, *Brennstoff-Chem.*, **18,** 6–8 (1937).
39. W. Frank and J. Mönch, *Ann.*, **556,** 200–23 (1944).
40. H. A. Laitinen and J. S. Nelson, *Ind. Eng. Chem., Anal. Ed.*, **18,** 422–5 (1946).
41. R. F. Robey and H. K. Wiese, *Ind. Eng. Chem., Anal. Ed.*, **17,** 425–6 (1945).
42. R. A. Chapman and K. Mackay, *J. Am. Oil Chem. Soc.*, **26,** 360–3 (1949).
43. P. Devi and S. C. Ray, *Cur. Sci. (India)*, **19,** 243–4 (1950).
44. G. L. Hills and C. C. Thiel, *J. Dairy Res.*, **14,** 340–53 (1946).
45. C. Furmanek and K. Monikowski, *Rocz. Państw. Zakl. Hig.*, **1953,** 447–57.
46. M. Filamoni and A. J. Siesto, *Boll. Soc. Ital. Biol. Sper.*, **27,** 1096–8 (1951).
47. A. M. Siddiqi and A. I. Toppel, *Chem. Anal.*, **44,** 52 (1955).
48. P. Duboules, M. F. Monge-Hedde, and J. Fondarai, *Bull. Soc. Chim. Fr.*, **1947,** 900–1.
49. S. Hartmann and J. Glavind, *Acta Chem. Scand.*, **3,** 954–8 (1949).
50. M. I. Eiss and P. Giesecke, *Anal. Chem.*, **31,** 1558–60 (1959).
51. P. Dugan, *Anal. Chem.*, **33,** 696–8 (1961); *Ibid.*, 1630 (1961).

not appear to be a simple oxidation, but rather catalyzed by a free radical mechanism, probably involving the splitting off of $—N\genfrac{}{}{0pt}{}{CH_3}{CH_3}$ groups by active OH.

The calculated precision of the method is ±4.1% at the 95% confidence limit, based on eight determinations for each peroxide.

Table 25 summarizes the analytical methods discussed in this chapter.

7

Unsaturation

The approaches for the determination of carbon-to-carbon unsaturation include bromination, catalytic hydrogenation, addition of iodine monohalides (iodine number), ozonization, and epoxidation. A specific determination of acetylenic unsaturation involves hydration to the corresponding ketone and determination of the ketone. Specificity of the hydrogenation procedure for acetylenic unsaturation is obtained by the use of a special catalyst. There are also some specific reactions for ethylenic compounds where the double bond is in the α-β position to a functional group, generally of the carboxylic type.

The bromination approach has the advantage of simplicity and rapidity, but it also has the disadvantage of substitution by bromine of some of the hydrogens on the organic compound, along with addition to the unsaturated linkages. This frequently leads to high results. Another common interference is oxidation of some component or group in the sample, which also causes high results. The bromination reaction is as follows:

$$RCH{=}CHR_1 + Br_2 \rightarrow \underset{\displaystyle\;\;\;Br\;\;Br}{RCHCHR_1}$$

Iodide monohalides are more selective for the unsaturated linkages than is bromine, since iodine monohalides do not undergo substitution reactions as readily as does bromine.

$$RCH{=}CHR_1 + IX \rightarrow \underset{\displaystyle\;\;\;I\;\;\;X}{RCHCHR_1}$$

where X is chlorine or bromine. Substitution is not eliminated completely, however, and it is particularly bothersome with aromatic ring compounds. The iodine monohalides are also fairly good oxidizing agents; thus, as in the case with bromine, any oxidizable group in the sample will react and cause high results for unsaturation.

Hydrogenation is a clear addition reaction that is very specific for carbon-to-carbon double or triple bonds and is a reaction that can be carried out with practically all such unsaturated compounds.

$$RCH{=}CHR' + H_2 \rightarrow RCH_2CH_2R_1$$

The methods for using these reactions, however, are generally time-consuming and not easy to run. They require either special devices or care and patience in the running of the analysis, and usually both. For some systems, however, these methods are generally worth the trouble, since other methods cannot be applied. For example, where there is trace unsaturation in an organic medium, the halogen addition methods fall down, since the interferences are maximized and the group being determined is minimized. Also, for very sterically hindered unsaturated compounds such as $\underset{\text{Cl}}{\underset{|}{\text{RC}}}{=}\underset{\text{Cl}}{\underset{|}{\text{CR}_1}}$, the halogens have difficulty adding to the unsaturated bond, but hydrogen does not.

Ozonization methods are generally free of interferences resulting from substitution and steric effects, but they have the disadvantage of resulting in the formation of explosive ozonides in some cases. Also, ozone generators are expensive.

Epoxidation has been made practical by the recent commercial availability of the relatively stable reagent *m*-chloroperbenzoic acid, which is particularly useful for the determination of unsaturation in polymers.

Mercuric salts also add onto unsaturated linkages. The acetate salt is most commonly applied probably because of its solubility in organic media.

$$\text{RCH}{=}\text{CHR}' + \text{Hg(Ac)}_2 + \text{R}''\text{OH} \rightarrow \underset{\text{OR}''}{\underset{|}{\text{RCH}}}{-}\underset{\text{HgAc}}{\underset{|}{\text{CHR}'}} + \text{HAc}$$

where R″ can also be a hydrogen atom.

Bromination

There are several bromination reagents commonly used for determining unsaturation: free bromine in a solvent, bromine generated from a bromate-bromide reagent using acid, pyridine sulfate–dibromide reagent, and bromine generated electrically.

FREE BROMINE IN A SOLVENT

The approach involving free bromine in a solvent was the first of the bromination methods to be used to determine unsaturation, but it has the significant disadvantage of maintaining standard solutions of bromine in a solvent because of the loss of bromine by volatilization on standing or handling. The most generally applicable solvents are acetic acid, carbon tetrachloride, propylene carbonate, and aqueous bromide solutions (to form Br_3^- complex). These solvents do not substitute bromine and are not oxidized by bromine. Also, the solubility of bromine in these solvents is good, although volatilization losses can occur if precautions are not taken. Of these solvents, carbon tetrachloride and aqueous bromide are the most commonly used.

Bromine in Carbon Tetrachloride—Shell Method

[(*Adapted from A. Polgar and J. L. Jungnickel, in Organic Analysis, Vol. I, Edited by J. Mitchell et al., Wiley-Interscience, New York, 1956, pp. 237–8, Reprinted in Part*)]

The Shell method involves the reaction of the sample in the dark at ice temperature with a measured excess of bromine in carbon tetrachloride in the presence of water, following which the excess bromine is determined iodometrically. One or two additional determinations are made at increased reaction times, and the results are extrapolated to zero time if there is significant change in bromine absorption with time. The extrapolation procedure is primarily necessary with cracked petroleum distillates. It is based on the assumption that addition, although rapid at first, decreases in rate, and a slow substitution begins as the saturation of the double bonds of diverse reactivity is approached. If the addition is complete when the first value is obtained and the rate of side reactions remains constant, extrapolation to zero time should give the correct addition value.

The application of this method to a large number of pure olefins has given very close agreement with the theoretical values. The extrapolated values in general are slightly on the lower side. Typical results are shown in Table 1. The value for pinene is high. This compound appears to undergo rupture of the bridge ring in addition to saturation of the double bond.

REAGENT

For each liter of solution, 17.2 grams (5.5 ml) of bromine is dissolved in carbon tetrachloride. This reagent (0.2*N*) is standardized daily as follows. Carbon tetrachloride (25 ml) and water (100 ml) in a glass-stoppered flask are cooled in an ice bath for 10 minutes. Reagent (15 ml) is added, and the flask is stored in the

Table 1. Analytical Data by the Extrapolation Method:[a] (bromine in carbon tetrachloride)

Compound	Extent of Reaction, % of Theoretical
1-Pentene	101
2-Pentene	97
2-Methyl-2-butene	100
2,2-Dimethylpropene	100
1-Hexene	100
2,3-Dimethyl-2-butene	99
1-Heptene	99
4-Methyl-2-pentene	101
2,3,3-Trimethyl-1-butene	99
1-Octene	100
2,3,4-Trimethyl-2-pentene	96
Diisobutylene	101
1-Decene	99
Triisobutylene	99
1-Tetradecene	100
1-Hexadecene	97
1-Octadecene	99
Styrene	99
α-Methylstyrene	100
4-Phenyl-1-butene	97
Stilbene	ca. 60
Cyclohexene	99
3-Methylcyclohexene	99
Pinene	ca. 150
2,5-Dimethyl-1,5-hexadiene	101
d-Limonene	100
2-Methyl-1,3-butadiene	100
4-Methyl-1,3-pentadiene	54
2,3-Dimethyl-1,3-butadiene	54

[a] Shell Method Series, 221/50.

dark for 10 minutes. Potassium iodide solution (15 ml of 20%) is added, and the titration is made with standard 0.1*N* sodium thiosulfate solution to the starch end point.

PROCEDURE

Dissolve the sample in carbon tetrachloride (25 ml) in a glass-stoppered flask, add water (100 ml), and cool the flask in an ice bath for 10 minutes.

Place a light shield around the neck of the flask, add reagent in 65 to 70% excess, and store the flask (kept in the ice bath) in the dark. Exactly 10 minutes after the addition of the reagent, add 15 ml of 20% potassium iodide solution and titrate the liberated iodine with $0.1N$ sodium thiosulfate solution to the starch end point. Repeat the determination with 20 minutes' reaction time. If these two results differ by 2% or more, make a third determination with 30 minutes' reaction time, and extrapolate the results to zero time.

BROMINE IN PROPYLENE CARBONATE

Method of B. Kratochvil, P. K. Chattopadhyay, and R. D. Krause

[*Anal. Chem.*, ***48***, *568* (*1976*)]

The method involves the direct potentiometric titration of olefins with bromine in propylene carbonate (PC).

PROCEDURE

The procedure, titrant, and apparatus are those described for the titration of phenols (p. 59).

RESULTS AND DISCUSSION

Results for several olefinic compounds are shown in Table 2. The change in potential in the region of the end point for the titration is on

Table 2. Percentage Recoveries in Titrations of Several Olefins with Bromine in Propylene Carbonate

Compound	Recovery, at 25°C, %
Cyclohexene	99
Vinyl butyrate	95
1- and 2-Octene (mixture of isomers)	100
Allyl alcohol	113
Allyl amine	118
Allyl ether	109[a]

[a] Calculated on the basis of 2 moles of bromine consumed per mole of ether.

Table 3. Effects of Sample Size on Recovery of Oleic and Linoleic Acids by Direct Titration with Bromine in PC and Indirect Iodine Monochloride Method[a]

Compound	Sample Size, grams	Recovery by Bromine in PC,[b] %	Recovery by ICl,[c] %
Oleic acid	0.1	107	104
	0.3	106	111
	0.5–1.2	105.5 ±0.2	105
	0.5–1.2	107.1 ±0.2 (0°C)	—
	0.5–1.2	107.5, 107.8 (50°C)	—
	1.2–2.0	104	104
Linoleic acid	0.1	89.3	85[d]
	0.2	88.7	90
	0.35–0.4	84.2 ±0.2	86
	0.65	81.5	—

[a] Temperatures controlled at 25 ±1°C unless otherwise indicated. Individual values listed for one or two titrations, average and standard deviation for three or more.

[b] Sample dissolved in 2:1 PC—$CHCl_3$ mixture.

[c] Sample dissolved in 10 ml of carbon tetrachloride, 20 ml of 0.6*M* iodine monochloride added. Each value is the average of two titrations.

[d] Sample dissolved in 10 ml of carbon tetrachloride, 20 ml of 0.2*M* iodine monochloride added.

Table 4. Comparison of Iodine Values of Vegetable Oils Determined by Bromine in PC and Iodine Monochloride Methods

Oil[a]	Iodine Value[b]: Bromine in PC Method	Iodine Value[b]: ICl Method
Corn	145	131
Cottonseed	124	119
Linseed	168	168
Olive	91	87
Peanut	103	97
Rapeseed (low erucic acid variety)	125	118
Soybean	123	112
Safflower	143	139

[a] Commercial oils used as received.

[b] Calculated as equivalent number of grams of iodine that would have been consumed by 100 grams of oil. All values are the average of two titrations.

the order of 150 to 200 mV. Rates of reaction were rapid; titrations typically required only 5 to 6 minutes to complete without use of catalyst. The high recoveries obtained in several instances may be attributed to the occurrence of substitution as well as addition; substitution tends to take place to a greater extent in polar solvents than in nonpolar ones.

A study of the effect of temperature on the reaction stoichiometry of the compounds in Table 2 showed that bromine consumption generally increases at temperatures above or below 25°C. Below 25°C the rate of substitution becomes faster relative to the rate of addition.

Most oils and fats are not sufficiently soluble in PC to allow titration. The addition of 30% by volume chloroform provided sufficient solubility for all the systems studied without affecting potentiometric end point detection. Commercial chloroform contains about 0.75% ethanol as a preservative. Because ethanol undergoes bromine substitution readily, it must be removed from the chloroform solvent before use.

Linoleic acid, which has two double bonds per molecule, shows two potential breaks; the second break was used for all analytical calculations. Titrations of 10 mixtures of oleic and linoleic acids in varying ratios gave overall recoveries within 1% of the sums of the individual acids. The percentage recovery is independent of the rate of titrant delivery up to at least 2 ml per minute, but it does depend on sample size; smaller samples give higher recoveries. Recovery with sample size by the direct bromine in PC and indirect iodine monochloride (ICl) methods are compared in Table 3. Recoveries are similar for the two methods, and both give lower recoveries as the amount of sample taken is increased, although the direct bromine in PC method is affected less. Iodine values for a number of oils by direct bromination and by indirect iodine monochloride determination are compared in Table 4. The results for the bromine method tend to be about 5 to 10% higher than for the iodine monochloride method. However iodine numbers are useful primarily for relative comparisons among various oils and, for this purpose, a direct titration is more rapid and convenient.

Bromine in Aqueous Potassium Bromide

[*Adapted from A. Polgar and J. L. Jungnickel, in Organic Analysis, Vol. I, Edited by J. Mitchell et al., Wiley-Interscience, New York, 1956, pp. 240–3*]

An aqueous bromination reagent containing bromine and excess potassium bromide is easy to prepare and has displayed very satisfactory keeping qualities. Since the potassium bromide is in excess, all the bromine content is assumed to be in the loose Br_3^- form. This brominating agent has been found to be especially applicable to oxygen-containing olefinic compounds that react incompletely in some other halogenation methods. Phenol reacts quantitatively to give tribromophenol, but most saturated alcohols, ketones, aldehydes, esters, and acids do not interfere to any appreciable degree. The normal procedure involves the shaking of the sample in the presence of ice with a 65 to 70% excess of reagent for 20 minutes and back-titrating the unreacted bromine iodometrically with sodium thiosulfate. To avoid loss of bromine, the reaction is carried out in an evacuated bottle (Procedure A). With samples that react only with difficulty, the reaction is carried out at room temperature for 1 hour (Procedure B).

REAGENT

For each liter of solution, dissolve 35.8 grams of potassium bromide in about 100 ml of water and add 17.2 grams (5.5 ml) of bromine. Mix the solution until bromine is dissolved, then dilute to volume. This reagent (0.2 *N*) is standardized daily in an evacuated bromination bottle (Fig. 7.1) by drawing in 15 ml of reagent, 15 ml of 20% potassium iodide solution, and about 100 ml of water) through a piece of rubber tubing; then the vacuum is released, the stopper is removed, and the titration is made with standard 0.1*N* sodium thiosulfate to the starch end point.

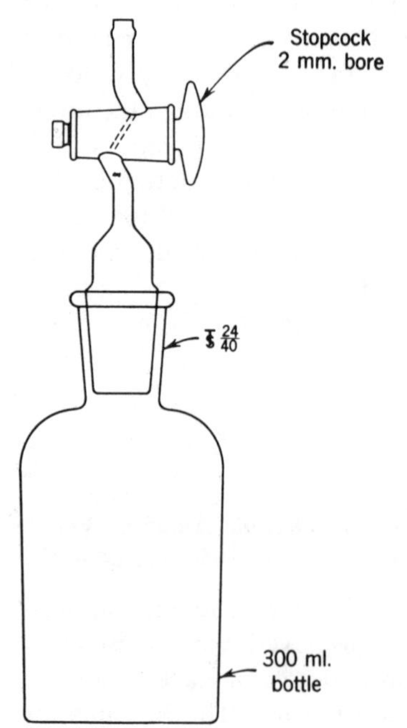

Fig. 7.1. Bromination bottle.

PROCEDURE A

Fill the bromination bottle two-thirds full with crushed ice; stopper, and evacuate. Introduce an aliquot of diluted sample (in alcohol or acetone) and the reagent (65 to 70% in excess), and shake the bottle for 20 minutes in a mechanical shaker. Add 15 ml of 20% potassium iodide solution, release the vacuum, and titrate the mixture with standard 0.1*N* sodium thiosulfate solution to the starch end point. If the result of a duplicate determination with 90% reagent excess differs by more than 1%, the method should be considered to be inapplicable to the sample.

PROCEDURE B

Follow Procedure A but with these differences: omit ice, use 80% reagent excess, and continue shaking for 1 hour.

Typical results obtained by these procedures are shown in Table 5.

Table 5. Results by the Aqueous Tribromide Method

Compound	Extent of Reaction, % of Theoretical	
	Procedure A	Procedure B
Allyl alcohol	99	—
β-Chloroallyl alcohol	98	100
Methallyl alcohol	102	—
Methylvinylcarbinol	97	—
Crotyl alcohol	100	—
4-Hydroxy-2-methyl-1-pentene	96	—
Acrolein	100	—
α-Chloroacrolein	10	92
Acrolein dimer	100	—
Methacrolein	100	—
Methyl isopropenyl ketone	99	—
Mesityl oxide	99	—
Acrolein diacetate	100	—
Methacrolein diacetate	99	—
1,1,3-Triallyloxypropane	101	—
Acrylic acid	86	97
Sodium acrylate	—	100
Methacrylic acid	99	99

Table 5 (*Continued*)

Compound	Extent of Reaction, % of Theoretical	
	Procedure A	Procedure B
Tetrahydrobenzoic acid	99	—
Vinyl acetate	100	—
Vinyl propionate	—	103
Vinyl butyrate	—	99
Vinyl crotonate	—	64
Vinyl oleate	—	100
Allyl acetate	98	—
Allyl propionate	100	—
Chloroallyl propionate	ca. 40	93
Allyl butyrate	101	—
Allyl caproate	98	—
Allyl caprylate	100	—
Allyl hexahydrobenzoate	99	—
Diallyl adipate	91	97
Diallyl phthalate	ca. 60	101
Allyl crotonate	77	100
Methyl α-chloroacrylate	<1	66
Ethyl acrylate	10	101
α-Allyl glycerol ether	100	—
Allyl glycidyl ether	100	—
Allyl 1,3-dichloro-2-propyl ether	99	—
Allyl sulfolanyl ether	100	—
Dimethallyl ether	86	101
Vinyl chloride	100	—
Dichloroethylene	—	36
1,1-Dichloro-1-propene	—	(low)
1,3-Dichloro-1-propene	(low)	96
2,3-Dichloro-1-propene	(low)	99
3-Chloro-1-butene	2	103
1-Chloro-2-butene	101	—
1,4-Dichloro-2-butene	80	97
Allyl N-ethyl carbamate	100	—
Dibutyl allyl phosphate	100	—
2,4-Dimethyl-3-sulfolene	98	100

BROMATE–BROMIDE

To avoid the volatilization problems involved in the use of free bromine, some investigators have resorted to the use of bromate-bromide solutions. This reagent is very stable, and the bromine is generated by adding acid at the time of analysis after all other reagent and sample handling is completed. This step eliminates the volatilization losses.

A bromate-bromide reagent can be used alone, or it can be used with a mercuric catalyst to accelerate the bromination. When the substitution reactions are too rapid for accurate analysis, the catalyst must be eliminated; also, the extrapolation approach of the preceding method can be used.

Method of H. L. Lucas and D. Pressman

[*Adapted from Ind. Eng. Chem., Anal. Ed.*, **10,** *140–2* (*1938*)]

REAGENTS

Bromate-bromide solution, 0.1*N*. (2.78 grams of potassium bromate and 15 grams of potassium bromide per liter of water solution).

Mercuric sulfate, 0.2*N*. (950 ml of water; 28 ml of concentrated sulfuric acid; 30 grams of mercuric sulfate).

CP Carbon tetrachloride.

CP Glacial acetic acid.

Potassium iodide, 2%.

Sulfuric acid, 6*N*.

Sodium thiosulfate, 0.05*N*.

Starch indicator solution.

Sodium chloride, 2*N*.

PROCEDURE

Dilute water-soluble samples to 0.08*N* in unsaturation and take a 25-ml aliquot (0.002 equivalent in unsaturation).

Dissolve hydrocarbon-soluble samples in carbon tetrachloride.

When volatile, weigh the sample in a sealed glass ampoule (see pp. 862–64) and place it in a volumetric flask of convenient volume to give a 0.08*N* solution in terms of unsaturation, along with enough carbon tetrachloride or water to cover the ampoule. Chill the volumetric flask in ice water, and break the ampoule with a glass rod. Rinse the rod with chilled carbon tetrachloride or water, and make the solution up to the mark with the same chilled solvent. To pipet a sample of this solution, do not apply a vacuum. Instead, equip the pipet with a two-hole rubber stopper to fit the volumetric flask, and a piece of glass tubing is inserted in the other hole. Then blow the solution into the pipet by compressed air or by mouth.

Introduce a calculated excess (10–15%) of 0.1*N* bromate-bromide solution (about 25 ml) into a 250- to 300-ml Erlenmeyer flask having a glass stopper equipped with a sealed-in three-way stopcock (Fig. 7.2). If the amount of unsaturation in the sample is unknown, make a preliminary analysis with a large excess of bromate-bromide solution. From this result, the desired excess can be calculated. The reason for avoiding the large excess is minimization of substitution, which leads to high results. If the excess is less than 10 to 15%, the addition proceeds so slowly toward the end of the bromination that the bromination may not be complete in the specified time.

After the addition of the bromate-bromide solution, evacuate the flask through tube *A* with a water aspirator, and add 5 ml of 6*N* sulfuric acid by means of the funnel attachment *B* on the stopper, allowing 2 or 3 minutes to elapse for the bromine to be liberated. Next, add 10 to 20 ml of 0.2*N* mercuric sulfate, followed by the sample containing solution

Fig. 7.2. Bromination apparatus.

(25 ml of the sample solution should be used if the concentration is $0.08N$ with respect to unsaturation). Use water or carbon tetrachloride, depending on which was used to dissolve the sample, to rinse the sample into the flask. Use a total of 15 ml of wash liquid in three portions. For samples dissolved in carbon tetrachloride, add 20 ml of glacial acetic acid. When the samples are dissolved in water, omit the acetic acid. Then wrap the flask in black cloth and shake for 7 minutes (time may vary with some samples). Add 15 ml of $2N$ sodium chloride and 15 ml of 20% potassium iodide and shake the flask for about 30 seconds. Break the vacuum and titrate the free iodine with $0.05N$ sodium thiosulfate, using a starch indicator. Run a blank under the same conditions as the samples, with about one-third the amount of bromate-bromide solution used in the analysis.

CALCULATIONS

$$\frac{\text{(Milliliters for blank minus milliliters for sample)} \times N \text{ thiosulfate} \times \text{mol. wt. of compound} \times 100}{\text{Grams of sample} \times 2000 \times B} = \% \text{ compound}$$

where

B = number of moles bromine absorbed by compound being determined

The molar ratio of mercuric ion to bromide ion should be greater than 1 if the mercuric salt is to have sufficient catalytic effect. Sodium chloride is necessary to liberate free bromine from its complex with mercuric sulfate. Acetic acid is necessary to solubilize the unsaturated compound in the water layer.

The 7-minute shaking is enough for most samples, but some samples require longer (see Table 6). Maleic and fumaric acids in water require 30 minutes in the presence of mercuric sulfate for complete reaction.

Cinnamic acid requires no mercuric sulfate for quantitative results; in fact mercuric sulfate is detrimental, since it causes substitution. Propiolic acid, dimethylbutadiene, and propargyl alcohol substitute bromine and yield high results when determined by this method. Because of varying amounts of substitution for different compounds, it is difficult to ascribe a general accuracy and precision value to this procedure.

Absorbed oxygen in solutions of alkynes and exposure to sunlight generally affect results significantly.

This procedure was tested with phenylacetylene, 1-heptyne, 1-pentyne, 1-hexyne, cyclohexene, 1-hexene, dichloroethylene, maleic acid, fumaric acid, 2-butyne-1,4-diol, anethole.

Table 6. Effects of Variations in the Analytical Procedure on Results

Compound	$HgSO_4/Br^-$, Ratio	HOAc, ml.	$KBrO_3$ Excess, %	Time, min.	Error, %
	Mercury-Bromide Ratio				
2-Heptyne	2.4	20	20	7	−3.2
2-Heptyne	1.2	20	20	7	−4.1
2-Heptyne	0.0	20	20	7	−23.0
	Acetic Acid				
Phenylacetylene	1.1	25	10	7	−3.0
Phenylacetylene	1.1	0	10	7	−23.0
1-Hexyne	1.1	25	10	5	−0.9
1-Hexyne	1.1	0	20	5	−10.0
	Excess Bromate				
1-Pentyne	1.0	0	25	20	+12.0
1-Pentyne	1.7	0	12	20	+7.5
1-Pentyne	1.7	0	5	20	+5.5
	Time of Bromination				
1-Heptyne	1.5	15	15	60	−2.1
1-Heptyne	1.5	15	15	30	+2.7
1-Heptyne	1.5	15	15	15	+2.7
Phenylacetylene	1.1	20	10	30	+3.7
Phenylacetylene	1.1	20	10	5	0.0

Interfering substances consist of phenols and amines, which substitute bromine, and also hydrazines, some aldehydes, and other materials that are oxidized by bromine.

ELECTRICALLY GENERATED BROMINE

In a more exotic method for determining unsaturation, the bromine is generated electrically at a constant rate in the reaction system. This prevents large excesses of bromine from being present at any time. Since the addition reaction of bromine onto the unsaturated bond is much faster than the substitution or oxidation reactions, this approach generally provides a more accurate analysis than the foregoing methods. Where large numbers of analyses have to be run, the application of this method can be justified quite readily.

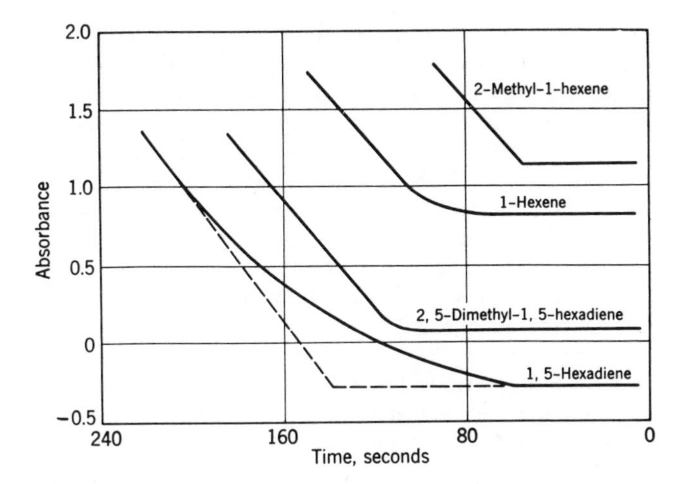

Fig. 7.3. Typical titration curves for determination of various olefins.

Because bromination is not an instantaneous reaction, the bromine is not all consumed as soon as it is generated; hence an electrometric end point is not possible. To circumvent this problem, the solution is viewed with a spectrophotometer to follow the bromine color in the reaction system. In the initial stages of the reaction, the bromine is consumed rather rapidly; hence the color remains low as bromine is continuously generated at a constant rate. As the unsaturated linkages become saturated, the bromine concentration in the reaction system builds up, and the color increases. Once the compound is completely saturated, the bromine builds up sharply, since the substitution reactions are slow. Figure 7.3 plots absorbance versus time yields curves. Extrapolation yields the value for the bromination of the unsaturated groups.

Method of J. W. Miller and D. D. DeFord

[*Adapted from Anal. Chem.*, ***29,*** *475–9* (*1957*), *Reprinted in Part*]

APPARATUS

A recording spectrophotometer capable of covering the region of 360 nm. The instrument should yield plots of absorbance versus time. A Cary Model 11 was used in the original work of Miller and DeFord. Miller and DeFord modified the cell holder as shown below to accommodate a 100-ml beaker in which to carry out the electrolysis.

CELL HOLDER (Fig. 7.4). The base platform *A* was made of $\frac{1}{2}$-in. aluminum, 10.0×11.5 cm. The bottom of base *A* was notched so that when the regular cell holder was removed, the base of the new cell holder would fit tightly into position on the platform pegs of the spectrophotometer sample cell compartment. Four holes were drilled to a depth of $\frac{1}{8}$ in. on the top of the platform base, the centers being 1.2 cm from each edge. These were threaded with an 8-32 tap. Four brass rods, each 10 cm long and $\frac{1}{8}$ in. in diameter, were threaded on each end and then screwed into these four holes.

The adapter *B* for the beaker was constructed of $\frac{1}{8}$-in. Bakelite with a hole cut exactly in the center to fit a 100-ml tall-form beaker. The adapter was 10.5×11.5 cm. Four holes were drilled 1.4 cm from each edge at the corners to allow the adapter to be set over the four brass rods, which were attached to the platform base. Each brass rod had a piece of brass tubing placed over it to hold the adapter 7 cm above the top of the platform base. Shorter pieces of tubing were placed over the rods after the adapter was in position, and these were held tightly in place by four nuts. The whole assembly was painted flat black. Although

Fig. 7.4. Titration cell holder and sample cell compartment with electrodes and stirrer attached.

only a 100-ml beaker was used in this work, titration vessels of other sizes could be accommodated by using appropriate adapters.

SAMPLE CELL COMPARTMENT COVER. The cover was identical with the standard cover on the Cary spectrophotometer except for three holes drilled near the center. The center hole held the stirring shaft of the motor, which was mounted on the cover, and the other two holes permitted the introduction of the generation electrodes. A glass paddle stirring rod was attached to the motor by means of a No. 1, one-hole rubber stopper. The stirrer was long enough to reach just above the light path. By pushing the rubber stopper firmly against the cover, room light was prevented from entering the sample compartment. A small rheostat was attached to the motor to vary the speed of stirring. In the final work with the apparatus, a large hole was drilled in the cover to allow the introduction of sample without removing the whole cover. This hole was closed satisfactorily by a No. 6 rubber stopper.

CONSTANT CURRENT SOURCE. A Kay-Lab Meter Calibrator, Model No. M10A10 was used as a source of constant current. This instrument is capable of delivering from 0.1 to 100.0 mA with an accuracy of 5 parts in 10,000.

ELECTRODES. The anode used in all determinations was a platinum foil electrode (1.0 × 1.4 cm); the cathode was a platinum wire, which was placed in a compartment isolated from the bulk of the solution by means of a 1-cm sintered-glass disk. The cathode compartment was tapered at the top to fit through a $\frac{3}{16}$-in. hole in the cover.

REAGENTS

All chemicals employed were CP or reagent grade. The arsenite samples were aliquots of standard solutions prepared in the conventional way from weighed quantities of arsenic trioxide. Standard solutions of the olefins were prepared by weighing a known amount of sample into a volumetric flask and diluting to the mark with carbon tetrachloride. The standard solutions were prepared so that 0.500 ml of the sample solution would take up 13 mg of bromine.

The generator electrolyte was essentially a nonaqueous mixture similar to that employed by Sweetser and Bricker (1) for the spectrophotometric titration of olefins with a standard bromate-bromide solution. A stock solution of the generator electrolyte was prepared by mixing 646 ml of glacial acetic acid, 256 ml of methanol, 16 ml of concentrated aqueous hydrochloric acid, and 30 ml of 40% (by weight) of aqueous potassium bromide. A 15% (by weight) mercuric chloride solution in methanol was prepared as a catalyst.

PROCEDURE

Allow the spectrophotometer and constant current source to warm up 15 to 20 minutes before beginning the titrations. Select a wavelength of

1. P. B. Sweetser and C. E. Bricker, *Anal. Chem.*, **24,** 1107 (1952).

360 nm when mercuric chloride is employed as a catalyst. This is the shortest wavelength that can be used in the presence of mercuric chloride, which has a high absorbance in the ultraviolet. Select a rate of bromine generation so that the total time of generation will be at least 250 seconds. When the standard solutions are prepared as described previously, use a current of 50.00 mA.

Place the cell holder in the sample cell compartment, and attach the stirring paddle and electrodes to the cover. Fill the cathode compartment with 1*N* hydrochloric acid (the nonaqueous solvent runs out too rapidly). Fill a standard quartz sample cell with distilled water and place in the reference beam; then add 95 ml of generator electrolyte and 8 ml of the mercuric chloride solution to the electrolysis beaker. Dry the sides of the beaker and polish with lens tissue. Add the olefin sample to the generator electrolyte by means of an appropriate sized washout-type micropipet, and place the beaker in the cell compartment. Set the cover in place and connect the electrodes to the current source.

With the stirring motor running, move the cell position shaft to a position where the absorbance reading is a minimum. This indicates that the beaker is centered in the light beam. By means of the balance control, set the recorder pen at any convenient position near the bottom of the absorbance scale.

When all these adjustments have been made, turn on the current and an electric timer simultaneously. Approximately 100 seconds before the expected end point, turn on the recorder chart drive. Note the exact reading of the time at the moment the chart drive was started. Continue the titration until the recorder has traced a line several inches long after the end point has passed. (The typical titration curves obtained are shown in Fig. 7.3).

Before each set of titrations, run a blank on the supporting electrolyte.

CALCULATION

The end point was obtained by extrapolation of the two linear portions of the recorded titration curve to the point of intersection. The sum of the time indicated on the time axis of the recorder chart plus that indicated by the timer at the moment recording was begun gave the number of seconds required for the titration. The time required for the blank was subtracted to obtain the net titration time. The number of milligrams of bromine generated was calculated from eq. 1.

$$\text{Milligrams of bromine} = \frac{[\text{time (seconds)}]\,[\text{current (amperes)}]\,(79.916)}{96.5} \tag{1}$$

The factor 79.916 is the atomic weight of bromine. The bromine number was calculated by use of eq. 2.

$$\text{Bromine no.} = \frac{\text{Milligrams of bromine} \times 100}{\text{Sample weight (mg)}} \tag{2}$$

The results of the coulometric titration of 13 olefins of various types are given in Table 7. The weight of olefin in each sample varied from 3.4 to 11.2 mg for compounds having the largest and the smallest bromine numbers, respectively.

Table 7. Summary of Bromination Results

Compound	Source and Purity of Olefins	Coulometric Procedure				ASTM Method, Average Error, %	Lucas-Pressman* Procedure, Average Error, %
		Bromine No.		Average Error, %	Standard Deviation		
		Theory	Found				
1-Hexene[a]	Phillips, 99+	190.0	185.4	−2.42	0.28(5)[b]	+1.16	+9.11
2-Methyl-2-butene	Phillips, 99+	227.9	221.6	−2.76	0.90(5)	+7.77	+4.78
4-Methyl-*cis*-2-pentene	Phillips, 95+	190.0	184.4	−2.95	1.02(5)	+0.37	+4.32
4-Methyl-*trans*-2-pentene	Phillips, 95+	190.0	179.4	−5.58	1.26(5)	−1.05	+5.53
2-Methyl-1-heptene	Peninsular Chem Research, b.p., 119–21°C	142.4	134.5	−5.55	0.89(5)	+9.69	+8.50
2,4,4-Trimethyl-2-pentene	Phillips, 95+	142.4	137.6	−3.23	0.33(5)	+7.94	+44.3
2,6-Dimethyl-1-heptene	Peninsular Chem Research, b p , 141–3°C	126.6	120.0	−5.21	0.35(5)	+9.95	+12.56
1-Phenyl-2-butene	Phillips, 99+	120.9	115.1	−4.80	1.06(5)	−3.72	+1.08
Cyclohexene[a]	Eastman	194.6	193.1	−0.77	0.20(5)	+2.36	+9.71
4-Vinyl-1-cyclohexene	Phillips, 99+	295.4	284.5	−3.69	3.24(4)	−0.14	+4.43
1,5-Hexadiene	Peninsular Chem Research, b.p., 58–60°C	389.2	342.9	−11.90	0.65(5)	−11.61	−1.41
2,5-Dimethyl-1,5-hexadiene	Peninsular Chem Research, Unknown	290.0	242.4	−16.41	0.66(5)	−6.97	+3.72
2-Methyl-1,3-butadiene	Phillips, 99+	469.2[c]	253.4	−45.99	1.94(5)	−43.48	+2.90
		234.6[d]	253.4	+8.00	1.94(5)	+13.04	+105.8

[a] Purified by silica gel chromatography (ASTM D 1158-55T).
[b] Numbers in parentheses are number of individual titrations used to calculate average bromine number and standard deviation.
[c] Bromine number for addition of 2 moles of bromine.
[d] Bromine number for addition of 1 mole of bromine.
* See pp. 383–386.

DISCUSSION

The efficiency of bromine generation in the generator electrolyte was checked by the titration of arsenic(III). The results of these titrations both with and without the mercuric chloride catalyst are given in Table 8. A wavelength of 320 nm was used when no catalyst was present. These results indicate that coulometric bromine titrations with spectrophotometric detection of the end point can be performed satisfactorily in this solvent. Mercuric chloride has no effect on the results obtained.

Because several of the olefins had been stored for over one year, their purity was uncertain and the results have little meaning. To evaluate the

Table 8. Check of Current Efficiency

Number of Runs	Current, Ma.	Meq. of As_2O_3 Taken	Meq. of As_2O_3 Found	Error, %	Standard Deviation	Catalyst
8	50.00	0.3254	0.3235	−0.58	0.49	None
10	50.00	0.1302	0.1292	−0.77	1.01	None
8	50.00	0.3254	0.3251	−0.09	0.12	8 ml. $HgCl_2$
10	50.00	0.1302	0.1295	−0.54	0.54	8 ml. $HgCl_2$

coulometric procedure, each of the olefins was titrated by two standard procedures. The first was that of Lucas and Pressman (2) (see pp. 383–386). This procedure uses a reaction time and an excess of bromine, which were found to give the best average results for a large number of unsaturated compounds. However certain types of compounds, particularly highly branched olefins, consume too much bromine because of the contribution of substitution reactions. For this reason the method cannot be expected to give good results with all types of unsaturated compounds.

The second method was that recommended by ASTM for the determination of bromine numbers (3). This method also uses empirically adjusted reaction conditions. Since the method has been tested on a large number of olefins, it should give more accurate results than the Lucas-Pressman method. Only highly branched olefins are reported to give irregular results with this method (3).

The results obtained for each of the 13 olefins by these two standard procedures are summarized in Table 7. Each result is the average of at least five individual determinations, and is given as the average error (in terms of per cent) from the theoretical bromine number.

The results shown in Table 7, column 6, indicate the precision that can be obtained in the coulometric determination of bromine numbers. All the olefin types except tetrasubstituted ethylenes were represented, and the precision was not determined by the olefin type. All the results obtained indicate that the samples were less than 100% pure. If substitution reactions had been taking place, the results would have indicated a purity greater than 100%.

In the hands of Miller and De Ford, the method of Lucas and Pressman gave results consistently higher than the theoretical bromine number for the pure compounds. The high results are probably caused by the fact that substitution reactions are taking place at the same time as the main

2. H. J. Lucas and D. Pressman, *Ind. Eng. Chem., Anal. Ed.*, **10,** 140 (1938).
3. American Society Testing and Materials, *Standards*, Pt. V, D 1158–55T, p. 595.

addition reaction. This side reaction is undoubtedly caused by the long contact time (7 minutes) of the sample with the excess bromine.

In the coulometric procedure there were two compounds for which the standard deviation was greater than 1.26%: 4-vinyl-1-cyclohexene and 2-methyl-1,3-butadiene. The extreme rounding in the titration curve of 4-vinyl-1-cyclohexene did not allow the extrapolation to be carried out with precision. The tendency of 2-methyl-1,3-butadiene to add either 1 or 2 moles of bromine did not allow a set of titrations to be carried out with precision. The break in the titration curve was sharp, but the end point showed a slight tendency to wander. If the results for these two compounds are omitted, the average standard deviation of the coulometric procedure is 0.69.

FACTORS AFFECTING TITRATION CURVES

The three factors that affect the sharpness of the break in the photometric titration curve are the rate of bromine generation, the wavelength used for the absorbance measurements, and the rate of reaction with the olefins. The greater the rate of bromine generation, the sharper will be the break in the titration curve. If the wavelength is varied toward shorter wavelengths, the sharpness of the break increases. As the rate of bromine generation and the wavelength were kept constant throughout all the analyses, only the rate of addition of bromine to olefins determined the shape of the titration curve.

Olefins with terminal unsaturation and with no branching at the double bond showed considerable rounding in the vicinity of the end point—for example, 1,5-hexadiene and 1-hexene (Fig. 7.3). The exact end point was easily found by extrapolation of the two linear portions of the curve. The rounding can be attributed to the slowness of the reaction of this type of olefin under the experimental conditions. Branching at the 2-position of compounds with terminal unsaturation causes a very sharp break with little round in the curve. This effect is shown in Fig. 7.3 by 2,5-dimethyl-1,5-hexadiene and 2-methyl-1-hexene. The strong electron-repelling nature of alkyl groups tends to increase the negative charge density on the double bond, and this facilitates the attack of the doubly positively charged bromine–mercuric ion complex; this is presumably the rate-determining step. Further branching of terminal double bonds or the reaction of internal double bonds did not change the shape of the titration curve. All the compounds titrated except 1-hexene, 1,5-hexadiene, and 4-vinyl-1-cyclo-hexene have titration curves nearly identical with that of 2,5-dimethyl-1,5-hexadiene. 4-Vinyl-1-cyclohexene shows even more rounding than does 1,5-hexadiene.

Even though all the results obtained by the coulometric procedure were low when compared with the theoretical bromine numbers, this error is not a real error. The apparent low results represent only the deviation of the samples from 100% purity. The facts that the "addition reaction" is complete, that insignificant abnormal reactions occur, and that olefins form peroxides under usual storage conditions, indicate that the coulometric method gives a true measure of the actual olefinic content of the sample.

Iodine Number Methods

Iodine number is the common designation for the determination of unsaturation via the addition of iodine monohalides. There are two common iodine monohalides used—iodine monochloride and iodine monobromide. Both methods cover approximately the same range of unsaturated compounds; however each method has particular advantages. It is the authors' opinion that the iodine monochloride method is faster, but the iodine monobromide method circumvents the reagent difficulties encountered with iodine monochloride. (This view is not shared by all users of these methods.) There are as many proponents to the iodine monochloride method as there are opponents. It suffices to say that both methods do work and are widely used; the reader can use both and judge for himself.

IODINE MONOCHLORIDE

Wijs Method

[*Reprinted with Permission from American Society for Testing and Materials, Standards, 1961, Part 10, D-460, pp. 954–6*]

REAGENTS

POTASSIUM DICHROMATE, STANDARD SOLUTION, 0.1*N*. Dissolve 4.903 grams of potassium dichromate ($K_2Cr_2O_7$) in water and dilute to 1 liter at the temperature at which titrations are to be made. *Note.* Occasionally potassium dichromate is found containing sodium dichromate ($Na_2Cr_2O_7$), although this is rare. If the character of the potassium compound is not certain, the purity can be ascertained by titration against freshly resublimed iodine. However this is usually unnecessary.

POTASSIUM IODIDE SOLUTION, 150 GRAMS PER LITER. Dissolve 150 grams of potassium iodide (KI) in water and dilute to 1 liter.

SODIUM THIOSULFATE, STANDARD SOLUTION, 0.1*N*. Dissolve 24.8 grams of sodium thiosulfate ($Na_2S_2O_3 \cdot 5H_2O$) in freshly boiled water and dilute to 1 liter at the temperature at which the titrations are to be made. To standardize, place 40 ml of potassium dichromate to which has been added 10 ml of the solution of potassium iodide in a glass-stoppered flask, add 5 ml of concentrated hydrochloric acid (sp. gr. 1:19), dilute with 100 ml of water, and allow the sodium thiosulfate solution to flow slowly into the flask until the yellow color of the liquid has almost disappeared. Add a few drops of the starch paste, and while shaking constantly, continue to add the sodium thiosulfate solution until the blue color just disappears.

STARCH PASTE. Boil 1 gram of starch in 2000 ml of water for 10 minutes and cool to room temperature. *Note.* An improved starch solution may be prepared by autoclaving 2 grams of starch and 6 grams of boric acid dissolved in 200 ml of water at 15 lb pressure for 15 minutes. This solution has good keeping qualities.

WIJS IODINE SOLUTION. Dissolve 13.0 grams of resublimed iodine in 1 liter of glacial acetic acid and pass in washed and dried chlorine gas until the original thiosulfate titration of the solution is not quite doubled. There should be no more than a slight excess of iodine, and no excess of chlorine. When the solution is made from iodine and chlorine, this point can be ascertained by not quite doubling the titration. For preparation of the Wijs solution, use glacial acetic acid of 99.0 to 99.5% strength. For glacial acids of somewhat lower strength, freezing and centrifuging or draining, as a means of purification, is recommended. Preserve the solution in amber, glass-stoppered bottles, sealed with paraffin until ready for use. Mark on the bottles the date on which the solution is prepared; do not use Wijs solution that is more than 30 days old. *Note.* For preparation of the solution, McIlhiney (4) gives the following details. The preparation of the iodine monochloride solution presents no great difficulty, but it must be done with care and accuracy in order to obtain satisfactory results. There must be in the solution no appreciable excess either of iodine or more particularly of chlorine, over that required to form the monochloride. This condition is most satisfactorily attained by dissolving in the whole of the acetic acid to be used the requisite quantity of iodine, using a gentle heat to assist the solution, if it is found necessary; then setting aside a small portion of this solution, while pure and dry chlorine is passed into the remainder until the halogen content of the whole solution is doubled. Ordinarily, it will be found that by passing the chlorine into the main part of the solution until the characteristic color of free iodine has just been discharged, there will be a slight excess of chlorine which is corrected by the addition of the requisite amount of the unchlorinated portion until all free chlorine has been destroyed. A slight excess of iodine does little or no harm, but an excess of chlorine must be avoided.

4. McIlhiney et al., "Report of the Sub-Committee on Shellac Analysis," *J. Am. Chem. Soc.*, **29**, 1222 (1907).

PROCEDURE

Weigh accurately from 0.10 to 0.50 gram (depending on the iodine number) of the sample prepared into a clean, dry, 450-ml (16-oz) glass-stoppered bottle containing 15 to 20 ml of carbon tetrachloride. Add 25 ml of the iodine solution from a pipet, allowing each sample to drain for the same length of time. The excess of iodine should be from 50 to 60% of the amount added, that is, from 100 to 150% of the amount absorbed. Let the bottle stand in a dark place for 30 minutes at a temperature of 25 ±2°C, then add 20 ml of potassium iodide solution and 100 ml of water. Titrate the iodine with 0.1*N* sodium thiosulfate solution, added gradually while shaking constantly, until the yellow color of the solution has almost disappeared. Add a few drops of starch paste and continue titration until the blue color has entirely disappeared. Toward the end of the reaction, stopper the bottle and shake vigorously, so that any iodine remaining in solution in the carbon tetrachloride may be taken up by the potassium iodide solution. Make two determinations on blanks, employing the same procedure as used for the sample except that no sample should be used in the blanks. Slight variations in temperature quite appreciably affect the titer of the iodine solution, since acetic acid has a high coefficient of expansion. It is therefore essential that the blanks and determinations on the sample be made at the same time.

CALCULATION

Calculate the iodine number of the sample tested (centigrams of iodine absorbed by 1 gram of sample; (i.e., percentage of iodine absorbed), as follows:

$$\text{Iodine value} = \frac{(B-A)N \times 12.69}{C}$$

where

A = milliliters of $Na_2S_2O_3$ solution required for titration of the sample
B = milliliters of $Na_2S_2O_3$ solution required for titration of the blank
C = grams of sample used
N = normality of the $Na_2S_2O_3$ solution

IODINE MONOBROMIDE

Hanus Method

[*As Described in F. D. Snell and F. M. Biffen, Commercial Methods of Analysis, McGraw-Hill, New York, 1944, pp. 345–6, 719*]

REAGENTS

HANUS IODINE MONOBROMIDE SOLUTION. Dissolve 13.6 grams of CP iodine in 825 ml of glacial acetic acid by warming and stirring. Cool the solution and pipet out 25 ml; dilute to about 200 ml and titrate with 0.1*N* thiosulfate.

Add 3 ml of CP bromine from a buret to 200 ml of glacial acetic acid, mix well, and pipet out 5 ml. Dilute to about 150 ml with water, and add 10 ml of 15% potassium iodide solution. Titrate the liberated iodine with 0.1*N* sodium thiosulfate. The titration for the 5 ml of bromine solution should be approximately 80% of the titration of the 25 ml of the iodine solution.

The amount of bromine solution to be added to the remaining 800 ml of iodine solution is calculated as follows:

$$800 \times \frac{\text{Titration of iodine solution}/25}{\text{Titration of bromine solution}/5}$$

After mixing, dilute the solution to 1 liter with acetic acid, and store in a glass-stoppered amber bottle. A blank should be run with each determination or with each set of determinations, if more than one sample is run at one time.

Standard sodium thiosulfate, 0.1*N*.

Potassium iodide solution, 15%.

Starch indicator solution.

PROCEDURE

Take a sample of such a size that titration of the sample solution will be at least 60% that of the blank. If the sample titration comes to less than 60% of the blank, not enough reagent is present for complete reaction and the analysis should be repeated with a smaller sample.

Dissolve the sample in chloroform or carbon tetrachloride using a 250-ml iodine flask. If warming is necessary to dissolve the sample, the solution should be cooled to room temperature before the Hanus solution is added. Put the same volume of solvent in a separate flask. Into each of these flasks, pipet 25 ml of Hanus solution, and shake the flasks to ensure homogeneity. Allow the samples to stand for exactly 30 minutes with occasional shaking. After that time, add 50 to 100 ml of water 10 ml of 15% potassium iodide solution. Titrate the liberated iodine with 0.1*N* thiosulfate until the iodine color has almost disappeared. Add 1 ml of starch indicator, and continue titration until the blue color is discharged. The flask should be agitated quite vigorously when the reaction is near the end point to ensure extraction of all the iodine from the organic layer.

CALCULATIONS

Milliliters of thiosulfate for blank

minus milliliters of thiosulfate for sample $= A$

$$\frac{A \times N \text{ thiosulfate} \times 126.9 \times 100}{\text{Grams of sample} \times 1000} = \% \text{ iodine} = \frac{\text{centigrams of iodine}}{\text{grams of sample}}$$

(The calculation is given in these arbitrary units because this procedure is used mostly on hydrocarbons, fatty acids, and esters, where no clear-cut compound exists but a mixture of unsaturated compounds is present. There is no conclusive molecular weight that can be used in these cases. When definite compounds are being determined, the molecular weight of the compound can be substituted for the 126.9, and this divided by 2 for each double bond present, since 2 equivalents of iodine are involved per double bond. The equation will then yield the percentage of compound.)

The foregoing procedure was found to operate very well on unsaturated hydrocarbons, fatty acids, esters, vinyl esters, and some unsaturated alcohols. There are compounds for which the procedure does not operate as cleanly. Trouble can sometimes be noted by a continual fading back of the end point.

In addition to the substitution problems mentioned at the beginning of this procedure, any compound that is readily oxidized will also give erroneous results.

In spite of the previously mentioned drawbacks, the procedure still has a relatively wide applicability and a rather high precision and accuracy ±1–2%) when none of the interferences are present.

It was found advisable in the work to check each new batch of Hanus solution against a standard to ensure proper preparation. The investigators used corn oil because of its stability and because it is so readily adapted to this procedure.

Hydrogenation Methods

There are a multitude of apparatus for determining unsaturated compounds by the addition of hydrogen. The authors have selected five, since each of these covers one or more of the various attributes of an analytical method.

RAPID, SIMPLE QUANTITATIVE METHOD

The following method has been used in the laboratory of S. Siggia for many years and is quite satisfactory if a precision and accuracy of ±5% is acceptable. The main advantage of the method is simplicity and ease of operation.

PROCEDURE

Add about 3 ml of solvent and about 0.5 gram of catalyst to the hydrogenation vessel. The amount of catalyst is not critical. If a larger amount of catalyst is used, the hydrogenation of the compound proceeds more rapidly. The time required to saturate the catalyst, however, is much longer, so that no time is saved. When a small amount of catalyst is used, the sample hydrogenates more slowly, but the time is saved in saturating the catalyst.

Put a weighed sample that will consume about 0.0002 mole of hydrogen in the cup indicated by *A* in Fig. 7.5. If the sample is a solid, weigh it in a small aluminum boat or envelope. Place the boat in the cup. Weigh liquid samples in small glass receptacles or in gelatin capsules. (When using capsules, water must be used as the solvent.) All joints and stopcocks should be well greased to prevent hydrogen leakage.

Flush the system with a slow stream of hydrogen for 3 to 5 minutes, introducing the hydrogen through stopcock at *B*, which is in position as indicated; the mercury level is as near the stopcock as possible. Allow the hydrogen to escape through stopcock *C*. After 3 to 5 minutes close stopcock *C* and lower the mercury to a position well below the calibrations on the buret. Then turn stopcock *B* to position *D* and disconnect the hydrogen.

Apply a pressure of about 3 cm Hg and start the stirrer. The stirring is brought about by a rotating magnet under the reaction flask with a small glass-covered iron paddle inside the flask. These magnetic stirrers can be purchased or they can be improvised by attaching a magnet to the shaft of a stirring motor. The more vigorous the agitation, the more rapidly will the catalyst be saturated and the sample hydrogenated. When the mercury level ceases to rise, it signifies that the catalyst is saturated. Then raise the leveling bulb and slowly open stopcock *C*, allowing hydrogen to escape until the mercury level is within the calibrations on the gas buret. Close the stopcock again and allow about 5 to 10 minutes for the apparatus to come to equilibrium. Note the buret reading and the temperature. Then allow the sample to fall into the solvent by turning *A*.

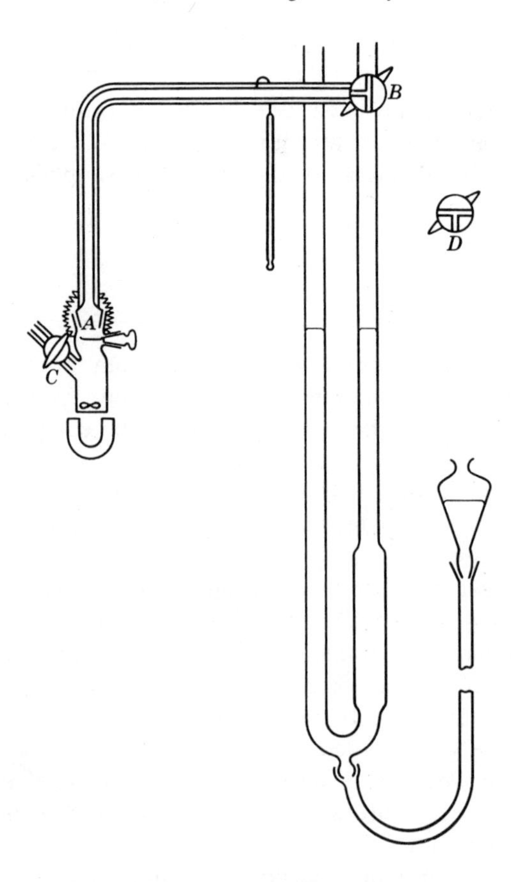

Fig. 7.5. Quantitative hydrogenation apparatus.

Apply about 3 cm pressure as agitation is continued, until the level of the mercury ceases to rise. Take the buret reading and again apply 3 cm pressure for 10 minutes to ensure complete reaction. Note the temperature.

For accurate results, the volume of the apparatus is needed to correct hydrogen-volume readings when the temperature at the end of an analysis is different from the temperature at the beginning. If this is not done, the volume readings will be in error because of expansion or contraction of the hydrogen in the free space of the apparatus. The volume of the apparatus is measured by completely filling it (up to the top mark on the buret) with water and then weighing the water. The correction is made by adding the volume of the apparatus to the volume between the top mark and the level of the mercury at the end of the hydrogenation minus the volume of solvent used. This equals the volume of free space at the end of the hydrogenation. This volume is corrected for temperature changes that may have occurred during the determination by the following equation:

$$\frac{V_1}{T_1} = \frac{V_2}{T_2}$$

where

$$T = \text{absolute temperature}$$

The difference between the volume before the temperature correction and the volume after the correction is the volume change due to expansion or contraction of the gas in the apparatus. This volume change is either added or subtracted from the volume of hydrogen consumed as read on the burets, depending on whether the final temperature was lower or higher than the initial temperature.

Low-boiling solvents are to be avoided, since temperature changes affect the vapor pressure of the solvent. Since the low-boiling solvents exhibit significant variations in vapor pressure over the range of temperatures encountered in the room, significant errors can result in the volume measurements.

In this procedure a 7-ml buret is used. Results have an accuracy and precision of ±5%. If a 50-ml gas buret is used and the sample is 5 to 8 times larger (the remainder of the apparatus remaining exactly the same), the results will be accurate and precise to ±1 to 2%.

Oxygen should be completely eliminated from the apparatus, since it will consume hydrogen. Any other materials that can be reduced with hydrogen will, of course, interfere.

CALCULATIONS

$$\frac{V_t}{T°C + 273} = \frac{V_0}{273}$$

$$\frac{V_0/22{,}400}{\text{Grams of sample/mol. wt. of sample}} = \frac{\text{moles of hydrogen consumed}}{\text{mole of sample}}$$

where

V_t = volume of hydrogen consumed (corrected for any temperature fluctuations)
V_0 = volume of hydrogen converted to 0°C
T°C + 273 = temperature of the experiment

and the temperature at 0°C = 273

Materials used to test the procedure were cinnamic acid, maleic acid, *n*-butyl, and *n*-propyl vinyl ethers, 2-butyne-1,4-diol, propargyl alcohol, and methyl acrylate.

CAUTION

All Raney nickel should be stored under alcohol (preferably a high-boiling alcohol). Dry Raney nickel will react with atmospheric oxygen, emitting sparks that can ignite any surrounding combustible material. The presence of hydrogen increases the hazards. None of the catalysts should be allowed to dry in the apparatus, on the desk top, or on the implements used to introduce catalyst into the apparatus.

Adam's catalyst when dry will glow in the presence of hydrogen and oxygen. Hydrogen is adsorbed on the catalyst and reacts with oxygen, emitting much heat. When the catalyst is put in the apparatus, it should be completely wet with solvent. Any dry particles of catalyst will glow as soon as the hydrogen is introduced, causing either the solvent vapors to flash or the hydrogen-air mixture to explode.

Method of N. Clausson-Kaas and F. Limborg

[*Acta Chem. Scand.*, **1,** *884–8* (*1947*)]

This method has a higher accuracy and precision (±1%) than the previous method. Also, temperature effects are eliminated, since the entire equipment is thermostated. The disadvantages of the equipment are that it is not composed of simple parts and must be either purchased or made by glassblowing; also, the hydrogen pressures that can be applied are slight and hence the hydrogenations proceed rather slowly.

Hydrogenation takes place in the vessel I (Fig. 7.6), which communicates with a compensation vessel II by a manometer. During hydrogenation, the drop of pressure in I is compensated by the addition of mercury from a buret, and in this way the manometer is kept at zero. The amount of hydrogen h (in milliliters at 0°/760 mm Hg) consumed by the substances given by eq. 3

$$h = \frac{273}{760}(B - e)\frac{a}{T} \tag{3}$$

where a is the volume of added mercury in milliliters, T the absolute temperature, B the atmospheric pressure in millimeters of mercury, and e the vapor pressure of the solvent in millimeters of mercury at $T°$.

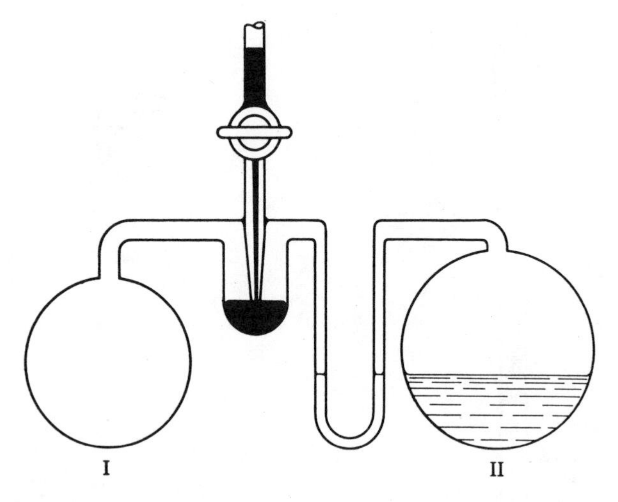

Fig. 7.6. Principle of the method: vessel I, hydrogenation; vessel II, compensation.

The foregoing principle was introduced by Smith (5), and later employed by Slotta and Blanke (6), Jackson and Jones (7), Breitschneider and Burger (8), and Prater and Haagen-Smit (9). The advantage of the method as compared to other methods has been reviewed by Breitschneider and Burger [*cf.* also Pfeil (10)].

The method described here uses a modification of the apparatus of Breitschneider and Burger. Magnetic stirring has been employed to avoid the use of the fragile glass coil that connects the hydrogenation vessel with the manometer; the air is washed out by a stream of hydrogen without evacuating the entire apparatus; this makes it possible to dispense with an arrangement to remove the manometer liquid. The new apparatus is simpler and easier to handle, whereas the obtainable accuracy remains the same.

APPARATUS

Figure 7.7 gives a detailed design of the apparatus (11). The total volume of the parts on either side of the manometer is about 70 ml and the two volumes should be of the same size within 2 to 3 ml.

5. J. H. C. Smith, *J. Biol. Chem.*, **96,** 35 (1932).
6. K. H. Slotta and E. Blanke, *J. Prakt. Chem.*, **143,** 3 (1935).
7. H. Jackson and R. N. Jones, *J. Chem Soc.*, 895 (1936).
8. H. Breitschneider and G. Burger, *Chem. Fabrik*, **10,** 124 (1937).
9. A. N. Prater and A. J. Haagen-Smit, *Ind. Eng. Chem., Anal. Ed.*, **12,** 705 (1940).
10. E. Pfeil, *Angew, Chem.*, **54,** 161 (1941).
11. The apparatus is manufactured by *Dansk Glasapparatur v/ Angelo Jensen,* Vesterbrogade 126 A, Copenhagen.

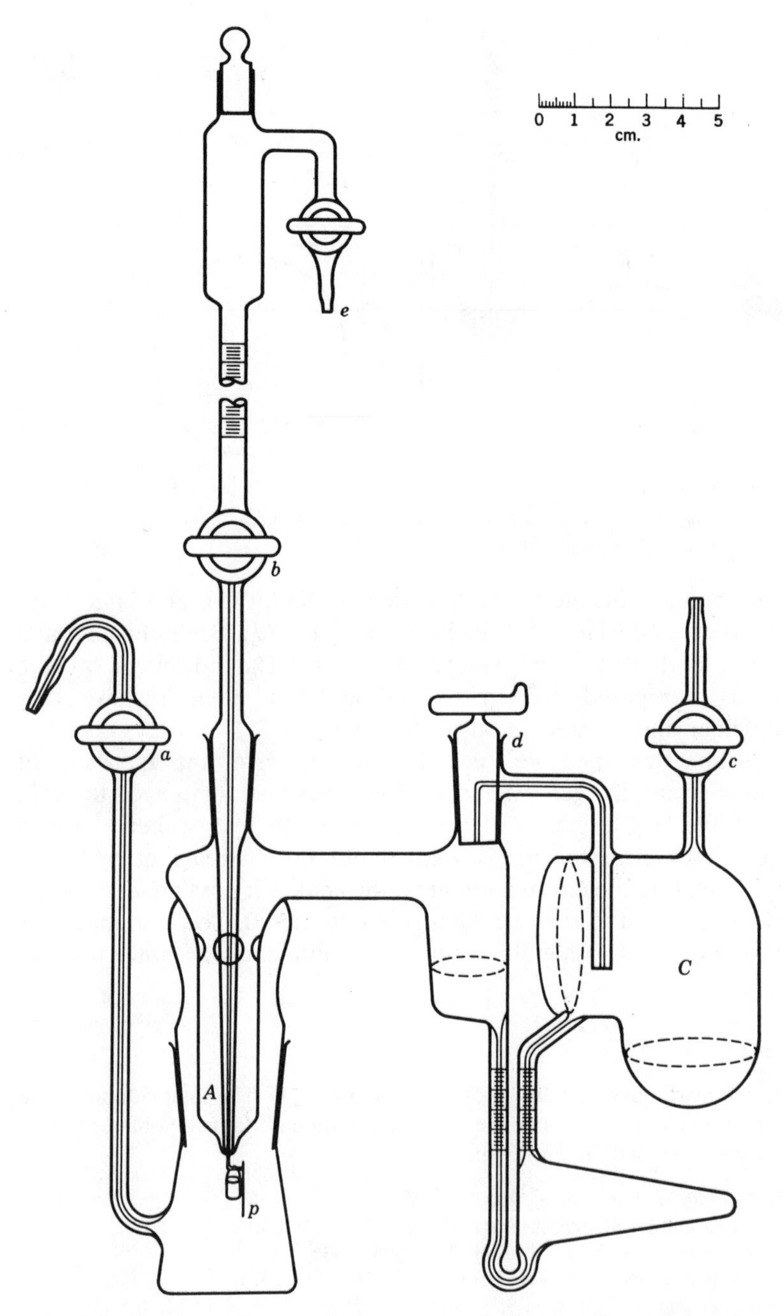

Fig. 7.7. Apparatus: *A*, reservoir for mercury; *C*, compensation vessel; *a*, *b*, *c*, *d*, and *e*, cocks; *p*, platinum beaker.

The graduation of the manometer must not necessarily be very accurate inasmuch as it only used to adjust the pressure difference to zero. A 2-cm long graduation in millimeters is very convenient.

The buret is about 40 cm long and contains 5 ml. Every milliliter is graduated in 50 parts; of course a longer and more accurate buret may be employed equally well. The tip of the buret is so fine that 1 ml of mercury flows through it in about 1 minute. Therefore it is not necessary to adjust the flow of mercury with the cock *b*.

A 25-mm long iron rod is attached to a platinum beaker *p* so that the beaker may be hooked off and dropped into the solvent by a magnet. The rod is wrapped in a platinum sheet, soldered with gold, to avoid corrosion of the iron by the solvent.

The whole apparatus is supported by a clamp just beneath the cock *b* and immersed into a bath filled with water. The water in the constant-temperature bath is stirred efficiently to assure equality of temperature throughout the bath. The liquid in the hydrogenation vessel is agitated by a Teflon-jacketed iron rod, which is rotated by a revolving magnet placed under and outside the glass constant-temperature bath.

Before use the apparatus is cleaned, and the cocks are carefully greased with Apiezon. Mercury is filled into *A* and is sucked up into the buret by applying suction to *e*. When the mercury has reached the upper mark of the buret, *b* is closed and suction turned off. There should now be only a small amount of mercury left in *A*, just enough to cover the tip of the end of the buret.

The apparatus including the constant-temperature bath and magnetic stirrer can be purchased from Dr. Hans Hosli, Bischofszell, Switzerland.

PROCEDURE

Fill the solvent into *C* (about 2 ml) and into the manometer by disconnecting the cock *d*. Place the catalyst and 2 ml of the solvent in the hydrogenation vessel. Weigh the substance to be analyzed into the platinum beaker, which eventually is suspended on the glass hook. Then connect the hydrogenation vessel with the main part of the apparatus and remove the air in the whole apparatus by leading a stream of hydrogen through *c* with *d* and *a* open.*

When all air has been removed, first close *a* and, shortly afterwards, *c*. In this way the pressure inside the apparatus will become slightly higher than that of the atmosphere. Start stirring to hydrogenate the catalyst. At the same time the hydrogen becomes saturated with the vapours of the solvent. When the *catalyst* is perfectly hydrogenated, the manometer will remain on level if *d* is closed. Now relieve the excess pressure by opening

* For the usual precautions concerning the employed reagents, and so on, see the monograph of Pregl on quantitative organic microanalysis (5th ed., Vienna, 1947.)

a for a moment (*d* must be open). Register the temperature and the atmospheric pressure. Close *d* and commence hydrogenation by dropping the beaker with the substance into the solvent.

Absorption of hydrogen causes the left part of the manometer to be filled with the manometer liquid.* When a reading of the consumption of hydrogen is desired, stop stirring and let mercury into *A* until the manometer again stands on level.† Hydrogenation is continued till no decrease of pressure is observed and the manometer remains on level.

Sorbic acid (*M* Mol. Wt. = 112.06) was employed to test the apparatus. The sample was alternately sublimated in vacuum (twice) and recrystallized from water (three times). The results of five hydrogenations in 2 ml of alcohol with Adam's catalyst are cited in Table 9.

Table 9

Sample mg.	PtO_2 mg.	*B* mm.	e mm.	*T*°	*a* after 90 Minutes ml.	*a* after 5, 10, 15, 20 and 40 Hours, Respectively	*a* Calculation	Error, %
8931	13	771.7	38.0	273 + 17.7	3952	3952	3955	0.1
8726	10	771.2	39.6	273 + 18.4	3882	3883	3885	0.1
8583	8	773.8	42.3	273 + 19.5	3835	3835	3836	0.0
8845	5	764.4	39.0	273 + 18.1	3961	3961	3967	0.1
3451	4	764.8	38.2	273 + 17.8	1541	1542	1544	0.1

The experiments demonstrate that the error is less than 0.2% even after hydrogenation for 40 hours.

Method of C. W. Gould and H. Drake

[*Reprinted in Part from Anal. Chem.*, ***23***, *1157–60* (*1951*)]

This method has the advantage of using hydrogen under relatively high pressure (up to 5 atm), hence the reactions are rapid. Also, the method has an accuracy and precision of ±1%. But the required apparatus is not simple and must be constructed because it cannot be purchased. The method utilizes 0.5 to 12.0 mM of hydrogen for optimum results.

* The volumes of the different parts of the apparatus are chosen so that 5 ml of hydrogen, that is, the total content of the buret, may be consumed without causing the solvent to flow into *A*. Thus it is possible to leave the apparatus unattended during hydrogenation, if only the equivalent weight of the substance is to be determined.

† Temperature fluctuations of the order of one degree may cause a small pressure difference because of volumetric changes of the solvent and the glass. If a very accurate determination is desired, the temperature of the water bath should therefore be kept constant within about 0.5°C.

APPARATUS

Figure 7.8 shows that the arrangement is like many described previously (12–16), except that a metal syringe replaces the double manometer and mercury leveling bulb, a sensitive Bourdon gauge serves as the pressure indicator, and a long stainless steel capillary is used to connect the reactor with the measuring system.

VALVES AND MANIFOLD. Connections are made with fittings having the standard $\frac{1}{8}$-in. pipe thread. 'Lunkenheimer' stainless steel needle valves are installed to resist leakage of hydrogen from the manifold.

SYRINGE. The syringe consists of a water-jacketed cylinder 7 in. (17.5 cm) long and 1.180-in. in bore, and a piston moved by a screw 18 threads per inch and 0.625 in. in diameter. Each turn of the screw displaces 1.00 ml. On the circumference of a wheel attached to the screw are engraved 10 divisions, each equivalent to 0.10 ml. A sectional drawing (Fig. 7.9) shows the construction of the syringe. The cord packing in the piston is clamped tightly between washers held by a cap bolt. To prevent leakage of hydrogen through and around this packing, a thick mixture of gear oil (Texaco Thuban 250) and fine graphite provides a satisfactory seal. At 5 atm, no leakage of hydrogen, can be detected in 24 hours.

GAUGE. An Ashcroft laboratory test gauge, 0 to 120 psi in 0.5-psi divisions, was calibrated up to 5 atm with a 10-foot open mercury manometer, and a new dial was made to read directly in absolute atmospheres. Any deviation from 1 atm

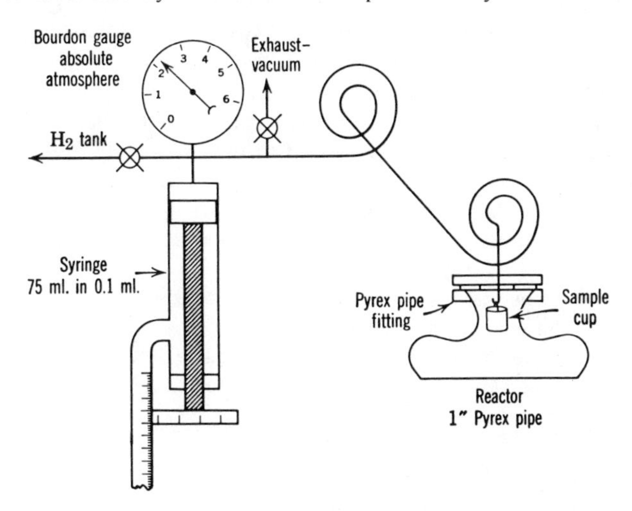

Fig. 7.8. Diagram of hydrogenation apparatus.

12. H. Jackson, *J. Soc. Chem. Ind.*, **57,** 97T (1938).
13. I. B. Johns and E. J. Seiferle, *Ind. Eng. Chem.*, *Anal. Ed.*, **13,** 341 (1941).
14. L. M. Joshel, *Ibid.*, **15,** 590 (1943).
15. C. L. Ogg and F. J. Cooper, *Anal. Chem.*, **21,** 1400 (1949).
16. H. E. Zaugg and W. M. Lauer, *Anal. Chem.*, **20,** 1022 (1948).

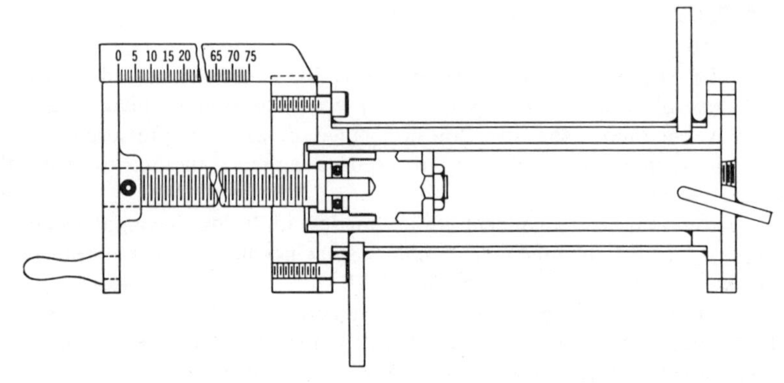

Fig. 7.9. Syringe.

caused by barometric change is applied as a correction to the initial and final gauge readings.

CONNECTION BETWEEN MANIFOLD AND REACTOR. Since the reactor is rocked by a shaker, a flexible connection is necessary. This is a 10-foot length of stainless steel tubing, 0.125 in. in outside diameter (1.25 mm in bore), with two coiled sections as shown in Fig. 7.8.

This tube is important in preventing solvent vapors from diffusing back into the syringe and gauge.

REACTOR. The reactor was made by the Corning Glass Works from standard heavy-walled 1-in. pipe according to the dimensions in Fig. 7.10. The open end is the regular Corning borosilicate glass pipe ending.

The reactor is connected to the connecting tube by bolting the standard borosilicate glass fitting to a flange brazed to the connecting tube.

From a hook on the inlet tube hangs a sample cup made from a lipless 5-ml borosilicate glass beaker. The inlet tube opens to the side, rather than to the bottom, to avoid blowing material from the sample cup when hydrogen enters the reactor.

The gasket is of two layers. Next to the glass face is a layer of 0.125-in. Teflon; next to the Teflon is smooth $\frac{1}{16}$-in. neoprene.

SHAKER. A Bodine motor, Type NC-1-12 RH, 1/50 hp, with a 30:1 reduction gear, furnishes power at 57 rpm which is transferred to a Fisher clamp by a wheel,

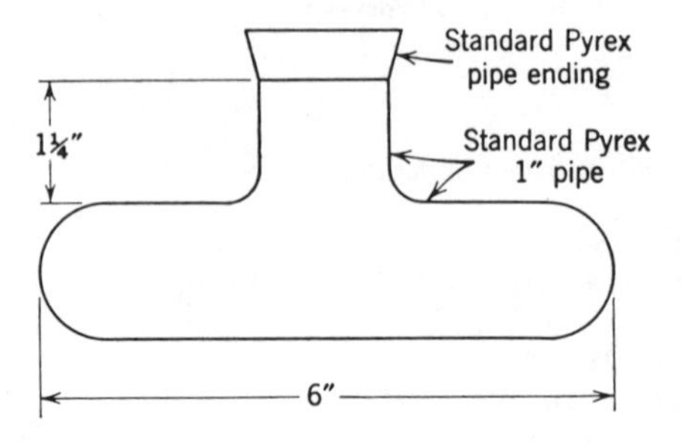

Fig. 7.10. Reactor.

connecting arm, and crank arm (Fig. 7.11). A slot in the crank arm permits changes in the amplitude of the rocking motion.

TEMPERATURE CONTROL. A centrifugal pump circulates water at 2.3 liters per minute between the jacket of the syringe and a constant-temperature bath (0.1°C control by mercury regulator), in which the reactor is immersed as far as the shaker clamp will allow.

PROCEDURE

Add a water slurry containing about 1.3 grams of fresh Raney nickel to 40 ml of solvent in the reactor. Hang the sample cup containing a weighed sample equivalent to 0.5 to 12.0 mM of hydrogen on the hook attached to the reactor inlet tube.

Lightly oil the gaskets with Texaco Thuban 250 gear oil and place on the flange of the reactor, then carefully lower the sample cup and inlet tube into the neck of the reactor, and bolt the flanges together. A 10-lb torque on the end of a 6-in. wrench is more than enough to tighten the bolts and prevent hydrogen leakage.

With the syringe set at a reading of 70 and the hydrogen value closed, evacuate the apparatus through the outlet valve, which was connected to the vacuum line. Then, without delay, close the outlet valve and slowly admit hydrogen until the gauge reading is 5 atm absolute. Repeat this purging operation twice more to eliminate air from the system. On the last filling, turn the syringe screw slowly to readings near 0.0 as hydrogen is admitted and the gauge pressure is increasing. This avoids any possibility of drawing solvent vapors from the reactor into the syringe and gauge.

Start the circulating pump, adjust the bath temperature, and start the shaker. Open the hydrogen valve as necessary to give a gauge reading of

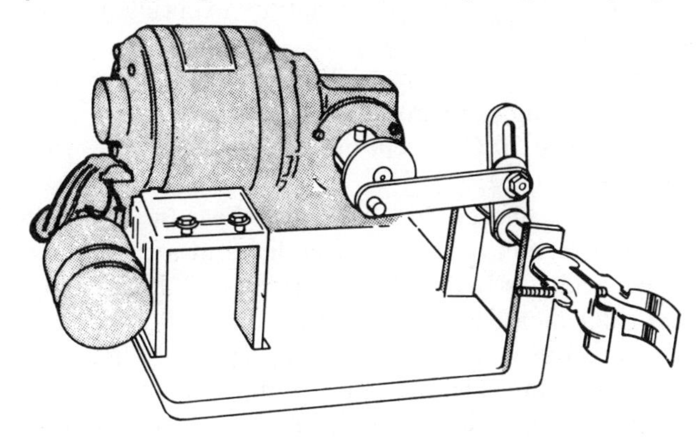

Fig. 7.11. Shaker.

5.00 atm (or any other desired working pressure above atmospheric), and then close. Then continue shaking for 5 minutes while saturating the catalyst and solvent with hydrogen; maintain the pressure by adjusting the syringe. After this equilibration, continue shaking for 5 minutes more to let the operator look for leaks or changes in bath temperature.

Make bath temperature and syringe readings at the exact desired working pressure. Stop the shaker, unclamp the reactor, and tip it so that the sample cup and its contents fall into the catalyst-solvent slurry. Then clamp the reactor as before and resume shaking.

Since hydrogen is used, turn the syringe screw to maintain the original pressure. If necessary, the syringe can be refilled with hydrogen at a slight sacrifice in accuracy; stop the shaker during such refilling so that hydrogen uptake is very slow.

Near the end of the hydrogenation, unclamp the reactor and tip it so that solvent can wash down any sample particles adhering in the neck of the reactor. Then resume shaking.

When the original pressure is maintained without further syringe adjustment and the temperature is constant at its original reading, record the syringe reading and use it in the calculation

$$\text{Millimoles of } H_2 = \frac{P(V_2 - V_1)}{0.08206T}$$

where P is the gauge reading in atmospheres, corrected for changes in barometric pressure, V_2 and V_1 are the final and initial syringe readings in milliliters, and T is the absolute temperature of the bath.

RESULTS

Data from several typical hydrogenation experiments are given in Table 10.

In these experiments 1.3 grams of Raney nickel and 40 ml of solvent were used. The working pressure in all cases was 5.00 atm, and the bath temperature was 25 to 30°C, except in the cases indicated, where temperature near 0°C were used.

Raney nickel was chosen for this work because of Whitmore and Revukas's demonstration (17) of its usefulness in azo splitting.

The cinnamic acid was J. T. Baker's purified U.S. Pharmacopoeia grade recrystallized from methanol. It melted sharply at 134°C corrected. The dye samples were purified until chromatographically homogeneous and recrystallized to constant melting point from methanol or ethanol.

17. W. F. Whitmore and A. J. Revukas, *J. Am. Chem. Soc.*, **59,** 1500 (1937); **62,** 1687 (1940).

Table 10. Tests of Hydrogenation Apparatus: Catalyst, 1.3 grams of Fresh Raney Nickel; Solvent, 40 ml; Pressure, 5.00 atm; Temperature, Room Temperature or Ice Temperature

Test Substance	Weight, grams	Solvent	Time, min.	Hydrogen: Theory, mM	Hydrogen: Found, mM	Error, %
Cinnamic acid	0.4947	EtOH	25[a]	3.34	3.39	+1.5
	0.9942	EtOH	20	6.71	6.69	−0.3
	1.0788	EtOH	30	7.28	7.30	+0.3
	1.0426	BuOH	20	7.04	7.09	+0.7
	0.9964	MeOH	30	6.72	6.71	−0.15
	1.0002	DMF[b]	20	6.75	6.76	+0.15
2,4-Dinitrodiphenylamine	0.3091	EtOH	105[a]	7.15	7.04	−1.5
	0.2612	EtOH	66	6.05	5.97	−1.3
	0.3941	EtOH	45	9.12	9.05	−0.8
	0.3160	EtOH	40	7.31	7.30	−0.1
	0.3307	BuOH	60	7.65	7.64	−0.1
	0.3505	BuOH	60	8.11	8.12	+0.1
Competitive azo dye I, mol. wt. 300.3, one nitro and one azo	0.4172	EtOH	90[a]	6.95	6.92	−0.4
	0.4283	EtOH	168[a]	7.13	7.12	−0.1
	0.3906	EtOH	20	6.50	6.45	−0.8
	0.3355	BuOH	30	5.59	5.53	−1.1
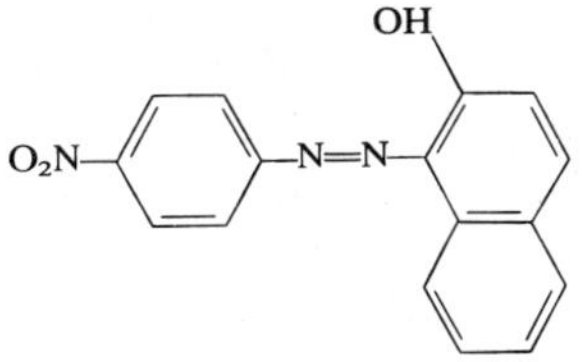 O_2N—⟨ring⟩—N=N—⟨naphthalene ring with OH⟩	0.2983	EtOH	180	5.80	5.82	+0.3
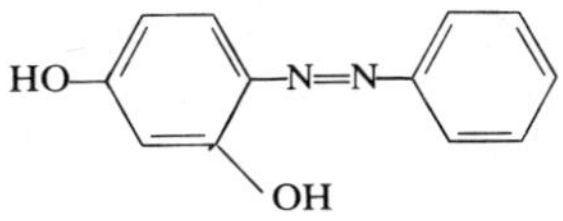 HO—⟨ring with OH⟩—N=N—⟨ring⟩	0.9953	DMF[b]	35	9.29	9.23	−0.6
Competitive azo dye II, mol. wt. 428.4, one nitro and one azo	2.003	EtOH	90	23.35	22.85 (Syringe refilled)	−2.1
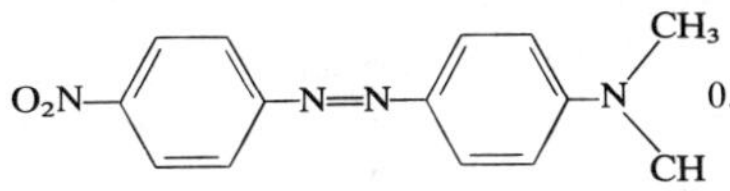 O_2N—⟨ring⟩—N=N—⟨ring⟩—N(CH_3)(CH)	0.2024	DMF[b]	60	3.74	3.72	−0.5

[a] Ice temperature.
[b] Dimethylformamide.

The solvents were of reagent grade.

The hydrogen was purchased from the Air Reduction Co. (manufactured by the Paschall Oxygen Co., Philadelphia). It had a purity above 99.5%, as stated by the manufacturer and confirmed by Gould and Drake.

DISCUSSION

From the results in Table 10 it can be calculated that standard deviation of the errors is 0.7, excluding the next to the last case where the syringe was refilled. There is no correlation between the errors and either the temperature or the solvent employed. The error in the case of cinnamic acid, for example, is no worse with methanol as the solvent than with *n*-butyl alcohol. If a correction for solvent vapor pressure were required, the errors caused by neglecting it would be −3.2 and −0.17% for methanol and butanol, respectively (for reductions at 25°C and 5 atms).

The explanation for this lack of dependence on vapor pressure is believed to be that the 10-foot connecting tube between manifold and reactor serves as an effective barrier against diffusion of solvent vapor back into the syringe-gauge system, particularly during a hydrogenation, when there is a counterflow of gas to the reactor.

To estimate the diffusion rate of solvent through the connecting tube, the following experiment was carried out.

One end of a 2-foot length of borosilicate glass capillary, 1.25 mm in bore, was drawn out to a fine tip. The other end was fastened through a 50-ml filter flask as shown in Fig. 7.12. Hydrogen gas was passed through the apparatus and forced through the capillary. The fine tip was dipped in methanol, and a column of solvent allowed to rise in the capillary until the meniscus was 20 mm above the drawn-down section; then the tip was removed and quickly sealed in the Bunsen flame. In this way, the solvent was sealed in the capillary under the hydrogen.

Finally, the capillary was mounted in a vertical position, and hydrogen was passed slowly (0.5 ml per minute) through the filter flask for several days. The average rate at which the meniscus receded was 2.5 mm per day at 25°C; this rate was not changed measurably during an additional 24 hours with an increased hydrogen flow of 15 ml per minute.

In a control experiment, the capillary was evacuated, whereupon the methanol meniscus receded at a rate of 3 mm per minute. The slow diffusion of methanol through hydrogen was thereby shown not to be a result of limitation by the extent of evaporation surface.

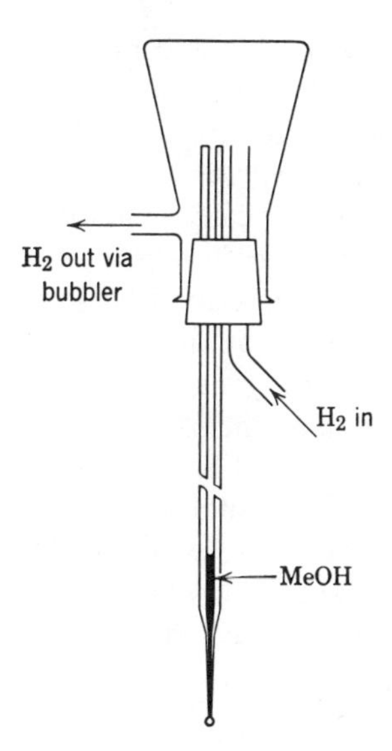

Fig. 7.12. Arrangement for diffusion experiments.

These results suggest that in this case correction for solvent vapor pressure is not justified because no significant amount of vapor can enter the measuring system under the specified operating conditions.

In the literature on laboratory hydrogenation at constant pressure, most experimenters who made a correction for solvent vapor had previously saturated their hydrogen before introducing it into their apparatus.

The closest analogy to the apparatus of Gould and Drake is described by Jackson (12), who used a flexible glass helix and other connections totaling 3.4 meters of tubing 5 mm in inside diameter (estimated) between measuring system and reactor. He did not saturate his hydrogen with solvent, but he did make a correction for its vapor pressure. This correction may be justified because solvent vapor could reach his measuring system through a smaller diffusion barrier than that of Gould and Drake, at 1 atm hydrogen pressure, during the 3 to 4 hours that Jackson allowed for temperature equilibration.

The authors of this textbook have used the procedure above and have found it to operate for chloroacrylates, 1,4-butynediol, hexene, traces of vinyl monomers in hexane, and unsaturation in natural oils.

Method of Miller and DeFord

[*Adapted from Anal. Chem.*, ***30,*** *295–8* (*1958*), *Reprinted in Part*]

The method of Miller and DeFord involves the use of electrically generated hydrogen. Again, as in the methods of Clausson-Kaas and

Gould and Drake, the apparatus is not standard equipment and must be constructed by the analyst. The apparatus is not complex, however, and it has the advantage that it operates more or less automatically. The accuracy and precision obtainable are of the order of ±2 to 3%.

Manegold and Peters (18) were the first to use electrically generated hydrogen in place of the conventional gas buret. Its use for the determination of relative rates of hydrogenation has been described by Farrington and Sawyer (19). The new apparatus is a modification of that of Manegold and Peters (18), which embodied a cell for generating hydrogen electrically, a gas coulometer to measure the amount of electricity consumed, and a reaction vessel. This apparatus measured both the rate of reaction as indicated by the current intensity and the total amount of gas consumed as shown by the coulometer, but the current has to be adjusted manually during a run, so that the rate of generation equaled the rate of disappearance of hydrogen. The apparatus was constructed to measure large volumes of gas (liters) and was too complex for easy reproduction and maintenance.

These disadvantages were overcome by automatic electrolysis, which regulated the pressure and varied the electrolysis current so that rates of consumption and generation of hydrogen were equal.

Figure 7.13 is a schematic diagram of the apparatus. The sample, the solvent, and the catalyst are contained in the reaction flask *F*. The flask is connected to the U-shaped electrolysis cell T_2 by the gas manifold and drying tube T_1. The entire system is closed to the atmosphere by stopcocks and by placing generator electrolyte in T_2.

The catalyst is prereduced with hydrogen; when reduction is complete, the increase in the hydrogen pressure in the system causes the liquid to be pushed away from the electrode. The contact between the electrode and electrolyte is broken, and the electrolysis current is automatically shut off. The pressure in the system exceeds the atmospheric pressure by an amount that corresponds to the difference in liquid levels in the U-tube arms. When the sample is introduced into the solvent, reaction takes place and the hydrogen pressure in the system decreases, causing the liquid to make contact with the electrode. The rate of hydrogen generation is adjusted to equal the initial rate of hydrogen consumption. After this setting is made, the hydrogenation is automatic.

Some degree of current control is achieved by employing a tapered electrode, so that its depth of immersion in the electrolyte governs the current flowing. As hydrogenation proceeds, the rate of hydrogen uptake decreases, and the liquid level falls, slowing down the rate of hydrogen

18. E. Manegold and F. Peters, *Kolloid-Z.*, **85,** 310 (1938).
19. P. S. Farrington and D. T. J. Sawyer, *J. Am. Chem. Soc.*, **78,** 5536 (1956).

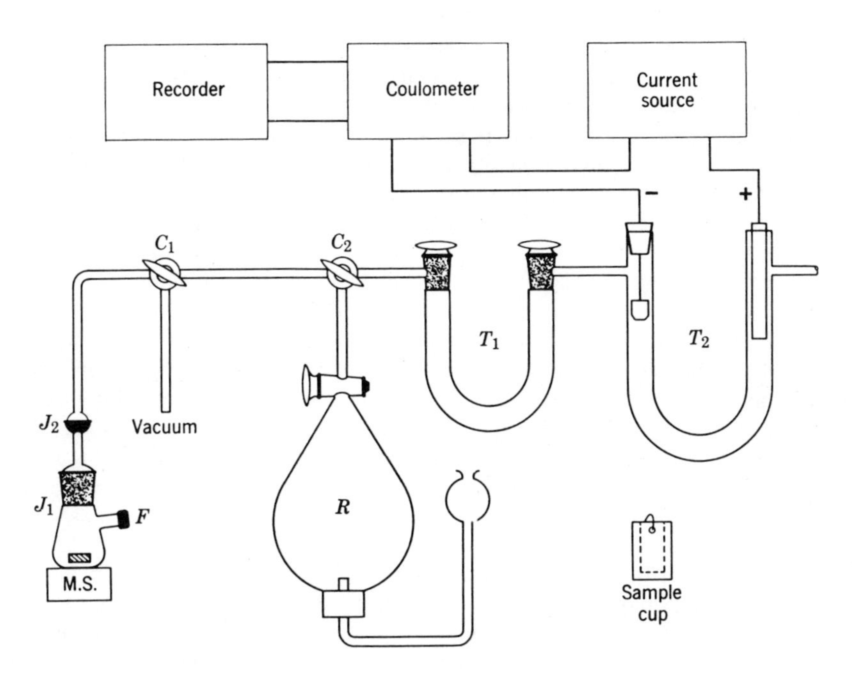

Fig. 7.13. Hydrogenation apparatus of Miller and DeFord.

generation. If the rate of generation exceeds the rate of consumption because of a slow hydrogenation reaction, the instrument will cycle on and off. During the off-cycle the pressure in the system slowly drops until the liquid makes contact with the electrode again. When no hydrogen is generated for a given length of time, the reaction is assumed to be complete.

A stepping recorder was placed in the coulometer circuit, so that the number of milliequivalents of hydrogen consumed was plotted against time. After the initial adjustments, the hydrogenation proceeds to completion automatically with no operator attention. The results are then calculated from the coulometer reading as registered on the recorder.

APPARATUS

The reaction vessel *F* was constructed from a 25-ml Erlenmeyer flask to which was added a standard-taper 14/20 joint, J_1. A short length of 6-mm tubing served as a side arm for introduction of sample. A female standard-taper 14/20 joint, sealed to a 3-in. length of 1-mm capillary tubing, connected the reaction vessel to the rest of the system. The ball and socket joint J_2 was placed above the reaction flask to give flexibility to the system. The three-way 2-mm capillary stopcock C_1 allowed the reaction flask to be evacuated before hydrogenation was begun. A water aspirator was satisfactory for evacuation.

A hydrogen reservoir *R*, a 500-ml separatory funnel filled with mineral oil and attached to the system by the three-way 2-mm capillary stopcock C_2, served as a source of hydrogen for the flushing procedure. The absorption tube T_1, filled with 8-mesh calcium chloride, dried the hydrogen before it passed into the reacting system. The generator cell T_2 was a U-shaped drying tube closed on the left by a No. 0 rubber stopper through which extended the platinum generating cathode (1.0×1.4 cm). The cathode was tapered so that the width at the bottom was one-half that at the top. A rod of reagent grade zinc served as the generator anode. The individual components were joined with 2-mm capillary tubing with butt joints of Tygon tubing between components. Such construction allowed easy cleaning of the apparatus and made it flexible and less subject to breakage. The total volume of gas in the apparatus, including the electrolysis cell, drying tube, connecting tubing, and empty reaction flask, was 51.5 ml. To reduce errors arising from changes in temperature or pressure, this gas volume should be kept as small as possible.

The side arm of the flask was closed by a rubber serum bottle stopper. Liquid samples were injected by a hypodermic syringe, which served as a weight buret. For solid samples a cup (shown in Fig. 7.13) was constructed of $\frac{1}{4}$-in. aluminum rod, $\frac{7}{16}$ in. long. A hole $\frac{3}{16}$ in. in diameter was drilled into the center of the rod to a depth of $\frac{3}{8}$ in. Two small holes were drilled opposite each other at the top and a wire loop of Chromel A wire was attached. The cup was hung from a hook made from a needle bent at the end and inserted through a serum bottle stopper. The cup was dropped into the catalyst-solvent mixture by a half-turn of the stopper. The hydrogenation mixture was agitated by a $\frac{5}{8}$-in. magnetic stirring bar placed in the reaction flask.

The current source was a Sargent-Slomin Electroanalyzer, chosen because low-voltage current was desired to prevent arcing between the electrode and the liquid when contact was broken. Two direct-reading coulometers were used during the work. The first (20) was based on the principle of charging a condenser to a given voltage with the generation current and counting the number of times it was charged during a run. The second used an integrating motor (Model 120T-3, Summers Gyroscopy Co., Santa Monica, Calif.) to measure the amount of electricity passed during a run. Miller and De Ford described the design and operation of these coulometers in subsequent publications.

REAGENTS

CP Concentrated sulfuric acid was used to prepare the 6*N* acid.

Platinum oxide catalyst and 10% palladium-on-charcoal were obtained from the American Platinum Works, Newark, N.J.

20. D. D. DeFord and C. E. Toren, Northwestern University, Evanston, Ill., unpublished results, 1955.

The purity of several of the compounds used was unknown. The following materials were used as received from the manufacturer: cinnamic acid and acetophenone (Eastman Kodak Co.), and phenyl propyl ketone (Matheson, Coleman, and Bell Division, Matheson Co., East Rutherford, N.J.). The low results obtained for these compounds were undoubtedly due to lack of purity.

Reagent grade benzene was dried with sodium wire before use. The naphthalene had been purified by being crystallized 4 times from each of the following solvents: glacial acetic acid, acetone, and ethyl alcohol; it was further purified by refluxing for 39 hours in ethyl alcohol over Raney nickel. *p*-Nitrophenol was purified by recrystallization from $1N$ sulfuric acid. Its melting point was 112 to 113°C.

PROCEDURE

By electrolyzing $6N$ sulfuric acid, hydrogen gas is produced at the cathode. With all the stopcocks open to the atmosphere, allow hydrogen to flow through the system 20 minutes before any determinations are carried out. During the flushing and filling operations, adjust the current to its maximum (1 ampere) by increasing the applied potential to full value. Then fill the reservoir with electrically generated hydrogen. While flushing the apparatus, weigh the catalyst into the reaction flask and add 5 ml of solvent. Place a weight of solid sample corresponding to a hydrogen uptake of between 2 and 22 ml in the aluminum cup. Hang the cup on the hook after the solvent and catalyst have been added. High-boiling liquids can also be weighed into the cup with no loss when the reaction vessel is evacuated. Make up liquid samples of high volatility as standard solutions in the hydrogenation solvent, so that 1 ml of the solution takes up the required amount of hydrogen. Fill a 1-ml tuberculin syringe (Yale No. 1 YT, Becton, Dickinson, and Co., Rutherford, N.J.) to the 1-ml mark with solution and weigh. After injection of exactly 1 ml of sample solution, reweigh the syringe, and calculate the sample size from the loss in weight.

Attach the reaction flask containing catalyst solvent and magnetic stirring bar to the apparatus. Turn stopcock C_2 so that only the hydrogen reservoir R is connected to the reaction vessel, and alternately evacuate and fill the reaction flask with hydrogen by proper adjustment of stopcock C_1. After five evacuations, turn both stopcocks so that the reaction vessel is connected directly to the generator cell. Set the magnetic stirring motor at a speed that allows rapid hydrogen transfer across the gas-liquid interface. Reduce the catalyst until the coulometer indicates no hydrogen uptake for 10 minutes. Turn on the recorder and reset the register on the coulometer to zero. Then introduce the sample, either by dropping the aluminum cup or by injecting 1 ml of the sample solution into the solvent. Note the temperature and barometric pressure.

For the hydrogenation of aromatic nuclei, use glacial acetic acid as the solvent and platinum oxide as the catalyst. Most of the other reductions are carried out in 95% ethyl alcohol; 10% palladium-on-charcoal is the catalyst. The catalyst usually weighs 10 to 20% as much as the sample. This amount is sufficient to permit the reaction to go to completion in less than an hour.

CALCULATION OF RESULTS

The number of counts on the coulometer was used to calculate the results. As the coulometers had been previously calibrated in terms of milliequivalents per count, the number of milliequivalents of hydrogen generated was the product of the number of counts and the calibration factor. This result was compared to the theoretical number of milliequivalents of hydrogen. The results were then calculated as the hydrogen number, defined in eq. 4 as

$$\text{Hydrogen number} = \frac{n(201.6)}{\text{mol. wt.}} \tag{4}$$

where n is the number of double bonds. The experimental hydrogen number was calculated by substituting the proper values in eq. 5.

$$\text{Hydrogen number} = \frac{(\text{Milliequivalents of } H_2)(100.8)}{\text{Weight of sample (mg)}} \tag{5}$$

If the sample was added as a standard solution, the number of milliequivalents of hydrogen generated did not represent actual hydrogen uptake. When 1 ml of solution was added, the liquid level in the generator cell was pushed a corresponding distance below the cathode. Even though hydrogen was consumed by the sample, no counts were registered until slightly more than 1 ml of hydrogen had been used. For this reason it was necessary to add to the number of milliequivalents obtained from the coulometer the number of milliequivalents of hydrogen equivalent to the volume of sample added. The volume was corrected to standard pressure and temperature before it was used in any calculations. The total number of milliequivalents taken up by the sample was then the sum of those from the coulometer reading and those from the volume of sample added.

RESULTS AND DISCUSSION

The accuracy and precision expected for the hydrogenation of a variety of compounds are shown in Tables 11 to 13. The standard deviation is

Table 11. Hydrogenation of Cinnamic Acid

Number of Determinations	Weight Range Sample, mg.	Hydrogen Number Found	Error, %	Standard Deviation, %	Maximum Deviation from Mean	H_2 Consumed ml.
4	72.6–74.4	1.342[a]	−1.47	0.58	0.011[b]	11
4	36.2–37.5	1.339	−1.69	1.98	0.034	5.5
4	18.6–19.6	1.330	−2.35	1.59	0.029	2.7
4	10.2–10.9	1.324	−2.79	1.14	0.019	1.6
4	5.2–5.7	1.343	−1.39	3.18	0.052	0.84

[a] Theoretical hydrogen number is 1.362.
[b] In terms of hydrogen number.

Table 12. Coulometric Hydrogenation of Unsaturated Compounds

Compound	Number of Determinations	Weight Range Sample, mg.	Hydrogen Number		Error, %	Standard Deviation, %	Maximum Deviation from Mean[a]
			Theory	Found			
Benzene	11	14.9–15.4	7.754	7.578	−2.27	3.26	0.152
Naphthalene	6	20.6–23.3	7.865	7.671	−2.47	1.44	0.173
p-Nitrophenol	8	47.2–54.3	4.348	4.273	−1.72	1.74	0.127
Acetophenone	5	72.5–74.2	1.678	1.638	−2.38	1.04	0.024
Phenyl propyl ketone	7	57.2–62.4	1.361	1.330	−2.28	3.75	0.090

[a] In terms of hydrogen number.

Table 13. Hydrogenation of Fumaric Acid[a]

Number of Determinations	Weight Range Sample, mg.	Hydrogen Number		Error, %	Standard Deviation, %
		Theory	Found		
3	51.5–53.9	1.737	1.741	+0.23	2.18
6	25.7–27.2	1.737	1.711	−1.50	1.72
10	12.0–13.0	1.737	1.741	+0.23	4.82

[a] Practical grade material was recrystallized from 1*N* hydrochloric acid and vacuum dried at 50°C before use.

less than 2% in all but four cases. The poor precision for benzene was probably due to partial loss of sample during introduction into the reaction flask. The average of three determinations of benzene on a hydrogenation apparatus identical to that described by Park, Planck, and Dollear (21) was −2.22%. This result agrees with that obtained on the new apparatus. The extrapolation procedure used to find the number of coulometer counts for the reduction of phenyl propyl ketone accounts for the high standard deviation of this compound. Typical rate curves are shown in Fig. 7.14.

21. F. C. Park, R. W. Planck, and F. G. Dollear, *J. Am. Oil Chem. Soc.*, **29,** 227 (1952).

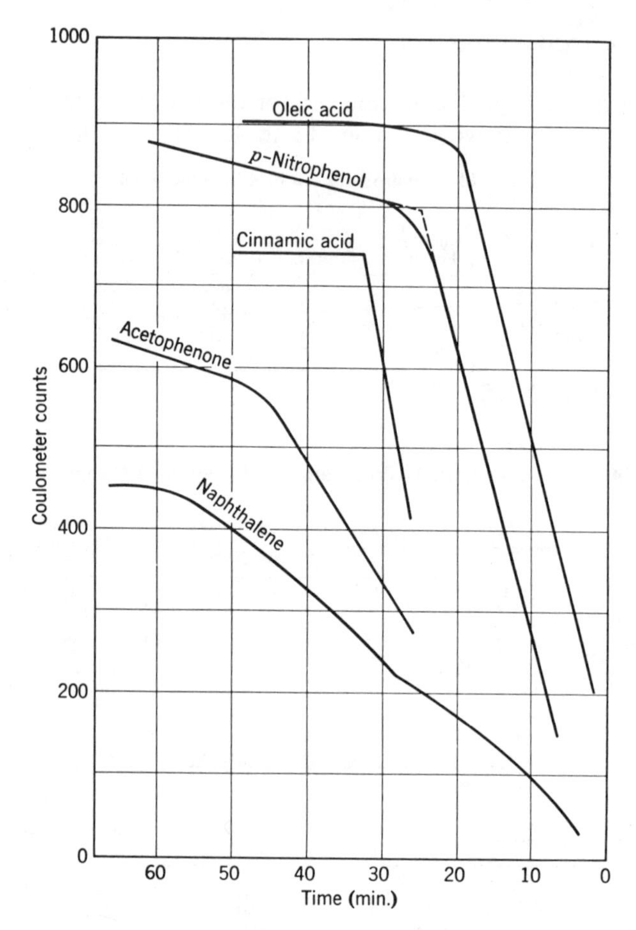

Fig. 7.14. Typical rate curves.

When the nitro group had been completely reduced, the uptake of hydrogen continued at a slow rate. The difference in the rate of the main reduction and the slow reaction was large enough so that the end point could be determined by extrapolation of the two linear portions of the curves. The end points for the hydrogenation of the two aromatic ketones, acetophenone, and phenyl propyl ketone, also were found by extrapolation. The curve after the break was linear only at an optimum catalyst sample ratio of 5%; higher ratios gave a rounded curve, which could not be used for extrapolation. In such cases the optimum concentration of catalyst must be determined before unknowns are analyzed.

Table 11 indicates the effect of sample size on the determination of cinnamic acid (average results). In three of the five weight ranges, at least one of the four determinations gave a high result. It was possible to hydrogenate microsamples with little loss in accuracy. The precision decreased slightly as the sample size decreased; this may be the result of the effect of temperature on the liquid level in the cathode compartment.

It was possible to reduce acetophenone rapidly to ethylbenzene, with no break in the rate curve corresponding to the alcohol intermediate, by using a 3:1 sample-catalyst ratio. The consumption of hydrogen stopped abruptly upon completion of this reaction. Two such hydrogenations gave errors of −4.41 and 3.75%, respectively.

Naphthalene might be expected to give low results because of its slow rate of hydrogenation. The low results on *p*-nitrophenol are unexplained.

Three main factors affect the accuracy of the hydrogenations: temperature, pressure, and surface tension. With a dead volume of 46.5 ml when the reaction flask is filled with 5 ml of solvent, a 1°C change in temperature during a hydrogenation is equivalent to 0.16 ml of gas. The final result may be high or low, depending on the direction of the temperature change. For relatively large samples that require on the order of 15 ml of hydrogen, this error is only 1%, but for every small sample, the error may be as large as 20%. In the laboratory of Miller and De Ford, temperature fluctuations were small or negligible during a determination and the accuracy and precision were good. During the summer months the room temperature varied by as much as 10°C during the day, but fluctuations during a run must have been small. In many situations some form of temperature control would have to be maintained.

The effect of temperature fluctuation could be lessened by decreasing the dead volume of the apparatus and by using a smaller reaction flask and no drying tube. The drying tube accounted for more than half the dead volume. The use of 85% sulfuric acid permitted elimination of the drying tube, since the vapor pressure of water is negligible over it. Because the low conductivity of 85% acid limited the current to a few milliamperes, however, only microsamples could be analyzed.

Since one arm of the generator cell is open to the atmosphere, changes in atmospheric pressure also produce errors. A pressure change of 1 mm Hg is equivalent to approximately 0.06 ml of hydrogen. It is possible that a pressure increase could be compensated by a temperature increase, so that no error would be introduced. It is unlikely that pressure would change by 1 mm Hg during a half-hour determination; thus the error from this source would, in general, be small.

Because of surface tension effects, the liquid level in the cathode arm of the generator cell at the moment when contact with the cathode is broken is about 1 mm lower than the level at which contact is reestablished. With the cell used, this difference in height corresponds to a volume of about

0.13 ml. For this reason there is a dead zone corresponding to this volume of hydrogen. The dead zone corresponds to a possible error of 1% when approximately 15 ml of hydrogen is consumed. The percentage of error increases as the volume of hydrogen uptake decreases. If desired, the dead zone could be reduced to a negligible value by use of a cathode compartment of smaller cross section or more elaborate manostats.

The apparatus should be useful for rate studies, since the rate curves are automatically plotted. Extremely small changes in rates are easily noted from the recorded curves. Naphthalene shows a small break at 40% reaction: this is because of the difference in reaction rates of naphthalene and Tetralin (22, 23). The end point of a hydrogenation can be determined by the sharp break in the recorded curve. The apparatus can be used to study rates only when the rate of reaction is slower than the rate of transfer of hydrogen across the gas-liquid interface. The rate of transfer has been shown to be dependent on stirring (24, 25). For rate studies, the reaction vessel described by Vandenheuvel (24) should be the most satisfactory. Because the hydrogenation rate curves were used only to locate the end points, the reliability of these curves was not studied. The curves for a compound such as naphthalene showed differences from determination to determination, undoubtedly due to the effect of agitation.

The low results are what might be expected for the compounds used, and no known systematic error exists in the system. A thermostated apparatus would eliminate one large source of error.

HYDROGEN FROM SODIUM BOROHYDRIDE

Method of C. A. Brown, C. S. Shanti, and H. C. Brown

[*Adapted from Anal. Chem.*, ***39,*** *823* (*1967*)]

The *in situ* preparation of highly active hydrogenation catalysts (26, 27) has been combined with an automatic valve (28, 29) for the generation of hydrogen from sodium borohydride to provide a convenient technique for the determination of unsaturation through quantitative hydrogenation.

22. R. H. Baker and R. D. Schuetz, *J. Am. Chem. Soc.*, **69,** 1250 (1947).
23. R. D. Schuetz, Ph.D. thesis, Northwestern University, Evanston, Ill., 1947.
24. F. A. Vandenheuvel, *Anal. Chem.*, **24,** 847 (1952).
25. F. A. Vandenheuvel, *Ibid.*, **28,** 362 (1956).
26. H. C. Brown and C. A. Brown, *J. Am. Chem. Soc.*, **84,** 1494, 2827 (1962).
27. H. C. Brown and C. A. Brown, *Tetrahedron, Suppl.*, **8,** Pt. I, 149 (1966).
28. C. A. Brown and H. C. Brown, *J. Am. Chem. Soc.*, **84,** 2829 (1962).
29. C. A. Brown and H. C. Brown, *J. Org. Chem.*, **31,** 3989 (1966).

In the recommended procedure, a standardized borohydride solution is used to measure the hydrogen uptake. Because the buret contains only liquid, it need not be thermostated. The only requirements for accuracy are that the hydrogenator vessel be thermostated and that the pressures at the beginning and end of the reaction be the same. The former is assured by a small constant-temperature bath, the latter by the analytical procedure. The method has been used to hydrogenate from 0.2 to 2.0 mM of compound with an accuracy of 1%. Analyses require about 3 minutes each, 20 minutes being sufficient for triplicate analyses, including the preparation of the catalyst.

APPARATUS

The hydrogenator (Fig. 7.15) consists of a 65-ml flask (containing a Teflon-coated stirring bar) with a 24/40 ground joint and a hydrogenator valve (also with a 24/40 joint) having a mercury well and injection port, which is fastened to a mercury bubbler by means of an 18/9 O-ring joint. A precision-bore 5-ml buret with a reservoir and a three-way stopcock is attached to the valve by means of a 12/30 ground joint. Mercury in the central well serves as a sealant and control liquid. A drop in the pressure of the system below atmospheric causes borohydride solution to be sucked from the buret tip, through the mercury, out of the vent holes just below the 12/30 joint, and into the flask. Acid in the flask and the borohydride react to produce hydrogen, the increase in hydrogen pressure causing

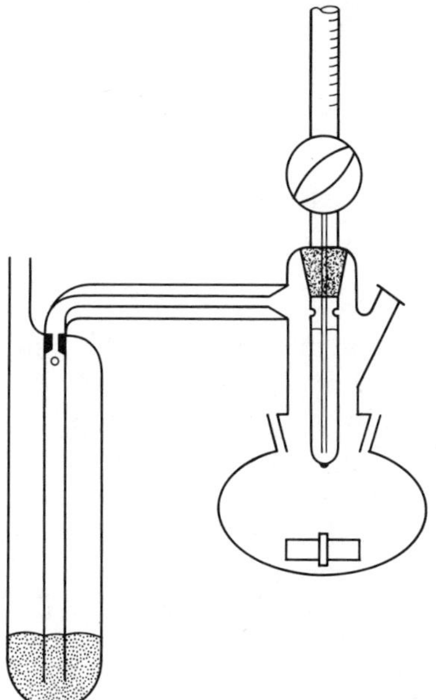

Fig. 7.15. Analytical hydrogenator (hydroanalyzer) used by C. A. Brown et al.

the flow of borohydride solution to cease. In this manner the addition of borohydride solution proceeds smoothly and automatically until reduction is complete. The catalyst is generated *in situ* via the reaction of sodium borohydride and platinum metal in the presence of activated carbon. This provides a highly active and reproducible catalyst as well as one already saturated with hydrogen and ready for use. As many as 12 replicate analyses may be run with one batch of catalyst. A commercial model of this apparatus is available from Delmar Scientific Laboratories (317 Madison St., Maywood, Ill. 60154).

SOLUTIONS

Sodium borohydride, 3.95 grams (Metal Hydrides, Inc., 98%) was dissolved in 100 ml of diglyme or triglyme (di- or trimethylene glycol dimethyl ether, Ansul Chemical Co.). For most purposes technical grade diglyme and triglyme from freshly opened tins are satisfactory. However if these ethers have been exposed to the atmosphere for some time, they must be made anhydrous by the addition of a small quantity of lithium aluminum hydride and distillation under vacuum (diglyme, b.p. 62–63°C per 15 mm, n^{20}D 1.4078; triglyme, b.p. 108°C per 15 mm, n^{20}D 1.4233). The solution was filtered or decanted from any separated solids. It was then standardized by injection of 10.0 ml into 30 ml of 50% aqueous acetic acid and measurement of the gas evolved with a wet test meter. The molarity was calculated as follows:

$$M_{NaBH_4} = \frac{V-v}{896} \times \frac{P-25}{760} \times \frac{273}{T+273}$$

where

V = total volume of gas evolved (milliliters)

v = volume of sodium borohydride solution injected (milliliters)

P = atmospheric pressure (mm Hg)

T = temperature (°C).

Alternatively, the sodium borohydride solution may be standardized by the acid titration method. In this case, 10.0 ml of standard hydrochloric acid (0.15M) and 5 ml of water are mixed in a flask, and 1.00 ml of sodium borohydride solution (approximately 1.0M) is added. The flask is covered loosely and swirled for a minute. The solution is then titrated with standard sodium hydroxide solution (0.10M). After the addition of about 4 ml of base, 1 or 2 drops of saturated methyl red in ethanol is added and the titration is continued to the end point. The molarity of the sodium borohydride solution is calculated as follows:

$$M_{NaBH_4} = \frac{V_A N_A - V_B N_B}{V}$$

where

N_A, N_B = normalities of standard acid and base solutions

V_A, V_B = volumes (milliliters) of standard acid and base solutions

V = volume of sodium borohydride (milliliters).

The standard sodium borohydride solution (1.0*M*) was diluted with isopropyl alcohol to give 0.10 or 0.050*M* solutions. (Isopropyl alcohol free from acid or acetone is satisfactory. If necessary, it may be refluxed over sodium borohydride for 30 minutes and then distilled). These solutions may be standardized by the acid titration method above (with 10.0 ml instead of 1.0 ml of 1.0*M* solution) or by hydrogenation of 1.0 mM of 1-octene (Phillips Petroleum Co., 99%) by the procedure described below. From the volume of sodium borohydride solution used, the molarity is calculated from the following equation:

$$M_{NaBH_4} = \frac{1}{4V}\left[1.00 - \frac{0.0160(V_O)(P-40)}{T+273}\right]$$

where

V = volume of sodium borohydride used

V_O = V + volume of 1-octene solution injected (milliliters)

P = atmospheric pressure (mm Hg)

T = temperature of the water bath (°C).

These solutions are fairly stable and do not deteriorate appreciably over several weeks.

Chloroplatinic acid (40% platinum by weight), 1.0 gram, was dissolved in 40 ml of isopropyl alcohol to make a 0.05*M* solution.

All the olefins were accurately weighed into volumetric flasks and diluted with isopropyl alcohol to make exactly 1.00*M* solutions. Very volatile materials such as isoprene were cooled to 0°C, then weighed, diluted with chilled isopropyl alcohol, warmed to room temperature, and diluted to volume. Oils (2–5 grams of sample) were weighed into 25-ml volumetric flasks, and the volume was made up with isopropyl alcohol or ethyl acetate.

HYDROGENATION PROCEDURE

Place a Teflon-covered stirring bar, 0.25 gram of Darco K-B activated carbon, and 1.0 ml of 0.05*M* chloroplatinic acid solution in the 65-ml hydroanalyzer flask. Fill the hydrogenator valve with mercury to the etched mark. Attach the hydrogenation flask to the valve by means of the ground joint, and place the flask in a water bath supported on a magnetic stirrer. Fill the buret and the reservoir with $0.10M$ (or $0.05M$) sodium borohydride solution, and mount on the hydrogenator valve by means of the glass joint.

Start vigorous stirring and inject 15 ml of $0.10M$ sodium borohydride into the flask, taking care in this and other injections that the point of the syringe needle is against the central well of the valve. After a few minutes inject 1.0 ml of $12N$ hydrochloric acid into the flask (wash the syringe immediately, or corrosion products from the needle and hub will poison the catalyst in subsequent runs), and equilibrate the system with the bath for 5 minutes. Inject a 1.00-ml portion of the $1.00M$ olefin solution in isopropyl alcohol into the flask and open the buret stopcock immediately. Adjust the stirring so that not more than 2.0 ml of borohydride solution is used per minute. On completion of the reaction, the flow of sodium borohydride automatically ceases. Ignore this initial reading of the volume of borohydride solution. (If only a small amount of sample is available for analysis, a standard solution of 1-octene may be used for the preliminary reaction). Enough sample should be taken for this initial run to take up an amount of hydrogen approximately equal to that utilized in the actual determinations. Stop the stirring and refill the buret. The system is now ready for sample analyses. Inject the desired volume of olefin into the flask, start the stirring and open the buret stopcock. Note the volume of borohydride consumed. Stop the stirring and close the stopcock. This process of injection of samples can be repeated until the flask becomes full. In general, triplicate consecutive results are quite sufficient. For 0.5 to 2.0 mM of olefin, use $0.100M$ borohydride solution. Use $0.050M$ solution for smaller quantities up to 0.2 mM.

CALCULATIONS

Calculate the amount of hydrogen absorbed in millimoles by the following equation:

$$\text{Millimoles of hydrogen} = 4V(M_{NaBH_4} + \text{FSD}$$

where

V = volume of standard borohydride solution used (milliliters)
M_{NaBH_4} = molarity of standard borohydride solution
FSD (free space displacement)
= millimoles of hydrogen in the free space of the hydroanalyzer displaced by the olefin solution injected and by the borohydride solution added.

$$\text{FSD} = \frac{0.0160(V_0)(P-40)}{T+273}$$

where

V_O = volume of olefin or olefin solution and sodium borohydride solution added
P = pressure (mm Hg),
T = water bath temperature (°C).

The iodine value of the sample may be calculated from the following equation:

$$\text{Iodine value} = \frac{25.38 \times \text{Millimoles of hydrogen}}{\text{Grams of sample used}}$$

RESULTS AND DISCUSSION

The results are summarized in Table 14. In general, the individual measurements fell within a range of ±1.0 to 1.5% of the average, although frequently the measurements were grouped much better. Accuracy was also ±1.0 to 1.5%.

A study of the range of the hydroanalyzer was made, and the results appear in Table 15. A range of ±1.0 to 1.5% was found in measurements even on the smallest sample. Accuracy was about ±2% on the lower levels of sample size.

Table 14. Analyses of Various Unsaturates

Compound	Olefin Found,[a] mM
1-Octene	0.999 ±0.013
1-Dodecene	1.001 ±0.009
2,4,4-Trimethylpentene-1	1.001 ±0.009
2,4,4-Trimethylpentene-2	1.003 ±0.011
4-Methylcyclohexene	0.995 ±0.009
Cyclooctene	0.987 ±0.013
Ethyl oleate	0.986 ±0.017
α-Pinene	0.993 ±0.007
β-Pinene	1.003 ±0.001
Isoprene	0.972 ±0.004
1,3-Cyclooctadiene	1.005 ±0.002
1,5-Cyclooctadiene	0.982 ±0.004
D-Limonene	0.993 ±0.006
1,5,9-Cyclodecatriene	1.006 ±0.006

[a] Amount present was 1.00 mM in all cases.

Table 15. Test of Range of Measurement of the Hydroanalyzer

1-Octene (mM)	Sodium Borohydride Solution	Number of Replicates	Average Volume of 0.100*M* Solution, ml	Standard Deviations, ml	FSD[a]	Olefin Found, mM
2.00	0.100*M*	7	4.42	0.017	0.245	2.013 ±0.01
1.00	0.100*M*	6	2.18	0.018	0.121	0.993 ±0.007
0.50	0.050*M*	7	2.05	0.0084	0.0972	0.507 ±0.002
0.30	0.050*M*	4	1.21	0.0057	0.0578	0.301 ±0.001
0.20	0.050*M*	5	0.82	0.0050	0.0389	0.203 ±0.001

[a] Millimoles of hydrogen in the free space displaced by the olefin solution injected and the borohydride solution added.

The analytical results for several samples of representative vegetable oils are given in Table 16. The analyses were precise within a range of ±1.5%.

Table 16. Analysis of Oils

Oil Used	Solvent for Oil	Number of Replicates	Iodine Value	
			Found	Literature
Corn	Isopropanol	6	123 ±0.60	111–128
Olive	Isopropanol	6	79.01 ±0.60	79–88
Cottonseed	Isopropanol	7	111.8 ±1.60	99–113
Tung	Ethyl acetate	5	219.5 ±1.00	163–171[a]

[a] Hydrogenation values up to 240.

Finally, the hard-to-reduce $\Delta^{9,10}$-octalin was analyzed. Successive replicate analysis required 6, 24, and 40 hours. Nonetheless, 0.975 mM was found, an error only 2.5% low. The range of measurements was ±0.5%.

This technique was reduced to the microscale (30), for which a special apparatus is available.

Mercuric Acetate Methods

Mercuric acetate adds to olefinic unsaturation as follows:

$$\mathrm{>C{=}C<} + \mathrm{Hg(Ac)_2} + \mathrm{ROH} \rightarrow \mathrm{>\underset{OR}{\underset{|}{C}}{-}\underset{HgAc}{\underset{|}{C}}<} + \mathrm{HAc}$$

where R can be H or an alkyl group. Marquardt and Luce (31) first carried out this reaction in 40% aqueous dioxane. The excess mercuric acetate is converted to the oxide using sodium hydroxide and the oxide is

30. C. A. Brown, *Anal. Chem.*, **39,** 1882 (1967).
31. R. P. Marquardt and E. N. Luce, *Anal. Chem.*, **20,** 751 (1948).

reduced to metallic mercury using boiling hydrogen peroxide. The solution is then acidified with nitric acid and titrated with thiocyanate. The thiocyanate pulls the mercury from the addition product, thus forming mercuric thiocyanate and liberating the original unsaturated compound. To overcome low results, Marquardt and Luce (32, 33) then carried out the reaction in methanol. The excess of mercuric acetate was measured by addition of acetone and a known excess of caustic. The caustic neutralizes the acetic acid formed in the mercuric acetate addition reaction and also takes part in the formation of a soluble complex between the mercuric ion and the acetone.

$$3HgAc_2 + 2\ \begin{array}{l} CH_3 \diagdown \\ \qquad C{=}O \\ CH_3 \diagup \end{array} + 6NaOH \rightarrow$$

$$\begin{array}{ccccc} & & Hg & & \\ & \diagup & & \diagdown & \\ HOHgCH & & & & CHHgOH \\ | & & & & | \\ HO \diagdown & & & & \diagup OH \\ C & & & & C \\ H_3C \diagup & \diagdown & O & \diagup & \diagdown CH_3 \end{array} + 6NaAc + 3H_2O$$

Potassium iodide is then added, which converts the mercuric addition product of the olefin to the iodide and also breaks up the mercuric addition product of the acetone, liberating the caustic once more.

$$\begin{array}{ccccc} & & Hg & & \\ & \diagup & & \diagdown & \\ HOHgCH & & & & CHHgOH \\ | & & & & | \\ HO \diagdown & & & & \diagup OH \\ C & & & & C \\ CH_3 \diagup & \diagdown & O & \diagup & \diagdown CH_3 \end{array} + 12KI + 3H_2O \rightarrow$$

$$6KOH + 3K_2HgI_4 + 2\ \begin{array}{l} CH_3 \diagdown \\ \qquad C{=}O \\ CH_3 \diagup \end{array}$$

The alkalinity is then titrated to the phenolphthalein end point.

A blank run is made on this reagent. The difference between the caustic titrated in the blank and that in the sample is a measure of the

32. R. P. Marquardt and E. N. Luce, *Ibid.*, **21,** 1194 (1949).
33. R. P. Marquardt and E. N. Luce, *Ibid.*, **22,** 363 (1950).

acetic acid formed on the formation of the mercuric addition product of the olefin.

Martin (34) has a system whereby he destroys the excess mercuric acetate by adding sodium chloride. The chloride complexes the mercuric ion, and this permits direct titration with alkali of the acetic acid formed in the reaction. Johnson and Fletcher (35) use sodium bromide to replace the chloride.

Das (36) uses a known amount of mercuric acetate in his reaction and determines the excess mercuric acetate by titration with standard hydrochloric acid. The mercuric acetate consumes 2 moles of acid,

$$Hg(Ac)_2 + 2HCl \rightarrow HgCl_2 + 2HAc$$

but the addition product consumes only one.

$$\underset{OCH_3}{\overset{\diagdown\quad}{\underset{\diagup |}{C}}}\!\!-\!\!\underset{HgAc}{\overset{\quad\diagup}{\underset{| \diagdown}{C}}} + HCl \rightarrow \underset{OCH_3}{\overset{\diagdown\quad}{\underset{\diagup |}{C}}}\!\!-\!\!\underset{HgCl}{\overset{\quad\diagup}{\underset{| \diagdown}{C}}} + HAc$$

The amount of hydrochloric acid consumed over what is consumed by the mercuric acetate reagent alone is a measure of the olefin.

Method of R. W. Martin

[*Adapted from Anal. Chem.*, **21,** *921* (*1949*), *Reprinted in Part*]

REAGENTS

Mercuric acetate, Merck's CP reagent grade. Must be low in free acetic acid content.

Synthetic methanol. The reagent should be substantially free of acids or aldehydes.

Sodium chloride. Commercial grades of sodium chloride are satisfactory. The aqueous salt solution should be saturated with sodium chloride, then filtered and made neutral to phenolphthalein.

Sodium hydroxide, 0.1*N* solution. It was carbonate-free and standardized against Baker's CP reagent grade benzoic acid.

Vinyl monomers. The unsaturated compounds were obtained from various sources. Where their purity was not indicated by the supplier, they were carefully

34. R. W. Martin, *Anal. Chem.*, **21,** 921 (1959).
35. J. B. Johnson and J. P. Fletcher, *Anal. Chem.*, **31,** 1563 (1949).
36. M. N. Das, *Anal. Chem.*, **26,** 1086 (1954).

redistilled and only a very narrow middle cut was used. In most cases the purified materials boiled over a 1°C range, and never over more than a 3°C range.

Sodium nitrate, CP reagent grade, is used as a saturated solution in methanol.

PROCEDURE

Accurately weigh approximately 4 meq. of the ethylenic compound in a weighing bottle and transfer the weighing bottle and contents to a 500-ml Erlenmeyer flask containing 20 to 25 ml of carbon tetrachloride. Empty the contents of the weighing bottle into the solvent. Add 4.00 grams of mercuric acetate and 30 ml of methanol. If the ethylenic compound reacts slowly with the reagents, the use of 30 ml of a saturated solution of sodium nitrate in methanol in place of the pure methanol will increase the rate of reaction. Stopper the flask, swirl the contents, and warm slightly, if necessary, to dissolve the mercuric acetate. Allow the reaction to proceed for 10 to 15 minutes; then add 75 ml of neutral saturated sodium chloride solution and 50 to 100 ml of water. Add 20 drops of phenolphthalein solution and titrate to the first pink end point with standard 0.1*N* sodium hydroxide. Shake the reaction mixture vigorously during the titration, to ensure complete removal of the acetic acid from the carbon tetrachloride layer.

A blank should be run immediately after mixing the reagents, omitting only the unsaturated compound. If the blank is allowed to stand too long, its titer has a tendency to increase slowly. Each milliequivalent of sodium hydroxide consumed in the titration, after subtraction of the blank, represents one milliequivalent of ethylenic group. The analyses reported in Tables 17 and 19 were obtained using the foregoing procedure; the factors reported in Table 18 were obtained by modifying the procedure as indicated.

Table 17. Analysis of Styrene and Styrene Derivatives

Compound	% Purity	Determined by	% Found, This Method
Dow styrene	99.8	Supplier	99.85, 99.75
Monsanto styrene	99.8	Melting point	99.80, 99.66
Styrene solution	2.02	Dilution of 99.8% styrene	2.06
Styrene solution	1.45	Dilution of 99.8% styrene	1.50
Divinyl- and ethylvinyl-benzene	142.2	Method of Marquardt and Luce	142.3, 142.9[a]

[a] Calculated as ethylvinylbenzene.

Table 18. Factors Affecting Reaction

Compound	Ethylene: $Hg(OAc)_2$	Temperature, °C	Time, Min.	Catalyst	% Reaction
Styrene	1:1	25	10	O	95.4, 95.6
	1:1	45	10	O	97.0
	1:1.25	25	10	O	99.8
	1:1.50	25	10	O	99.9, 100.6
	1:1	25	120	O	99.1, 99.8
	1:1	25	240	O	100.1
	1:1	25	15	$NaNO_3$	99.7, 100.5
Methyl methacrylate	1:2	25	10	O	1.3
	1:2	25	420	O	6.8
	1:2	25	10	$NaNO_3$	4.0
Allyl alcohol	1:1	25	10	O	98.1
	1:1.5	25	15	O	98.7
2-Vinylpyridine	1:1	25	30	O	19.8
	1:2	25	30	O	36.1
	1:3	25	30	O	47.4
	1:2	25	30	$NaNO_3$	51.8
	1:2	25	960	$NaNO_3$	97.0
	1:3	25	1200	$NaNO_3$	97.3

Table 19. Analysis of Unsaturated Compounds

Compound	% Purity	Purification	% Found, This Method
Methyl methacrylate	99[a]	Redistillation	1.3
Diallyl phthalate	—	Redistillation	94.5
Diethyl itaconate	—	Redistillation	Trace
Methyl acrylate	98[a]	—	43.5
Vinyl acetate	99[a]	—	201
Vinyl benzoate	—	3°C boiling range	187
Diethyl maleate	—	1°C boiling range	4.0
Acrylonitrile	—	1°C boiling range	3.4
Allyl alcohol	—	1°C boiling range	99.48, 98.71
Crotyl alcohol	—	1°C boiling range	101.5
β-Chloroallyl alcohol	—	2°C boiling range	15
N-Vinylcarbazole	—	1°C melting point range	99.4, 99.7
2-Vinylpyridine	92.8[b]	—	39.8
Diallyl ether	—	1°C boiling range	99.9
Divinyl ether	96.5[a]	Anesthetic grade	96.8

[a] As given by commercial supplier.

[b] Determined by acid titration using $Fe(OH)_3$ indicator.

FACTORS INFLUENCING REACTION

When the elements of methoxy mercuric acetate add to an ethylenic double bond, a number of factors influence the rate and extent of the reaction. The effect of some of these factors can be seen from an inspection of the data in Table 18, where variations in the regular procedure were employed. Increasing the temperature and adding sodium nitrate (37) accelerates the reaction. An excess of mercuric acetate not only increases the rate of reaction but also serves to force the reaction to completion, the excess necessary depending on the particular ethylenic compound. In practice 1 gram of mercuric acetate was used for approximately 1 meq. of ethylenic compound. If this amount of mercuric acetate was not sufficient to give complete reaction in 15 minutes, the method was considered impractical where that particular compound is concerned. As indicated in Table 18, longer reaction periods give more complete reaction but may result in some inaccuracies because of an increase in the blank.

The structure of the compound is the ultimate factor in determining the extent, rate, and type of reaction occurring when an ethylenic compound is treated with mercuric acetate in methanol. Although this method deals primarily with the analysis of styrene and styrene derivatives, Table 19 shows the results of one or two determinations on 15 other ethylenic compounds. These exploratory results indicate that the procedure will give satisfactory results with allyl and crotyl alcohol, certain allyl ethers and esters, certain vinyl ethers, and vinylcarbazole. Unsatisfactory results were obtained with acrylate, methacrylate, itaconate, and maleate esters, as well as with acrylonitrile and vinylpyridine. Vinyl acetate and vinyl benzoate have results approximately twice what was expected.

Of major importance in the use of this procedure is the quality of the mercuric acetate. This reagent should not only be low in free acid but the entire supply should be thoroughly blended, so that each portion used for an analysis has the same acid content. If this is not the case, checks will be poor. If the mercuric acetate contains appreciable amounts of acid, it is often advisable to place it in a vacuum desiccator for 2 or 3 hours and then blend carefully before use. The bottle containing this reagent should be tightly stoppered except when in use.

Some difficulty was experienced in detecting the end point. However after a few titrations and the addition of somewhat more than the usual amount of phenolphthalein the end point gave little trouble.

The procedure described has been in use for the rapid routine estimation of styrene and divinyl- and ethylvinylbenzene samples. It has been

37. G. F. Wright, *J. Am. Chem. Soc.*, **57,** 1994 (1935).

entirely satisfactory and in general gave results checking within a few tenths of 1%. Typical data on styrene and ethylvinyl and divinylbenzene appear in Table 17.

Using sodium bromide instead of the chloride, Johnson and Fletcher (38) have obtained the results shown in Table 20. This method is described in Chapter 10 pp. 518–20. The method was devised for vinyl ethers but was tried on other olefins at the same time.

Method of M. N. Das

[*Adapted from Anal. Chem.*, **26,** *1086* (*1954*), *Reprinted in Part*]

Das has found (39) that mercuric acetate can be very accurately estimated by Palit's method (40) of glycolic titration using propylene glycol–chloroform (1:1) as the titration medium and thymol blue as indicator (see pp. 166–67). A standard solution of hydrochloric acid in the same solvent medium as above is used for titration, and the end point (yellow to pink) is extremely sharp. Mercuric chloride may be precipitated out in the course of the titration, but it does not interfere with the detection of the correct end point. Perchloric acid cannot be used for titration of mercuric acetate, and hydrochloric acid must be used.

When titrating with hydrochloric acid, not only does the excess unreacted mercuric acetate react with the acid, but the mercury addition product also takes up 1 equivalent of acid as follows:

$$\underset{OCH_3}{\overset{\diagdown}{\underset{\diagup|}{C}}}\text{—}\underset{Hg\cdot O\cdot CO\cdot CH_3}{\overset{\diagup}{\underset{|\diagdown}{C}}} + HCl \rightarrow \underset{OCH_3}{\overset{\diagdown}{\underset{\diagup|}{C}}}\text{—}\underset{HgCl}{\overset{\diagup}{\underset{|\diagdown}{C}}} + CH_3\cdot CO\cdot OH \qquad (6)$$

Hence the difference between the milliequivalents of mercuric acetate taken and the milliequivalents of acid used in titration directly gives the amount of unsaturation in millimoles.

REAGENTS

MERCURIC ACETATE, APPROXIMATELY $0.25N$ SOLUTION IN METHANOL. Dissolve about 20 grams of mercuric acetate in 500 ml of methanol; add 1 ml of glacial

38. J. B. Johnson and J. P. Fletcher, *Anal. Chem.*, **31,** 1563–4 (1959).
39. M. N. Das, *J. Indian Chem. Soc.*, **31,** 39 (1954).
40. S. R. Palit, *Ind. Eng. Chem., Anal. Ed.*, **18,** 246 (1946).

Table 20. Determination of Purity of Unsaturated Compounds by Modified Mercuric Acetate Procedure

Compound	Purity, Wt. %	Average Deviation	Number of Determinations	Minute Time, Min.	Temperature, °C
Allyl acetate	98.8	0.0	2	60	25
Allyl acetone	98.8	0.1	2	20	−10
Allyl alcohol	99.1	0.1	4	1	25
2-Allyl-3-methyl-2-cyclopenten-4-ol-1-one	100.1	0.0	2	10	−10
Butyl chrysanthemum monocarboxylate	95.4	0.0	3	60	25
2-Chloro-1-propenyl butyl ether	97.4	0.4	2	15	25
Cyclohexene	98.5	0.1	4	1	25
Dichlorostyrene	99.7	0.2	3	120	25
2,5-Dimethyl-1,5-hexadiene	94.1	0.1	3	15	25
3,4-Epoxy-1-butene	97.6	0.0	2	60	25
2-Ethoxy-3,4-dihydropyran	97.0	0.1	4	30	0
2-Ethoxy-5-methyl-3,4-dihydropyran	97.1	0.2	4	30	0
3-Ethoxy-4-propyl-5-ethyl-3,4-dihydropyran	96.2	0.0	4	30	25
2-Formyl-3,4-dihydropyran	97.0	0.0	4	30	25
3-Hydroxy-8-nonen-2,5-dione	98.9	0.2	3	10	−10
Methallyl chloride	97.9	0.1	4	15	25
2-Methoxy-3-ethyl-3,4-dihydropyran	100.0	0.0	7	30	25
2-Methoxy-3-ethyl-4-methyl-3,4-dihydropyran	96.1	0.0	6	30	25
2-Methoxy-3-methyl-3,4-dihydropyran	101.5	0.1	6	30	25
2-Methoxy-3-methyl-4-propyl-5-ethyl-3,4-dihydropyran	97.2	0.1	2	120	25
4-Methyl-3,4-dihydropyran	98.7	0.0	2	15	−10
4-Methyl-1-pentene	98.2	0.0	2	30	25
α-Methylstyrene	99.2	0.1	4	5	−10
Styrene	99.6	0.1	2	10	25
Vinyl acetate[a]	99.0	0.1	4	10	25
N-Vinylpiperidone	99.1	0.0	2	10	25
N-Vinylpyrrolidone	97.0	0.3	10	10	25

[a] Each mole of vinyl acetate results in consumption of two equivalents of KOH due to ease of saponification of reaction product.

acetic acid, and filter. Standardize the solution by titration in propylene glycol–chloroform (1:1) solvent medium with 0.1 N glycolic hydrochloric acid, using thymol blue as indicator.

HYDROCHLORIC ACID, 0.1N SOLUTION IN PROPYLENE GLYCOL–CHLOROFORM (1:1). Prepare by adding 8 to 9 ml of concentrated hydrochloric acid in 1 liter of the mixed solvent. The acid is standardized as follows. Accurately weigh out about 0.2 gram of mercuric oxide (analytical reagent) and dissolve in 5 ml of glacial acetic acid by gently heating, then evaporate almost to dryness. Take up the residue with 25 ml of propylene glycol–chloroform and titrate with hydrochloric acid, using thymol blue as indicator (1 ml of 0.1N hydrochloric acid 0.01083 gram of mercuric oxide).

MIXED SOLVENT. Prepare by mixing propylene glycol and chloroform in equal volumes. The solvents should be neutral to thymol blue.

Thymol blue, 0.2% solution in ethyl alcohol.

PROCEDURE

Treat about 2 mM of the unsaturated compound with 20 to 25 ml of the mercuric acetate solution in a glass-stoppered bottle or conical flask and allow to stand at room temperature (about 30°C) for 10 to 30 minutes. Then dilute the reaction mixture with about 25 ml of propylene glycol–chloroform (1:1) and titrate the sample with glycolic hydrochloric acid, using thymol blue as indicator. At the end point, the indicator shows a sharp color change from yellow to pink.

To test the accuracy of this method, the samples of the unsaturated compounds were simultaneously estimated by the alkalimetric method of Marquardt and Luce (41); the analytical data are represented in Table 21.

Table 21. Analytical Data

Compound	Time of Reaction, Min.	% Found by Marquardt-Luce Method	% Found by Present Method
Styrene	5–10	99.4, 99.7	99.2, 99.6
Cyclohexene	10–15	96.7, 96.3	96.6, 96.9
Allyl alcohol	30	97.7	98.2, 98.2
Allyl acetate	30	88.1, 76.3	97.4, 96.8
Vinyl acetate	30	33.2, 28.4	96.9, 97.5, 97.0

CALCULATION

$\frac{(a-b)\times N}{1000}\times M =$ Amount of unsaturate compound in grams where

41. R. P. Marquardt and E. N. Luce, *Anal. Chem.*, **21**, 1194 (1949).

a = the acid titer milliliters for blank (mercuric acetate solution); b = acid titer for the sample; N = normality of the acid; M = molecular weight of the ethylenic compound.

DISCUSSION

The method seems to be particularly useful for analysis of some unsaturated esters like vinyl acetate and allyl acetate for which the methods of Marquardt and Luce (41, 42) and of Martin (43) (see pp. 430–34) are not suitable, since under the conditions used by them, the esters may undergo hydrolysis and lead to erroneous results. With vinyl acetate and vinyl benzoate Martin (43) obtained about twice the theoretical value. This arises from the fact that his method measures not only the acetic acid liberated in the course of the mercury addition reaction, but also an equivalent amount of acid liberated as a result of the hydrolysis of the vinyl esters. The alkalimetric method of Marquardt and Luce gives low and erratic results with vinyl acetate and allyl acetate for the same reason. Methyl acrylate and methyl methacrylate cannot be estimated with mercuric acetate by any of these methods, since they do not quantitatively react with mercuric acetate under the conditions employed.

Mercury addition products are, in general, unstable toward halogen acids. In some cases, they are so very susceptible to acidity that titration with hydrochloric acid is not possible. This has been observed with α-methylstyrene, cinnamic acid, and diisobutylene, where the addition products undergo quick decomposition in the course of titration with hydrochloric acid. In other cases that have been studied, the decomposition is very slow, and at the end point, the acid color of the indicator persists for several minutes, so that no difficulty is involved in detecting the correct end point.

From the Method of R. P. Marquart and E. N. Luce

[*Anal. Chem.*, ***39,*** *1655* (*1967*)]

The preceding mercuric acetate addition procedure of Martin was studied and changed to simplify it and to increase its accuracy.

REAGENTS

MERCURIC ACETATE, APPROXIMATELY 0.20M IN METHANOL. Dissolve 65.0 grams of analytical reagent grade mercuric acetate and 1.0 ml of glacial

42. R. P. Marquardt and E. N. Luce, *Ibid.*, **20,** 751 (1948).
43. R. W. Martin, *Ibid.*, **21,** 921 (1949).

acetic acid in approximately 800 ml of ACS grade methanol in a 1000-ml volumetric flask. Gentle warming of the contents of the flask on a steam bath helps to dissolve the mercuric acetate. Cool the solution to room temperature and dilute it to 1000 ml with methanol. Filter any solids, if necessary, and keep the filtrate in an amber bottle.

A neutral saturated aqueous solution of sodium chloride.

REGULAR PROCEDURE

Pipet 50.0 ml of 0.20*M* mercuric acetate solution into a 50-ml Erlenmeyer flask. Add an accurately weighed sample containing less than 4.0 meq. of the unsaturated compound, and mix the sample with the mercuric acetate solution. Let the contents of the flask stand for 10 minutes or longer, as necessary for quantitative addition of the mercuric acetate to the compound being analyzed. Add 25 ml of carbon tetrachloride and 150 ml of water. Stopper the flask and mix the contents well by shaking. Add 20 or more drops of phenolphthalein and 50 ml of saturated sodium chloride solution; immediately shake the contents of the flask vigorously for 3 to 5 seconds and titrate the acidity with 0.1*N* sodium hydroxide solution. Stopper the flask and shake the contents well to extract the remaining acid from the carbon tetrachloride. Let stand, and after the liquids have separated, continue the titration of the acidity in the upper aqueous layer to a definite pink end point. Shake the contents of the flask a third time and finish the titration when the liquids have separated.

Subtract the number of milliliters of 0.1*N* sodium hydroxide representing the acidity of the reagent blank (see below) from the number of milliliters of 0.1*N* sodium hydroxide used for the analysis to obtain the net titration of the acid produced by the addition of mercuric acetate to the sample. Calculate unsaturation in terms of the compound being analyzed.

DETERMINATION OF THE REAGENT BLANK

Pipet 50 ml of 0.20*M* mercuric acetate solution into a 500-ml Erlenmeyer flask. Add 50 ml of saturated sodium chloride solution, 25 ml of carbon tetrachloride, and 150 ml of water. Shake the contents of the flask well. Add 20 or more drops of phenolphthalein indicator and titrate the acidity of the reagents in the same manner as described in the regular procedure. Be careful not to overtitrate.

MODIFIED PROCEDURE FOR ALLYL ALCOHOL, METHALLYL ALCOHOL, AND SO ON

The modified procedure and the determination of the acidity of the reagents are the same as for the regular procedure, but use 100 ml instead of 150 ml of water and 100 ml instead of 50 ml of saturated sodium chloride solution.

CALCULATION

$$\text{Unsaturation, as \% compound} = \frac{\text{Net milliliters of } 0.1N \text{ NaOH} \times F \times 100}{\text{Weight of sample}}$$

where

$$F = \frac{\text{Molecular weight}}{10{,}000 \times \text{Number of double bonds}}$$

RESULTS AND DISCUSSION

Data are given in Table 22.

Table 22. Determination of Unsaturation with Mercuric Acetate

Compound	Number of Determinations	Reaction Time, min.	Found, %	Standard Deviation
Styrene A, 99.50%[a]	8	10	99.46	0.08
Styrene B, 99.75%[a]	4	10	99.69	0.06
Styrene C, 99.83%[a]	8	10	99.76	0.08
α-Methylstyrene A, 99.58%[a]	4	15	99.68	0.10
α-Methylstyrene B, 99.81%[a]	4	15	99.74	0.07
α-Methylstyrene C, 99.44%	4	15	99.47	0.06
Cyclohexene, 99.9%[a]	8	30	99.36	0.09
Allyl alcohol[b]	4	15	99.16	0.07
Vinyl acetate, practical grade[b]	4	15	99.88	0.26
Methallyl alcohol, b.p. 66.0°C[b], 100 mm Hg	4	15	99.78	0.15

[a] The purities were determined by cryoscopic means.
[b] The purities were not determined.

The mercuric acetate in methanol solution is sensitive to ordinary light, which promotes the formation of insoluble mercurous acetate. The solution is stable if it is kept in an amber bottle and stored in a dark place. If, on long standing, mercurous acetate does precipitate, it should be filtered before the solution is used.

Do not add carbon tetrachloride to the solution in the 500-ml flask until the addition of mercuric acetate to the olefinic compound is complete. The efficiency of the reaction, best in methanol, is lowered when carbon tetrachloride is present.

The mercuric acetate addition compound in acid solution is slowly decomposed by excess sodium chloride. Carbon tetrachloride is added to dissolve most of the chloride salt of the addition compound, thus protecting it from the action of the excess sodium chloride.

As the acidity of the reagent blank is determined, the acidity of the mercuric acetate solution is low enough to cause mercuric oxide to appear if the carbon tetrachloride and the water are added prior to the addition of the saturated sodium chloride solution. Acidic material is then produced, which increases the acidity of the reagents. Therefore, the sodium chloride is added first.

The titration of acidity in the presence of the excess mercury can be influenced within narrow limits by varying the amount of sodium chloride present in the solution. When the regular procedure for styrene is used for determining the unsaturation in allyl alcohol, methallyl alcohol, and so on, and probably other vinyl compounds of this type, the analytical results will be too high. The amount of $0.1N$ sodium hydroxide solution required to titrate the acidity is reduced by the larger volume of saturated sodium chloride solution recommended in the modified procedure, giving analytical results that are more nearly correct for these compounds.

For vinyl acetate and similar esters, the acid produced by hydrolysis is titrated also. Therefore the factor is based on one-half the molecular weight of the compound.

If free acid is present in the sample, it can be determined separately and the correction made in the calculation of unsaturation.

OZONIZATION

Ozonization is the basis for methods that are generally applicable for the determination of double bonds. Ozone acts on unsaturated compounds to give ozonides that break down to carbonyl, carboxyl, or other oxygenated cleavage derivatives.

$$\rangle C{=}C\langle \; + O_3 \rightarrow \; \rangle C{-}C\langle \text{ (bridged by } O_3)$$

Such methods are advantageous in that interferences resulting from substitution and steric and substituent effects are not significant. Disadvantages are the explosive nature of some ozonides and the difficulties and expense of fabrication or purchase of suitable electrolytic ozone generators.

Boer and Kooyman devised a method (44) based on ozone consumption as a measure of olefinic unsaturation. An oxygen stream containing a constant ozone concentration is passed through the sample solution, and the time required to realize complete ozonization is determined by indicator decoloration or by the appearance of iodine liberated from potassium iodide by unused ozone in the escaping gas stream. Significant improvement was obtained by use of commercial ozone-generating equipment (45).

Subsequently a method has been developed that eliminates the problems of ozone control and end point detection associated with the previous methods without sacrifice of accuracy. An outline of this method follows.

Method of K. F. Guenther, G. Sosnovsky, and R. Brunier

[*Anal. Chem.*, **36,** *2508* (*1964*)]

The ozone stream produced by the ozonator is divided into two equal parts. One stream is passed through the olefin sample and then into a potassium iodide solution. The other stream goes directly into another potassium iodide solution. The amount of ozone consumed is calculated from the difference in the titration values for the iodine liberated in the two potassium iodide solutions.

APPARATUS

The ozone stream was produced by a noncommercial ozonator at a rate of approximately 3.4 mg per minute and introduced into the apparatus shown in Fig. 7.16. In the apparatus the ozone stream is divided into two equal parts A_1 and A_2 by the use of two precision valves V_1 and V_2 (Ideal No. 2512A, Ideal Aerosmith, Cheyenne, Wyo.). The flow rate of A_1 and A_2 is controlled by two calibrated precision flowmeters F_1 and F_2 (Precision Bore Flowmeter Tube No. O8F-1/16-20-4/36 55 19008, Fischer and Porter, Warminster, Pa.). Stream A_1 is passed first into vessel C, containing the olefin sample dissolved in 18 ml of chloroform, and

44. H. Boer and E. C. Kooyman, *Anal. Chim. Acta*, **5,** 550 (1951).
45. A. Maggiolo and A. L. Tumolo, *J. Am. Oil Chem. Soc.*, **38,** 279 (1961).

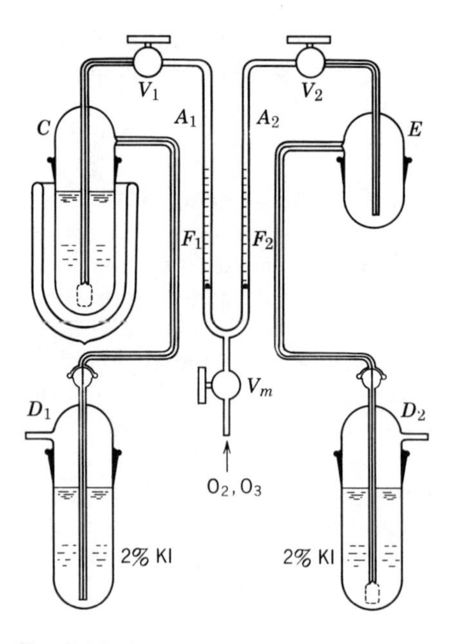

Fig. 7.16. Apparatus for determination of unsaturation in olefins by titration with ozone.

then into vessel D_1, containing 50 ml of 2% neutral aqueous potassium iodide solution. Stream A_2 is first passed through an empty vessel E. The volume of vessel E is equal to the volume above the liquid in vessel C. Then stream A_2 is passed into vessel D_2, also containing 50 ml of 2% neutral potassium iodide solution. The total gas volumes above the liquids in lines A_1 and A_2 are approximately equal.

PROCEDURE

Weigh a 5-mM sample of the olefin into a 50-ml volumetric flask and dilute to volume with chloroform. Pipet a 5-ml aliquot into vessel C and dilute with chloroform to a total volume of 18 ml. Connect the vessel to the apparatus (Fig. 7.16) and cool to $-30° \pm 2°C$ by immersion in a Dewar flask containing acetone and chips of carbon dioxide. Place vessels D_1 and D_2, each containing 50 ml of 2% neutral aqueous potassium iodide, in the apparatus. Turn on the ozone stream, approximately 50 ml per minute, by the main valve V_m. Preset V_1 and V_2 to avoid having to make major adjustments during an experiment.

Iodine is immediately liberated in vessel D_2. Terminate the ozonization when a yellow color is visible in vessel D_1.

Acidify the contents of D_1 and D_2 with 15 ml of dilute sulfuric acid; and titrate the iodine with 0.1*N* sodium thiosulfate solution, using starch indicator.

CALCULATION

Calculate the ozone consumption from the difference in the titration of the contents of D_1 and D_2 as follows:

$$\text{Milligrams of } O_3 = [(\text{ml}_{D_2} - \text{ml}_{D_1})2.4] - F$$

where ml_{D_1} and ml_{D_2} are the volumes of thiosulfate solution used to titrate the contents of vessels D_1 and D_2, respectively, and F is the empirical correction factor needed to account for the solubility of ozone in the chloroform, which is 0.25 mg of ozone under the conditions recommended.

RESULTS AND DISCUSSION

Table 23 gives results for a series of determinations of unsaturated compounds.

Table 23. Determination of Unsaturation of Olefins with Ozone

	Olefin, mg		Ozone Value[a]		Error;
Olefin	Added	Found[b]	Theory	Found	Relative, %
n-Octene-1[c]	53.12	53.18	42.86	43.02	0.4
n-Octene-1[d]	32.50	32.32	42.86	43.04	0.4
n-Octene-1[d]	55.77	54.67	42.86	42.44	1.0
n-Decene-1[d]	74.89	73.82	34.29	34.14	0.4
n-Dodecene-1[d]	82.16	81.45	28.57	28.61	0.1
n-Tetradecene-1[d]	100.30	98.69	24.49	24.34	0.6
n-Hexadecene-1[d]	115.20	111.39	21.43	20.93	2.3

[a] Amount of ozone (in grams) consumed by 100 grams of olefin.
[b] Mean values of eight replicate experiments.
[c] 99.73% Research grade, Philips Petroleum Co.
[d] 99% Development chemicals, Gulf Oil Corp.

In this procedure changes in the ozone concentration during the determination do not affect the results, and it is not mecessary to know the exact ozone concentration, in contrast to the conventional analytical procedure. Other advantages are that no warm-up time is required for the ozonator, and the detection of the end point of the reaction is not critical.

An olefin can be ozonized at temperatures ranging from room temperature to the temperature of acetone–dry ice mixtures. A temperature of −20°C is used in most conventional procedures. At room temperature the solubility of ozone in chloroform should be negligible. Because of the risk of degradative reactions of certain olefins and of the solvent, however, a temperature of −30°C was chosen. This temperature was maintained within ±2°C. A solubility factor of 0.25 mg of ozone was established and was used in all calculations.

Method of M. M. Smits and D. Hoefman

[*Adapted in Part from Anal. Chem.*, **44**, *1688* (*1972*)]

Although more elaborate, this arrangement for the quantitative determination of olefinic unsaturation by measurements of ozone consumption has advantages in speed and convenience over the other ozonometric methods.

A mixture of oxygen and ozone of constant composition is passed at constant flow rate through the sample, to which a color indicator has been added. Immediately after the ozone has converted the olefins present, the color of the indicator changes, marking the end point. The olefin content of the sample is then calculated by comparison of its reaction time with that of a calibration standard compound. The flow rate is controlled by a precision valve and a rotameter. Ozone is supplied by a commercial dry ozone generator at a very constant output. The color change of the indicator is measured by a photometric device connected to a recorder. The absorption of the solution drops rapidly during discoloration of the indicator, resulting in a change in resistance of a photocell that is recorded continuously. The time required for the ozonolysis can be defined accurately from the curve obtained.

APPARATUS

The ozonolysis equipment is assembled as in Fig. 7.17 and includes a high-pressure flow regulator *a* (Brooks Instrument Division, Emerson Electric Co., Hatfield, Pa. 19440), and the ozone generator *b* with a built-in rotameter *c*. A suitable instrument is that manufactured by Fischer Labortechnik (53 Bonn-Bad Godesberg, Heerstrasse 35-37, Germany). On the front panel a calibration curve is sketched, giving ozone output in grams per hour versus oxygen flow in

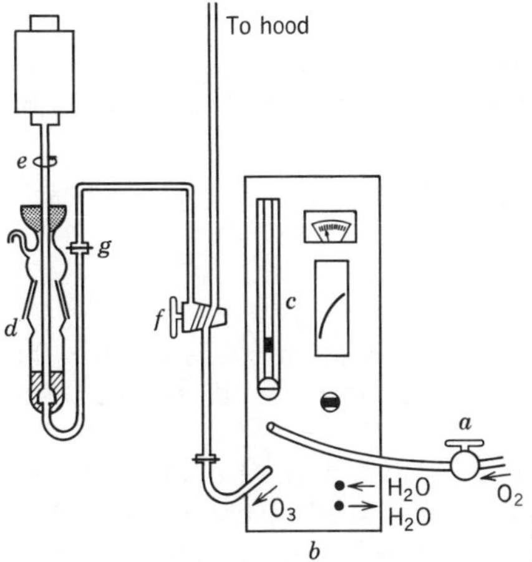

Fig. 7.17. Assembly of apparatus for ozonolysis.

liters per hour. Since ozone is highly toxic and may cause severe irritation of the respiratory tract and eyes, the instrument should be placed under a well-ventilated hood. The reaction vessel *d* consists of two parts connected by ground-glass joints held in position by springs. The stirrer *e* is bell shaped to ensure thorough mixing of gas and liquid. It rotates at approximately 1200 rpm.

The photoelectric circuit is illustrated in Fig. 7.18. Light source *E* and light-dependent resistor (LDR) R_1 both belong to a photoelectric relay (N.V. Instrumentenfabriek H. M. Smitt, Midellaan 3-5, Bilthoven, the Netherlands). Before entering the reaction vessel, the light beam passes through an interference filter, which has its maximum transmission in the same region as the maximum absorbance of the indicator solution (i.e., at about 550 nm). Care must be taken that the light path is not interrupted by the stirrer.

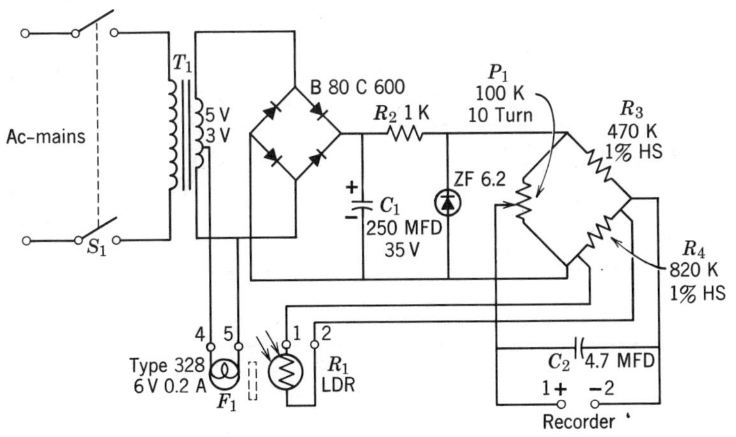

Fig. 7.18. Photodetector for titration.

REAGENT

The indicator is Rouge Organol B. S. from Cie Française des Matières Colorantes (15 Boulevard de l'Amiral-Brieux, Paris 16e). It is used as a 0.1% by weight solution in chloroform.

PROCEDURE

Allow the ozone generator to stabilize for 30 minutes before the determination is carried out. The gas escapes into the hood through valve *f*.

Weigh an amount of sample containing 1 to 2 meq. of olefins into the reaction vessel *d* (Fig. 7.17), and dilute with 40 ml of chloroform. Add 1 ml of indicator solution. Connect the reaction vessel to the apparatus via the joint *q*, and start the stirrer. Introduce ozone into the reaction mixture through valve *f*, and at the same time start the recorder. Continue the ozonization until the color of the indicator fades. On the recorder chart the S-shaped curve levels off, as in Fig. 7.19. Close valve *f* to shut off the ozone supply.

Apply the same procedure to a solution of pure olefin, (e.g., *n*-hexadecene-1) to serve as a calibration standard. Make blank determinations in the same way with pure chloroform, but omit the olefin sample.

Adjust the flow rate and ozone concentration of the oxygen stream to obtain the ozonation of 1 mM of hexadecene in about 10 minutes. Read the end point as illustrated in Fig. 7.19. Measure the distance *t* from the start to the deflection point corresponding to the reaction time.

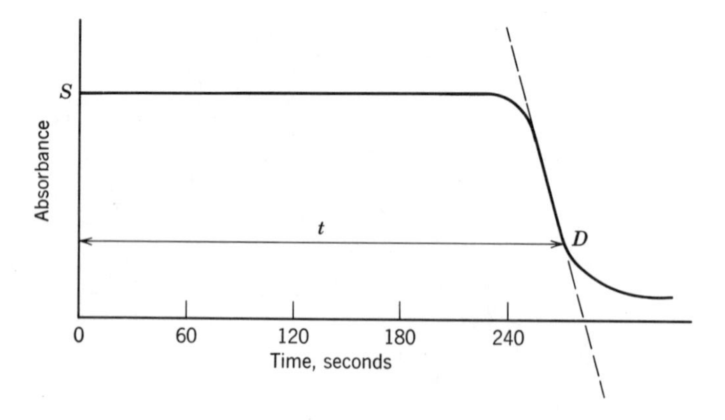

Fig. 7.19. End point determination in ozonolysis.

CALCULATION

$$\text{Ozone supply} = \frac{W_c\ 1000}{(t_c - t_0)E_c)} \text{ millimoles per second}$$

$$\text{Ozone number (milliequivalents of olefin per gram)} = \frac{t_s - t_0}{W_s} \text{ ozone supply}$$

where t is the measured reaction time, W, the sample weight in grams, and the subscripts s, c and 0 refer to sample, calibration standard, and

Table 24. Ozonometric Determination of Unsaturation of Olefins

Olefin[a]	Sample Weight, gram	Ozone Number[b] Found	Mean Value	Theory
3-Heptene	0.1663	10.195	10.25	10.20
	0.1822	10.299		
	0.1975	10.252		
1-Octene	0.1733	8.909	8.88	8.93
	0.2029	8.884		
	0.2423	8.835		
2-Methylheptene-2	0.1462	9.146	9.29	8.93
	0.2168	9.316		
	0.2472	9.418		
2,4,4-Trimethylpentene-1	0.1355	8.969	8.98	8.93
	0.2128	8.954		
	0.2389	9.019		
2,3,4-Trimethylpentene-2	0.1496	8.674	8.68	8.93
	0.2098	8.675		
	0.2380	8.688		
4-Methylpentadiene-1,3	0.0681	25.74	25.4	24.4[c]
	0.0686	25.06		
1,3-Cyclohexadiene	0.0710	27.63	26.7	25.0[c]
	0.0790	25.81		
2,5-Dimethylhexadiene-2,4	0.0717	18.31	18.3	18.2[c]
	0.0723	18.20		
Di(cyclopentadiene)	0.0925	15.84	15.6	15.3[d]
	0.0951	15.39		
Hex-*trans*-2-enal		10.26		10.2

[a] All the olefins are commercially available. Purity, as determined by gas-liquid chromatography, exceeds 99%.

[b] Milligram equivalent of ozone absorbed per gram of olefin.

[c] Calculated for two double bonds per molecule.

[d] Per molecule of dimer.

blank, respectively; E_c is the equivalent weight of the calibration standard.

RESULTS AND DISCUSSION

The method has been tested successfully on a wide variety of olefins (Table 24) as well as on low and high molecular weight technical samples (Table 25). It shows no major discrepancies from the theoretical values for all olefins tested, including terminal substituted olefins. Accurate results are obtained also with conjugated dienes and unsaturated aldehydes (Table 24).

The differences between the values obtained for different sample sizes do not exceed ±0.02 meq. per gram for ozone numbers less than 0.5 meq. per gram, and ±5% relative for ozone numbers greater than 0.5 meq. per gram. The detection limit is approximately 0.01 meq. per gram.

Some sulfur- and nitrogen-containing compounds, as well as aromatics, may absorb ozone in varying amounts, depending on their structure (Table 26).

Table 25. Ozonometric Determination of Unsaturation in Olefinic Samples

Sample	Sample Weight, gram	Ozone Number [a]		
		Found	Mean Value	Calculated
n-Eicosene[b]	0.3730	3.55	3.58	3.57
	0.4763	3.61		
Propylene tetramer	0.0742	5.94	5.95	5.95[c]
	0.0773	5.97		
	0.1537	5.98		
	0.1556	5.91		
	0.2239	6.01		
	0.2318	5.88		
Allylchloride in chlorobenzene I	0.3630	4.26	4.24	4.25[d]
	0.5053	4.19		
	0.5306	4.29		
II	0.2025	6.00	5.91	5.87[d]
	0.3265	5.89		
	0.4051	5.83		

[a] Milligram equivalent of ozone absorbed per gram of sample.

[b] Mixture of double-bond isomers.

[c] Calculated for one double bond per molecule of tetramer. Between 5.95 and 6.00 meq. per gram, determined by ultraviolet.

[d] Samples prepared by mixing weighed quantities of components.

Table 26. Ozonization of Nonolefinic Compounds

Compound	Ozone Number[a]	1000/M[b]
Cyclohexane	<0.01	11.9
Toluene	<0.01	10.9
Chlorobenzene	<0.01	8.9
Pyridine	<0.01	12.7
Acetonitrile	<0.01	24.4
8-Hydroxyquinoline	15.6	6.9
Naphthalene	6.3	7.8
Anthracene	13.7	5.6
Phenanthrene	6.2	5.6
1-Decanethiol	7.6	5.7

[a] Millimoles of ozone absorbed per gram of compound.

[b] Number of millimoles of compound per gram, for comparison.

Epoxidation with *m*-Chloroperbenzoic Acid

Adapted from P. Dreyfuss and J. P. Kennedy

[*Anal. Chem.*, **47,** 771 (1975)]

The epoxidation of olefins proceeds as follows:

$$\mathrm{>C{=}C<} + \mathrm{R{-}\overset{\overset{O\cdots H}{\|\quad\ |}}{C\quad\ O}}\ \text{(C–O–O ring closure through bottom O)} \rightarrow \mathrm{>\overset{O}{C{-}C}<} + \mathrm{RCOOH}$$

The instability of perbenzoic acid has prevented its use as a general analytical reagent. The stability and commerical availability of *m*-chloroperbenzoic acid makes the epoxidation reaction practical as a basis for the determination of unsaturation in olefins and in polymers.

REAGENTS

m-Chloroperbenzoic acid (technical grade, Aldrich Chemical Company) was purified by washing with a phosphate buffer, 67.1 grams of $Na_2HPO_4 \cdot 7H_2O$ and 34.0 grams of KH_2PO_4 in 1 liter of distilled water and then with distilled water.

After it was dried overnight with vacuum at room temperature, it was further dried to constant weight over P_2O_5 in a vacuum desiccator. Recovery was 85%.

The 0.6*M* solutions of the perbenzoic acid were stored in a freezer.

GENERAL PROCEDURE FOR OLEFINS

Dissolve the olefin (0.2–0.6 ml, 0.8–3 mM) in 10 ml of methylene chloride in a 250-ml Erlenmeyer flask. Add 5 ml of 0.6*M* solution of *m*-chloroperbenzoic acid in methylene chloride and let stand at room temperature until complete reaction is obtained (0.08–20 hours) with occasional swirling. Add 10 ml of 10% potassium iodide solution, 7 ml of glacial acetic acid, and 40 ml of distilled water. Titrate the liberated iodine with standard 0.1*N* sodium thiosulfate solution to the starch end point.

PROCEDURE FOR POLYMERS

Add 0.2 to 0.5 gram of polymer to 5 ml of heptane in a 250-ml Erlenmeyer flask. Cover the flask tightly with aluminum foil and shake overnight on a wrist-action shaker to dissolve the polymer. Add 20 ml of methylene chloride and 5 ml of 0.6*M* *m*-chloroperbenzoic acid in methylene chloride, and stir magnetically until complete epoxidation is obtained. Carry out the titration in the manner used for olefins.

RESULTS AND DISCUSSION

The particular olefins studied were chosen to serve as the basis for an exploration of the rate of epoxidation of unsaturation in Nordel and in butyl rubbers. These polymers contain small amounts of random unsaturation. The unsaturation arises as a result of the copolymerization of 1,4-hexadiene and isoprene in Nordel and in butyl rubbers, respectively.

The rate of epoxidation of the model olefins by *m*-chloroperbenzoic acid is rapid and essentially quantitative. The reaction is more rapid in solvents of higher dielectric constant. The order of reactivity is

2,4,4-trimethyl-2-pentene ≥ 2-hexene > 2,4,4-trimethyl-1-pentene > 1-decene

Increasing substitution of the olefin carbon leads to increased reactivity, and 1,2-disubstitution seems to be more effective than 1,1-disubstitution.

Reaction rates were especially rapid for both 2-hexene and 2,2,4-trimethyl-2-pentene, the models chosen for the copolymers. Epoxidation for these compounds was complete in 5 to 10 minutes.

The average deviation of duplicate analyses of the olefins was 0.003.

Analytical resuts for two butyl rubbers and one Nordel are given in Table 27. The variation in the results arises both from the usual errors in titration and from the random unsaturation in the polymers. Therefore

Table 27. Results of Epoxidation Studies on Polymers

		Unsaturation[b], mole %		
Polymer	C=C[a] %	Found	Literature	Other[c]
Butyl LM430, crude	Av 1.95	Av 4.56	4.2	—
	Exptl 1.65	Exptl 3.85		
	1.52	3.55		
	1.77	4.12		
	1.98	4.62		
	2.85	6.65		
Butyl LM430, purified	Av 2.08	Av 4.18	4.2	—
	Exptl 1.79	Exptl 4.18		
	2.34	5.45		
	1.92	4.48		
	2.28	5.31		
Butyl 268, purified	Av 0.69	Av 1.63	1.5–2.0	1.61
	Exptl 0.69	Exptl 1.62		
	0.76	1.68		
	0.64	1.49		
Nordel 1440, purified	Av 1.19			
	Exptl 1.13			
	1.38			
	1.39			
	1.97			
	0.85			

[a] % C=C = [(meq. consumed) (24.04) (100)]/[2(sample weight) (1000)].

[b] Mole % unsaturation = moles of copolymerized isoprene per 100 moles of isobutylene = [(equivalent weight isoprene) × (milliequivalents consumed) × (molecular weight of isoprene) × 100]/[2(1000) × (grams of polymer) × (molecular weight of isoprene)]. The equivalent weight of isoprene equals the molecular weight of isoprene, the final equation is mole % = [milliequivalents consumed) × (molecular weight of isobutylene) × 100]/[2(1000) × (grams polymer)].

[c] Determined by iodine number.

some of the variations observed may represent real differences in the samples. Like the model compounds, the polymers react very rapidly with *m*-chloroperbenzoic acid. There was no change in the determined unsaturation after 5 minutes of reaction, although the reaction was followed for more than 2 hours. Nevertheless, a 60-minute reaction period is recommended. When the literature data are available for comparison, the agreement of the observed unsaturation with the literature value is found to be excellent.

Determination of α,β-Unsaturated Compounds

α,β-Unsaturated compounds can be determined by any of the preceding reactions. These compounds undergo two other addition reactions—addition of sodium bisulfite and of morpholine—which are more characteristic and are often faster and/or easier to carry out than those described. Of the two methods, the bisulfite method is generally the more precise, but the morpholine method is subject to less interferences.

SODIUM BISULFITE METHOD

Adapted from F. E. Critchfield and J. B. Johnson

[*Reprinted in Part from Anal. Chem.*, **28,** *73–6* (*1956*)]

In this reaction a substituted sodium sulfonate is formed (46) with a corresponding decrease in acidity according to the equation

$$NaHSO_3 + CH_2{=}CH{-}X \rightarrow NaO_3S{-}CH_2CH_2{-}X \qquad (7)$$

where X is any strong electron-attracting group. This reaction has been utilized commercially in the manufacture of anionic surface-active agents (47). It has also been used as an analytical procedure by Rosenthaler (48) for the determination of maleic acid; but under the conditions Rosenthaler established, he was unable to determine fumaric acid. The bisulfite reaction was also used by the Siggia and Maxcy (49) method to determine aldehydes (see pp. 100–115).

46. R. T. E. Schenck and I. Danishefsky, *J. Org. Chem.*, **16,** 1683 (1951).
47. C. R. Caryl, *Ind. Eng. Chem.*, **33,** 731 (1941).
48. L. Rosenthaler, *Pharm. Acta Helv.*, **17,** 196 (1942).
49. S. Siggia and W. Maxcy, *Ind. Eng. Chem., Anal. Ed.*, **19,** 1023 (1947).

In the method of analysis presented here, a known excess of sulfuric acid is added to the sodium sulfite reaction mixture. The decrease in acidity as determined by titration with standard sodium hydroxide to alizarin yellow R-xylene cyanol FF mixed indicator is a direct measure of the unsaturated compound originally present. Under these conditions, sodium sulfite is neutral and a large excess can be used. For maximum reactivity a saturated or nearly saturated solution of the reagent is recommended.

REAGENTS

Isopropyl alcohol, anhydrous commerical grade, Carbide and Carbon Chemical Company.

SULFURIC ACID, APPROXIMATELY 1*N*. Dissolve 49 grams of reagent grade sulfuric acid in 1000 ml of water.

SODIUM SULFITE, APPROXIMATELY 2*M*. Dissolve 252 grams of anhydrous sodium sulfite in 100 ml of distilled water. This reagent should be prepared fresh at least once a week.

ALIZARIN YELLOW R-XYLENE CYANOL FF MIXED INDICATOR. Dissolve 0.1 gram of alizarin yellow R and 0.06 gram of xylene cyanol FF in 100 ml of distilled water.

PROCEDURE

To make all sample and blank determinations in duplicate, introduce into each of a sufficient number of glass-stoppered Erlenmeyer flasks, 25 ml of the sodium sulfite reagent by means of a graduated cylinder. For reaction at 98°C, use heat-resistant pressure bottles. Pipet exactly 25.0 ml of 1*N* sulfuric acid into each fiask and add the amount of isopropyl alcohol specified in Table 28. Purge the flasks with nitrogen and then stopper. Reserve two of the flasks for blanks. Into each of the other flasks add an amount of sample that contains not more than 15.0 meq. of unsaturated compound. For dilute solutions, the samples may be pipetted and the weight calculated from the specific gravity. Allow the samples to react under the conditions specified in Table 28. If the reaction is carried out at 98°C, allow the pressure bottles to cool to room temperature before the bottles are uncapped, If specified in Table 28, place the samples and blanks in a −10°C bath for 10 minutes. Add 5 or 6 drops of the alizarin yellow R–xylene cyanol mixed indicator and titrate the samples and blanks with 0.5*N* sodium hydroxide just to the disappearance of the green color.

Table 28. Reaction Conditions for Determination of α,β-Unsaturated Compounds by Sodium Sulfite Reagent

Compound	Reaction Conditions: Temperature, °C	Time, Min.
Acrylic acid	98[a]	15 to 50
Acrylonitrile	25	5 to 30
Crotonic acid	98[a]	60 to 120
Diethyl fumarate	25[a,b]	15 to 90
Diethyl maleate	25[b,c]	60 to 90
Ethyl acrylate	25[b]	30 to 60
Maleic acid	98[a]	15 to 90
Methyl acrylate	25[b]	15 to 60

[a] Use 15.0 ml. of isopropyl alcohol as a cosolvent.

[b] Place samples and blank in a −10°C bath for 10 minutes before titration.

[c] Place on mechanical shaker for 15 minutes.

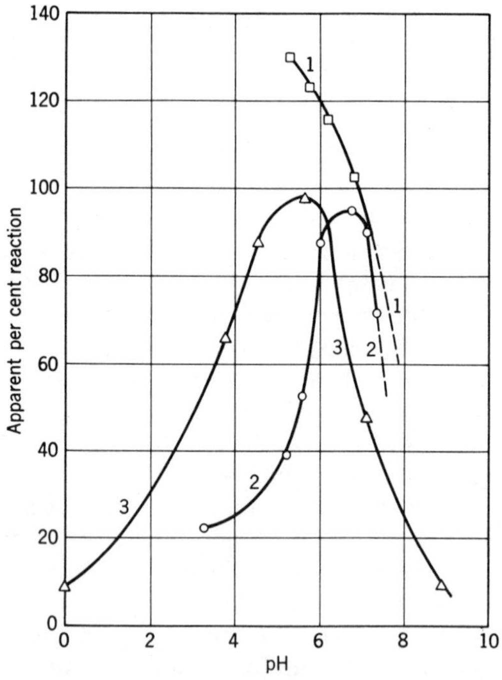

Fig. 7.20. Effect of pH on rate of addition of sodium sulfite α,β-unsaturated compounds: 1, ethyl crotonate at room temperature; 2, ethyl crotonate at 100°C; 3, crotonic acid at 100°C (46).

For unsaturated acids, it is necessary to determine the acidity of the sample independently by titration with standard sodium hydroxide to phenolphthalein indicator. The procedure for the determination of unsaturation in these acids is identical to that just described, except that the sample titration is usually larger than that of the blank.

DISCUSSION AND RESULTS

The rate of addition of sodium sulfite to α,β-unsaturated compounds is appreciably affected by the pH of the reagent. Curve 3, (Fig. 7.20), illustrates this effect on the reaction with crotonic acid according to Schenck and Danishefsky (46). A similar result is obtained for the reaction with ethyl crotonate at 100°C, curve 2. It is evident from these curves that the optimum range for maximum reactivity is pH 5.5 to 6.5. For ethyl crotonate at room temperature (curve 1), the apparent conversion over this range of pH is above 100%, indicating the occurrence of secondary reactions.

Two of the secondary reactions which are believed to occur between sodium sulfite and α,β-unsaturated compounds are given by the equations

$$NaHSO_3 \rightleftharpoons NaOH + SO_2 \qquad (8)$$

$$CH_2{=}\overset{\displaystyle H}{\overset{|}{C}}{-}X + SO_2 \underset{\Delta}{\rightleftharpoons} \underset{\diagdown\ SO_2\ \diagup}{CH_2{-}CH{-}X}$$

$$SO_3^{--} + \text{oxidant} \rightarrow [SO_3\cdot]^- + \text{oxidant}^- \qquad (9)$$

$$CH_2{=}CH{-}X + [SO_3\cdot]^- \rightarrow {}^-O_3S{-}CH_2{-}CH{-}X$$

$${}^-O_3S{-}CH_2{-}\dot{C}H{-}X + HSO_3^- \rightarrow {}^-O_3SCH_2CH_2{-}X + [SO_3\cdot]^-$$

where X is any strong electron-attracting group. The occurrence of the reaction shown in eq. 8 gives high results, because this method is based on an acidimetric determination. The net effect of the reaction is an apparent decrease of two equivalents of acidity. This is shown graphically in curve 1 of Fig. 7.20. At pH 5.3 and at room temperature the apparent addition of sodium sulfite to ethyl crotonate is 130%. At 100°C (curve 2), this reaction apparently does not occur, presumably because of the thermal instability of the cyclic sulfone reaction product. Also, compounds that are appreciably soluble in the reagent do not undergo this reaction, suggesting that the reaction takes place only at the gas-liquid interface. This reaction is favored at low values of pH, presumably because of the increase of free sulfur dioxide in the gas phase. In some cases this side reaction can be minimized by the addition of a cosolvent such as isopropyl alcohol. Using the reagent as specified, approximately 20.0 ml of cosolvent can be tolerated before phase separation occurs. Although no literature confirmation of the reaction of sulfur dioxide with α,β-unsaturated compounds has been found, the reaction with butadienes to form cyclic sulfones has been reported (50).

The secondary reaction shown in eq. 9 is similar to the principal nucleophilic reaction in that the same reaction product is fomed. However this reaction proceeds by a free-radical mechanism and is not restricted to conjugated systems (51). The occurrence of this reaction therefore prohibits the use of this reagent for the determination of α,β-unsaturation in the presence of olefinic-type unsaturation. As would be expected, this reaction is inhibited by the presence of hydroquinone and the absence of peroxides and molecular oxygen (51).

Secondary reactions other than those involving the carbon-to-carbon double bond are known to occur with this reagent. Among these are the

Table 29. Analysis of Substantially Pure α,β-Unsaturated Compounds

	Average Purity, wt. %	
Compound	Sodium Sulfite Method[a]	Other
Acrylic acid	98.4 ± 0.2 (5)	98.7[b]
Acrylonitrile	98.1[c]	—
Crotonic acid	98.7 ± 0.1 (3)	99.1[d]
Diethyl fumarate	99.9 ± 0.1 (8)	99.9[e]
Diethyl maleate	98.6 ± 0.1 (5)	98.6[e]
Ethyl acrylate	99.2 ± 0.1 (6)	99.0[e]
Maleic acid	99.2 ± 0.2 (5)	99.0[d]
Methyl acrylate	98.4 ± 0.1 (7)	98.3[e]

[a] Figures in parentheses represent number of determinations.

[b] Modified Kaufmann bromination of potassium salt.

[c] Standard deviation for eight degrees of freedom is 0.09.

[d] Acidity titration.

[e] Saponification.

50. H. J. Backer, J. Strating, and C. M. H. Kool, *Rec. Trav. Chim. Pays-Bas.*, **58,** 778 (1939).
51. M. S. Kharasch, E. M. May, and F. R. Mayo, *J. Org. Chem.*, **3,** 175 (1938).

Table 30. Selection of Reagent for Determination of α,β-Unsaturated Compounds of Type $R'{\equiv}\overset{H}{\overset{|}{C}}{\equiv}\overset{R}{\overset{|}{C}}{\equiv}X$

X	R	R′	Method[a]
—C≡N	H	H	A, B
	Alkyl	H	A
	H	Alkyl	A
	Alkyl	Alkyl	C
—COOH	H	H	A, B
	Alkyl	H	A[b], B[b]
	H	Alkyl	A[b], B
	Alkyl	Alkyl	C
	H	—COOH	B
—COONa	—	—	C[c]
—COOR″	H	H	A, B[d]
	Alkyl	H	A
	H	Alkyl	A
	Alkyl	Alkyl	C
	H	—COOR″	A[c], B[d]
$—CONH_2$	H	H	A
	H	Alkyl	A[b]
	Alkyl	H	A[b]
	Alkyl	Alkyl	C
—COH	—	—	C[c]
—COR″	—	—	C[e]

[a] A, Morpholine method (53) (see pp. 458–467); B, sodium sulfite method; C, modified Kaufmann bromination method (54).
[b] No experimental data; reaction predicted.
[c] Independent of R and R′.
[d] Only if R″ is methyl or ethyl.
[e] Use conductometric end point determination.

substitution reactions involving a carbon-to-halogen bond, addition to carbonyls, and saponification of esters. Saponification does not occur at room temperature when the specified reagent, at pH 6.2, is used. However, at 100°C saponification does take place to an appreciable extent. For ethyl crotonate (Fig. 7.20, curve 2), an apparent 95% reaction is obtained at this temperature. This low result can be attributed to the saponification of 5.0% of this compound. Esters that fail to react quantitatively with the reagent at room temperature cannot be determined by this method.

Among the other reactions that are known to take place with this reagent are the addition to epoxides (52), reaction with strong oxidizing and reducing agents, and reaction with aldehyde. Since the method is based on a measurement of the decrease in acidity, a correction must be applied for the presence of mineral and organic acids and inorganic and most organic bases.

Table 29 lists a number of compounds that have been successfully determined by the sulfite method.

Table 30 is a guide for the selection of the method best suited for the determination of a particular compound from a consideration of its structure.

MORPHOLINE METHOD

Adapted from F. E. Critchfield, G. L. Funk, and J. B. Johnson

[*Reprinted in Part from Anal. Chem.*, ***28,*** *76* (*1956*)]

In the method presented here an excess of morpholine, a secondary amine, reacts in the presence of acetic acid catalyst with the unsaturated compound according to the following equation:

$$O\begin{matrix} \diagup CH_2CH_2 \diagdown \\ \diagdown CH_2CH_2 \diagup \end{matrix}N\text{—}H + R\text{—}\overset{H}{\overset{|}{C}}\!=\!\overset{H}{\overset{|}{C}}\text{—}X \xrightarrow{CH_3COOH} O\begin{matrix} \diagup CH_2CH_2 \diagdown \\ \diagdown CH_2CH_2 \diagup \end{matrix}N\text{—}\underset{R}{\underset{|}{\overset{H}{\overset{|}{C}}}}\text{—}CH_2\text{—}X$$

where X is any strong electron-attracting group. After the reaction has been completed, the excess morpholine is acetylated with acetic anhydride in acetonitrile medium to form the corresponding amide and acetic acid. The tertiary amine formed in the reaction is titrated with standard alcoholic hydrochloric acid to methyl orange–xylene cyanol mixed indicator. Under these conditions the amide and acetic acid are neutral, and the amount of tertiary amine formed is a direct measure of the α,β-unsaturated compound originally present.

52. J. D. Swan, *Anal. Chem.*, **26,** 878 (1954).
53. F. E. Critchfield, G. L. Funk, and J. B. Johnson, *Anal. Chem.*, **28,** 76 (1956).
54. H. P. Kaufmann and O. O. Kornmann, *Z. Unters.-Lebensm.*, **51,** 3 (1926).

APPARATUS

All potentiometric titrations are performed using a Leeds & Northrup pH meter equipped with glass and calomel electrodes. Conductometric titrations are made using a Model RCM 15 Serfass direct-reading conductance bridge. With this instrument, a dip-type conductivity cell with platinized-platinum electrodes having a cell constant of approximately 0.1 cm^{-1} is used.

REAGENTS

Acetontrile, commercial grade, Carbide and Carbon Chemicals Company.

ACETIC ACID SOLUTION. Mix 1 part of glacial acetic acid, Grasselli reagent grade, with 1 part of distilled water.

Acetic anhydride, commerical grade, Carbide and Carbon Chemicals.

Methanol, anhydrous, commerical grade, Carbide and Carbon Chemicals.

Morpholine, anhydrous, commercial grade, Carbide and Carbon Chemicals.

STANDARD 0.5*N* HYDROCHLORIC ACID IN METHANOL. Transfer 85 ml of 6*N* hydrochloric acid to a 1000-ml volumetric flask and dilute to volume with methanol. Standardize by titrating exactly 40.0 ml of the solution with standard 0.5*N* sodium hydroxide using phenolphthalein indicator. This titrant can best be dispensed from an automatic buret assembly and should be standardized daily. For the change of normality with increase in temperature use $\Delta N/°C = -0.0005$.

METHYL ORANGE–XYLENE CYANOL MIXED INDICATOR. Dissolve 0.15 gram of methyl orange and 0.08 gram of xylene cyanol FF in 100 ml of distilled water.

PROCEDURE

INDICATOR METHOD. By means of a graduated cylinder or dispensing buret add 10 ml of morpholine to each of two 250-ml glass-stoppered Erlenmeyer flasks. For reactions at 98 ±2°C, use heat-resistant pressure bottles. Reserve one of the flasks as a blank. Into the other flask introduce an amount of sample containing not more than 23 meq. of the unsaturated compound. For substantially pure material, weigh to the nearest 0.1 mg. For dilute solutions, the sample may be added by means of a pipet and the sample weight calculated from the specific gravity. To each flask add 7.0 ml of acetic acid solution, unless otherwise specified, and the amount of cosolvent specified in Table 31. Allow both the sample and the blank to stand for the time and at the temperature specified in Table 31. If elevated temperatures are used, carefully cool the pressure bottles to room temperature. Add 50 ml of acetonitrile to each flask by means of a graduated cylinder. While constantly swirling the flask, add

Table 31. Reaction Conditions for α,β-Unsaturated Compounds by Morpholine Reaction

Compound	Reaction Conditions	
	Time, Min.	Temperature, °C
Acrylamide	5 to 120	25
Acrylic acid	15 to 120[a]	98
Acrylonitrile	5 to 60	25
Allyl cyanide (3-butenenitrile)	30 to 60	98
Butyl acrylate	5 to 60	25
Diethyl fumarate	30 to 90[b]	25
Di(2-ethylhexyl) maleate	45 to 90[b,c]	25
Diethyl maleate	30 to 90[c]	25
Ethyl acrylate	5 to 60	25
2-Ethylbutyl acrylate	5 to 30	25
Ethyl crotonate	15 to 60	98
2-Ethylhexyl acrylate	5 to 60	25
Methacrylonitrile	120 to 240	98
Methyl acrylate	5 to 60	25
Methyl methacrylate	40 to 80	98

[a] Use 10 ml. methanol cosolvent.
[b] Use conductometric titration procedure.
[c] Use 40 ml. methanol cosolvent and 2 ml. acetic acid solution.

20 ml of acetic anhydride to both the sample and blank from a suitable graduated cylinder or dispensing buret and stopper. Allow to cool to room temperature. Add 5 or 6 drops of methyl orange–xylene cyanol mixed indicator and titrate with standard 0.5*N* methanolic hydrochloric acid to the disappearance of the green color. A Fisher titrating light or similar device greatly facilitates the selection of the end point.

CONDUCTOMETRIC METHOD. Follow the foregoing procedure, except limit the sample to less than 10.0 meq. of unsaturated compound. Titrate the blank using the indicator method. Quantitatively transfer the contents of the sample flask to a 250-ml tall-form beaker. Immerse the conductivity cell in the solution and add methanol to cover the electrodes. Titrate with standard 0.5*N* hydrochloric acid in methanol, taking conductance measurements at three or four points on each side of the anticipated end point. The end point is selected from a graphical plot of these data.

DISCUSSION

ACID CATALYSIS. The esters of acrylic acid react so readily with morpholine that these compounds can be determined without the addition of catalytic substances. As shown in Fig. 7.21, curve 3, the reaction with ethyl crotonate proceeds only with difficulty at elevated temperatures. In the presence of a catalytic amount of hydrochloric acid (curve 2), this reaction is quantitative in 60 minutes at 98°C. Acetic acid is even more effective (curve 1) and does not interfere in the subsequent determination of the tertiary amine reaction product.

Figure 7.22 shows the effect of acetic acid on the reaction rate of morpholine with methyl methacrylate. From this curve it can be seen that by the addition of 3 ml of catalyst solution containing 50% acetic acid to the reaction mixture a threefold increase is effected in the rate of reaction. At concentrations of acid greater than 55%, morpholinium acetate is precipitated, with subsequent depletion of the reagent. In the procedure finally adopted, 7.0 ml of 50% acetic acid has been specified, because in most cases the maximum rate of reaction is obtained with this volume of 50% acetic acid.

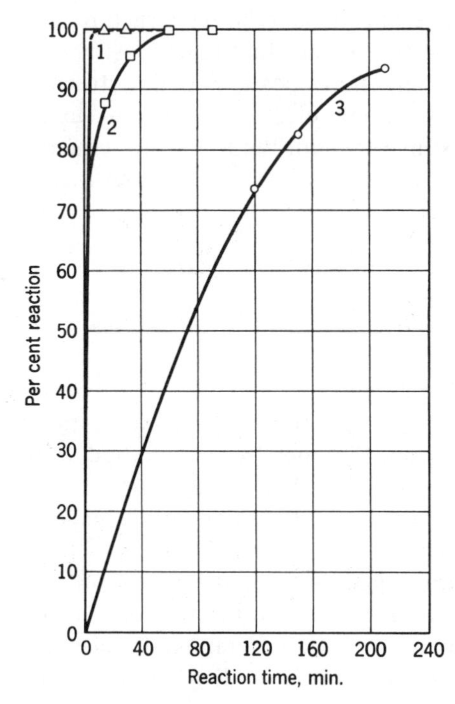

Fig. 7.21. Effect of acid catalysts on reaction of morpholine with ethyl crotonate: 1, acetic acid catalyst; 2, hydrochloric acid catalyst; 3, no catalyst.

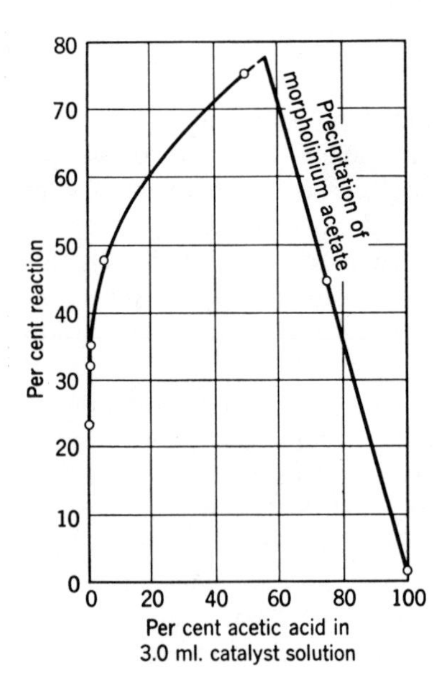

Fig. 7.22. Effect of acetic acid concentration on reaction of morpholine with methyl methacrylate. Reaction time is 15 minutes at room temperature.

EFFECT OF STRUCTURE ON REACTIVITY. From the data obtained in this investigation, several important generalizations relating structure and reactivity can be made. In general the compounds that react most readily with the reagent are acrylic-type compounds possessing the structure H_2—C=C(H)—*X*, where *X* is any strong electron-attracting (meta-orienting) group. Of the compounds investigated, the following sequence of reactivity with morpholine, as a function of the electron-attracting group of acrylic compounds, was found,

$$-\mathrm{C{\equiv}N} \gtreqless -\overset{\overset{\displaystyle O}{\|}}{\mathrm{C}}-\mathrm{OR} \gg -\overset{\overset{\displaystyle O}{\|}}{\mathrm{C}}-\mathrm{OH} > -\overset{\overset{\displaystyle O}{\|}}{\mathrm{C}}-\mathrm{R'} \geqq -\overset{\overset{\displaystyle O}{\|}}{\mathrm{C}}-\mathrm{ONa}$$

where R is alkyl and R′ is alkyl or hydrogen.

Attempts to determine the sodium or potassium salts of α,β-unsaturated acids and α,β-unsaturated aldehydes and ketones by this method were unsuccessful. These compounds, which are relatively unreactive with morpholine, react smoothly and completely with brominating reagents such as the modified Kaufmann reagent.

The substitution of alkyl groups on either the α or β carbon atoms of acrylic compounds has a marked retarding effect on the rate of reaction with morpholine. For the determination of these compounds, elevated temperatures must be employed. The effect of these alkyl groups can be attributed to their electron-releasing tendencies. The net effect of α- or β-alkyl substitution is an increase of electron density about the β carbon atom, thereby decreasing the possibility of attack by a nucleophilic reagent. For the same reason, there is an increase in the susceptibility of attack by electrophilic reagents such as bromine. The retarding effect of alkyl substituents on the reactivity of α,β-unsaturated compounds with morpholine is as follows:

$$\alpha,\beta \gg \alpha > \beta$$

Of the compounds investigated that possessed both α and β substituents, no detectable reaction was observed, even at elevated temperatures. As was pointed out, these compounds react more readily with brominating reagents, and many can be determined in this manner.

Certain β,γ-unsaturated compounds, such as allyl cyanide (Table 32), are isomerized to the corresponding α,β-unsaturated compound under

Table 32. Analysis of Substantially Pure Unsaturated Compounds

Compound	Average Purity,[a] wt. %	
	Morpholine Method	Other
Acrylamide	100.0 ± 0.1 (5)	—
Acrylic acid	98.6 ± 0.1 (4)	98.7[b]
Acrylonitrile	98.3[c]	—
Allyl cyanide	98.3 ± 0.1 (4)	98.3[d]
Butyl acrylate	99.8 ± 0.1 (4)	99.8[e]
Diethyl fumarate	99.5 ± 0.2[b] (2)	99.9[e]
Di(2-ethylhexyl) maleate	99.7 ± 0.2[f] (2)	100.0[e]
Diethyl maleate	98.4 ± 0.2[f] (2)	98.6[e]
Ethyl acrylate	99.2 ± 0.2 (4)	99.0[e]
2-Ethylbutyl acrylate	99.6 ± 0.1 (4)	99.5[e]
Ethyl crotonate	100.0 ± 0.1 (3)	99.9[e]
2-Ethylhexyl acrylate	99.0 ± 0.1 (4)	99.0[e]
Methacrylonitrile	97.9 ± 0.2 (3)	—
Methyl acrylate	98.7 ± 0.2 (4)	98.3[e]
Methyl methacrylate	97.8 ± 0.1 (3)	98.8[e]

[a] Figures in parentheses represent number of determinations.

[b] Bromination of potassium salt.

[c] Standard deviation for five degrees of freedom is 0.11.

[d] Bromination by a modified Kaufmann procedure.

[e] Saponification.

[f] Conductometric endpoint.

the conditions of the reaction and can be determined by this method. Allyl cyanide is the only compound investigated that can be determined by both the usual brominating reagents and by reaction with morpholine.

Table 32 lists 15 compounds that have been analyzed successfully by the morpholine method.

RATIO OF MORPHOLINE TO UNSATURATED COMPOUNDS. Figure 7.23 shows the effect of excess reagent on the reactivity of morpholine with acrylonitrile at a fixed acetic acid to morpholine mole ratio of 0.5:1 and at a fixed reaction time of 5 minutes at room temperature. From this curve it is apparent that a reagent-reactant mole ratio of at least 2.5:1 is necessary to obtain quantitative reaction in the shortest length of time. To provide a reasonable amount of reaction medium and a margin of safety, a mole ratio of 5:1 is used.

EFFECT OF TERTIARY AMINE STRENGTH. The potentiometric titration curves in Fig. 7.24 indicate that the sharpness of the equivalence point is appreciably affected by the nature of the tertiary amine formed in this reaction. The effect is particularly accentuated in the case of the reaction product of morpholine with maleic and fumaric esters. As shown in curve

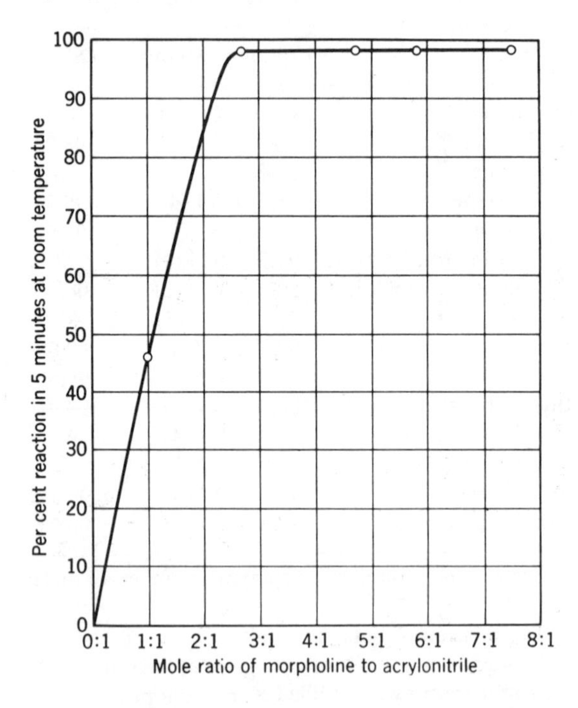

Fig. 7.23. Effect of mole ratio on reaction of morpholine with acrylonitrile.

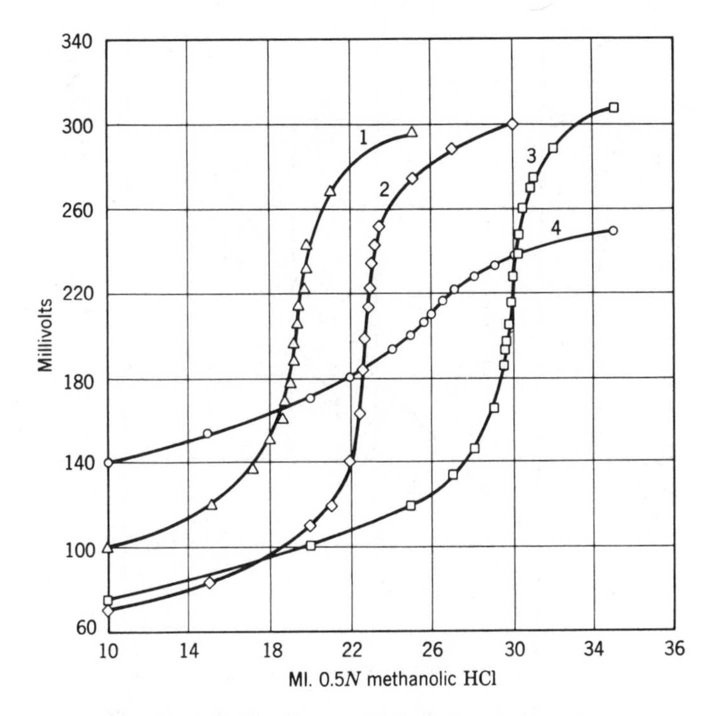

Fig. 7.24. Potentiometric titration curves of tertiary amines formed from reaction of morpholine α,β-unsaturated compounds: 1, acrylonitrile; 2, methyl methacrylate; 3, acrylic acid; 4, diethyl maleate.

4, this amine is too weak to be determined by either indicator or potentiometric methods. The decreased basicity of these amines can be attributed to the fact that the tertiary nitrogen is alpha to a strong electron-attracting group. Although an acidic solvent, such as glacial acetic acid, enhances the basicity of these amines, it also enhances the basicity of amides to the extent that they interfere in the titration. These weak tertiary amines can be determined, however, by employing a conductometric titration procedure. Figure 7.25 shows the titration of the amines formed by the reaction of morpholine with diethylfumarate and di(2-ethylhexyl) maleate. This method of end point determination can be employed here, because straight lines are obtained on each side of the equivalence point. In the selection of the end point only these lines are considered, and the points in the vicinity of the end point are ignored. With this procedure it is necessary to use a smaller sample size, so that the net titration does not exceed 20 ml of titrant. Because a separate curve must be plotted for each determination, this method is not readily adaptable to routine determinations. In the hands of an experienced

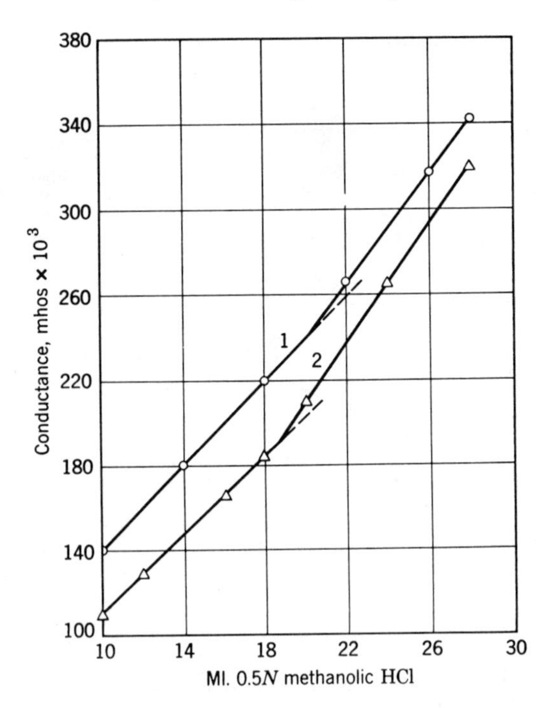

Fig. 7.25. Conductometric titration curves of tertiary amines formed from reaction of morpholine with diethyl fumerate and di(2-ethylhexyl) maleate: 1, diethyl fumerate; 2, di(2-ethylhexyl) maleate.

operator, however, results within ±0.2% have been obtained in the determination of the purity of maleic and fumaric esters.

INTERFERENCES

Because this method is based on a nonaqueous titration of the tertiary amine formed in the reaction, it is subject to interference from acid and basic constituents present in the sample. Acids with ionization constants in water greater than 2×10^{-2}, tertiary amines, and inorganic bases titrate quantitatively under these conditions and a correction may be applied.

Most epoxides react quantitatively with the reagent to form tertiary amines, which also are basic under the conditions of the titration. A method for the determination of epoxides, based on this reaction, has been developed in these laboratories.

Large quantities of aldehydes, ketones, and anhydrides may interfere by depleting the reagent. Organic halides react with morpholine to liberate halogen acids and therefore cannot be tolerated.

Most compounds with unsaturation not conjugated to a strong electron-attracting group (isolated unsaturation) and most α,β-unsaturated compounds substituted in both the α and β positions do not react.

There is another reaction system specifically for acetylenic compounds. This reaction system makes possible the determination of acetylenic compounds in the presence of ethylenic compounds. It involves the reaction of acetylenic compounds and water or alcohols in the presence of mercury catalyst to form the corresponding ketone, in the case of water, or ketal, in the case of alcohol. The reactions are shown in the method described below.

Acetylenic Unsaturation

The methods involving bromination, iodine number, and hydrogenation can be applied to acetylene compounds as well as to ethylenic compounds, except that 2 moles of reagent is added. For the halogen reaction systems, the second mole adds much more slowly than the first and occasionally adds slowly enough to cause difficulty in analysis. This behavior results from the halogen atoms being relatively large, and once one mole of halogen has added to a triple bond, the added halogen atoms sterically hinder the addition of further halogens to the remaining double bond. Hence in using the halogenation methods for determining acetylenic compounds, the reaction time factor should be considered.

There is also a slower rate of reaction of the second mole of hydrogen in the hydrogenation methods. However the difference in rate between the addition of the first and second moles of hydrogen in the methods described previously is so small that it causes no problems in obtaining complete saturation. But now it has been shown that under the influence of a special catalyst, zinc-deactivated palladium–calcium carbonate, hydrogenation stops after the addition of the first mole of hydrogen in many cases. This is the basis for a quantitative method for the selective hydrogenation of acetylenic bonds.

Hydration Method of S. Siggia, C. R. Stahl, and R. Reinhardt

[*Reprinted in Part from Anal. Chem.*, ***28,*** *1481–3* (*1956*)]

Wagner, Goldstein, and Peters (55) describe a method for determining mono- and dialkyl acetylenes of four or five carbon atoms by reaction

55. C. D. Wagner, T. Goldstein, and E. D. Peters, *Anal. Chem.*, **19,** 103–5 (1947).

with methanol, using mercuric oxide and boron trifluoride as catalyst. The acetylenic compound is thus converted to the ketal. The ketal is distilled into hydroxylamine hydrochloride reagent, which hydrolyzes the ketals to the ketones and then forms the oxime of the ketone. The reactions used in this method are as follows:

$$RC{\equiv}CR' + 2MeOH \xrightarrow[BF_3]{HgO} RC(OMe)_2CH_2R'$$

$$RC(OMe)_2CH_2R' + H_2O \rightarrow R\overset{\overset{\displaystyle O}{\|}}{C}CH_2R' + 2MeOH$$

$$R\overset{\overset{\displaystyle O}{\|}}{C}CH_2R' + NH_2OH{\cdot}HCl \rightarrow R\overset{\overset{\displaystyle NOH}{\|}}{C}CH_2R' + H_2O + HCl$$

The hydrochloric acid formed in the last reaction is titrated, and the amount of acetylenic compound present in the sample is calculated from this value. The results obtained by this method are about 92% of the theoretical values. This procedure could not be used for the acetylenic compounds under investigation because of the high boiling points of the acetals formed, which made distillation impossible, the instability of the acetals, which resulted in decompositon at the distillation temperatures, or the low accuracy inherent in the method. No distillation is required in the procedure described below.

Koulkes (56) used the reaction between the acetylenic triple bond and mercuric acetate to determine several disubstituted acetylenic compounds. The mercuric acetate presumably adds onto the triple bond, and the excess of the acetate is determined by addition of sodium chloride and titration of the acetic acid liberated. This approach is fast, but impurities that react with mercuric acetate interfere: ethylenic compounds, some of which also add mercuric acetate, and inorganic and some organic halides, which complex with the mercuric ion. Some organic compounds such as carboxylates and sulfonates, form precipitates, and others are readily oxidized by the mercuric ion.

In the procedure described, the hydration reaction is used, with a mercuric catalyst in a strongly acidic medium, and the ketone

$$RC{\equiv}CR' + H_2O \xrightarrow[H_2SO_4]{HgSO_4} R\overset{\overset{\displaystyle O}{\|}}{C}CH_2R$$

formed is a measure of the acetylenic material. This approach is subject to less interference than the mercuric acetate addition method, since the

56. M. Koulkes, *Bull. Soc. Chim. Fr.*, **1953,** 402–4.

hydration will proceed if a portion of the catalyst is consumed by other components. The accuracy of this system is, in general, 100 ±2%, and a variety of acetylenic compounds are determinable (Table 33).

Table 33. Determination of Acetylenic Compounds

Compound	Reflux Time, min.		Mole Used	Mole Found	% Recovery	% Purity	Other Method
	Hydration	Oximation					
Butynediol[a]	15	15	0.01398	0.01339	95.8	98.9 ± 1	A
	30	30	0.01398	0.01391	99.5	—	—
	30	30	0.01398	0.01381	98.8	—	—
	30	30	0.00702	0.00682	97.2	—	—
	30	45	0.01404	0.01415	100.8	—	—
	30	45	0.01404	0.01422	101.3	—	—
	30	45	0.00702	0.00700	99.7	—	—
	45	60	0.01271	0.01277	100.5	—	—
	60	60	0.01398	0.01395	99.8	—	—
	120	60	0.01398	0.01388	99.3	—	—
2-Propyne-1-ol[b]	30	18 hr. at room temp.	0.01551	0.01473	95.0	99.0 ± 1	A
	30	30	0.01551	0.01387	89.4	—	—
	30	30-min. reflux + 18 hr. at room temp.	0.01551	0.01526	98.4	—	—
	30	60	0.01551	0.01517	97.8	—	—
	30	60	0.01551	0.01512	97.5	—	—
1-Butyne-3-ol[b]	60	60	0.01321	0.01125	85.2	84.1 ± 1	B
	60	60	0.00881	0.00732	83.1	—	—
3-Butyne-1-ol[b]	30	30	0.01477	0.01317	89.2	93.8 ± 1	B
	60	60	0.01507	0.01376	91.3	—	—
	60	60	0.01507	0.01414	93.8	—	—
	60	60	0.01484	0.01355	91.3	—	—
	60	60	0.01507	0.01364	90.5	—	—
	60	60	0.01507	0.01402	93.0	—	—
Ethynylcyclohexanol[c]	30	30	0.01381	0.01280	92.7	—[d]	—
	60	60	0.00810	0.00781	96.4	—	—
	60	60	0.00810	0.00793	97.9	—	—
Phenylacetylene[c]	60	—[f]	0.0001985	0.0001915	96.5	97.2 ± 1	B
	60	—[f]	0.0005958	0.0005744	96.4	—	—
1-Hexyne[c]	75	60	0.01223	0.01088	89.0	89.1 ± 1	C
	90	60	—	0.01083	88.6	—	—
3-Hexyne[c]	75[g]	60	0.01384	0.01235	89.2	91.5 ± 1	C
	75	60	—	0.01263	91.3	—	—
3-Octyne[c]	90[g]	60	0.00936	0.00875	93.5	93.0 ± 1	D
	150	60	—	0.00868	92.7	—	—
3-Octyne-1-ol[c]	90[g]	60	0.00806	0.00678	84.1	87.5 ± 1	C
	150	60	—	0.00675	83.7	—	—

A. Acetylation method (for alcohols) (57); see pp. 12–14.
B. Acetylenic hydrogen method (58); see pp. 508–509.
C. Bromination method (59); see pp. 383–387.
D. Hydrogenation method (60); see pp. 406–413.
[a] Recrystallized from ethyl acetate.
[b] Laboratory samples distilled once through helix-packed column.
[c] Analyzed as purchased from Farchan Research Laboratory.
[d] Could not be determined by method A, B, C, or D.
[e] Analyzed as purchased from Eastman Kodak.
[f] Dinitrophenylhydrazine method used because acetophenone could not be measured by oximation method.
[g] 20 ml extra methanol used in hydration step because of insolubility of compounds.

57. C. L. Ogg, W. L. Porter, and C. O. Willits, *Anal. Chem.*, **17,** 394 (1945).
58. S. Siggia and J. G. Hanna, *Anal. Chem.*, **21,** 1469 (1949).
59. H. J. Lucas and D. Pressman, *Ind. Eng. Chem., Anal. Ed.*, **10,** 140–2 (1938).
60. C. W. Gould and H. J. Drake, *Anal. Chem.*, **23,** 1157 (1951).

Acidic or alkaline impurities in the sample do not interfere, since the system is neutralized prior to oximation. Ethylenic unsaturated compounds do not interfere because they do not form carbonyl compounds under the conditions of the reaction. The only interferences that can be envisioned are from carbonyl compounds or carbonyl-forming compounds such as acetals or vinyl ethers. However these can be determined by running the oximation analysis alone on a sample without first running the hydration; this should yield just the carbonyl compound. The hydration analysis should then yield the total of carbonyl compound and acetylenic compound; the acetylenic component can then be determined by subtraction. Some aldehydes are not stable (oxidize) with mercuric ion, and samples containing large amounts of aldehydes or acetals should be examined thoroughly to make sure that the corrections applied are valid.

A potentiometric titration is used to measure the hydrochloric acid liberated in the oximation step. The end points are not sharp, but, in general, the precision is within ±2% and sometimes within ±1%.

Acetylenic compounds with substituents on the carbons adjacent to the acetylenic linkage

$$(R_1\underset{\displaystyle R_3}{\underset{|}{C}}HC{\equiv}C\underset{\displaystyle R_4}{\underset{|}{C}}HR_2)$$

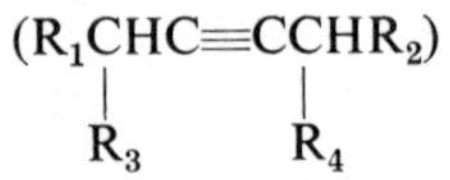

do not hydrate rapidly, owing to the hindrance caused by substituents. An analysis

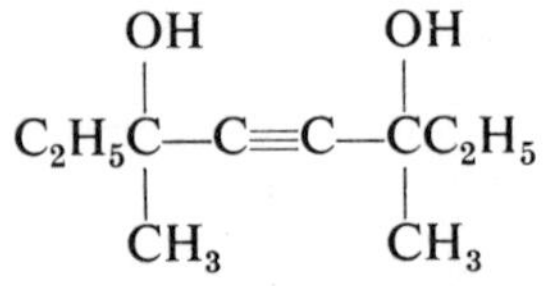

$$C_2H_5\overset{\displaystyle OH}{\overset{|}{\underset{\displaystyle CH_3}{\underset{|}{C}}}}{-}C{\equiv}C{-}\overset{\displaystyle OH}{\overset{|}{\underset{\displaystyle CH_3}{\underset{|}{C}}}}C_2H_5$$

was attempted by this method, but conversions of only about 50% were obtained under the conditions described; 70% conversions were obtained when the hydration time was doubled. Unfortunately, not enough compounds could be obtained with substituents adjacent to the acetylenic linkage to permit comparison of the rates of hydration with the type of substituent.

Attempts were made to analyze acetylenic bromine compounds of the type $RC{\equiv}CCH_2Br$. The bromine atoms on these compounds are so labile, however, that they are removed by the mercuric catalyst to form the mercuric bromine complex, and the catalytic action of the mercuric ion is much decreased; this type of compound cannot be determined by this method. The chloride compounds of the same structure will probably behave in the same manner, but this has not been tested.

For phenylacetylene, the ketone formed on hydration is acetophenone. This ketone is one of the very few that cannot be determined by oximation in an aqueous or partially aqueous medium, because of the equilibrium that is present in the oximation system. Water is a product of the oximation reaction, and an aqueous system keeps the reaction from going to completion. For phenylacetylene, after hydration any excess mercuric ion is removed by bubbling hydrogen sulfide through the solution and filtering off the mercuric sulfide. Then the acetophenone in the solution is determined by the 2,4-dinitrophenylhydrazine method of Iddles and Jackson (61) (see p. 114). The same technique should be applicable to other acetylenic compounds yielding ketones that are not readily determined by oximation. This technique is applicable to samples containing small amounts of acetylenic compounds, since the 2,4-dinitrophenylhydrazine method requires only 4×10^{-4} mole of ketone for optimum operating conditions. This precipitation approach is less applicable to the hydroxyacetylenic compounds because of the solubilizing effect of the hydroxyl groups.

Before the 2,4-dinitrophenylhydrazine approach was tried on the acetylenic compounds, blanks were run, in which all the steps were included. This was to make sure that no extraneous precipitate formed on addition of the hydrazine, which would affect the results. Known samples of acetophenone were run through the entire procedure to establish the conditions for complete recovery of the ketone. Then the acetylenic compounds were used.

REAGENTS

Hydroxylamine hydrochloride ($0.5N$), in 1:1 methanol-water.

The catalyst is made from 0.5 gram of mercuric sulfate, 2 ml of sulfuric acid, and 63.4 ml of water.

ALCOHOLIC SODIUM HYDROXIDE, $1.0N$. Dissolve sodium hydroxide in as little water as possible. Filter off the sodium carbonate, and dilute the solution with methanol to the desired volume. This solution need not be standardized.

Aqueous sodium hydroxide, $0.5N$ (standard).

APPARATUS

Glass and calomel electrodes, with a Model H-2 Beckman pH meter.

61. H. A. Iddles and C. E. Jackson, *Ind. Eng. Chem., Anal. Ed.*, **6,** 454 (1934).

PROCEDURE

Dissolve a sample containing 0.05 to 0.20 mole of acetylenic compound in methanol and dilute to 100 ml in a volumetric flask; use 10-ml aliquots for the determinations. Add 10 ml of sample solution to 20 ml of catalyst in a 200-ml three-necked flask connected to a reflux condenser. Insert glass stoppers in the two unused necks of the flask. Reflux the mixture for 1 hour and then cool in ice with the condenser still attached. After cooling the condenser, wash it with 10 ml of 1:1 methanol-water and drain. Disconnect the flask from the condenser and insert glass-calomel electrodes into the flask through the two side necks. Just neutralize the acid (pH 7) with 1.0N alcoholic sodium hydroxide.

Add 50 ml of hydroxylamine hydrochloride, reflux the mixture for 1 hour, and cool in ice; wash the condenser with equal parts of methanol and water. Transfer the mixture to a 400-ml beaker, using 50 ml of 1:1 methanol-water to wash the flask. Leave as much of the solid residue as possible in the flask during transfer.

Titrate the liberated hydrochloric acid potentiometrically with standard 0.5N sodium hydroxide, using the glass and calomel electrodes. Determine the end point from a plot of milliliters of reagent versus pH.

If carbonyl compounds are present in the sample, they should be determined using the hydroxylamine hydrochloride analysis (see discussion above).

2,4-DINITROPHENYLHYDRAZONE METHOD

REAGENTS

Catalyst as described.

A saturated solution of 2,4-dinitrophenylhydrazine at 0°C in 2N hydrochloric acid.

PROCEDURE

Dissolve a sample in methanol and dilute to 100 ml, so that a 10-ml aliquot contains approximately 4×10^{-4} mole. Add 10 ml of sample to 20 ml of mercuric sulfate–sulfuric acid catalyst and reflux for 1 hour in a three-necked flask with glass stoppers in the two unused necks. After the hydration reaction period, cool the flask in ice with the condenser attached, and wash the condenser with 10 ml of 1:1 methanol-water. At this point there is a white precipitate in the flask, which does not appear to affect the results.

With the condenser still in position, pass hydrogen sulfide into the solution to precipitate mercury as the sulfide. When this reaction is complete (5–10 minutes), filter off the sulfide through a Whatman No. 30 filter paper and wash the flask and paper with a 1:1 solution of methanol-water.

To the filtrate add 50 ml of 2,4-dinitrophenylhydrazine solution, and allow the mixture to stand for 30 to 60 minutes. Warm the resulting solution on a hot plate, stirring constantly to coagulate the precipitate. When the supernatant liquid is clear, filter off the precipitate through a Gooch crucible with an asbestos mat, wash with water, dry at 100°C, and weigh. If the resultant hydrazone exhibits a significant solubility with the alcohol present (this must be predetermined), boil the solution for a few minutes to remove as much alcohol as possible before filtration. Acetylenic compounds containing hydroxyl groups usually cannot be determined by this method, because of the solubilizing effects of these groups.

TRACE AMOUNT OF ACETYLENES

The preceding method has been extended to the 5- to 500-ppm range for C_4–C_5 acetylenes in hydrocarbons by a spectrophotometric measurement of the 2,4-dinitrophenylhydrazones of the carbonyl hydration products.

Method of M. W. Scoggins and H. A. Price

[*Anal. Chem.*, ***35,*** *48* (*1963*)]

REAGENTS AND APPARATUS

The catalyst is 30 grams of mercuric sulfate dissolved in 630 ml of water containing 20 ml of sulfuric acid.

The 2,4-dinitrophenylhydrazine reagent was a saturated solution in 3*M* sulfuric acid.

Measurements were made in 1-cm cell with a Beckman DU spectrophotometer.

PROCEDURE

Dilute a quantity of sample containing not more than 100 μg of acetylenes to 10 ml with spectrograde cyclohexane in a 40-ml screw-cap vial equipped with a polyethylene gasket in the cap. Make a blank determination simultaneously with 10 ml of cyclohexane. Add 10 ml of catalyst solution and allow the solutions to react for 1 hour at room temperature, with continuous mixing.

Add approximately 0.5 gram of sodium chloride to remove the mercuric ion from the solution as the slightly ionized mercuric chloride. A white precipitate appears at this point if the sample contains an appreciable amount of olefins. If the precipitate is not removed, results will be low. Remove the precipitate as follows:

Centrifuge the vials until the solid mass settles to the bottom of the vial. Carefully withdraw 5 ml of each of the two phases and transfer them to another vial. Add 5 ml of cyclohexane to bring the organic phase back to its original volume. Proceed now as if the solid phase had not formed.

Table 34. Total Acetylene by Hydration

	Acetylene		
Compound[a]	Added, ppm	Found, ppm	Recovery, %
Butyne-2	4.4	4.1, 3.9, 4.1	91.0
	6.4	6.2, 6.2, 6.0	98.4
	22.1	23.8, 20.5, 22.6	100.8
	66.0	67.0, 68.0, 67.0	101.8
	132	120, 114, 116	88.8
Butyne-2, Pentyne-1	502	475, 500, 475	96.2
Average			96.1

[a] The solvent in all cases was cyclohexane.

Table 35. Effects of Aromatics and Olefins

		Acetylene, ppm		
Compound	Solvent	Added	Found, Average[a]	Range
Butyne-2	Cyclohexane, hexene-1	26.5	25.8	1.5
Butyne-2, Pentyne-2	Cyclohexane, hexene-1	26.5	27.0	0.0
Butyne-2	Cyclohexane, benzene	13.0	12.3	1.1

[a] Average of at least three determinations.

Add 10 ml of 2,4-dinitrophenylhydrazine reagent solution and allow the reaction to proceed for 30 minutes at room temperature with continuous mixing. Allow the two phases to separate, and withdraw portions of the cyclohexane phase of the sample and blank. Measure the absorbance of the sample versus the blank in 1-cm cells at 340 nm. Determine the acetylene concentration from a previously prepared calibration curve.

CALIBRATION CURVE. Prepare a solution of methyl ethyl ketone in cyclohexane to contain 10 μg of ketone per milliliter of solution. Transfer aliquots of the solution to 40-ml screw-cap vials and dilute to 10 ml. Add 10 ml of the dinitrophenylhydrazine reagent solution and continue as outlined in the procedure above. Plot absorbance versus the corresponding carbonyl concentrations. In using the curve, obtain the acetylene concentration from the carbonyl concentration by multiplying by the proper molecular weight ratios.

DISCUSSION AND RESULTS

Recovery results are summarized in Table 34.

Results in Table 35 show the effect of aromatics and olefins on the method. Olefins in concentrations up to 5% do not interfere if the solid olefin–mercuric chloride complex is removed from the solution by withdrawing an aliquot of each phase after the addition of sodium chloride and centrifugation. When this step was omitted, the olefin-acetylene blends of Table 35 were low by about 13 ppm. An aliquot of both phases is required because carbonyls containing less than five carbon atoms are almost completely miscible with aqueous solutions. Benzene in concentrations up to 10% does not interfere seriously. Isoprene interferes because it yields a yellow precipitate with the mercuric reagent that is soluble in cyclohexane.

SELECTIVE HYDROGENATION

From the Method of W. Merz and K. Müller

[*Z. Anal. Chem.*, ***237***, *264* (*1968*)]

APPARATUS

One of the arrangements for the hydrogenation methods described previously (pp. 402–406) is suitable.

Table 36. Hydrogenation of Acetylene Derivatives with Zinc-Deactivated Palladium–Calcium Carbonate Catalyst

Compound	Solvent				
	Water	Aqueous Ammonia	Aqueous Ethanol amine	Aqueous Potassium Hydroxide	Acetic Acid
Phenylacetylene	—	—	—	1.0	2.0
Propargyl alcohol	1.1, 1.0	1.2	1.2	1.0	2.0
3-Methyl-1-butyn-3-ol	1.0, 1.0, 0.9	1.0	1.0	1.0	2.0, 2.1, 1.9
3-Methyl-1-pentyn-3-ol	1.0, 0.9, 0.9	1.1	1.0	1.0	2.0
4-Ethyl-1-octyn-3-ol	1.1	1.0	1.0	1.0	2.0
1-Ethynylcyclohexanol	1.0, 1.0	1.0	0.9	1.0	2.0, 2.0
2-Methyl-1-ethynyl cyclohexanol	1.0, 1.0 0.9	1.0	1.0	1.0	2.0, 2.0, 1.9
2-Butyn-1,4-diol	1.9, 1.0, 1.0	1.0, 1.0, 1.0	1.0	1.0	2.0, 2.0
2-Methyl-3-pentyn-2,5-diol	1.0	1.0	1.0	1.0	2.0
Dehydronerolidol	1.0, 1.0, 1.0	1.2	1.1	1.0	2.0
Ethynylionol	1.1, 1.0	0.9, 1.0	1.0	1.1	2.0, 2.0, 1.9
Dehydroisophytol	1.1, 1.0	1.9	0.9	1.0	2.2, 2.0, 2.0

Table 37. Hydrogenation of Compounds Containing C=C, C=O, and C≡N Groups with Zinc-Deactivated Palladium–Calcium Carbonate Catalyst

Compound	Solvent				
	Water	Aqueous Ammonia	Aqueous Ethanol amine	Aqueous Potassium Hydroxide	Acetic Acid
Styrol	—	—	—	0	1.0
Cyclohexa-1,3-diene	0	—	—	0	0.6
2-Butene-1,4-diol	0	0	0	0	1.0, 1.0
Cinnamic acid	0	—	—	—	1.0, 1.0
Methyl acrylate	—	—	—	1.0	—
Acrylamide	—	1.0	—	—	—
Cyclohexanone	0	0	—	—	0
Acrylonitrile	—	1.0	1.0	1.0	1.0
Acetonitrile	—	—	—	—	0
Adiponitrile	—	—	—	—	0

REAGENTS

ZINC-DEACTIVATED PALLADIUM–CALCIUM CARBONATE CATALYST. Stir a palladium–calcium carbonate catalyst with aqueous zinc acetate solution and boil for 1 to 2 hours. Filter the catalyst and boil with water. Again filter the catalyst, wash with water, and dry. The catalyst should contain 0.6 to 0.7% palladium and about 9% zinc.

SOLVENTS. Water; aqueous ammonia (5 μg per milliliter); aqueous ethanolamine (20 μg per milliliter); aqueous potassium hydroxide (16.5 μg per milliliter).

PROCEDURE

Use 20 to 25 mg of catalyst for about 5-mg samples for microhydrogenation and about 200 ml catalyst for 50 to 100-mg samples for macrohydrogenation.

RESULTS AND DISCUSSION

Experimental results obtained by Merz and Müller expressed as moles of used hydrogen per mole of acetylenic bonds are given in Table 36. Results for the method on some olefinic compounds, nitriles; and a ketone to indicate selectivity appear in Table 37.

With water as the solvent, acetylenic bonds accept one mole of hydrogen. The addition of traces of alkaline substances significantly increases the rate of hydrogenation. Two moles of hydrogen is added in acetic acid solvent. The carbon-to-nitrogen triple bond was not hydrogenated, as shown by the examples of acetonitrile and adiponitrile. Some double bonds sufficiently activated by adjacent groups such as those in acrylonitrile, methyl acrylate, and acrylamide, are hydrogenated in aqueous medium. However the double bonds of styrols, butenediol, and cinnamic acid were not hydrogenated. The CO group in cyclohexanone and its acid derivatives likewise were not hydrogenated.

8

Active Hydrogen

An active hydrogen atom may be defined as a hydrogen atom that is attached to any atom except a carbon atom. Hence alcohols, amines, and amides (primary and secondary only), carboxylic and sulfonic acids, mercaptans, and sulfonamides (primary and secondary only) can be considered to contain active hydrogens.

Even though these active hydrogen atoms are in radically different groupings, they have one set of reactions in common. They will react with Grignard reagent to liberate the corresponding hydrocarbon. Methyl

$$\text{A—H} + \text{RMgX} \rightarrow \text{A—MgX} + \text{RH}$$

Grignards are generally used for analytical purposes so that the hydrocarbon formed is methane, which is easily measured gasometrically.

Lithium aluminum hydride also is used for measuring active hydrogens. In this case, hydrogen is given off by the reaction and this is measured.

$$4\text{A—H} + \text{LiAlH}_4 \rightarrow \text{LiAlA}_4 + 4\text{H}_2$$

Diazomethane also reacts with active hydrogens liberating nitrogen. This reaction has never achieved much utility from an analytical standpoint,

$$\text{A—H} + \text{CH}_2\text{N}_2 \rightarrow \text{A—CH}_3 + \text{N}_2$$

however, so it is not discussed.

Active hydrogen analysis as a whole has lost much of its significance in the past 15 to 20 years, since many of the functional groups of which the active hydrogen is a part can readily be determined by other methods. These other methods not only have higher specificity but are generally easier to operate, more accurate, and more precise. For example, all the functional compounds mentioned in the first paragraph of this chapter are better and more easily measured by the methods given in this text for the respective groups than by any àctive hydrogen method.

The active hydrogen methods are no more accurate than ±3%, hence can be considered only quantitative estimations, usuable for determining the number of active hydrogens per molecule. The methods lack specificity because so many compounds, including water, contain active hydrogen.

The active hydrogen method has only one remaining valuable application. This is the determination of a small amount of active hydrogen compound in the presence of a large amount of nonactive hydrogen components, that is, traces of alcohol in hydrocarbons. Since the analysis is gasometric, very small quantities can be detected. The main reason for continuing to include active hydrogen methods in this book is their historical significance. They enjoyed a high degree of popularity 25 years ago.

Grignard Reagent Approach

The use of Grignard reagent for measuring active hydrogen was initiated by Tschugaeff (1) but was developed into a firm analytical method by Zerewitinoff (2). The apparatus used to measure the methane evolved has undergone many modifications. The one shown below is a potpourri the authors have derived from the apparatus described in the literature. It works as well as any other and is simpler than most. The controlling factor in this analysis is not the apparatus so much as the sensitivity of the reagents used; that is, water and even most stopcock greases will react with the Grignard and other reagents.

The Grignard method was also once used to measure functional groups that add Grignard reagent as well as to measure active hydrogen.

$$RCHO + CH_3MgI \rightarrow RCH(CH_3)OMgI$$

$$R_1C(=O)R_2 + CH_3MgI \rightarrow R_1C(OMgI)(CH_3)R_2$$

$$RC{\equiv}N + CH_3MgI \rightarrow RC(CH_3){=}NMgI$$

$$RNC + CH_3MgI \rightarrow RN{=}C(CH_3){-}MgI$$

$$RC(=O)X + 2CH_3MgI \rightarrow RC(CH_3)_2{-}OMgI$$

1. L. Tschugaeff, *Ber.*, **35,** 3912 (1902).
2. T. Zerewitinoff, *Ber.*, **40,** 2023 (1907); **41,** 2233 (1908).

$$RC(=O)OR + 2CH_3MgI \rightarrow RC(CH_3)_2{-}OMgI$$

$$RCH_2X + CH_3MgI \rightarrow RCH_2CH_3$$

In these cases, excess Grignard is added and the excess is decomposed with aniline. The difference between the methane collected from the reagent alone and from the reagent after reaction with the sample is a measure of the group adding the Grignard. Because of the nonspecificity of the Grignard reagent and the cumbersomeness of the method, however, this approach is not recommended, especially in light of the relatively simple, accurate, and precise existing methods to measure the same functional groups.

Method Described by Niederl and Niederl

[*Adapted with Permission from Micromethods of Quantitative Organic Analyis, 2nd ed., Wiley, New York, 1942, pp. 263–72. The apparatus has been modified to simplify handling and to increase the accuracy of the method.*]

REAGENTS

METHYLMAGNESIUM IODIDE, 0.4 TO 0.5*M*. In a 250-ml round-bottomed distilling flask, equipped with an inlet tube (sealed in about an inch from the neck) and a ground-glass joint with reflux condenser to fit, place about 0.6 gram of magnesium turnings. To the magnesium turnings, add 2.5 ml of methyl iodide and 50 ml of freshly distilled *n*-amyl ether (the ether should be distilled from and stored over metallic sodium). Pass nitrogen through the system at a rate of about one bubble per second by means of the inlet tube on the reaction flask. The nitrogen should be of 99.9% purity (Siefert nitrogen).

The Grignard reaction is started by dropping a crystal of iodine into the reaction mixture and warming on a water bath. Should the Grignard reaction still fail to start, put 2 to 3 ml of absolute diethyl ether in a small test tube with a few magnesium turnings, and add 2 to 3 drops of methyl iodide. The reaction starts rapidly in the diethyl ether. As soon as the reaction in the test tube is proceeding at a vigorous rate, add the contents of the test tube, while still reacting, to the reaction mixture in the flask. This procedure usually induces the reaction to begin.

If, even after this priming, the reaction does not begin, another method can be employed. Prepare the reaction mixture as above, but use 25 ml of absolute diethyl ether instead of the amyl ether. This reaction begins immediately. Allow it to proceed to completion; then add 50 ml of amyl ether by means of the condenser, and boil off the ethyl ether as described below. This method of preparing the Grignard reagent is seldom needed, however. The priming of the reaction with the iodine crystal or with some methylmagnesium iodide in ethyl ether is usually enough to get the reaction started.

Once the reaction is started in amyl ether, allow it to proceed for 2 hours while heating gently on a steam bath or until the reaction has appreciably subsided. Then turn off the steam, and stop the flow of water in the condenser and the flow of nitrogen. The condenser outlet is connected by means of a rubber tube to a vacuum pump or an efficient aspirator (both equipped with dry ice traps to trap the vapors coming over). A pinch clamp is attached to the rubber tube leading to the source of vacuum, to regulate the vacuum. Turn on the vacuum and slowly open the pinch clamp, causing the contents of the flask to boil. As soon as the pinch clamp is fully opened, resume the nitrogen flow at a rate of about one bubble per second for about 15 minutes. This procedure removes any unreacted methyl iodide and any ethyl ether that may have been added to initiate the reaction. Little amyl ether is lost during the evaporation.

After the evaporation, carefully remove the rubber tube from the condenser to the source of vacuum, while allowing the apparatus to fill up with nitrogen. Then decant the methylmagnesium iodide quickly through a funnel containing a glass wool plug into a long-stemmed glass ampoule of the type used for liquid bromine, acetaldehyde, acetyl chloride, and so on. Seal the top of the ampoule with a soft rubber stopper. The ideal stopper is a serum cap such as is used on ampoules of serum and certain medicinals. The stopper has to be of such a nature that a hypodermic needle can be inserted through it easily. The reagent should be about 0.4 to 0.5 *M*. The strength of the reagent can be ascertained by adding excess 0.1*N* hydrochloric acid to a 5 ml sample of reagent. Titrate the excess of acid with 0.1*N* sodium hydroxide, using phenolphthalein as an indicator.

SOLVENTS

Anethole, pyridine, and xylene have been found to be convenient solvents. They should be redistilled and should be thoroughly dry.

APPARATUS

In the apparatus illustrated in Fig. 8.1, *A* is a reaction vessel of about 10-ml capacity, *B* a three-way stopcock, *C* a rubber serum cap, *D* a 2-mm capillary, *E* a 7-ml gas buret, *F* a three-way stopcock, and *G* a thermometer.

PROCEDURE

Weigh a sample containing about 0.0001 to 0.00015 equivalent of active hydrogen into the reaction vessel *A* and dissolve it in 1.0 ml of one

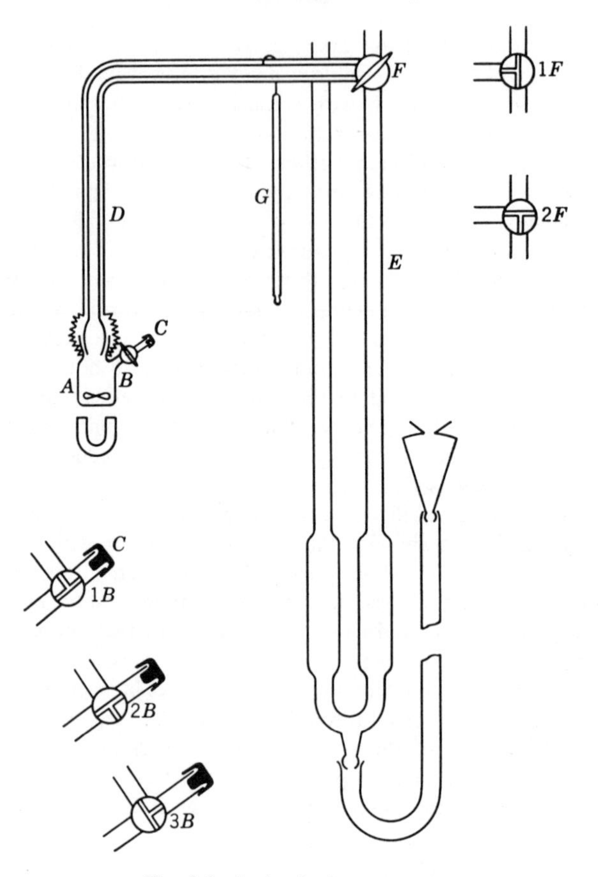

Fig. 8.1. Active hydrogen apparatus.

of the solvents mentioned. The solvent should be measured with a pipet. Drop a glass-encased iron paddle into the solution. All the apparatus should be thoroughly dry. It is best to rinse the reaction flask with acetone and then with ethyl ether after each determination; the stopcock should be taken out of the apparatus and all grease removed. Then dry all the parts in air until all the ether has evaporated and put the parts in an oven at 130°C for at least 15 minutes before the next experiment.

After the sample and solvent have been added to the reaction flask, grease the top edge of the ground-glass joint and join the reaction flask to the rest of the apparatus; the top edge of the ground joint must be completely sealed by the grease. The stopcock on the reaction flask should be lubricated with a minimum of grease. There should be no grease in the bore of the stopcock. Pass through the system for 5 minutes the nitrogen that has been bubbled through concentrated sulfuric acid.

Introduce the nitrogen through stopcock *F*, which is in position 1*F*, and keep the mercury in the gas buret as close to the stopcock as possible. The nitrogen escapes through stopcock *B*, which is in position 1*B*. While the nitrogen is still flowing, slowly lower the mercury level to the beginning of the calibrations and put the serum cap over the end of the exit tube on the reaction flask at *C*. Then turn stopcock *B* to position 2*B*; turn stopcock *F* to position 2*F*, and discontinue the flow of nitrogen.

Use a hypodermic syringe of about 2-ml capacity with a wide bore needle to introduce the Grignard reagent. Insert the needle through the serum cap on the ampoule containing the Grignard reagent and tilt the ampoule until the reagent solution comes in contact with the needle. Fill the hypodermic with reagent; then invert it, thus driving the air out, until reagent comes through the needle tip. With the mercury levels equalized and the initial reading recorded, quickly insert the needle through the serum cap at *C* on the reaction flask and pass the needle through the bore of the stopcock into the body of the flask. Inject reagent to the amount of 0.5 to 1.0 ml (or an amount in excess of the theoretical amount required) into the solution. Carefully note and record the temperature of the apparatus and the volume of reagent added. Withdraw the needle carefully, and as soon as the tip is past stopcock *B*, turn the stopcock to position 3*B*. Lower the mercury level, and immerse the reaction flask in boiling water for 15 minutes. Then remove the water bath and attach the magnetic stirrer. Stir the reaction mixture until the mercury level becomes constant. Again note the temperature along with the mercury level reading.

Make a blank determination on the solvent used for the analysis to determine how much methane is liberated by the solvent. Analyze the blank by the same method as the sample. It is advisable to use about 5 ml of solvent. Divide the result by 5 to get the blank per milliliter of solvent. If the number of groups that *add* Grignard reagent is to be found, the strength of the reagent has to be determined. It can be done together with the blank. Use 1 ml of solvent, add a known amount of reagent, and determine the blank. (For accurate work, the blank should not be greater than 0.4 ml of methane per milliliter of solvent.) Inject a measured amount of aniline (about 1 ml) into the reaction flask in the same manner as the reagent (each time an injection of fluid is made into the reaction flask, the mercury levels should be equalized to prevent leaks), and stir the contents until the mercury level becomes constant. The methane liberated by the solvent plus the methane liberated by the aniline corresponds to the strength of the Grignard reagent.

In analyzing a sample that adds Grignard reagent, add the reagent to the sample and determine the active hydrogen as described. After the mercury level is read, decompose the excess Grignard reagent with aniline as described. After the level of the mercury has become constant, read the level along with the temperature.

All parts of the apparatus can be cleaned of precipitated magnesium salts by using dilute hydrochloric acid.

For accurate work, measure the volume of the entire apparatus by filling all parts of the apparatus with water and weighing the water (see Hydrogenation, p. 400). This is to correct for any temperature changes that occur between buret readings. Any temperature changes cause errors in volumes because of expansion or contraction of the gas in the free space in the apparatus. After each buret reading, note the temperature. If the temperature is different from the initial buret reading, correct the readings to the temperature of the initial buret reading before the calculations are carried out. (See section on Hydrogenation, p. 401, for the method of correcting for temperature changes.)

CALCULATIONS

$$\frac{I \times 273 \times \text{Mol. wt. of compound}}{(T+273) \times 22{,}400 \times \text{Grams of sample}} = \frac{\text{atoms of active hydrogen}}{\text{mole of sample}}$$

$$\frac{K \times 273 \times \text{Mol. wt. of compound}}{(T+273) \times 22{,}400 \times \text{Grams of sample}} = \frac{\text{moles of Grignard reagent added to the sample}}{\text{mole of sample}}$$

where

A = volume of Grignard reagent added

B = blank on the solvent

C = volume of aniline added

D = initial buret reading

E = buret reading after reaction with the Grignard reagent (corrected for temperature as indicated above)

F = buret reading after reaction of aniline with excess Grignard reagent (corrected for any temperature change)

G = total methane that would be evolved by the amount of Grignard reagent used, as calculated from the strength of the Grignard reagent predetermined as described above

T = temperature of the experiment

$E - D$ = volume change recorded = H

$H - B - A$ = volume of methane liberated by active hydrogen = I

$F - E - C$ = volume of methane evolved by the excess Grignard reagent (reaction with aniline) = J

$G - J$ = volume of methane corresponding to reagent added onto the sample = K

The foregoing procedure requires some practice to operate successfully. There are several sources of error. All pieces of the apparatus must be thoroughly dry as must the sample and the solvent. Oxygen must be excluded from the apparatus. Temperatures should be observed and corrections applied to buret readings, since a degree change in temperature causes variation of about 0.08 ml in the buret reading; and if 4 ml of methane were collected, this change in temperature would cause a 2% error.

The Grignard reagent as stored for this procedure is stable for a month or more, provided the stopper is air-tight.

The following compounds were used to test this procedure for determining active hydrogen and for determining compounds that react with Grignard reagent: aniline, water, acetophenone, amyl alcohol, *tert*-butylphenol, benzaldehyde, and salicylaldehyde.

Modifications by J. F. Lees and R. T. Lobeck

[*Adapted from Analyst,* ***88,*** *782 (1963)*]

A study was made of the Grignard reagent method to find modifications that would produce more accurate and precise results.

To eliminate completely errors due to solubility of methane, the ideal practice would be to use pure methane as the inert gas. Both the reagent and the solvent could then be saturated with methane before the reaction begins. The use of methane in this manner is not practical in some cases, and it is desirable to have a method for solubility corrections when nitrogen is used as the inert gas. After reaction, because of the methane generated, there is a decrease in the partial pressure of the nitrogen and some of the dissolved nitrogen is liberated from the reaction mixture. These solubility effects make it necessary to keep constant for all determinations the volume of solvent for the sample, the volume of the Grignard reagent added, and the volume of nitrogen in the system.

Therefore the blank value that is the increase in volume observed on mixing solvent and Grignard reagent alone is the sum of two values: (1) the volume of methane generated by traccs of active hydrogen, if any, in the solvent, and (2) the volume of vapor produced by the reaction mixture in the presence of a constant volume of nitrogen. After reaction of a sample, the total volume of the system increases, and there is a corresponding increase in the volume of vapor produced by the reaction mixture.

The volume of the system and all these other effects are temperature dependent, and the temperature of the system must be accurately controlled. The reaction flask, the manometer, and the gas buret in this study were regulated at 25 ±0.1°C by water pumped through jackets from a constant-temperature bath.

An equation was derived that related the true volume of methane to the observed volume, the blank value, and the correction terms for the solubility and vapor pressure effects. To obtain values for these correction terms, however, the coefficients of solubility of pure methane and of nitrogen in the reaction mixture, and also the vapor pressure of the reaction mixture, were required. Since these values are not readily available, the equation was simplified and the correction values were represented by a single term V_c, as follows;

$$V_t - (V_o - B) = V_c$$

where V_t is the true volume of methane generated by the sample, V_o is the observed volume increase after the reaction, and B is the blank volume.

For a known weight of a pure standard substance, if the reaction is assumed to be stoichiometric, V_t can be calculated; from the value $V_o - B$ obtained experimentally, the corresponding value of the correction V_c can be calculated by the equation given. Thus by varying the weight of standard substance, a series of values of $V_o - B$ and V_c can be obtained and $V_o - B$ can be plotted against V_c. Then, from the value of $V_o - B$ obtained for the sample, the corresponding correction V_c may be obtained from the graph. Then V_t for the sample is given by

$$V_t = (V_o - B) + V_c$$

With such corrections applied, the results in Table 1 were obtained. The accuracy of the graphical correction procedure is ±1.5% of the theoretical value. It is possible to determine hydroxyl percentages down to 0.04%.

Table 1. Grignard Determination Results with Graphical Corrections

Compound	Average Purity,[a] %
Tetramethyldisiloxane-1,3-diol	99.9 (20)
Diphenylsilanediol	100.0 (20)
Diphenyldimethyldisiloxane	101.1 (9)
Triphenylsilanol	99.8 (5)
Tetraphenyldisiloxane-1,3-diol	100.5 (2)
α-Naphthol	101.7 (2)
Resorcinol	99.7 (2)
Phenol	101.1 (2)
Thymol	100.7 (2)
Salicylaldehyde	99.9 (2)
Vanillin[b]	59.4 (3)[c]
Benzoic acid	100.5 (4)
Glacial acetic acid[b]	89.9 (2)
n-Butanol	99.3 (3)
2-Octanol	100.3 (3)
Aniline	100.5 (2)
Cyclohexylamine[d]	101.7 (2)
Di-*n*-propylamine[e]	99.8 (2)

[a] Figures in parenthesis indicate number of determinations.
[b] Precipitate formed on reaction. The formation of insoluble product usually leads to low results, presumably because of occlusion of unreacted sample.
[c] After heating reaction mixture at 100°C for 5 minutes.
[d] Reaction time, 15 minutes.
[e] Reaction time, 60 minutes.

Lithium Aluminum Hydride Approach

Lithium aluminum hydride, which has been used instead of Grignard reagent, possesses a few advantages. Namely, it can be purchased readily, whereas the Grignard reagents must be prepared; also, where the hydride is concerned, hydrogen is the gas evolved, and this has less solubility in the solvents used than the methane liberated by the Grignard. The sensitivity and lack of selectivity of the hydride are as undesirable as for the Grignard, however.

REAGENT

A 0.1 to 0.2*N* solution of lithium aluminum hydride in any suitable ether in which both the reagent and sample are soluble. The higher boiling ethers, namely, di-*n*-butyl, di-*n*-amyl, and dimethyl ethers of ethylene glycol or diethylene glycol, are recommended. Dioxane and tetrahydrofuran can be used, however, if the solubility of the sample so dictates.

All solvents used must be distilled from lithium aluminum hydride solutions to ensure lack of active hydrogen.

The reagent and solvents are best kept in ampoules sealed with serum caps to protect against the atmosphere.

PROCEDURE

The procedure and apparatus for lithium aluminum hydride is identical to that for the Grignard reagent. The only difference is that only ether solvents can be used for the samples when lithium aluminum hydride is the reagent, since the reagent is soluble in these solvents alone.

9

Acetylenic Hydrogen

$$HC{\equiv}CR$$

A hydrogen atom on an acetylenic carbon atom is easily replaced with certain metal atoms. These reactions are very rapid and appear to be almost ionic. The acetylenic hydrogen atom exhibits no acidity in itself, however, hence is probably not ionic in character.

The metals that replace the acetylenic hydrogen atom the most readily are silver, copper(I), and mercury. The metal derivatives are quite insoluble, and precipitation is generally quantitative unless solubilizing groups such as —OH, COONa, $—SO_3Na$ are present on the same molecule. The metal derivatives also tend to be explosive when dry; however their sensitivity decreases as the molecular weight of the compound increases.

The replacement of acetylenic hydrogen atoms with silver is the reaction most commonly used to determine these types of compounds, since

$$2AgNO_3 + HC{\equiv}CR \rightarrow AgC{\equiv}CR{\cdot}AgNO_3 + HNO_3$$

silver is easily measured quantitatively and is a common analytical reagent. The six methods of applying the silver replacement reaction are described below.

Method A consists of using ammoniacal silver nitrate and determining excess silver. Method B is used when aldehydes are present in the sample. It is similar to Method A except that sodium acetate is used instead of ammonia.

A basic reaction medium helps pull the reaction to completion by removing the acid formed in the reaction. Ammonia is a stronger base than sodium acetate; however ammoniacal silver can oxidize aldehydes quite readily, and interference in the analysis will result if aldehydes are present in the sample. The oxidation of aldehydes is minimized when sodium acetate is used.

Methods C and D also employ the reaction with silver nitrate, only here no alkaline agent is added; instead, the acid formed in the replacement reaction is titrated. The reaction actually is proceeding as titrant is added. These methods are preferred, since the metal derivatives need not be removed from the reaction system. Method D is so arranged that neither aldehydes nor halogens interfere. It utilizes a high concentration of silver reagent relative to sample; hence enough reagent is available for reaction with all components. Also, since the final analysis is based on the amount of strong acid liberated, the relatively weak carboxylic acid formed on oxidation of aldehydes does not interfere, and the precipitation of halide ions does not alter the acidity of the system.

Methods E and F are particularly useful for water-insoluble samples. Silver perchlorate in methanol is used as the reagent in Method E, and titration of the liberated acid is done with a weak amine that eliminates interferences from readily saponified esters. Pyridine, used as the solvent in Method F, offers advantages because of its basic and complexing properties.

Copper and mercury rarely serve for determining acetylenic hydrogen compounds. They are used only when the silver methods cannot be.

Silver Methods

Method A

The reaction above (p. 489) will not take place if a solution of silver nitrate as such is added to the acetylenic compound. An alkaline material such as ammonia has to be present to remove the nitric acid formed in the reaction. Without the ammonia, the reaction would take place only to a slight degree, if at all, since acids react with the silver acetylides to liberate the free acetylenic compound.

All the acetylenic compounds tried quantitatively add one more silver atom than there are acetylenic hydrogen atoms. This has also been found to be true by other investigators (1). This method is reproducible to about ±0.5%.

REAGENTS

Standard silver nitrate, $0.1N$.
Ammonium thiocyanate, $0.1N$.
CP Concentrated ammonium hydroxide.
CP Concentrated nitric acid.
Approximately $6N$ sodium hydroxide.
Approximately $1M$ ammonium hydroxide.
Saturated ferric alum indicator solution.

1. For instance, V. J. Altieri, *Gas Analysis and Testing of Gaseous Materials*, American Gas Association, Inc., New York, 1945, p. 329.

PROCEDURE

Into a 250-ml glass-stoppered Erlenmeyer flask pipet 25 ml of standard 0.1N silver nitrate. Then add 3 ml of 6N sodium hydroxide, followed by concentrated ammonia until the precipitate just dissolves; then add 2 ml of excess ammonia. In a glass ampoule, weigh a sample containing about 0.0008 mole of acetylenic hydrogen (see pp. 862–4) and place it in the Erlenmeyer flask, together with the ammoniacal silver nitrate. Add some glass beads to aid in breaking the ampoule. The stopper should be greased if the sample is very volatile. Stopper the flask and shake to break the ampoule. Continue shaking for 3 to 5 minutes to assure complete reaction. Open the flask and quantitatively transfer the contents to a 50-ml centrifuge tube. Rinse the flask three times with 3-ml portions of 1N ammonium hydroxide and add the washings to the centrifuge tube. Not all the precipitate need be removed from the flask. Then centrifuge the mixture, and carefully decant the supernatant liquid into a 250-ml Erlenmeyer flask, taking care not to get any of the precipitate into the flask. If particles of the precipitate persist in remaining at the surface of the liquid, the liquid can be decanted into the flask through a funnel containing a plug of glass wool. The precipitate in the funnel should not be allowed to dry for the reason given in Cautions (p. 494). Wash the contents of the centrifuge tube three times with about 10-ml portions of 1N ammonium hydroxide, centrifuging each time. Add the washings to the flask with the filtrate, taking care not to get any particles of precipitate into the filtrate.

Then carefully neutralize the filtrate with concentrated nitric acid and add a few milliliters of excess nitric acid. Titrate the solution with standard 0.1N ammonium thiocyanate, using ferric alum indicator.

CALCULATIONS

Milliliters of NH_4CNS for $AgNO_3$ alone minus milliliters NH_4CNS for sample $= A$

$$\frac{A \times N\ NH_4CNS \times \text{Mol. wt. of compound} \times 100}{\text{Grams of sample} \times 1000 \times B} = \%\ \text{acetylenic compound}$$

where B is the number of acetylenic hydrogen atoms per molecule plus 1 (see equation, p. 489).

The foregoing procedure was tested with propargyl alcohol and solutions of acetylene in methyl vinyl ether. Aldehydes or other materials that reduce silver ion will interfere. Cyanides will interfere because of complex ion formation, and sulfides will interfere because of formation of the insoluble silver sulfides.

Gaseous Samples

In gas samples in which acetylene or any gaseous acetylenic compound is to be determined, the method is essentially the same as the silver method just described except that a suitable apparatus is needed. In most cases, an ordinary Orsat gas analysis apparatus can be used for these analyses. The method described here, however, has a much higher degree of accuracy and is capable of determining much smaller amounts of acetylenic compound than the Orsat apparatus. This procedure is accurate and precise to ±0.5%.

REAGENTS

Aqueous standard silver nitrate, 0.1*N*.
Standard ammonium thiocyanate, 0.05*N*.
Concentrated ammonium hydroxide.
Sodium hydroxide, 6*N*.
Concentrated nitric acid.
Saturated ferric alum indicator.

APPARATUS

Gaseous samples are analyzed using the apparatus depicted in Fig. 9.1.

PROCEDURE

Introduce the gas sample into the 100-ml gas buret *A* through *B*. In a 50-ml volumetric flask, place 25 ml of 0.1*N* standard silver nitrate. To this add 3 ml of approximately 6*N* sodium hydroxide, followed by concentrated ammonium hydroxide, until the silver oxide precipitate just dissolves. Then dilute the solution to the mark. Into receiver *E* pipet 25 ml of this solution. Receiver *E*, a tube of about 1.5 cm diameter with a plug of glass wool at the bottom is packed two-thirds full with 6-mm, glass beads.

Allow the gas sample to stand for about 10 minutes in the gas buret for thermal equilibrium to be established; then read the volume and the

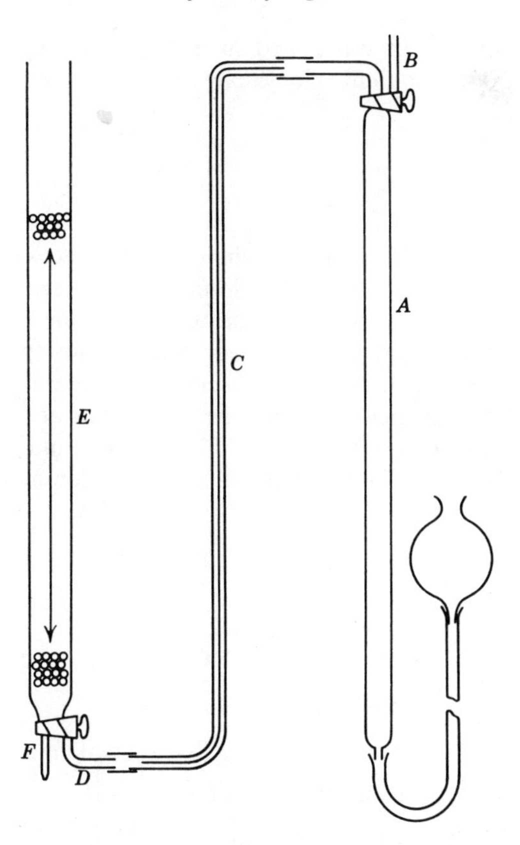

Fig. 9.1. Acetylenic hydrogen apparatus for gaseous samples.

temperature. Raise the leveling bulb, putting the sample under about 10 cm pressure, and turn the stopcock on the gas buret to allow the gas in the buret to escape through *C*. Then open the stopcock on the receiver to permit the gas into the receiver. The rate of flow of sample should be about one bubble per second. Raise the mercury level as the gas flows out of the gas buret. Force the mercury over through *C* and *D* until it just appears in the receiver. This is to ensure sweeping all the gas sample into the receiver. Drain out the contents of the receiver by means of *F* into a 250-ml Erlenmeyer flask through a funnel with a glass wool plug to catch any silver acetylide particles that might come out of the receiver. There should be no particles of the silver acetylide in the Erlenmeyer flask. The contents of the receiver should be transferred quantitatively by rinsing with water and adding the washings to the Erlenmeyer flask through the funnel. Acidify the contents of the flask with concentrated nitric acid and add a few milliliters of nitric acid in excess. Titrate the solution with $0.05N$ standard ammonium thiocyanate solution, ferric alum indicator.

CAUTION

1. Silver acetylides are explosive when dry. All pieces of apparatus that contain silver acetylides should be rinsed in nitric acid, to dissolve the acetylides, then rinsed with water. The receiver in the foregoing analysis should be cleaned in the same manner; then, if the apparatus is not in use, fill the receiver with dilute nitric acid in case there are any undissolved silver acetylide particles between the glass beads or on the glass wool.
2. Ammoniacal silver nitrate, when allowed to stand, can form explosive compounds. Therefore dispose of all excess ammoniacal silver nitrate.

CALCULATIONS

Milliliters of $AgNO_3$ (ammon.) $\times N$ $AgNO_3$ (ammon.) $= A \times N$ NH_4CNS

A minus milliliters of NH_4CNS used in titration

= milliliters NH_4CNS equivalent to silver consumed by acetylenic compound $= B$

$$\frac{B \times N\,\mathrm{NH_4CNS} \times \text{mol. wt. of compound}}{1000C} = \text{grams of acetylenic compound} = D$$

where C is the number of silver atoms added to the acetylenic compound plus 1 (see equation, p. 489).

In the case of solid or liquid samples,

$$\frac{D \times 100}{\text{Grams of sample}} = \%\ \text{acetylenic compound}$$

To obtain the percentage of acetylenic compound in a gaseous sample, the density of the sample should be known at a certain temperature, in which case the volume of sample is converted to the volume at the temperature at which the density is known, by means of the equation $PV/T = P_1V_1/T_1$. Then, if the density is known, the volume of sample is converted to grams of sample, and this value is substituted in the foregoing equation to calculate percentage of acetylenic compound.

If the density of the sample is not known, the result can be obtained in terms of volume percentage. The grams of acetylenic compound *D* is converted by using the density (obtainable in chemical or engineering handbook) and

$$\frac{\text{Volume of acetylenic compound} \times 100}{\text{Volume of sample}}$$

$$= \% \text{ acetylenic compound by volume}$$

When tested with gaseous mixtures of acetylene, air, and carbon dioxide, this procedure was found to be accurate to ±0.5%. The interfering substances in this procedure are the same as those in the preceding one.

Method B

The next procedure is of service when aldehydes are present with the acetylenic compound. In the two preceding methods aldehydes interfere seriously because of their reducing action on silver ions. This procedure employs sodium acetate instead of ammonia to remove the nitric acid formed by the reaction of the silver nitrate with the acetylenic compound. The aldehydes reduce the silver ion much more slowly when sodium acetate is used as the basic medium instead of ammonium hydroxide.

REAGENTS

Standard 0.1*N* silver nitrate.
Standard 0.1*N* ammonium thiocyanate.
Sodium acetate solution 2%.
Ferric alum indicator.
Concentrated nitric acid.

PROCEDURE

Into a 250-ml Erlenmeyer flask, introduce 50 ml of 0.1*N* silver nitrate. Add to this 50 ml of 2% sodium acetate solution. Cool the solution in ice and add the sample, which should contain about 0.002 equivalent of acetylenic compound. Filter the resultant precipitate with suction by using a sintered-glass funnel, and thoroughly wash it with a minimum of chilled 0.01% sodium acetate. The precipitate should be covered with liquid at all times, since it decomposes when exposed to air.

Acidify the filtrate with 5 to 6 ml of concentrated nitric acid and titrate with $0.1N$ standard ammonium thiocyanate, ferric alum indicator being used.

Large amounts of aldehydes will reduce the silver ions relatively rapidly and consequently will interfere. However this difficulty can be surmounted by cooling the silver nitrate–sodium acetate solution thoroughly before adding the sample. Then, after the sample has been added, filter off the resultant precipitate immediately into a filtration flask containing 5 to 6 ml of concentrated nitric acid. In this way, the filtrate is acidified as soon as it passes through the funnel, and any reaction between the silver ions and the aldehydes is prevented. The filter should be as coarse as possible, to hasten the filtration. The reaction between the aldehyde and the silver ions is sufficiently slow, especially when the solutions are cold, that the filtration of the silver salt of the acetylenic compound can be achieved before any significant interference of the aldehyde can be noted.

CAUTION

See previous procedures.

CALCULATIONS

Most acetylenic compounds consume one more atom of silver than the number of acetylenic hydrogen atoms:

$$HC\equiv CR + 2AgNO_3 \rightarrow AgC\equiv CR\cdot AgNO_3 + HNO_3$$

$$\text{Milliliters of } NH_4CNS \text{ equivalent to 50 ml of } AgNO_3 = A$$

$$A \text{ minus milliliters } NH_4CNS \text{ used to titrate sample} = B$$

$$\frac{B \times N\ NH_4CNS \times \text{Mol. wt. of compound} \times 100}{\text{Grams of sample} \times 1000 \times C} = \%\text{ acetylenic compound}$$

$$C = \text{number of acetylenic hydrogen atoms plus 1}$$

Method C

This method involves the reaction of the acetylenic compound with silver nitrate, the solution being alcoholic to deal with water-insoluble samples. The acid liberated by the reaction is titrated with standard alkali:

$$HC{\equiv}CR + 2AgNO_3 \rightarrow AgC{\equiv}CR \cdot AgNO_3 + HNO_3$$

Method of the Texas Company

(*As described by V. J. Altieri, Gas Analysis and Testing of Gaseous Materials, American Gas Association, New York, 1945, pp. 330–2. Adapted with Permission of the American Gas Association, Mr. Altieri, and the Texas Company*)

REAGENTS

Sodium hydroxide, 0.02*N* for samples containing less than 1% acetylenic compound.

Sodium hydroxide, 0.1*N*, for samples containing more than 1% acetylenic compound.

SILVER NITRATE SOLUTION. Dissolve 100 grams of silver nitrate in water and dilute to 1 liter (store in dark bottle). For the working solution, dilute 35 ml of the aqueous silver nitrate solution to 140 ml with 95% ethanol.

INDICATOR. Alcoholic solution composed of a mixture of 0.1% methyl red and 0.05% methylene blue. Store in a dark dropping bottle.

PROCEDURE

In a 250-ml Erlenmeyer flask, place 50 ml of silver nitrate solution together with the sample, which should contain 0.002 equivalent of acetylenic hydrogen if the 0.1*N* sodium hydroxide is to be used in the final titration, or 0.0004 equivalent of acetylenic compound if the 0.02*N* sodium hydroxide is to be used.

Add about 6 drops of indicator and titrate the contents of the flask with the sodium hydroxide solution of appropriate strength. The end point is the change of the red color of the indicator to a dull yellow.

For gaseous samples, the procedure described on pages 492–95 can be used, except that the silver nitrate can be put in the receiver and the ultimate analysis made as just described instead of determining excess silver.

This procedure is accurate and precise to ±0.5 to 1.0%. It was tested with 3-butyne-1-ol, ethynyl-cyclohexanol, 1-hexyne, and 3-butyne-2-ol. Any acidic or basic materials in the sample will interfere; they should be determined on a separate sample, and the proper correction applied to the acetylenic hydrogen result. Aldehydes or other reducing substances interfere by reducing the silver ions to metallic silver, which darkens the solution so much that the end point is completely obscured. Also, the acids formed on the oxidation of the aldehyde would interfere even if the metallic silver formed did not obscure the end point.

CAUTION

Silver salts of acetylenic compounds are explosive when dry. Destroy these salts by dissolving in dilute nitric acid, or by making the solution ammoniacal and dissolving the silver acetylide with 5% potassium cyanide (Hood). The resulting solution is poured into 5% ferrous sulfate to destroy the excess cyanide.

Method D—Adapted from L. Barnes, Jr., and L. J. Molinini

[*Anal. Chem.*, **27**, *1025–7* (*1955*)]

REAGENTS

Silver nitrate, 2.0 to 3.5*M* aqueous solution.

Silver perchlorate, 2.0 to 3.5*M* aqueous solution. Anhydrous or hydrated silver perchlorate is obtainable from G. Frederick Smith Chemical Company, Columbus, Ohio.

Sodium hydroxide, 0.1*N*, carbonate-free.

Methyl purple indicator solution. Obtainable from Fleisher Chemical Company, Benjamin Franklin Station, Washington, D.C.

PROCEDURE

LIQUID AND SOLID SAMPLES. Place 40 ml of silver nitrate or silver perchlorate reagent solution in a 250-ml Erlenmeyer flask and add 3 or 4 drops (amount critical) of methyl purple indicator. Neutralize with 0.1*N* base or acid. Add a weighed amount of sample equivalent to 2.0 to 3.5 meq. of acetylenic hydrogen and titrate the liberated acid with 0.1*N* base to the green color of the indicator as viewed by transmitted light. Acidic or basic impurities and/or functional groups, if present, may be determined separately and appropriate corrections applied.

GASEOUS SAMPLES. Bring a known volume of the gas into intimate contact with a suitable quantity of the concentrated silver reagent and titrate the liberated acid with standard base.

Equipment and manipulative technique for handling gas samples may take any one of many forms. However the following apparatus and procedure have proved to be eminently satisfactory. The manifold illustrated in Fig. 9.2 provides means for the transfer of gas from a cylinder under pressure directly into calibrated sample tubes convenient for analysis.

With the main cylinder valve closed but the needle valve open, and the gas sample tubes open to the manifold, evacuate the system. Close the large stopcock *D*, thus isolating the system from the vacuum pump. If the

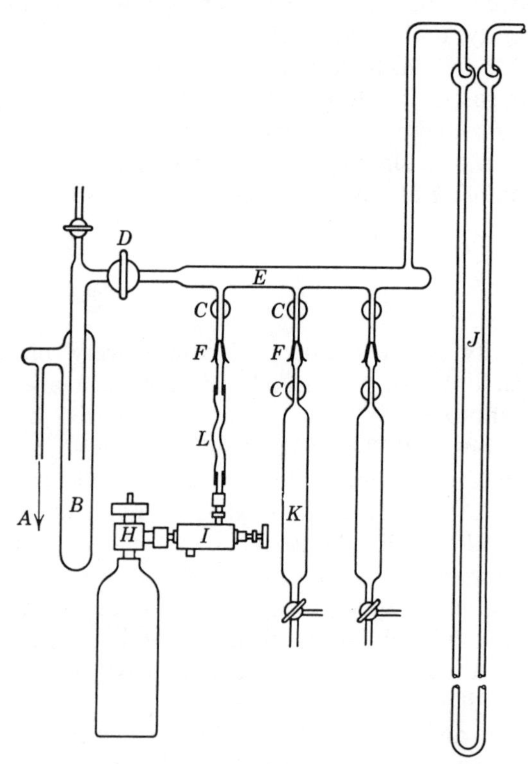

Fig. 9.2. Gas-sampling manifold: *A*, to vacuum pump; *B*, trap (4×31 cm); *C*, stopcock, straight bore, 4 mm; *D*, stopcock, straight bore, 10 mm; *E*, manifold (3×40 cm); *F*, joint, standard taper 12/38; *C*, three-way stopcock, 2 mm; *H*, main cylinder valve; *I*, needle valve; *J*, open-end mercury manometer; *K*, gas sample tube, approximately 100-ml capacity (3.3×18 cm), *L*, flexible pressure tubing.

system is free from leaks, as indicated by a steady manometer reading, close the needle valve and open the main cylinder valve. Open the needle valve slowly, and fill the sample tubes to slightly above atmospheric pressure. Close the stopcock on both gas sample tubes and remove the tubes from the manifold. Mount the tubes vertically on an apparatus support stand, allow 5 minutes for temperature equilibration, and open the stopcocks momentarily to adjust sample to atmospheric pressure. Record temperature and barometric pressure. By means of rubber tubing, connect a leveling bulb to the lower (three-way) stopcock of the sample tube. Place a suitable quantity of the concentrated silver reagent solution in the leveling bulb and displace the air from the rubber tubing. Open the three-way stopcock and allow the reagent solution to enter the sample tube. Shake gently, drain, and wash contents into a suitable vessel, and titrate with $0.1N$ sodium hydroxide. The volume of distilled water used for washing the gas sample tube should be held at a minimum to prevent dilution, hence precluding acetylide precipitation.

By modification of sampling technique, the apparatus (Fig. 9.2) may be used for the sampling of gases under pressures less than atmospheric.

RESULTS

The applicability of the method to various acetylenic compounds is indicated in Table 1. All compounds listed were analyzed as received without purification. Some of the compounds tested had very limited water solubility, but, with the exception of 3-methyl-1-nonyn-3-ol, when added directly to the concentrated silver solution, all reacted rapidly and quantitatively to produce a clear, single-phase system. When the latter compound was added directly to the concentrated silver solution, an acetylide precipitated that would not dissolve completely, and analytical results were erratic. Upon weighing the compound into 5 to 10 ml of 95% ethyl alcohol, then adding the silver solution to the alcoholic one, no precipitation of acetylides was observed and analytical results were reproducible and quantitative.

The acetylenic amines listed in Table 1, which formed insoluble acetylides, were titrated (without preneutralization of the amino group) with the aid of a pH meter equipped with standard glass and external calomel electrodes connected by a sodium nitrate salt bridge. Their inclusion in Table 1 demonstrates the versatility of the method.

The effects of halogens and aldehydes are shown in Table 2. Halogens, in the amounts present, did not interfere. The visual end point was readily observed after the precipitated silver halide had settled. Samples containing halogens have also been analyzed by filtering off the precipitated

Table 1. Application of Method to Various Compounds

Compound[a]	Purity, %	
	$AgNO_3$ Reagent	$AgClO_4$ Reagent
Alcohols		
1-Phenyl-2-propyn-1-ol	99.4	99.2
2-Propyn-1-ol[b]	98.5	98.8
2-Methyl-3-butyn-2-ol	99.0, 99.2	—
3-Methyl-1-pentyn-3-ol	99.3, 99.5	99.5
3,4-Dimethyl-1-pentyn-3-ol	99.1	99.5
3-Ethyl-1-pentyn-3-ol	98.5, 98.5	98.5
1-Ethynylcyclopentanol	97.4, 97.7	97.9
3-Methyl-1-hexyn-3-ol	98.2, 98.7	98.8
1-Ethynylcyclohexanol	97.5, 97.9	97.8
2,6-Dimethyl-4-ethynyl-heptan-4-ol	98.0, 98.0	—
3-Methyl-1-nonyn-3-ol[c]	96.3, 96.3	96.5, 96.6
Hydrocarbons		
Acetylene[d]	99.82 ± 0.32	99.85 ± 0.15
1-Propyne	99.3, 99.4	—
1-Butyne[e]	98.9, 98.4	—
2-Methylbuten-3-yne	99.8, 99.6	—
1-Hexyne[f]	96.1, 96.3	—
Carboxylic Acid		
Propiolic acid[g]	98.0, 98.4	—
Amines		
3-Aminopropyne[h]	99.5	—
Di(2-propynyl)amine[i]	99.4	—
Miscellaneous		
Sodium acetylide[j]	97.4, 97.2	—

[a] Technical grade samples of reasonable purity; no attempts made at further purification.

[b] Boiling range 113 to 115°C.

[c] 3.2% Ketone present by hydroxylamine hydrochloride analysis.

[d] Average analysis of 14 samples ($AgNO_3$) and 4 samples ($AgClO_4$) from same cylinder of acetylene on acetone-free basis. Hempel absorption analysis of 24 samples from same cylinder, using bromine solution as absorbent, indicates 0.38 ±0.08% nonacetylenics (i.e., 99.62% acetylene).

[e] 1.3% Nonacetylenics present by Orsat analysis.

[f] 95.8% Assay by potassium mercuric iodide [J. G. Hanna and S. Siggia, *Anal. Chem.*, **21,** 1469 (1949)]; see p. 509, this book.

[g] 98.5% Assay by titration of carboxylic acid group.

[h] 98.6% Assay by nonaqueous titration of amino group. [C. W. Pifer and E. G. Wollish, *Anal. Chem.*, **24,** 300 (1952)].

[i] 99.3% Assay by nonaqueous titration of amino group.

[j] 2.1% Nonacetylenics present.

Table 2. Effects of Halogens and Aldehydes on Determination of Acetylenic Hydrogen

Compound	Taken, meq.	Found, meq.	Halogen or Aldehyde		Reagent	Error, %
			Present	meq.		
3-Methyl-1-pentyn-3-ol	3.180	3.180	Iodide	6.0	$AgNO_3$	0.0
	3.390	3.390	Chloride	3.0	$AgNO_3$	0.0
	2.667	2.671	Bromide	3.0	$AgNO_3$	+0.1
	1.542	1.543	Butyraldehyde	0.544	$AgNO_3$	+0.1
	2.358	2.358	Butyraldehyde	0.833	$AgClO_4$	0.0
2-Methyl-3-butyn-2-ol	2.654	2.640	Formaldehyde	0.698	$AgNO_3$	−0.5
	2.730	2.724	Formaldehyde	0.718	$AgClO_4$	−0.2
1-Hexyne	3.313	3.300	Salicylaldehyde	0.688	$AgNO_3$	−0.4
	3.430	3.414	Salicylaldehyde	0.713	$AgNO_3$	−0.5

silver halide through a coarse paper, followed by titration of the liberated acid in the filtrate. Although the metathesis of silver acetylides by halides has been reported (2), no such interference would likely be involved here, since the relatively high concentration of silver ion precludes the possibility of excess (free) halide ions. Presumably the aldehydes do not interfere because the pH of the solution is never maintained at a high value long enough to reduce silver ion. Apparently local excess concentrations of base are without effect in this respect.

DISCUSSION

Certain distinct advantages are associated with the use of concentrated silver solutions for the determination of acetylenic hydrogen. The general absence of precipitated acetylides facilitates detection of the end point and is particularly desirable in the analysis of gaseous samples, where precipitates are likely to obstruct volumetric apparatus, clog stopcocks, and so on. The limited number of reagents required, coupled with the fact that filtration is unnecessary, makes the method extremely rapid. In addition, certain compounds of very limited water solubility, such as 1-butyne and 1-hexyne, may be analyzed directly without prior addition of a solubilizing agent.

The principal objection to the method lies in the high cost of the silver reagent solutions. However careful selection of compounds to be analyzed and reclamation of waste silver solutions help to reduce the cost. The concentration of silver ion necessary to prevent acetylide precipita-

2. R. F. Robey, B. E. Hudson, Jr., and H. K. Wiese, *Anal. Chem.*, **24,** 1080–2 (1952).

tion is much less for some compounds than for others. 2-Methyl-3-butyn-2-ol and 3-methyl-1-pentyn-3-ol are analyzed routinely by the addition of 1.5 meq. of sample to 100 ml of 0.3*M* aqueous silver nitrate and subsequent titration of the liberated acidity with 0.05*N* alkali. Under the described conditions no acetylide precipitation is observed. On the other hand, most hydrocarbons, such as acetylene and 1-hexyne, whose acetylides would precipitate from dilute silver solutions, are analyzed utilizing concentrated (2.0–3.5*M*) solutions of silver nitrate or silver perchlorate as reagents.

The amount of methyl purple indicator added to the solution to be titrated is critical: 3 or 4 drops from the conventional dropping-type indicator bottle is sufficient when silver nitrate is used; larger amounts of indicator tend to obscure the end point. When silver perchlorate is employed, 4 or 5 drops of indicator may be used. For many of the compounds tested, a more distinct end point was obtained with silver perchlorate than with silver nitrate. The effects of the high salt concentration in the proposed concentrated silver solutions on the indicator appear to have a negligible influence on the accuracy of the method. The selection of appropriate indicators is limited to those which change color in the pH range of 4 to 6. At higher pH values, hydrated silver oxide starts to precipitate. Bromocresol purple, bromocresol green, methyl red, and methyl red–methylene blue indicators have been utilized, however methyl purple is preferred by most analysts.

The concentration of sodium hydroxide used to titrate the liberated acidity is not extremely critical. Normalities between 0.05 and 0.2 have proved satisfactory, although 0.1*N* seems to be ideal for most compounds. More concentrated solutions—that is, 0.5*N*—set up local excesses of base so large that the locally precipitated silver oxide redissolves with considerable difficulty. This lengthens the time required for the analysis and also produces a condition that may be more favorable for interference by aldehydes. Base solutions that are too dilute do not produce a sharp color change at the end point and may cause some acetylide precipitation.

Method E—Adapted from the Method of L. Barnes, Jr.

[*Anal. Chem.*, ***31,*** *405* (*1959*)]

REAGENTS

SILVER PERCHLORATE, 1*M*. Dissolve 104 grams of anhydrous silver perchlorate in anhydrous methanol and dilute to 500 ml. Store in a polyethylene bottle.

STANDARD 0.1*N* TRIS(HYDROXYMETHYL)AMINOETHANE (THAM). Dissolve 12.15 grams of THAM in methanol with mechanical stirring and dilute to 1 liter. Filter any remaining insolubles. To standardize, dilute 40 ml of this solution with 200 ml of water and titrate with aqueous acid to the methyl purple end point.
SCREENED THYMOL BLUE INDICATOR SOLUTION. Dissolve 0.100 gram of thymol blue and 0.025 gram of alphazurine in 100 ml of methanol. Store in a brown bottle. Prepare fresh every week, since the solution decomposes on standing. Alphazurine is also known as Alphazurine A or Patent Blue A.
STANDARD 0.1*N* PERCHLORIC ACID. Dissolve 8.5 ml of 72% perchloric acid in 1 liter of methanol. Standardize against THAM using screened thymol blue as indicator.

PROCEDURE

Add 3 drops of screened thymol blue indicator to 10 ml of 1*M* methanolic silver perchlorate contained in a 50-ml beaker. Neutralize any free perchloric acid with 0.1*N* THAM to the green of the indicator. Weigh the sample containing 1 to 3 meq. of acetylenic hydrogen and place in a 250-ml Erlenmeyer flask containing 5 to 10 ml of neutral methanol and 5 to 10 drops of indicator. Pour the neutralized silver perchlorate into the sample flask and titrate the liberated perchloric acid to a permanent green end point with 0.1*N* THAM.

For colored samples, make a potentiometric titration with a glass-indicating electrode–double-junction reference electrode system. Replace the solution in the double-junction electrode with 3*M* sodium perchlorate in 2:1 methanol-water solution.

If the sample contains a strong acid (indicator purple), neutralize with 0.1*N* THAM to the green of the indicator prior to the addition of silver perchlorate. If strong bases such as alkali hydroxides are present (indicator blue), neutralize with 0.1*N* perchloric acid to the purple of the indicator before adding the silver perchlorate. If amines with $K_B \gtrless 1\times10^{-6}$ are present, the indicator will remain green and should be neutralized with perchloric acid prior to the addition of silver perchlorate. Sulfuric acid is a special interference; THAM neutralizes only one of its hydrogens, the other is released by silver perchlorate. Also, sulfuric acid can esterify methanol rapidly with loss of hydrogen ions. Therefore sulfuric acid should be precipitated by the addition of an excess of neutral barium perchlorate dissolved in methanol. The liberated perchloric acid is then neutralized, and the addition of silver perchlorate and titration with THAM are performed without filtration of the precipitated barium sulfate.

DISCUSSION AND RESULTS

The method is particularly useful for water-insoluble samples. Because the strong bases normally used as titrants readily saponify many esters, the weak amine THAM was used for this method. Acetylide precipitation rarely occurs in this system. Water causes low results, but for practical purposes, 0.5 to 1.0 gram can be present without serious effect. There are no interferences from halogens or aldehydes. Table 3 gives results obtained by this method.

Table 3. Application to Various Compounds

	Purity, %	
Compound	Nonaqueous Silver Perchlorate	Aqueous Silver Nitrate[a]
2-Propyn-1-ol	98.1	98.0
2-Methyl-3-butyn-2-ol	99.0	98.9
3-Phenyl-1-butyn-3-ol	99.7	99.1
3-Methyl-1-pentyn-3-ol	98.8	99.0
3-Methyl-1-hexyn-3-ol	100.3	—
3,5-Dimethyl-1-hexyn-3-ol	98.1	98.6
2-Ethyl-1-heptyn-3-ol	100.1	99.5
3-Ethyl-5-methyl-1-heptyn-3-ol	98.0	98.3
3,6-Dimethyl-3-hydroxy-1-heptyne	99.5	100.0
4-Ethyl-1-octyn-2-ol	100.3	99.7
3-Methyl-1-nonyn-3-ol	100.1	—
Acetylene in acetone	1.19	1.18
2-Methyl-1-butene-3-yne	99.7, 99.1	98.5
1-Pentyne	84.9	85.3
1-Hexyne	97.5, 97.9	98.6
1,6-Heptadiyne	92.6, 92.7	93.3

[a] Method D (Barnes and Molini, p. 498).

Method F—Adapted from the Method of M. Miocque and J. A. Gautier

[*Bull. Soc. Chim. Fr., 467* (*1958*)]

REAGENTS

SILVER NITRATE IN PYRIDINE. Dissolve 17 grams of silver nitrate in pyridine and make to 100 ml with pyridine.

Table 4. Results by the Silver Nitrate in Pyridine Method

Compound	Found, %: Silver Nitrate in Pyridine	Found, %: Method of Siggia[a]
Propargyl alcohol	*b*	
1-Pentyn-3-ol	91.4	91.3
3-Methyl-1-pentyn-3-ol	99.2	98.8
3-Methyl-1-butyn-3-ol	95.5	95.5
Ethynylcyclohexanol	99.8	99.1
Phenylethynylcarbinol	*b*	
3-Methyl-1-pentyn-3-ol-carbamate	100.0	99.6
3-Methyl-1-butyn-3-ol-*p*-nitrobenzoate	99.4	101–105
3-Methyl-1-pentyn-3-ol-*p*-nitrobenzoate	99.3	109–112
3-Methyl-1-pentyn-3-ol-hydrogen phthalate	99.5	
Propargyl bromide	*b*	
6-Chloro-1-hexyne	99	*b*
11-Chloro-1-undecyne	89.5	
Dimethylpropargylamine	97.3	*b*
5-(2-Pyridyl)-1-hexyne	89	*b*
5-(2-Pyridyl)-1-hexyne picrate	97.5	*b*
7-(2-Pyridyl)-1-heptyne	93.4	*b*
7-(2-Pyridyl)-1-heptyne hydrochloride	98.5	*b*
7-(2-Pyridyl)-1-heptyne picrate	98.7	*b*
12-(2-Pyridyl)-1-dodecyne	92	*b*
12-(2-Pyridyl)-1-dodecyne hydrochloride	97	*b*
12-(2-Pyridyl)-1-dodecyne picrate	96.3	*b*
4-(4-Pyridyl)-1-butyne	97.5	*b*
4-(4-Pyridyl)-1-butyne picrate	98.7	*b*
12-(4-Pyridyl)-1-dodecyne	91	*b*
12-(4-Pyridyl)-1-dodecyne picrate	95.5	*b*
7-(4-Pyridyl)-1-heptyne	97	*b*
7-(4-Pyridyl)-1-heptyne picrate	96.3	*b*

[a] See page 490.
[b] Determination unsuccessful.

Thymolphthalein indicator, 1% in absolute alcohol.
Sodium methoxide in methanol, standard 0.1*N* solution.

PROCEDURE

Neutralize about 5 ml of the silver nitrate solution to a pale blue color of the indicator with 0.1*N* sodium methoxide solution. Add the sample, which causes decoloration of the solution. Titrate with standard 0.1*N* sodium methoxide solution to the return of the pale blue color. Occasionally a yellow color appears during the titration, but this usually does not affect the sharpness of the end point, although the final color is then green instead of blue.

RESULTS AND DISCUSSION

The scope and accuracy of the method are suggested by the data in Table 4.

The method fails for propargyl alcohol, propargyl bromide, and phenylethynylcarbinol because a brown coloration develops rapidly when these substances are present.

Cuprous Method

The second principal method involves reaction of cuprous chloride with the acetylenic compound in a pyridine solution. The hydrochloric acid liberated is titrated with standard alkali.

$$Cu_2Cl_2 + 2HC{\equiv}C{-}R \xrightarrow{\text{pyridine}} 2CuC{\equiv}C{-}R + 2HCl$$

Pyridine is a weak base that combines with the liberated hydrochloric acid so that the reaction will proceed to completion. The hydrochloric acid can be titrated with sodium hydroxide, a stronger base than pyridine.

This method is less accurate and precise than the silver methods just described; the reproducibility of this procedure is ±2%. The method can be used, however, for samples containing impurities that interfere with the silver methods.

REAGENTS

CUPROUS CHLORIDE, APPROXIMATELY 0.1*N*. In a solution of 150 ml of pyridine and 150 ml of water, dissolve 10 grams of cuprous chloride. Dilute the resulting solution to 1 liter with distilled water. Fresh solutions should be prepared each day because of the instability of the reagent toward oxygen.

Standard sodium hydroxide, 0.5*N*.

PROCEDURE

Dissolve a sample containing about 0.005 equivalent of acetylenic compound in a minimum of water (pyridine can be used if the sample is insoluble in water), and add 100 ml of cuprous reagent to the sample. A precipitate of the cuprous derivative of the acetylenic compound will result. Then titrate the solution with 0.5*N* sodium hydroxide by using a pH meter with the standard calomel and glass electrodes. An indicator cannot be used because of the intensity of color of the solution and the presence of the colored precipitate. A plot of pH versus milliliters of reagent will reveal the end point accurately. However the pH change at the end point is large enough to be visible from the pH readings recorded; a plot is necessary only for higher accuracy.

CALCULATIONS

$$\frac{\text{Milliliters of NaOH} \times N\ \text{NaOH} \times \text{mol. wt. of compound} \times 100}{\text{Grams of sample} \times 1000 \times A} = \%\ \text{acetylenic compound}$$

where A is the number of cuprous ions that add to the acetylenic compound.

This procedure was tested by using propargyl alcohol. Interfering substances consist of any acidic or alkaline material that may be present. Aldehydes do not interfere with this method as they do with the silver methods A and B.

Mercuric Method

The disadvantages of using silver ion are mainly attributable to the interference of any halides, cyanides, sulfides, and traces (down to 0.01%) of aldehydes in the sample. These interferences consume silver ion, halides, and sulfides by precipitation of the silver salts, the cyanides by complex ion formation, and the aldehydes by reduction of the silver ion to metallic silver. Traces of aldehydes affect the determination of the nitric acid liberated by the silver nitrate, for the metallic silver formed obscures the end point completely. Small amounts of aldehydes (up to 0.5% as formaldehyde) can be tolerated in the procedure below. Because aldehydes are oxidized by the potassium mercuric iodide, however, amounts larger than 0.5% will begin to affect the acetylenic hydrogen results.

For qualitative identification purposes, Shriner and Fuson (3) recommend the use of potassium mercuric iodide to form the mercuric derivative of those compounds which contain the grouping —C≡CH, according to the formula

$$2RC{\equiv}CH + K_2HgI_4 + 2KOH \rightarrow (R{-}C{\equiv}C{-})_2Hg + 4KI + 2H_2O$$

The procedure described in this section uses this reaction for the quantitative estimation of acetylenic compounds by the determination of the hydroxide consumed.

Adapted from the Procedure of Hanna and Siggia

[*Anal. Chem.*, **21,** *1469* (*1949*)]

REAGENTS

POTASSIUM MERCURIC IODIDE REAGENT. Dissolve 50 grams of mercuric iodide in 250 ml of 20% potassium iodide solution.

CP Methyl alcohol.

Standard 0.5*N* sulfuric acid solution.

Standard 0.5*N* sodium hydroxide solution.

PROCEDURE

Add 50 ml of the potassium mercuric iodide reagent to 100 ml of CP methyl alcohol in a 250-ml Erlenmeyer flask. Add the sample containing 0.010 to 0.015 mole of acetylenic hydrogen to the flask, and pipet 50 ml of standard 0.5*N* sodium hydroxide solution into the mixture. Titrate the excess sodium hydroxide immediately to the phenolphthalein end point with standard 0.5*N* sulfuric acid.

Used to test this procedure were propargyl alcohol, 1-hexyne, 3-butyn-1-ol, and acetylene.

3. R. L. Shriner and R. C. Fuson, *Systematic Identification of Organic Compounds*, Wiley, New York, 1948, p. 200.

10

Acetals, Ketals, and Vinyl Ethers

$$\mathrm{R{-}CH(OR')(OR'')},\quad \mathrm{R{-}C(OR')(OR''){-}R'''},\quad \text{and}\quad \mathrm{ROCH{=}CH_2}$$

Acetals, ketals, and vinyl ethers are grouped together because they are all forms of combined carbonyl compounds. The acetals and vinyl ethers are combined forms of aldehydes, and the ketals are combined forms of ketones. Most of the methods of analysis of these materials are based on hydrolysis of these compounds back to the parent carbonyl compound and determination of this carbonyl compound. There are specific reactions of the vinyl ethers that are not common to acetals or ketals (Methods, pp. 515–20).

Hydroxylamine Hydrochloride Method for Acetals, Ketals, and Vinyl Ethers

The original source of this very generally used procedure cannot be traced. The method is similar to those described earlier (pp. 95–100), except that water is used as the solvent. The procedure involves hydrolysis of acetals, vinyl ethers, or ketals, with hydroxylamine hydrochloride used as the acid catalyst and also as the agent to remove the resulting aldehyde or ketone by forming the oxime; this pulls the hydrolysis reaction to completion. The liberated hydrochloric acid (see equation, p. 95) is titrated with standard alkali.

REAGENTS

Aqueous hydroxylamine hydrochloride solution, $0.5M$. The pH of the solution should be adjusted to 3.1 with diluted alkali solution, a pH meter with the standard glass and calomel electrodes being used.

Standard $0.5N$ sodium hydroxide.

PROCEDURE

Weigh a sample containing about 0.012 mole of acetal, vinyl ether, or ketal into a 250-ml beaker or flask containing 50 ml of hydroxylamine reagent.

For water-insoluble samples, methanol can be added to the reagent to bring about solution. However this should be kept at a minimum because of the slowing effect it has on the hydrolysis.

Allow the mixture to stand until the reaction is complete. Some compounds react completely in 30 minutes at room temperature. Others hydrolyze so slowly that the reaction mixture has to be refluxed for some hours. Most acetals and vinyl ethers will hydrolyze in less than 2 hours at room temperature.

Ketals usually take a longer time. Chlorodimethylacetal requires 4 hours of reflux.

After hydrolysis, titrate the reaction mixture with 0.5*N* sodium hydroxide, using a pH meter, with the standard calomel and glass electrodes. The end point can be detected from the recorded pH and milliliter data. If the break is too gradual, however, a curve of pH versus milliliters should be plotted. The pH break at the end point is too gradual for indicators to be used.

If much methanol has been added to dissolve the sample, at least an equal volume of water should be added before the titration is begun. This produces a sharper pH change at the endpoint.

If the sample is acidic or basic, bring it to pH 3.1 before adding it to the hydroxylamine reagent. Use pH 3.1 because the end point break is obtained in the vicinity of this pH.

Used to test this procedure were dimethylacetal, dibutylacetal, chlorodimethylacetal, dihydroxyethylacetal; methyl, ethyl, *n*-propyl, *n*-butyl, and isobutyl vinyl ethers, as well as 2,2-dimethoxy-1,4-dihydroxybutane.

The procedure is usually accurate to better than ±1%.

Bisulfite Method for Acetals and Vinyl Ethers

The analysis described below involves the hydrolysis of the acetal and the determination of the aldehyde formed by the bisulfite method for aldehydes as described previously (pp. 100–107).

$$RCH(OR_{(1)})(OR_{(2)}) + H_2O \xrightarrow{H^+} RCHO + R_{(1)}OH + R_{(2)}OH$$

Methods for determining acetals in which the acetal was hydrolyzed and the aldehyde determined have been described. The drawbacks in these procedures are due, in most cases, to the shortcomings of the aldehyde procedure used. (These were discussed in the section on aldehydes pp. 95–160.)

Since all the reactions involved in this analysis are carried out in the same closed vessel, there is less danger of loss of aldehyde by volatilization. The reagents used are stable, and there are no end point or equilibrium difficulties to harass the operator. Also there is no need for distilling the aldehyde to determine it; the aldehyde is determined in the hydrolysis reaction mixture. The procedure is accurate and precise, on the average, to ±0.3%. The same procedure can be used to determine alkyl vinyl ethers, since these compounds also hydrolyze readily to yield acetaldehyde.

$$ROCH{=}CH_2 + H_2O \xrightarrow{H^+} CH_3CHO + ROH$$

Mixtures of acetals and alkyl vinyl ethers can be determined by first hydrolyzing the sample and determining the total aldehyde produced. The foregoing vinyl ether is determined by the iodimetric procedure described in the next section. The difference between the two results will yield the amount of acetal present.

Adapted from the Procedure of Siggia

[*Ind. Eng. Chem., Anal. Ed.*, ***19***, *1025* (*1947*)]

Figure 10.1 illustrates the acetyl–vinyl ether apparatus. The top portion *A* consists of a 500-ml Erlenmeyer flask with a 29/26 standard-taper ground-glass female joint blown onto the bottom, with glass stoppers to fit. To the mouth of the flask is attached a 6-mm stopcock, and onto the stopcock is sealed a 29/26 standard-taper ground-glass male joint.

The bottom portion of the apparatus *B* is a 500-ml Erlenmeyer flask with a 29/26 standard-taper female joint sealed onto the mouth.

REAGENTS

Standard sulfuric acid, 1*N*.
Standard sodium hydroxide, 1*N*.
Approximately 1*M* sodium sulfite prepared as described on p. 105.

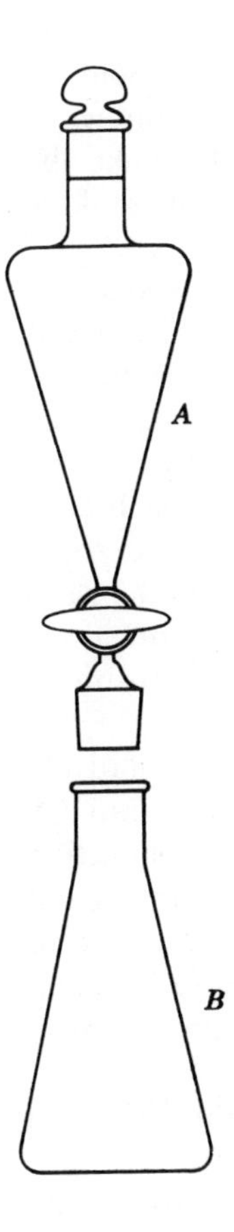

Fig. 10.1. Acetal–vinyl ether apparatus.

PROCEDURE

Weigh a sample of 0.02 to 0.04 mole of acetal or alkyl vinyl ether in a sealed thin-walled ampoule (see pp. 862–64) and introduce it into section *A* of the apparatus (Fig. 10.1), together with 50 ml of standard 1*N* sulfuric acid. Include about 10 to 20 ml of glass beads to aid in breaking the ampoule. Stopper the apparatus, the stopper being well greased to prevent leakage of any volatile aldehyde. If the aldehyde produced is acetaldehyde, totally immerse the flask in a mixture of ice and water for 5 minutes to avoid building up pressure when an ampoule is broken. If the flask is not chilled, the stoppers sometimes are forced open or leaks are created in the stopcock by the pressure.

After the flask is chilled, shake it vigorously to break the ampoule containing the sample; continue shaking for 15 minutes to ensure complete hydrolysis of the sample.

For most acetals and vinyl ethers, 15 minutes of shaking is sufficient for hydrolysis; but for some, for example, di-*n*-butyl acetal, 30 minutes is needed to ensure complete hydrolysis. Use a Kahn shaker to shake the sample. This type of shaker clamps the stopper on the flask while shaking is in progress.

When hydrolysis is complete, attach section *A* of the apparatus, containing the hydrolyzed sample, to section *B*, which contains 250 ml of 1*M* sodium sulfite. The joint connecting the two sections should be greased to prevent loss of low-boiling aldehydes. Then open the stopcock in section *A* and shake the acid solution containing the aldehyde down into the sulfite solution. The acid and sulfite form bisulfite, which consumes the aldehyde. After all the acid solution is in section *B*, shake some of it back into section *A* to react with the aldehyde contained on the walls and on the glass beads. Vigorously shake the whole apparatus for a few minutes to assure removal of all the aldehyde vapor from the atmosphere in both sections of the apparatus.

Separate the sections of the apparatus and rinse them quantitatively into an 800- to 1000-ml beaker. Titrate the solution with standard 1*N* sodium hydroxide. Use a pH meter, such as is described in connection with aldehyde determinations (p. 103). As described earlier, plotting pH versus milliliters of sodium hydroxide is the most accurate method of determining the end point. However if a large enough sample is taken, titrating to a definite pH for the aldehyde does not incur any very significant error.

Determine any free aldehyde in the sample (as on pp. 100–107), and subtract the value from the figure obtained for the acetal.

In the determination of acetals, the hydrolysis step depends on the effectiveness of shaking and the ease of hydrolysis of the particular acetal as far as time necessary for hydrolysis is concerned. The shaking is important, since the samples are immiscible with water; and the more vigorous the shaking, the more sample surface is in contact with acid. The procedure as described calls for 15 minutes of shaking for complete hydrolysis. However, this time interval is used to ensure complete hydrolysis for all samples with a rather leisurely shaking of the sample with the acid.

The hydrolysis can be completed in 5 minutes if the shaking is vigorous, and a good mixing of sample with the acid is maintained. If the shaking is too leisurely, or if the flask is not shaken at all, the hydrolysis may not be complete even after 15 minutes.

The nature of the sample has a great deal to do with the time required for hydrolysis. Dimethyl acetal hydrolyzes very easily. However di-*n*-butyl acetal requires more vigorous treatment and should be shaken vigorously with the acid for 15 minutes to ensure complete hydrolysis. If agitation is moderate, shaking should proceed for 30 minutes.

The following compounds were used to test this procedure: dimethyl, diethyl, di-*n*-propyl, diisopropyl, di-*n*-butyl, and diisobutyl acetal, and methyl, ethyl, propyl, *n*-butyl, and isobutyl vinyl ether.

Iodimetric Method Specific for Vinyl Ethers

Adapted from S. Siggia and R. L. Edsberg

[*Reprinted in Part from Anal. Chem.*, **20,** *762–3* (*1948*)]

The iodometric method described is by far the simplest method yet devised that will still yield results of a high degree of precision and accuracy. The reaction time is 10 minutes, and the entire analysis is carried out in one vessel. The end point is very sharp to the naked eye, and no color comparisons or electrical indicating devices are necessary.

The reaction between the iodine and the vinyl ether requires the presence of an alcohol, since the end product of the reaction is the corresponding iodoacetal.

$$R{-}OCH{=}CH_2 + I_2 + CH_3OH \rightarrow \underset{\displaystyle OCH_3}{\underset{|}{ROCH}}{-}CH_2I + HI$$

The mechanism of the reaction was established in several ways. On omission of the alcohol, no quantitative reaction between the ether and the iodine occurred. The pH of the solution was measured before and after the sample was introduced; the pH dropped considerably on introduction of the ether, indicating the formation of hydriodic acid. To substantiate these qualitative observations, several reactions were run using *n*-butyl vinyl ether. The final reaction mixtures were combined and extracted with carbon tetrachloride. The carbon tetrachloride extracts were distilled and a product was obtained whose carbon, hydrogen, and iodine analysis corresponded to the methyl butyl iodoacetal.

The impurities usually associated with vinyl alkyl ethers—namely, alcohols, acetaldehyde, acetals, acetylene, and water— do not interfere under the conditions of the iodometric analysis. This procedure is an excellent method for determining vinyl alkyl ethers in the presence of acetals and acetaldehyde. Because all previous methods for vinyl alkyl ethers depended on acid hydrolysis and determination of the resulting acetaldehyde, they would react with the acetal along with the ether. Since the ultimate analysis was for acetaldehyde, samples containing acetaldehyde presented difficulty.

Acid impurities in the sample will not interfere unless they lower the pH of the analysis solution below 2. In such an acid solution the hydriodic acid is sensitive to atmospheric oxygen, and oxidation to free iodine results. Because the sample is titrated to the disappearance of the free iodine color, the oxidation of the hydriodic acid obscures the end point by causing the color to reappear very rapidly.

Alkaline impurities interfere in the analysis by causing the reaction to take a different course. In alkaline solution the ethers react with the iodine, as does acetaldehyde, to form iodoform. Samples containing free alkali or too much free acid should be neutralized before analysis.

Acetaldehyde shows no interference whatsoever as long as the sample contains no free alkali. When a sample containing approximately 0.060 gram of ether and 2.0 gram of acetaldehyde was analyzed, there was no interference, even in the presence of such a large amount of acetaldehyde.

PROCEDURE

Accurately weigh vinyl ether samples by sealing 0.001 to 0.002 mole in small glass ampoules. Blow bulbs of soft glass to about 1.25 cm (0.5 in.) diameter, having a capillary stem of about 10 cm (4 in.) (see pp. 882–884). Weigh an ampoule and then fill it by dipping the capillary below the surface of the vinyl ether, cooling the bulb with a dry ice–acetone cooling mixture until the desired amount of ether has entered the bulb. Seal the ampoule by touching the end of the capillary to a small flame, while keeping the bulb in the cooling mixture. Leave the ampoule to dry, and allow it to come to room temperature before reweighing.

The reaction vessel used is a 500-ml wide-mouthed glass-stoppered bottle. Place about 25 ml of glass beads in the bottle, and add 50 ml of standard iodine (0.1*N*) by pipet. To this add 50 ml of CP methanol and the vinyl ether sample. Grease the stopper well to prevent any leaks, and shake the bottle violently to enable the glass beads to crush the sample ampoule thoroughly. After the ampoule is crushed, place the bottle on a mechanical shaker for 10 minutes. Shaking is very important to complete the reaction with methyl vinyl ether which, because of its low boiling point, will be present largely in the vapor phase. Since the higher boiling ethers will be in solution at room temperature, complete reaction may take place without continuous shaking for 10 minutes.

After the shaking, rinse the stopper and side walls of the bottle and titrate the contents of the bottle with standard 0.1*N* thiosulfate. Inasmuch as starch indicator in the presence of methanol is not satisfactory, take the end point as the disappearance of the last trace of iodine color. On standing a few minutes after titrating, the solution may regain a slight iodine color, but in all cases the first disappearance of iodine is the correct end point. If the sample size is large, the excess iodine titration will be small and the iodine color will return more rapidly. If a smaller sample is used, there will be a large excess of iodine left to titrate and the iodine color will return much more slowly. The iodine color returns so slowly that it does not noticeably affect the determination of the end point.

It is very important to use CP methanol in the foregoing reaction. As a check, it was the custom of Siggia and Edsberg to run a blank titration of 50 ml of standard iodine in 50 ml of CP methanol. If the iodine titration in the presence of the methanol differs from the calculated titer, a correction for the blank must be applied. From the titration, the excess iodine present in the reaction mixture is determined. Subtracting this excess from the original 50 ml of iodine added yields the iodine used in the reaction. The percentage of vinyl ether is then calculated as follows:

$$\frac{\text{Milliliters of } I_2 \text{ used in reaction}}{1000} \times \text{normality of } I_2 \times \frac{\text{mol. wt. of ether}}{2} \times \frac{100}{\text{wt. of sample}} = \%\ \text{vinyl ether}$$

DISCUSSION

The method was found to work for almost all the vinyl ethers tried (see Table 1). Lauryl vinyl ether and octadecyl vinyl ether could not be determined by this method because of their limited solubility in the reagents used. Attempts to analyze these samples by eliminating water from the system entirely, using alcoholic solutions of iodine and solutions of iodine in carbon tetrachloride, gave very unsatisfactory results. Apparently water is necessary to obtain a clean-cut reaction. The use of the standard aqueous solution of iodine provides sufficient water to obtain a good reaction, and the aqueous iodine is a very stable reagent.

The end point in this analysis is very sharp. Starch is not used because of its inactivity in an alcoholic medium; however, the disappearance of the iodine color can be very easily detected (within one drop).

Table 1. Analysis of Vinyl Alkyl Ethers

Sample	Analysis by Hydrolysis and Bisulfite Addition, %	Iodometric Analysis, %
Methyl vinyl ether	96.7	96.51
		96.48
		97.02
Methyl vinyl ether	[a]	99.61
		99.01
Ethyl vinyl ether	[a]	99.03
		99.75
		99.51
		99.72
n-Butyl vinyl ether	98.7	98.45
		98.85
		98.81
Isobutyl vinyl ether	[a]	99.34
		99.35
		99.25

[a] Ether was purified by washing five times with alkaline water (pH = 8) to remove acetaldehyde and alcohol; cooled to −50°C and filtered from any ice which had separated; and finally distilled from sodium.

Mercuric Acetate Method for Vinyl Ethers

Mercuric acetate has been applied to determining olefin compounds (see pp. 428–40). Vinyl ethers have a double bond, which makes them suitable for this analysis. The reaction as used for olefins, however, could not be applied to vinyl ethers until Johnson and Fletcher found that for these compounds, a low temperature was needed.

Methods of J. B. Johnson and J. P. Fletcher

[*Reprinted in Part from Anal. Chem.*, **31,** *1563–4 (1959*]

REAGENTS

MERCURIC ACETATE. Approximately 0.12*M* solution in anhydrous, reagent grade methanol. Dissolve 40 grams of mercuric acetate (reagent grade) in sufficient methanol to make 1 liter of solution. Stabilize the reagent by the addition of 3 to 8 drops of glacial acetic acid. Filter the reagent before using. When used in the procedure, 50 ml of the reagent should have a titration of from 1 to 10 ml of 0.1*N* potassium hydroxide.

Standard potassium hydroxide, 0.1*N* solution in methanol.

Phenolphthalein indicator, 1.0% methanolic solution.

Sodium bromide, reagent grade crystals.

PROCEDURE

Pipet 50 ml of the mercuric acetate reagent into each of two 250-ml glass-stoppered Erlenmeyer flasks. If a sealed glass ampoule is specified, use heat-resistant pressure bottles containing a few pieces of 8-mm glass rod. Cool the contents of the flasks between −10 and −15°C. (A bath of chipped ice and methanol can be maintained below −10°C for more than an hour without difficulty.) Reserve one of the flasks for a blank determination. Into the other flask introduce an amount of sample containing from 3.0 to 4.0 meq. of vinyl ether. Allow both the sample and the blank to stand in the bath at −10°C or lower for 10 minutes. To each flask add 2 to 4 grams of sodium bromide and swirl the contents to effect solution. Add approximately 1 ml of the phenolphthalein indicator and titrate immediately with standard 0.1*N* methanolic potassium hydroxide to a pink end point. Do not permit the temperature of the solution to exceed 15°C during the titration. Because the method is based on an acidimetric titration, take the usual precautions to avoid interference from carbon dioxide.

Reaction time and temperatures for other compounds are shown in Table 20, Chapter 7.

DISCUSSION

The vinyl ethers and certain other compounds react with mercuric acetate in methanol to form mercury addition compounds that are unstable at room temperature. These compounds can be determined quantitatively if the solution temperature is maintained below −10°C during the reaction and is prevented from exceeding 15°C during the titration step. Vinyl ethyl ether may also be determined by the procedures of both Martin (1) and Das (2) (see pp. 430–37) if these conditions of temperature are observed. Presumably other vinyl ethers may also be determined

1. R. W. Martin, *Anal. Chem.*, **21,** 921 (1949).
2. M. N. Das, *Anal. Chem.*, **26,** 1086 (1954).

by these two procedures if the temperature is controlled. The vinyl ethers that have been quantitatively determined by the procedure described here are listed in Table 2.

Table 2. Vinyl Ethers Determined by Modified Mercuric Acetate Procedure

Compound	Purity, wt. %	Average Deviation	Number of Determinations	Compound	Purity, wt. %	Average Deviation	Number of Determinations
1-Butenyl methyl ether[a]	97.7	0.0	2	Vinyl 2-chloroethyl ether	97.3	0.0	2
Divinyl Carbitol	98.4	0.1	2	Vinyl ethyl ether[a]	98.9	0.2	14
1-Propenyl ethyl ether	97.4	0.4	2	Vinyl hexyl Carbitol	100.1	0.1	2
Vinyl allyl ether	99.0	0.0	2	Vinyl isobutyl ether[a]	98.6	0.2	2
Vinyl butyl Cellosolve	100.0	0.1	2	Vinyl propyl ether	97.0	0.3	2
Vinyl butyl ether[a]	98.9	0.2	3	Vinyl tetradecyl ether	95.2	0.1	2
Vinyl (2-butylmercapto)ethyl ether	100.0	0.1	2	Vinyl undecyl ether	96.2	0.1	2
Vinyl Carbitol	100.8	0.1	2				

[a] Use a sealed glass ampoule or an aliquot from a methanolic dilution of the sample.

Allyl acetate and certain other unsaturated esters are not hydrolyzed under the nonaqueous conditions of this procedure, whereas vinyl acetate is saponified quantitatively. α-Methylstyrene, a compound whose mercury addition product is unstable toward halogen acids (2), can be determined by this procedure.

LIMITATIONS AND INTERFERENCES

Like the other methoxymercuration methods, this procedure is most suitable for the determination of unsaturated compounds containing a terminal double bond or an internal double bond with a cis configuration. α,β-Unsaturated acids, aldehydes, esters, ketones, and nitriles do not react quantitatively under the conditions employed. Inorganic salts, especially halides, must be absent from the reaction mixture.

Inasmuch as the method is based on an acidimetric titration, the sample must be neutral to phenolphthalein or a suitable correction applied. Care must also be taken to exclude carbon dioxide, which titrates as an acid.

General Method for Traces of Acetals, Ketals, and Vinyl Ethers

The trace methods described earlier (pp. 148–60) for carbonyl compounds can generally be applied to these compounds. The tests involving acidic reagents such as 2,4-dinitrophenylhydrazine hydrochloride (pp. 148–52) can be applied directly, since the acidity of the reagent will rapidly convert these compounds to the free carbonyl compound, which will in turn react with reagent. For the nonacidic reagents, the acetals, ketals, or vinyl ethers can be hydrolyzed first with a dilute solution of mineral acid to liberate the carbonyl compound. The carbonyl test can then be applied to the hydrolyzed system.

Method for Traces of Acetals of Acetaldehyde and for Traces of Vinyl Ethers

Adapted from the Method of M. C. Bowman, M. Beroza, and F. Acree, Jr.

[*Reprinted in Part from Anal. Chem.*, **33,** *1053–5* (*1961*)]

The method is a modification of the one employed by Barker and Summerson (3) for lactic acid in biological materials, by Stotz (4) for acetaldehyde in blood, and by Giang and Smith (5) for metaldehyde (tetramer of acetaldehyde) in plant material. The compound is hydrolyzed with sulfuric acid; the acetaldehyde generated is distilled into aqueous sodium bisulfite and reacted with *p*-phenylphenol in the presence of concentrated sulfuric acid and cupric ion. A violet color having an absorption maximum at 572 nm (Fig. 10.2) is produced.

The analysis is simple to run, it is accurate to within 3% in the microgram range, and results are reproducible.

APPARATUS

Absorbance measurements were made with a Beckman Model DU spectrophotometer in square Corex cells having a 1-cm light path.

3. S. B. Barker and W. H. Summerson, *J. Biol. Chem.*, **138,** 535 (1941).
4. Elmer Stotz, *J. Biol. Chem.*, **148,** 585 (1943).
5. P. A. Giang and F. F. Smith, *J. Agr. Food Chem.*, **4,** 623 (1956).

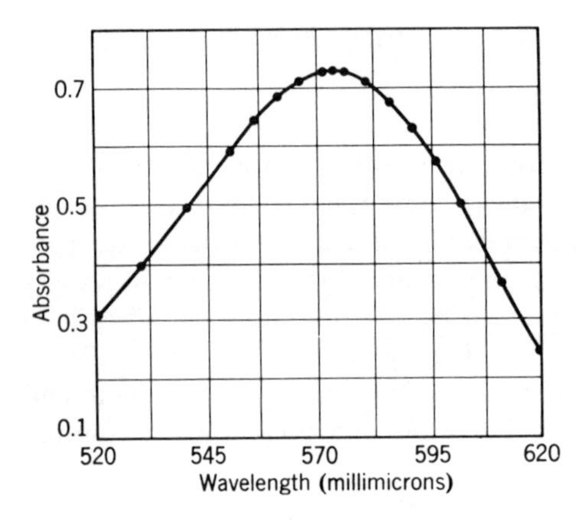

Fig. 10.2. Absorption curve of color.

The reaction apparatus, described by Giang and Schechter (6), is an all-glass system with two vertical, parallel condensers. One permits refluxing and the other distillation in the same apparatus without transfer. The reflux condenser is at least 15 cm long attached to the reaction flask. The second condenser (25 cm long) is attached to the reflux condenser by a suitable adapter, such that condensate runs into a suitable collection vessel joined to the second condenser by means of a ground-glass joint.

REAGENTS

The sulfuric acid–cupric sulfate, *p*-phenylphenol, and standard metaldehyde reagents were prepared as follows (5).

SULFURIC ACID–CUPRIC SULFATE. Add 5 ml of 5% cupric sulfate to 500 ml of sulfuric acid and mix well.

p-PHENYLPHENOL. Dissolve 1 gram crystalline material in 25 ml of hot 2*N* sodium hydroxide. Add 75 ml of water. Store in brown bottle.

METALDEHYDE. Recrystallize from ether and ethyl alcohol.

PROCEDURE

PREPARATION OF STANDARD CURVE. Attach a test tube (15×150 mm) containing 5 ml of 2% sodium bisulfite solution (freshly prepared) to the reaction apparatus so that the tip of the exit tube from the second condenser extends to the bottom of the solution. Partly immerse the test tube in an ice bath during reflux and distillation. Chill in an ice bath 10 ml of 10% aqueous sulfuric acid (v/v) in a 50-ml distilling flask containing 1

6. P. A. Giang and M. S. Schechter, *J. Agr. Food Chem.*, **6,** 845 (1958).

to 2 mg of Carborundum powder. Add the appropriate volume of standard metaldehyde in chloroform solution and attach it to the apparatus. Reflux the mixture for 15 minutes while water maintained at 5°C is passed through both condensers. Then drain the water from the reflux condenser and distill the mixture until 3 to 4 ml remains. Disconnect the apparatus, and wash both condensers and the delivery tube with small portions of cold water, which are combined with the distillate. Separate the chloroform and extract with three 10-ml portions of cold water, which are combined with the aqueous distillate and diluted to 50 ml in a volumetric flask. Extract the solution with 5 ml of redistilled hexane to remove traces of chloroform.

Mix 1 ml of the aqueous solution by swirling, with 8 ml of cold sulfuric acid–cupric sulfate reagent in a 15 × 150 mm test tube partly immersed in an ice bath. Add the *p*-phenylphenol reagent (0.2 ml) in a similar manner; then remove the tube from the ice bath and allow to stand in the dark at room temperature for 1 hour. Heat the tube in a water bath at 100°C for 90 seconds, then return it to the dark for about 30 minutes to adjust to room temperature. With distilled water in the reference cell, determine the absorbance at 572 nm. Prepare a standard curve by plotting concentration against the absorbance corrected for the blank of the reagents carried through the entire process. The color reaction conformed to Beer's law, and 1 μg of acetaldehyde produced an absorbance of about 0.150.

ANALYSIS OF SAMPLES. Prepare chloroform solutions of the samples for analysis by diluting 200 μl of each to 50 ml in volumetric flasks. Calculate the weights of the samples in these solutions from the volumes and the specific gravities of the samples. This procedure serves to minimize any error resulting from the high volatility of some of the compounds. Then dilute each solution with chloroform to contain 100 to 200 μg of combined acetaldehyde per milliliter. Analyze both 1- and 2-ml aliquots of each solution by the method used for the preparation of the standard curve.

RESULTS AND DISCUSSION

Determinations of acetals and vinyl ethers were first attempted by the method of Giang and Smith (5) for the determination of metaldehyde, but the low and erratic yields of acetaldehyde indicated that a more vigorous hydrolysis was necessary. Yields were increased by using 10% sulfuric acid for hydrolysis instead of the 0.4% concentration employed by Giang and Smith. Even with the higher acid concentration, quantitative yields were obtained only after the reaction mixture had been refluxed for 15 minutes prior to distillation. The introduction of cold water in the condenser during the reflux period was especially necessary with highly volatile materials, to prevent their loss by distillation before the hydrolysis could be effected.

Confirming the experience of Stotz (4), it was found that 8 parts of acid to 1 part of water is probably the most practical proportion for color development, no color being obtained with acid-water ratios of 2:1 or less. No better substitute for *p*-phenylphenol was found, although 2,4-dinitro-6-phenylphenol, *p*-hydroxyazobenzene, *o*-phenylphenol, 2-hydroxy-5(α-methylbenzyl)biphenyl, 2-chloro-4-phenylphenol, *m*-nitrophenol, 4-phenylazo-*m*-cresol-1-naphthol, and -2-naphthol, and *p*-nitrosophenol were tried. Only 2-chloro-4-phenylphenol produces color, but it was about 60% as strong as that obtained with *p*-phenylphenol.

The data in Table 3 show that the color fades after development, but any error in fading is insignificant when absorbance readings are made at the interval designated in the procedure. The absorbance blank differs with each preparation, but it is usually 0.030 to 0.040. It is therefore advisable to run a blank with each set of analyses.

Table 3. Stability of Color at 25°C

Hours after First Reading	Absorbance Loss, %	
	In Dark	In Artificial Light
2	0.6	3.7
6	6.9	10.0
18	13.4	26.5
26	17.6	33.2
50	30.7	49.2

The results of determining a number of vinyl ethers and acetals of acetaldehyde are seen in Table 4. In all instances the values found for acetaldehyde were within 3% of the theoretical.

The method has been used to determine the persistence of Sesamex residues on glass (Table 5). The presence of Sevin (1.75%) or pyrethrins (1.75%), the insecticides that Sesamex synergizes (7, 8), did not interfere with the determination of the synergist in chloroform or in deodorized kerosene solution.

7. M. E. Eldefrawi et al., *Science*, **129,** 898 (1959).
8. J. H. Fales, O. F. Bodenstein, and M. Beroza, *Soap and Chem. Specialties*, **33** (2), 79 (1957).

Table 4. Recovery of Acetaldehyde from Compounds Containing Combined Acetaldehyde Groups

Compound	Amount, Added, μg.	Acetaldehyde Equivalent, μg.	Acetaldehyde Recovered	
			μg.	%
Ether				
2-Butoxyethyl vinyl	343	105	106	101
	686	210	207	99
Butyl vinyl	312	137	136	99
	624	274	274	100
Isobutyl vinyl	306	134	138	103
	612	268	265	99
2-Ethylhexyl vinyl	644	181	181	100
	1288	363	364	100
Ethyl vinyl	300	183	181	99
	600	366	363	99
Methoxyethyl vinyl	358	154	151	98
	716	308	301	98
Acetaldehyde				
Diisobutyl acetal	656	166	168	101
	1312	332	331	100
2-(2-Ethoxyethoxy)-ethyl 3,4-methylenedi-oxyphenyl acetal (sesamex)	904	134	135	101
	1808	267	275	103

Table 5. Persistence of Sesamex Residues on Glass

Days after Application	Recovery, mg./sq. cm.	
	Applied in Chloroform	Applied in Deodorized Kerosene
0	0.55	0.55
3	0.50	0.50
7	0.50	0.50
10	0.47	0.47
14	0.47	0.44

$$CH_2{=}CH{-}O{-}R \xrightarrow[H+]{HOH} CH_3{-}CH\begin{matrix}OH\\OR\end{matrix} \xrightarrow[H+]{HOH}$$

$$\left[CH_3CH\begin{matrix}OH\\OH\end{matrix}\right] \xrightarrow{-H_2O} CH_3CHO$$

The method is potentially valuable for the structure determination of compounds with combined acetaldehyde groups, but little work has been done on this problem. This potential may be realized when the present analysis of combined acetaldehyde groups is compared with that of combined formaldehyde groups, a method that has been more thoroughly studied (9–11). In the formaldehyde analysis, mineral acids liberate formaldehyde from methylene groups attached to oxygen, nitrogen, and sulfur: for example, $—OCH_2O—$, $—NCH_2N—$, $—SCH_2S—$, $—SCH_2N—$, $—N{=}CH_2$ (9). Acetals of acetaldehyde behave in a like manner when CH_3CH OR hydrolyzes with acid to give CH_3CH OH, which is hydrated acetaldehyde. Vinyl ethers hydrolyze to hemiacetals, which cleave further to give acetaldehyde:

$$CH_3CH\begin{matrix}OR\\OR\end{matrix}$$

$$CH_3CH\begin{matrix}OH\\OH\end{matrix}$$

Similarly, it would be expected that $CH_3CH<$, if attached like the above-mentioned methylene to nitrogen or sulfur, would yield acetaldehyde on acid hydrolysis.

It is not anticipated that all compounds with combined acetaldehyde groups should always be determined by the procedure reported here.

9. M. Beroza, *Anal. Chem.*, **26,** 1970 (1954).
10. C. E. Bricker and H. R. Johnson, *Ind. Eng. Chem., Anal. Ed.*, **17,** 400 (1945).
11. C. E. Bricker and W. A. Vail, *Anal. Chem.*, **22,** 720 (1950).

Determination of combined formaldehyde groups (9) showed that more color could often be produced by varying such conditions as heating time and amounts of reagents. Such differences would be expected to apply also to the determination of combined acetaldehyde groups.

Another potential use is for the determination of end ethylidene ($CH_3—CH=$) groups in a manner analogous to the ingenious procedure of Bricker and Roberts (12) for determining end methylene groups. The procedure would involve oxidation of the double bond to a vicinal pair of hydroxyl groups, which could then be split by periodic acid and the resultant acetaldehyde distilled and determined:

$$CH_3CH{=}CH{-}R \xrightarrow{KMnO_4} CH_3\overset{OH}{\overset{|}{C}}H\overset{OH}{\overset{/}{C}}H{-}R \xrightarrow{HIO_4} CH_3CHO + RCHO$$

This type of analysis has already been carried out by Huggins and Miller (13), who determined 1,2-propanediol by oxidizing it with periodic acid to get acetaldehyde and formaldehyde; the acetaldehyde was selectively aerated, collected in bisulfite, and determined colorimetrically.

To use the method for structural determinations would require full knowledge of interferences, and some of this information is available. Barker and Summerson (3) listed 71 compounds that do not interfere when present in 50 to 100 times the amount of the lactic acid for which they were analyzing. They also listed a number that do, and gave means of overcoming interferences. Their analysis is similar to this one. Ethanol, acetylmethylcarbinol, acetone, and 2,3-butylene glycol do not interfere (3, 4). Diacetyl develops a green color in the analysis, which may be destroyed by preliminary treatment with periodic acid (4). Lactic and pyruvic acids interfere and undoubtedly can be removed by preliminary treatment with an ion-exchange resin in a manner similar to that of Markus (14), who removed interfering cations. Henry et al. (15) reported that acetic acid gives a color with *p*-phenylphenol; it could also be removed with an ion-exchange resin.

The distillation step lends specificity to the analysis by separating the highly volatile acetaldehyde (b.p. 21°C) from the nonvolatile interferences. Thus Barker and Summerson (3), Stotz (4), Westerfield (16), and Giang and Smith (5) were able to determine acetaldehyde in biological materials without interferences. However in determining metaldehyde in plant material, Giang and Smith washed the chloroform extract of the

12. C. E. Bricker and K. H. Roberts, *Anal. Chem.*, **21,** 1331 (1949).
13. G. C. Huggins and O. N. Miller, *J. Biol. Chem.*, **221,** 377 (1956).
14. R. L. Markus, *Arch. Biochem.*, **29,** 159 (1950).
15. R. J. Henry, *J. Lab. Clin. Med.*, **33,** 241 (1948).
16. W. W. Westerfield, *J. Lab. Clin. Med.*, **30,** 1076 (1945).

plant with a 2% sodium bisulfite solution to remove free acetaldehyde and related materials. By the exercise of this precaution, their blank was kept low. The separation of acetaldehyde from formaldehyde (gives green color in test) by aeration has been mentioned (15).

As in most highly sensitive methods, care must be taken to avoid contamination. Lubricating grease, chromic acid (from cleaning solution), perspiration, even the analyst's breath in blowing out a pipet have been reported to interfere with the acetaldehyde determination (3–5).

11

Amino Groups

Titration, amide formation, bromination, and reaction with nitrous acid are the reactions most commonly used to determine amino groups. Titration is by far the most common method used. Almost any amine can be titrated either in water or in certain organic solvents. The aliphatic amines are usually basic enough to be titrated directly in aqueous solutions using standard acids. They have dissociation constants ranging from 10^{-3} to 10^{-6}. The aromatic or other weakly basic amines cannot be titrated in water but can be successfully titrated in solvents to be discussed later. The latter amines have dissociation constants ranging from 10^{-9} to 10^{-12}.

As in the case of the carboxylic acids, the basicity of the amines varies little with structure (except aromatic vs. aliphatic) or with position in a homologous series. Hence unsubstituted aliphatic amines have approximately the same dissociation constant (see Table 1). Also, unsubstituted

Table 1. Dissociation Values for Aliphatic Amines[a]

Amine	pKa
Methyl	10.6
Ethyl	10.7
Propyl	10.6
n-Butyl	10.6
n-Amyl	10.6
n-Hexyl	10.6
n-Heptyl	10.7
Heptadecyl	10.6
Isopropyl	10.7
Isobutyl	10.4
tert-Butyl	10.8
Isoamyl	10.6
Triethyl	11.0

[a] E. A. Braude and F. C. Nachod, *Determination of Organic Structures by Physical Methods*, Academic Press, New York, 1955, p. 573.

aromatic amines have approximately the same dissociation constant (see Table 2). Again as in the case of carboxylic acids, however, the basic strength of the amine varies with substituents and with the position of the substituent (Tables 3 and 4.)

The alicyclic amines vary in basicity depending on their character. Alicyclic amines of an aromatic character such as pyridine and pyrrole are weakly basic. Pyrrole (pK_a 0.4) is an acid. The saturated alicyclic amines such as piperidine and pyrrolidine are more strongly basic. Piperidine

Table 2. Dissociation Values for Aromatic Amines[a]

Amine	pKa
Aniline	4.6
o-Methyl aniline	4.4
m-Methyl aniline	4.7
p-Methyl aniline	5.1
o-Phenyl aniline	3.8
m-Phenyl aniline	4.2
p-Phenyl aniline	4.3
Naphthylamine	3.9
β-Naphthylamine	4.1

[a] E. A. Braude and F. C. Nachod, *Determination of Organic Structures by Physical Methods,* Academic Press, New York, 1955, pp. 590, 599.

Table 3. Dissociation Values of Substituted Aliphatic Amines[a]

Substituent*	pKa					
	NH_3	Methyl Amine	Ethyl Amine	Propyl Amine	Butyl Amine	Pentyl Amine
None	9.3	10.6	10.7	10.6	10.6	10.6
RO_2C—	—	7.8	9.1	9.7	10.2	10.4
HO—	6.0	—	9.5	—	—	—
C_6H_5—	4.6	9.4	9.8	10.2	10.4	10.5
H_2N—	8.1	—	10.0	10.7	10.8	11.1
HOOC—	—	2.3	3.6	4.2	4.3	4.4

* On position on opposite end of the molecule to the amino group.

[a] E. A. Braude and F. C. Nachod, *Determination of Organic Structures by Physical Means,* Academic Press, New York, 1955, pp. 575, 579.

Table 4. Dissociation Values for Substituted Anilines[a]

Substituent	pKa		
	Ortho	Meta	Para
None	4.6	—	—
O_2N—	0.3	2.5	1.0
HOOC—	2.0	3.1	2.3
CH_3OOC—	2.2	3.6	2.3
HO—	4.7	4.2	5.5
Br—	2.6	3.5	3.9
Cl—	2.6	3.3	3.8
F—	3.0	3.4	4.5
CH_3O	4.5	4.2	5.3
C_6H_5—	3.8	4.2	4.3
H_2N—	4.5	4.9	6.1

[a] E. A. Braude and F. C. Nachod, *Determination of Organic Structures by Physical Means*, Academic Press, New York, 1955, p. 590.

(pK_a 11.1) is more strongly basic than many aliphatic amines. Pyrrolidine (pK_a 2.9), although more basic than pyrrole, is not quite as basic as aniline (pK_a 4.6). Substitution on these compounds also markedly affects their basicity (Table 5). Note again that when a substituent affects the basic strength of the amino group, the position of substitution makes a

Table 5. Dissociation Values for Pyridine and Substituted Pyridines[a]

Substituent	pKa		
	2 Position	3 Position	4 Position
None	5.2	—	—
CH_3—	6.0	5.7	6.0
C_2H_5—	5.8	5.7	6.0
Isopropyl—	5.8	5.7	6.0
t-Butyl—	5.8	5.8	5.8
F	−0.4	3.0	—
Cl	0.7	2.8	—
Br	0.9	2.8	—
I	1.8	3.3	—
OH	0.8	4.7	3.1
NH_2	6.7	5.8	9.0

[a] E. A. Braude and F. C. Nachod, *Determination of Organic Structures by Physical Methods*, Academic Press, New York, 1955, p. 597.

difference as to its effect. The substituents that affect basicity generally exert this effect greatest when they are in the 2-position. The $—NH_2$ group is an exception which the data indicate shows its greatest effect in the 4- position. This is comparable in aniline (Table 4), where the $—NH_2$ in the para position exerts the greatest effect.

In addition to titration, reactions with anhydrides can be used to determine primary and secondary amines.

$$\begin{matrix} RNH_2 \\ + \\ R_2NH \end{matrix} \quad (R'CO)_2O \rightarrow \begin{matrix} R'C(=O)NHR + R'COOH \\ R'C(=O)NR_2 + R'COOH \end{matrix}$$

The anhydride methods applicable to alcohols (pp. 12–31) are also applicable to amines. The reactions of the anhydrides with amines are more rapid than those with alcohols. Although the titration methods for amines are preferred because of their speed and simplicity, the anhydride methods can be readily used to determine primary and/or secondary amines in the presence of tertiary amines.

Aromatic amines have reactions specific to them which can be used for analysis. For example, primary aromatic amines can be diazotized with nitrous acid which can be measured.

$$C_6H_5NH_2 \cdot HCl + HONO \rightarrow C_6H_5N{\equiv}NCl + 2H_2O$$

Nitrous acid is also used to determine aliphatic, primary amines. The reaction proceeds as follows and the evolved nitrogen is measured.

$$RCH_2NH_2 + HONO \rightarrow RCH_2OH + N_2 + H_2O$$

In addition, primary, secondary, and tertiary aromatic amines can be brominated much like phenols. This reaction can sometimes be applied for analytical purposes. The bromination proceeds in the unoccupied ortho and para positions.

$$C_6H_5NH_2 + 3Br_2 \rightarrow 2,4,6\text{-}Br_3C_6H_2NH_2 + 3HBr$$

Some aromatic amines will couple with diazonium compounds, and this also is the basis of an analytical method.

$$\text{NH}_2\text{C}_6\text{H}_4\text{NH}_2 + \text{ClN}{\equiv}\text{NC}_6\text{H}_5 \xrightarrow{\text{base}} (\text{NH}_2)_2\text{C}_6\text{H}_3\text{—N}{=}\text{NC}_6\text{H}_5 + \text{HCl}$$

In addition to these systems of analysis for amines, there are other specialized reactions whose main application is to the analysis of individual amines in mixtures of amines. These methods are described later (pp. 567–633).

Titration Methods

Amines can be titrated in almost any solvent in which they can be dissolved. The only undesirable solvents for amines are the basic solvents, since these solvents depress the basicity of the amino groups and poor end points result.

There are several aspects to be considered in selecting a solvent for titrating amines:

1. The solvent must dissolve the sample containing the amine.
2. The solvent should be such that the basicity of the amine is apparent enough for accurate and precise titration.
3. The solvent must not react with the amine.
4. If more than one amino group is present, a solvent may be used that will resolve the different basicities of the several amino groups.

Amines can generally be titrated in a wide range of solvents including acetic acid, dioxane, ketones, alcohols, nitriles, ethers, glycols, nitromethane, and mixtures of these solvents with each other and with hydrocarbons. Hence aspect 1 is easily satisfied; a solvent is generally available that will dissolve the sample and in which the amine is determinable.

As to aspect 2, water is a suitable solvent for aliphatic amines since these are relatively strong bases as indicated on p. 529 and in Table 1. The more weakly basic amines, however, such as aromatics, alicyclics, alkaloids, and nitro- or carboxy-substituted aliphatic amines, do not give sharp enough end points in aqueous media for accurate and precise analysis. The acidic solvents generally are preferred for accurate and precise determination of amines, especially the more weakly basic amines. These solvents accentuate the basicity more markedly than do the more acidic solvents. Hence acetic acid is a very popular solvent for amino group determination. Nitromethane is also used (1), but it has no advantage over acetic acid except that it may dissolve some materials not completely soluble in acetic acid. Phenol has also been used (2), but only because the polyamide (nylon) polymers being titrated for residual amino groups are not soluble in other more convenient solvents.

For aspect 3, the solvents for determining amines must be carefully selected since some solvents will react with certain amines. For example, acetic anhydride is a very good solvent for titrating very weak bases (3); however the anhydride reacts with primary and secondary amines, which is not desirable. Even though the amide reaction products of this system are still titratable, the end points are not as good as those obtainable for amines in other solvents. Ketone solvents should be avoided for primary amines, especially aromatic primary amines; since ketones react with primary amines to yield the corresponding imines, which are more weakly basic than the original amines.

$$\begin{matrix} R_1 \\ \diagdown \\ \end{matrix} C{=}O + RNH_2 \rightarrow C{=}NR + H_2O \quad (R_1R_2C{=}O + RNH_2 \rightarrow R_1R_2C{=}NR + H_2O)$$

Care must also be taken with ether solvents such as dioxane and tetrahydrofuran, since these often contain aldehydes as impurities. These aldehydes form the corresponding imines with primary amines as shown in the preceding equation, with the same effect. The ether solvents should be purified by the method used for dioxane, as indicated below, if primary amines are to be determined (p. 550).

If amino groups of different basic strengths are present (item 4 above), glacial acetic acid is to be avoided. Although this solvent intensifies the basic strengths, it also has a "leveling" effect whereby all the amino groups exhibit the same basic strength. Hence in glacial acetic acid, different amino groups usually are all titrated together and cannot be differentiated; even mixtures of aromatic and aliphatic amines that differ widely in basicity in water, the basicities in acetic acid do not vary much

1. C. Streuli, *Anal. Chem.*, **31,** 1652–4 (1959).
2. H. J. Stolten, *Private Communication.*
3. D. C. Wimer, *Anal. Chem.*, **30,** 77–80 (1958).

at all. Water is one of the best solvents for resolving the basicities of the different amino groups. It has been indicated that aliphatic amines can be readily titrated in water but that aromatic amines cannot be titrated at all. In aqueous strong salt solutions, weak bases such as aniline can be titrated, and yet some of the discriminating features of water are maintained (pp. 537–44). In glycol solvents or mixtures of glycols with other solvents (see Methods below) mixtures of amino groups can readily be resolved and determined. Figure 11.1 shows an extreme case of the titration curve of a compound that contains four amino groups of comparable basicities. This titration curve shows the titration of triethylenetetramine in 1:1 glycol-isopropanol; three inflection points are visible (4). Acetonitrile is another solvent recommended for resolving amines. Of lesser significance in this regard are alcohols and dioxane.

WATER SYSTEMS

Water Alone—Adapted from E. F. Hillenbrand, Jr., and C. A. Pentz

[*Reprinted in Part from Organic Analysis, Vol. III, Edited by J. Mitchell, Jr., et al., Wiley-Interscience, New York, 1956, pp. 192–3.*

Certain water-soluble amines exhibit sufficiently strong alkaline dissociation in aqueous solution to ensure a satisfactory end point using either indicator or potentiometric means. These compounds react stoichiometrically with hydrochloric acid in water to form the corresponding amine

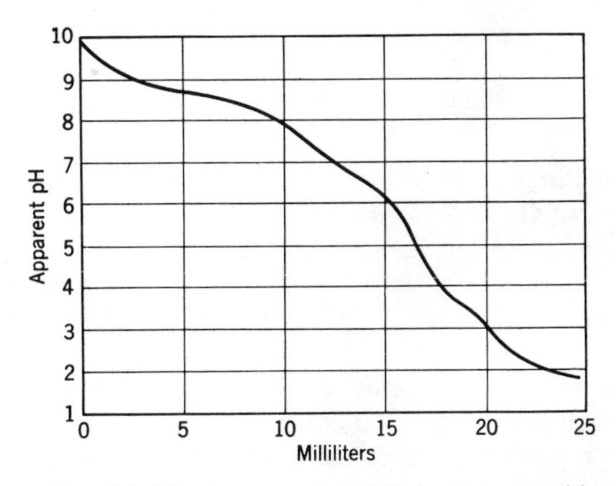

Fig. 11.1. Titration curve for triethylenetetramine (4).

4. S. Siggia and H. J. Stolten, *An Introduction to Modern Organic Analysis*, Wiley-Interscience, New York, 1956, p. 69.

hydrochloride. The amount of standard hydrochloric acid required for the titration is a direct measure of the amine originally present.

REAGENT

MIXED INDICATOR. Prepare separate solutions of 0.1% bromocresol green and 0.1% methyl red in methanol. Mix, using 5 parts of the bromocresol green solution and 1 part of the methyl red solution. Prepare fresh solutions of each indicator and a new mixture every two weeks.

PROCEDURE

Transfer 50 ml of water to each of a sufficient number of 250-ml glass-stoppered Erlenmeyer flasks to make all sample determinations in duplicate. Add 6 to 8 drops of bromocresol green–methyl red mixed indicator and make neutral by the dropwise addition of hydrochloric acid to the disappearance of the green color. Into each flask introduce an amount of sample containing 3 to 4 meq. of amine. Swirl the flask to effect solution. Titrate with standard 0.1*N* hydrochloric acid to the disappearance of the green color. If, because of sample color, it is desirable to titrate potentiometrically, consult Table 6 for the pH at the end point for several representative amines. Glass-calomel electrodes can be used.

Calculate the amount of amine present:

$$\frac{A \times N \times \text{E.W.}}{\text{Grams of sample} \times 10} = \text{amine, \% by wt.}$$

where A is milliliters of N normal hydrochloric acid required and E.W. is the equivalent weight of the amine.

Table 6 shows the proper sample size for a number of compounds for which this procedure was found satisfactory. Also included are some ionization constants K_b and the pH of the end point if a potentiometric titration is required. Of course, the potentiometric data can be plotted and the end point calculated. The sample size is based on the amount of pure amine equivalent to 25 to 40 ml of 0.1*N* hydrochloric acid.

Although the foregoing procedure is not as generally applicable as the titration of amines with perchloric acid in acetic acid medium (see pp. 545–6), the method is often used as a control of the quality of commercially available materials.

Table 6. Titration of Amines with Hydrochloric Acid in Water

Compound	K_b	Sample, grams[a]	Endpoint, pH
Butylamine	4.1×10^{-4b}	0.18 to 0.29	6.4
N-Butyldiethanolamine	5.4×10^{-6c}	0.40 to 0.65	5.4
Diethanolamine	6.0×10^{-6c}	0.26 to 0.42	5.4
Diethylamine	1.3×10^{-3b}	0.20 to 0.28	—
Diisopropanolamine	2.0×10^{-5c}	0.36 to 0.44	—
Diisopropylamine	3.7×10^{-4c}	0.30 to 0.38	—
Dimethylethanolamine	1.3×10^{-5c}	0.26 to 0.34	—
Ethylamine	5.6×10^{-4b}	0.16 to 0.22[d]	—
Ethylenediamine	8.5×10^{-5b}	0.08 to 0.14	4.5
Hexylamine	1.3×10^{-4c}	0.25 to 0.40	6.3
N-(2-Hydroxyethyl)morpholine	6.5×10^{-8c}	0.33 to 0.52	4.5
Isopropylamine	4.3×10^{-4b}	0.18 to 0.22	—
N-Methylmorpholine	1.7×10^{-7c}	0.25 to 0.40	4.6
Monoethanolamine	2.8×10^{-5b}	0.15 to 0.25	6.1
Morpholine	3.3×10^{-6c}	0.22 to 0.35	5.1
γ-Picoline	6.3×10^{-9c}	0.23 to 0.37	4.0
Piperazine	$K_1 = 6.4 \times 10^{-5b}$ $K_2 = 3.7 \times 10^{-9b}$	0.11 to 0.17	4.0
Triethanolamine	3.1×10^{-7c}	0.37 to 0.60	5.0
Triethylamine	5.7×10^{-4b}	0.36 to 0.44	—
Triethylene glycol 3-aminopropyl ether	6.6×10^{-5c}	0.52 to 0.83	6.0

[a] Weigh the samples to the nearest 0.1 mg. in the most convenient manner.
[b] From Lange, *Handbook of Chemistry*, Handbook Publishers, Sandusky, Ohio.
[c] From Carbide and Carbon Chemicals Co., unpublished data.
[d] Use a sealed-glass ampule.

Aqueous Strong Salt Solutions—Adapted from F. Critchfield and J. B. Johnson

[*Reprinted in Part from Anal. Chem.*, **30**, *1247–9* (*1958*)]

Neutral salts enhance the potentiometric break for the titration of weak bases with aqueous mineral acids. This effect and some if its analytical applications are discussed here.

APPARATUS AND REAGENTS

All potentiometric titrations were performed using a Leeds & Northrup line-operated pH meter equipped with glass and calomel electrodes.

All salts were Baker's Analyzed Reagents, J. T. Baker Chemical Company.

DISCUSSION

The ability of a neutral salt to enhance the potentiometric break in the acidimetric titration of a weak base is demonstrated in Fig. 11.2. Curve 1 is the potentiometric titration of aniline in water with $0.5N$ hydrochloric acid. Obviously, the amine is too weak ($K = 3.8 \times 10^{-10}$) to be titrated satisfactorily under these conditions. The same titration in $7M$ sodium iodide gives a potentiometric break that is satisfactory for precise analytical measurements.

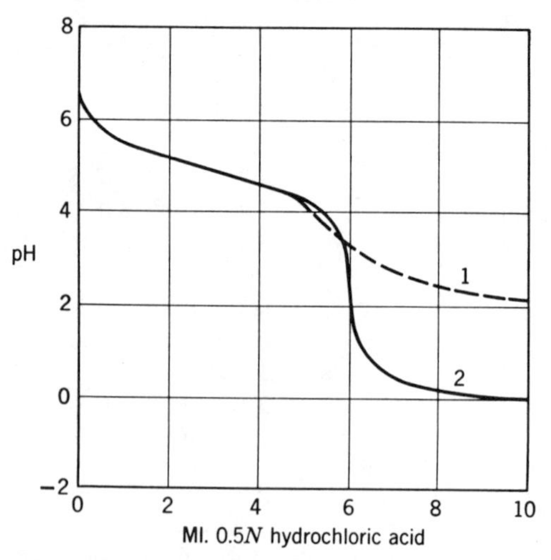

Fig. 11.2. Potentiometric titration of aniline: 1, in water; 2, in $7M$ aqueous sodium iodide.

The potentiometric curves in Fig. 11.3 were obtained for the titration of aniline in solutions containing various concentrations of sodium iodide. For sake of clarity, the curves have been displaced along the abscissa, and only the end point portions are shown. These curves show that the

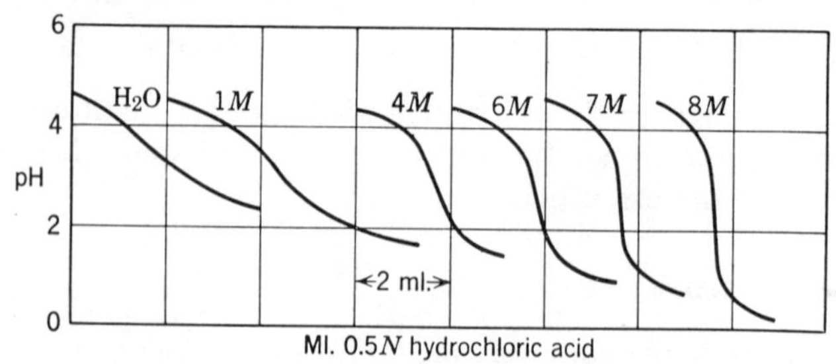

Fig. 11.3. Potentiometric titration of aniline in aqueous sodium iodide solutions.

enhancement of the potentiometric break by neutral salts is a definite function of the salt concentration and is noticeable at a sodium iodide concentration of 1*M*. For this series of curves the sharpest potentiometric break is obtained at a concentration of 8*M* sodium iodide.

An examination of the curves in Fig. 11.2 shows that the initial portions of the curves for the titration of aniline in water and in 7*M* sodium iodide are superimposed. This suggests that the pH of weak bases is independent of salt concentration. After aniline is neutralized, the pH of the titration solution is dependent on the amount of excess hydrochloric acid present and is independent of aniline hydrochloride. The curve in Fig. 11.4 shows that the pH of 0.0192*M* hydrochloric acid decreases linearly with sodium iodide concentration. At a concentration of 7*M* the solution has a pH of zero. An aqueous solution containing the same concentration of hydrogen ion has a pH of 1.8. An inspection of curve 2, Fig. 11.2, shows that, in the titration of aniline in 7*M* sodium iodide, the addition of 4.0 ml of excess hydrochloric acid, which makes the solution 0.0192*M*, lowers the pH of the titration solution to −0.10. This value corresponds with the value of zero predicted by the curve in Fig. 11.4.

Although the explanation of the enhancement of the potentiometric break by neutral salts is unknown, the mechanism must be associated with the decrease of pH of mineral acids by neutral salts, because the magnitude of this decrease shows up in the potentiometric break.

Any salt of a fairly strong base and a strong acid will enhance the potentiometric break for the titration of weak bases. Among the effective salts are sodium chloride, lithium chloride, sodium iodide, and calcium chloride. Salts of strong bases and weak acids inhibit the potentiometric break, as would be expected because they are appreciably basic. Sodium sulfate normally considered a salt of a strong acid and a strong base, also inhibits the break in the titration of aniline. This suggests that the acid from which the salt is derived must have an ionization constant greater than 1×10^{-2} in order to exhibit this effect.

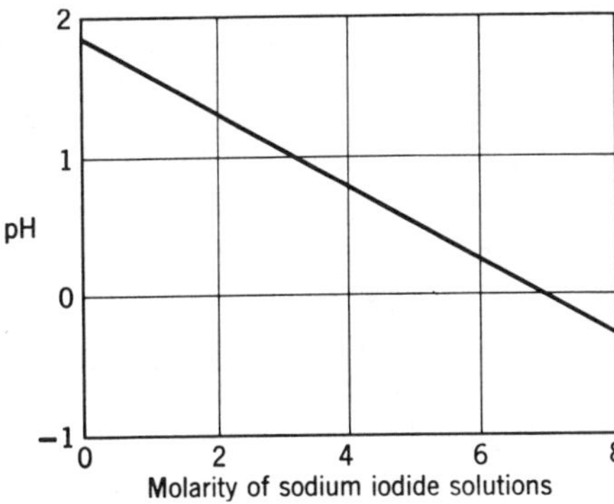

Fig. 11.4. Effect of sodium iodide concentration on pH of 0.0192*M* hydrochloric acid.

INDICATORS

In general, the pH at which indicators change color in strong solutions of neutral salts is the same as in water. Table 7 lists the approximate pH range at which certain indicators change color in water, $8M$ lithium chloride, and $4.5M$ calcium chloride. These indicators were selected to cover the pH range from 1.5 to 7.5 and are recommended for use in strong salt solutions. Certain indicators are subject to salt effects that shift the pH of their color transition. Among these are bromophenol blue, bromocresol green and phenolphthalein. The indicators listed in Table 7 are free from salt effects and behave in salt solutions as they do in water.

Table 7. Behavior of Indicators in Strong Salt Solutions

Indicator	Color Change, Acid to Base	Medium[a]	Approximate pH Range
Thymol blue	Red to orange	Lithium chloride	1.5 to 2.7
		Calcium chloride	1.5 to 3.0
		Water	1.9 to 2.6
M-Alka Ver	Red to green	Lithium chloride	3.0 to 4.0
		Calcium chloride	2.9 to 4.2
		Water	3.3 to 4.2
Methyl orange	Red to orange	Lithium chloride	4.1 to 5.5
		Calcium chloride	4.0 to 5.0
		Water	4.1 to 5.0
Methyl red	Red to yellow	Lithium chloride	5.6 to 6.4
		Calcium chloride	5.4 to 6.4
		Water	5.4 to 6.4
Bromothymol blue	Yellow to blue	Lithium chloride	6.2 to 7.5
		Calcium chloride	6.1 to 7.4
		Water	6.0 to 7.6

[a] All salt solutions are aqueous. Lithium chloride, $8M$; calcium chloride, $4.5M$.

The data in Table 8 show the purities of some weak bases as obtained by titration with aqueous $0.5N$ hydrochloric acid in various salt solution media using indicators specified in Table 7. Comparison with the purities obtained by titration in acetic acid medium using perchloric acid as the titrant and crystal violet indicator shows generally good agreement. The indicator method is limited by the lack of indicators that can be used below pH 2.0. Thymol blue cannot be used for amines with an ionization constant less than 1×10^{-11}.

For example, the ionization constant of the third amino nitrogen of

Table 8. Titration of Weak Bases in Strong Salt Solutions

Compound	Ionization Constant	Medium	Indicator	Purity, % by wt.	
				Salt Solution	Other[a]
Diethyl-amine	1.3×10^{-3}	Calcium chloride, 4.5M	Methyl red	98.0	98.2
Triethanol-amine	4.5×10^{-7}	Sodium chloride, 6M	Methyl orange	102.1	102.6
γ-Picoline	1.1×10^{-8}	Lithium chloride, 8M	Thymol blue	95.6	95.4
Pyridine	1.7×10^{-9}	Lithium chloride, 8M	Thymol blue	99.1	99.6
Aniline	3.8×10^{-10}	Lithium chloride, 8M	Thymol blue	99.8	99.4
Diethylene-triamine	4.7×10^{-11}[b]	Lithium chloride, 8M	Thymol blue	99.3	99.4
Sodium acetate	—	Sodium iodide, 8M	Thymol blue	100.0	99.4

[a] By titration in acetic medium with perchloric acid using crystal violet indicator.

[b] Corresponds to ionization of third amino nitrogen (K_3).

diethylenetriamine is 4.7×10^{-11}. This compound can be titrated satisfactorily using thymol blue. The ionization constant of the fourth amino nitrogen of triethylenetetramine is approximately 1×10^{-12}; this group cannot be titrated in salt solutions using thymol blue. Possibly other indicators will eventually be found that can be used below pH 1.5.

DIFFERENTIATING TITRATIONS

The curves in Fig. 11.5 show the potentiometric titration of diethylenetriamine in water and in 6M sodium iodide. The first break obtained in the aqueous titration corresponds to the neutralization of two of the amino nitrogens of the molecule. After two of the amino nitrogens have been neutralized, the third nitrogen is too weak ($K_3 = 4.7 \times 10^{-11}$) to be titrated in water, and only a small inflection can be observed in curve 1 for this end point. In 6M sodium iodide two sharp breaks are obtained that correspond to the neutralization of the second and third amino nitrogens of this compound.

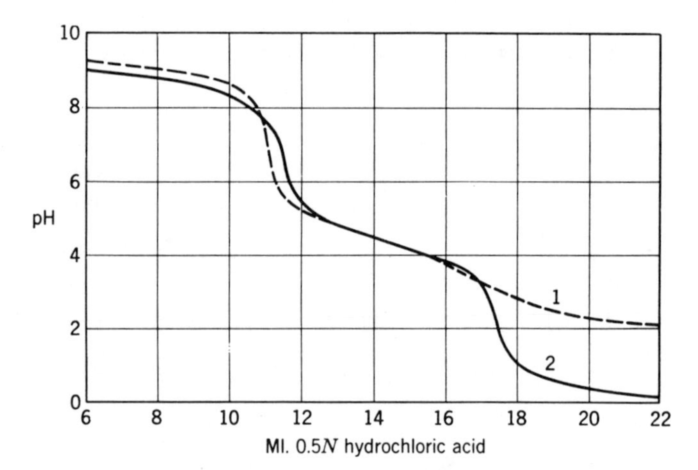

Fig. 11.5. Potentiometric titration of diethylenetriamine: 1, in water; 2, in $6M$ aqueous sodium iodide.

The curves in Fig. 11.6 show the potentiometric titration of triethylenetetramine in water and in $8M$ sodium iodide. The first inflection that occurs in the aqueous titration corresponds to the neutralization of two of the amino nitrogens. The second inflection corresponds to the neutralization of three nitrogens. After three of the amino nitrogens have been neutralized, the fourth is too weak ($K_4 = 1 \times 10^{-12}$) to be titrated in water.

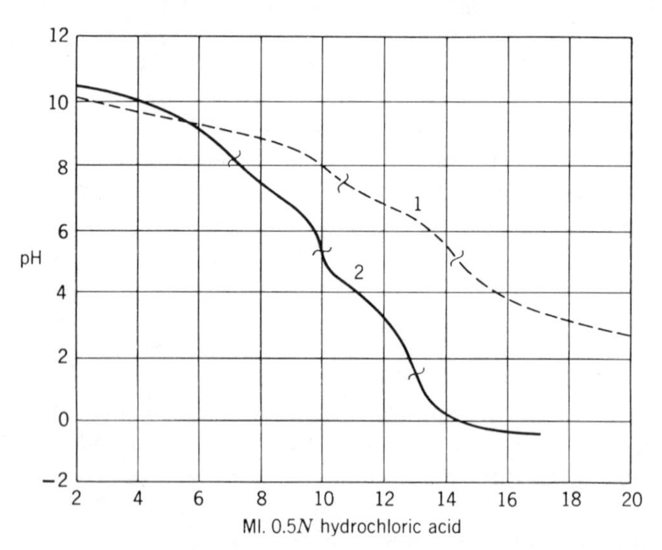

Fig. 11.6. Potentiometric titration of triethylenetetramine. 1, in water; 2, in $8M$ aqueous sodium iodide.

In $8M$ sodium iodide three inflections are obtained that correspond to the neutralization of two, three, and four nitrogens.*

Titrations in strong salt solutions should find considerable applicability in the analysis of polyfunctional amines such as the ethyleneamines. The titration of the individual amino nitrogens of compounds of this class is difficult to obtain. In water, as shown in Figs. 11.5 and 11.6, breaks can be obtained for the neutralization of one-half, two-thirds, or three-fourths of these molecules; however the breaks for the total neutralization are usually obscure. Total neutralization values can be obtained for these compounds in acetic acid or other acidic solvents, but acidic solvents cannot be used to differentiate the individual amino nitrogens.

Nondissociating compounds such as acetonitrile (5) or methyl isobutyl ketone (6) are usually good differentiating solvents. When diethylenetriamine or similar compounds are titrated in these media, insoluble salts form prior to complete neutralization. This limits the use of these differentiating solvents for the analysis of polyfunctional amines. A comparison of the curves in Figs. 11.5 and 11.6 shows that the differentiating powers of water are essentially unaffected by the addition of neutral salts. The only effect obtained is an enhancement of the sharpness of the potentiometric break for the weaker amino nitrogens. In this respect, titrations of weak amines in salt solutions appear to have certain advantages over titrations in other media.

The potentiometric titration curves in Fig. 11.7 show the application of

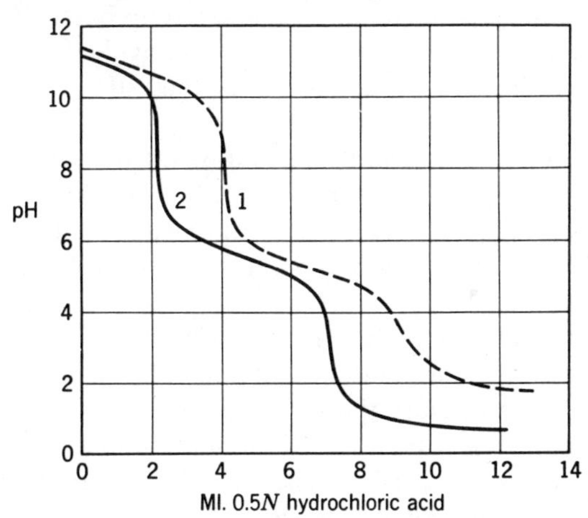

Fig. 11.7. Potentiometric titration of triethylamine-pyridine mixtures: 1, in water; 2, in $6M$ aqueous sodium chloride.

* Compare this titration to that in Fig. 11.1, where the same compound is titrated in 1:1 glycol-isopropanol with the four amino groups titrating.

this technique to the differentiation of a mixture of triethylamine and pyridine. As indicated by curve 1, water is a good medium for the titration of strong amines in the presence of weak amines but cannot be used for the reverse determination. Because neutral salts enhance the potentiometric break for weak amines without affecting the differentiating powers of water, strong salt solutions are applicable to the resolution of mixtures of weak and strong amines. Titrations of this type can also be done in nondissociating solvents such as acetonitrile (5) and methyl isobutyl ketone (6). In most cases comparable results should be obtained by these methods. In cases where the amines form insoluble salts in nonaqueous media, titrations in salt solutions may have certain advantages.

SCOPE

Strong solutions of neutral salts enhance the potentiometric break for the titration of most bases. Because of the lack of suitable indicators, the indicator method cannot be used satisfactorily for amines with ionization constant less than 1×10^{-11}. The potentiometric method cannot be used for amines with ionization constants less than 1×10^{-12}. The method is not applicable to the titration of amino acids but can be used for the titration of certain salts of weak acids and strong bases. The method has unique applicability for the titration of the individual amino nitrogen of polyfunctional amines and is applicable to the differentiating titration of bases in general.

The volume of aqueous titrant that can be introduced into the titration medium is dependent on the amount of dilution that can be tolerated without affecting the potentiometric break. Usually a 10 to 15% dilution will not appreciably affect the end point if the salt solution is originally 7 to 8*M*.

Organic bases such as aniline, which have low water solubility, are even less soluble in strong salt solutions. In some cases these amines can be titrated into solution because their salts are normally more soluble. Small amounts of methanol can be used to solubilize amines without precipitating excessive quantities of salts. When cosolvents are necessary, lithium chloride is usually the salt of choice because of its organic solubility, particularly in methanol.

5. J. S. Fritz, *Anal. Chem.*, **25,** 407 (1953).
6. D. B. Bruss and G. E. A. Wyld, *Anal. Chem.*, **29,** 232 (1957).

ACETIC ACID SYSTEM

Adapted from J. Fritz

[*Anal. Chem.*, **22**, *1028–9* (*1950*)]

In glacial acetic acid the titration curves are so accentuated that even good indicator end points can be obtained.

REAGENTS

Perchloric acid, 0.1*N*. Dissolve about 8.5 ml of 70% perchloric acid in 1 liter of glacial acetic acid. Add 15 ml of acetic anhydride *cautiously* in small portions, and allow to stand overnight.

Sodium acetate solution, 0.1*N*, to standardize the perchloric acid. Dissolve a weighed portion (about 0.53 gram) of dried sodium carbonate in enough acetic acid to make 100 ml of solution. Potassium acid phthalate makes a very good and also convenient standard.

Methyl violet indicator or naphthol benzein, 0.25% solution in acetic acid.

PROCEDURE

The regular acid-base titration is the procedure used here except that 25 to 50 ml of glacial acetic acid is used as a solvent. The methyl violet indicator changes from the violet to a green, and the naphthol benzein goes from a yellow to a green. Either indicator can be used. This solvent can be used for a potentiometric titration, using the standard pH meter with glass and calomel electrodes. Another set of electrodes can be used for this system, the glass electrode as indicator electrode and a silver wire with a thin coating of silver chloride as the reference electrode.

To test this procedure, aniline, pyridine, *N*-ethyl aniline, *N*,*N*-diethyl aniline, α-naphthylamine, and quinoline were used. Good indicator end points were obtained for these compounds. Fritz also used brucine and benzylamine. The procedure is generally applicable to weak bases with dissociation constants down to 10^{-10}. The end points obtained in acetic acid are generally sharper than those obtained in nonacid solvents, and accuracy and precision of $\pm 0.3\%$ can easily be obtained.

As indicated on page 534, acetic acid is highly recommended for sharp end points but not for resolving mixtures of amines, since all will generally titrate together with one end point.

GLYCOL SOLVENTS AND SOLVENT MIXTURES

The procedure of S. Palit for determining carboxylic salts (pp. 166–8) can also be used for titrating amines. The glycol systems do not yield as sharp end points as the acetic acid system does, but they do yield much better end points for the weaker amines than does water. The glycol solvent systems generally have a high discrimination for different amines of relatively small differences in basicity (see discussion, p. 534).

KETONE SOLVENTS

Methyl Isobutyl Ketone and Methyl Ethyl Ketone—Adapted from D. B. Bruss and G. E. A. Wyld

[*Anal. Chem.*, ***29***, *232–5* (*1957*)]

REAGENTS

Perchloric, acid 0.2*N*, in dioxane. The appropriate amounts of 70 to 72% perchloric acid are dissolved in dioxane that has been refluxed over potassium.

PROCEDURE

In the desired ketone, dissolve a sample containing an amount of amine to give an optimum size titration with the titration devices used. Titrate the solution with the perchloric acid reagent using a glass indicator electrode and a calomel (sleeve-type) electrode. The titrations are best carried out potentiometrically.

ACETONE

Acetone can be used as a solvent for titrating amines using the same acid titrant as previously for the other ketone solvents, or the acid titrants used in the glycol-solvent system above can be used. The ketones must be avoided in titrating primary amines for the reasons given earlier (p. 534).

DIOXANE SYSTEM

Adopted from J. Fritz

[*Reprinted in Part from Anal. Chem.*, ***22***, *578–9* (*1959*)]

Dioxane serves as an excellent solvent for the titration of most organic bases. The titrating acid is a solution of perchloric acid in dioxane, which may be kept for several weeks with little change in titer. Modified methyl orange (xylene cyanol) or methyl red serves as indicator, giving sharp end points. The titration of most bases may thus be quickly and conveniently carried out.

NITROGEN HETEROCYCLIC BASES

Pyridine ($K_b = 1.4 \times 10^{-9}$), although too weak to be titrated accurately in water, can be easily determined by titration in dioxane to the modified methyl orange end point. During the titration, pyridine perchlorate comes down as a slight soluble white precipitate. This precipitate in no way interferes with the end point—in fact, it is chiefly responsible for the sharpness of the end point because the pyridinium ion is thus effectively removed from solution.

2,2′-Bipyridine and 1,10-phenanthroline may be titrated as monoacid bases in dioxane to the modified methyl orange end point or in ethyl ether to the methyl red end point. The perchlorate salt of these bases precipitates during the titration. Brucine ($K_1 = 9 \times 10^{-7}$, $K_2 = 2 \times 10^{-12}$) is insufficiently soluble in water to be titrated directly, but in dioxane it may be conveniently titrated as a monoacid base. Brucine monoperchlorate is precipitated. 2,6-Lutidine gives an extremely sharp end point with modified methyl orange; an emulsion is formed during the titration. Hexamethylenetetramine ($K_b = 8 \times 10^{-10}$) can also be successfully titrated in dioxane with modified methyl orange indicator. It is usually difficult to get hexamethylenetetramine completely in solution without using excessive amounts of dioxane. If the titration is carried out slowly with efficient stirring, however, accurate results are obtained even if solution of the original sample was not complete.

Quantitative results for the titrations above are given in Table 9.

Table 9. Titration of Heterocyclic Bases with Perchloric Acid in Dioxane

Base	Weight Taken, grams	$HClO_4$ Used, ml.	$HClO_4$, N	Purity, %
Pyridine	0.1987	24.87	0.0994	98.39
	0.1483	18.64		98.81
	0.2373	29.70		98.39
	0.1622	20.30		98.39
				Av. 98.50
2,6-Lutidine	0.2179	20.00	0.0992	97.58
	0.2613	23.98		97.57
				Av. 97.58
2,2′-Bipyridine	0.2105	13.65	0.0988	100.01
	0.3000	19.43		99.89
	0.3167	20.50		99.83
	0.3175	20.59		100.02
				Av. 99.94
1,10-Phenanthroline	0.3596	18.20	0.0994	99.67
	0.4168	21.03		99.36
	0.3700	18.70		99.53
				Av. 99.51
Brucine	0.4700	12.20	0.0976	99.88
	0.7590	19.72		99.97
	0.7486	19.47		100.07
				Av. 99.97

AMINES

The titration of aniline in dioxane was attempted, but both the modified methyl orange and methyl red end points are very poor. No precipitate appears during the titration. Aliphatic amines can be successfully titrated in dioxane, even though the perchlorate salt of the amine does not precipitate. As an example, benzylamine was titrated in both dioxane and ethyl ether, using as indicators modified methyl orange and methyl red, respectively. The results obtained by this titration agree with those obtained by titrating benzylamine in water with aqueous perchloric acid (Table 10).

Table 10. Titration of Benzylamine with Perchloric Acid in Dioxane and Water

Solvent	Weight Taken, grams	$HClO_4$ Used, ml.	$HClO_4$, N	Purity, %
Water	0.2075	19.48	0.0977	98.31
	0.2330	21.83		98.11
			Av.	98.21
Dioxane	0.2342	21.55	0.0996	98.24
	0.2561	23.52		98.06
Ether	0.2262	20.86		98.46
			Av.	98.25

INTERFERENCES

Water and alcohols interfere with all the previously described titrations. The indicator blank with both modified methyl orange and methyl red in 90% dioxane–10% water, for example, amounts to several milliliters of 0.1*N* perchloric acid. The conversion of these indicators to their acid colors is very gradual. Alcohols interfere in a similar manner to water, but not to such a great extent. Ketones, aldehydes, hydrocarbons, nitrobenzene, and most carboxylic acids do not interfere. Table 11 gives the results obtained by titrating hexamethylenetetramine in the presence of various impurities.

Table 11. Titration of Hexamethylenetetramine in the Presence of 0.5 to 1 gram of Added Impurities

Weight Taken, grams	Impurity	$HClO_4$ Used, ml.	$HClO_4$, N	Purity, %
0.2716	None	19.93	0.0972	99.99
0.3271	$CH_3COCH(CH_3)_2$	24.00	0.0972	99.98
0.2464	None	18.05	0.0972	99.82
0.2771	*tert*-Butyl alcohol	20.50	0.0972	100.80
0.2862	Nitrobenzene	21.01	0.0972	100.03

PROCEDURE

Weigh a sample of the proper size and dissolve it in dioxane or ethyl ether (25–50 ml of solvent usually suffices). Add 2 drops of indicator (modified methyl orange or methyl red for dioxane; methyl red for ether) and titrate the solution with perchloric acid. One drop or less of $0.1N$ acid is usually sufficient to give a sharp color change at the end point. In titrations where a precipitate is formed, the use of a magnetic stirrer is recommended but not required.

Prepare the perchloric acid solution by dissolving approximately 8.4 ml of 72% perchloric acid in 1 liter of dioxane.* This solution is standardized against diphenylguanidine. Diphenylguanidine is an excellent primary standard because it is readily available, easily purified, nonhygroscopic, easily soluble in dioxane and ether, has a high equivalent weight, and is a strong base.

DISCUSSION

The samples of 2,2′-bipyridine, brucine, and hexamethylenetetramine were known to be of very high purity. The fact that their purity was in each case very close to 100%, as determined by titration in dioxane, indicates that this method is capable of great accuracy. This would also seem to confirm the advisability of using diphenylguanidine as a primary standard.

The ionization of most acids and bases dissolved in dioxane and other solvents with low dielectric constants is very slight. (An exception would be solvents possessing pronounced acid or basic properties. The ionization of acids in pyridine, e.g., would be comparatively great.) In such media, therefore, an acid-base titration involving ions would be expected to proceed in a sluggish fashion to give a very poor end point. All the bases that have been titrated successfully in dioxane, however, are neutral and do not react as ions. A strong acid HA will react readily with such a neutral base

$$C_5H_5N + HA \rightarrow C_5H_5NH^+ + A^-$$

* *Authors' note.* Dioxane often contains significant amounts of aldehyde impurity because of peroxide formation and subsequent decomposition. This aldehyde affects the titration of primary amines, since it forms the Schiff base with these compounds, noticeably reducing this basicity. The dioxane should be purified by distillation from lithium aluminum hydride. An alternative purification consists of allowing the dioxane to stand for several days over solid sodium hydroxide. The purified dioxane should be stored over sodium hydroxide.

Even with acids as strong as methanesulfonic acid, this titration is poor because the anion liberated acts as a base in dioxane and partially reverses the reaction. With perchloric acid, however, a sharp end point is obtained because the perchloric ion possesses only extremely weak basic properties and the perchlorate ion is largely removed from solution in most titrations by the formation of a slightly soluble perchlorate salt.

NITROMETHANE SYSTEM

Adapted from C. Streuli

[*Reprinted in Part from Anal. Chem.*, ***31,*** *1652–4* (*1959*)]

Nitromethane, as well as acetonitrile, is an eminently practical medium for the titration of weak organic bases. A very weak acid, nitromethane has an extensive potential range, neither levels nor reacts with most solutes, and has a dielectric constant of nearly 40, which allows potentiometric measurements to be made easily. The insolubility of most salts in nitromethane and the effect of water on the potential range are its principal disadvantages. Fritz and Fulda (7) have shown the uses of nitromethane as a differential medium, but the solvent mixture they used contained 20% of acetic anhydride.

Because of the utility of nitromethane and the limited data for this solvent, the work described was initiated. The compounds selected for use in this work were those that offered a variety of molecular organic structure and pK_a range consistent with basic properties. Primary, secondary, and tertiary amines, heterocyclics, ureas, amides, guanidine derivatives, and imidates have been included.

REAGENTS

The nitromethane was obtained from the Fisher Scientific Company. It was reagent grade material and was used without further purification.

Standard 0.05*N* perchloric acid solutions were prepared by diluting 4.2 ml of 72% acid to 1 liter with nitromethane. Solutions were stored in closed brown bottles and were standardized against potassium hydrogen phthalate dissolved in acetic acid (8) at weekly intervals. The standard solution appears to be stable for at least a month.

Most of the materials titrated were Eastman Kodak White Label grade. The cyanoethyl compounds and imidates were research samples prepared and purified in these laboratories.

7. J. S. Fritz and M. O. Fulda, *Anal. Chem.*, **25,** 1837 (1953).
8. H. LeMarie and H. J. Lucas, *J. Am. Chem. Soc.*, **73,** 5198 (1951).

APPARATUS

All titrations were performed using a Precision-Dow Recordomatic titrator. Glass and aqueous sleeve calomel electrodes were employed to measure potentials during titrations. Solvents were agitated with a magnetic stirrer.

PROCEDURE

Dissolve 1 mM of material in nitromethane and dilute to 100.0 ml in a volumetric flask. Then dilute an aliquot of 25.0 ml of this solution to 100 ml with nitromethane and titrate it. Run three aliquots on each compound and determine blanks on each batch of nitromethane.

RESULTS AND DISCUSSION

All titration curves were corrected for blanks and plotted as potential against per cent neutralization. Several examples of the plots appear in Fig. 11.8. The titration curves illustrated include examples of amines, ureas, and amides.

No leveling of base strength was observed, even for amines as strong as piperidine. Because nitromethane does have a labile hydrogen, leveling probably does occur, but only in a more basic region. All the amines, heterocyclics, and guanidine derivatives titrated gave curves qualitatively similar to those illustrated for these types of compound. Between 20 and 80% neutralization, the titration curve for those bases with pK_a (H_2O) values less than 8 are essentially linear with a slope of 1.1 ±0.1 mV per per cent neutralization. It can also be seen that pyridine, which in water is a weaker base than either *N,N*-diethylaniline or *N*-ethyl-*N*-methylaniline, shows a base strength greater than the latter compound when dissolved in nitromethane. Urea is also considerably stronger than diphenylamine in this solvent, which is contrary to literature pK_a data (8, 9).

The shape of the titration curve for urea is interesting, being much steeper than the curve for amines. The slope between 20 and 80% neutralization is 2.3 mV per per cent neutralization. This type of curve is characteristic of all the ureas as well as mono- and unsubstituted amides. Disubstituted amides show titration curves similar to those for amines. This type of behavior was not noted in either acetic acid (9) or acetic anhydride (10), but is characteristic for phenols in both acetone (11) and

9. N. F. Hall, *J. Am. Chem. Soc.*, **52,** 5115 (1930).
10. C. A. Streuli, *Anal. Chem.*, **30,** 997 (1958).
11. J. S. Fritz and J. J. Yamamura, *Anal. Chem.*, **29,** 1079 (1957).

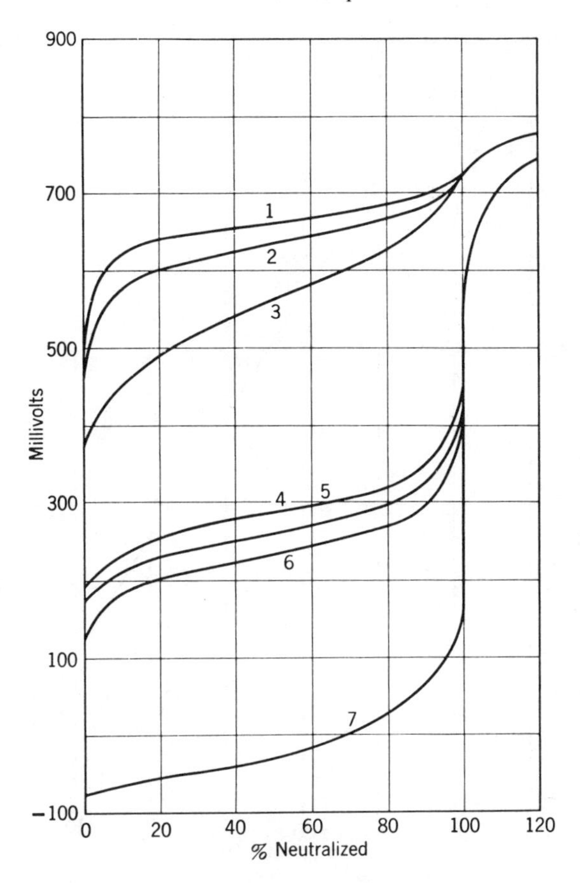

Fig. 11.8. Titration of nitrogen bases in nitromethane: 1, diphenylamine; 2, *N*-methylacetanilide; 3, urea; 4, *N*-ethyl-*N*-methylaniline; 5, pyridine; 6, *N,N*-diethylaniline; 7, *N,N'*-diphenylguanidine.

pyridine (12). It is probably because of intermolecular hydrogen bonding between amide or urea molecules rather than hydrogen bonding between solute and solvent. Disubstituted amides with no available hydrogen atoms behave like the amines as exemplified by *N*-methylacetanilide (Fig. 11.8).

Results were quantitative for all the compounds titrated, with the exception of some very weak amides. In the latter cases the potential change is so slow that it is difficult to locate an end point.

Figure 11.9 illustrates the relation between pK_a (H_2O) and the half-neutralization potential (ΔHNP) (CH_3NO_2) for the compounds titrated.

12. C. A. Streuli and R. R. Miron, *Anal. Chem.*, **30**, 1978 (1958).

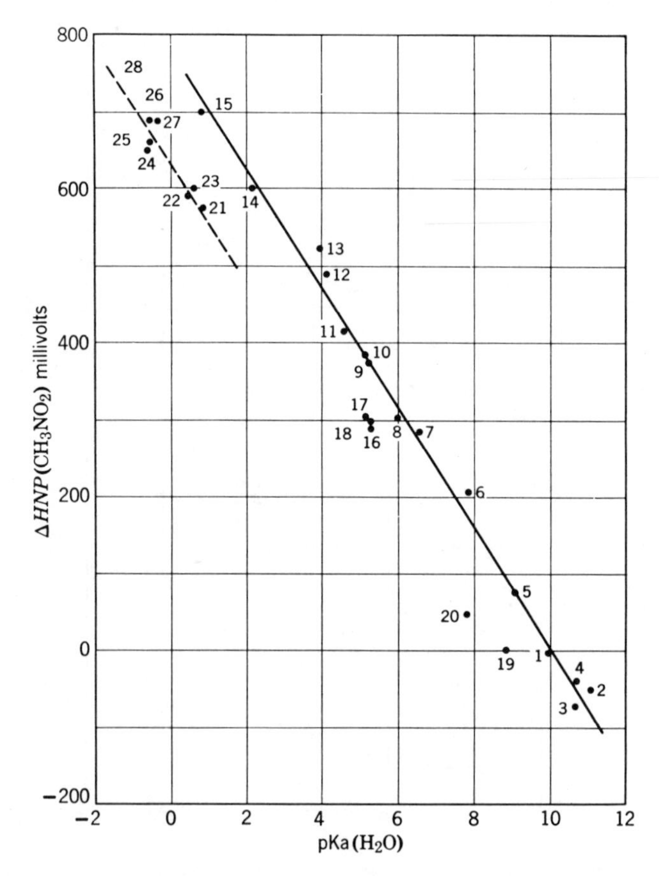

Fig. 11.9. Basic strength of nitrogen compounds in nitromethane and in water.

Pertinent data are given in Table 12. ΔHNP Values were calculated by algebraically subtracting the half-neutralization potential (*HNP*) for a *N,N′*-diphenylguanidine sample run on the same day from the *HNP* value of the compound being tested. This is necessary because of day-to-day shifts in liquid junction potentials of the electrodes. ΔHNP Values are reproducible within 5 to 6 mV, whereas the *HNP* value of a compound using this electrode system may shift 100 mV over several days' time. *N,N′*-Diphenylguanidine was chosen as a reference standard because of its high base strength, purity, and availability. Using this compound as a reference, all titration curves and ΔHNP values stand in correct relative relation to one another.

Two linear relations between pK_a (H_2O) values and ΔHNP (CH_3NO_2) values are apparent in Fig. 11.9. The main sequence is determined by the amines, the shorter sequence by amides and ureas. Heterocyclic nitrogen compounds and hydroxylated amines do not fit either sequence. Essentially this illustrates that compounds containing the grouping $\mathrm{>C(=O)-N<}$ as well as heterocyclics and hydroxy amines are, relative to amines, stronger bases in nitromethane than they are in water. This type of behavior was not noted for amides and ureas, in either acetic acid (9) or acetic anhydride (10), but probably holds true for such solvents as acetonitrile or acetone. It is again most probably related to hydrogen bonding between solute and solvent molecules in the various solvents.

The straight line drawn for the amine sequence is the least-squares solution of the data for all the amines exclusive of the heterocyclic and hydroxy amines. The equation relating pK_a (H_2O) and ΔHNP (CH_3NO_2) values for the amine sequence is

$$pK_a\,(H_2O) = 10.12 - 0.0129 \times \Delta HNP\,(CH_3NO_2) \qquad (1)$$

The least-squares solution for all the amide and urea data in Table 12 is given by the equation

$$pK_a\,(H_2O) = 8.26 - 0.0120 \times \Delta HNP\,(CH_3NO_2) \qquad (2)$$

and is shown by the broken line in Fig. 11.9. Calculated pK_a (H_2O) values using ΔHNP (CH_3NO_2) data and the appropriate equation are listed in third column of Table 12; deviations between literature and calculated values are given in column 4. The standard deviation for all the data is 0.27 pK_a unit exclusive of the heterocyclics and hydroxy amines. Both equations fail badly for these compounds, which show strengths intermediate between the two main classes of compounds.

The two lines are essentially parallel with slopes of 78 mV per pK_a unit. Hall (9) assumed a slope of 59 mV per pK_a unit for this relation in acetic acid. In acetic anhydride this value is 51 mV per pK_a unit (10). On this basis, nitromethane should have greater resolving power than the anhydride.

A number of other molecular species with basic characteristics but unknown or poorly defined pK_a values were also titrated. Typical titration curves are given in Fig. 11.10 and ΔHNP (CH_3NO_2) values in Table 13. Values of pK_a have been calculated from these data using either eq. 1 or 2 depending on the molecular structure of the solute. In the cases of the imidates and the phosphine, eq. 1 was used, since the titration behavior of these compounds was similar to that of amines. Assuming that this

Table 12. Basicity Data for Amines, Amides, and Ureas in Nitromethane and Water

No.	Compound	ΔHNP	pK_a Literature	Calculated	ΔpK_a	Reference
1	*N,N'*-Diphenylguanidine	0	10.00	10.12	0.12	*a*
2	Piperidine	−47	11.09	10.72	−0.37	*a*
3	Triethylamine	−69	10.64	11.01	0.37	*b*
4	Tri-*n*-butylamine	−37	10.67	10.60	−0.07	*a*
5	*N,N*-Dimethylbenzylamine	78	9.02	9.12	0.10	*b*
6	2-Cyanothylamine	204	7.86	7.49	−0.37	*c*
7	*N,N*-Diethylaniline	286	6.52	6.43	−0.09	*a*
8	*N*-Ethyl-*N*-methylaniline	304	5.99	6.20	0.21	*a*
9	*N,N*-Dimethylaniline	374	5.21	5.30	0.09	*d*
10	Bis-2-cyanoethylamine	386	5.14	5.14	0.00	*c*
11	*o*-Phenylenediamine	412	4.52	4.81	0.29	*b*
12	2-Naphthylamine	490	4.11	3.80	−0.31	*a*
13	1-Naphthylamine	522	3.92	3.39	−0.53	*a*
14	Anthranilic acid	602	2.15	2.36	0.20	*b*
15	Diphenylamine	701	0.85	1.08	0.23	*a*
16	Pyridine	286	5.30	(6.43)	1.13	*a*
17	Quinoline	305	5.06	(6.19)	1.13	*d*
18	Isoquinoline	299	5.30	(6.26)	0.96	*d*
19	Bis-2-hydroxyethylamine	4	8.87	(10.07)	1.20	*c*
20	Tris-2-hydroxyethylamine	49	7.82	(9.49)	1.67	*e*
21	Methylurea	576	0.90	0.79	−0.11	*a*
22	Urea	590	0.50	0.61	0.11	*e*
23	Caffeine	600	0.61	0.48	−0.13	*b*
24	*N*-*n*-Propylacetanilide	650	−0.60	−0.17	−0.43	*a*
25	*N*-Methylacetanilide	663	−0.50	−0.34	−0.16	*a*
26	Acetamide	691	−0.48	−0.70	0.22	*a*
27	Phenylurea	691	−0.30	−0.70	0.40	*a*
28	Acetanilide	757	−1.70	−1.56	−0.14	*e*

[a] Ref. 9.

[b] N. A. Lange, *Handbook of Chemistry*, 4th ed., Handbook Publishers, Sandusky, Ohio, 1941, pp. 1220–1.

[c] C. A. Streuli and S. Sandler, unpublished data.

[d] I. M. Kolthoff and N. H. Furman, *Potentiometric Titrations*, Wiley, New York, 1926, p. 329.

[e] Ref. 8.

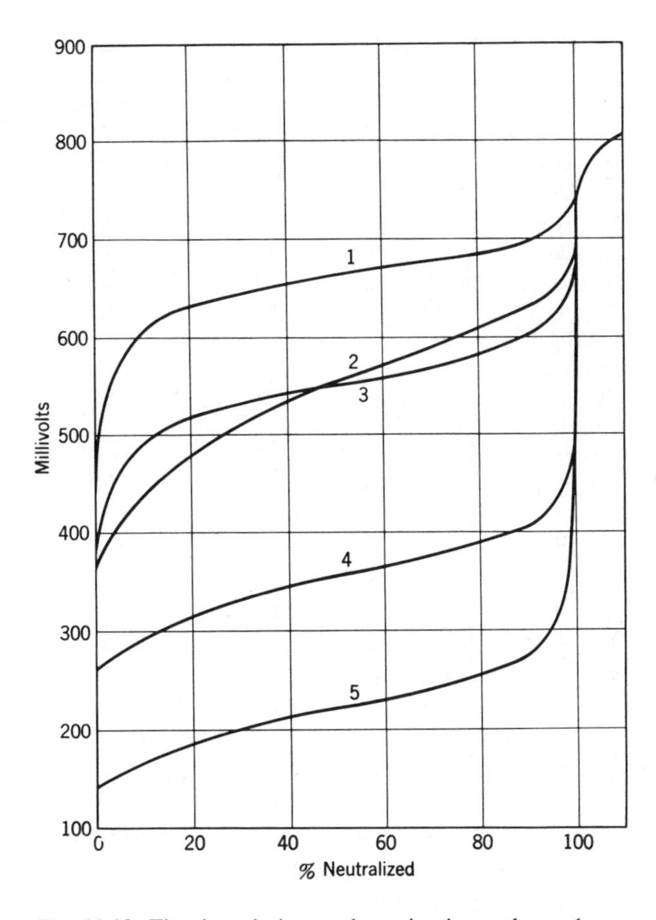

Fig. 11.10. Titration of nitrogen bases in nitromethane: 1, acrylamide; 2, 2-pyrrolidinone; 3, tris-2-cyanoethylamine; 4, ethyl *p*-nitrobenzimidate; 5, ethyl benzimidate.

equation applies in these cases, the 90% confidence limits for these pK_a values are $\pm 0.4\,pK_a$ unit. Nitromethane makes a convenient titration medium for imidates that are hydrolyzed in water to esters.

Very weak bases such as dimethylsulfoxide, tetrahydrofuran, and acetonitrile show no titratable basicity in this solvent. This is most probably due to interference by water in the titrant.

It is possible to resolve mixtures of amines in the main sequence that differ in ΔHNP values by approximately 180 mV. The theoretical limit of resolution should be about 100 mV. However this has not been achieved with any of the compounds tested.

Table 13. Titration of Various Organic Bases in Nitromethane

Compound	ΔHNP (CH_3NO_2)	Calculated pKa
2-Pyrrolidinone	699	−0.8
N-Methyl-2-pyrrolidinone	655	−0.2
Acrylamide	657	−0.3
Nitrilotrispropionitrile	568	2.8
Triphenylphosphine	573	2.7
Ethylbenzimidate	308	6.2
Ethylacetimidate	198	7.6
Ethylcarbethoxyacetimidate	385	5.2
Trichloroacetimidate	779	0.1
p-Nitrobenzimidate	441	4.4
Ethyl α-chloropropionimidate	403	4.9

CONCLUSIONS

The foregoing data, with the studies made in acetic acid (9) and acetic anhydride (10), illustrate that whereas the relative basicity within a class of compounds is not greatly affected by changes in titration medium, basicity relations between various classes such as amides and amines are affected to a considerable degree. This change appears to be related to the hydrogen bonding characteristics of the solvent.

An unknown pK_a value may be calculated from nitromethane data if the class of the compound is considered; pK_a values so obtained are probably reliable within 0.5 unit.

Resolution of mixtures of compounds can be predicted on the basis of pK_a data or nitromethane data if the shape of the titration curve and structure of the solute are considered. Mixtures of amines are more readily resolved than mixtures of amides or ureas because of the shape of the titration curve.

Acylation Methods

Any of the anhydride methods using either acetic, phthalic, or pyromellitic anhydrides shown for determining alcohols (pp. 9–31) can be used to determine amines. The amines react much more rapidly than do the alcohols, generally in less than half the time. The procedures otherwise remain the same as those used for the alcohols.

Mitchell, Hawkins, and Smith (13) devised an acylation method using acetic anhydride whereby the excess anhydride was determined by adding excess water and titrating the excess with Karl Fischer reagent. This acylation approach has no advantages over the more direct methods.

Since only primary and secondary amines react with anhydrides, only these types of amines can be determined. Tertiary amines do not interfere in the analysis for the primary and secondary species. The method applies very widely for all types of primary and secondary amines including aliphatic, aromatic, and alicyclic. Hydroxy groups interfere in the analysis, of course, since they acylate, but samples containing hydroxy compounds can be handled as described earlier (pp. 40–1). Aldehydes would normally interfere if acetic anhydride is used (see pp. 10–12), however aldehydes are not common in amine systems. Of course, if aldehydes were present in a primary amine system, it would react to yield the imine.

Diazotization and Nitrosation Procedures

No specific reference can be given for any general method employing nitrous acid. The following procedure is a conglomeration of many known procedures.

The procedure will mainly determine primary and secondary amino groups (see reactions 3–6). However some tertiary amines will react with nitrous acid to form the nitrite salt, but these usually cannot be determined.

$$C_6H_5NH_2HCl + HONO \rightarrow [C_6H_5\overset{+}{N}{\equiv}N]Cl^- + 2H_2O \qquad (3)$$

$$C_6H_5NHCH_3 + HONO \rightarrow C_6H_5N(NO)CH_3 + H_2O \qquad (4)$$

$$RNH_2 + HONO \rightarrow ROH + N_2 + H_2O \qquad (5)$$

$$R_2NH + HONO \rightarrow R_2NNO + H_2O \qquad (6)$$

(5), (6): where R is aliphatic

13. J. Mitchell, W. Hawkins, and D. M. Smith, *J. Am. Chem. Soc.*, **66,** 782–4 (1944).

Reaction 5 proceeds as written only when R is methyl or ethyl. When R is a higher alkyl group the reaction does often proceed quantitatively, but the end product is not the alcohol expected. For instance, if propyl amine is reacted with nitrous acid, both *n*-propyl and isopropyl alcohol result.

AROMATIC AMINES

REAGENTS

Sodium nitrite, 1*N*, 0.5*N*, 0.1*N*.
Starch iodide paper.
Concentrated hydrochloric acid.
Glacial acetic acid.
Potassium bromide solution, 25%.

PROCEDURE

Into a 1-liter beaker, weigh a sample of amine large enough to give a 20-ml titration with nitrite of a certain strength and dissolve in about 500 ml of water, 30 ml of concentrated hydrochloric acid, and 25 or 50 ml of 25% potassium bromide. (The amount of potassium bromide used depends on the ease of nitrosation of the amine.) Potassium bromide can be eliminated if the amine reacts rapidly. If the sample is not soluble in dilute acid, the sample can be dissolved in a few milliliters of glacial acetic acid, followed by addition of the water and hydrochloric acid. The amine, in such cases, generally remains in solution.

The required strength of sodium nitrite depends on several factors: ease of nitrosation, clarity of end point, instability of the nitrous acid. The 1*N* sodium nitrite will give a better end point, but the danger of losing nitrous acid is much greater, since more is formed on each addition than is formed with the more dilute sodium nitrite. When the more dilute sodium nitrite is used, a poorer end point results, but the chances of loss of nitrous acid are lower.

Immerse the beaker containing the sample to within 1 in. of the rim in chopped ice and water until the temperature is about 5°C. This is to lower chances for loss of nitrous acid. Then, with the tip of the buret well under the surface of the solution (see Fig. 11.11), add the nitrite standard solution at a rate depending on how rapidly the amine consumes the

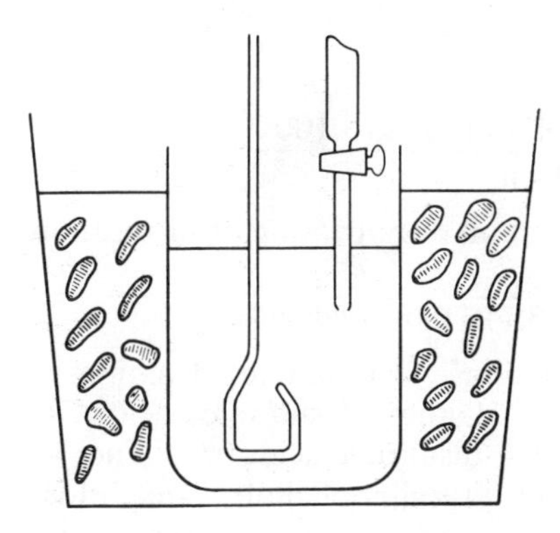

Fig. 11.11

nitrous acid. There should never be a large excess of nitrite present, since this causes loss of nitrous acid. At first, add the nitrite in small increments and test the solution by dipping a strip of starch-iodide paper into the solution; the paper turns the blue-black of iodine in starch if nitrous acid is present. If the amine consumes the nitrous acid rapidly, add the nitrite more rapidly. If the amine nitrosates slowly, add the nitrite more slowly. The purpose of the potassium bromide is to catalyze the nitrosation; it should be added whenever the nitrite is consumed too slowly. If the amine nitrosates too slowly, the end point is very hard to detect. The nitrite is added in smaller increments as the end point is approached. This decreases the chance of overtitration.

The end point is the point at which the blue-black color is produced on the starch-iodide paper when the solution has stood for quite a time after the addition of nitrite. As the end point is approached, the nitrite will be consumed more slowly; thus a sufficient length of time (depending on the rate at which the particular amine nitrosates) should be allowed before testing for the end point. An excess of nitrite gives the blue-black color on the paper immediately on immersion in the solution. In some cases, the starch-iodide paper, when dipped into the solution, does not darken immediately but darkens slowly on exposure to air. This should not be construed as indicating excess nitrite, since the nitrite produces the color immediately. Colored amines and precipitates will hamper the viewing of any color on the paper. However the presence of excess nitrite still can be shown by the solution that diffuses up the indicator paper, showing a blue-black portion above the mark indicating the depth of the immersion of the paper in the solution.

Standardize the sodium nitrite by reaction with sulfanilic acid, which can be purchased in a pure state. The procedure is that described earlier.

CALCULATIONS

$$\frac{\text{Milliliters of nitrite} \times N\ NO_2^- \times \text{mol. wt. of compound} \times 100}{\text{Grams of sample} \times 1000 \times n}$$
$$= \%\ \text{compound based on diazotization or nitrosation}$$

where

n is the number of amino groups per molecule.

Amines tested were aniline, 4-nitroaniline; p-toluidine; 1-amino-2-naphthol-4-sulfonic acid; 2,6-toluylene diamine-4-sulfonic acid; 4-aminoazobenzene; 1-amino-8-naphthol-3-6-disulfonic acid; benzidine; 4-4′-diaminodiphenylamine-2′-sulfonic acid; 4-aminophenol; 1-naphthylamine; o-aminodiphenyl.

Secondary amines tested were acetoacetanilide,

$$[C_6H_5NHC(=O)CH_2C(=O)CH_3] + HONO \rightarrow [C_6H_5N(NO)-C(=O)CH_2C(=O)CH_3] + H_2O$$

N-ethyl-1-naphthylamine, diphenylamine, N-methyl- and N-ethylaniline.

Interfering substances consist mainly of compounds that are readily reduced by nitrous acid and also compounds that contain active methylene groups. The compounds with active methylene groups will nitrosate and can be determined by this procedure. The pyrazolones are such compounds.

ALIPHATIC PRIMARY AMINES (VAN SLYKE METHOD)

The reaction between aliphatic primary amines and nitrous acid as shown earlier (p. 532) is not as clean-cut as it appears. There are some unexplained variations that occur, which makes the approach limited in its scope. Because of a lack of complete stoichiometry, the method cannot be used for assaying amines. The results will generally be low because of some nitrogen-containing by-products remaining in the systems. The method excels, however, for determining small quantities of amines in complex media. The reaction produces a gas (nitrogen, see p. 532), and this gas is measured; hence minute quantities of amines can be detected. In addition, the reaction is fairly specific. The main application of this method is in the analysis of biological materials (see pp. 586–92 for method).

Diamines

SCHIFF BASE PRECIPITATION

Method of V. J. Farrugia and P. C. Christopher

[*Reprinted in Part from Anal. Chem.*, **36**, *435* (*1964*)]

To determine the diamine content of mixtures containing the diamines in the presence of large quantities of dibasic acids, amino acids, and monofunctional amines without separation, a gravimetric procedure was used by which the diamine is precipitated as a Schiff base (salicylaldimine) by salicylaldehyde.

Huggins and Drinkard (14) reported the determination of ethylenediamine in the presence of large amounts of hydroxyalkylamines, ethanolamine, and *n*-hydroxyethylenediamine. The following method consists of slight modifications of the Huggins and Drinkard procedure to make it more general.

REAGENTS

SALICYLALDEHYDE SOLUTION. Add 10 ml of salicylaldehyde to 200 ml of water containing 5.0 grams of sodium hydroxide and stir rapidly on a magnetic stirrer until all the oil is dissolved. If a precipitate forms, filter through a medium porosity sintered-glass filter. Prepare fresh daily.

PROCEDURE

Weigh a sample containing 0.10 to 0.20 gram of the diamine and dissolve it in 300 ml of deionized water in a 400-ml beaker. Add 50 ml of the basic salicylaldehyde solution with vigorous stirring. Allow to stand for 30 minutes with occasional stirring. Adjust to pH 8.3 (pH meter) by the dropwise addition of 27% aqueous sulfuric acid (by weight). Adjust samples containing monofunctional amines to pH 11.0 in the same manner. Allow to stand at ambient temperature for 1 hour. Collect the precipitated salicylaldimine in a tared Gooch crucible containing an asbestos mat. Dry the washed precipitate in a 70°C oven for 5 hours, cool, and weigh.

14. D. Huggins and W. C. Drinkard, *Anal. Chem.*, **34**, 1756 (1962).

Table 14. Recovery of Diamines at pH 8.3

Compound	Sample, grams	Recovery,[a] %
Hexamethylenediamine dihydrochloride	0.0050	96.73
	0.1003	97.55
	0.1200	98.63
	0.2000	98.65
	0.2360	97.87
	0.2380	97.74
p-Phenylenediamine	0.1227	99.22
	0.1278	99.41
Hydrazine sulfate	0.0987	94.00
	0.1221	94.50
	0.1366	93.50
Ethylenediamine	0.1275	97.37
	0.1438	97.65
Benzidine dihydrochloride	0.1100	100.40
	0.2000	100.39

[a] Quotient of observed and theoretical grams of imine yield expressed as percentages.

Table 15. Results of Determinations of Hexamethylenediamine Dihydrochloride and Ethylenediamine in Synthetic Mixtures Containing Adipic and Amino Acids

Amino Acid	Amino Acid Taken %	Adipic Acid Taken %	Diamine, %		Standard Deviation
			Taken	Found	
ε-Amino-*n*-caproic acid hydrochloride	0.05	40.0	59.5[a]	59.34[b]	±0.53
	1.0	40.0	59.0[a]	58.93[b]	±0.51
	10.0	40.0	50.0[a]	50.43[b]	±0.44
Glycine	30.0	40.0	30.0[a]	30.64[b]	±0.19
Anthranilic acid	30.0	40.0	30.0[a]	29.98[b]	±0.05
DL-Aspartic acid	30.0	40.0	30.0[a]	29.95[b]	±0.05
Glycine	30.65	38.87	30.48[c]	30.28[b]	±0.05
	30.70	38.82	31.09[c]	30.98[b]	±0.03
Glycine	31.04	38.21	30.75[d]	30.59[e]	±0.09
Taurine	27.98	43.70	28.61[d]	28.14[b]	±0.08

[a] Hexamethylenediamine dihydrochloride.
[b] Based on 98.0% recovery.
[c] Ethylenediamine.
[d] Benzidine.
[e] Based on 100.4% recovery.

Table 16. Results of Determinations[a] of Hexamethylenediamine Dihydrochloride in Synthetic Mixtures Containing Adipic Acid and Amines

Amine	Amine Taken, %	Adipic Acid Taken, %	Diamine, %		Standard Deviation
			Taken	Found	
Aniline	38.72	34.00	27.28	27.22	±0.10
n-Propylamine	38.00	33.66	28.34	27.11	±0.50
o-Toluidine	30.37	43.20	26.43	26.59	±0.25
Aminoethylpiperazine	33.75	40.00	26.25	26.13	±0.30

[a] Precipitation at pH 11.0.

RESULTS AND DISCUSSION

Recoveries of synthetic mixtures are shown in Tables 14 to 16. The data reported represent triplicate runs. In the case of 0.12 gram of hexamethylenediamine dihydrochloride (Table 14), they are representative of a replicate of nine runs. Interference was not encountered from either adipic acid or the amino acids; ε-amino-*n*-caproic acid hydrochloride, glycine, anthranilic acid, or DL-aspartic acid.

Only in the case of *n*-hexylamine was there precipitation. The adducts of the monofunctional amines, *n*-propyl, *o*-toluidine, aminoethylpiperazine, and aniline did not precipitate.

Bromination (Aromatic Amines Only)

Aromatic amines will substitute bromine on the ring as shown in the equation on page 532. The bromination methods for determining phenols (pp. 57–9), can be used for amines as well. Aliphatic amines do not undergo the reaction but will interfere to some degree, since the aliphatic amines are prone to oxidation by bromine.

Results obtained by Krause and Kratochvil (p. 566) by the titrations of several aromatic amines with bromine in propylene carbonate, with and without the addition of excess pyridine, appear in Table 17, and their analytical data are summarized in Table 18.

The bromination of aromatic amines proceeds only in the unoccupied ortho and para positions on the ring. The bromination method does not enjoy wide popularity because there are so many interferences owing to the oxidizing and substitution properties of bromine.

Table 17. Titrations of Aromatic Amines with Bromine in Propylene Carbonate

	First End Point			Second End Point		
Compound Titrated	Bromine-Amine Ratio	Approximate Break, mV	Midpoint Potential, mV	Bromine-Amine Ratio	Approximate Break, mV	Midpoint Potential, mV
p-Nitroaniline	1.07	50	+150	—	None	—
Anthranilic acid	2.00[a]	50	+110	—	None	—
Anthranilic acid[b]	1.4–1.6	150	−150	4.1	50	+10
Aniline	0.6–0.8	50	−250	2.00[a]	250	+50
Aniline[b]	1.0–1.1	150	−325	4.00[a]	100	−30
m-Phenylenediamine	1.14	500	−300	3.00[a]	50	+100
m-Phenylenediamine[b]	2.6	300	−400	6.00[a]	50	0
p-Toluidine	0.47–0.52	150	−300	1.7–1.8	300	0
p-Toluidine[b]	0.9–1.1	100	−275	4.00[a]	150	0
p-Phenetidine	0.46–0.47	200	−350	1.8	200	+50
p-Phenetidine[b]	1.00	200	−350	—	None	—

[a] End point recommended for analytical use.
[b] Excess pyridine present.

Table 18. Analysis of Aromatic Amines by Titration with Bromine in Propylene Carbonate

Compound Titrated	End Point Ratio Bromine-Amine	Approximate Sample Size, mM	Average Analysis, %	Relative Standard Deviations, ppt
Aniline	2	0.6	99.0 (5)	1
Aniline[a]	4	0.3	99.5 (4)	6
Anthranilic acid	2	0.6	99.4 (4)	8
Anthranilic acid[a]	4	0.3	103.9 (5)	6
p-Toluidine	4	0.3	99.5 (4)	4

[a] Excess pyridine present.
[b] Figures in parentheses indicate number of titrations.

Coupling (Aromatic Amines Only)

Some aromatic amines will couple as do phenols with diazonium compounds as shown in the equation on page 62. These amines are generally the polyfunctional aromatic amines. The method is not widely used because of its limited range and also because of the complexity of the coupling methods. This method is described in detail in the section on phenols pp. 62–6. The coupling method is very specific, however, and occasions calling for such specificity do arise.

Determination of Amines in Mixtures

PRIMARY, SECONDARY, AND TERTIARY AMINE MIXTURES: SCHIFF BASE AND ACETYLATION APPROACH

Wagner, Brown, and Peters (15) devised a system for determination of primary, secondary, and tertiary aliphatic amines. Their system could not be applied to aromatic mixtures because of the much weaker basic properties of the aromatic amines. The methods described below utilize the same reactions as those devised by Wagner, Brown, and Peters, but the reaction media and techniques used make possible the utilization of the analysis system for the determination of aromatic amine mixtures as well as aliphatic systems. Results appear in Table 19.

Table 19

System	Primary		Secondary		Tertiary	
	% Calculated	% Found	% Calculated	% Found	% Calculated	% Found
Aniline, ethylaniline, and diethyl aniline	33.29	33.05	31.89	31.87	34.79	34.90
	74.95	75.41	12.02	11.29	13.02	12.96
	9.70	10.52	43.30	42.04	47.00	47.32
Aniline, methylaniline, and dimethylaniline	33.39	32.86	33.34	32.61	33.24	34.00
	74.97	74.57	12.50	12.44	12.58	12.52
	9.77	10.43	45.13	45.27	45.05	45.42
α-Naphthylamine, ethyl-α-Naphthylamine, and diethyl-1-α-naphthylamine	34.16	33.34	32.96	33.78	32.88	32.50
	9.89	10.14	45.07	44.53	45.04	44.74
α-Naphthylamine and dimethyl-α-naphthylamine	49.96	49.38	[a]	—	46.44	46.59

[a] Secondary amines to this system could not be obtained.

15. C. D. Wagner, R. H. Brown, and E. D. Peters, *J. Am. Chem. Soc.*, **69,** 2609–14 (1947).

Mitchell, Hawkins, and Smith (16) proposed determining tertiary amines by determining first the sum of primary and secondary amines by acetylation and determining excess anhydride by aquametric means; the tertiary amine was obtained by determining total base and subtracting the sum of primary and secondary amines. In aromatic systems the amines involved are so weakly basic that titration of total base for determination of tertiary amines is impossible by ordinary means. Also, the acetylation procedure used for determining primary plus secondary amines cannot be used when alcohols are present with the amines. The procedure has an unnecessary step in the addition of excess water and determination of the water by Karl Fischer reagent. Simply determining the excess anhydride after acetylation by titration with sodium hydroxide has been found to yield very good results.

Hawkins, Smith, and Mitchell (17) (see pp. 601–4) also devised a procedure for determining primary amines in the presence of secondary and tertiary amines. The sample is reacted with benzaldehyde, and the water liberated is determined by Karl Fischer reagent. This procedure involves the use of hydrogen cyanide, which necessitates special handling and is more time-consuming than the procedures to be described.

The method for determining amine mixtures described in this section can be outlined as follows. The tertiary amine is determined by adding acetic anhydride directly to a weighed sample. After a short time the acetylated mixture is dissolved in 1:1 ethylene glycol–isopropanol, and the tertiary amine (which is not affected by the anhydride) can be titrated using standard hydrochloric acid. The glycol-isopropanol solvent is used to accentuate the titration of the tertiary aromatic amines, which are weak bases. It was first proposed by Palit (18) to make possible the titration of weak bases that could not be determined accurately by titration in aqueous solution (see pp. 166–7).

The primary amine in the mixture is determined by titrating the total base in the sample in the 1:1 ethylene glycol–isopropanol solvent. To a separate sample, salicylaldehyde is added to remove the primary amine by the Schiff reaction; the remaining base in the sample is then titrated. The difference between the two titrations will yield the primary amine content.

The secondary amine content is determined by taking the titration value after addition of salicylaldehyde; this value is tertiary amine plus secondary amine. By subtracting the value obtained for the tertiary amine as described previously, the amount of secondary amine can be determined.

16. J. Mitchell, Jr., W. Hawkins, and D. M. Smith, *Ibid.*, **782**.
17. W. Hawkins, D. M. Smith, and J. Mitchell, Jr., *J. Am. Chem. Soc.*, **66,** 1662 (1944).
18. S. Palit, *Ind. Eng. Chem.* (*Anal. Ed.*), **18,** 246–51 (1946).

Procedure of Siggia, Hanna, and Kervenski

[*Anal. Chem.*, **22**, *1295* (*1950*)]

REAGENTS

Ethylene glycol–isopropanol mixture, 1:1.

Standard 1*N* hydrochloric acid in ethylene glycol–isopropanol mixture. Dilute 96 ml of concentrated hydrochloric acid to 1 liter with equal parts of ethylene glycol and isopropanol.

CP Acetic anhydride.

Salicylaldehyde (from bisulfite addition compound).

PROCEDURE A

TOTAL AMINES

In a weighing bottle, accurately weigh a sample containing approximately 0.02 mole of total amines. Wash the contents of the weighing bottle into a 150-ml beaker with 1:1 ethylene glycol–isopropanol mixture. Add ethylene glycol–isopropanol mixture until the volume is approximately 50 ml. Use a pH meter to indicate the apparent pH after each addition of acid as the sample is titrated with 1*N* hydrochloric acid prepared in the ethylene glycol–isopropanol mixture. Determine the neutralization point by plotting the apparent pH against milliliters of acid.

$$\frac{\text{Milliliters of HCl} \times \text{N}}{\text{Grams of sample} \times 1000} = \frac{\text{moles of total amines}}{\text{gram of sample}}$$

PROCEDURE B

SECONDARY PLUS TERTIARY AMINES

In a weighing bottle, weigh a sample containing approximately 0.02 mole total of secondary and tertiary amines. Wash the contents of the weighing bottle into a 150-ml beaker with 1:1 ethylene glycol–isopropanol mixture. Add ethylene glycol–isopropanol mixture until the volume is approximately 50 ml. Add 5 ml of salicylaldehyde (more should be used if the amount of primary amine is larger than 0.035 mole). Stir the mixture thoroughly and allow to stand at room temperature for 30 minutes. Use a pH meter to indicate the apparent pH after each addition of acid as the sample is titrated with 1*N* hydrochloric acid prepared in the ethylene glycol–isopropanol mixture. Determine the neutralization point by plotting apparent pH against milliliters of acid.

$$\frac{\text{Milliliters of HCl} \times \text{N}}{\text{Grams of sample} \times 1000} = \frac{\text{moles of secondary plus tertiary amine}}{\text{gram of sample}}$$

In the case of the aliphatic amines, two breaks are usually obtained in the titration curve after the addition of the salicylaldehyde. This is because the Schiff bases of the aliphatic amines still have a noticeable basicity, although it is not so great as that of the original primary amine or that of the secondary and tertiary amines. The first break in the curve, therefore, is the secondary plus the tertiary amine content of the sample. The difference between the first and second breaks represents the primary amine content of the sample.

PROCEDURE C

TERTIARY AMINES

Accurately weigh a sample that contains approximately 0.02 mole of tertiary amine in a 20 × 150 mm test tube and cool by placing it in a beaker of ice. Slowly add 10 ml of acetic anhydride while swirling the test tube. Allow the test tube and contents to stand for 15 minutes at room temperature. Quantitatively transfer the contents from the test tube into a 150-ml beaker by washing with 1 : 1 ethylene glycol–isopropanol mixture.

Add ethylene glycol–isopropanol mixture until the volume is approximately 50 ml. Use a pH meter to indicate the apparent pH after each addition of acid as the sample is titrated with 1*N* hydrochloric acid prepared in the ethylene glycol–isopropanol mixture. Determine the neutralization point by plotting the apparent pH against milliliter of acid.

$$\frac{\text{Milliliters of HCl} \times N}{\text{Grams of sample} \times 1000} = \frac{\text{moles of tertiary amines}}{\text{gram of sample}}$$

CALCULATIONS

PRIMARY AMINE

$$\frac{\text{Moles of total amine}}{\text{Gram}} - \frac{\text{moles of secondary} + \text{tertiary amine}}{\text{gram}} = \frac{\text{moles of primary amine}}{\text{gram}}$$

$$\frac{\text{Moles of primary amine}}{\text{Gram}} \times \text{mol. wt. of primary amine} \times 100 = \%\ \text{primary amine}$$

SECONDARY AMINE

$$\frac{\text{Moles of secondary amine} + \text{tertiary amine}}{\text{Gram}} - \frac{\text{moles of tertiary amine}}{\text{gram}} = \frac{\text{moles of secondary amine}}{\text{gram}}$$

$$\frac{\text{Moles of secondary amine}}{\text{Gram}} \times \text{mol. wt. of secondary amine} \times 100 = \%\ \text{secondary amine}$$

TERTIARY AMINE

$$\frac{\text{Moles of tertiary amine}}{\text{Gram}} \times \text{mol. wt. of tertiary amine} \times 100 = \%\ \text{tertiary amine}$$

Aliphatic and alicyclic systems that have been used with this method of analysis include the butyl, dibutyl, and tributyl amines; lauryl amine, *N*-methyl- and *N,N*-dimethyl lauryl amines; cyclohexylamine, *N*-methyl- and *N,N*-dimethyl cyclohexylamines; piperidine, and *N*-ethyl piperidine.

One of the desirable features of this system of analysis is the small number of interferences. First of all, for a substance to be an interference, it has to be basic. In the determination of tertiary amines, most basic impurities are neutralized by the acetic anhydride, thus are removed. In the primary amine determination, it is the decrease in basicity on addition of salicylaldehyde that is measured. Any basic impurity is figured in both titrations and does not affect the difference.

A basic impurity will affect only the determination of the secondary amine. If the alkaline impurity is strong enough, however, it can be determined by a differential titration in the presence of the aromatic amines, which are very weak bases. The secondary amine value can then be corrected. Ammonia present in mixtures of aniline, monoethyl, and diethyl aniline was handled very satisfactorily by the foregoing technique.

It was found impossible to determine *N,N*-di(β-hydroxyethyl) aniline by the tertiary amine procedure. On acetylation, the hydroxyl groups on the molecule were esterified, and this resulted in such a decrease in the basicity of the compound that it was no longer titrable even in the special solvent mixture.

It was also found that systems containing diphenylamine and triphenylamine could not be determined because these materials are too weakly basic to be titrated.

In the determination of secondary plus tertiary amines, the buffering action of the Schiff base formed after the reaction with salicylaldehyde is sometimes strong enough to decrease the sensitivity of the titration curve. In these cases it is advisable to make the plotted curve more sensitive by extending the pH scale of the graph over a greater length so that each division on the graph equals a smaller pH unit. In this way the break in the curve will be accentuated, and a more accurate reading is made possible.

The foregoing procedures provide an easily applicable way of determining primary, secondary, and tertiary amines when all are present in the sample. Combinations of these procedures can also be applied to determine mixtures when only two of the three types of amine are present. In the latter case, however, other techniques can sometimes be used to give the analysis for the two components more directly, more accurately, or faster.

The procedure for tertiary amines just described is the most efficient for these compounds when they are in the presence of primary or secondary amines. This procedure for primary amines is a very good method for determining these compounds in the presence of secondary amines. If the mixture to be analyzed consists of a primary and a tertiary amine, however, the primary amine is best determined by the acetylation procedure of Ogg, Porter, and Willits (19). The acetylation procedure is faster to carry out than the salicylaldehyde method. It must be kept in mind that alcohols will interfere with the acetylation method.

The acetylation method is also the best way of determining secondary amines in the presence of tertiary amines and is a more direct analysis than determining total amines and subtracting the tertiary amine.

System of F. Critchfield and J. B. Johnson

[*Reprinted in Part from Anal. Chem.,* ***28,*** *430–6 (1956)*]

In the methods presented here the reaction of primary and secondary amines with carbon disulfide to form dithiocarbamic acids has been utilized (20)

$$RNH_2 + CS_2 \rightleftharpoons R\text{—}NH\text{—}\overset{\overset{\displaystyle S}{\|}}{C}\text{—}SH \tag{7}$$

19. C. L. Ogg, W. L. Porter, and C. O. Willits, *Ind. Eng. Chem., Anal. Ed.*, **17,** 397 (1945); see also pp. 12–14 of this book.
20. A. W. Hofman, *Ber.*, **1,** 169 (1868).

and

$$R_2NH + CS_2 \rightleftharpoons R_2N-\overset{\overset{\large S}{\|}}{C}-SH \quad (8)$$

It has been found that, under certain conditions, these dithiocarbamic acids can be titrated directly and quantitatively with standard sodium hydroxide. This principle has been utilized for the analysis of mixtures of aliphatic amines, inorganic base–amine, acid–amine, or carboxylic acid–anhydride.

DETERMINATION OF PRIMARY AND SECONDARY ALIPHATIC AMINES AND ANALYSIS OF AMINE–ACID AND AMINE–STRONG BASE MIXTURES

This section describes a specific method for the direct determination of the total primary and secondary amine content in the presence of tertiary amines. An excess of carbon disulfide is reacted with the primary or secondary amine in an essentially nonaqueous medium such as isopropyl alcohol or a pyridine–isopropyl alcohol mixture. The dithiocarbamic acid formed in the reaction is then titrated with standard sodium hydroxide using phenolphthalein indicator. Potentiometric curves for the titration of the reaction products of carbon disulfide and three amines are shown in Fig. 11.12. The method is unique in that primary and secondary amines are converted to acids and titrated with a base. This permits the determination of these amines in the presence of strong inorganic bases, ammonia, tertiary amines, and most acids.

APPARATUS AND REAGENTS

All potentiometric titrations were performed using a Leeds & Northrup line-operated pH meter equipped with glass and calomel electrodes.

Carbon disulfide, reagent grade.

Isopropyl alcohol, commercial grade, Carbide and Carbon Chemicals Co.

Pyridine, redistilled. This material should contain less than 0.0005 meq. of primary and secondary amines per gram. The amines are determined by the procedure described below.

Sodium hydroxide, 0.5*N*.

Hydrochloric acid, 0.5*N*.

Phenolphthalein, 1.0% pyridine solution.

Thymolphthalein, 1.0% pyridine solution.

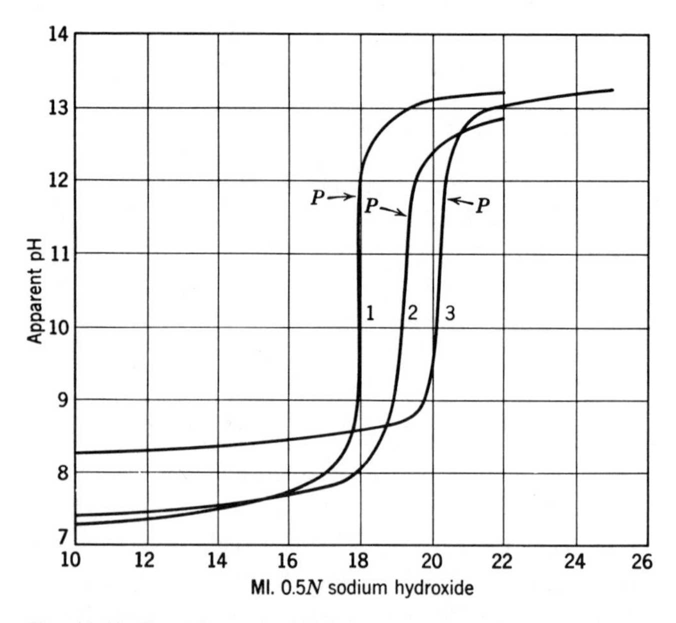

Fig. 11.12. Potentiometric titration curves of dithiocarbamic acids from reactions of carbon disulfide with primary and secondary amines: 1, 0.6771 gram of isopropanolamine, solvent *C*; 2, 0.2868 gram of ethylenediamine, solvent *A*; 3, 1.3085 grams of dibutylamine, solvent *B*; *P*, phenolphthalein end point.

PROCEDURE

To each of two 250-ml glass-stoppered Erlenmeyer flasks, add the solvent specified in Table 20. Reserve one of the flasks as a blank. Into the other flask introduce an amount of sample that contains not more than 15 meq. of primary or secondary amine. The sample aliquot should contain not more than 15 meq. of alkali or acid, and the total tertiary amine and ammonia content should not exceed 30 meq. If the sample contains alkali, it should be neutralized with standard 0.5*N* hydrochloric acid using thymolphthalein indicator. For samples that contain acids, neutralize with standard 0.5*N* sodium hydroxide. For samples void of alkalies or acids, use phenolphthalein indicator. If more than 2.0 meq. of ammonia is present in the sample aliquot, cool the contents of the flasks to approximately −10°C. Carbon dioxide from a dry ice–organic solvent bath interferes in the subsequent titration; therefore a brine bath is most convenient for this purpose.

By means of a pipet add 5 ml of carbon disulfide to each flask and swirl to effect solution. Titrate the contents of each flask with 0.5*N* sodium

Table 20. Analysis of Primary and Secondary Amines by Reaction with Carbon Disulfide

Compound	Solvent Composition	Average Purity, wt. %[a]	
		Carbon Disulfide Method	Other[b]
2-Aminoethylethanolamine	A	100.0 ± 0.0 (2)	100.0 ± 0.1 (2)[c]
N-Aminoethylmorpholine	A	98.6 ± 0.1 (4)	98.8 ± 0.1 (2)[c]
Butylamine	B	97.9 (1)	97.5 ± 0.1 (2)
Butylamine, secondary	C	96.4 ± 0.1 (2)	96.4 (1)
Dibutylamine	B	100.1 ± 0.0 (2)	99.7 ± 0.0 (2)
Diethanolamine	C	99.8 ± 0.1 (2)	99.8 (1)
Diethylamine	B	97.4 ± 0.1 (3)	97.4 (1)
Diethylenetriamine	A	92.0 ± 0.0 (2)[d]	98.1 (1)[c]
Di(2-ethylhexyl)amine	C	100.0 ± 0.1 (2)	99.5 ± 0.1 (2)
Dihexylamine	B	93.4 (1)	93.4 ± 0.1 (2)
Dimethylamine, aqueous	B	39.6 ± 0.0 (2)	39.6 (1)
2,6-Dimethylpiperazine	A	98.9 ± 0.1 (2)	99.5 (1)[c]
Ethanolamine	C	99.3 ± 0.2 (3)	99.2 ± 0.2 (2)
Ethylamine, aqueous	B	70.2[e]	70.2 ± 0.1 (2)
Ethylenediamine	A	99.3 ± 0.2 (2)	99.2 (1)
2-Ethylhexylamine	C	99.0 (1)	98.7 (1)
Hexylamine	B	98.9 (1)	99.1 ± 0.0 (2)
Isobutylamine	B	99.3 ± 0.1 (2)	98.8 (1)
Isopropanolamine	C	99.2 ± 0.1 (2)	99.2 ± 0.0 (2)
Isopropylamine	C	99.1 ± 0.2 (2)	99.2 (1)
Methylamine, aqueous	B	44.9 ± 0.1 (2)	45.0 (1)
Morpholine	A	99.3 ± 0.1 (4)	99.4 (1)
Propylenediamine	A	99.0 ± 0.0 (2)	98.9 (1)

[a] Figures in parentheses represent number of determinations.

[b] By titration in water with standard $0.5N$ hydrochloric acid using bromocresol green–methyl red mixed indicator, unless otherwise specified.

[c] By titration in glacial acetic acid with standard $0.1N$ perchloric acid using crystal violet indicator.

[d] Sample contains 0.34 meq. of tertiary amine per gram.

[e] Standard deviation for 8 degrees of freedom is 0.11.

A = 50 ml. of pyridine, 25 ml. of water, and 50 ml. of isopropyl alcohol.

B = 75 ml. of isopropyl alcohol. If more than 2.0 meq. of ammonia is present, add 25 ml. of pyridine.

C = 25 ml. of pyridine and 75 ml. of isopropyl alcohol.

hydroxide. For sample aliquots that contain more than 2.0 meq. of ammonia, conduct this titration below 0°C, by placing the flask in a 1000-ml beaker containing a mixture of crushed ice and methanol. Stir the contents of the flask by means of a magnetic stirrer to prevent a local excess of sodium hydroxide from accumulating in the titration medium. The end point selected should be the first definite pink color for phenolphthalein, or blue or blue-green for thymolphthalein. The color should be stable at least 1 minute.

For samples that contain alkalies, the amount of hydrochloric acid consumed in the first titration is a measure of the alkali. For samples that contain acids, the amount of sodium hydroxide necessary to neutralize the sample is a measure of the acid present. The amount of standard sodium hydroxide necessary to neutralize the sample after the addition of carbon disulfide is a direct measure of the primary and secondary amine content of the sample.

DISCUSSION

In the presence of a large excess of carbon disulfide, the reaction of primary and secondary amines to give dithiocarbamic acids is approximately 90 to 95% complete. In general, secondary amines are more reactive than primary amines. To adapt this reaction to a quantitative method of analysis, it is forced to completion by means of the sodium hydroxide titrant. The potentiometric titration curves in Fig. 11.12 were obtained by permitting the system to reach equilibrium before the addition of each increment of titrant. Usually equilibrium is established rapidly, except in the vicinity of the equivalence point.

An attempt was made to determine the dithiocarbamic acids by the addition of a measured excess of sodium hydroxide, and subsequent determination of the excess by titration with standard hydrochloric acid. Results obtained in this manner were erroneous because of the incompatibility of sodium hydroxide and carbon disulfide. In isopropyl alcohol medium and in the presence of excess sodium hydroxide, carbon disulfide reacts with the solvent to form a xanthate. Xanthate formation is even more pronounced when methyl or ethyl alcohol is substituted for isopropyl alcohol. Because of the tendency of isopropyl alcohol to react with carbon disulfide and sodium hydroxide, the end points obtained in this method are stable for only 4 or 5 minutes.

The reaction of carbon disulfide and primary and secondary amines cannot be forced to completion by heating because of the instability of the dithiocarbamic acids at elevated temperatures. In the case of alkyl amines, substituted ureas are formed by the evolution of hydrogen sulfide (20). Diamines, such as ethylenediamine, also liberate hydrogen sulfide but give polyalkylurea derivatives (21). Pyridine present in the reaction medium acts as a proton acceptor and tends to force the reaction of carbon disulfide with primary and secondary amines to completion.

Listed in Table 20 are several primary and secondary amines that have been determined successfully by this method. In each case the purity obtained by the carbon disulfide method is compared to the purity obtained by another acid-base titration method. Also listed are the solvent mixtures recommended for the determination of these amines. Solvent *A*, which contains 50 ml of pyridine, 25 ml of water, and 50 ml of isopropyl alcohol, is used for the determination of amines that give dithiocarbamic acids insoluble in isopropyl alcohol. Solvent *B*, isopropyl alcohol, is used for reactive amines that give soluble reaction products. Solvent *C*, which contains 75 ml of isopropyl alcohol and 25 ml of pyridine, is recommended for unreactive amines that give soluble reaction products. Of the primary and secondary amines investigated, only the aromatic amines and aliphatic amines that are highly branched in the 2-position, such as tertiary butylamine and diisopropylamine, do not react quantitatively under the conditions of the method.

Although the reaction of primary and secondary amines with carbon disulfide is specific, there is a slight tendency for ammonia to react with the reagent. The extent of the reaction is shown in Table 21, which gives data for the determination of butylamine in the presence of ammonia. It can be seen that at temperatures below 0°C the interference is negligible. At room temperature, however, the reaction is appreciable. In general, the reaction is more pronounced in pyridine medium than in isopropyl alcohol. For this reason pyridine alone is not used as a solvent for the reaction. Even in the absence of ammonia, it is necessary to use either isopropyl alcohol or water as a cosolvent to solubilize the sodium hydroxide in the titrant. As stated previously, methanol cannot be used because of xanthate formation. The maximum amount of ammonia that can be tolerated is given under Procedure.

Tertiary amines do not react with carbon disulfide. In the method described here, large quantities of pyridine (a tertiary amine) have been used as a cosolvent. The extent of interference from tertiary amines is determined by the basicity of the amine in the titration medium. Under the conditions of the method, as much as 30 meq. of most tertiary amines will not interfere. Tertiary amines and ammonia are more basic in isopropyl alcohol than in pyridine–isopropyl alcohol mixtures. When large

21. A. L. Carpenter, S. Coldfield, and D. L. Wilson, U.S. Patent 2,566,717 (September 4, 1951).

Table 21. Determination of Butylamine in Aqueous Ammonia Solutions

(Extent of ammonia interference)

Sample[a]	Added		Butylamine Found	Deviation
	Ammonia	Butylamine		
Ammonia-butylamine	27.00	6.73	6.79[b]	+0.06
	23.50	18.85	18.95[b]	+0.10
	19.40	32.60	33.10[b]	+0.50
Ammonia	29.00	0.00	0.00[b]	0.00
	2.90	0.00	0.00[c]	0.00
	5.80	0.00	0.28[c]	+0.28
	29.00	0.00	0.73[c]	+0.73

[a] All values are per cent by weight.
[b] Reaction with carbon disulfide and subsequent titration at $<0°C$.
[c] Reaction with carbon disulfide and subsequent titration at 25°C.

quantities of tertiary amines or ammonia are present, it is sometimes necessary to add sufficient pyridine to suppress the basicity of these compounds.

Because the method is based on an alkalimetric titration, compounds that are acidic or basic under the conditions of the titration will interfere. Acids with ionization constants greater than approximately 1×10^{-7} in water and bases with ionization constants greater than about 1×10^{-2} interfere quantitatively. Figure 11.13 plots a potentiometric titration of an aqueous mixture of sodium hydroxide and morpholine. The curve was obtained by first titrating the sodium hydroxide with standard hydrochloric acid. After this equivalence point was obtained, carbon disulfide was added and the dithiocarbamic acid formed by reaction with morpholine was titrated with standard sodium hydroxide. The amount of morpholine present in the sample was calculated from the total amount of sodium hydroxide titrant added minus the hydrochloric acid titrant added in excess of the first equivalence point. Table 22 shows data for the analysis of known mixtures of morpholine and sodium hydroxide. These data were obtained in the manner just described, except that thymolphthalein indicator was used, and the titration of sodium hydroxide with hydrochloric acid was not extended beyond the end point. Although other methods are available for the analysis of mixtures of this type, the method described is of merit because a single sample is sufficient for both determinations.

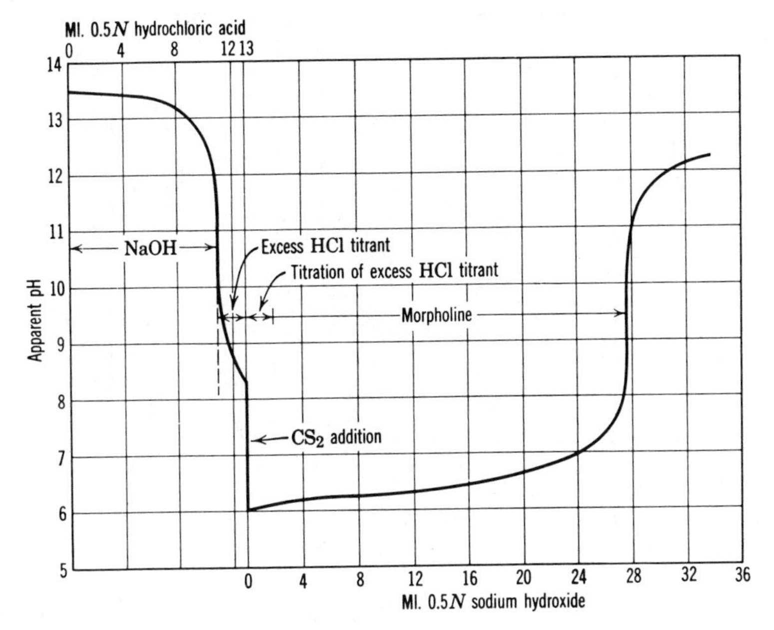

Fig. 11.13. Potentiometric titration curve for determination of a mixture containing 3.46% sodium hydroxide and 18.5% morpholine.

Mixtures of acids and primary and/or secondary amines can be analyzed by a procedure similar to that described. Figure 11.14 shows a potentiometric titration of a mixture of morpholine and hydrochloric acid. The hydrochloric acid present in the sample was titrated with standard sodium hydroxide. After the end point was obtained, carbon disulfide was added and the titration of the dithiocarbamic acid was performed. The

Table 22. Analysis of Aqueous Morpholine–Sodium Hydroxide Mixtures of Known Composition by Carbon Disulfide Reaction

Added[a]		Found[a]		Deviation[a]	
Morpholine	NaOH	Morpholine	NaOH	Morpholine	NaOH
34.2	2.79	34.0	2.80	+0.20	+0.01
16.0	3.56	16.0	3.57	0.00	+0.01
31.9	2.89	32.0	2.94	+0.10	+0.05
15.9	3.57	16.0	3.55	+0.10	−0.02
3.60	4.04	4.67	4.02	+0.07	−0.02
0.48	4.22	0.50	4.19	+0.02	−0.03

[a] All values are per cent by weight.

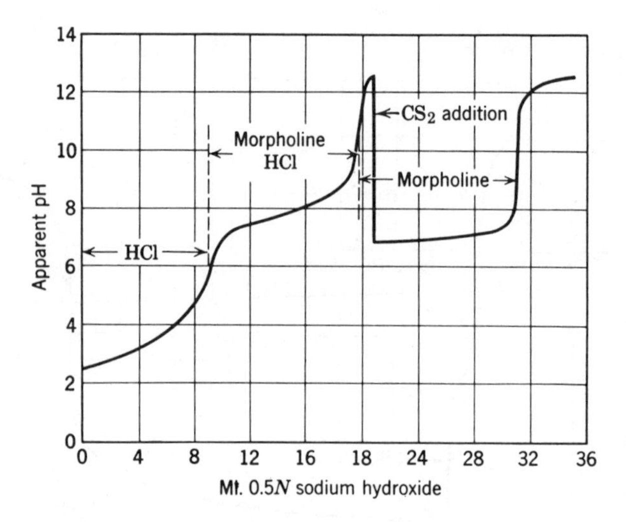

Fig. 11.14. Potentiometric titration curve for determination of a mixture containing 3.39% hydrochloric acid and 4.56% morpholine.

data in Table 23, for the analysis of morpholine–hydrochloric acid mixtures and dibutylamine–acetic acid mixtures, were obtained in this manner, except that thymolphthalein indicator was used. This method should be of considerable advantage for the analysis of amine-acid mixtures because the use of a single sample makes the method rapid and convenient.

By combining the method presented here with a total base and a tertiary amine determination, it is possible to determine indirectly the

Table 23. Analysis of Aqueous Amine–Acid Mixtures of Known Composition by Carbon Disulfide Reaction

Sample	Added[a]		Found[a]		Deviation[a]	
	Acid	Amine	Acid	Amine	Acid	Amine
Morpholine-hydrochloric	2.32	34.50	2.39	33.90	+0.07	−0.60
	2.92	17.60	2.97	16.40	+0.05	−0.20
	3.37	4.81	3.40	4.78	+0.03	−0.03
	3.51	0.84	3.52	0.85	+0.01	+0.01
Dibutylamine-acetic acid	38.80	61.20	38.50	61.80	−0.30	+0.60
	10.20	89.80	10.20	89.80	0.00	0.00
	87.60	12.40	87.70	12.60	+0.10	+0.20

[a] All values are per cent by weight.

amount of ammonia present in a particular sample. A further discussion is given below.

ANALYSIS OF MIXTURES OF AMMONIA AND PRIMARY, SECONDARY, AND TERTIARY ALIPHATIC AMINES

The analysis of mixtures of amines and ammonia is a difficult analytical problem that confronts any manufacturer or consumer of amines. For the complete resolution of amine mixtures, a method of analysis must be available for the specific determination of either primary or secondary amine. The salicylaldehyde method of Siggia, Hanna, and Kervenski (22, see pp. 569–72) and the benzaldehyde procedure of Hawkins, Smith, and Mitchell (23) have been used for the determination of secondary and tertiary amines in the presence of primary amines. Ammonia interferes in these methods and must be removed.

A specific method for primary and secondary amines has just been described in which the amines are reacted with carbon disulfide and the dithiocarbamic acids formed are titrated with standard sodium hydroxide. Ammonia and tertiary amines do not interfere in this method. In this method primary amines are reacted with 2-ethylhexaldehyde to form the corresponding imine.

$$\mathrm{RNH_2} + \mathrm{(C_4H_9)(C_2H_5)CH{-}CH{=}O} \rightarrow \mathrm{R{-}N{=}CH{-}CH(C_4H_9)(C_2H_5)} + \mathrm{H_2O} \qquad (9)$$

Secondary amines do not react with the aldehyde and are converted to the corresponding dithiocarbamic acids, as in eq. 8. The imines from the primary amines do not react with carbon disulfide to form acids and therefore do not interfere. The dithiocarbamic acids are titrated, at reduced temperature and in organic medium, with standard sodium hydroxide using phenolphthalein indicator. The imines formed according to eq. 9, ammonia, and tertiary amines, are not basic under these conditions and do not interfere.

REAGENTS

Carbon disulfide, reagent grade.

Isopropyl alcohol, Carbide and Carbon Chemicals Company, commercial grade.

22. S. Siggia, J. G. Hanna, and I. R. Kervenski, *Anal. Chem.*, **22,** 1295 (1950).
23. W. Hawkins, D. M. Smith, and J. Mitchell, Jr., *J. Am. Chem. Soc.*, **66,** 1662 (1944).

Pyridine, redistilled. This material should not contain more than 0.0002 meq. of primary and secondary amines per gram.

2-Ethylhexaldehyde, 50% by volume pyridine solution containing approximately 0.5% phenyl-1-naphthylamine inhibitor. The 2-ethylhexaldehyde used was Carbide and Carbon Chemicals Company commercial grade. The reagent was stored in brown bottles and discarded when the blank described below exceeded 0.5 ml.

Phenolphthalein, 1.0% pyridine solution.

Sodium hydroxide, 0.5*N*.

PROCEDURE

Pipet 10.0 ml of the 2-ethylhexaldehyde solution into each of two 250-ml glass-stoppered Erlenmeyer flasks. If specified in Table 24, add 50 ml of isopropyl alcohol to each flask. Reserve one of the flasks as a blank. Into the other flask introduce an amount of sample that contains not more than 13 meq. of secondary amine. The combined ammonia and tertiary amine content of the sample aliquot should not exceed 30 meq. The primary amine content of the sample aliquot should be less than 16 meq. Allow the sample and blank to stand at room temperature for 5 minutes and add the solvent mixture specified in Table 24. Cool the contents of the flask to $-10 \pm 2°C$ using a suitable bath. Do not use dry ice–organic solvent baths because carbon dioxide interferes in the titration. Remove the flasks from the bath and add 5 ml of carbon disulfide by means of a pipet. Add 5 or 6 drops of phenolphthalein indicator to each flask, and immediately titrate with standard 0.5*N* sodium hydroxide at below 0°C. This is best done by placing the flask in a 1000-ml beaker containing a slurry of crushed ice and methanol. Stir the contents of the flask by means of a magnetic stirrer. The end point selected should be the first definite pink color that is stable for at least 1 minute.

DISCUSSION

It has already been mentioned that most aliphatic secondary amines react with carbon disulfide; under the conditions of the first method described, they can be quantitatively titrated as the dithiocarbamic acids. Aromatic amines and highly branched amines such as diisopropylamine cannot be titrated in this manner. The limiting factor in this method for the determination of secondary amines in the presence of primary amines is the ability to destroy the primary amine quantitatively. Several aldehydes were investigated for this purpose. Any aldehyde suitable for use

Table 24. Determination of Secondary Amines by Use of the 2-Ethylhexaldehyde–Carbon Disulfide Method

Compound	Solvent Mixture	Purity, wt. %[a]	
		Carbon Disulfide Method	Other[b]
Dibutylamine	B	99.6 ± 0.0 (2)	99.7 ± 0.0 (2)
Diethylamine	B	96.8[c]	97.4 (1)
Di(2-ethylhexyl)amine	C	98.6 ± 0.1 (2)	99.5 ± 0.1 (2)
Dihexylamine	B	92.7 (1)	93.4 ± 0.0 (2)
Dimethylamine, aqueous	D	38.6 ± 0.05 (3)	39.6 (1)
2,6-Dimethylpiperazine	A	98.8 ± 0.05 (2)	99.5 (1)[d]
Morpholine	A	99.3 ± 0.1 (2)	99.4 (1)

[a] Figures in parentheses represent number of determinations.

[b] By titration with standard hydrochloric acid to bromocresol green–methyl red mixed indicator unless otherwise indicated.

[c] Standard deviation for 6 degrees of freedom is 0.16.

[d] By titration with standard perchloric acid in acetic acid using crystal violet indicator.

A = 50 ml. of pyridine, 25 ml. of water, and 50 ml. of isopropyl alcohol.

B = 75 ml. of isopropyl alcohol. If more than 2 meq. of ammonia is present, add sufficient pyridine to suppress its basicity.

C = 75 ml. of isopropyl alcohol and 25 ml. of pyridine.

D = Reaction with 2-ethylhexaldehyde in presence of 50 ml. of isopropyl alcohol. Add 25 ml. isopropyl alcohol before reaction with carbon disulfide.

in this method must have the following characteristics: (1) resistance to sodium hydroxide, (2) resistance to oxidation, (3) ability to react with primary amines to form imines, and (4) failure to react with secondary amines. Salicylaldehyde meets these requirements satisfactorily but cannot be used in the method because it is acid to phenolphthalein. Benzaldehyde satisfies conditions 1, 2, and 4, but is easily oxidized to benzoic acid, thus causing unstable blanks. Lower aliphatic aldehydes, such as formaldehyde, acetaldehyde, propionaldehyde, and butyraldehyde, tend to react with secondary amines. Of all the aldehydes investigated, 2-ethylhexaldehyde was found to be best suited for use in this method (isobutyraldehyde is less effective than ethylhexaldehyde for reaction of primary amine). The aldehyde is sufficiently resistant to sodium hydroxide titrant. Although autoxidation occurs to a slight extent, it can be inhibited by the addition of phenyl-1-naphthylamine to the aldehyde reagent. In one instance an apparent reaction of 2-ethylhexaldehyde with a secondary amine was noted, giving low results. For the determination of dimethylamine in the presence of methylamine, it was necessary to carry out the reaction with 2-ethylhexaldehyde in 50 ml of isopropyl alcohol to inhibit the reaction with dimethylamine. In all other cases the reaction takes place with 10 ml of 50% by volume of the aldehyde in pyridine.

2-Ethylhexaldehyde reacts quantitatively with most simple aliphatic primary amines to form the corresponding imine. A 100% excess of the aldehyde is sufficient to force this reaction to completion. Using a 50% excess of the aldehyde, high results were obtained for the determination of dimethylamine in the presence of methylamine, thus indicating an insufficient excess of aldehyde reagent. When a 10-ml aliquot of the reagent is used, which contains approximately 32 meq. of aldehyde, no more than 16 meq. of primary amine should be present in the sample aliquot used for analysis. This aldehyde does not react quantitatively with primary alcohol amines, aromatic amines, primary amines highly branched in the 2-position such as *tert*-butylamine and isopropylamine, and polyamines such as ethylenediamine. When the amines above are present, they interfere in the determination of secondary amines by this method.

Table 24 lists several secondary amines that have been successfully determined by this method; only the secondary amines, whose corresponding primary amines react quantitatively with 2-ethylhexaldehyde, are given. Secondary amines for which no corresponding primary amine exists are also included. The purity obtained by the 2-ethylhexaldehyde–carbon disulfide method is compared with that obtained by a total base titration. The data obtained by the former method are usually slightly lower than those obtained by the latter; this is no doubt caused by the presence of small amounts of primary and tertiary amines in these samples. Also included in Table 24 is the solvent mixture recommended for each amine. The effect of the solvent on the reaction of these amines with carbon disulfide has already been discussed.

Data are listed in Table 25 for the analysis of several mixtures of primary and secondary amines of known composition. In a few cases both the primary and the secondary amine contents of the samples were obtained. For these analyses the total primary and secondary amine contents were determined by titration of the corresponding dithiocarbamic acids of these amines. Secondary amines were determined in these samples by the procedure described in this section, and the primary amine content was obtained by difference.

By using this method for the specific determination of secondary amines and the carbon disulfide method for primary and secondary amines, combined with a total base and a tertiary amine determination, it is possible to resolve certain amine mixtures completely. Listed in Table

Table 25. Analysis of Known Mixtures of Primary and Secondary Amines by Use of the 2-Ethylhexaldehyde–Carbon Disulfide Method

Mixture	Added, wt. %		Found, wt. %[a]		Deviation	
	Primary	Secondary	Primary	Secondary	Primary	Secondary
Methylamine	19.4	22.3	19.4 ± 0.03 (2)	22.3 ± 0.06 (5)	0.0	0.0
Hexylamines	25.6	65.6	26.0	64.8	+0.4	−0.8
2-Ethylhexyl-	79.8	19.4	—	19.1	—	−0.3
amines	39.0	60.0	—	60.4	—	+0.4
	49.6	49.9	—	50.2	—	+0.3
Butylamines	47.4	51.8	47.1 ± 0.05 (3)	51.8 ± 0.05 (3)	−0.3	0.0
	61.2	38.5	—	61.0	—	−0.2
	28.7	69.5	—	28.9	—	+0.2
	69.6	30.2	—	69.4	—	−0.2

[a] Single determinations unless otherwise indicated by figures in parentheses.

26 are data on the analysis of known mixtures of ammonia and primary, secondary, and tertiary amines. The data were obtained as described above. For the determination of the tertiary amine content, the samples were reacted with acetic anhydride in methanol; the tertiary amine was then titrated with alcoholic hydrochloric acid using methyl yellow–methylene blue mixed indicator. The total amine content was determined

Table 26. Analysis of Known Amine Mixtures by the Use of the 2-Ethylhexaldehyde–Carbon Disulfide Method

Mixture	Compound	wt. %		Deviation
		Added	Found	
No. 1	Ammonia	5.5	5.2	−0.3
	Butylamine	19.8	19.9	+0.1
	Dibutylamine	20.9	21.1	+0.2
	Tributylamine	16.6	16.7	+0.1
No. 2	Ammonia	0.9	0.7	−0.2
	Butylamine	25.7	26.0	+0.3
	Dibutylamine	38.8	38.9	+0.1
	Tributylamine	8.1	8.3	+0.2
No. 3	Ammonia	1.8	1.0	−0.8
	Ethylamine	26.0	26.4	+0.4
	Diethylamine	43.8	43.2	−0.6
	Triethylamine	21.6	22.3	+0.7

by titration in water with standard hydrochloric acid and bromocresol green–methyl red mixed indicator.

Because this method is based on an alkalimetric titration of a dithiocarbamic acid, materials that are acidic or basic under the condition of the titration interfere. The maximum amounts of ammonia and tertiary amines that can be tolerated are given under Procedure in the section on the 2-ethylhexaldehyde–carbon disulfide method. A more detailed discussion of interferences is in the section on the determination of primary and secondary amines.

This method should be of considerable use for the analysis of mixtures of simple aliphatic amines and for the determination of purities of secondary amines. Mixtures of amines and ammonia can be analyzed without resorting to a specific removal or determination of ammonia. Even in the presence of large concentrations of ammonia, certain amine mixtures can be completely resolved. The lower limit of determination of ammonia, using the two carbon disulfide methods with a tertiary amine and a total base determination, is limited because ammonia is obtained by difference.

DETERMINATION OF PRIMARY AMINES IN THE PRESENCE OF SECONDARY AND TERTIARY AMINES

All the methods except the Van Slyke method (see below) are based in one way or another on the Schiff base reaction

$$RNH_2 + R_1CHO \rightarrow R_1C{=}NR + H_2O$$

In the salicylaldehyde methods (pp. 567–73, 593–4) the change in the basicity of the primary amine after reaction is measured. In the 2,4-pentanedione method (pp. 594–600) an acidic aldehyde is used as reagent and the excess of acid is measured. In the carbon disulfide–ethylhexaldehyde system (pp. 581–86), the Schiff base reaction is used to remove the primary amine from the reaction system to permit the use of carbon disulfide. In the benzaldehyde method (pp. 601–4) the measurement of water in the reaction above is used for analytical purposes. In the copper-salicylaldehyde method (pp. 604–12), a colored copper complex is formed with the Schiff base and the color is measured.

Van Slyke Method—Adapted from E. F. Hillenbrand, Jr., and C. A. Pentz

[*Reprinted in Part from Organic Analysis, Vol. 3, Edited by J. Mitchell, Jr., et al., Wiley-Interscience, New York, 1956, pp. 178–53*]

The first quantitative adaption of the nitrous acid reaction with aliphatic amino groups was presented by Van Slyke (24) in 1911. Although it

24. D. D. Van Slyke, *J. Biol. Chem.*, **9**, 185 (1911).

was used primarily in protein studies to determine amino groups in amino acids, the method may be extended to many primary amines.

Nitrous acid reacts with certain primary amines and ammonia to liberate a quantitative volume of nitrogen. Any other volatile reaction products are removed by scrubbing through the proper solvents. The low molecular weight ethers formed as reaction products from methyl- and ethylamine are water soluble enough to be removed by scrubbing through distilled water. As the molecular weight of the amine increases, however, the volatile reaction products must be scrubbed through solvents such as tributyl phosphate or dimethoxytetraglycol (Table 27). The volume of nitrogen, which is measured in a gas buret, is a direct measure of the primary amine or ammonia originally present.

REAGENTS

Cylinder nitrogen, Linde or equivalent, convenient for purging.

Sodium nitrite, 25% aqueous solution. Prepare fresh daily.

Glacial acetic acid, Grasselli reagent grade or equivalent.

ALKALINE POTASSIUM PERMANGANATE. Dissolve 25 grams of CP sodium hydroxide and 60 grams of potassium permanganate in 1 liter of distilled water.

APPARATUS

Assemble the apparatus as shown in Fig. 11.15. Make certain that all stopcocks are carefully greased and all connections vapor-tight. The stopcock at the base of the buret C should be open. Transfer sufficient alkaline potassium permanganate to the autobubbler pipet B to fill one bulb, and add 30 ml additional. If auxiliary scrubbing is required, transfer sufficient liquid to the auxiliary autobubbler pipet B' to fill one bulb, and add 30 ml additional. If not required, set stopcock R' to bypass the scrubber completely.

Set the stopcocks as follows: X and Y to connect sample tube D with reaction chamber A; R to connect pipet B with both sides of the manifold; R' to close pipet B' from the header; S and T to connect buret C with both the vent and the nitrogen supply. When B and B' are closed from the system, stopcocks R and R' should be positioned so that the manifold is open from stopcock Y to stopcock S. Raise the mercury bottles F and G sufficiently to just fill the reaction chamber A and the gas buret C with mercury. Rinse the reaction chamber A with two 50-ml portions of distilled water. Introduce the water into sample tube D, and lower mercury bottle F to draw the water into the chamber. Set X to connect A with the sewer vent E and raise mercury bottle F to force the water to the sewer. After the rinses, chamber A should be full of mercury and stopcocks X and Y should be closed.

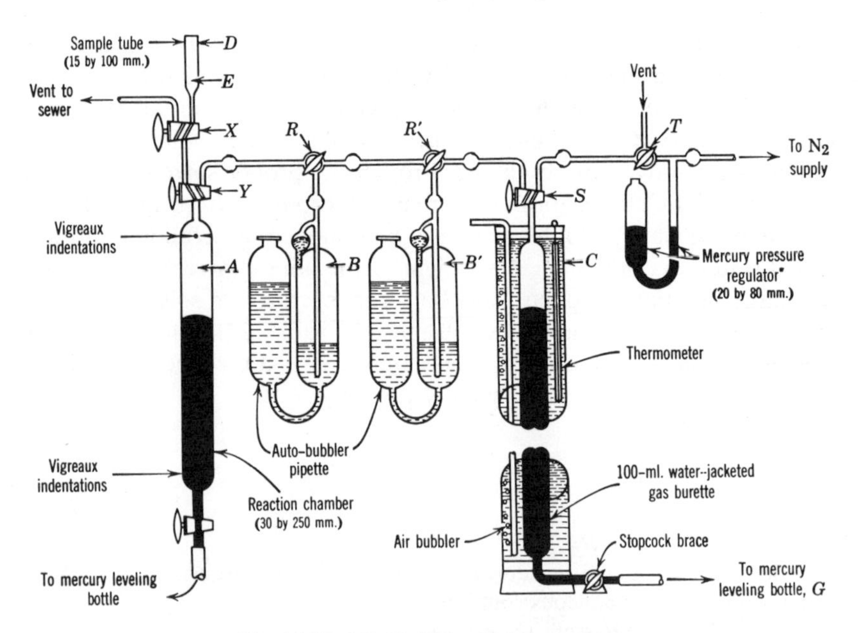

Fig. 11.15. Modified Van Slyke apparatus.

Set stopcock *S* to connect buret *C* to the manifold and carefully adjust the height of mercury bottle *G* until the permanganate solution in *B* rises to the inlet tube just to the side arm containing the check valve. Turn stopcock *R* to close *B* from the manifold. If auxiliary scrubbing is required, set stopcock *R'* to connect pipet *B'* with buret *C* and carefully adjust the height of the mercury in bottle *G* until the liquid in *B'* rises in the inlet tube just to the side arm containing the check valve. Turn stopcock *R'* to close *B'* from the manifold.

Set *S* to connect buret *C* with the vent-nitrogen header and raise bottle *G* to fill the buret with mercury. Turn on a gentle stream of nitrogen, set *T* to bypass the vent, and slowly lower bottle *G* to draw 100 ml of the nitrogen into buret *C*. Turn stopcock *S* to connect the buret with the manifold and *T* to permit the nitrogen to pass out the vent. Set *Y* to connect reaction chamber *A* with the manifold, lower bottle *F*, and raise bottle *G* to force nitrogen from buret *C* into *A*.

Turn *X* and *Y* to connect *A* with sample tube *D*, and raise bottle *F* until mercury rises just above *X*, forcing the nitrogen out through *D*. Repeat this operation twice to purge the system thoroughly, after which stopcocks *X* and *Y* should be closed. The right bore of stopcock *Y* should be full of mercury.

Draw 100 ml of nitrogen into buret *C* and turn off the nitrogen. Slowly raise bottle *G* until the mercury in *C* is at the 50-ml mark. Vent the excess nitrogen in buret *C* through buret *A* out sample tube *D*.

Hold bottle *G* at the top of buret *C* and set stopcock *R* to connect pipet *B* with *C*, forcing the nitrogen into *B*. Lower bottle *G* slowly to draw the gas back into buret C. Repeat this scrubbing operation twice, and close B from the manifold. If auxiliary scrubbing is required, scrub the nitrogen through pipet B' by proper manipulation of stopcock R', as directed for B and R. Close B' from the manifold. Read the volume of gas in buret C by leveling the mercury column.

Scrub the gas twice again through the permanganate solution and read the volume of gas in the buret. If this volume is lower than that read previously, the system has a leak and the apparatus must be reassembled and prepared as described earlier. Set stopcock S to connect buret C with the vent-nitrogen header and raise bottle G until the mercury in C is between the 18- and 25-ml marks, venting the excess gas through stopcock T. Immediately turn stopcock S to connect C with the manifold and read the initial volume of gas in the buret by leveling the mercury column.

Van Slyke equipment also is available from commercial sources.

PROCEDURE

Run a blank on the apparatus and reagent as described below, but omit the addition of sample. After preparation of the apparatus as described earlier, the stopcocks should be as follows: R and R' with pipet B and B' closed from the manifold; S with buret C connected with the manifold; T so that vent-nitrogen header is vented; X and Y closed, with Y having its right bore (side toward the manifold) full of mercury.

Into the sample tube D introduce an aliquot of the sample sufficient to evolve not more than 75 ml of nitrogen. Turn X and Y to connect sample tube D with reaction chamber A and lower bottle F to draw the sample into A until the level of the sample is just at the top of stopcock X. Close the stopcocks. Add 10 ml of the sodium nitrite solution into sample tube D from a graduate. Turn X and Y to connect D with A and lower bottle F to draw the solution into the reaction chamber until the level is just at the top of stopcock X. Again close the stopcocks.

Add, in order, 5 ml of distilled water and 10 ml of the glacial acetic acid to sample tube D. Keep stopcocks X and Y *closed* and lower bottle F so the mercury level is quite near the bottom of the reaction chamber.

PRECAUTION. *Perform the following step slowly and carefully because the ensuing reaction may be somewhat violent if a considerable amount of primary amine is present. The sudden evolution of a large volume of nitrogen may develop sufficient pressure to force the gas out through stopcock* Y. Open stopcock X to connect sample tube D with Y and slowly turn stopcock Y to connect X with A, allowing the acid to be drawn into the reaction chamber until the level of the acid is just at the top of stopcock Y. *Close Y immediately.*

Raise and lower bottle F to provide thorough mixing of the sample and reactants, and allow the reaction to proceed for 30 minutes. If necessary, allow gas to flow into pipet *B* during the reaction. Level the mercury column in buret *C* and record the initial volume of gas in the buret. Except for slight changes due to temperature and pressure, this volume should be the same as that read in preparation of the apparatus. Record the barometric pressure and the temperature indicated by the thermometer in the water jacket.

Open stopcock *Y* to connect reaction chamber *A* with the manifold and slowly raise bottle *F* until the level of the liquid in the reaction chamber just reaches the bottom of *Y*, forcing the vapor from *A* into buret *C*. With bottle *G* at the top of buret *C*, turn stopcock *R* to connect *C* with pipet *B* and lower *G* until the permanganate solution in *B* rises in the inlet tube just to the side arm containing the check valve. Scrub the gas through the permanganate solution until a constant volume is obtained in the buret.

If auxiliary scrubbing is required, scrub the gas in buret *C* through pipet B' by proper manipulation of stopcock R'. Scrub the gas through the permanganate solution again until the constant volume is obtained. If it is necessary to repeat the auxiliary scrubbing, always make a final pass through the permanganate solution. Level the mercury column and record the final volume of gas in *C*. Turn stopcocks *X* and *Y* to connect chamber *A* with the sewer vent, and raise bottle *F* to force the used reagents through vent *E* to the sewer.

ADDITIONAL INSTRUCTIONS

Never make more than three determinations without changing the alkaline potassium permanganate scrubber solution. The autobubbler pipets and any other equipment that becomes fouled by the permanganate solution may be effectively cleaned by washing with a solution of 2% oxalic acid to which has been added 2 to 3% by weight sulfuric acid. Ordinarily the only cleaning of the reaction chamber necessary is rinsing with distilled water. If sufficient care is not exercised in the manipulation of the leveling bottles, however, the entire assembly may become fouled, necessitating complete dismantling for cleaning.

At all times keep a drop of distilled water floating on the mercury in the gas buret so that the theoretical correction for the vapor pressure of water is applicable.

In the manipulation of the mercury leveling bottles it is best not to make too sudden or extreme changes in the level of the mercury because this brings unnecessary strain to bear on points of possible weakness.

CALCULATIONS

Correct the net volume evolved by the blank to standard conditions:

$$\frac{P_b - P_v}{760} \times \frac{273}{273 + t_b} \times (B_f - B_i) = B$$

where B = milliliters of gas, at standard conditions, evolved by the blank
B_f = final milliliters of gas in buret
B_i = initial milliliters of gas in buret
P_b = atmospheric pressure (mm Hg)
P_v = vapor pressure of water at t_b°C (mm Hg)
t_b = temperature of water jacket (°C).

Correct the net volume evolved by the sample to standard conditions:

$$\frac{P_a - P_w}{760} \times \frac{273}{273 + t_a} \times (A_f - A_i) = A$$

where A = milliliters of gas, at standard conditions, evolved by the sample,
A_f = final milliliters of gas in buret
A_i = initial milliliters of gas in buret
P_a = atmospheric pressure (mm Hg)
P_w = vapor pressure of water at t_a°C (mm Hg)
t_a = temperature of water jacket (°C).

Calculate the primary amine as milliequivalents:

$$\frac{(A - B)}{22.4} = C$$

where A = milliliters of gas, at standard conditions, evolved by the sample
B = milliliters of gas, at standard conditions, evolved by the blank
C = primary amine (meq.).

Calculate the primary amine as per cent by weight:

$$\frac{C \times \text{E.W.}}{\text{Grams of sample} \times 10} = \text{primary amine, per cent by weight}$$

where C is milliequivalents of primary amine and E.W. is the equivalent weight of the amine.

Table 27 shows the proper sample size and reaction conditions for each of several compounds for which this method has been found satisfactory.

The sample size is calculated from the amount of pure compound theoretically equivalent to not more than 75 ml of nitrogen at standard conditions. Because of the relatively small amounts involved, the original samples should be diluted so that a 5- or 10-ml aliquot contains the proper amount of sample.

Table 27 also lists the proper auxiliary scrubber liquid for each compound. This liquid, if required, is placed in pipet *B′* and used as described.

The Van Slyke procedure is applicable to the determination of aliphatic primary amines and alcoholamines in the presence of the corresponding secondary and tertiary amines. Ammonia reacts quantitatively with the reagent and, if present, may be determined independently by a suitable procedure, such as the cobaltinitrite method (25), and a suitable correction applied. The procedure is not applicable to the determination of amino nitrogen in alcoholic solution because the alcohol reacts with nitrous acid reagent to form alkyl nitrites, which are difficult to remove, and also depletes the excess nitrous acid reagent, thus resulting in incomplete amine reaction. This method has not been found suitable for the determination of complex aliphatic amines or the higher simple aliphatic amines.

Table 27. Primary Amines by the Van Slyke Method

Compound	Auxiliary Scrubber Liquid	Maximum Sample[a]
Monomethylamine	Distilled water	0.078
Monoethylamine	Distilled water	0.12
Monoisopropylamine	Tributyl phosphate	0.15
Monobutylamine	Dimethoxytetraglycol	0.19
Monoisobutylamine	Tributyl phosphate	0.19
Monoethanolamine	—	0.16
Monoisopropanolamine	—	0.20
Ammonia	—	0.044

[a] Use an aqueous dilution such that a 5 or 10 ml. aliquot contains the amount indicated.

The first and most often used application of the Van Slyke procedure is the determination of amino nitrogen in amino acids and proteins (25, 26).

25. D. D. Van Slyke, *Ibid.*, **12,** 275 (1912).
26. E. G. Wollish, in *Acid-Base Titrations in Nonaqueous Solvents*, J. S. Fritz, Ed., G. Frederick Smith, Columbus, Ohio, 1952.

SALICYLALDEHYDE METHODS

Potentiometric Method

See pages 567–73.

Indicator Method—Adapted from F. Critchfield and J. B. Johnson

[*Reprinted in Part from Anal. Chem.*, ***29,*** *957–8* (*1957*)]

REAGENTS

Salicylaldehyde, reagent grade.
Dioxane, commercial grade, Carbide and Carbon Chemicals Company.
Chloroform, Mallinckrodt Chemical Works analytical reagent.
PERCHLORIC ACID, 0.5*N* SOLUTION IN DIOXANE. Shake the dioxane used to prepare this solution overnight with Amberlite IRA-100 ion-exchange resin (26). Prepare the solution by diluting 70 grams of 70 to 72% perchloric acid to 1 liter with the special dioxane. Standardize against potassium acid phthalate.
Bromocresol green indicator, 0.5% solution in methanol.
Congo red indicator, 0.1% solution in methanol.

PROCEDURE

Add the volume of chloroform specified in Table 28 to a 250-ml glass-stoppered Erlenmeyer flask, followed by the specified volume of salicylaldehyde. Add 4 to 6 drops of the bromocresol green indicator, and neutralize just to the end point with perchloric acid in dioxane or alcoholic potassium hydroxide. Usually the latter step can be omitted because the blank on the reagent is very small. Weigh into the flask an amount of sample calculated to contain no more than 12.5 meq. of primary amine and 12.5 meq. of secondary plus tertiary amine. If the solution becomes turbid because of the separation of the water of reaction, add the minimum quantity of dioxane necessary to effect solution. Allow the sample to react for 15 minutes at room temperature before titrating it with standard perchloric acid in dioxane just to the disappearance of the green color. Record this titration and level the buret at zero. Add the volume of dioxane specified in Table 28 and 8 to 10 drops of the Congo red indicator. Titrate the sample with the perchloric acid solution until it is pure green.

The amount of perchloric acid required for the first titration is a measure of the secondary plus tertiary amine content of the sample. The amount required for the second titration is a measure of the primary amine content.

Table 28. Reaction Conditions for Determination of Primary, Secondary, and/or Tertiary Amines

	Reagent and Solvent Systems		
Amine Mixtures	Chloroform, ml.	Salicylaldehyde, ml.	Dioxane, ml.[a]
Butylamines	25	5	75
Ethylamines	25	5	75
2-Ethylhexylamines	25	5	75
Isopropylamines	50	10	50
Hexylamines	25	5	25
Methylamines	25	5	75
Propylamines	25	5	75

[a] Add dioxane after titration to bromocresol green indicator endpoint.

2,4-Pentanedione Method—Adapted from F. Critchfield and J. B. Johnson

[*Reprinted in Part from Anal. Chem.*, ***29,*** *1174–6 (1957)*]

The enol form of 2,4-pentanedione (acetylacetone) reacts with primary amines in much the same manner as salicylaldehyde:

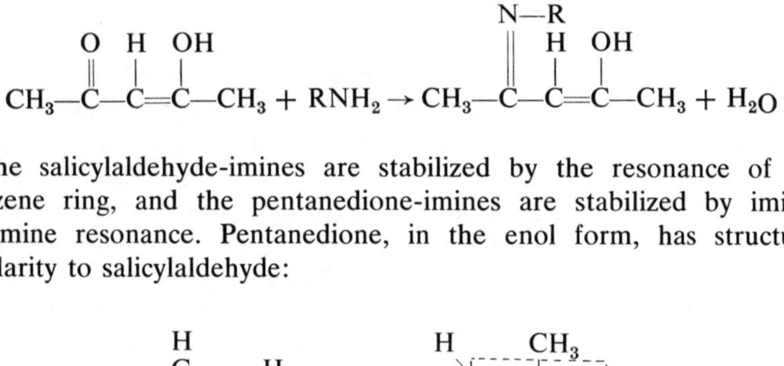

$$CH_3-\overset{\overset{\displaystyle O}{\|}}{C}-\overset{\overset{\displaystyle H}{|}}{C}=\overset{\overset{\displaystyle OH}{|}}{C}-CH_3 + RNH_2 \rightarrow CH_3-\overset{\overset{\displaystyle N-R}{\|}}{C}-\overset{\overset{\displaystyle H}{|}}{C}=\overset{\overset{\displaystyle OH}{|}}{C}-CH_3 + H_2O$$

The salicylaldehyde-imines are stabilized by the resonance of the benzene ring, and the pentanedione-imines are stabilized by imine-ketamine resonance. Pentanedione, in the enol form, has structural similarity to salicylaldehyde:

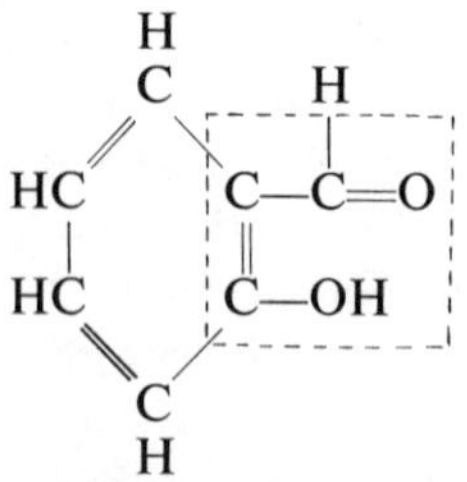

Salicylaldehyde

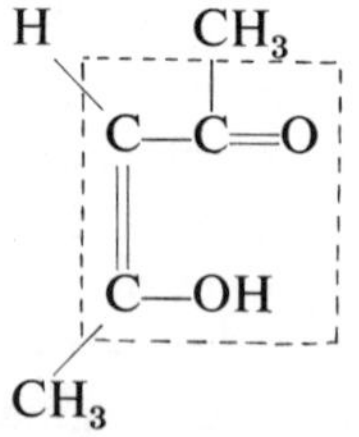

Pentanedione

In the structural formulas the portion of the compounds enclosed within the dotted lines are identical. Other compounds with this type of structure should also react readily with primary amines and may be of interest for the determination of these compounds. Another compound of this class, ethylacetoacetate, was investigated but is not as reactive as pentanedione.

Pentanedione is a weak acid and can be titrated in pyridine medium with sodium methylate in pyridine by using either phenolphthalein or thymolphthalein indicator. The reaction products of pentanedione and a primary amine are neutral under these conditions. Therefore, by reaction of a sample with a measured excess of pentanedione in pyridine medium, the primary amine content can be determined by titrating the unreacted pentanedione. Fritz (27) has discussed the titration of enols in other media.

REAGENTS

Methanol, anhydrous, Union Carbide Chemicals Company.

2,4-PENTANEDIONE. Commercial grade, Union Carbide Chemicals Company. Distill under atmospheric pressure and at 3:1 reflux, using a column having at least 10 theoretical plates. Collect a heart fraction consisting of approximately one-third of the charge.

Pyridine, freshly redistilled.

2,4-PENTANEDIONE, APPROXIMATELY 2.5N. Dilute 260 ml of the redistilled pentanedione to 1 liter with the freshly distilled pyridine.

SODIUM METHYLATE, 0.5N SOLUTION IN PYRIDINE. Prepare 3N sodium methylate by dissolving 163 grams of dry sodium methylate in sufficient methanol to make 1 liter of solution. Transfer 167 ml of the 3N sodium methylate and 40 ml of methanol to a 1-liter volumetric flask; dilute to volume with redistilled pyridine. Standardize this solution against Bureau of Standards benzoic acid, using pyridine as solvent and thymolphthalein indicator. For the change in normality with temperature, use ΔN per °C = −0.0005. The reagent readily absorbs carbon dioxide from the air and is best preserved in a bottle equipped with a 50-ml automatic buret. All vents open to the air should be protected with Ascarite tubes.

Thymolphthalein indicator, 1.0% solution in pyridine.

PROCEDURE

Pipet exactly 10 ml of the 2.5N pentanedione reagent into each of two 250-ml glass-stoppered Erlenmeyer flasks. For reactions at 98 ±2°C use heat-resistant pressure bottles. Reserve one of the flasks for a blank determination. Into the other flask introduce an amount of sample that

27. J. S. Fritz, *Anal. Chem.*, **29**, 674 (1952).

contains 10.0 to 15.0 meq. of primary amine. Allow both the sample and blank to stand for the time and at the temperature specified in Table 29. If elevated temperatures are used, carefully cool the pressure bottles to room temperature at the end of the reaction time. To each flask add approximately 1 ml of the thymolphthalein indicator and titrate with the standard 0.5*N* sodium methylate in pyridine to the appearance of the first definite blue color. The difference between a blank and sample titration is a measure of primary amine.

DISCUSSION

REACTION CONDITIONS. Most primary aliphatic amines react quantitatively in pyridine medium with an excess of pentanedione to form imines. Reaction conditions are listed in Table 29 for a number of compounds that can be determined by this method. The reaction of pentanedione with the compounds listed is quantitative over the time intervals and at the temperatures indicated.

An 80% excess of pentanedione must be maintained during the reaction to obtain quantitative results. The sample size specified in the procedure ensures this excess.

A reagent concentration of 2.5*N* pentanedione provides quantitative results for most amines in 15 to 60 minutes at room temperature. If weaker concentrations are employed (0.5*N* pentanedione and 0.1*N* sodium methylate), longer reaction times are required for quantitative reaction.

AMINO ACIDS. Pentanedione reacts readily with the sodium salt of primary amino acids to form the corresponding imines. The reaction of pentanedione with the acid form of these compounds is very slow. Evidently the formation of zwitterions by these compounds inactivates the amino groups.

Amino acids can be determined by converting them to the sodium salt prior to reaction by addition of a calculated equivalent amount of sodium methylate in pyridine. The correct amount can be determined by titration in ethylenediamine medium, using benzopurpurin 4B indicator. The color change for the indicator in this titration is from pink to blue. Pyridine medium cannot be used for the titration because of solubility difficulties. Reaction conditions are shown in Table 29 for three compounds of this class.

RESULTS

Purity data are listed in Table 30 for 23 primary amines and amino acids that can be determined by the pentanedione method. Data for most

Table 29. Reaction Conditions for Determination of Primary Aliphatic Amines by Reaction with Pentanedione

Compound	Temperature, °C	Time, min.
Alanine	98	15 to 90[a]
β-Alanine	98	15 to 90[a]
Aminoethylethanolamine	25	30 to 60
N-Aminoethylpiperazine	0	30 to 60
N-Aminopropylmorpholine	25	15 to 60
Ammonia	25	30 to 120
Benzylamine	25	15 to 120
Butylamine	25	15 to 60
Diethylenetriamine	25	15 to 60
Ethanolamine	25	15 to 60[b]
Ethylamine	25	15 to 60
Ethylenediamine	25	30 to 90
2-Ethylhexylamine	25	15 to 60
Glycine	98	60 to 120[a]
Hexylamine	25	15 to 60
Isobutylamine	25	15 to 60
Isopropanolamine (1-amino-2-propanol)	25	15 to 60[c]
Isopropylamine	98	15 to 60
Methylamine	25	15 to 60
Propylamine	25	30 to 60
Propylenediamine	25	60 to 120
Tetraethylenepentamine	25	90 to 120
Triethylenetetramine	25	90 to 120

[a] Dissolve sample in 10 ml. of water, add a calculated equivalent amount of standard 0.5*N* sodium methylate to neutralize sample, and add 75 ml. of pyridine before titration.

[b] In presence of over 50% diethanolamine sample should contain 10 to 15 meq. of primary amine and reaction time should be limited to 20 to 30 minutes.

[c] In presence of over 50% diisopropanolamine sample should contain 10 to 15 meq. of primary amine and reaction time should be limited to 40 to 60 minutes.

Table 30. Determination of Purity of Primary Aliphatic Amines by Reaction with Pentanedione

Compound	Average Purity, wt. %	
	Pentanedione Method[a]	Other
Alanine	97.0 ± 0.0 (4)	99.7[b]
β-Alanine	99.8 ± 0.0 (2)	99.7[b]
Aminoethylethanolamine	98.2 ± 0.2 (2)	99.0[c]
N-Aminoethylpiperazine	101.4 ± 0.1 (2)	98.8[c]
N-Aminopropylmorpholine	96.1 ± 0.0 (3)	97.3[c]
Ammonia, aqueous	27.5 ± 0.0 (2)	27.6[d]
Benzylamine	95.5 ± 0.1 (4)	98.5[c]
Butylamine	97.0 ± 0.0 (3)	97.5[e]
Diethylenetriamine	98.6 ± 0.2 (3)	99.1[c]
Ethanolamine	98.5 ± 0.0 (2)	99.8[d]
Ethylamine, aqueous	67.9 ± 0.2 (3)	70.0[e]
Ethylenediamine	99.2 ± 0.2 (3)	99.1[c]
2-Ethylhexylamine	97.1 ± 0.1 (4)	97.2[e]
Glycine	98.3 ± 0.1 (2)	99.3[b]
Hexylamine	99.1 ± 0.0 (3)	99.2[e]
Isobutylamine	96.9 ± 0.1 (3)	97.7[d]
Isopropanolamine	97.6 ± 0.3 (3)	98.8[d]
Isopropylamine	99.0 ± 0.1 (2)	98.8[e]
Methylamine, aqueous	37.4 ± 0.1 (3)	37.5[e]
Propylamine, aqueous	45.6 ± 0.1 (2)	45.5[d]
Propylenediamine	98.2 ± 0.1 (3)	98.0[c]
Tetraethylenepentamine	98.4 ± 0.2 (2)	95.0[c]
Triethylenetetramine	96.5 ± 0.3 (2)	93.2[c]

[a] Figures in parentheses represent number of determinations.

[b] By titration in ethylenediamine medium with sodium methylate in pyridine using benzopurpurin 4B indicator.

[c] By titration with perchloric acid in glacial acetic acid medium using crystal violet indicator.

[d] By titration with aqueous hydrochloric acid using bromocresol green–methyl red mixed indicator.

[e] By indicator salicylaldehyde method (28); see pages 593–4, this book.

of the primary amines are compared with the total basicities of these compounds. Data for the simple aliphatic amines are compared with purities obtained by the indicator salicylaldehyde method (28). The data for the amino acids are compared with total acidities. The purities of the primary amines tend to be lower than the total basicities, probably because of the presence of small amounts of secondary and tertiary amines in these compounds.

Table 31 gives data for the determination of known concentrations of primary amines in the presence of secondary amines. These data show that an accuracy within ±0.1% can be obtained for the determination of primary amines in the presence of secondary amines by this method.

Table 31. Determination of Primary Aliphatic Amines in the Presence of Other Amines

Primary Amine	Matrix	Added, wt. %	Found, wt. %	Deviation
Ethanolamine	Diethanolamine	69.9	70.0	+0.1
		49.6	49.5	−0.1
		17.8	17.7	−0.1
		7.1	7.4	+0.3
	Triethanolamine	54.4	54.5	+0.1
		36.8	36.8	0.0
2-Ethylhexylamine	Di(2-ethylhexyl)amine	74.6	74.5	−0.1
		14.6	14.5	−0.1
Isopropanolamine	Diisopropanolamine 1,1′-iminodi (2-propanol)	13.9	13.8	−0.1
Isopropylamine	Diisopropylamine	58.1	58.0	−0.1
Ethylenediamine	Piperazine	27.7	27.8	+0.1
β-Alanine	Nicotinic acid	86.8	87.4	+0.6
		30.2	30.5	+0.3

INTERFERENCES

Most acids, and bases with ionization constants greater than 1×10^{-2}, will interfere in the method. Usually corrections can be made if these substances are present.

Because the method is based on a nonaqueous titration, large quantities of water tend to interfere with the selection of the equivalence point. Usually 10% water can be tolerated in the titration medium

28. F. E. Critchfield and J. B. Johnson, *Anal. Chem.*, **29,** 957 (1957); see pages 593–4, this book.

without serious interference. Large quantities of other solvents such as alcohols, ketones, tertiary amines, esters, and nitriles can be present without affecting the end point.

Secondary alcoholamines, such as diethanolamine and diisopropanolamine, tend to react under the conditions of the method. A similar reaction with salicylaldehyde was discussed by Wagner, Brown, and Peters (29). This secondary reaction can be minimized by adjusting the sample size so that at least 10 meq. of primary amine is present for reaction. In this way the effective concentration of pentanedione is reduced by reaction with primary amine, and interference from secondary amines is negligible. Because of this interference, the determination of primary alcoholamines in the presence of secondary alcoholamines is limited to samples containing more than 5.0% of the former. Data are shown in Table 31 for the determination of mixtures of these amines.

Heterocyclic secondary amines, such as piperazine and morpholine, react slowly with pentanedione. Usually this interference can be inhibited by conducting the reaction at 0°C. The procedure recommended for eliminating the interference of alcoholamines can also be applied to the heterocyclic amines.

SCOPE

The method is applicable to the determination of the purity of most of the aliphatic primary amines investigated. Aromatic amines, such as aniline, do not react appreciably under the conditions of the method. Highly branched primary amines, such as tertiary butylamine, do not react quantitatively with pentanedione. Tertiary butylamine can be determined by the salicylaldehyde method of Johnson and Funk (30).

The pentanedione method is applicable to the determination of the following classes of primary amines that cannot be determined by the salicylaldehyde method of Johnson and Funk: ethyleneamines, alcoholamines, compounds containing a heterocyclic secondary amino nitrogen, and amino acids.

In the pentanedione method ammonia interferes quantitatively; a correction can be applied by determining the ammonia independently.

Carbon Disulfide–Ethylhexaldehyde System

See pages 581–86.

29. C. D. Wagner, R. H. Brown, and E. D. Peters, *J. Am. Chem. Soc.*, **69,** 2611 (1947).
30. J. B. Johnson and G. L. Funk, *Anal. Chem.*, **28,** 1977 (1956).

Benzaldehyde Method—Adapted from W. Hawkins, D. M. Smith, and J. Mitchell, Jr.

[*Reprinted in Part from J. Am. Chem. Soc.*, ***66****, 1662–3 (1944)*]

A technique has been developed based on the Schiff type reaction:

$$RNH_2 + C_6H_5CHO \rightarrow RN{=}CHC_6H_5 + H_2O$$

The water formed from the rapid quantitative reaction in the presence of pyridine between the amine and aldehyde is titrated with Karl Fischer reagent, after the excess aldehyde has been removed by means of the cyanhydrin reaction (31).

This procedure is applicable to aliphatic, alicyclic, and aromatic primary amines, and also amino alcohols that do not contain a secondary amino nitrogen group. Heterocyclic secondary amines interfere.

EXPERIMENTAL

ANALYTICAL PROCEDURE. Weigh the sample, containing up to 0.1 equivalent of primary amine, into a 100-ml volumetric flask about one-third filled with dry pyridine. After dilution to the mark with more pyridine, transfer a 10-ml portion to a 250-ml glass-stoppered volumetric flask. Add 3 ml of benzaldehyde,* stopper the flask and, together with a blank, place it in a 60°C bath. After 30 minutes remove the flasks and allow to cool spontaneously to room temperature. Transfer the flasks to a well-ventilated hood. Add about 0.2 gram of dry sodium cyanide and 30 ml of 6% hydrogen cyanide in pyridine. Shake the flasks vigorously for about 1 minute;† then set aside in the hood for 45 minutes. At the end of this time, titrate the mixture with Fischer reagent‡ to the usual visual end point. This titration should be fairly rapid. The first sharp end point should be taken, for occasionally some fading may be observed after standing a few minutes.

The water found after correction for that present in the blank and original sample is equivalent to the amount of primary amine in the sample. Free water is best obtained by titrating the original sample in acetic acid solution.

* Either freshly distilled or acid-free benzaldehyde inhibited with about 0.1% hydroquinone is preferred. Results with benzylamine were only 95% quantitative using benzaldehyde containing about 10% benzoic acid.

† This shaking is required to initiate the cyanhydrin reaction, since the sodium cyanide catalyst is insoluble in the pyridine.

‡ Preferably in a well-ventilated hood, since the mixture contains excess hydrogen cyanide.

31. W. M. D. Bryant, J. Mitchell, Jr., and D. M. Smith, *J. Am. Chem. Soc.*, **62,** 3504 (1940).

ANALYTICAL RESULTS

A group of 17 widely different primary and primary-secondary amines was analyzed by the method (Table 32). With the exception of *p*-bromoaniline, which was recrystallized from chloroform, the trade products were used without further purification. The precision and accuracy were usually within ±0.2%.

Table 32 Analytical Data for Primary Amines

Substance	% Amine (Other Methods)	Analysis (Using Fischer Reagent) %		
		Amine	Water	Total
Propylenediamine[b]		(2)[a] 86.3 ± 0.3	14.1	100.4
n-Butylamine		(6) 98.2 ± 0.2	1.7	99.7
Diethylenetriamine[b]		(4) 96.7 ± 0.1	1.3	98.0
Hexamethylenediamine[c]		(4) 97.2 ± 0.1	2.7	99.9
Decamethylenediamine[c]		(2) 97.0 ± 0.5		
Laurylamine[c]	98.1[g]	(2) 98.2 ± 0.2	0.6	98.8
Cyclohexylamine	100.2[g]	(4) 99.7 ± 0.2	0.3	100.0
2-Aminomethylcyclopentylamine		(2) 94.2 ± 0.3	5.4	99.0
o-Aminodicyclohexyl[d]		(2) 94.2 ± 0.2	0.1	94.3
Aniline[e]	98.8[h]	(2) 98.8 ± 0.1	0.1	98.9
m-Aminophenol		(2) 99.5 ± 0.1	0.0	99.5
p-Bromoaniline	99.4[h]	(6) 94.8 ± 0.3	0.0	94.8
p-Phenylenediamine	99.6[h]	(4) 99.9 ± 0.0	0.0	99.9
Benzylamine	92.8[g]	(5) 92.5 ± 0.2	0.7	92.9
Toluidine[c]		(2) 97.1 ± 0.3	0.0	97.1
o-Aminodiphenyl[d]		(2) 92.6 ± 0.2		
β-Naphthylamine[f]	97.3[h]	(4) 97.2 ± 0.1	0.0	97.2

[a] Figures in parentheses represent number of individual determinations.
[b] Carbide and Carbon Chemical Company, 85%.
[c] Du Pont.
[d] Monsanto.
[e] Arthur H. Thomas Company.
[f] J. T. Baker; all others Eastman Kodak Company Chemicals.
[g] By titration to bromphenol blue.
[g] By acetylation according to J. Mitchell, Jr., W. Hawkins, and D. M. Smith, *J. Am. Chem. Soc.*, **66,** 782 (1944).

Results for *p*-bromoaniline were precise but consistently about 5% low as compared with the value obtained by the acetylation method. Urea and methyl urea reacted to the extent of about 10% on the basis of one mole of water formed per mole. This value would be doubled if the disubstitution product were formed (32).

Amino alcohols containing only primary amine groups react quantitatively. This reaction, presumably, might follow one of two courses, either normal imine formation or condensation involving both the amine and hydroxyl to form a substituted oxazine-type compound (33). In this case the exact mechanism of the reaction was not established, since by either route one mole of water is formed from each mole of amino alcohol present in the sample. Experimental data obtained on several amino alcohols are given in Table 33.

Table 33. Analytical Data for Primary Amino Alcohols

Substance	Found, wt. %
Monoethanolamine[a]	100.1 ± 0.2
2-Amino-2-methylpropanol	91.1 ± 0.2
2-Amino-2-methyl-1,3-propandiol	91.6
2-Amino-1-butanol	88.9 ± 0.2
Tris-(hydroxymethylaminomethane)	96.1 ± 0.2

[a] Purified Carbide and Carbon chemical; all others Commercial Solvents materials.

Aliphatic and aromatic secondary amines and tertiary amines do not interfere to any appreciable extent. Negative results were obtained on diisobutylamine, methylaniline, diphenylamine, carbazole, triethanolamine and triisopropanolamine. Values of 1.5 and 2.0%, respectively, on diethyl and di-*n*-butylamine may have been due to primary amine impurity.

INTERFERING SUBSTANCES. Amino alcohols having a secondary amine group usually react nearly quantitatively. Following the conditions of the general procedure, diethanolamine gave one mole of water per mole of amine; diisopropanolamine and hydroxyethylethylenediamine gave 0.9 and 1.8 moles of water, respectively. A compound boiling at 160°C (10 mm) was isolated after reaction of diethanolamine with benzaldehyde. Elemental analysis indicated the compound $C_{11}H_{15}NO_2$. The molecular weight found was 180, as compared with 193 calculated. The presence of both tertiary amino nitrogen and some free hydroxyl suggested that possibly this type of amino alcohol also condenses with benzaldehyde to form a heterocyclic ring compound.

32. Schiff, *Ann.*, **291,** 368, 370, 371 (1896).
33. Kohn [*Monatsh.*, **26,** 956 (1905)] isolated 4,6,6-trimethyl-2-phenyltetrahydrometoxazine after treating benzaldehyde with 2-hydroxy-2-methylpentylamine-4.

Heterocyclic amines react with benzaldehyde to form the *N,N′*-benzaldimines, eliminating one mole of water for every 2 moles of amine. *N,N′*-Benzaldipiperidine, melting at 83°C, was isolated as the product from the reaction of piperidine with benzaldehyde, in the molar proportions used in the analytical procedure. The literature values for the melting point of this material vary from 78 to 81°C (34–36). Apparently both piperazine and morpholine react in the same manner, for in both cases the ratio of water found to amine added was 1:2.

Copper-Salicylaldehyde Method—Adapted from F. Critchfield and J. B. Johnson

[*Reprinted in Part from Anal. Chem.*, ***28,*** *436–40* (*1958*)]

The qualitative test of Duke (37) was investigated as a possible quantitative procedure. The basis of this test is the reaction of a primary amine with 5-nitrosalicylaldehyde. The imine formed in this reaction reacts with nickel ion in the presence of triethanolamine to form an insoluble compound.

In attempting to adapt the reagent specified by Duke to a quantitative basis, it was found that nickel-5-nitrosalicylaldehyde derivatives of the primary amines were insoluble in most solvents and could not be determined colorimetrically. By substituting copper for nickel and salicylaldehyde for the corresponding 5-nitro compound, derivatives soluble in certain organic solvents were obtained. A reagent containing salicylaldehyde, cupric acetate, and triethanolamine was prepared in methanol. With this reagent, primary amines give a soluble, colored reaction product that has an absorption maximum at 445 nm and follows Beer's law. Secondary amines interfere with this reagent because they also form a colored reaction product. Because of this interference an aqueous reagent was prepared, in which most of the reaction products of primary amines are insoluble. Reaction products of primary amines can be extracted into isopropyl ether or benzene; when this technique is used, however, the colored reaction products have no absorption maximum in the visible portion of the spectrum.

An attempt was made to utilize the procedure in spite of this limitation by measuring the absorbance at 430 nm, which is on the shoulder of the curve. A calibration curve, which was linear except near the origin, was obtained by this procedure. The addition of 0.01% ethanolamine to

34. Laun, *Ber.*, **17,** 678 (1884).
35. Ehrenberg, *J. Prakt. Chem.* [2] **36,** 130 (1887).
36. Lachowicz, *Monatsh.*, **9,** 698 (1888).
37. F. R. Duke, *Ind. Eng. Chem., Anal. Ed.*, **17,** 196 (1945).

the triethanolamine used in the reagent gave a calibration curve that followed Beer's law over all ranges of concentration. However secondary and tertiary amines cause a shift in the absorption curve. Because of this effect, it was necessary to develop a color that possesses an absorption maximum in the visible portion of the spectrum independent of secondary or tertiary amines.

An aliquot of the benzene layer was removed and the copper present made to react with diethyldithiocarbamic acid (38). The absorbance at 430 nm is a direct measure of copper and, therefore, an indirect measure of primary amine. Fair results were obtained by this method for the determination of primary amines in the presence of secondary and tertiary amines. However there were three serious disadvantages: (1) the amount of monoethanolamine present in the triethanolamine used in the reagent determined the slope of the calibration curve, (2) low results were obtained for the analysis of alcoholamines in the presence of the corresponding secondary and tertiary amines, and (3) the presence of monoethanolamine in the reagent caused the reagent blank to have a high absorbance.

Because of these disadvantages, several organic solvents were investigated for the extraction of the copper complexes. The use of 1-hexanol eliminated the necessity of adding monoethanolamine to the reagent. A superior color stability was also imparted by substituting bis(2-hydroxyethyl)dithiocarbamic acid for diethyldithiocarbamic acid as the reagent for copper (39).

EFFECT OF TRIETHANOLAMINE CONCENTRATION. Triethanolamine is used in the reagent to act as a proton acceptor to shift the equilibrium reaction between cupric chloride and the imine. A minimum of 15.0% by volume of the amine is necessary to give quantitative reaction. The minimum amount of triethanolamine is specified in the reagent to allow for the presence of secondary and tertiary amines in the sample.

EFFECT OF SALICYLALDEHYDE CONCENTRATION. The compatibility of cupric ion and salicylaldehyde is limited by the formation of an insoluble complex. Fortunately, in dilute solutions and in the presence of triethanolamine, this chelate formation is minimized. A maximum of 0.5% by volume of salicylaldehyde can be tolerated in the reagent without appreciable reaction with cupric ion. This concentration has been specified in the reagent.

EFFECT OF COPPER CONCENTRATION. The concentration of cupric ion in the reagent is fairly critical, as shown in Fig. 11.16. Curve 1 is a plot of the absorbance obtained for 0.372 mg of ethanolamine as a function of the

38. L. A. Haddock and N. Evers, *Analyst*, **57**, 495 (1932).
39. W. C. Woelfel, *Anal. Chem.*, **20**, 722 (1948).

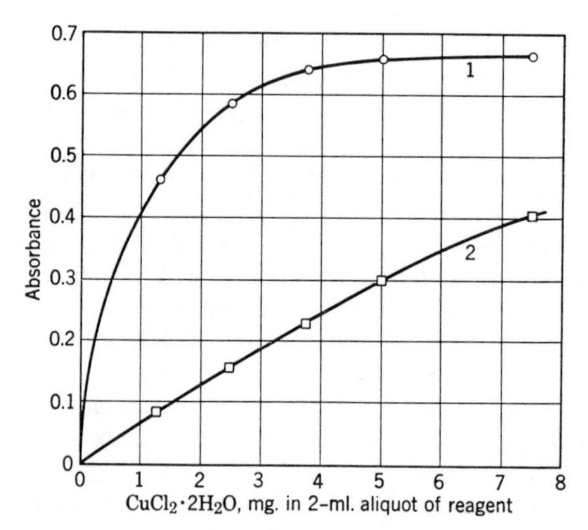

Fig. 11.16. Effect of copper concentration on reaction of copper-salicylaldehyde reagent with ethanolamine: 1, absorbance for 0.372 mg of ethanolamine; 2, absorbance of reagent blank versus methanol.

concentration of cupric chloride dihydrate in 2.0 ml of reagent. The concentrations of triethanolamine and salicylaldehyde are fixed at 15.0 and 0.5% by volume, respectively. From this curve it can be seen that at least 5.0 mg of cupric chloride dihydrate is necessary to obtain maximum reaction. Curve 2 shows the effect of cupric ion on the absorbance of the reagent blank compared to methanol. It is apparent that a minimum concentration of cupric ion is desired to obtain a minimum absorbance of the reagent blank. For this reason 5 mg of cupric chloride dihydrate in a 2-ml aliquot of the reagent is specified. This corresponds to a cupric chloride concentration of 0.25% in the reagent.

EFFECT OF QUALITY OF TRIETHANOLAMINE. Figure 11.17 shows calibration curves obtained for ethanolamine using a highly purified grade of triethanolamine (curve 1) and commercial triethanolamine (curve 3), demonstrating that the quality of triethanolamine used in the reagent determines the slope of the calibration curve. This is not desirable because a separate calibration curve would be required for each lot of triethanolamine used in the reagent. This effect is no doubt caused by the presence of chelating agents in commercial triethanolamine, which effectively decrease the cupric ion concentration of the reagent, thus decreasing the sensitivity of the method as shown by curve 1 in Fig. 11.16. To establish the theoretical calibration curve for a primary amine, a calibration curve (curve 2, Fig. 11.17) was obtained for copper acetate, employing the relative volumes used in the amine procedure. For these curves a

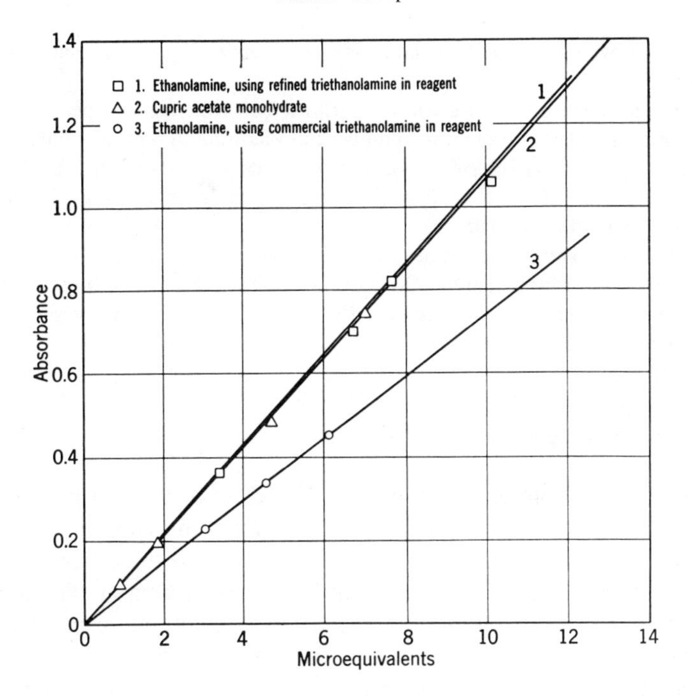

Fig. 11.17. Effect of triethanolamine quality on slope of calibration curve for ethanolamine.

plot of absorbance versus microequivalents was made. Within the experimental error of the determinations, curves 1 and 2 are identical. Therefore a refined grade of triethanolamine is necessary for the preparation of the reagent, if a reproducible calibration curve is desired. The method of preparing a suitable grade of triethanolamine is described next.

REAGENTS

Cupric chloride dihydrate.

1-Hexanol, Carbide and Carbon Chemicals Company.

Salicylaldehyde, reagent grade.

TRIETHANOLAMINE. Carbide and Carbon Chemicals Company. Distill 98% material under 1- to 2-mm pressure, using a column 6 in. long and 30 mm in diameter, packed with 2-mm glass beads and heated by means of resistance wire. Use a 3-liter, round-bottomed distillation flask fitted with a thermometer well. Stir the contents of the flask by means of a magnetic stirrer, and do not allow the kettle temperature to exceed 185°C during the distillation. An absorbance of 0.65 ±0.02 for 0.372 mg of ethanolamine should be obtained by the procedure described below, when this material is used to prepare the copper-salicylaldehyde reagent.

COPPER–SALICYLALDEHYDE REAGENT. Into a 100-ml glass-stoppered graduated cylinder, measure 15.0 ml of redistilled triethanolamine, 0.5 ml of salicylaldehyde, and 0.25 gram of cupric chloride dihydrate. Dilute to 100 ml with distilled water and mix the contents. This reagent is stable for at least a month; however the reagent blank increases with age.

BIS(2-HYDROXYETHYL)DITHIOCARBAMIC ACID REAGENT. Prepare a 2% by volume solution of carbon disulfide in methanol and a 5% by volume solution of diethanolamine in methanol. Prepare the reagent fresh daily by mixing equal volumes of the two components.

CALIBRATION CURVE

Prepare a dilution of the pure compound in distilled water so that a 5-ml aliquot contains not more than the maximum sample size given in Table 34. To each of five 25-ml glass-stoppered graduated cylinders, add 2 ml of copper-salicylaldehyde reagent by means of a pipet. Transfer 1.0-, 2.0-, 3.0-, and 5.0-ml aliquots of the dilution above to respective 25-ml graduated cylinders, reserving one as the blank. Measure the absorbance of each standard at 430 nm, using 1-cm cells and a suitable spectrophotometer.

PROCEDURE

Add 2.0 ml of copper-salicylaldehyde reagent from a transfer pipet to each of two 25-ml glass-stoppered graduated cylinders. Reserve one of the cylinders as a blank, and into the other measure an amount of sample calculated to contain not more than the maximum amount of primary amine listed in Table 34. The sample must not contain more than 0.01 mg of ammonia or 0.5 gram of secondary and tertiary amine. For samples of less than 0.1 gram, use an aliquot of a suitable aqueous dilution. Dilute the contents of each graduate to the 10-ml mark with distilled water, stopper, and mix thoroughly. Allow the sample to react under the conditions specified in Table 34.

When the reaction is complete, add sufficient 1-hexanol to bring the total volume of liquid to 25 ml. Stopper the cylinders, shake vigorously 15 or 20 times, and allow the layers to separate. Add 5 ml of bis(2-hydroxyethyl)dithiocarbamic acid reagent to each of two additional 25-ml glass-stoppered graduated cylinders. In this step it is important that the graduated cylinders and stoppers be clean and void of any metallic ions,

Table 34. Reaction Conditions for Determination of Primary Amines by Copper-Salicylaldehyde Method

Compound	Primary Amine, mg., Maximum	Time, min.[a]
Aminoethylethanolamine	1.10	30 to 60
N-Aminoethylmorpholine	0.85	15 to 60
Amylamine	1.20	15 to 45
Butylamine	0.70	15 to 60
Ethanolamine	0.50	15 to 60
Ethylamine	0.53	15 to 60
2-Ethylhexylamine	1.40[b]	15 to 60
Hexylamine	1.10	15 to 60
Isoamylamine	1.10	15 to 45
Isobutanolamine	0.60	60 to 120
Isobutylamine	0.90	15 to 60
Isopropanolamine	0.60	15 to 60
Methylamine	0.30	15 to 60
Propylamine	0.73	15 to 60
Propylenediamine	0.42	10 to 20[c]

[a] Reaction time at 20 to 30°C unless otherwise specified.

[b] Make dilutions using a 10% solution of methanol.

[c] Perform reaction at 98 ± 2°C. Use 50-ml. glass-stoppered graduated cylinders; do not stopper during reaction.

which react with this reagent. Pipet 5.0 ml of the hexanol layer from the graduated cylinders in which the reaction was performed to the graduates containing the dithiocarbamic acid reagent. Add the hexanol dropwise to prevent the material from clinging to the walls of the pipet. Dilute the contents of each cylinder to the 25-ml mark with methanol, stopper, and mix the contents. Measure the absorbance of the sample versus the blank at 430 nm, using 1-cm cells. Read the concentration of primary amine from the calibration curve.

DISCUSSION AND RESULTS

A reaction rate study and a calibration curve were obtained for each primary amine investigated. Table 34 gives the recommended reaction conditions for primary amines to which this method has been applied. In all cases the maximum sample size is based on the amount of pure primary amine corresponding to an absorbance of 0.9 under the conditions of the method. Figure 11.18 shows the calibration curves obtained

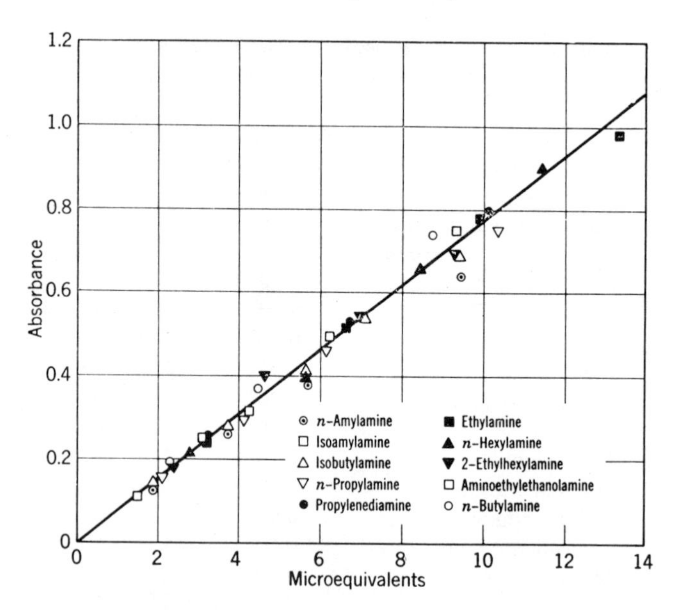

Fig. 11.18. Calibration curve for 10 primary amines.

for 10 primary amines. For this curve absorbance is plotted against microequivalents of primary amine. Using such a plot, all primary amines should fall on the same calibration curve, provided (1) the ratio of amine to copper is constant, (2) a quantitative extraction of the complex from water is obtained, or (3) the distribution coefficient is the same for each compound. Figure 11.18 shows that, within certain limits, one calibration curve suffices for these 10 primary amines. Of all the compounds investigated, five primary amines did not fall on this curve. Figure 11.19 shows the calibration curves obtained for these compounds, three of which are alcoholamines. The sensitivity is greater for these compounds than for the other amines investigated. The calibration curves obtained for ethanolamine and isopropanolamine correspond to the theoretical curve for a 1:1 mole ratio of copper to amine, as in Fig. 11.17. This is not in agreement with the findings of other investigators, who report a 2:1 ratio of amine to copper (40). Evidently in dilute solution the formation of the 1:1 complex is favored. In no case was a calibration curve obtained that corresponded to the 2:1 complex.

A few primary amines do not react quantitatively with this reagent. In general, compounds that react incompletely can be divided into three classes: (1) aromatic amines such as aniline; (2) compounds that contain

40. A. E. Martell and M. Calvin, *Chemistry of the Metal Chelate Compounds*, Prentice-Hall, Englewood Cliffs, N.J., 1953, p. 273.

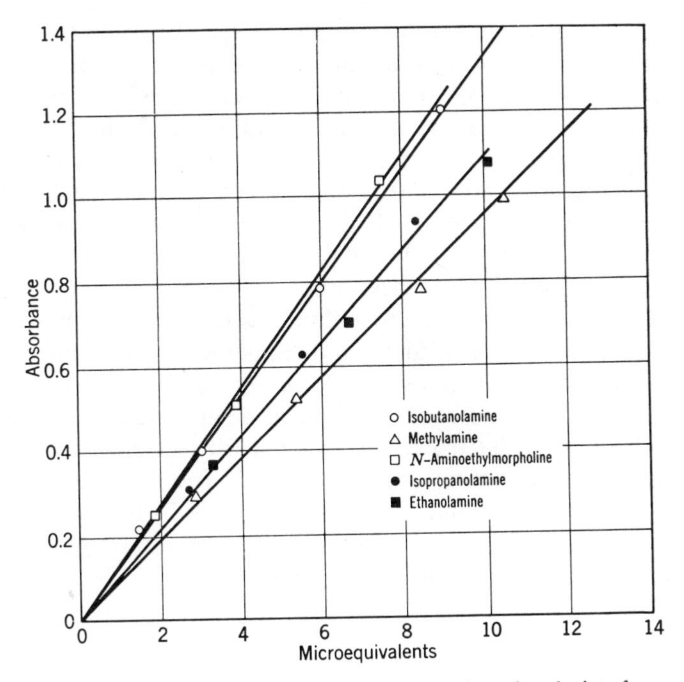

Fig. 11.19. Calibration curves for primary amines that deviate from curve in Fig. 11.18.

more than one primary amine group—for example, ethylenediamine and diethylenetriamine; and (3) primary amines that are branched in the 2-position, such as tertiary and secondary butylamine and isopropylamine. Propylenediamine and 2-ethylhexylamine are exceptions to this generalization.

Ammonia interferes if more than 0.01 mg is present in the sample aliquot. This amount of ammonia corresponds to an absorbance of 0.03, which is within the experimental error of the method. Attempts to determine ammonia by this method were unsuccessful because the reaction apparently is not quantitative. The combined secondary and tertiary amine contents of the sample aliquot must not exceed 0.5 gram. Greater amounts of secondary or tertiary amines tend to solubilize the copper complex in the aqueous layer. Strong oxidizing or reducing agents interfere by depleting the reagent. Compounds that form copper complexes soluble in 1-hexanol give high results; those forming water-soluble complexes give low results.

This method has been successfully applied to the analysis of several amine mixtures. To date the method has replaced the Van Slyke method (pp. 586–92) in many cases for the control of plant processes with a considerable saving of time and laboratory costs. Tables 35 to 37 present data for the analysis of several mixtures of primary, secondary, and tertiary amines. Table 35 shows the determination of several amine mixtures of known primary amine content. In Tables 36 and 37, analyses of several plant process samples are given. In each case, the primary amine content was determined by the copper-salicylaldehyde and Van Slyke methods. The data in Table 36 were obtained by an experienced operator; those in Table 37 were obtained in a routine control laboratory. For the determination of primary amines in the presence of secondary and tertiary amines, an accuracy within 5% can be anticipated by this method. In some cases Van Slyke and copper-salicylaldehyde comparative results are substantially better (Table 37), while in others they are somewhat poorer. Because of the limited sample size, primary amine contents of less than 0.01% in the presence of secondary and tertiary amines cannot be determined by this method. The method was successfully applied to the determination of ethanolamine hydrochloride without prior neutralization of the sample. The absorbance of the hydrochloride is proportionally the same as the free amine.

DETERMINATION OF SECONDARY AMINES IN THE PRESENCE OF PRIMARY AND TERTIARY AMINES

Secondary amines in mixtures with primary amines can be determined by determining the total primary and secondary amines acidimetrically (see pp. 533–44) or by the esterification methods (p. 558). Then the primary amines can be determined by any of the methods described previously. The secondary amines are then determined by difference. Methods already described (pp. 567–8, 586) specifically discuss these indirect approaches.

Secondary amines in mixtures with tertiary amines can be determined directly by the esterification methods (p. 558). In addition, the secondary amine can be determined by difference by measuring the tertiary amines by any of the methods shown on pages 567–633 and determining the total amine content acidimetrically.

The discussions on pages 567–586 cover the determination of secondary amines in the presence of both primary and tertiary amines. Two additional direct methods exist for secondary amines in the presence of either or both other types of amine. These methods are based on the reaction of the secondary amine and carbon disulfide as illustrated on page 572. If

Table 35. Determination of Primary Amines in the Presence of Secondary and Tertiary Amines

Sample	Primary Amine, wt. %		
	Added	Found[a]	Difference
Isopropanolamine in 2,5-dimethylpyrazine	0.12	0.12 (2)	0.00
	0.57	0.57 (2)	0.00
Ethanolamine in diethanolamine	2.35	2.16 (1)	−0.19
	0.36	0.32 (2)	−0.05
	0.12	0.13 (2)	+0.01
	0.09	0.13 (2)	+0.07
	2.4	2.5 (1)	+0.1
Ethanolamine in triethanolamine	0.31	0.31 (1)	0.00
	0.10	0.10 (1)	0.00
	0.98	0.90 (1)	−0.08
Ethanolamine in di- and triethanolamine	47.6	47.6 (2)	0.0
	21.3	21.2 (2)	+0.1
	29.2	28.7 (4)	−0.5
Butylamine in dibutylamine	1.07	1.14 (1)	+0.07
	0.56	0.53 (1)	−0.03
	0.22	0.15 (1)	−0.07
	0.49	0.54 (1)	+0.05

[a] Figures in parentheses represent number of determinations.

Table 36. Determination of Ethanolamine in the Presence of Di- and Triethanolamines

Sample Number	Ethanolamine, wt. %		
	Van Slyke Method	Copper-Salicylaldehyde Method	Difference
1	11.5	12.1	+0.6
2	12.9	13.0	+0.1
3	12.5	12.5	0.0
4	11.7	12.2	+0.5
5	13.1	12.5	−0.6
6	10.8	11.1	+0.3
7	13.1	12.1	−1.0
		Average difference	±0.5

Table 37. Control Laboratory Determination of Ethanolamine in the Presence of Di- and Triethanolamine

Sample	Operator	Ethanolamine, wt. %		
		Van Slyke Method	Copper-Salicylaldehyde Method	Difference
1	1	42.2	41.8	−0.4
2	1	39.1	41.2	+2.1
2	2	39.5	40.1	+0.6
3	1	38.3	36.6	−1.7
3	2	42.2	42.4	+0.2
4	1	34.3	35.4	+1.1
4	2	36.4	37.2	+0.8
5	2	14.4	14.5	+0.1
			Average difference	±0.9

the reaction is performed in the presence of copper or nickel in aqueous solution, formation of the copper or nickel dialkyldithiocarbamate results.

$$2\left(\begin{matrix}R \\ & NH \\ R'\end{matrix}\right) + CS_2 + M^{2+} \longrightarrow \left(\begin{matrix}R \\ & N-\underset{\substack{\| \\ S}}{C}-S \\ R'\end{matrix}\right)_2 M + 2H^+$$

Spectrophotometric analysis of the copper complex and atomic absorption analysis of the nickel in the nickel complex are the end determinations for the quantitative procedures.

Dithiocarbamic Acid Colorimetric Method—Adapted from G. R. Umbreit

[*Reprinted in Part from Anal. Chem.*, **33,** *1572–3* (*1961*)]

The reaction of primary and secondary amines with carbon disulfide has been utilized by Critchfield and Johnson (41) (pp. 572–81). The resulting dithiocarbamic acids are titrated with base. Tertiary amines do not react, and primary amines are masked by imine formation with 2-ethylhexaldehyde to make the method selective for secondary amines. Katcher and Voroshilova (42) determined dimethylamine by titration of

41. F. E. Critchfield and J. B. Johnson, *Anal. Chem.*, **28,** 430–6 (1956).
42. E. Katcher and M. Voroshilova, *Anilinokras. Prom.*, **4,** 39–41 (1934); *Chem. Abstr.*, **28,** 3689_7 (1934).

the dithiocarbamate with standard copper sulfate solution. Several investigators (43–45) have used colorimetric measurement of the copper dithiocarbamate complex but have restricted their investigations to the determination of dimethylamine in various systems.

The method described here defines conditions for more general applicability of the colorimetric methods mentioned earlier for dimethylamine. Specific application is made to diethylamine and *N*-methylaniline. Application to a number of other secondary amines is indicated. Data on the analysis of primary-secondary and tertiary-secondary amine mixtures are presented.

REAGENTS

CARBON DISULFIDE–PYRIDINE–ISOPROPYL ALCOHOL. Accurately measure and mix 35 ml of carbon disulfide, 25 ml of pyridine, and 65 ml of isopropyl alcohol. When stored in a glass-stoppered reagent bottle, the solution is usable for at least 2 months.

CUPRIC CHLORIDE SOLUTION, 0.0013*M*. Dissolve 0.1 to 0.12 gram of aqueous cupric chloride in 250 ml of water and dilute to 500 ml with pyridine.

None of the volume or weight measurements is excessively critical as long as standards are used for comparison when any reagent solution is replaced.

PROCEDURE

Transfer 1 ml of the sample solution to a 15×150 mm glass-stoppered test tube or other suitable reaction vessel. Add 4 ml of the carbon disulfide–pyridine–isopropyl alcohol reagent and 2 ml of the cupric chloride reagent. Agitate the mixture; then allow it to stand for 5 to 20 minutes at room temperature. (Some of the carbon disulfide settles out during this period.) Then add 3.0 ml of acetic acid (10% volume of glacial acetic acid in water) and 3.0 ml of benzene. Agitate the mixture by inversion several times, and allow the phases to separate. From the upper (organic) phase, remove 4.0 ml and dilute to 5 ml with isopropyl alcohol. After this solution has been standing for 60 to 90 minutes (diethylamine) or 20 minutes (*N*-methylaniline), measure the absorbance.

43. H. C. Dowden, *Biochem. J.*, **32,** 455–9 (1938).
44. L. Nebbia and F. Guerrieri, *Chim. Ind.* (*Milan*), **35, 896** (1953); *Chem. Abstr.*, **48,** 3869*c* (1954).
45. E. L. Stanley, H. Baum, and J. L. Gove, *Anal. Chem.*, **23,** 1779–82 (1951).

RESULTS AND DISCUSSION

Table 38 summarizes recovery data for diethylamine and *N*-methylaniline, which were chosen as representative of aliphatic- and aromatic-substituted secondary amines. The responses in both cases are linear over the concentration ranges indicated.

Table 38. Analysis of Secondary Amines

Taken, μg.	Found, μg.	Recovery, %
	Diethylamine	
11.3	11.0	97.3
22.7	22.4	98.7
45.4	45.4	100.0
68.0	67.3	99.0
90.7	90.5	99.8
	N-Methylaniline	
36.2	35.3	97.6
72.4	72.2	99.8
108.6	108.6	100.0
	Av.	99.1 ± 1.0

The application of this method to primary-secondary and tertiary-secondary amine mixtures is summarized in Table 39. Tertiary amines do not react with carbon disulfide. Thus minute fractions of secondary amines in tertiary amines can be analyzed accurately. Primary amines constitute a positive interference. The primary amine dithiocarbamate complexes have an absorption maximum significantly removed from that of the secondary amines (Table 40). In addition, their molar absorptivity is significantly smaller. For these reasons, primary amines present in approximately equivalent molar quantities will result in an error of 1% or less in the determination of secondary amines. Because the primary amine dithiocarbamate complexes are much less soluble in both phases, however, they cannot normally be tolerated in amounts greater than 100 μg per sample. Larger amounts of primary amines result in a haze that interferes with the spectrophotometric measurements.

Compounds tested specifically that do not react under the specified conditions are pyridine, triethylamine, tributylamine, *N,N*-dimethylaniline, 1,3-diphenylguanidine, diphenylamine, 3-carbonyl indoles, and ammonia. The secondary amines that do not react are all highly conjugated.

Table 39. Determination of Mixtures

$C_6H_5NHCH_3$ Taken, μg.	$C_6H_5NHCH_3$ Found, μg.	Error, μg.
N-Methylaniline in N,N-Dimethylaniline[a]		
18.2	19.0	+0.8
36.4	35.4	−1.0
N-Methylaniline + Aniline[b]		
18.2	19.7	+1.5
36.4	37.0	+0.6
54.6	55.3	+0.7
Et_2NH	Et_2NH	
Diethylamine in Triethylamine[c]		
11.3	11.5	+2.0
22.7	22.9	+0.2
34.0	33.7	−0.3
Diethylamine + Ethylamine[d]		
11.3	13.3	+2.0
22.7	23.3	+0.6
34.0	35.1	+1.1

[a] 4.779 mg. of $C_6H_5N(CH_3)_2$ taken.
[b] 100.3 μg. of $C_6H_5NH_2$ taken.
[c] 3.615 mg. of Et_3N taken.
[d] 64.8 μg. of $EtNH_2$ taken.

Table 40. Reacting Compounds

Compound	Wave-length of Absorption max., mμ	Time to Reach Maximum Absorbance, min.	ϵ'[a], $\times 10^2$
Piperidine	440	30	5.37
Di-*n*-butylamine	440	30	7.08
N-Methylaniline	445	20	5.36
Diethylamine	440	60	9.67
Isoleucine	360		
n-Butylamine	350, 430[b]		
Aniline	355, 430[b]		
Neomycin B	390		

[a] Based on concentration in original sample aliquot.
[b] Secondary amine impurity is suspected.

EFFECTS OF VARIABLES

The reaction medium, prior to extraction, provides a sufficient excess of reactants to permit rapid formation of the dithiocarbamates and subsequent complexation with copper. Some amines may require additional reaction time. However after extraction there is a significant period of time during which the absorbance increases. As indicated in Table 41,

Table 41. Response as a Function of Time After Extraction

Sample	Absorbance at Time, *t*, after Extraction, min.						
	10	15	20	25	40	120	180
N-Methylaniline[a]	0.335	0.350	0.360	0.352	—	—	—
Diethylamine[b]	—	—	0.225	—	0.240	0.270	0.269

[a] 72.4 μg. measured at 445 mμ.
[b] 22.7 μg. measured at 440 mμ.

this time varies with the amine in question. This problem is believed the result of a reversible equilibrium between 1:1 and 1:2 amine-to-copper compounds as in eq. 10.

$$2R_1R_2N{-}\overset{\overset{\displaystyle S}{\|}}{C}{-}SCuAc \rightleftharpoons (R_1R_2{-}\overset{\overset{\displaystyle S}{\|}}{C}{-}S)_2Cu + Cu(Ac)_2 \qquad (10)$$

The 1:1 compound would appear to be preferred in the aqueous phase where a significant excess of copper is available, while in the organic phase the 2:1 compound is preferred. The time required to attain maximum absorbance should be a measure of the rate of attainment of this equilibrium in the organic phase. The response curves do not become linear until this maximum absorbance is obtained.

The necessity of adding acetic acid to accomplish the extraction supports this hypothesis. If water only is substituted as a diluent in place of the acetic acid solution before extraction, the response curve becomes nonlinear, and color remains partially in the aqueous phase. The use of greater amounts of acetic acid results in the formation of haze in the organic layer.

Variations in ionic strength of the sample solution using both potassium chloride and sodium acetate from 0.01 to 0.24*M* have no effect.

The formation of the dithiocarbamic acids in the reaction mixture requires the free base of the amine. Although pyridine is a weaker base than most of the amines tested here, the large excess that is present in the reaction mixture is sufficient to drive the amine-amine salt equilibrium in the direction of the free amine that is removed by dithiocarbamate formation. A sample solution that is strongly acidic should be neutralized with sodium hydroxide or ammonia prior to analysis. Amines that are very strong bases may require addition of sodium hydroxide to attain the necessary fraction of free base. In some cases increased reaction time is required. This must be determined individually for each amine.

Atomic Absorption Method of P. J. Oles and S. Siggia

[*Reprinted in Part from Anal. Chem.*, ***45,*** *2150* (*1973*)]

APPARATUS

Absorbances were measured at 232.0 nm with a Perkin-Elmer atomic absorption spectrophotometer. All filtering was done with a medium-frit (10–15 μm) borosilicate glass funnel.

REAGENTS

AMMONIACAL NICKEL REAGENT (46). Dissolve 200 g of ammonium acetate and 5.0 g of $NiCl_2 \cdot 6H_2O$ in 300 ml of water in a 1-liter bottle. Add 100 g of sodium hydroxide in 200 ml of water and 200 ml of ammonium hydroxide (sp. gr., 0.90). Dilute to approximately 1 liter with distilled, deionized water.

NICKEL DI-*n*-BUTYLDITHIOCARBAMATE FOR CALIBRATION CURVE. Prepare according to the experimental procedure, recrystallize twice from acetone and dry *in vacuo* at room temperature for 48 hours.

PROCEDURE

Add 0.50 to 1.0 ml of sample containing 1 to 4 μM of secondary amine to a 6-in. test tube. Add 1.0 ml of the ammoniacal nickel reagent and approximately 0.025 to 0.05 ml of carbon disulfide. Mix the contents of the test tube thoroughly and place in a 30 to 35°C water bath for 60 to 75 minutes, mixing occasionally. Transfer the contents of the test tube to a medium-fritted glass funnel and apply suction. Rinse the test tube three times with water, each time allowing the rinsings to pass through the filter. Place a clean 125-ml suction flask below the filter and add 1 ml of warm benzene-acetone (1:1) to the reaction test tube to ensure the

46. E. L. Stanley, J. Baum, and J. L. Grove, *Anal. Chem.*, **23,** 1779 (1951).

dissolution of any adhering precipitate. Transfer this solvent to the funnel and apply suction. Then rinse the funnel with two portions of warm benzene-acetone. Remove the organic solvents with the aid of suction and a hot water bath. Add 4 ml of equal parts of concentrated nitric and hydrochloric acids to the flask, and allow digestion to take place at 100°C for 15 to 20 minutes. Transfer the contents of the flask to a 10-ml volumetric flask, dilute to volume with rinsings from the flask, and analyze the resulting solution for nickel content by atomic absorption spectrophotometry.

For the calibration curve, remove the organic solvent from aliquots of a prepared standard solution of the nickel di-*n*-butyldithiocarbamate in acetone, digest and dilute as described previously. The standard solution of nickel di-*n*-butyldithiocarbamate in acetone was found to be stable for at least 4 months.

RESULTS AND DISCUSSION

The results obtained for nine secondary amines appear in Table 42.

Table 42. Determination of Secondary Amines as Their Nickel Dialkyldithiocarbamates

	Secondary Amines		
Compound	Taken, μM	Recovery, %	Relative Standard Deviation,[a] %
Diethylamine	2.43	99.5	±1.4 (5)
N-Ethyl-*n*-butylamine	1.83	98.9	±1.1 (6)
Di-*n*-butylamine	1.48	98.7	±1.1 (6)
Di-*n*-hexylamine	2.13	99.6	±2.4 (6)
Di-*n*-octylamine	1.90	97.4	±3.0 (5)
N-Butyl-*n*-dodecylamine	2.52	99.6	±3.4 (5)
Di-*n*-dodecylamine	2.20	100.7	±2.9 (5)
Dicyclohexylamine	2.55	97.6	±3.9 (5)
Piperidine	2.52	99.6	±2.6 (5)

[a] Figures in parentheses indicate the number of determinations.

Although extensive xanthate formation occurs in those samples containing alcohol, no interference in the determination is observed.

The practical detection limit of this procedure is approximately 0.30 μM of secondary amine per milliliter of solution. Results in Table 43

Table 43. Determination of Secondary Amines at the ppm level

Compound	Secondary Amines Taken, μM	Secondary Amines Recovery, %	Relative Standard Deviation,[a] %
Diethylamine	0.486	86	±7.3 (6)
Di-*n*-butylamine	0.296	90	±12 (6)

[a] Figures in parentheses indicate number of determinations.

are representative of the precision and accuracy obtainable at this level of concentration.

The only aromatic amines studied, *N*-methylaniline, failed to react quantitatively under a wide range of conditions.

DETERMINATION OF TERTIARY AMINES IN THE PRESENCE OF PRIMARY AND SECONDARY AMINES

Methods given previously (pp. 567–86) indicate procedures for determining tertiary amines in mixtures with either or both primary and secondary amines. An additional direct method exists which is a modification of the foregoing acetylation approaches (pp. 570–3) for measuring small quantities of tertiary amines.

Method for Small Amounts of Tertiary Amines—Adapted from J. Ruch and F. E. Critchfield

[*Reprinted in Part from Anal. Chem.*, **33**, *1569–72* (*1961*)]

The basic technique for a satisfactory titrimetric determination of tertiary (3°) amines in the presence of primary (1°) and secondary (2°) amines is embodied in the method of Blumrich and Bandel (47), further extended by Wagner, Brown, and Peters (48). Both 1° and 2° amines are reacted with acetic anhydride in acetic acid to produce amides. The 3° amine is titrated with perchloric acid in acetic acid. A more satisfactory method, used in these laboratories, replaces acetic acid with methanol as the solvent (49). Both these procedures are relatively insensitive and are

47. K. Blumrich and G. Bandel, *Angew. Chem.*, **54,** 374 (1941).
48. C. D. Wagner, R. H. Brown, and E. D. Peters, *J. Am. Chem. Soc.*, **69,** 2609 (1947).
49. E. F. Hillenbrand, Jr., and C. A. Pentz, in *Organic Analysis*, Vol. III, J. Mitchell, Jr., et al., Eds., Wiley-Interscience, New York, 1956, p. 175.

likely to fail when the tertiary amine concentration is less than 0.5%. Furthermore, the procedures cannot be modified either by reducing the titrant normality or by increasing the sample size without losing definition of the titration curve because of a buffering action of the amide reaction product.

Various solvents were evaluated by acetylating 5 ml of diethylamine for 30 minutes with 20 ml of acetic anhydride in 100 ml of the solvent. Triethylamine was determined by potentiometric titration with $0.01N$ perchloric acid in Methyl Cellosolve, and the shapes of these titration curves were compared (Fig. 11.20).

Methyl Cellosolve is definitely superior to 2-propanol, methanol, or acetonitrile, as shown by the larger and sharper potentiometric break in that solvent. 2-Propanol and glycols also are inferior because they yield a higher solvent blank. Finally, in nonhydroxylic solvents (e.g., ethyl acetate, acetonitrile, and chloroform), the usefulness of the method is somewhat restricted because acetic anhydride tends to attack the indicators found suitable for visual detection of the endpoint.

Although Methyl Cellosolve is a hydroxylic compound, this solvent does not compete seriously with 1° and 2° amines in the acetylation reaction. Under the conditions previously specified, all amines reacted quantitatively within 30 minutes to form the corresponding amide. In nonhydroxylic solvents, acetylation is complete within 15 minutes; before titration, however, the composition of the medium must be converted to at least two-thirds Methyl Cellosolve by addition. Unless time is a critical factor, or the amine is difficult to acetylate, Methyl Cellosolve is generally recommended because it is a suitable medium for both the acetylation and titration.

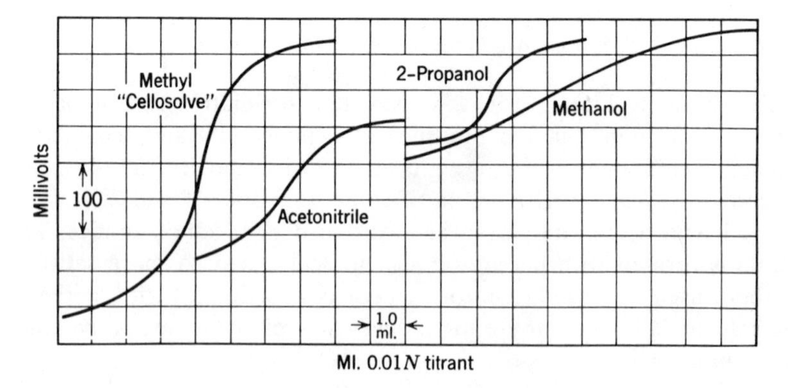

Fig. 11.20. Comparison of media for determination of triethylamine in diethylamine.

REAGENTS

PERCHLORIC ACID, $0.01N$. Dilute 0.8 ml of 70% perchloric acid to 1 liter with Methyl Cellosolve (Union Carbide Chemicals Co.). Standardize by titration of tris(hydroxymethyl)aminomethane (primary standard grade) dissolved in Methyl Cellosolve.

Thymol blue indicator, 0.3% solution in dimethylformamide.

Congo red indicator, 0.1% solution in methanol.

PROCEDURE

Transfer approximately 100 ml of Methyl Cellosolve into two flasks and reserve one for a blank determination. Into the other flask introduce from 5 to 7 grams of sample (for pure amines) containing not more than 0.5 meq. of tertiary amine. Add 20 ml of acetic anhydride (99%) to each flask, swirl gently, and allow the blank and sample to stand for 30 minutes at room temperature.

Titrate the contents of each flask with standard $0.01N$ perchloric acid. Determine the end point potentiometrically or by an indicator (see Table 44). The difference between blank and sample titrations is a measure of the tertiary amine content.

RESULTS

AMINE PURITIES. An accepted general method for determining amine purities is the perchloric acid procedure, whereby the total basicity in acetic acid medium is determined by titration with a standard solution of perchloric acid in acetic acid. Each of seven tertiary amines, with basic ionization constants ranging from 1×10^{-4} to 1×10^{-10}, was titrated to the end point of a suitable indicator with $0.1N$ solutions of perchloric acid in acetic acid and Methyl Cellosolve, respectively. The results of these titrations are summarized in Table 44.

Inspection of the total basicity data reveals a good correlation between the results for a given amine, irrespective of the medium. Methyl Cellosolve, unlike acetic acid, is not a base-leveling solvent; however amines of widely varying basic strengths can be titrated quantitatively in this medium. Indicator choices for visual detection of the end points in the titration of various amines in Methyl Cellosolve are discussed subsequently.

DETERMINATION OF SMALL AMOUNTS OF 3° AMINES IN 1° AND 2° AMINES. Using the method previously described, various 1° and 2° amines were acetylated for 30 minutes in Methyl Cellosolve medium, and the residual 3°

Table 44. Comparison of Total Basicity for Tertiary Amines

Compound	Purity, wt. %	
	$HClO_4$–HOAc	$HClO_4$–Methyl Cellosolve
N,N-Dimethylethanolamine	99.8	99.8
	99.8	99.8
N-Ethylmorpholine	99.9	100.0
	99.7	100.1
Triethanolamine	102.3	101.8
	101.7	102.0
Triethylamine	99.3	100.0
	99.5	99.8
γ-Picoline	93.6	93.6
	93.9	94.1
Pyridine	100.0	99.7
	100.0	99.3
Dimethylaniline	99.8	99.4
	99.6	

amine content was titrated with $0.01N$ perchloric acid to thymol blue indicator. The average results are shown in Table 45. The values are reproducible and are independent of sample size.

RECOVERY DATA FOR 3° AMINES IN 1° AND 2° AMINES. Four 1° and 2° amines were selected from the group previously shown to be low in 3° amine content. To each of these four 1° and 2° amines was added its 3° amine counterpart. The recovery data for these small increments of 3° amine are shown in Table 46. Quantitative recoveries of added 3° amine in the 0.01 to 0.20% range demonstrate the validity and accuracy of the procedure.

DISCUSSION

COMPARISON OF METHODS. A sample of diethylamine was acetylated and analyzed for triethylamine by the hydrochloric acid–methanol, perchloric acid–acetic acids and perchloric acid-Methyl Cellosolve systems. Sample-size studies using titrant-solvent combinations provide conclusive data

Table 45. Tertiary Amine Content of Various Primary and Secondary Amines

1° or 2° Amine Analyzed	3° Amine Found, wt. %
Ethylamine	Triethylamine, 0.20
Diethylamine	Triethylamine, 0.15
Monoethanolamine	Triethanolamine, 0.04 and 0.03
Monoisopropanolamine	Triisopropanolamine, 0.007
Morpholine	as NaOH, 0.004
Aniline	as NaOH, 0.002

concerning the relative merits of these systems for determining 3° amines in the presence of 1° and 2° amines.

No titration curves were obtained for the perchloric acid–acetic acid system because the large amounts of amide formed by acetylation of diethylamine are sufficiently basic to interfere in glacial acetic acid, thus precluding any titration of triethylamine.

Table 46. Recovery Data for 3° Amines in 1° and 2° Amines

1° or 2° Amine Analyzed	3° Amine Sought	Per cent		
		Added	Found	Recovery
Diethylamine	Triethylamine	0.052	0.052	100
		0.082	0.082	100
		0.149	0.151	101
Monoethanolamine	Triethanolamine	0.204	0.190	93
		0.204	0.200	99
		0.204	0.192	94
		0.204	0.195	96
Morpholine	N-Ethylmorpholine	0.010	0.008	80
		0.020	0.018	90
		0.030	0.027	90
		0.040	0.037	92
		0.050	0.047	93
Aniline	N,N-Dimethylaniline	0.012	0.011	95
		0.012	0.011	95
		0.024	0.023	96
		0.024	0.023	96

The titration curves obtained by the other two systems offer a significant contrast (Fig. 11.21). These sample-size studies reveal the effect of amide concentration on the shape of the titration curve. The hydrochloric acid–methanol titration curve becomes progressively flatter with increasing sample size and rapidly loses its definition until it fails beyond a 5-ml sample size, In contrast, the perchloric acid–Methyl Cellosolve titration system is not seriously affected by the amide concentration. Visual end points to thymol blue indicator give the same result for sample sizes of 5, 10, and 15 ml. These experiments demonstrate the superiority of the perchloric acid–Methyl Cellosolve system over existing titrimetric procedures for determining small amounts of 3° amines in 1° and 2° amines.

CHOICE OF INDICATOR. Earlier, discussions of the determination of tertiary amines by the perchloric acid–Methyl Cellosolve method have not treated fully the means of end point detection; however, the inflection points in the potentiometric curves for the amines listed in Table 44 are matched by the color transitions of certain indicators. The indicator color changes, in turn, can be correlated with the basic strengths (in water) of these amines, so that a suitable indicator can be chosen for titrating a given amine by knowing the approximate basic ionization constant (K_b) of that amine.

Table 47 illustrates the dependence of indicator choice on the basic

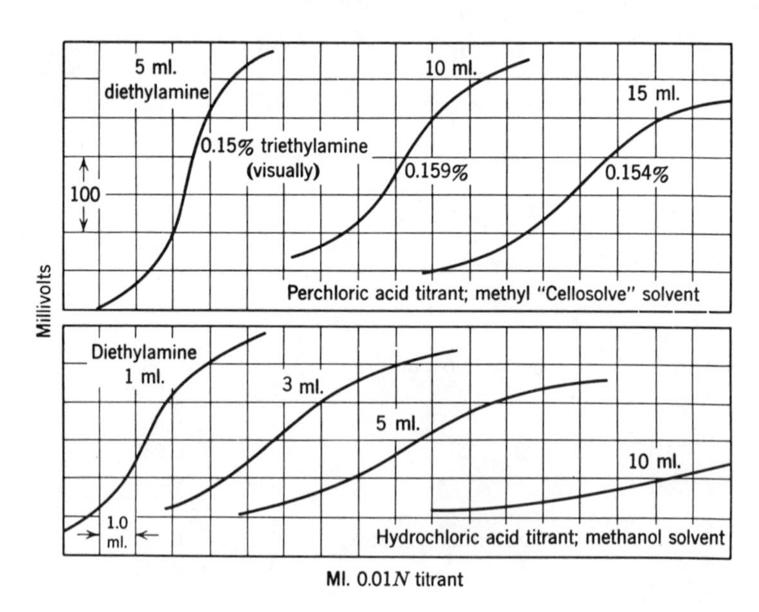

Fig. 11.21. Determination of triethylamine in diethylamine. Comparison of titrant-solvent pairs.

Table 47. Correlation of Indicator Choice with Amine Strength

Tertiary Amine	Ionization Constant in Water	Suitable Indicators[a]
Triethylamine	5.7×10^{-4}	*A*
N,N-Dimethylethanolamine	1.6×10^{-5}	*A*
Triethanolamine	4.5×10^{-7}	*A*
N-Ethylmorpholine	3.1×10^{-7}	*A*
γ-Picoline	1.1×10^{-8}	*B, C*
Pyridine	1.7×10^{-9}	*B, C*
Dimethylaniline	$\sim 10^{-10}$	*C*

[a] *A*. Thymol blue (alone or screened with Xylene Cyanol FF). *B*. Methyl yellow-methylene blue. *C*. Congo red.

ionization constant of the amine in question. Amines whose ionization constants are less than 1×10^{-10} cannot be successfully differentiated from amides by this method.

The choice of an indicator for titrating a given amine is not solely dependent on the basic strength of the amine. Other factors to be considered are the composition of the medium (all Methyl Cellosolve or mixed solvent, acetic anhydride present or absent); the normality of the titrant (for 0.1–0.5N perchloric acid, use thymol blue screened with Xylene Cyanol FF; for 0.01N titrant, thymol blue alone is preferred); the concentration of both titrated and acetylated species (excessively large amounts of titrated and/or acetylated amines tend to produce buffered end points). In case of doubt, the indicator requirement for a specific application should be determined experimentally.

LIMITATIONS AND INTERFERENCES

Amines with ionization constants less than 1×10^{-10} should not interfere. For example, the direct titration of *N*-phenylpiperazine in Methyl Cellosolve gave a quantitative result based on the secondary nitrogen ($K_b = 1 \times 10^{-6}$) without interference from the tertiary nitrogen ($K_b = 1 \times 10^{-12}$). Acetylation of the secondary nitrogen completely destroyed the basic character of the molecule under the conditions of solvent and titrant. The *N*-phenyl nitrogen, a weak tertiary amine, contributed no basicity and again did not interfere.

Compounds that do interfere include stronger amines that do not acetylate, such as the poly(ethyleneamines) and organic acid salts of inorganic bases. The latter compounds include the alkali metal and alkaline earth salts, which interfere quantitatively.

ANOMALOUS BEHAVIORS OF PYRIDINE AND TRIETHANOLAMINE. Although pyridine behaves normally during a titration in Methyl Cellosolve alone, in the presence of anhydride a very poor potentiometric end point is obtained by titrating with $0.01N$ perchloric acid. However some improvement is gained by using $0.1N$ acid. (This behavior is atypical because some amines that are weaker bases than pyridine yield good results with $0.01N$ titrant.) In addition, thymol blue indicator is destroyed in this pyridine-anhydride mixture and must be added at the conclusion of the acetylation period rather than at the beginning. Presumably, acetyl pyridinium acetate is formed by the action of pyridine and acetic anhydride. This organic species reacts with the indicator and is itself a weaker base than the parent amine, pyridine. Pyridine can be determined more satisfactorily by this procedure by titrating with $0.1N$ acid to a potentiometric end point; however the increased titrant normality limits the sensitivity.

Triethanolamine also exhibits a marked deviation from the behavior of other amines tested. The hydroxyl groups of triethanolamine react with acetic anhydride on standing. The extent of reaction, as well as the number of product species, varies with the reaction time.

Figure 11.22 shows this effect of reaction time on the acetylation of triethanolamine. Within a 30-minute acetylation time, a normal titration curve is obtained, whereas the inflection point is completely obscured in

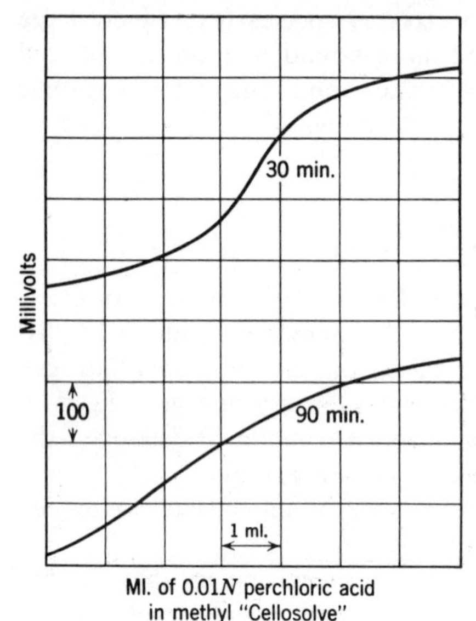

Fig. 11.22. Acetylation of triethanolamine in Methyl Cellosolve.

90 minutes. Evidently, acetylation of the three hydroxyl groups successively increases the chain length of the original triethanolamine molecule with acetoxy groups, whose electron-withdrawing tendencies progressively diminish the basic character of the nitrogen atom. Titration of these various basic moieties produces a result similar to that obtained by titrating an amine that contains small amounts of other amines of varying basic strengths; that is, the titration curve does not yield a well-defined inflection point. Dimethylethanolamine does not behave in this manner because only one acetoxy group can be added to the molecule, and this effect is insufficient to overcome the contributions to basicity of the electron-releasing *N*-methyl groups. The triethanolamine problem can be avoided by restricting the acetylation time to 30 minutes.

DETERMINATION OF PRIMARY AND SECONDARY AMINES ALONE AND IN MIXTURES WITH TERTIARY AMINES: THE ISOTHIOCYANATE REACTION

Procedures based on the determination of amine consumption in the amine-isothiocyanate reaction have been available for some time for the analysis of isothiocyanates (p. 694). The more recent development of a quantitative iodate titration for the thiourea product of this reaction (50) affords, on the other hand, a basis for the determination of the amine reactant.

Method of B. C. Verma and S. Kumar

[*Reprinted in Part from Analyst,* ***99****, 498* (*1974*)]

The method consists in treating the amine with an excess of phenyl isothiocyanate in dimethylformamide and titrating the substituted thiourea formed with potassium iodate in a sulfuric acid medium at room temperature. Excess isothiocyanate does not interfere in the titration of thioureas. Tertiary amines can be determined in the same aliquot. After the primary and secondary amines have been converted to the corresponding thioureas, the unreacted tertiary amines are titrated conductimetrically with trichloroacetic acid and the potassium iodate titration of the substituted thioureas is then made.

50. B. Singh and B. C. Verma, *Z. Anal. Chem.*, **194,** 112 (1963).

APPARATUS

A bright platinum wire indicator electrode and a saturated calomel reference electrode were used for the potentiometric titrations.

Acid-base conductimetric titrations were formed with a Philips PR9500 conductivity bridge that operated at a frequency of 50 Hz. The cell electrodes consisted of two rigidly held square plates (approximately 0.8 cm^2) of platinized platinum, facing each other at a distance of 1 cm.

REAGENTS

DIMETHYLFORMAMIDE, COMMERCIAL GRADE, This solvent was allowed to stand over sodium carbonate for 2 days; then it was decanted and distilled with the 148.5 to 149.5°C fraction collected.

Potassium iodate solution, 0.05*N*.

TRICHLOROACETIC ACID, 1.0*N*. The solution was standardized by conductimetric titration of anhydrous sodium acetate.

PROCEDURE FOR PRIMARY AND SECONDARY AMINES

Place an aliquot of the amine sample in dimethylformamide in a glass-stoppered flask and add 3 to 5 ml of phenyl isothiocyanate (an approximately 0.3*N* solution in dimethylformamide). Add solvent until a 10-ml volume of total solution is obtained. Stopper the flask, swirl to mix the reactants, and allow to stand for 10 minutes. Then add sufficient water and sulfuric acid to obtain 100 ml of solution 2.0 to 2.5*N* in sulfuric acid. Cool to room temperature (25°C) and titrate with 0.05*N* potassium iodate solution to the appearance of a distinct permanent yellow color. If preferred, add 0.2 ml of a 1% aqueous solution of amylose, in which case, the solution becomes blue at the end point. Titrated potentiometrically, a sharp potential change is observed at the equivalence point.

DETERMINATION OF PRIMARY AND SECONDARY AMINES AND TERTIARY AMINES IN THE PRESENCE OF EACH OTHER

Place an aliquot of the sample dissolved in dimethylformamide in a glass-stoppered flask containing an excess (7–10 ml of approximately 0.3*N* solution) of phenyl isothiocyanate in dimethylformamide solution. Make the volume to 20 ml with the solvent. Swirl to mix, and set aside for 10 minutes. Mix the solution with 40 to 45 ml of approximately 1.0*N* acetic acid, cool to room temperature (25°C), and titrate conductimetrically with standard 1.0*N* trichloroacetic acid solution. To the same solution, then add sufficient water and sulfuric acid to produce a volume of 125 ml that is 2.0 to 2.5*N* in sulfuric acid. Cool to room temperature and titrate potentiometrically with standard 0.05*N* potassium iodate solution.

The volume of standard acid used in the acidimetric titration corresponds to the tertiary amine content; the volume of potassium iodate solution gives the amount of di- and/or trisubstituted thiourea and, consequently, the amount of primary and/or secondary amine present in the sample.

RESULTS AND DISCUSSION

Results for primary and secondary amines are recorded in Table 48.

Table 48. Potassium Iodate Determination of Primary and Secondary Amines

	Primary Amines, Amount Found[a], mg		Secondary Amines, Amount Found,[b] mg	
Compound	Visual Method[c]	Potentiometric Method[c]	Visual Method[c]	Potentiometric Method[c]
Ethylamine	9.94 ±0.072	9.98 ±0.036	39.85 ±0.082	39.88 ±0.056
n-Propylamine	10.04 ±0.034	10.03 ±0.028	40.18 ±0.063	40.16 ±0.046
Isopropylamine	9.97 ±0.053	10.02 ±0.051	40.28 ±0.061	40.20 ±0.025
n-Butylamine	10.01 ±0.056	9.99 ±0.041	39.72 ±0.092	39.81 ±0.057
Isobutylamine	9.96 ±0.068	10.02 ±0.038	40.28 ±0.061	40.24 ±0.047
Diethylamine	9.95 ±0.076	9.95 ±0.062	40.10 ±0.075	40.02 ±0.039
Pyrrolidine	9.97 ±0.063	9.98 ±0.058	40.26 ±0.056	40.21 ±0.026
Piperidine	10.06 ±0.075	10.05 ±0.046	39.80 ±0.078	39.82 ±0.030

[a] Amount taken, 10 mg.
[b] Amount taken, 40 mg.
[c] Mean of six determinations, ±standard deviation.

The results of the analysis of various ethylamine-triethylamine and diethylamine-triethylamine mixtures appear in Tables 49 and 50, respectively.

Amides (acetamide, urea, salicylamide, and nicotinamide), imides (phthalimide and succinimide), Schiff bases (*N*-*p*-chlorobenzylideneaniline and *N*-cinnamylideneanisidine), and tertiary amines (triethylamine, pyridine, α-picoline, quinoline, and isoquinoline) do not

Table 49. Analysis of Mixtures of Ethylamine and Triethylamine

Ethylamine, mg		Triethylamine, mg		Ratio of Ethylamine to Triethylamine
Amount in Mixture	Amount Found[a]	Amount in Mixture	Amount Found[a]	
20.00	20.12 ±0.084	20.00	20.08 ±0.126	1:1
20.00	20.14 ±0.094	40.00	39.90 ±0.132	1:2
20.00	19.90 ±0.097	60.00	60.32 ±0.094	1:3
20.00	20.06 ±0.078	80.00	79.60 ±0.127	1:4
40.00	39.85 ±0.076	20.00	19.92 ±0.088	2:1
60.00	59.72 ±0.082	20.00	20.05 ±0.095	3:1
80.00	80.36 ±0.088	20.00	20.16 ±0.096	4:1

[a] Mean of six determinations; ±standard deviation.

cause any interference, even when present in up to a fivefold excess in the determination of primary or secondary amines by the procedure. Thiourea, thiosemicarbazide, thioacetamide, phenylhydrazine, xanthates, and organic isocyanates, however, do interfere.

The method could not be extended to the determination of aromatic amines because of their extremely slow reactions.

Amylose was not suitable in the visual end point determination of piperdine and pyrrolidine.

Table 50. Analysis of Mixtures of Diethylamine and Triethylamine

Diethylamine, mg		Triethylamine, mg		Ratio of Diethylamine to Triethylamine
Amount in Mixture	Amount Found[a]	Amount in Mixture	Amount Found[a]	
20.00	19.92 ±0.083	20.00	20.10 ±0.124	1:1
20.00	20.10 ±0.081	40.00	40.14 ±0.156	1:2
20.00	19.88 ±0.096	60.00	59.58 ±0.141	1:3
20.00	19.96 ±0.055	80.00	80.56 ±0.104	1:4
40.00	40.22 ±0.087	20.00	19.86 ±0.088	2:1
60.00	60.35 ±0.073	20.00	19.90 ±0.114	3:1
80.00	79.55 ±0.082	20.00	20.08 ±0.096	4:1

[a] Mean of six determinations, ±standard deviation.

GENERAL MIXTURES OF AMINO COMPOUNDS

We have already seen methods to distinguish the different types of amines (pp. 567–633). There is an additional approach that not only distinguishes primary from secondary amines but also resolves mixtures of primary amines and mixtures of secondary amines, including isomeric and homologous compounds. This method, which also can resolve two amino groups on the same molecule, involves the use of the differences in the reaction rates of the different amino compounds with phenyl isothiocyanate. The approach and method are described in Chapter 25 (pp. 837–40) which deals with the use of differential reaction rates to analyze mixtures of organic compounds containing the same functional group.

Determination of Small Quantities of Amines

AZEOTROPIC DISTILLATION FOR CONCENTRATION AND DETERMINATION

Amines lend themselves well to two-phase azeotropic distillation with certain very polar solvents. Water, for example, works very well for the steam distillation of amines. Kjeldahl distillation can be applied to many amines for separations and concentration from a sample. This procedure permits taking extremely large samples, even several kilograms, and separation of a few milligrams of amine. The amine distillate, as in the regular Kjeldahl analysis, is caught in standard mineral acid and the excess acid is back-titrated; or the amine can be caught in boric acid and then the amine can be titrated directly with mineral acid.

High-boiling amines do not steam-distill too efficiently; however ethylene glycol distillation is possible here. Dodecylamine, hexadecylamine, octadecylamine, and *N*-methyl octadecylamine have been successfully distilled and determined by using ethylene glycol instead of water.

The procedures used distillation and determination of amines by distillation and titration of the distillate, as shown in the section on carboxylic acid amides (p. 183). In this procedure, Siggia and Stahl reduced amides to amines. They then azeotropically distilled the amine, using water or ethylene glycol as dictated by the sample; finally, they titrated the distillates.

PRIMARY ALIPHATIC AMINES ONLY

The Van Slyke method (pp. 586–92) is very desirable for determining small amounts of primary amines. Since the measurement is based on the measurement of gaseous nitrogen, very low amounts of amines are determinable. The exact limit depends on the amine and the other materials contained in the sample; based on calculations, however, 1 ml of nitrogen is equivalent to 0.0447 mM of nitrogen, which is equivalent to 0.0894 mM of primary amino groups.

PRIMARY AROMATIC AMINES ONLY

The primary aromatic amines are specifically and sensitively determinable by diazotization of the amine to the corresponding diazonium compound, then coupling this diazo compound with a phenolic or other amino compound(s).

Adapted from Method of F. J. Bandelin and C. R. Kemp

[*Ind. Eng. Chem., Anal. Ed.*, **18**, *470* (*1946*)]

REAGENTS

Sulfuric acid, 4*N*.
N-(1-Napthoyl) ethylene diamine, 0.1%.
Sodium nitrite, 0.1%.
Ethyl alcohol, 0.1%.

PROCEDURE

Dissolve or extract the amine sample with 5 ml of 4*N* sulfuric acid. To the solution or extract, add 1 ml of 0.1% sodium nitrite and allow to stand for 3 minutes. The aryl amine is now as the diazonium salt. To this solution add 5 ml of 95% ethyl alcohol and allow the mixture to stand for 2 minutes. Add to this 1 ml of 0.1% *N*-(1-naphthoyl)ethylenediamine. The color generally develops rapidly at this stage, but often an alkaline agent must be added to bring about coupling. The subsequent discussion under Diazonium Salts (pp. 685–87) covers alkaline agents and their use for coupling reactions. The color is then measured by visual comparison with standards or the comparison can be made spectrophotometrically.

AUTHORS' NOTE. Phenols and naphthols can be used in place of the *N*-(1-naphthoyl)ethylenediamine. This follows the diazotization step, and the procedure is as described under Diazonium Salts.

PRIMARY ALIPHATIC AND AROMATIC AMINES

Reaction with Orthoquinones

Orthoquinones give colored compounds with primary amino compounds (51). The orthoquinone used for analysis is sodium 1,2-naphthoquinone-4 sulfonate (52):

$$\text{(1,2-naphthoquinone-4-)}SO_3N_2 + H_2NR \xrightarrow{OH^-} \text{(1,2-naphthoquinone-4-)}{=}NR$$

Combined Method of E. G. Frame, J. A. Russell, and A. E. Whilhelmi and D. H. Rosenblatt, P. Hlinka, and J. Epstein

[*J. Biol. Chem., 1949,* ***255,*** *(1943)*]
Adapted from Pharmaceutical Analysis, Edited by T. Higuchi and E. Brochman–Hanssen, Wiley-Interscience, New York, 1961, p. 427, Reprinted in Part.

PROCEDURE A

Treat a 5-ml sample containing 8 to 30 μg pf amino nitrogen with 1 drop of 0.25% alcoholic phenolphthalein, and add 0.1*N* sodium hydroxide dropwise until a pink color is attained. Add 1 ml of pH 9.3 borate buffer plus 1 ml of freshly prepared 0.5% sodium β-naphthoquinone-4-sulfonate; mix the solution, and place it in a boiling water bath for 10 minutes, then in an ice bath for 5 minutes. Add 1 drop of 4% sodium lauryl sulfate, followed by 1 ml of a mixture of 3 parts of 1.5*N* hydrochloric acid, 1 part of glacial acetic acid, and 4 parts of 0.15*M* formaldehyde. Stir the mixture and add 1 ml of 0.1*N* sodium thiosulfate. Then

51. O. Folin and H. Wu, *J. Biol. Chem.*, **51,** 377 (1922).
52. M. E. Auerbach, *Drug Standards*, **20,** 165 (1952).

dilute the mixture to 15 ml with distilled water and mix, allow to stand for 10 to 30 minutes, and measure the absorbance at 480 to 490 nm.

A technique for avoiding interferences in the older methods was introduced by Rosenblatt et al. They noted that dilute aqueous solutions of *n*-butylamine reacted with β-naphthoquinone-4-sulfonate in pH 10.3 phosphate buffer to give a reddish dye, extractable with chloroform for measurement at 450 nm. The product with ethanolamine was not extractable with chloroform, but did shake out with isoamylalcohol for measurement at 420 nm. The extraction removed the products from excess reagent and decomposition products, and increased the sensitivity of the reaction. Ammonia reacted with the reagent, but the dye formed was not extractable by chloroform. They examined several solvents for various amines and showed that it was possible to choose a solvent for a particular amine to eliminate interference by other basic compounds present.

PROCEDURE B

Mix a 50-ml portion of an aqueous amine solution (1–4 ppm) with 10 ml of 0.138% potassium 1,2-naphthoquinone-4-sulfonate and 1 ml of pH 10.3 phosphate buffer in a 125-ml glass-stoppered Erlenmeyer flask. After a reaction period of 1 minute, add 10 ml of chloroform, introduce a Teflon-covered stirring bar, stopper the flask, and vigorously stir the contents electromagnetically for 20 minutes. After the phases have separated, pipet out the chloroform layer and measure the absorbance at 450 nm against a chloroform blank. Determine the concentration from a standard curve similarly prepared.

Reactions with Aldehydes

Certain amines condense with various aldehydes in strongly acid media to give products that are oxidizable to give a color. Among the many aldehydes that have been shown to react are *p*-dimethylaminobenzaldehyde, vanillin, formaldehyde, benzaldehyde, salicylaldehyde, piperonal, paraldehyde, *p*-acetylaminobenzaldehyde, *m*- and *p*-nitrobenzaldehyde, *m*-aminobenzaldehyde, and metaldehyde. The most common oxidant used is atmospheric oxygen, but the process has been hastened by the addition of hydrogen peroxide, nitrites, nitrates, ferric ion, and several other metal ion catalysts (53).

53. T. Higuchi and J. Bodin, *Pharmaceutical Analysis*, T. Higuchi and E. Brochmann-Hanssen, Eds. Wiley-Interscience, New York, 1961, p. 425.

Of the foregoing aldehydes, best results have been obtained with *p*-N-dimethylbenzaldehyde.

Method of C. Menzie

[*Adapted from Anal. Chem.*, **28,** *1321–2 (1956), Reprinted in Part*]

The reaction of *p*-*N*-dimethylaminobenzaldehyde with indoles and pyrroles has been used qualitatively and quantitatively for many years (54–58). The reaction with aromatic amines to give Schiff bases is equally well documented (59). Not so well known, the Wasicky reaction, utilizing 94% aqueous sulfuric acid solution, gives a color reaction with alkaloids (60) and purines (61, 63). In 1944 Werner (63) reinvestigated the reaction with nitrogen compounds and found that in a more dilute aqueous system "aromatic compounds react in the presence of mineral acid, provided the —NH_2 group is directly attached to the benzene nucleus. . . . No reaction occurs with (*i*) aliphatic amines and amino acids, (*ii*) *N*-substituted aromatic amines, (*iii*) heterocyclic amino compounds, or (*iv*) amino derivatives of the cycloparaffins—e.g., cyclohexylamine." More recently, Burmistrov (64) isolated as picrates, from toluene solutions, the reaction products of secondary aromatic amines with Ehrlich's reagent.

Menzie has reinvestigated this reaction and extended it with some modification. It has been possible to obtain color formation with every class of nitrogen compounds tried.

PROCEDURE

To several milligrams of sample, add an equal amount of Ehrlich's reagent (*p*-*N*-dimethylaminobenzaldehyde). Add 0.3 ml of toluene and then 0.02 ml of concentrated sulfuric acid. Allow to stand for about 1 minute. Agitate, add 1.0 ml of ethyl alcohol, and mix thoroughly.

54. F. Blumenthal, *Biochem. Z.*, **19,** 521 (1909).
55. G. O. Burr and R. A. Gortner, *J. Am. Chem. Soc.*, **46,** 1224–46 (1924).
56. E. Fischer, *Ber.*, **19,** 2988 (1886).
57. W. Frieber, *Centr. Bakteriol, Parasitenk.*, **87,** 254–77 (1922).
58. C. Renz and K. Loew, *Ber.*, **36,** 4326 (1903).
59. *Chem. Rev.*, **26,** 324–7 (1940).
60. R. Wasicky, *Z. Anal. Chem.*, **54,** 393–5 (1915).
61. Raymond-Hamlet, *Bull. Sci. Pharmacol.*, **33,** 447–56 (1926).
62. Raymond-Hamet, *Ibid.*, **33,** 518–25 (1926).
63. A. E. A. Werner, *Sci. Proc. Roy. Dublin Soc.*, **23,** 214–21 (1944).
64. S. I. Burmistrov, *J. Gen. Chem.*, **19,** 1511–14 (1949).

On addition of sulfuric acid to the toluene suspensions of the amino acids, a yellow color was obtained. Upon addition of ethyl alcohol to these solutions, the colors noted in Table 51 developed. There were five

Table 51. Color Reaction of Amino Acids and Related Compounds

α-Alanine	Pinkish purple	Histidine	Orange
β-Alanine	Pale orange	Hydroxyproline	Orange
α-Aminobutyric acid	Pale orange	Isoleucine	Pale orange
Arginine	Orange	Leucine	Almost colorless
Aspartic acid	Orange	Lysine	Pale purple
Citrulline	Pale orange	Methionine	Reddish
Creatine	Pale purple	Methionine sulfoxide	Orangish yellow
Creatinine	Pale orange	Norleucine	Almost colorless
Cysteine	Intense red	Norvaline	Pale orange
Cystine	Pale yellow	Ornithine	Pale orange
Dihydroxyphenylalanine	Pale purple	Phenylalanine	Pale orange
Diiodotyrosine	Pale purple	β-Phenylserine	Pale purple
Djenkolic acid	Intense red	Proline	Bright orange
Ethionine	Almost colorless	Sarcosine	Pale purple
Glutamic acid	Pale orange	Serine	Pale orange
Glutamine	Pale orange	Taurine	Pale orange
Glutathione	Bright orange	Threonine	Orange
Glycine	Orange-yellow	Allothreonine	Pale purple
Homocysteine	Intense red	Tyrosine	Pale orange
Homocystine	Orange	Tryptophan	Orange solution and blue precipitate
Homoserine	Reddish	Valine	Pale orange

exceptions to this: the color shown in Table 51 for cysteine, homocysteine, dihydroxyphenylalanine, djenkolic acid, and tryptophan was developed in toluene. Addition of ethyl alcohol effected no change.

Tables 52 and 53 indicate the color reactions for both steps of the test: acid toluene solution and acid-toulene plus ethyl alcohol.

When toluene is omitted, no color development occurs with the amino acids (except tryptophan) nor with such compounds as octadecylamine, the benzo-(*f*)-quinolines, purines, pyrimidines, or quinine. Fleig (65), and later van Urk (66), found that when *p*-*N*-dimethylaminobenzaldehyde reacted with various nonnitrogen compounds and heterocyclics, color

65. C. Fleig. *Bull. Soc. Chim. Fr.* (4) **3,** 1038–45 (1908).
66. H. W. van Urk, *Pharm. Weekbl.*, **66,** 101–8 (1929).

Table 52. Color Reaction of Nitrogen Heterocyclics

	Acid-Toluene Only	Acid-Toluene and Ethyl Alcohol
Piperidine	Reddish	Orange-yellow
Pyridine	Light purple	Colorless
Pyridoxine	Yellow	Pale purple
Quinidine . HCl	Yellow	Light purple
Quinine	Amber	Reddish
Quinoline	Yellow	Yellow
Benzo(*f*)quinoline	Red	Pale purple and precipitate
3-Methylbenzo-(*f*)quinoline	Yellow	Pale purple and precipitate
Caffeine	Yellow	Red-orange
Adenine	Blue to lavender upon standing	Red-purple ring
Adenylic acid	Blue to brownish upon standing	Trace or no color
Guanine	Light purple	Ring
Guanylic acid	Reddish	Trace of ring
Guanosine	Reddish	No ring
Cytosine	Blue-purple	No ring
Cytidine	Dark lavender purple	Reddish ring
Cytidylic acid	Pale lavender purple	Trace or no ring
Uridine	Light lavender	Trace of ring
Xanthine	Purple color, dissipated by agitation	No ring

Table 53. Miscellaneous Nitrogen Compounds

	Acid-Toluene	Acid-Toluene Plus Ethyl Alcohol
Ethanolamine	Yellow	Yellow
Octadecylamine	Dark amber	Purple
Urea	Yellow	Orange-yellow
Biuret	Yellow	Pale orange
Methylurea	Yellow	Orange-yellow
Phenylurea	Orange	Orange-yellow
Hippuric acid	Yellow	Pale purple
Diphenylamine	Brownish	Yellow-green
Benzylamine	Yellowish	Reddish purple, dissipated by agitation

Twenty-one primary aromatic amines, three containing naphthalene ring systems, were also tested and all gave positive color reactions.

developed. In both cases, however, it was necessary to heat the reaction mixtures for varying periods of time or even evaporate and then redissolve in water.

When the modification indicated is employed, Ehrlich's reagent will react at room temperature with every class of nitrogen compounds tried. Although this behavior precludes its use in identification of a class of compounds, under the modified conditions used, differentiation within a class of compounds is possible. In one case, color differentiation between the stereoisomers quinine and quinidine has been shown (Table 52).

SPECTROFLUOROMETRIC DETERMINATION OF PRIMARY AMINES

Ninhydrin (1,2,3-indantrione hydrate) and α-amino acids react to form aldehyde, carbon dioxide, ammonia, and a purple product.

$$C_6H_4(CO)_2C{=}(OH)_2 + RCH(NH_2)COOH \rightarrow RCHO + CO_2 + C_6H_4(CO)_2CH{-}OH + NH_3$$

Measurement of the carbon dioxide produced provides the basis for a quantitative method for amino acids (67, 68). The production and measurement of the colored product has been used for many years for the detection and determination of amino acids, amines and peptides (69). The ninhydrin color reaction has been widely adapted for automatic analysis of the amino acid composition of proteins.

Sensitivities 10 to 100 times higher were obtained when it was found that ninhydrin can yield highly fluorescent products with amine-containing compounds (70, 71). In the assay of phenylalanine, the phenylacetaldehyde formed on interaction with ninhydrin combined with additional ninhydrin and primary amine to yield the highly fluorescent ternary product. The structure of this product was subsequently elucidated (72), and a novel reagent was then synthesized based on this study

67. D. D. Van Slyke, R. T. Dillion, D. A. MacFadyer, and P. Hamilton, *J. Biol. Chem.*, **141,** 627 (1941).
68. D. D. Van Slyke, D. A. MacFadyer, and P. Hamilton, *Ibid.*, **141,** 671 (1941).
69. R. West, *J. Chem. Educ.*, **42,** 386 (1965).
70. K. Samejima, W. Dairman, and S. Udenfriend, *Anal. Biochem.*, **42,** 222 (1971).
71. K. Samejima, W. Dairman, J. Stone, and S. Udenfriend, *Ibid.*, **42,** 237 (1971).
72. M. Weigele, J. F. Blount, J. P. Tengi, R. C. Czaijkowski, and W. Leimgruber, *J. Am. Chem. Soc.*, **94,** 4052 (1972).

(73). This reagent, 4-phenylspiro[furan-2(3*H*), 1′-phthalan]-3,3′-dione, given the trivial name fluorescamine, acts directly with primary amines to form the same fluorophors (390-nm excitation, 475-nm emission) as are generated by the ninhydrin-phenylacetaldehyde reaction (74).

Fluorescamine

RNH_2

Fluorophor

Ninhydrin Phenylacetaldehyde

Efficient fluorogenic reactions have been observed with a large variety of aliphatic and aromatic primary amines, including amino acids, catecholamines, sulfonamides, and antibiotics.

The following procedure, developed specifically for the determination of drugs containing primary aromatic or aliphatic amino groups, can be used as a model for the analysis of primary amines in general.

Method of J. A. F. deSilva and N. Strojny

[*Reprinted in Part from Anal. Chem.*, **47,** *714 (1975)*]

APPARATUS

Fluorescence measurements were made with a Farrand Mark I Spectrofluorometer equipped with a 150-watt Hanovia xenon arc energy source

73. M. Weigele, S. L. DeBernardo, J. P. Tengi, and W. Leimgruber, *Ibid.*, **94,** 5927 (1972).
74. S. Udenfriend, S. Stein, P. Bohlen, W. Dairman, W. Leimgruber, and M. Weigele, *Science*, **178,** 871 (1972).

and an RCA I-P-28 photomultiplier. The instrument sensitivity was adjusted each day to a constant energy using a Pyrex rod as a reference standard.

REAGENTS AND SOLUTIONS

FLUORESCAMINE. Dissolve 100 mg of fluorescamine in 100 ml of anhydrous reagent grade acetone and "age" prior to use by allowing to stand at room temperature for 24 hours.

Molar solutions of phosphoric acid, monobasic potassium phosphate, and dibasic potassium phosphate were prepared in the pH ranges desired, using a pH meter.

PROCEDURE

Dissolve the sample in either methanol or water. Transfer an aliquot (0.1 ml) equivalent to 10 μg of compound to a 15-ml tube. Add 15 ml of buffer and mix the contents. Add 0.1 ml of the fluorescamine solution (100 μg of fluorescamine) and mix the contents again. Allow to stand for 15 minutes, and scan the solution on the spectrofluorometer. Dilute as necessary to obtain readings on-scale.

RESULTS AND DISCUSSION

Fluorescamine and its hydrolysis products are nonfluorescent; therefore they do not interfere with the quantitation of the derivative formed. The data in Table 54 indicate its applicability to both primary aromatic amines and aliphatic primary amines of widely differing chemical structures.

The main fluorophor formed is an acidic compound. Extraction of the fluorophor and its determination in an organic solvent was investigated. It was noted that in general, irrespective of the pH of optimal reactivity for the respective compounds (Table 54), the fluorophor was quantitatively extractable ($>80\%$) at pH 5.0 to 5.5 into ethyl acetate, and its fluorescence was also measured optimally in this solvent.

The quantitation sensitivity of some of the compounds can be improved significantly by this procedure (Table 55). In each case, the reaction was carried out at the pH of optimal reactivity, as given in Table 54; the aqueous medium was then titrated to pH 5.0 with 1.0*M* H_3PO_4 and the fluorophor extracted directly into 10 ml of ethyl acetate in which its fluorescence was measured at the respective excitation-emission maxima (Table 54).

Table 54. Luminescence Properties of 1 μ*M* Solutions of Fluorescamine Derivatives of Aliphatic and Aromatic Primary Amines

Compound	Optimal Reaction, pH	Excitation-Emission Maxima, nm	Sensitivity Limit, ng/ml	Upper Limit of Linear Range, μg/ml	Fluorescence Intensity, arbitrary units
Amphetamine sulfate	9.3–9.4	395/490	1000	30	20
Phenylpropanol amine	9.3–9.4	395/490	1000	30	20
p-Aminobenzoic acid	3.0–4.5	405/500	3	3	500
Procaine hydrochloride	3.0–4.5	405/495	3	3	1100
p-Aminosalicylic acid	2.0–3.0	405/495	200	10	100
Sulfanilamide	3.0–4.5	400/495	10	1	1000
Sulfadoxidine	3.0–4.5	400/495	10	1	1000
Sulfamethoxazole	3.0–4.5	400/495	10	1	900
Isosulfisoxazole	3.0–4.5	400/495	10	1	600
Sulfisoxazole	3.0–4.5	400/495	10	1	500
Sulfadiazine	3.0–4.5	400/495		1	500
7-Aminoclonazepam	5.5–9.3	412/505	100	20	100
7-Amino-3-hydroxy-clonazepam	5.5–9.3	412/505	100	20	100

Table 55. Luminescence Properties of 1 μ*M* Solutions of Fluorescamine Derivatives Following Extraction at pH 5.0 into Ethyl Acetate

Compound	Excitation-Emission Maxima in Ethyl Acetate, nm	Sensitivity Limit, ng/ml	Fluorescence Intensity, arbitrary units
Amphetamine sulfate	395/475	300	200
Phenylpropanol amine	395/475	300	300
p-Aminobenzoic acid	410/485	1	1450
Procaine hydrochloride	405/485	3	630
p-Aminosalicylic acid	405/490	30	500
Sulfanilamide	410/495	3	1320
Sulfadoxidine	405/485	3	1620
Sulfamethoxazole	405/485	3	1350
Isosulfisoxazole	405/485	3	1440
Sulfisoxazole	405/485	3	1430
Sulfadiazine	405/485	3	1630
7-Aminoclonazepam	405/500	30	350
7-Amino-3-hydroxy-clonazepam	405/500	30	330
L-DOPA[a]	390/470	1000	60
Dopamine[a]	395/475	1000	80
1-Adamantanamine[a] (amantadine)	395/475	1000	60

[a] These compounds were only measurable after extraction into ethyl acetate.

Halogenating agents cause secondary amino acids to undergo oxidative decarboxylation to produce imines, which are then hydrolyzed to primary amines. This technique therefore makes it possible to determine secondary amino acids by the fluorescamine method also (75). *N*-Chlorosuccinimide was used as the halogenating agent for proline: 1-ml aliquots of 4×10^{-6} to $4\times10^{-5}M$ solutions of proline, or hydroxyproline, respectively, at pH 2 with 1 ml of $4\times10^{-4}M$ aqueous *N*-chlorosuccinimide, 1 ml of 2% sodium bicarbonate, and 1 ml of $2\times10^{-3}M$ fluorescamine in acetone. Fluorescence was measured 2 minutes after addition of the last reagent. Optimal fluorescence from sarcosine was generated by the use of bromine water ($2\times10^{-3}M$) in place of *N*-chlorosuccinimide.

ALIPHATIC AND AROMATIC PRIMARY, SECONDARY, AND TERTIARY AMINES

Method of J. P. Rawat and J. P. Singh

[*Reprinted in Part from Anal. Chem.*, **47,** *738* (*1975*)]

Small amounts of the amino compounds, acetyl chloride, and ferric ion react to produce a greenish-violet complex. The use of this reaction for the determination of amines is described.

APPARATUS

A Bausch & Lomb Spectronic-20 was used for the spectrophotometric measurements.

PROCEDURE

Place 50 μg to 10 mg of the amino compound in a test tube and add a small amount of gelatin, 1 ml of a 1% (v/v) aqueous solution of acetyl chloride, and 2 ml of 5% aqueous ferric nitrate solution. Heat on a water bath at 65°C for 20 minutes. Make the volume to 25 ml with conductivity water, and read the absorbance at the optimum wavelength (550 nm).

75. M. Weigele, S. DeBernardo, and W. Leimgruber, *Biochem. Biophys. Res. Commun.*, **50,** 352 (1973).

RESULTS AND DISCUSSION

Each amino compound requires the preparation of a separate calibration curve. Reproducible calibration curves were obtained for aniline, diphenylamine, pyridine, diethylaniline, phenylenediamine, *p*-toluidine, methylamine, ethylamine, isopropylamine, amylamine, 1,3-diaminopropane, morpholine, and piperidine. The standard deviation for 10 different measurements of 205 μg of aniline was 4.50 μg, and the maximum error was ±3%, which is in the spectrophotometric error range.

Certain organic compounds were added to aniline, and it was found that nitrobenzene, acetic acid, bromobenzene, and benzaldehyde have no effect on the intensity of the color of the complex. Phenol, acetic anhydride, and benzamide interfere.

12

Imino Groups

Imino groups are alkaline and can be titrated; however their basicity is much lower than that of the parent amines (see discussion on pp. 567–70). Imines can be titrated directly in nonaqueous media. Also, these compounds can be hydrolyzed back to the parent carbonyl compound, and this can be determined.

Nonaqueous Titration Methods

Adapted from S. Freeman

[*Reprinted in Part from Anal. Chem.*, ***25,*** *1750–1* (*1953*)]

APPARATUS

Beckman pH meter (Model G) or similar titrator.
Glass electrode, Beckman Catalog No. 1190–42.
Calomel electrode, Beckman Catalog No. 1170.

REAGENTS

Acetic acid, reagent grade, glacial.
Acetic anhydride, reagent grade, 90 to 95%.
Acetonitrile, Eastern Chemical Company, or equivalent.
p-Dioxane, Carbide and Carbon Chemicals Company, or equivalent.
Methyl violet, 0.1% solution in glacial acetic acid.

Perchloric acid in acetic acid. Dissolve 8.5 ml of 72% perchloric acid in 100 ml of glacial acetic acid; add 22 ml of acetic anhydride, and let stand overnight. Dilute to 1 liter with glacial acetic acid.

Perchloric acid in *p*-dioxane, 0.01*N*. Dissolve 0.85 ml of 72% perchloric acid in 1 liter of *p*-dioxane.

Perchloric acid in *p*-dioxane, 0.1*N*. Dissolve 8.5 ml of 72% perchloric acid in 1 liter of *p*-dioxane.

Potassium acid phthalate, primary standard grade.

PROCEDURE A

Dissolve the sample containing 3 meq (accurately weighed) of Schiff base in 50 ml of acetic acid. Titrate potentiometrically with 0.1*N* perchloric acid in acetic acid using the millivolt scale of the pH meter or methyl violet indicator to a blue-green end point.

PROCEDURE B (1)

Dissolve the sample containing approximately 0.3 meq of total Schiff base and amine in 25 ml of chloroform and titrate potentiometrically with 0.01*N* perchloric acid in dioxane. Standardize the titrant against purified diphenylguanidine in chloroform, using the pH meter.

PROCEDURE C (2)

Dissolve the sample containing about 3 meq of total Schiff base and amine in 50 ml of acetonitrile and titrate potentiometrically with 0.1*N* perchloric acid in dioxane. Run a 50-ml blank on each batch of acetonitrile.

With the exception of N^4-*p*-methoxybenzylidinesulfathiazole, the potential change at the equivalence point using Procedure A was so great (70–100 mV per 0.1 ml of volumetric solution) than no record of the potential change was necessary. For the sulfathiazole Schiff base, the potential change was on the order of 10 to 15 mV per 0.1 ml of titrant, and its exact determination required recording to note its position. However when Procedure C was employed for this substance, a much greater break in the titration curve occurred, which obviated the need to plot points.

SCOPE

Ten Schiff bases were prepared, and the analytical data were compiled by titrimetric, nitrogen, and dinitrophenylhydrazone methods of analyses (Table 1). Purity values obtained by the latter procedure vary between 0.4 and 0.9% higher than those of the other two methods, Shoppee (5) noted that the *p*-nitrophenylhydrazones of halogen and methoxy-substituted benzaldehydes appeared to occlude small amounts of impurities, and he reported analytical results on these classes of compounds

1. C. W. Pifer, E. G. Wollish, and M. Schmall, *Anal. Chem.*, **25,** 310 (1953).
2. J. S. Fritz, *Anal. Chem.*, **25,** 407 (1953).

Table 1. Analytical Results on Schiff Bases by Three Different Methods

Schiff Base	Reference	Procedure A, %	Nitrogen, %	2,4-Dinitrophenylhydrazone, %
N-*n*-Butylidine-*n*-butylamine	(3)	99.6	99.6	—
N-Benzylidine-*n*-butylamine	(4)	99.5	99.6	100.0
N-Benzylidine-*n*-hexylamine	—	99.4	99.4	100.2
N-*p*-Chlorobenzylidine-*n*-hexylamine	—	100.0	99.9	100.9
N-*p*-Methoxybenzylidinebenzylamine	(5)	98.9	98.9	99.8
N-*p*-Methoxychlorobenzylidinebenzylamine	(6)	99.4	99.5	100.1
N-Benzylidineaniline	(7)	99.7	99.8	100.2
N-*n*-Butylidineaniline	(8)	99.0	99.1	—
1,2-Bis(benzylidineamino)-ethane	(9)	99.7	99.8	100.1
N^4-*p*-Methoxybenzylidinesulfathiazole	(10)	97.8	97.6	98.3

between 100.5 and 104%. There is excellent agreement between the titrimetric and nitrogen assay methods. The results of assays using methyl violet indicator lie within 0.1% of those obtained potentiometrically (Procedure A).

No differentiation between Schiff bases and the parent aliphatic or aromatic amines could be accomplished by procedure A (Fig. 12.1). This is probably due to the leveling effect of acetic acid (2) (see p. 534). Employing the technique of Pifer, Wollish, and Schmall (Procedure B), aliphatic amines could be quantitatively distinguished from their respective Schiff bases (Fig. 12.2). Only by the method of Fritz (Procedure C); was it possible to effect a quantitative differentiation between aniline and

3. W. S. Emmerson, S. M. Hess, and F. C. Uhle, *J. Am. Chem., Soc.*, **63,** 872 (1941).
4. C. W. Stein and A. R. Day, *J. Am. Chem. Soc.*, **64,** 2569 (1942).
5. C. W. Shoppee, *J. Chem. Soc.*, **1932,** 696.
6. C. W. Shoppee, *J. Chem. Soc.*, **1931,** 1225.
7. A. H. Blatt, Editor, *Organic Syntheses*, 2nd ed., Coll. Vol. I, Wiley, New York, 1941, p. 80.
8. M. S. Kharasch, I. Richlin, and F. R. Mayo, *J. Am. Chem. Soc.*, **62,** 494 (1940).
9. G. Loeb, *Rec. Trav. Chim. Pays. Bas*, **55,** 859 (1936).
10. R. N. Castle, *J. Am. Pharm. Assoc., Sci. Educ.*, **50,** 162 (1951).

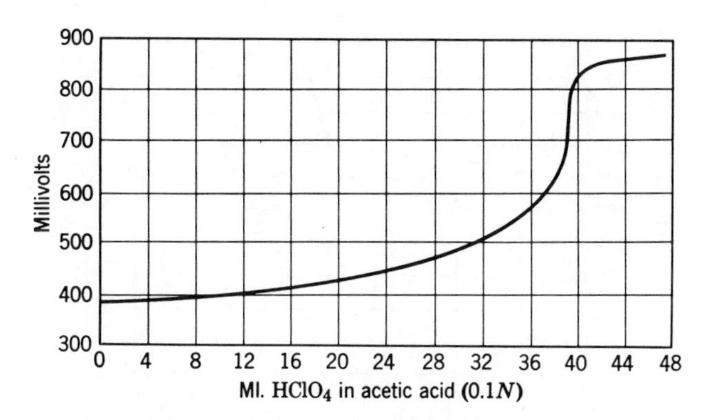

Fig. 12.1. Titration of 1 : 1 mixture of aniline and *N*-benzilidine-aniline in acetic acid.

N-benzylidineaniline (Fig. 12.3). By means of Procedure B and C, amine impurities occurring in Schiff bases in concentrations as low as 1% can be determined by simply increasing the sample size (see p. 534 for further discussion of the leveling effect of solvents; also see pp. 567–70 for discussion for relative basicities of amines and their Schiff bases).

Only one inflection was observed in the titration curve of *N*-butylidine-aniline when following Procedures A through C. This compound is reported to contain both an azomethine and a secondary amine groupings (8). In addition, equimolar mixtures of aniline and *N*-butylidineaniline

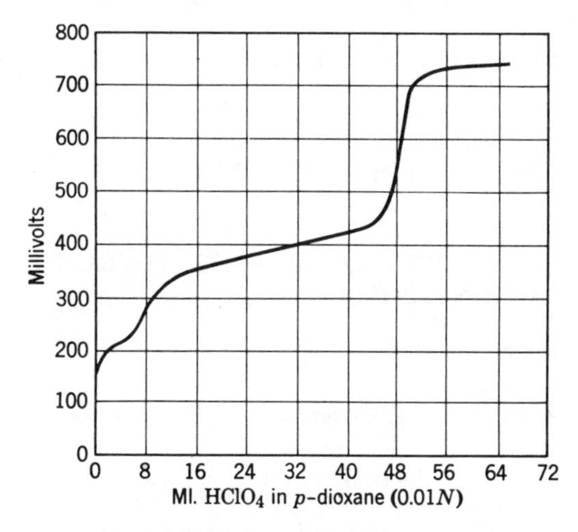

Fig. 12.2. Titration of *N*-benzilidine-*n*-hexylamine containing 10% *n*-hexylamine in chloroform.

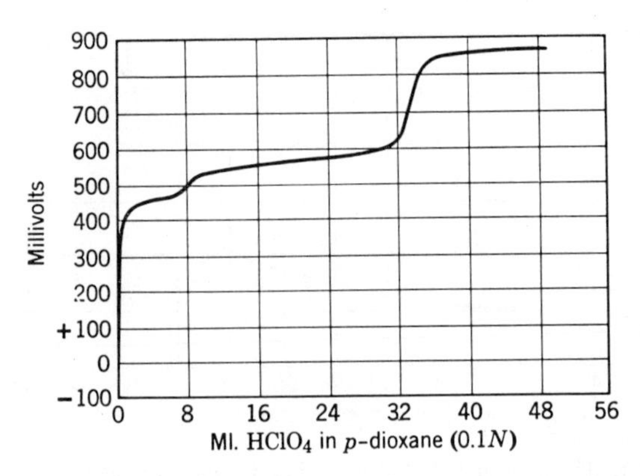

Fig. 12.3. Titration of *N*-benzylidineaniline containing 10% aniline in acetonitrile.

and sulfathiazole and N^4-*p*-methoxybenzylidinesulfathiazole yielded only one break in their titration curves. These results were duplicated when acetic acid was substituted for dioxane as solvent for the perchloric acid (2) using Procedure C.

Hydrolysis Methods

$$\begin{matrix} R \\ \quad\diagdown \\ \quad\quad C{=}NR_2 \\ \quad\diagup \\ R_1 \end{matrix} \xrightarrow{H_2O} \begin{matrix} R \\ \quad\diagdown \\ \quad\quad C{=}O \\ \quad\diagup \\ R_1 \end{matrix} + R_2NH_2$$

(where R_1 and R_2 can be hydrogen)

2,4-DINITROPHENYL HYDRAZONE PRECIPITATION OF CARBONYL PORTION OF IMINE

Method of S. Freeman

[*Reprinted in Part from Anal. Chem.*, ***25,*** *1750–1 (1953)*]

PRECIPITATION OF 2,4-DINITROPHENYLHYDRAZONES

To about 1 meq (accurately weighed) of Schiff base in 50 ml of ethyl alcohol, add 50% excess of a 1% solution of 2,4-dinitrophenylhydrazine in $2N$ hydrochloric acid. The precipitate forms immediately and is allowed to remain overnight at room temperature. It is then filtered off, washed with $2N$ hydrochloric acid, and dried to constant weight.

The compounds run by this method were listed in Table 1. The results by the hydrogen method are compared to those obtained by nonaqueous titration.

This analysis is more selective than the titration method because it is a function of the carbonyl portion of the Schiff base. Basicity is a common property, hence the titration is not as selective. The titration methods, however, are much faster than this gravimetric approach.

BISULFITE METHOD

This approach also measures the carbonyl portion of the imine by hydrolysis and bisulfite addition product formation with the resultant carbonyl compound.

Adapted from E. F. Hillenbrand, Jr., and C. A. Pentz

[*Reprinted in Part from Organic Analysis, Vol. III, Edited by J. Mitchell, Jr., et al., Wiley-Interscience, New York, 1956, pp. 194–6*]

REAGENTS

Sulfuric acid, approximately $0.6N$.
Sodium bisulfite, $0.2N$ aqueous solution.
Methyl red indicator, 0.2% aqueous solution.
Starch indicator, 1.0% aqueous solution.
Standard $0.1N$ iodine.

PROCEDURE

Add 25 ml of distilled water to each of four 500-ml glass-stoppered Erlenmeyer flasks. Reserve two of the flasks for a blank determination. Into each of the other flasks introduce an amount of sample containing 1 to 2 meq of imine. Place the flasks in an ice bath at 0 to 5°C for 10 minutes. Add 2 drops of the methyl red indicator and neutralize by adding $6N$ sulfuric acid from a buret. Swirl the flasks sufficiently to make sure the sample is completely neutralized, but do not add an excess of acid. Return the flasks to the ice bath and allow them to stand for 10 minutes. Introduce 25 ml of 0.2N sodium bisulfite into each flask using a pressure-type pipet. Keep the tip of the pipet just at the surface of the solution. Space the addition of the reagent to allow equal reaction time for all blanks and samples. Remove the flasks from the ice bath, swirl the contents gently, and allow them to stand at room temperature for exactly 30 minutes. Swirl the flasks occasionally during this time. Replace the flasks in the ice bath for exactly 5 minutes. Remove the flasks from the ice bath and add approximately 25 grams of crushed ice. Add 2 ml of the starch indicator and titrate immediately with 0.1N iodine to the first appearance of a blue end point.

CALCULATION

$$\frac{(B-A)N \times \text{E.W.}}{\text{Grams of sample} \times 10} = \text{imine, wt. \%}$$

where A = milliliters of N normal iodine required for the sample
B = milliliters of N normal iodine required for the blank
E.W. = equivalent weight of imine

Compounds for which this procedure was found to give satisfactory results include ethylimine polymer; butylimine; *N*-butylidenebutylamine; *N*-decylidenedecylamine; amylimine; *N*-amylideneamylamine; *N*-(2-butyloctylidene)-3-butyloctylamine; *N*-ethylideneaniline; and *N*-butylideneaniline. In cases in which solubility difficulties occur, the procedure is modified to use a 15:10 mixture of isopropanol and water as the solvent medium.

HYDROXYLAMINE METHOD

Adapted from E. F. Hillenbrand, Jr.

[*Reprinted in Part from Organic Analysis, Vol. III, Edited by J. Mitchell, Jr., et al., Wiley-Interscience, New York, 1956, pp. 194–6*]

When the aldehyde released by the neutralization (hydrolysis) reaction cannot be determined using sodium bisulfite reagent, the free hydroxylamine reaction may often be applied. To use this procedure, add the sample to 15 ml of isopropanol and 10 ml of water. Add 1 ml of 0.04% solution of bromophenol blue indicator in methanol, and nearly neutralized the sample with 6N hydrochloric acid; the final neutralization is completed with 0.5N hydrochloric acid. To both a blank and the sample, add 35 ml of 0.5*N* hydroxylamine hydrochloride that has been neutralized with 0.5*N* sodium hydroxide to bromophenol blue indicator. Introduce exactly 3 ml of 0.5*N* hydrochloric acid, and allow the blank and the sample to stand at room temperature for 60 minutes. Then titrate both blank and sample with standard 0.5*N* sodium hydroxide to a blue-green end point.

This method has been applied successfully to the following imines and *N*-substituted imines: ethylimine; *N*-isopropylideneisopropylamine; butylimine; *N*-butylidenebutylamine; 2-ethylbutylimine; 2-ethylhexylimine; *N*-(2-ethylbutylidene)-2-ethylbutylamine; and α-methylbenzylimine.

13

Titanous, Chromous, and Ferrous Reductions

(—N=N—; —NO_2; —NO; —NHNH—; —N≡N—)

These groups—azo, nitro, nitroso (or *N*-oxide), hydrazo, or diazonium groups, respectively—can be readily reduced using titanous or chromous ions. Titanous ion is the most used reducing agent but probably only because it has the longer history. Chromous ion has all the earmarks of being just as widely applicable, although the work to date reports its use only for nitro, azo, and nitroso groups. The iron(II) systems offer some advantages of selectivity for nitro and nitroso compounds.

The reductions proceed as shown below.

$$RN{=}NR_1 + 4(H) \rightarrow RNH_2 + R_1NH_2$$

$$RNO_2 + 6(H) \rightarrow RNH_2 + 2H_2O$$

$$RNO + 4(H) \rightarrow RNH_2 + H_2O$$

$$RNHNHR_1 + 2(H) \rightarrow RNH_2 + R_1NH_2$$

$$[RN^+{\equiv}N]X^{-*} + 6(H) \rightarrow RNH_2 + NH_4X$$

The systems must be handled in the absence of oxygen, since the reagent ions are readily oxidized. This is the only drawback of the approaches; however, the disadvantage is minimized with the proper apparatus and handling, and the methods are quite readily applicable with an attainable accuracy and precision of about ±1%.

The titanous method to be shown has been used for many years in the laboratories of the authors.

Titanous Reduction

REAGENTS

TITANOUS CHLORIDE. Boil 1 liter of water containing 100 ml of concentrated hydrochloric acid to remove dissolved oxygen. Cool this solution under nitrogen,

* X^- is usually halide or bisulfate ion.

and add to it 560 ml of 20% titanous chloride ($TiCl_3$ is sold in 20% solution). Dilute the solution to 2 liters with freshly boiled and cooled distilled water. Store under hydrogen- or oxygen-free nitrogen in an apparatus as described in Fig. 13.1, or in any other storage buret that permits storage with inert atmosphere.

Ammonium thiocyanate, 10% solution.

FERRIC AMMONIUM SULFATE SOLUTION. Dissolve 282 grams of ferrous ammonium sulfate in 600 ml of water, and add 50 ml of concentrated sulfuric acid. To the still warm solution, slowly add 80 to 100 ml of Perhydrol (30% hydrogen peroxide), with stirring. Boil off the excess peroxide, cool the solution and dilute to 4 liters. To standardize the ferric ammonium sulfate solution, place 20 ml of the solution in a 250-ml iodine flask together with 4 ml of 4*N* hydrochloric acid. Then add 2 grams of potassium iodide and allow the solution to stand for 5 minutes. Titrate the liberated iodine with 0.1*N* sodium thiosulfate to the starch end point.

APPARATUS

Storage buret for titanous chloride (Fig. 13.1).

The buret, as indicated in Fig. 13.1, is used for Karl Fischer reagent: Since this reagent must be protected from water vapor, Ascarite tubes are shown in the diagram. Titanous chloride solution must be protected against oxygen, however; thus the Ascarite tubes do not apply in this case. For titanous chloride solution, a hydrogen generator (a Kipp generator) is attached to the vent from the storage bulb. In this manner, a continuous hydrogen atmosphere is maintained above the solution.

Attached to the vent of the measuring buret section of the apparatus is a small stopcock.

When the apparatus is charged with titanous chloride solution, the atmosphere above the solution is purged of air by turning the buret stopcock so that the measuring buret is open to the atmosphere at the tip. The hydrogen from the generator will then flow through the bypass tube from the reservoir, to the buret, and on to the outside through the buret tip. This operation will purge the buret as well as the atmosphere above the titanous chloride solution. In filling the measuring buret, the stopcock is turned to connect the buret to the reservoir. Then the stopcock on the vent of the buret is opened, and the buret will fill with reagent. The reagent entering the buret will force out the hydrogen through the stopcock at the vent. As soon as the buret is filled to the desired level, both the stopcock at the vent and at the tip of the buret are closed. To introduce the reagent to the reaction vessel, the stopcock at the buret tip is opened to the reaction vessel containing the sample (through which oxygen-free nitrogen is bubbling). The hydrogen pressure in the apparatus will cause the solution to flow out of the buret.

Reaction flask (Fig. 13.2).

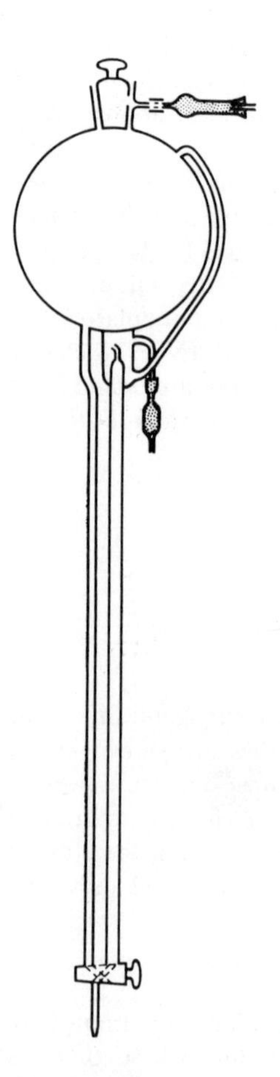

Fig. 13.1. Storage buret for titanous chloride.

PROCEDURE

Place about 0.005 equivalent of sample in a 500-ml Erlenmeyer flask and dissolve in water or in glacial acetic acid if the sample is water insoluble. Then equip the flask as in Fig. 13.2 to keep the reaction under nitrogen. Pass oxygen-free nitrogen through the sample for 5 to 10 minutes. Next add 25 ml of titanous chloride solution, 30 ml of concentrated hydrochloric acid, and 2 ml of hydrofluoric acid. Boil the solution for 5 minutes (a longer time may be needed for some samples). A condenser should be attached to the reaction flask for volatile samples. With the nitrogen still bubbling through the solution, cool the flask in ice, and add 10 ml of 10% ammonium thiocyanate. Titrate the solution with the ferric ammonium sulfate solution to a red end point, keeping the nitrogen bubbling through the solution.

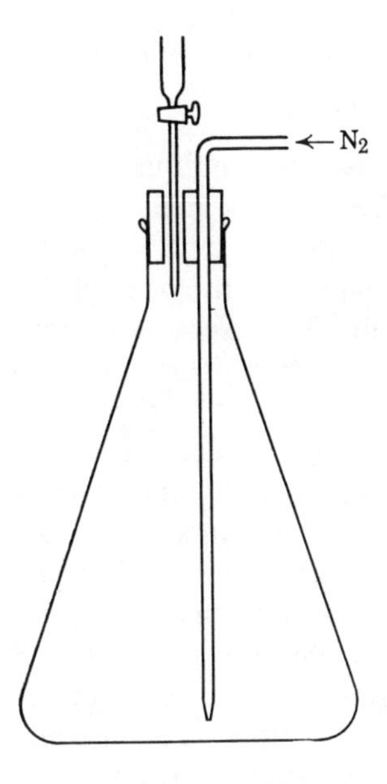

Fig. 13.2. Reaction flask.

A blank should be run on the titanous chloride for each set of reductions. Use the same solvent for the blank as is used for the sample.

CALCULATIONS

$$\frac{\text{(Milliliters of FAS blank minus milliliters of FAS sample)} \times N\ \text{FAS} \times \text{mol. wt. of compound} \times 100}{\text{Grams of sample} \times 1000 \times A} = \%\ \text{compound}$$

where A = number of hydrogen atoms consumed per molecule of compound

FAS = ferric ammonium sulfate solution

The preceding procedure was used successfully to analyze 2,4-dinitrophenylhydrazine; *p*-nitraniline; 1,3,5-trinitrobenzene; benzene-azodiphenylamine; *p*-nitrobenzeneazoresorcinol; hydrazobenzene. The average accuracy and reproducibility are better than ±2%. With special care to keep oxygen out of the reagents and the reaction mixture, results better than ±1% can be obtained.

Oxygen is the most troublesome interference in this procedure; however any reducible substance will usually interfere. Ethylenic or acetylenic compounds generally are not reduced by titanous chloride.

Chromous Reduction

Adapted from R. S. Bottei and N. H. Furman

[*Reprinted in Part from Anal. Chem.*, **27**, *1182–4* (*1955*)]

Chromous salts have not been as extensively used for the quantitative analysis of reducible organic compounds as have titanous salts. Someya (1) reduced *p*-nitroaniline, picric acid, and *p*-nitrophenol with an excess of chromous chloride solution, which was prepared by the incomplete reduction of chromic chloride by amalgamated zinc. The excess chromous chloride was titrated with standard ferric alum solution. Terent'ev and Goryacheva (2) had titrated quinone, azobenzene, and *m*- and *p*-nitroaniline directly, using methyl red as an indicator. Their precision for the determination of azobenzene was very poor. They prepared their chromous solution by dissolving chromous acetate in hydrochloric acid. Both these methods necessitated the frequent standardization of the chromous solution. Lingane and Pecsok (3) have shown that it is relatively easy to prepare and maintain a standard chromous solution of exactly determinate strength. Since chromous solutions are stronger reducing agents than titanous solutions, and their reactions are generally faster than titanous salts in that the reductions are usually carried out at room temperature, it was decided to reinvestigate the use of chromous chloride for organic analysis.

This section presents the results of the use of this reagent for the determination of *o*-nitrobenzoic acid, 2,4,6-trinitrobenzoic acid, 2,4,6-trinitroresorcinol, 2,4-dinitrophenylhydrazine, nitroguanidine, *p*-nitrobenzeneazoresorcinol, nitroso-R salt, anthraquinone 2,7-disodium sulfonate, and the monopotassium salt of acetylene dicarboxylic acid. All are quantitatively reduced, the nitrogen-containing compounds to the corresponding amines (rupture of the N—N link in the azo compound), the anthraquinone to the corresponding anthrahydroquinone, and the acetylenic compound to the corresponding ethylenic compound.

EXPERIMENTAL

APPARATUS

The titration cell was a tall-form 200-ml electrolytic beaker, covered by a rubber stopper provided with a gas inlet tube, a saturated calomel reference

1. K. Someya, *Z. Anorg. Allgem. Chem.*, **169,** 293 (1928).
2. A. P. Terent'ev and G. S. Goryacheva, *Uche, Zap.* (*Wiss. Ber. Mosk. Staats-Univ.*), **3,** 277 (1934).
3. J. J. Lingane and R. L. Pecsok, *Anal. Chem.*, **20,** 425 (1948).

electrode, a platinum indicator electrode, a thermometer, if the reaction was to be carried out at an elevated temperature, and openings for a gas outlet and the delivery tips of two burets. If any of the openings were not used, they were closed by means of corks. If the solution was to be heated, a beaker encircled with asbestos-covered heating wire was used. The temperature was controlled by regulating the current flowing through the heating wire by means of a Variac.

The solutions were stirred with a magnetic stirrer.

The end point of the reaction was determined potentiometrically by measuring the voltage change by means of a Leeds & Northrup line-operated pH meter, Model 7664. Since the titrant could be added in very small increments (about 0.02 ml, if the stopcock is spun rapidly and the buret tip has a rather small opening), it was not necessary to plot the voltage change. The voltage change at the end point in the back-titration of excess chromous chloride with ferric alum solution is about 500 mV, whereas the voltage change at the end point in the direct titration of the anthraquinone salts is only about 175 mV.

The end point in the back-titration of chromous solution with ferric alum solution can also be determined using the derivative polarographic end point (4), in which case the saturated calomel reference electrode is replaced by a second platinum electrode. The pair of platinum electrodes are polarized by a small constant current of about 2 μA.

The apparatus used for storing and dispensing standard chromous solution was the same as that utilized by Lingane and Pecsok (3), except that a 2-liter storage flask was used in place of the 1-liter one. In this way enough chromous solution was available for about 35 separate determinations.

REAGENTS

STANDARD 0.1000*N* SOLUTION OF CHROMOUS CHLORIDE IN 0.1*N* HYDROCHLORIC ACID. Prepared directly in the storage flask by the procedure described by Lingane and Pecsok (3). Fill the 2-liter storage flask (Fig. 13.3) about two-thirds full with amalgamated mossy zinc, and add about 1 liter of 0.1000*M* chromic chloride in 0.1*N* hydrochloric acid. The reduction is usually allowed to proceed overnight. Store the solution under pure hydrogen obtained from a Kipp generator, the hydrogen being freed from oxygen by passage through a bubble tower containing chromous chloride solution in 1*N* sulfuric acid in contact with amalgamated zinc. Standardize the chromous solution against standard cupric solution in 6*N* hydrochloric acid as recommended by Lingane and Pecsok (3). Prepare the standard cupric solution from copper sulfate pentahydrate and standardize electrogravimetrically as described by Willard and Furman (5).

BAKER AND ADAMSON'S "REAGENT QUALITY" ZINC. Amalgamate with about 2% mercury by shaking it for about 10 minutes in a mercuric chloride solution in

4. C. N. Reilley, W. D. Cooke, and N. H. Furman, *Anal. Chem.*, **23**, 1223 (1951).
5. H. H. Willard and N. H. Furman, *Elementary Quantitative Analysis*, 3rd ed., Van Nostrand, New York, 1940, p. 44.

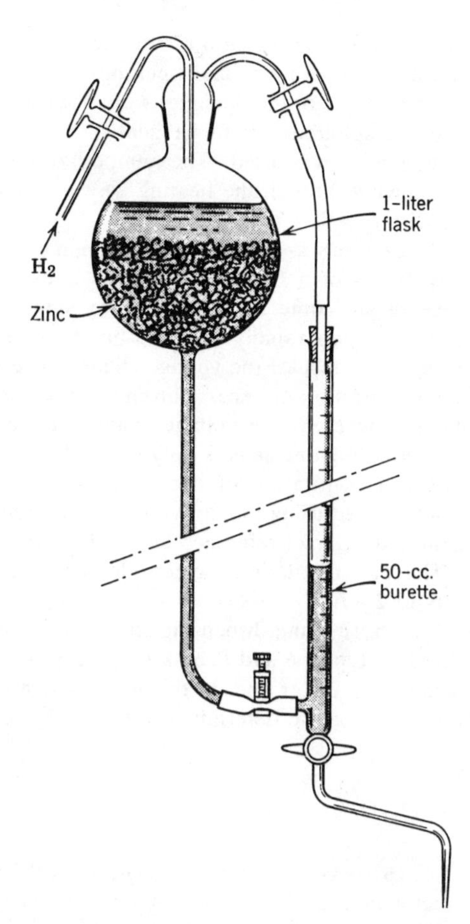

Fig. 13.3. Apparatus for preparation, storage, and dispensing of standard chromous sulfate solutions.

dilute hydrochloric acid. At first, a solution of mercuric nitrate in dilute nitric acid was used to amalgamate the zinc; however the chromous solution obtained by using this amalgamated zinc always had a normality about 2 to 3% too low, although the amalgam was washed thoroughly before use.

APPROXIMATELY $0.1N$ FERRIC ALUM SOLUTION, ACIDIFIED WITH SULFURIC ACID (TO $1N$). Use this in the back-titration of excess chromous chloride solution. Free the solution of oxygen by passing nitrogen through it for about 15 minutes; free the nitrogen of oxygen in the usual manner. Standardize the solution either by titrating with standard permanganate (6) portions of the solution, which have been reduced by amalgamated zinc in a Jones reductor, or by titrating aliquots with standard chromous solution.

6. *Ibid.*, pp. 230–1.

Standard solutions of the organic compounds investigated were generally prepared by dissolving a known weight of the purest commercially available material in water, or in glacial acetic acid, if the substance was insoluble in water. The sample of nitroguanidine was recrystallized from water two times and air dried. One of the solutions of the monopotassium salt of acetylene dicarboxylic acid was prepared from material synthesized according to the method of Moureu and Bongrand (7).

PROCEDURE

Except where otherwise specified, the following procedure was employed. Pipet suitable aliquots of the solution to be analyzed into a titration vessel containing about 15 ml of water. Add 10 ml of concentrated hydrochloric acid, and adjust the initial volume of the solution to about 50 ml by adding water. Bubble carbon dioxide, freed from traces of oxygen by passage through acidified chromous solution in contact with amalgamated zinc, through the solution for about 10 minutes. At the end of this period, pass the carbon dioxide over the surface of the solution. Add an excess of $0.1000N$ chromous chloride solution and generally allow the solution to stand about 1 to 2 minutes, depending on the rate of reaction, before back-titrating with standard ferric alum solutions. The titrations were performed at room temperature.

Since platinum is a catalyst for the decomposition of chromous ion by hydrogen ion, keep the platinum electrode out of the solution until it is time to back-titrate, at which time lower it into the solution. Follow a similar procedure with the saturated calomel electrode.

In the determination of *p*-nitrobenzeneazoresorcinol, 25 ml of concentrated hydrochloric acid has to be used in place of the usual 10 ml; otherwise a precipitate forms.

The samples of nitroguanidine and of the monopotassium salt of acetylene dicarboxylic acid are not prepared in acid medium. Add the aliquots to enough water to give an initial volume of about 50 ml. In the presence of hydrochloric or sulfuric acid, the results are very low. In the case of nitroguanidine, the use of a citrate-buffered solution does not improve the results appreciably, and in addition, the end point response is sluggish as compared with a nonbuffered solution. In these determinations it is not necessary to allow the solution to stand for several minutes before back-titrating the excess chromous solution.

The determination of 2,4,6-trinitrobenzoic acid must be performed at an elevated temperature. At room temperature in either hydrochloric or sulfuric acid solution only about 70% reduction is obtained, whereas in a

7. C. Mouceay and J. C. Bongnand, *Ann. Chim.*, **14**, 47 (1970).

citrate-buffered solution, the reduction is increased to about 80%. The procedure that was finally adopted involved heating a hydrochloric acid solution of the sample to 85°C and allowing it to cool to 55°C before back-titrating the excess chromous chloride.

Run a blank using the same solvent conditions employed for the samples on the chromous chloride for each set of reductions.

Titrate the anthraquinone-2,7-disodium sulfonate as well as an impure sample of 1-nitro-anthraquinone-5-sodium sulfonate, directly in a hydrochloric acid medium. The reaction at the end point is slow; therefore allow a 5-minute wait before making the final reading. Heating does not improve the end point response.

RESULTS AND DISCUSSION

Table 1 contains the results that were obtained for the compounds investigated in this study, which indicates that satisfactory results can be obtained.

The minimum excess of chromous chloride to be added to a given sample, to obtain satisfactory results, varied with the nature of the compound and generally depended on its concentration in the solution. Therefore a systematic investigation was necessary to carry out the percentage reduction for a given added excess at a particular concentration of the sample. In general, for 10 ml of an approximately $0.1N$ solution an excess of 200 to 250% was sufficient. With smaller samples the required excess may be the same as or slightly greater than with samples of a higher concentration, or it may be considerably greater as in the case of the monopotassium salt of acetylene dicarboxylic acid or 2,4-dinitrophenylhydrazine.

The indirect determination of anthraquinone-2,7-disodium sulfonate using the general procedure invariably produced low results. This was because of the very easy oxidation of the anthrahydroquinone by ferric ion. However the indirect determination of 1-nitroanthraquinone 5-sodium sulfonate gave the same results as the direct titration. Since the sample was impure, these data are not presented in Table 1. The first appreciable voltage change (about 200 mV) was taken to be the end point for the back-titration. After this break, there was a gradual voltage change with added ferric solution, and finally another substantial break corresponding to the oxidation of the anthrahydroquinone. This was observed qualitatively in that the reddish-brown of the anthrahydroquinone eventually gave way to the characteristic green of the chromic ion.

Table 1. Analysis of Organic Compounds Using Chromous Chloride

	Milli-equivalents Taken	Minimum Excess, %	Milli-equivalents Found	Number of Determinations
o-Nitrobenzoic acid	0.634_5	200	0.635 ± 0.004	4
	0.903	200	0.905 ± 0.003	3
	1.269	200	1.264 ± 0.004	4
	1.806	200	1.808 ± 0.008	3
2,4,6-Trinitrobenzoic acid	0.796_5	300	0.795 ± 0.007	4
	1.593	200	1.596 ± 0.019	5
2,4,6-Trinitroresorcinol	0.511	300	0.511 ± 0.005	6
	0.635	300	0.630 ± 0.007	5
	1.022	250	1.011 ± 0.005	8
	1.270	250	1.259 ± 0.005	4
2,4-Dinitrophenyl hydrazine	0.528_5	450	0.425 ± 0.004	8
	0.857	250	0.851 ± 0.007	5
Nitroguanidine	0.452	200	0.448 ± 0.003	3
	0.501	250	0.495 ± 0.002	4
	0.904	200	0.890 ± 0.002	3
	1.002	250	0.990 ± 0.008	10
p-Nitrobenzeneazoresorcinol	0.369	250	0.371 ± 0.002	4
	0.738	250	0.731 ± 0.003	6
	0.500	250	0.493 ± 0.002	3
	1.000	250	0.995 ± 0.001	4
Nitroso R salt	0.425	300	0.423 ± 0.004	4
	0.537	300	0.536 ± 0.001	3
	0.850	300	0.850 ± 0.004	5
	1.074	250	1.074 ± 0.002	4
Monopotassium salt of acetylene dicarboxylic acid	0.517[a]	400	0.514 ± 0.004	6
	0.527_5	400	0.528 ± 0.005	5
	0.661_5	400	0.662 ± 0.002	3
	1.034[a]	250	1.025 ± 0.003	4
	1.055	250	1.055 ± 0.001	3
	1.323	250	1.325 ± 0.002	3
Anthraquinone-2,7-disodium sulfonate	1.000	—	1.006 ± 0.001	3
	1.187	—	1.194 ± 0.000	4
	2.000	—	2.015 ± 0.002	4
	2.374	—	2.396 ± 0.002	4

[a] Material synthesized according to procedure of Moureu and Bongrand (7).

Ferrous Reductions of Nitro and Nitroso Groups

From the Method of W. I. Awad, S. S. M. Hassan, and M. T. M. Zaki

[*Anal. Chem.*, **44**, *911* (*1972*)]

Because the reducing power of iron(II) is less than that of titanium(III) or of chromium(II), the ferrous ion offers the advantage of less interference than other reducing agents suffer in the determinations of nitro and nitroso compounds. As an example, 8 equivalents of tin(III) is required per mole of nitrophenylhydrazine, 2 for the hydrazine and 6 for the nitro group, but only 6 equivalents of iron(II) is needed in acid medium. The activation of one nitro group on the hydrazine function is not sufficient to permit reduction with iron(II).

Also reduction in alkaline medium is possible with iron(II), which offers advantages in some situations. Other reductants, such as titanium(III), chromium(II). vanadium(II), and tin(II), are used only in acid media.

APPARATUS

An electrolytic reduction automatic microburet was used for the preparation, storage, and use of titanium(III) solution. The arrangement illustrated on page 655 with a microburet can be used.

PROCEDURE

REDUCTION IN ACID MEDIUM. To determine the nitro group, weigh 3 to 5 mg of the sample into the reaction flask. Dissolve the sample in acetone and add 20 ml of 11N hydrochloric acid. Sweep the air from the flask with carbon dioxide or nitrogen for 5 minutes at the rate of 100 bubbles per minute. Follow with the addition of 1.5 to 2 grams of ferrous ammonium sulfate. Boil for 15 minutes, cool, and add 1 ml of 10% ammonium thiocyanate solution. Titrate with 0.04N titanium(III) sulfate solution to the disappearance of the red color of the ferric thiocyanate. Carry a blank experiment through the procedure. Calculate the nitro group percentage based on consumption of 6 equivalents of iron(II) per mole of nitro group.

To determine the nitroso group, follow the same procedure but with 15 ml of 11N hydrochloric acid, 1 gram of iron ferrous ammonium sulfate, and a 17-minute boiling period. Calculate the nitroso group percentage based on the consumption of 4 equivalents of iron(II) per mole of nitroso group.

REDUCTION IN ALKALINE MEDIUM. Weigh 3 to 5 mg of the nitro or nitroso sample into the reaction flask. Dissolve the sample in ethanol and add 30 ml of 10% sodium hydroxide solution. Sweep the air from the flask with carbon dioxide or nitrogen for 5 minutes at the rate of 100 bubbles per minute. Add 1 ml of 1*N* ferrous ammonium sulfate solution. Boil 10 minutes, cool, acidify with concentrated hydrochloric acid (about 20 ml), and add 1 ml of 10% ammonium thiocyanate solution. Titrate with standard 0.04*N* titanium(III) sulfate solution until the red color disappears. Carry out a blank under the same experimental conditions.

Calculate the percentage of nitro or nitroso group based on the consumption of 6 and 4 equivalents of iron(II) per mole of the nitro and nitroso group, respectively.

RESULTS AND DISCUSSION

The nitro group in organic compounds is satisfactorily analyzed by reduction with ferrous ammonium sulfate in strong hydrochloric acid solution. The reduction of mononitro aromatic compounds substituted with electron-attracting groups [e.g., COOH, Cl, Br, COOR, NO_2, NH_3, $AsO(OH)_2$] shows an average recovery of 99.3%, with an average error of ±0.2%. Di- and polynitro aromatic compounds show an average recovery of 99.4%; the average error is ±0.4%. However nitro hydrocarbons and nitro compounds substituted with electron-repelling groups (e.g., OH, OCH_3, CH_3) are not reduced quantitatively. The average recovery is 76.5%, with an absolute error of 8.9%.

The reduction of nitro compounds in alkaline medium is successful for nitro hydrocarbons and mononitro aromatic compounds substituted with both electron-attracting and electron-repelling groups. The average recovery is 99.9%, and the mean absolute error is ±0.1%. *p*-Nitrophenylhydrazine is the only nitro compound that is not reduced quantitatively. The absolute error for this compound is −12% and the average recovery is 60%. Di- and polynitro aromatic compounds show an average recovery of 75.2%, with an average absolute error of −12.9%.

Nitroso compounds are satisfactorily analyzed with iron(II) in acidic and alkaline media; the average recoveries are 99.6 and 99.4%, respectively.

Very little interference is observed with other nitrogen functions. Some reducible nonnitrogeneous groups that are occasionally present in organic compounds (e.g., ketonic, aldehydic, carboxylic, sulfonic, arsonic, phosphonic, as well as ethylenic and acetylenic bonds) are not reduced by iron(II) in either acidic or alkaline media.

Results obtained by this method are shown in Table 2.

Table 2. Microdetermination of Nitro and Nitroso Groups in Some Aromatic Compounds by Reduction with Iron(II) in Acidic and Alkaline Media

Sample	Recovery,[a] %	
	Acid Reduction	Alkaline Reduction
Mononitro compounds		
p-Nitrobenzoic acid	99.4	99.4
3-Nitrophthalic acid	99.5	99.7
m-Nitroethylbenzoate	99.1	99.7
p-Nitroaniline	99.0	99.8
2-Chlroro-5-nitroaniline	99.7	99.8
3-Chloro-4-nitrophenylacetic acid	98.8	99.9
p-Bromonitrobenzene	99.6	99.6
p-Nitroacetophenone	99.5	99.7
m-Nitro(α-phenyl)cinnamic acid	98.7	99.9
3-Nitro-4-hydroxyphenylarsonic acid	100.2	100.0
p-Toluene-3-nitroarsonic acid	99.9	100.5
p-Nitrophenylhydrazine	99.8	60.1
1-Nitronaphthalene	85.1	99.6
3-Nitrodiphenyl	83.3	99.9
m-Nitrophenol	76.1	100.0
p-Nitrophenylacetic acid	91.8	100.0
1-Nitroanisole	53.0	99.9
p-Nitrotoluene	32.4	100.1
Dinitro compounds		
2,5-Dinitrobenzoic acid	99.9	82.1
m-Dinitrobenzene	99.7	81.7
Polynitro compounds		
Trinitrotoluene (TNT)	98.8	68.1
2,4,6-Trinitrobenzoic acid	99.7	77.8
Nitroso compounds		
1-Nitroso-2-naphthol	99.8	99.6
Nitroso-R salt	99.7	99.7
2-Nitroso-1-naphthol-4-sulfonic acid, sodium salt	99.3	99.0
p-Nitrosodimethylaniline hydrochloride	99.9	99.6
Nitrosobenzene	99.4	99.3
Research mononitroso sample ($C_{10}H_{20}N_7OCl$)	98.9	98.3

[a] Each value is the average of three determinations.

14

Hydrazines and Hydrazides

Hydrazines ($RNHNHR_1^*$)

Hydrazines can be analyzed chemically by several approaches. One of the most common is the titration with standard acid. Hydrazines are bases of much the same order of basicity as amines and can be titrated in the same manner in aqueous or nonaqueous media (see pp. 533–558). The one exception is that glacial acetic acid cannot be used as a solvent in the case of hydrazines, since these react with acetic acid to form the corresponding hydrazides.

$$CH_3COOH + RNHNH_2 \rightarrow CH_3C(=O)\text{—}NHNHR$$

The symmetrical disubstituted hydrazines ($R\text{—}NHNH\text{—}R_1$) may not suffer from this difficulty, but the authors have never attempted titration of these disubstituted compounds in glacial acetic acid. Ketone solvents must also be avoided for the titration of hydrazines due to hydrazone formation.

Hydrazines can also be readily oxidized using iodine, iodate, periodate, or cupric ion. The oxidation also forms the basis of analysis. Chapter 13 concerns the determination of hydrazines, mono- and disubstituted, using reduction via titanous chloride. This method applies to a wide range of hydrazines but suffers from a lack of specificity. As Chapter 13 indicates, many nitrogen-containing functional groups interfere (nitro, nitroso, azo, diazo groups).

Hydrazone formation with carbonyl compounds can also be used to determine hydrazines. These analytical methods are mainly colorimetric, but one volumetric method does exist (1). This method involves titration of the hydrazine with 0.2*N* benzaldehydesulfonic acid. The end point is followed by spot tests on filter paper to which an indicator is added.

TITRATION WITH ACIDS

See section on Amines in accordance with the foregoing discussion.

* R_1 can be H atom.

1. R. Meyer, *Z. Anal. Chem.*, **140**, 124–5 (1943).

OXIDATION METHODS

Oxidation with Iodine—Method Described by Joseph Rosin

$$NH_2NH_2{\cdot}H_2SO_4 + 2I_2 + 6NaHCO_3 \rightarrow 4NaI + Na_2SO_4 + 6CO_2 + 6H_2O + N_2$$

$$RNHNH_2{\cdot}H_2SO_4 + 2I_2 + 5NaHCO_3 \rightarrow RI + 3NaI + Na_2SO_4 + 5CO_2 + 5H_2O + N_2$$

(*Adapted with Permission from Reagent Chemicals and Standards*; *D. Van Nostrand, New York, 1939, p. 194*).

REAGENTS

Standard iodine, 0.1*N*.
Sodium bicarbonate.
Starch indicator.

PROCEDURE

Dissolve about 0.0005 to 0.001 mole of hydrazine in about 20 to 50 ml of water. Add about 1 gram of sodium bicarbonate and titrate the solution with 0.1*N* iodine, starch being used as an indicator.

$$\frac{\text{Milliliters of } I_2 \times N\, I_2 \times \text{mol. wt. of hydrazine} \times 100}{\text{Grams of sample} \times 4000} = \%\ \text{hydrazine}$$

The procedure above was described only for the determination of hydrazine sulfate. It was, however, found to work rather generally for water-soluble hydrazines.

Oxidation with Iodate—Direct Iodate Method Using Solvent Described by R. A. Pennman and L. F. Andrieth

[*Anal. Chem.*, **20,** *1058–61* (*1948*)]

$$N_2H_4 + KIO_3 + 2HCl \rightarrow KCl + ICl + N_2 + 3H_2O$$

Hydrazine may be titrated directly with standard iodate solution in the presence of concentrated hydrochloric acid (4*N* or above). Actually the reaction is in two steps: iodate is first reduced to I_2, then the latter is oxidized by additional iodate to iodine monochloride (ICl). The normality of the acid must be kept between 3 and 5.

A suitable aliquot is placed in a glass-stoppered flask and 20% more than an equal volume of concentrated hydrochloric acid is added, in addition to 10 ml of carbon tetrachloride. Standard *N*/10 iodate is added with frequent vigorous shaking until the aqueous layer changes from brownish to yellowish. Toward the end, add iodate dropwise and shake after each addition until the iodine disappears from the solvent layer. Acid normality at the end must be between 3 and 5:

$$1 \text{ ml of } N/10 \ KIO_3 \cong 0.000534 \text{ gram of } N_2H_4$$

Oxidation with Iodate—Direct Iodate Method with Internal Indicator Described by R. A. Pennman and L. F. Andrieth

[*Anal. Chem.*, ***20,*** *1058–61* (*1948*)]

This procedure is the same as given previously except that a water-soluble dye is used as internal indicator. Either Amaranth or Brilliant Ponceaux 5R (British Colour Index Numbers 184 and 185, respectively), may be used. National Aniline Company produces these under the names Wool Red-40F, and Brilliant Scarlet, 3R. Three to five drops of 0.2% aqueous solution is sufficient to give a distinct end point in 250 ml of solution. The dyes are not affected by hydrochloric acid, iodine, or iodine monochloride under conditions of the titration, but they are readily destroyed by a trace of iodate in 3 to 5*N* hydrochloric acid at temperatures above 30°C. Heat of dilution of the acid is usually sufficient to produce this temperature. Addition of the indicator is delayed until the end point is approached, as indicated by a lighter color. Here also, 1 ml of *N*/10 $KIO_3 \cong 0.000534$ gram N_2H_4.

The procedure used for sampling and determining anhydrous hydrazine for the assay is as follows.

Tare a cold weighing bottle containing 10 ml of water before adding the sample. (Cool with ice water to 18–25°C, to remove the heat of solution. Handling the weighing bottle or allowing the temperature to rise above 25°C will affect the accuracy of the results.) Use a 1-ml pipet that has been washed with water and methanol, then dried with dry nitrogen to withdraw the sample (pipet must not be rinsed with hydrazine), and allow it to drain into the weighing bottle with its tip against the side of the bottle, a few millimeters above the water. After replacing the top of the bottle, mix the sample and water by gently swirling the bottle. (Take care that the solution does not touch the ground glass.) Reweigh to determine weight of sample.

Transfer sample carefully to a 250-ml volumetric flask containing about 175 ml of water. Dilute to 250 ml. Do not shake flask to mix—just invert flask 6 or 7 times, slowly. From this flask take a 5-ml aliquot, add to 15 ml of water in a 250-ml Erlenmeyer flask, then add 30 ml of concentrated hydrochloric acid and titrate with standard iodate. When the brown changes to a light yellow, add 5 to 6 drops of National Aniline Wool Red-40F indicator and shake the solution vigorously after the addition of each drop of iodate solution. The red color of the solution fades through a pink and changes to a lemon yellow. This is the end point.

The method above was published for the analysis of hydrazine, but the authors of this book have found it applicable to a wide range of aliphatic hydrazines as well. Aromatic hydrazines have not been tried with this method.

Oxidation with Periodate—Adapted from A. Berka and J. Zyka

[*Chem. Listy*, ***50*** (*2*), *314–6* (*1956*)]

Potassium periodate is used for oxidative volumetric determination of the following hydrazine derivatives: hydrazide of isonicotinic acid, phenylhydrazine, semicarbazide, thiosemicarbazide. The determination was carried out in hydrochloric acid, and it was ascertained that the reduction of periodate to iodine monochloride proceeded depending on the concentration of hydrochloric acid and on the determination character, by the reaction

$$IO_4^- + 6e + 8H^+ = I^+ + 4H_2O \qquad (1)$$

REAGENTS AND APPARATUS

SOLUTION OF POTASSIUM PERIODATE, $0.01M$. Prepare from pure potassium metaperiodate, and determine its titer iodometrically. The solutions of smaller molarity were prepared by its direct dilution.

Potentiometric determinations were preformed with a platinum indicating electrode and a saturated calomel reference electrode. All given potentials are versus a saturated calomel electrode.

RESULTS

It was found that the reaction of investigated hydrazine derivatives proceeded to the reduction of periodate to iodine monochloride according to eq. 1 if the titration is done in $9N$ hydrochloric acid. Then 3 moles of corresponding derivative reacts with 2 moles of periodate. For phenylhydrazine and semicarbazide, the reaction proceeds quite rapidly, and the values of potential can be read after 30 seconds; for the remaining derivatives, especially those of low concentrations, the reaction is slower and the potential is steady after 2 to 5 minutes. Total volume is about 40 ml. Phenylhydrazine must be dissolved beforehand in 96% ethanol and thiosemicarbazide must be hydrolyzed by hydrochloric acid. The potential of the inflection point lies at about 700 mV for hydrazide of isonicotinic acid, 670 mV for thiosemicarbazide, 680 mV for semicarbazide, and 550 mV for phenylhydrazine. Results for some organic compounds titrated by periodate appear in Table 1.

Oxidation with Cupric Ion—Adapted from S. Siggia and L. J. Lohr

[*Anal. Chem.*, **21,** *1202* (*1949*)]

The procedure described here has a higher selectivity than the foregoing oxidation methods because a product of the reaction, nitrogen, is measured. Many organic compounds can be oxidized, but few liberate nitrogen.

In this determination, the aromatic hydrazines are oxidized to the corresponding diazonium salt.

$$C_6H_5NHNH_2\cdot HCl \xrightarrow{CuSO_4} \left[C_6H_5\overset{+}{N}\equiv N\right]Cl^- \xrightarrow[H_2O]{heat} C_6H_5OH + N_2 + HCl \qquad (2)$$

The diazonium salt is then decomposed and the liberated nitrogen is measured. The apparatus for this determination is the same as that used in the determination of diazonium salts (Fig. 15.1). In the case of the aliphatic hydrazines, the mechanism may be the same, although the diazonium derivatives are not known (very unstable). In any event, nitrogen is liberated by both aromatic and aliphatic types.

PROCEDURE

Weigh about 0.001 mole of hydrazine into the reaction flask and add a boiling chip. Attach the reaction flask to the condenser, using grease to

Table 1. Titration of some Organic Compounds by Periodate

Given, mg.	Found, mg.	Deviation, mg.	Given, mg.	Found, mg.	Deviation, mg.
Hydrazide of isonicotinic acid (2.057)[a]			Cysteine hydrochloride (11.03)		
1.37	1.35	−0.02	1.57	1.49	−0.08
6.85	6.58	−0.27	2.36	2.31	−0.05
13.71	13.47	−0.24	15.76	15.44	−0.32
27.42	26.94	−0.48	47.28	47.99	+0.71
Phenylhydrazine (1.622)[a]			Ascorbic acid (6.164)[a,b] (5.283)[a,c]		
1.37	1.45	−0.08	0.88	0.95	+0.07
4.11	4.13	−0.02	1.76	1.77	+0.01
13.72	13.29	−0.43	17.61	16.90	−0.71
27.44	27.07	−0.37	35.22	35.66	+0.44
Phenyl hydrazine chloride (2.168)[a]			Hydroquinone (3.854)[a,b] (3.303)[a,c]		
2.88	2.60	−0.28	5.50	5.61	+0.11
7.22	6.93	−0.29	11.01	10.70	−0.22
14.45	13.86	−0.59	22.02	22.13	+0.11
28.90	27.74	−1.16			
Semicarbazide hydrochloride (1.671)[a]			Thiourea (6.088)[a]		
5.57	5.68	−0.11	0.38	0.41	+0.03
11.14	11.11	−0.03	3.80	3.95	+0.15
22.28	21.97	−0.31	15.22	15.22	0
			22.83	22.83	0
Thiosemicarbazide (1.367)[a]			Thiosinamine (9.288)[a]		
4.55	4.50	−0.05	0.58	0.65	+0.07
9.11	9.02	−0.09	5.80	6.03	+0.23
18.22	18.03	−0.19	11.61	12.07	+0.46
			23.22	23.22	0

[a] 1 ml. 0.01 KIO_4 corresponds to 1 mg. substance.
[b] Calculated from equation $IO_4^- \rightarrow I_2$.
[c] Calculated from equation $IO_4^- \rightarrow I^+$.

make all the ground-glass joints gas-tight. Through the separatory funnel, add 40 ml of saturated copper sulfate solution, followed by 10 ml of concentrated sulfuric acid. Begin the stream of carbon dioxide and apply heat to the reaction vessel. Bring the contents of the reaction flask to boiling and continue to boil until the reaction is complete, which is indicated when the volume of nitrogen ceases to increase. The gas buret sometimes becomes warm during the reaction because of the hot gas bubbling through it. Allow the solution to cool to room temperature before taking the reading, or else the temperature recorded on the thermometer will be considerably lower than the actual temperature of the gas and the 50% potassium hydroxide. This can cause serious errors. The temperature and the barometric pressure are noted.

As in the procedure for determining diazonium salts, correct the volume for vapor pressure of the potassium hydroxide solution and the blank for the carbon dioxide (see p. 683).

CALCULATIONS

A = volume for nitrogen (corrected for vapour pressure of potassium hydroxide solution and the blank for the carbon dioxide at temperature T)

$$\frac{PA}{T+273} = \frac{A^1\,760}{273}$$

$$\frac{1}{22{,}400} = \frac{x}{A^1} \quad (x = \text{moles of nitrogen collected})$$

$$\frac{x \times \text{mol. wt. of hydrazine} \times 100}{\text{Grams of sample}} = \%\ \text{hydrazine}$$

A^1 = volume calculated to normal temperature and pressure

This method is reproducible to better than ±2%. It was tested by using phenylhydrazine, 2,4-dinitrophenylhydrazine, tolylhydrazine, and unsymmetrical dimethyl hydrazine.

REDUCTION METHOD

See Chapter 13 (pp. 654–7) on reductions using titanous chloride.

TRACE QUANTITIES OF HYDRAZINES

The colorimetric method for determining carbonyl compounds using 2,4-dinitrophenyl hydrazine (pp. 148–52) can be turned around, and hydrazines can be measured by adding a carbonyl compound. The chromatographic adsorption systems mentioned earlier (pp. 151–2) also apply here for concentrating or isolating the resultant hydrazones. The authors of this book have found cinnamaldehyde to be a useful reagent for forming hydrazones from hydrazines. But other aldehydes containing chromophoric groups can be used, such as *p*-dimethylamino benzaldehyde and the nitro benzaldehydes. If water solubility is desired, sulfonated aromatic aldehydes such as benzaldehyde *p*-sulfonic acid can be chosen. Also, salicylaldehyde can be used in such cases.

Hydrazides ($R\overset{\overset{O}{\|}}{C}NHNHR_1^*$)

These compounds can, in many cases, be determined much like hydrazines.

The hydrazides of the preceding structure, where R_1 is a hydrogen atom or an alkyl group, can generally be titrated as bases, although their basicity is much lower than that of the hydrazines. The relationship in basicity between hydrazide and hydrazine is much the same as between amide and amine. The acid-base titrations for amides (pp. 183–208) can generally be applied to hydrazides; however the analyst should be wary of the acetic anhydride solvent for determining hydrazides of the type $R\overset{\overset{O}{\|}}{C}NHNH_2$, since the unsubstituted end of the hydrazine may sometimes acetylate (primary amides do not acetylate at all). The glycol solvents mentioned for amines (p. 546) can often be used, and the glacial acetic acid system for amines (pp. 545–6) should also apply. The symmetrical diacyl hydrazines are generally not basic enough to titrate.

Hydrazones formed by the action of dichlorobenzaldehyde on primary acid hydrazides ($RCONHNH_2$) are acidic and can be titrated in pyridine with tetrabutylammonium hydroxide. Symmetrical secondary hydrazides (RCONHNHCOR) are acidic and can be titrated directly in basic solvents with tetrabutylammonium hydroxide.

Many hydrazides can be determined by the oxidation methods shown for hydrazines. No one has reported the use of iodine or iodate for determining hydrazides, although these methods might apply. The periodate method of Berka and Zyka (pp. 670–71) was applied successfully to semicarbazide, thiosemicarbazide, and the hydrazide of nicotinic acid (see

* R_1 can be another acyl group, an alkyl group; or a hydrogen atom.

Table 1). The cupric sulfate oxidation (pp. 671–73) for hydrazines has also been successfully applied to hydrazides even of the diacyl type.

In some stubborn cases, one can try the acid hydrolysis of the hydrazide to liberate the corresponding hydrazine

$$R\overset{O}{\overset{\|}{C}}NHNHR_1 + H_2O \xrightarrow{H+} RCOOH + R_1NHNH_2$$

The liberated hydrazine might then be determined oxidatively by the above methods.

The titanous reduction method for hydrazines can also be applied to hydrazides with essentially no changes.

FORMATION AND TITRATION OF ACIDIC HYDRAZONES WITH 2,4-DICHLOROBENZALDEHYDE

Adapted from the Method of A. A. Latour, E. J. Kuchar, and S. Siggia

[*Anal. Chem.*, ***36***, *2479* (*1964*)]

Primary hydrazides ($RCONHNH_2$, where R may be an alkyl or aryl group), and 2,4-dichlorobenzaldehyde react to form the hydrazone,

$$RCONHN{=}CH{-}C_6H_3Cl_2 \text{ (2,4-dichlorophenyl)}$$

which is subsequently titrated as an acid in pyridine with tetrabutylammonium hydroxide. 4-Amino-1,2,4-triazole types do not interfere. Symmetrical secondary impurities titrate at the same potential as the acid hydrazones and will cause high primary assay results if no corrections are made. The symmetrical secondary hydrazide content can be determined by a separate titration in aniline, however, and a correction applied to the hydrazone titration.

REAGENTS

TETRABUTYLAMMONIUM HYDROXIDE (TBAOH). Dilute 100 ml of 1*M* methanolic TBAOH (Southwestern Analytical Chemicals, Austin, Texas) to 2 liters with reagent grade benzene. Standardize by titration against benzoic acid in pyridine.

2,4-DICHLOROBENZALDEHYDE (2,4-DCBA)–*o*-CHLOROBENZOIC ACID (*o*-CBA)

SOLUTION. Dilute 22 ±1 grams of 2,4-DCBA (Eastman Kodak Practical Grade) in 100 ml of toluene. Transfer to a separatory funnel and extract the acidic impurity with 50 ml of 2% aqueous potassium carbonate. Use phenolphthalein to indicate the presence of excess carbonate. Wash the toluene phase with three 50-ml portions of distilled water. Filter the toluene phase through cotton in a 60° funnel into a 500-ml volumetric flask. Dissolve 1.6 ±0.1 grams of *o*-CBA (Matheson, Coleman, and Bell) in this solution and dilute to the mark with toluene.

ANILINE SOLUTION. Add 25 ml of Eastman White Label aniline to 25 ml of toluene.

APPARATUS

Precision-Dow Recordomatic Titrator or a pH meter with a glass-calomel electrode system. The calomel electrode was of the fiber type and was filled with a 0.1*M* methanolic tetrabutylammonium bromide solution. A sleeve-type calomel electrode was also used, in which the electrolyte was potassium chloride in methanol.

PROCEDURE

DETERMINATION OF PRIMARY AND SECONDARY SYMMETRICAL HYDRAZIDES. Weigh between 0.6 and 1.0 meq. of sample in a 150-ml beaker. Add 25 ml of toluene from a graduate and add by pipet 10.0 ml of 2,4-DCBA–*o*-CBA solution; cover with a watch glass and boil gently on low heat to a volume of about 10 to 15 ml. Then pipet 1 ml of 50% aniline solution, add 10 ml of toluene by graduate, and boil to about 10 to 15 ml. Cool, add 50 ml of pyridine, and titrate with the 0.05*N* TBAOH using the automatic titrator. The first potentiometric inflection is due to *o*-chlorobenzoic acid. This inflection occurs around 400 to 500 mV. The last inflection is due to hydrazone and any secondary hydrazide that may be present. This end point occurs around 850 mV. Take the blank through the same procedure, omitting only the sample.

CALCULATION

Milliequivalents per gram, primary + secondary hydrazide

$$= \frac{\text{Sample meq. minus blank meq.}}{\text{Gram of sample}}$$

DETERMINATION OF SYMMETRICAL SECONDARY HYDRAZIDES. Weigh between 1 and 2 grams of sample into a 250-ml beaker. Add 100 ml of aniline and dissolve. Heat (low heat and in a hood), if necessary to dissolve, taking care to prevent carbon dioxide absorption. If the sample is not readily soluble in aniline, a solvent mixture of 1:1 isopropanol-toluene can be used to dissolve the sample and the aniline then added. Titrate using the automatic titrator with 0.05*N* TBAOH. Any secondary hydrazide should titrate between 400 and 700 mV. Titrate a blank of the reagents along with the sample.

CALCULATIONS

Milliequivalents per gram of secondary hydrazide

$$= \frac{\text{Sample meq. minus blank meq.}}{\text{Grams of sample}}$$

% Primary hydrazide =

$$[(\text{meq./gram of primary} + \text{secondary}) \text{ minus meq. of secondary}] \times \text{meq. wt.} \times 100$$

DISCUSSION AND RESULTS

The procedure for secondary hydrazide content was tested on synthetic mixtures of primary and secondary stearic hydrazide and benzhydrazide. These data are presented in Table 2, along with recovery data obtained

Table 2. Analysis of Synthetic Mixtures of Primary and Secondary Hydrazides

Hydrazide	Added, %	Found, %
Distearic hydrazide	11.9	11.4
	16.7	16.8
	12.3	12.6
	10.5	10.8
Stearic hydrazide	79.9	79.1
	80.0	79.5
	79.1	79.2
	80.0	80.2
Dibenzhydrazide	10.4	10.4
	14.2	14.1
	19.6	19.5
	23.1	23.1
Benzhydrazide	82.4	82.5
	72.6	73.0

for the primary hydrazide in the presence of the secondary. Pyridine and aniline were used as the solvents in the titration of distearic hydrazide in the presence of primary stearic hydrazide; however aniline was required as the solvent for the other hydrazides tested because high and erratic results for the secondary hydrazide were obtained on synthetic mixtures when pyridine was used.

The data in Table 3 were obtained with the outlined procedure and have been corrected for any symmetrical hydrazide. Solubility problems arose with several compounds. The phthalic hydrazides are not soluble in hot toluene, but they are soluble in a solution of hot isopropanol-toluene. Isopropanol can be added to dissolve the sample without adverse effect. Carbohydrazide and thiocarbohydrazide were dissolved in a minimum of water (about 25 ml), isopropanol was added, and then toluene. The mixture was then heated to boiling, and additional toluene was added until there was no evidence of water, as shown by the absence of two phases. Prolonged heating (up to 30 minutes) at the temperature of boiling toluene does not affect the hydrazone. Although the hydrazino groups of carbohydrazide and the aldehyde react, the resulting hydrazone titrates as a monofunctional compound, that is, 1 equivalent per mole. The primary dihydrazides tested titrated as difunctional compounds, with

Table 3. Primary Acid Hydrazide Content of Aliphatic and Aromatic Hydrazides

Compound	Primary Hydrazide,[a] %
Acetic hydrazide	100.1
Benzhydrazide	100.0
Cyclopropanecarboxylic hydrazide	99.2
Carbohydrazide	100.5
2-Hydrazinobenzothiazole	97.6
Isonicotinic acid hydrazide	94.0
Isophthalic dihydrazide	100.0
Lauric hydrazide	100.8
Monomethylterephthalic hydrazide	88.4
m-Nitrobenzhydrazide	100.5
Propionic hydrazide	87.4
Salicylhydrazide	99.5
Stearic hydrazide	99.3
Terephthalic hydrazide	99.9
Thiocarbohydrazide	97.9

[a] Average of at least three determinations.

two inflections observed. As expected, the aromatic hydrazides form stronger acid hydrazones than do the aliphatic ones. In fact, the aromatic primary hydrazides are titratable as is, in pyridine; however side reactions occur and prevent direct determination in this way.

The standard deviation for this method is 0.69%; it was obtained by analyzing recrystallized benzhydrazide.

15

Diazonium Salts

Diazonium compounds can be determined in several ways. These compounds will couple with phenols under alkaline conditions (see pp. 62–66 for equation). In addition, diazonium compounds decompose readily, yielding stoichiometric quantities of nitrogen that can be collected and measured volumetrically. Diazo compounds also can be quantitatively reduced with titanous chloride.

Decomposition followed by nitrogen collection is the most generally applicable method; it applies to a very wide range of diazonium compounds. The authors have never found a diazo compound that was not determinable by this method. The decomposition method also has a high selectivity, since few compounds decompose under the conditions of the decomposition reaction to liberate a gas. The accuracy and precision of this method are also generally the highest of the three approaches described, being in the order of ±1%.

The decomposition method also has an advantage in the low ranges of concentrations of diazos, since it is a gasometric method. A small quantity of diazo will yield a fairly large volume of nitrogen (1 mM of diazo yields 22.4 ml of nitrogen).

The coupling method also enjoys a high selectivity because few other materials couple with phenols or active methylene compounds (see pp. 62–66 for equation). The coupling methods are not the simplest to carry out, however, and experience is needed to accurately ascertain the end point. Because of the end point observation problem, the accuracy and precision of the coupling method are generally lower than for the decomposition method. The accuracy and precision values vary with the specific diazo, since the rates of coupling vary; this variation determines the accuracy of end point detection. The faster-reacting diazos yield the sharpest end points.

The titanous chloride reduction method is the least specific, since this reagent reduces almost all nitrogen functional groups except amines and amides. If a laboratory is equipped for titanous chloride titrations, however, and has systems to analyze for diazonium compounds to which the method can be applied, this titration is quite rapid.

Traces of diazonium salts are easily determined because the coupling products with phenols and naphthols are highly colored and are readily amenable to colorimetric measurement.

Nitrogen Evolution Measurement

$$\left[C_6H_5\overset{+}{N}{\equiv}N \right] Cl^- \xrightarrow[H_2O]{CuCl+HCl} C_6H_5OH + N_2 + HCl$$

APPARATUS

In the apparatus illustrated in Fig. 15.1, *A* is a 25-ml separatory funnel; *B*, a 100-ml nitrometer containing 50% potassium hydroxide; *C*, a 10-mm tubing; *D*, a 2-mm capillary; *G*, a 5-in. condenser; *H*, a bubbler; *I*, a thermometer; and *J*, a 150-ml flask.

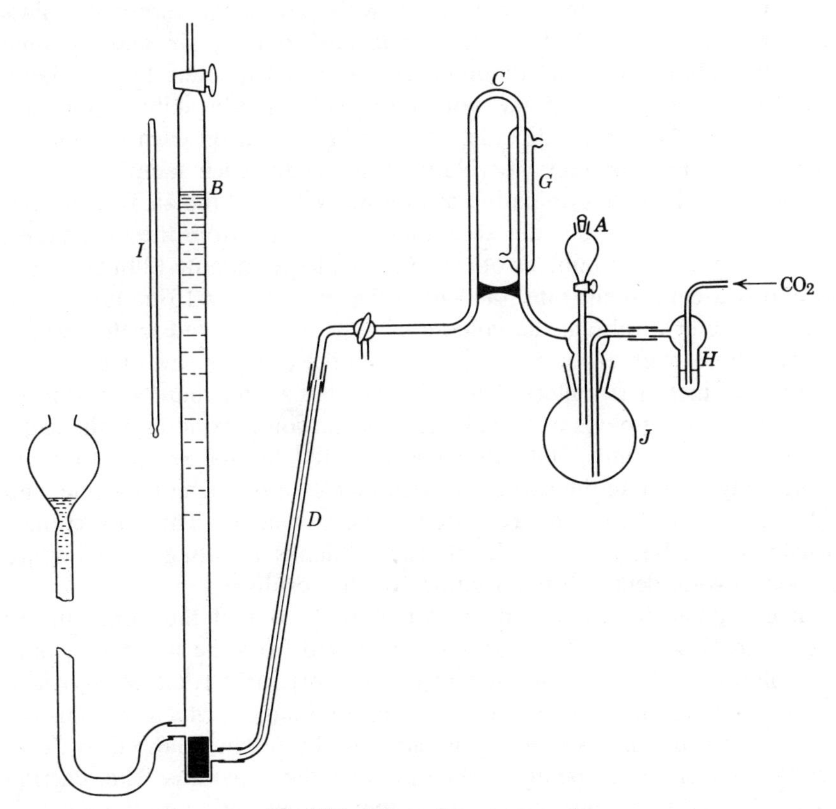

Fig. 15.1. Diazo nitrogen apparatus.

REAGENTS

Cuprous chloride.
Concentrated hydrochloric acid.

PROCEDURE

In about 15 ml of concentrated hydrochloric acid, dissolve 3 grams of cuprous chloride and place the resultant solution in the separatory funnel *A*. Weigh a sample containing about 0.001 mole of diazonium salt into the reaction flask. Attach the flask to the condenser, all joints being greased to prevent leakage. The nitrometer *B* contains potassium hydroxide solution (71.5 grams of potassium hydroxide per 100 ml of water) and a layer of mercury at the bottom. The level of the mercury extends about one-half inch above the entrance tube. The function of the mercury is to prevent clogging of the capillary *D* with potassium carbonate. Pass carbon dioxide through the apparatus to flush out the air, and continue until the bubbles reach a minimum size in the nitrometer. These minute bubbles are indicative of the potassium hydroxide-insoluble impurities contained in the carbon dioxide. Make a blank run on each cylinder of carbon dioxide to correct the volume readings for each sample.

Make the blank determination as follows. When all the air is swept out of the apparatus, remove the accumulated inert gas from the nitrometer, count the number of bubbles of carbon dioxide per minute (2 bubbles per second is a convenient rate), and note the time. Every 15 minutes, read the nitrometer to obtain the volume of inert gas collected in the period. This is done over a period of 2 to 3 hours. Take the average for the volume of inert gas collected per 15 minutes at that particular rate of flow. When a sample is analyzed, feed the carbon dioxide in at the same rate as for the blank. Note the time required for the sample to react completely, compute the carbon dioxide blank for that length of time, and subtract it from the volume recorded in the nitrometer. This correction is usually small, but it is significant. Large blanks resulting from impure carbon dioxide decrease the accuracy of the results.

In an actual determination the air is swept out of the apparatus as previously described. The inert gas is removed from the nitrometer, and the solution of cuprous chloride in hydrochloric acid is allowed to flow from the separatory funnel *A* into the reaction flask *J*, which contains the sample, and the time is noted. The cuprous chloride is followed by 50 to 75 ml of water, no air being allowed to enter the apparatus. The reaction flask is heated to boiling and is kept at that temperature until the reaction is complete, as indicated by the bubbles in the nitrometer reaching their minimum size. The heat is shut off, and the apparatus is allowed to stand for 5 to 10 minutes to reach temperature equilibrium. The carbon dioxide flow is continued throughout this period of standing. The volume of gas collected is read, the leveling bulb on the nitrometer being used to set the pressure of the gas in the nitrometer equal to atmospheric pressure. The temperature is read on thermometer *I*, and the barometric pressure and the time are noted. The volume of gas collected is corrected for the blank determination made on the carbon dioxide, the time required for the reaction being taken into account. The volume is also corrected for the vapor pressure of the potassium hydroxide solution using values from Table 1.

Table 1. Vapor Pressure of Potassium Hydroxide Solutions Containing 71.5 grams of Potassium Hydroxide per 100 grams of Water (International Critical Tables)

T°C	mm. Hg
15	4.1
20	5.6
25	7.4
30	9.6
35	12.7

CALCULATIONS

A = volume of gas, corrected for carbon dioxide blank

P = atmospheric pressure minus vapor pressure of the 50% potassium hydroxide solution

T = temperature recorded at the end of analysis

$$\frac{AP}{T+273} = \frac{A_1 760}{273}$$

A_1 = volume calculated to normal temperature and pressure

$$\frac{1}{22{,}400} = \frac{x}{A_1} \quad (x = \text{moles of nitrogen collected})$$

$$\frac{x \times \text{mol. wt. of diazonium compound} \times 100}{\text{Grams of sample}} = \%\ \text{diazonium compound}$$

The determination is accurate and precise to better than ±1%. The apparatus is very simple to operate and to clean. The time required for each analysis varies, depending on the stability of the particular diazonium salt being determined. However samples requiring more than an hour are rare. The compounds that follow were used to test this procedure: benzenediazonium chloride; and the following zinc chloride salts: *p*-dimethylaminobenzenediazonium salt, *p*-diphenylamine-diazonium salt, and *p*-diethylaminobenzenediazonium salt.

This method is suitable for determining hydrazines, which are first oxidized to diazonium salts, then determined as described previously (pp. 671–673).

Coupling Method

$$\left[C_6H_5\overset{+}{N}\equiv N\right]Cl^- + H_2C\begin{matrix} C(=O)-N(C_6H_5) \\ C(CH_3)=N \end{matrix} \rightarrow C_6H_5N{=}N{-}CH\begin{matrix} C(=O)-N(C_6H_5) \\ C(CH_3)=N \end{matrix} + HCl$$

$$\left[C_6H_5\overset{+}{N}\equiv N\right]Cl^- + C_6H_3(CH_3)(NH_2)_2 \rightarrow C_6H_5N{=}N{-}C_6H_2(CH_3)(NH_2)_2 + HCl$$

In this method, the procedure is identical with that used to standardize the diazonium solutions in the section on coupling procedures (pp. 62–66). In that section an aliquot of standard coupling agent is titrated with the unknown diazonium compound as a means of standardizing the diazonium solution. In this case the diazonium standard is known, and the ratio of coupler solution to diazonium solution is also known. An aliquot containing excess coupling agent is added to a water solution containing the weighed amount of unknown diazonium salt. The required amount of acetic acid–sodium acetate buffer is added. The excess coupler is than titrated with the standard diazonium solution. Full details can be obtained in Chapter 1.

Reduction Method Using Titanous Chloride

This method was described in detail earlier (pp. 654–657). The reduction generally proceeds past the hydrazine stage to the corresponding amines. Hence one mole of diazo generally consumes 6 equivalents of titanous ion.

Determination of Traces of Diazonium Compounds

Traces of diazonium compounds can be readily detected by coupling with phenols or naphthols. The approach is very sensitive, often extending to the tenths of 1 ppm range, and since the coupling reaction is very specific, the test is quite selective.

The method shown below is based on the chemistry of the coupling reaction and the solubility of the resultant dyes. It was used for many years in the laboratory of S. Siggia to determine diazonium compounds on paper for duplicating processes utilizing diazo salts. The method is written to cover a range of possible situations and to deal with as wide a range of compounds as possible.

REAGENTS

COUPLING REAGENTS

Aqueous R-salt solution (2-naphthol-3,6 disodium sulfonate), 1%.
Naphthol in acetone, 1%.
Aqueous phloroglucinol solution (1,3,5, trihydroxybenzene), 0.5%.

ALKALINE AGENTS

Aqueous sodium bicarbonate, 5%.
Aqueous sodium carbonate, 5%.
Aqueous or alcoholic sodium hydroxide, 5%.
Aqueous or alcoholic sodium acetate, 5%.

PROCEDURE

The diazonium compounds are generally water soluble, hence aqueous systems can be used. Occasionally, however, alcohol has to be used to dissolve the sample containing the diazo. Sometimes a diazo may be found that is only alcohol soluble. Other solvents might be used, but usually water or alcohol must be added because the alkaline agents shown previously are soluble either only in water or in both water and alcohol.

AQUEOUS SYSTEMS

In aqueous systems R-salt can be used as coupler. An adequate amount of coupling agent is added to react with all the diazo in the sample. Then an adequate amount of sodium bicarbonate or sodium acetate solution is added to fully develop the color. Color development is usually quite rapid, and 5 to 10 minutes' standing at room temperature is usually sufficient for complete reaction. The resultant dyes are generally water soluble; hence no precipitates form. The resultant colors are generally red and can be measured optically or spectrophotometrically in the standard manner.

Some diazo compounds couple rather slowly at the pH values resulting from sodium bicarbonate or sodium acetate (pH 7–8.5). In these cases, sodium carbonate can be used (pH 7–10). Sodium hydroxide results in still faster reactions, since it can yield a pH above 10. It should only be used as a last resort, however, since sodium hydroxide causes decomposition of the diazo to the corresponding phenol, liberating nitrogen. The decomposition is generally, but not always, slow enough to permit coupling before any significant amount of diazo is lost. Sodium carbonate can also cause some decomposition, but the rate is much slower than with caustic. Decomposition with sodium bicarbonate or acetate is very slow. Hence coupling rate has to be evaluated versus decomposition rate of the diazonium salt for the particular system at hand.

For aqueous systems, the R-salt couple is desired because of its solubility. But R-salt is not the most desirable from the standpoint of sensitivity. The phloroglucinol and naphthol generally produce deeper colors, hence are the best selections for detecting the smaller quantities of diazonium compounds. Both the latter couplers generally result in water-insoluble dyes. However the coupling reaction can be carried out as described earlier for R-salt, and the resultant dye can be extracted into benzene, petroleum ether, or carbon tetrachloride with the colorimetric determination made on the organic layer. Some of the naphthol or phloroglucinol dyes can be dissolved by adding sodium hydroxide to the system (after coupling to prevent decomposition of the diazo). The caustic forms the corresponding salt of the phenolic groups on the dye and renders them water soluble. Colorimetric analysis can then be made directly on the aqueous solution.

ALCOHOLIC SYSTEMS

In alcoholic systems, the sample is dissolved in alcohol, and either phloroglucinol or naphthol is used as coupler. Only alcoholic sodium acetate or alcoholic sodium hydroxide can be used as the alkaline agent. In general, the resultant dyes remain dissolved unless large amounts of diazo are present, in which case the dye may precipitate out of solution. The colors are generally red, purple, or brown and can readily be measured by optical or spectrophotometric means.

Mixtures of Diazonium Compounds

Mixtures of diazonium compounds can be analyzed utilizing the differences in their rates of decomposition. The sample mixture is decomposed as indicated in the nitrogen evolution method above. The rate of nitrogen evolution is measured. The results are plotted and evaluated as shown in Chapter 25 (pp. 848–50). Full details and experimental results on mixtures of known diazonium salts are also given in Chapter 25 (pp. 847–50).

16

Quaternary Ammonium Compounds

Hydroxides

Quaternary ammonium hydroxides are very strong bases, of the same order of basicity as sodium or potassium hydroxide. Hence these compounds are very readily titrated with standard acid by using indicators. The common contaminates in these hydroxides are the tertiary amines used to synthesize either the quaternary or the tertiary amine decomposition products. These amines are less basic than the hydroxide and can be titrated separately. In most solvents, two distinct end points are obtained, one for the quaternary and one for the tertiary base (see p. 534 for a discussion of resolution of basic compounds when titrating in the various available solvents).

Salts

As for any strong base, the quaternary ammonium salts of weak acids such as acetic acid are weakly basic and can be titrated as bases in the systems shown earlier (pp. 534–58) for determining carboxylic acid salts of the alkali metals.

The quaternary salts of the mineral acids are neutral and cannot be titrated as either acids or bases. However, many quaternary salts contain a fairly hydrophobic cation, for example, cetyl pyridinium bromide, tetrabutyl ammonium chloride, or cetyl trimethyl ammonium chloride. These can be determined titrimetrically using alkyl or aryl sulfates or sulfonates. This procedure is described in the section dealing with sulfonates (pp. 803–6). For sulfonates, a quaternary is used as titrant. To determine unknown quaternary salts, a known alkyl or aryl sulfate or sulfonate is used. Since there are no such sulfates or sulfonates that are primary standards, a practical form of these salts must be used and standardized by the Hyamine method (pp. 803–6). Two such compounds that perform well for determining quaternaries are nonyl phenoxy polyethoxy ammonium sulfate (mol. wt. 504, Alipal CO 436) and oleyl methyl tauride (Igepon T-77).*

* Both these compounds are obtainable from Antara Chemicals, 435 Hudson St., New York, N.Y.

This organic sulfate or sulfonate titration is quite selective for quaternaries but is limited to quaternary salts with a fairly hydrophobic cation, as indicated earlier. Also, the titration end points (p. 806) are not always as sharp as is desirable.

Quaternary halides can also be measured by reacting them with mercuric acetate. The mercuric ion complexes the halide ion, forming the quaternary acetate, which is basic and can be titrated in nonaqueous media. The mercuric acetate method has sharp end points and consequently a high accuracy and precision (generally better than ±1%). The method is not as specific as the preceding method, however; any halide salt that yields a basic acetate salt will interfere (i.e., sodium and potassium halides are interferences).

Titration Method Using Organic Sulfates or Sulfonates

This procedure is the same as that given in Chapter 22 (pp. 803–6) except that the quaternary salt is the unknown and the sulfate or sulfonate is the known. The details were outlined earlier.

Mercuric Acetate Method—Adapted from C. W. Pifer and E. G. Wollish

[Anal. Chem., ***24***, *300–6* (*1952*)]

REAGENTS

PERCHLORIC ACID IN DIOXANE OR GLACIAL ACETIC ACID, 0.1*N*. Dissolve 8.4 ml of 70 to 72% perchloric acid in 1 liter of solvent. Both the dioxane and the acetic acid solution can be standardized with potassium acid phthalate in glacial acetic acid solution.

Glacial acetic acid, CP.

MERCURIC ACETATE REAGENT. Dissolve 6 grams of CP mercuric acetate in 100 ml of hot glacial acid, and cool the solution.

PROCEDURE

Dissolve a sample containing about 0.003 equivalent of quaternary salt in about 150 ml of glacial acetic acid. If the quaternary is of the weakly basic variety (not a halide), titrate it directly with the standard perchloric acid solution. If the quaternary salt is a halide, add mercuric acetate solution so that no more than 2 moles of mercuric acetate is present per mole of quaternary; yet enough should be present to complete conversion of the quaternary halide to the acetate (1 mole of mercuric acetate can yield 2 quaternary acetates). Usually 10 ml of mercuric acetate solution is sufficient for samples containing up to 0.003 equivalent of quaternary. Then titrate the resultant solution.

Potentiometric titration using the glass-calomel electrode is the preferred method for end point detection. Crystal violet indicator has been reported as being a usable indicator (0.5 ml of a 0.1% solution of crystal violet in glacial acetic acid is used to indicate end point). However color matching of sample versus primary standard is often necessary to adjust the shade of sample end point to the shade of the primary standard end point.

Any free base (amine) in the quaternary halide sample should be determined by titrating a separate sample without addition of mercuric acetate. Also, any amine hydrohalide should be determined on a separate sample by titrating a separate sample either in glycol-isopropanol (1:1) or in acetone using $0.1N$ sodium hydroxide in methanol. The quaternary value as obtained after the mercuric acetate reaction should be corrected for these materials. Amines interfere because they are basic and consume perchloric acid, and their hydrohalides interfere because the mercuric acetate complexes the halides and forms the amine acetates, which usually consume perchloric acid.

CALCULATION

$$\frac{\text{Milliliters of perchloric acid} \times N \text{ acid} \times \text{mol. wt. of quaternary} \times 100}{\text{Grams of sample} \times 1000} = \%\ \text{quaternary salt}$$

Pifer and Wollish used this method on the following quaternary salts: choline chloride, neostigmine bromide, (3-hydroxyphenyl)-ethyl dimethyl ammonium chloride, cetylpyridinium chloride, tetraethyl ammonium bromide, (3-hydroxyphenyl)-ethyl dimethyl ammonium bromide, dimethyl carbamate of (2-hydroxy-3-cyclohexylbenzyl)-methyl piperidinium bromide, choline dihydrogen citrate, and choline dihydrogen tartrate. The authors of this book have also tried choline chloride, tetramethyl ammonium bromide, and tetrabutylammonium iodide.

Trace Quantities of Quarternary Ammonium Salts

Quaternary salts in dilute systems can be determined by the addition of a sulfonated dye. The common sulfonated indicators are usually used, and these react to form a colored compound that is soluble in hydrocarbon. The intensity of the color in the hydrocarbon layer is a measure of the quaternary. The principle of this method is the same as that used in the foregoing titration involving organic sulfates and sulfonates, namely, that the salt of a quaternary ion and a sulfonated organic compound is soluble in hydrocarbon solvents, provided both the quaternary ion and the sulfonate ion contain sufficient hydrophobic groups. In this case the sulfonate ion is a dye; hence the quaternary results in color being pulled into the hydrocarbon layer in proportion to the amount of quaternary present.

Method of M. E. Auerbach

[*Adapted from Ind. Eng. Chem., Anal. Ed.,* ***15****, 492–3 (1943); Ibid.,* ***16****, 739 (1944), Reprinted in Part*]

The quaternary ammonium salts named below form colored salts with dibromothymolsulfonephthalein (bromothymol blue) and with tetrabromophenolsulfonephthalein (bromophenol blue). These salts are readily extracted from alkaline aqueous solutions by a number of organic solvents, particularly the hydrocarbon solvents. Bromophenol blue, which Auerbach prefers to use, is itself insoluble in benzene, for example, either in its acid form or as the sodium salt; nor does it form salts extractable from alkaline solution with any of a large number of primary, secondary, and tertiary amines at the investigator's disposal, or with alkaloids. It

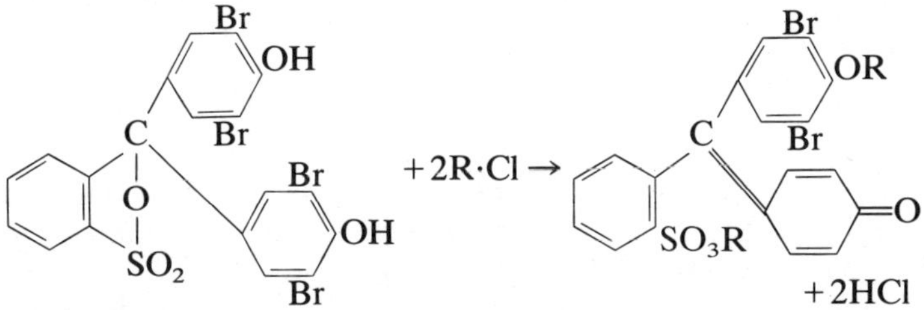

Br, OH, C, Br, O, SO$_2$, Br, OH, Br $+ 2R{\cdot}Cl \rightarrow$ Br, OR, C, Br, SO_3R, =O $+ 2HCl$

R = quaternary ammonium cation

seems probable that all cations of the composition $[R_1R_2R_3R_4N]^+$ where R_1, R_2, and R_3 are CH_3 or longer-chained alkyls and R_4 is (benzene ring) CH_2, C_4H_9, or longer-chained aryl-alkyl, will form salts with bromophenol blue that can be extracted from alkaline solution. It is not claimed that only quaternary amines respond to this test, but it is true that of 50 or 60 nonquaternary amines, all gave negative tests. These negative compounds include such typical amines as ethylamine, dimethylamine, diethanolamine, diethylaminoethanol, aniline, phenylenediamine, and ethylenediamine. It is especially significant that lauryl dimethylamine, which can be considered a possible impurity in alkyl benzyl dimethyl ammonium chloride, gives a negative test.

The method consists of forming the quaternary ammonium–dye salt in carbonate solution, extracting the color with benzene, and measuring in a photoelectric colorimeter the intensity of color so extracted. By means of a factor previously derived from a standard solution, the concentration of quaternary ammonium salt in the sample is readily calculated.

To illustrate the method under discussion, consider the assay of a commercial tinted alcohol-acetone solution containing 1 to 1000 ppm alkyl benzyl dimethylammonium chloride (referred to as Z):

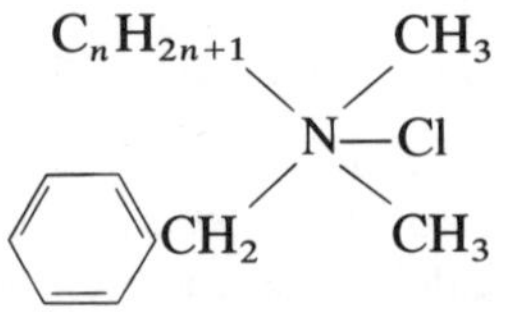

$$\begin{array}{ccc} C_nH_{2n+1} & & CH_3 \\ & \diagdown\ \diagup & \\ & N{-}Cl & \\ & \diagup\ \diagdown & \\ C_6H_5CH_2 & & CH_3 \end{array}$$

in which C_nH_{2n+1} represents a mixture of alkyl groups ranging from C_8H_{17} to $C_{18}H_{37}$. In spite of the apparently indeterminate constitution of this substance, the average molecular weight, because of rigidly standardized manufacturing conditions, is remarkably constant, being about 357.5. The molecular weight was determined by preparing the ferricyanide salt $R_3Fe(CN)_6$ (R = quaternary ammonium cation), igniting a portion of the salt to Fe_2O_3, and calculating back to the value of R.

PROCEDURE

In a 125-ml Squibb separatory funnel, take 50 ml of water containing 50 to 75 μg of the quaternary compound. Ordinary stopcock grease should be avoided. Starch-glycerol lubricant is satisfactory. Add 5 ml of 10% sodium carbonate solution, 1 ml of aqueous 0.04% bromophenol blue indicator solution, and exactly 10 ml of benzene. (The indicator solution should be prepared on the day it is to be used. Dissolve 40 mg of bromophenol blue powder in 100 ml of water containing 1 ml of 0.01 *N* sodium hydroxide.) Shake steadily for 2.5 to 3 minutes, let the layers separate roughly (20–30 seconds), and then swirl the funnel contents. Let stand for several minutes or until well separated. Rinse a 15-ml centrifuge tube with a portion of the lower aqueous layer, discard this layer entirely, and then run the colored benzene layer into the tube. Stopper the tube with a clean rubber diaphragm stopper and centrifuge for a few minutes at about 1000 rpm, if necessary to clarify. Transfer to a dry Klett-Summerson colorimeter tube, and read, using filter No. 60.

Under the specified test conditions (more precisely, in alkaline medium), the dye salts of alkaloids are not extractable. However these salts become extractable from acid solutions, and useful methods can be developed for the determination of minute amounts of various alkaloids in solution. In fact, such a method has already been found useful for the determination in urine of an alkaloidlike synthetic drug, the ethyl ester of methyl phenyl piperidine carboxylic acid (1).

Very small amounts of strychnine have been determined in a complex solution by extracting it as the yellow bromothymol blue salt in the presence of acetate buffer pH 5.0, using toluene as solvent. A somewhat similar method was described many years ago for quinine, using eosin as the dye anion (2, 3). This method is reliable to ±2 to 5%.

Fogh, Rasmussen, and Shadhauge (4) used a method similar to Auerbach's method and determined cetyl pyridinium chloride using bromocresol purple as the color reagent. They found that the quaternary adsorbed on glass and in the foam (if foam existed) constituted difficulties, which were minimized by using cuvettes coated with polymethylmethacrylate polymer.

1. R. A. Lehman and M. S. Aitken, *J. Lab. Clin. Med.*, **28,** 787 (1943).
2. R. O. Prudhomme, *Bull. Soc. Path. Exot.*, **31,** 929 (1938).
3. R. O. Prudhomme, *J. Pharm. Chim.*, **1,** 8 (1940).
4. J. Fogh, P. O. H. Rasmussen, and K. Shadhauge, *Anal. Chem.*, **26,** 392–5 (1954).

17

Isocyanates and Isothiocyanates

$$RN{=}C{=}O \quad \text{and} \quad RN{=}C{=}S$$

Both the isocyanates and the isothiocyanates compounds react readily with primary and secondary amines to yield the corresponding substituted ureas or thioureas. The equations for the reaction with primary amines are as follows:

$$RN{=}C{=}O + R_1NH_2 \rightarrow RNH\overset{\overset{\displaystyle O}{\|}}{C}NHR_1$$

$$RN{=}C{=}S + R_1NH_2 \rightarrow RNH\overset{\overset{\displaystyle S}{\|}}{C}NHR_1$$

The secondary amines react in the corresponding manner.

The reaction rate with both types of amines is very rapid, but in general the reaction with the primary amine is the more rapid. Rate becomes of significance, however, only with hindered isocyanates or isothiocyanates.

Primary Amine Method

Butylamine Method—Adapted from S. Siggia and J. G. Hanna

[*Anal. Chem.*, **20,** *1084* (*1948*)]

REAGENTS

n-BUTYLAMINE SOLUTION. Dilute 12.5 grams of *n*-butylamine to 500 ml with purified dioxane. To purify the dioxane allow it to stand over sodium hydroxide pellets and change the pellets each day until they no longer become brown. Or the dioxane can be purified more rapidly by distillation from sodium dispersion or after adding lithium aluminum hydride and then distilling. Store the dioxane over sodium hydroxide. The dioxane must be free of water, alcohol, and peroxide, to avoid the reaction of these compounds with the isocyanate or isothiocyanate. Also, the dioxane must be free of aldehydes, which will react with the butylamine reagent. The foregoing purification methods are quite adequate.

Standard sulfuric acid or standard hydrochloric acid, $0.1N$.

Methyl red indicator.

PROCEDURE

To a 100- or 150-ml glass-stoppered Erlenmeyer flask, charge 20 ml of *n*-butylamine reagent. Add the sample, which should contain about 0.002 mole of the isocyanate or isothiocyanate, to the reagent. Stopper the flask and allow it to stand at room temperature for 15 minutes. A very few samples may require heating for a few minutes to complete the addition of the *n*-butylamine to the isocyanate or isothiocyanate. All heating should be done under reflux to prevent loss of *n*-butylamine. Rinse the condenser with water. Titrate the contents of the flask with 0.1*N* acid, using methyl red as an indicator.

Run a blank on 20 ml of the reagent to determine its strength.

CALCULATIONS

Milliliters of acid for blank minus milliliters acid for sample $= A$

$$\frac{A \times N\,H_2SO_4 \times \text{mol. wt. of compound} \times 100}{\text{Grams of sample} \times 1000} = \%\ \text{compound}$$

The following compounds were used to test the foregoing procedure: phenyl and 1-naphthyl isocyanates, and methyl, ethyl, phenyl isothiocyanates and toluene diisocyanate (2,4- and 2,6-isomers).

Interferences in this procedure consist of any acidic or basic impurities that may be included in the sample. These impurities should be determined by titration before the analysis for the isocyanate or isothiocyanate is made; the final analysis should then be corrected for these impurities.

Thiocyanates will interfere, since these compounds will isomerize to the isothiocyanate on heating.

Representative results for pure compounds are summarized in Table 1.

Table 1. Determinations of Isocyanates and Isothiocyanates in Dioxane

Compound	Found,[a] %
α-Naphthyl isocyanate	98.6 ±0.2
Phenyl isocyanate	98.8 ±0.2
Phenyl isothiocyanate	100.2 ±0.5
Ethyl isothiocyanate[b]	98.6 ±0.1
Methyl isothiocyanate[b]	98.3 ±0.2

[a] Average value with standard deviation.
[b] After 45 minutes at room temperature.

Because dipolar solvents accelerate the rate of nucleophilic reactions, dimethylformamide was used as a solvent to shorten the time of the *n*-butylamine–organic isocyanate and isothiocyanate reactions.

Adapted from the Method of J. A. Vinson

[*Anal. Chem.*, ***41,*** *1661* (*1969*)]

REAGENTS

DIMETHYLFORMAMIDE, REAGENT GRADE. Store over molecular sieves.
n-BUTYLAMINE, COMMERCIAL GRADE. Distill from barium oxide and store over nitrogen in an amber bottle. Add 3 grams of the amine to a 100-ml volumetric flask and dilute to volume with dimethylformamide.

PROCEDURE

Add 5 ml of the *n*-butylamine solution to a 125-ml Erlenmeyer flask containing a Teflon stir bar and add 10 ml of dimethylformamide. Weigh approximately 1 mM of isocyanate or isothiocyanate in a glass micro weighing bottle fitted with a glass stopper. Drop the bottle into the flask so that the stopper separates from the bottle. Cork the flask and stir for a few seconds until the contents are mixed. Allow to stand 5 minutes at room temperature for aromatics and for 10 minutes for aliphatics. Add 50 ml of distilled water and cool the flask to room temperature with tap water or an ice bath. Add a few drops of methyl red indicator solution and titrate the excess amine to the pink end point with standard 0.1*N* hydrochloric acid.

DISCUSSION

Although dimethylsulfoxide was found to accelerate the reaction, the results obtained were high. Results for the determinations of isocyanates and isothiocyanates in dimethylformamide are found in Table 2.

The dimethylformamide method gave accuracy and precision equal to the ASTM method, (p. 697), with a significant reduction in reaction time; 5 minutes for the dimethylformamide method versus 1 hour for the ASTM method. The comparison results appear in Table 3.

Table 2. Determination of Isocyanates and Isothiocyanates in Dimethylformamide Solvent

Compound	Reaction Time, Min.	Reaction,[a] %
Allyl isothiocyanate	5	99.6 ±0.6
n-Butyl isocyanate	10	98.8 ±0.2
n-Butyl isothiocyanate	10	98.3 ±0.4
Phenyl isocyanate	5	99.1 ±0.1
Phenyl isothiocyanate	1	101.4 ±0.1
Naphthyl isocyanate	1	99.3 ±0.2
Naphthyl isothiocyanate	2	99.8 ±0.3
Tolylene-2,4-diisocyanate	5	98.7 ±0.4

[a] Average value with standard deviation.

Table 3. Determination of Polymeric Polyisocyanates

	% NCO	
Sample	ASTM	DMF[a]
A	30.7 ±0.1	30.9 ±0.1
B	31.2 ±0.1	31.1 ±0.0
C	30.8 ±0.0	31.1 ±0.1

[a] Reaction time, 5 minutes at room temperature.

Secondary Amine Methods

Dibutylamine Method—Adapted from American Society for Testing and Materials

[*Reprinted in Part from the ASTM Tests for Urethane Foam Raw Materials (D1638)–59T (1959), p. 3*]

The dibutylamine method was developed for determination of toluene diisocyanates in their various isomeric forms.

The foregoing primary amine methods have been tested on a wider range of isocyanates and isothiocyanates than has the secondary amine method, although the latter method may also apply.

REAGENTS

BROMOCRESOL GREEN INDICATOR SOLUTION. Triturate 0.100 gram of bromocresol green, indicator-grade powder, with 1.5 ml of 0.1*N* sodium hydroxide solution until solution of the bromocresol green is complete. Dilute to 100 ml with water.
DIBUTYLAMINE SOLUTION, 260 GRAMS PER LITER. Dilute 260 grams of dry dibutylamine to 1 liter with dry toluene.
HYDROCHLORIC ACID, 1*N*. Prepare 1*N* hydrochloric acid and standardize in any suitable manner, frequently and accurately enough so that the maximum error of the normality factor shall be not more than 0.001.
Isopropyl alcohol.
TOLUENE, DRY. Dry toluene conforming to the specifications for Nitration Grade Toluene (ASTM Designation: D 841) (1), or the equivalent, and dried by a drying agent. The Linde 4A Molecular Sieve, or equivalent, has been found satisfactory for this purpose.

PROCEDURE

To a 500-ml Erlenmeyer flask, rinsed successively with water, alcohol, and benzene, dried at 100°C, and cooled, add 40 ml of dry toluene. Accurately add 50 ml of the dibutylamine solution by pipet or buret and mix carefully.*

Transfer to the flask between 6.5 and 7.0 grams of the sample, weighed to the nearest 0.001 gram. Carefully swirl the flask while slowly adding the sample. If spattering is anticipated, cool the flask and contents in the cooling bath prior to the addition of the sample, after which cool until the heat of reaction has been dissipated. Add 10 ml of dry toluene, stopper the flask loosely, and allow the contents to come to room temperature. Wash down the sides of the flask with 10 ml of dry toluene. Stopper the flask loosely and allow to stand at room temperature for 15 minutes.

Add 225 ml of isopropyl alcohol from a 250-ml graduated cylinder and 0.8 ml of bromocresol green indicator solution from a graduated 1-ml pipet. Titrate with 1*N* hydrochloric acid solution in a 50- or 100-ml buret while swirling the flask contents to effect rapid mixing. Near the end point add the hydrochloric acid dropwise and shake the solution vigorously. Consider the end point reached with the disappearance of the blue color and the reappearance of a yellow color persists for at least 15 seconds.

1. 1958 Book of ASTM Standards, Part 8.

* Burets and pipets shall conform to National Bureau of Standards tolerances, as given in E. L. Peffer and G. C. Mulligan, "Testing of Glass Volumetric Apparatus," National Bureau of Standards *Circular C434*, Superintendent of Documents, Government Printing Office, Washington D.C., 1941.

Run a blank determination at the same time in exactly the same manner as above but omitting the sample.

CALCULATION

Calculate the assay percentage as follows:

$$\text{Assay, \%} = \frac{(B-S)NE}{1000W} \times 100$$

where

B = milliliters of hydrochloric acid required for titration of the blank,
S = milliliters of hydrochloric acid required for titration of the sample,
N = normality of the hydrochloric acid,
E = equivalent weight of the toluene-diisocyanate, in grams (87.08),
W = grams of sample used

Reactive halides, if present, interfere with the usual amine reagents. A study by Beazley, described below, of the reactivities of several amines relative to halides and isocyanates, revealed that dicyclohexylamine suffered very little interference of this type.

Adapted from the Method of P. M. Beazley

[*Anal. Chem.*, ***43,*** *148* (*1971*)]

REAGENTS

DICYCLOHEXYLAMINE. Dilute 50 ml of dicyclohexylamine, Eastman White Label 4627, to 1 liter with dry dimethylformamide.

PROCEDURE

Dissolve the sample containing 1 meq. of isocyanate in 5 ml of dry dimethylformamide. Pipet 10 ml of amine solution into the sample, and after 2 minutes add 40 ml of isopropanol and 8 drops of bromocresol green indicator. Titrate the unreacted amine with 0.1N hydrochloric acid until the indicator remains yellow for at least 15 seconds. Run a blank to determine the total amount of amine taken, and calculate the equivalent isocyanate content of the sample by subtraction of the amount of unreacted amine from the total amount of amine taken.

RESULTS AND DISCUSSION

Table 4 gives comparisons of the reactivities of different amines toward benzyl bromide. A portion of each amine in dimethylformamide was mixed with an equal portion of benzyl bromide in dimethylformamide at room temperature, and the reaction was quenched after 5 minutes.

Table 4. Reactivities of Different Amines Toward Benzyl Bromide

Amine	Benzyl Bromide Reacted, %
Piperidide	96
Dibutylamine	84
Diisopropylamine	10
Dicyclohexylamine	6

Results of the reaction of dicyclohexylamine and a mixture of benzyl bromide and tolylisocyanate (Table 5) indicate very little interference from the halide.

Table 5. Results of the Reaction of Dicyclohexylamine and a Mixture of Benzyl Bromide and Tolylisocyanate

Time, min.	Benzyl Bromide, mole %	*o*-Tolylisocyanate, mole %	
		Present	Found
1	38.1	61.9	60.6
2	38.1	61.9	61.4
4	38.1	61.9	62.2

Recoveries of pure isocyanates as determined by the recommended procedure are shown in Table 6.

Table 6. Recoveries of Pure Isocyanates

Compound	Isocyanate Found, %
2,4-Toluene diisocyanate	99.4
Phenyl isocyanate	98.9
Allyl isocyanate	97.2
n-Butyl isocyanate	87.4
Cyclohexyl isocyanate	91.2

Cyclohexyl isocyanate and *n*-butyl isocyanate fail to react completely within the 2 minutes allowed. The method is therefore applicable to isocyanate-halide mixtures, provided the halides are no more reactive than benzyl bromide and the isocyanates are at least as reactive as allyl isocyanate.

Piperidine Method—Adapted from R. Venkataraghavan and C. N. R. Rao

[*Reprinted in Part from Chemist-Analyst,* ***51****, 48–9* (*1962*)]

Piperidine was used in this work to determine isothiocyanates, but Stagg (2) had used it for determining isocyanates. The procedure as described works for both classes of compounds.

REAGENTS

Piperidine in anhydrous dioxane (store in a tightly sealed container), 0.1*M*.
Hydrochloric acid, 0.1*M*.
Sodium hydroxide, 0.0500*M*.
MIXED INDICATOR. Five parts 0.1% methyl red solution and 2 parts 0.1% methylene blue solution.

PROCEDURE

Add the sample of the organic isothiocyanate (about 0.1 meq. in 1 ml) in dioxane to a conical flask containing a known excess amount (5.00 ml) of the piperidine reagent. Shake the mixture vigorously for about 1 minute. Add a known, excess amount (10.0 ml) of 0.1*M* hydrochloric acid. Now titrate the excess of hydrochloric acid with 0.0500*M* sodium hydroxide, employing 5 drops of the mixed indicator to a very sharp end point marked by a color change of violet to green. Repeat the procedure, omitting the sample addition. The difference in the volumes of sodium hydroxide required in the two titrations corresponds to the amount of isothiocyanate present in the sample. If the sample contains free acidic or basic material, titrate a further portion without the addition of piperdine and then apply a correction.

2. H. E. Stagg, *Analyst,* **71,** 557 (1946).

RESULTS AND REMARKS

The recovery ranged from 99.2 to 99.8% in the application of the procedure to the following isothiocyanates: methyl, benzyl, phenyl, *p*-bromophenyl and *p*-methoxyphenyl. Organic thiocyanates do not react with piperidine, hence do not interfere. Thus it was found that methyl and benzyl isothiocyanates could be determined in the presence of the corresponding thiocyanates. The procedure may therefore be applied to the analysis of mixtures of thiocyanates and isothiocyanates or to study isomerization of the former.

DETERMINATION OF TRACE QUANTITIES OF ISOCYANATES

Kubitz (3) developed a method for traces of isocyanates in urethane-based polymers. The method was based on the *n*-butylamine reaction with the excess amine determined colorimetrically with malachite green. Mercali (4) determined toluene diisocyanate in the atmosphere by a procedure that depends on hydrolysis of the isocyanate to the corresponding diamine followed by diazotization of the diamine and coupling with *N*-(naphthyl)ethylenediamine to form a color that is measured spectrophotometrically. The method of Mercali was substantially modified by other workers (5, 6). Meddle, Radford, and Wood selected it as suitable for adaptation to a general procedure for aromatic isocyanates in air.

Adapted from the Methods of D. W. Meddle, D. W. Radford, and R. Wood, and D. W. Meddle and R. Wood

[*Analyst*, ***94***, *367* (*1969*); ***95***, *402* (*1970*)]

This method is applicable to the determination of aromatic isocyanates in air in the presence of primary aromatic amines. Samples of the test atmosphere are drawn simultaneously through two different absorbing solutions. One sample is collected in a dimethylformamide solution of 1,6-diaminohexane, and the other in a dimethylformamide–hydrochloric acid solution of 1,6-diaminohexane. The aromatic amine present or

3. K. A. Kubitz, *Anal. Chem.*, **29,** 814 (1957).
4. K. Mercali, *Anal. Chem.*, **29,** 552 (1957).
5. K. E. Grim and A. L. Linch, *Am. Ind. Hyg. Assoc. J.*, **25,** 285 (1964).
6. D. A. Reilly, *Analyst*, **88,** 732 (1963).

produced in each absorbing solution is diazotized and coupled with *N*-(1-naphthyl)ethylenediamine to form a colored complex, whose optical density is measured spectrophotometrically. The difference between the resulting optical densities gives a measure of the isocyanate in the air.

APPARATUS

ALL-GLASS ABSORBERS. These are the type shown in Fig. 17.1.
SAMPLING PUMP. Capable of drawing air through the apparatus at 2 liters per minute.
Colorimeter or spectrophotometer.

REAGENTS

DILUTED HYDROCHLORIC ACID. Dilute 15 ml of concentrated acid (sp. gr., 1.9 at 20°C) to 100 ml with distilled water.
Dimethylformamide solution of 1,6-diaminohexane (DAH), 70 μg per milliliter.
DIAZOTIZATION SOLUTION. Dissolve 3 grams of sodium nitrite and 5 grams of sodium bromide in water and dilute to 1 liter.
Aqueous sulfamic acid solution, 10%, w/v.

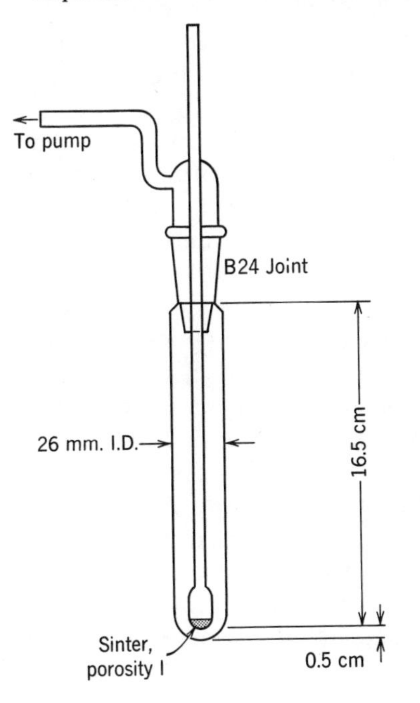

Fig. 17.1. All-glass absorber.

N-(1-NAPHTHYL)ETHYLENEDIAMINE SOLUTION. Dissolve 0.75 gram of *N*-(1-naphthyl)ethylenediamine dihydrochloride in water, add 2 ml of concentrated hydrochloric acid, and dilute to 100 ml with water. Prepare a fresh solution after 2 days.

STANDARD ISOCYANATE SOLUTION. Dissolve 30 mg of isocyanate in 100 ml of toluene.

STANDARD AROMATIC AMINE SOLUTION. Dissolve 30 mg of aromatic amine in 100 ml of toluene.

PROCEDURE

Add 3 ml of the dimethylformamide-DAH solution to each of two absorber tubes; then add 2 ml of hydrochloric acid solution to one of them. Insert the inlet tubes and mount the assembled absorbers side by side in a vertical position at the sampling site. Attach the pump to each absorber, and simultaneously draw 10 liters of the test atmosphere through each absorbing solution at the rate of 1 liter per minute. Leave the absorbing solutions for 10 minutes to allow any reactions to reach completion. To the absorber tube containing only dimethylformamide-DAH solution, add 2 ml of hydrochloric acid solution through the top of the inlet tube. Lift the inlet tube so that the sinters in the absorber tubes are clear of the respective solutions, and expel the liquid trapped in each domed sinter as completely as possible.

Ensure that the temperature of the absorbing solution is not above 20°C: then to each absorber tube add 0.5 ml of the diazotization solution. Shake the mixture and allow it to stand for 2 minutes. To each absorber tube add 0.5 ml of the sulfamic acid solution and shake until the effervescence has ceased; then 2 minutes later add 0.5 ml of *N*-(1-naphthyl)ethylenediamine solution to each tube and mix well. Allow the colors to develop for 10 minutes before measuring the optical density of each solution. Determine the optical density of each in a 20-mm cell against water as a reference. If a wide-band color filter is used with a colorimeter, the filter should be chosen so that it ensures as high an optical transmission as possible at each of the wavelengths of the absorbance maxima of the compounds being determined. Any arbitrary wavelength at or between those of absorbance maxima of the individual components of the mixture can be selected when a spectrophotometer is used; for example, for a mixture of 4,4′-methylenebis(*o*-chloroaniline) (MOCA) and naphthylene-1,5-diisocyanate (NDI), select a wavelength between 550 and 585 nm.

PREPARATION OF CALIBRATION CURVES

To a series of absorber tubes each containing 3 ml of the dimethylformamide-DAH solution and 2 ml of hydrochloric acid solution add 0.01, 0.02, 0.03, 0.04, and 0.05 ml of standard isocyanate solution covering the range of 0 to 15 μg from a micrometer syringe. Develop the color in each tube as indicated above. Measure the optical densities of the solutions in a spectrophotometer at the selected wavelength or in a colorimeter with a suitable wide-band color filter, in the same way as the samples. Plot micrograms of isocyanate against optical density. Using the standard amine solution, prepare a calibration curve of micrograms of aromatic amine against optical density, as described for the isocyanate.

DISCUSSION

Since there is considerable variation in the molecular weights of the isocyanates and amines likely to be found together in the atmosphere and also in the intensities of the colors they produce, it is difficult to give a general concentration range over which this method can be used. However, for a mixed tolylene-2,4-diisocyanate (TDI)-MOCA atmosphere when weight for weight, these compounds give similar color intensities; a total of up to 16 μg of mixed species can be separately identified into TDI and MOCA concentrations with an E.E.L. colorimeter fitted with a No. 625 Ilford filter and with 14 mm I.D. glass tubes for color measurements.

Over the range 0 to 15 μg, the colors produced by both TDI and MOCA with the proposed procedure exhibited a linear relationship between concentration and optical density, 15 μg of TDI and MOCA giving optical densities of 1.05 and 1.15, respectively, when measured in 2-cm cells on a Unicam SP600 spectrophotometer at 550 nm. Both TDI and MOCA produced colors with a maximum absorbance at 550 nm. Thus with the optical density of MOCA measured in the dimethylformamide-DAH absorbing solution and that of MOCA with TDI in the dimethylformamide-DAH–hydrochloric acid solution, the optical density of TDI alone was obtained by subtraction. The respective concentrations of MOCA and TDI were then determined by reference to the calibration curves. To check the feasibility of determining the individual concentrations of amine and isocyanate when the absorbance maxima differed, mixtures of MOCA and naphthylene-1,5-diisocyanate (NDI) were studied, the latter having a maximum absorbance at 585 nm. Aliquots of standard toluene solutions of NDI were added to each absorbing solution containing a known amount of MOCA. An arbitrary wavelength of 550 nm was chosen for reading the optical density of the absorbing solutions. Calibration curves had been prepared previously for each compound at this wavelength.

The results of this work showed that atmospheres containing mixtures of NDI and MOCA could be analyzed readily and the individual concentrations of the two species determined. Although 550 nm was found to be a convenient wavelength for the measurement of MOCA and NDI, any wavelength between their respective maximum absorptions at 550 to 585 nm could be used, since over this range each compound is measured with a relatively high sensitivity.

18

Mercaptans

Mercaptans analytically can be considered as monosubstituted hydrogen sulfide, and they have the reactions of such compounds. For example, metals replace the mercapto hydrogen

$$x\text{RSH} + \text{Me}^{+x} \rightarrow (\text{RS})_x\text{Me} + x\text{H}^+$$

The analysis can be carried out by titrating the mercaptan with a metal ion such as in the silver methods below. Also, mercaptans can be oxidized to the corresponding disulfides with iodine or cupric ion; these approaches also form the basis of analysis.

Silver Methods

The silver methods can be approached in three ways: (*a*) direct titration with silver ion; (*b*) addition of excess silver, precipitation of the mercaptide, and back-titration of the excess silver with a Volhard type titration; (*c*) addition of excess silver nitrate and titration of the acid liberated by the reaction.

The back-titration of the excess silver (1, 2) is the least desirable of the three approaches, since any material that precipitates with silver ion will interfere. The direct method does not suffer from this disadvantage, since the silver mercaptides are usually the least dissociated silver compound in the system and in a mixture the silver mercaptide precipitates first. It is thus indicated separately from the other components in the system. Also, the back-titration often yields high values because the precipitated mercaptides tend to occlude some of the excess silver ion, which is then not back-titrated. The procedure of Malisoff and Arding (2) tends to minimize this difficulty by using dilute solutions.

The direct potentiometric titration of mercaptans with silver nitrate is the most generally applicable method with the least interferences. The acidimetric method is also quite widely applicable.

1. P. Borgstrom and E. E. Reid, *Ind. Eng. Chem., Anal. Ed.*, **1,** 186 (1929).
2. W. M. Malisoff and C. E. Arding, *Ibid.*, **7,** 86 (1935).

POTENTIOMETRIC TITRATION

Potentiometric titration is a very simple, direct, and generally accurate and precise approach to determining mercaptans. The authors have successfully used this approach in almost any solvent system that will dissolve silver nitrate and the sample in question. An ordinary pH meter equipped with a silver indicating electrode and calomel reference electrode suffices. As low as 0.001 meq of mercaptan can often be measured using $0.001N$ silver nitrate solution. In general, an alkaline agent such as ammonia or sodium acetate aids the end point because of neutralization of the acid formed in the reaction

$$RSH + AgNO_3 \rightarrow RSAg + HNO_3$$

The acid tends to dissolve the mercaptide slightly and yield a small amount of silver ion.

The procedure of Tamele and Ryland is a formalized potentiometric titration method.

Method of M. W. Tamele and L. B. Ryland

[*Adapted from Anal. Chem.*, ***8****, 16–9 (1936), Reprinted in Part*]

The following procedure is based on precipitation of mercaptans with silver nitrate. By using enough alcohol to dissolve the sample and by titrating with an alcoholic solution of silver nitrate, the separation of phases, the formation of emulsions, and the resulting adsorption are completely eliminated. Furthermore, to avoid an excess of silver nitrate at the end of precipitation, the end point is determined potentiometrically with a silver electrode indicator. The method thus becomes applicable to colored solutions, since the selection of the end point is not dependent on a color change of the indicator. Finally because of the very low solubility of silver mercaptides, approximately equal to that of silver iodide, the method becomes applicable in the presence of substances that normally react with silver nitrate but form compounds more soluble than the silver mercaptides. The danger of the simultaneous precipitation of common impurities is thus minimized.

The influence of a number of substances likely to interfere with the suggested procedure was studied, especially those occurring naturally in petroleum products. Many were found to have no influence on the accuracy of the results, but hydrogen sulfide and elementary sulfur interfere with the procedure (see pp. 711–13 for a method to deal with these interferences).

APPARATUS

A suitable simple arrangement for potentiometric titration of mercaptans is shown in Fig. 18.1.

The cell consists of a silver half-cell sensitive to changes in silver-ion concentration and a mercury half-cell, *R*, used as the reference electrode. The silver half-cell consists of a silver electrode immersed in the beaker containing 50 cc of 0.1*N* sodium acetate in 96% ethyl alcohol. The mercury half-cell and the bridge are filled with the same solution.

The silver electrode is a polished silver wire of about 2 mm diameter, and the mercury electrode is a layer of mercury about 3 to 4 cm in diameter. The cell is represented by the diagram

$-$Ag	0.1*N* sodium acetate in alcohol	0.1*N* sodium acetate in alcohol	Hg+

The emf of the cell is reasonably constant, about -0.070 volt, the minus sign signifying that the silver wire is the negative electrode. In the absence of any generally accepted standard reference electrode in alcoholic solutions, the mercury half-cell is used as the standard and the potential of the silver electrode is considered to be equal to the numerical value of the emf of the cell. Variations

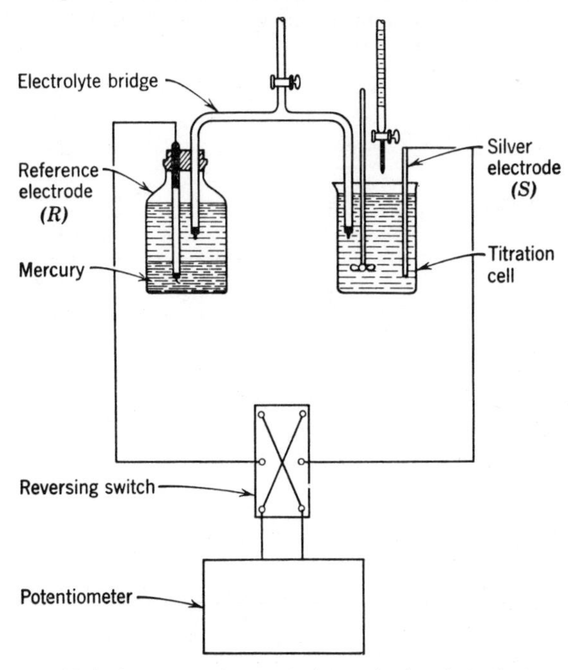

Fig. 18.1. Apparatus for potentiometric titration of mercaptans.

from this constant value may be caused by impurities on the silver wire. It is advisable to clean the electrode with a solution of potassium cyanide and then to wash it carefully with water.

Because of the somewhat high resistance of the cell, a reasonably sensitive potentiometer arrangement is required.

The sample is added to 50 cc of the alcoholic sodium acetate solution in the silver half-cell. The size of the sample may vary according to the solubility in alcohol of the solution examined and the amount of mercaptan present, and it should be such that about 10 to 15 cc of 0.01N silver nitrate solution is consumed. From 5 to 10 cc of a hydrocarbon sample usually is soluble in the cell liquid.

If the sample (free from hydrogen sulfide and elementary sulfur) contains mercaptan, the potential of the silver electrode will rise from −0.070 to about −0.380 volt.

Standard 0.01N solution of silver nitrate in isopropyl alcohol is added in small portions to the beaker with occasional stirring, and the values of the emf of the cell are recorded. A sudden drop of the potential of the silver electrode to less negative values occurs at the end point.

SELECTION OF END POINT. If titration curves are constructed from the data observed, they appear to be symmetrical, as would be expected, since the reaction involves two monovalent ions (3). The end point of titration is therefore at the point of inflection of the curve. If the results are not plotted, the values of $\Delta E/\Delta c$ are calculated and the end point is taken where this value is a maximum. A typical titration curve appears in Fig. 18.2.

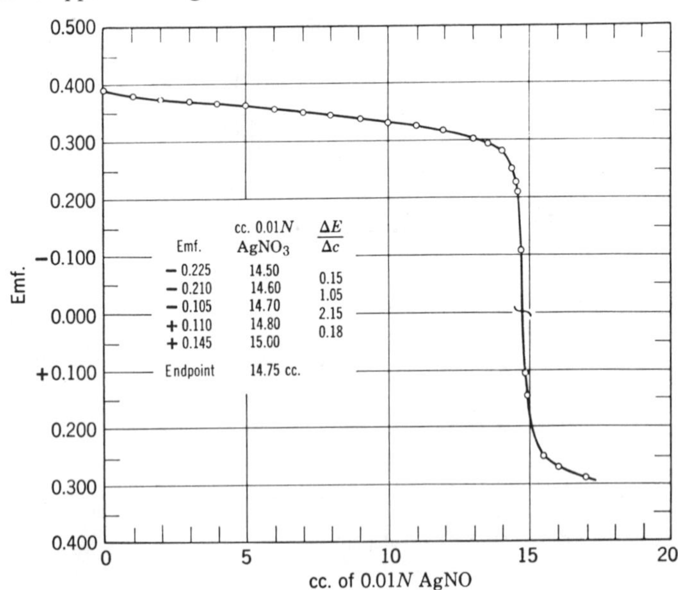

Fig. 18.2. Titration of n-butyl mercaptan in kerosene; 5-cc sample, 0.0259M.

3. I. M. Kolthoff and N. H. Furman, *Potentiometric Titrations*, Wiley, New York, 1926, pp. 70 ff.

REAGENTS

SODIUM ACETATE IN ALCOHOL. Approximately $0.1N$ solution of sodium acetate in 96% ethyl alcohol is used as a medium for titration. Ethyl alcohol denatured with benzene or gasoline was found to be suitable.

STANDARD SOLUTION OF SILVER NITRATE. Prepare a solution of silver nitrate of $0.01N$ strength in isopropyl alcohol containing about 9% of water by exact dilution of a $0.1N$ stock solution. The stock solution is stable for months but should preferably be kept in darkness. The dilute $0.01N$ solution should be prepared as needed, but it has been found to remain stable for several weeks. Ethyl alcohol is unsuitable for preparation of these solutions because acetaldehyde is slowly formed and fine silver powder is precipitated.

ISOPROPYL ALCOHOL. This must be free of aldehydic impurities. Purify commercial alcohol by dissolving 0.5 gram of silver nitrate in 1 liter of alcohol and exposing in a clear glass bottle to direct sunlight for several hours. Decant the alcohol from the precipitated silver, remove the excess silver nitrate with sodium chloride, and redistill the alcohol. The azeotropic mixture containing 9% of water can be used directly for preparation of solutions of silver nitrate, since this amount of water aids in the solution of silver nitrate.

INTERFERING SUBSTANCES AND CONDITIONS

INFLUENCE OF SOLVENT. The freedom of interference by the hydrocarbon solvent with the precipitation of silver mercaptides has been demonstrated by previous workers (4, 5). No disturbing influence of the solvent could be detected experimentally when titrations were performed in kerosene, cracked gasoline, amylene, acetone, and various aliphatic alcohols.

COMMON IMPURITIES. From a theoretical point of view, substances reacting with silver nitrate and forming compounds substantially more soluble than silver mercaptides should not interfere with the procedure. Since the solubility of silver mercaptides is very low and nearly equal to the solubility of silver iodide, the chances of encountering an interfering substance are small.

Possible interference by a number of substances likely to be encountered in practice was studied experimentally. The procedure consisted in the titration of a solution of mercaptan of known strength, in the presence of a measured amount of substance examined for possible interference. Table 1 shows the results of analyses of *n*-butyl mercaptan solutions in kerosene containing varying amounts of diethyl disulfide.

4. G. R. Bond, Jr., *Ind. Eng. Chem., Anal. Ed.*, **5,** 257 (1933).
5. W. M. Malisoff and E. M. Marks, *Ind. Eng. Chem.*, **23,** 1114 (1931).

Table 1. Titration of *n*-Butyl Mercaptan in the Presence of Diethyl Disulfide in Kerosene

Diethyl Disulfide Present, mole/l.	Mercaptan Present, mole/l.	Mercaptan Found, mole/l.
None	0.0297	0.0297
0.030	0.0297	0.0297
0.030	0.0297	0.0297
0.090	0.00425	0.00428

By similar experiments it was found that certain substances had no influence on the results when present in the cell in the amounts given in Table 2.

Table 2. Impurities Not Influencing Results

	%		%
Sulfur Compounds			
Ethyl sulfide	0.1	Ethyl sulfone	0.1
Ethyl disulfide	0.1–0.2	Butyl sulfone	0.1
Carbon disulfide	0.2	Sodium cymene sulfonate	0.1
Thiophene	0.1–0.6	β-Trithioacetaldehyde	0.7
Reducing Substances			
Formaldehyde	0.04–0.40	Acetaldehyde	0.04–0.40
Miscellaneous			
Nitrobenzene	2.0	Phenol	0.1
Fuchsin	0.004	Pyridine	10
Sodium oleate	0.1	Light nitrogen bases[a]	0.4
Sodium sulfate	1.4	Medium nitrogen bases[a]	0.4
Magnesium sulfate	0.5	Heavy nitrogen bases[a]	0.4
Aluminum sulfate	0.5	Purified naphthenic acids[a]	0.4
Sodium bicarbonate	0.1	Crude naphthenic acids[a]	0.4

[a] Separated from California crude oil.

Determinations were not influenced by the presence of chloride ion. The end point of mercaptan titration is reached long before the precipitation of the chloride ion begins. Figure 18.3 shows a titration of a solution of *n*-butyl mercaptan with and without chloride added.

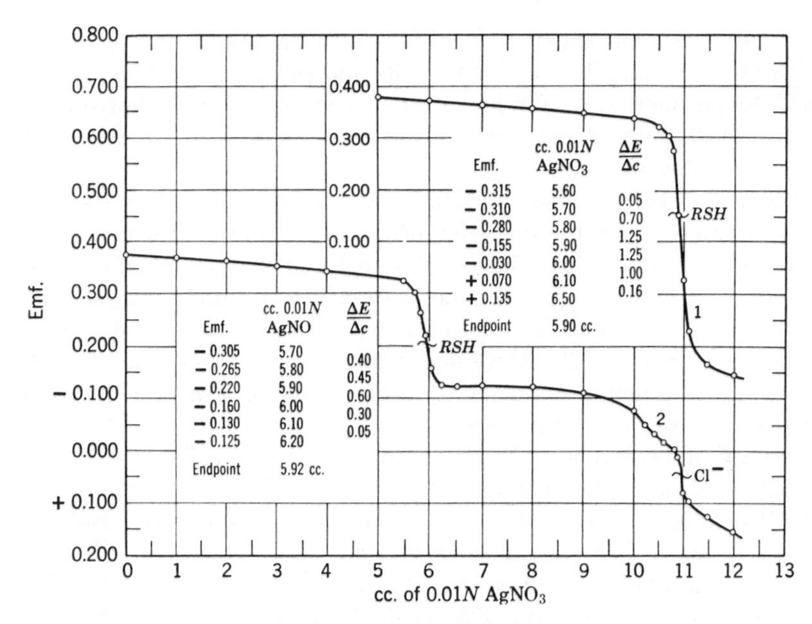

Fig. 18.3. Titration of *n*-butyl mercaptan in alcohol: 1, in the absence of chloride: 2, in the presence of chloride.

OXIDIZING SUBSTANCES, PEROXIDES. The determination of mercaptans in the presence of peroxides is often required, although this fact is not generally appreciated in the literature concerned with the determination of mercaptans in petroleum products. Occurrence of peroxide in gasolines is common, and the oxidation of mercaptans may proceed very slowly when both substances are present in concentration often found in gasolines (6). The present procedure gives reliable results in the presence of reasonable amounts of hydrogen peroxide, ethyl ether peroxide, and organic peroxides formed in cracked gasoline when exposed to ultraviolet light.

INFLUENCE OF CONCENTRATION

A number of mercaptans purchased from Eastman Kodak Company were analyzed. A weighed amount was dissolved in kerosene and the solution was titrated at various dilutions. In all cases the end point was reached at or near −0.100 volt. This value was found experimentally to coincide nearly with the end point of the precipitation of silver iodide in the same solution, from which it is concluded that the solubility of silver

6. J. A. C. Yule and C. P. Wilson, Jr., *Ind. Eng. Chem.*, **23,** 1254 (1931).

mercaptides is approximately equal to (somewhat lower than) the solubility of silver iodide. The variations with individual mercaptans were too small to be considered for analytical purposes. Table 3 shows the influence of dilution on the analyses and gives an estimate of the absolute accuracy of the method.

ACCURACY. The principles on which the method is based and the freedom of interference by many impurities allow the conclusion that the absolute accuracy is high. The samples analyzed were not chemically pure preparations, and therefore the purity calculated from the analyses would be expected to remain below 100%, as it actually did in all cases. To prove experimentally the absolute accuracy, pure mercaptans should be analyzed. An attempt was made to purify a sample of *n*-butyl mercaptan by fractionation in nitrogen atmosphere. The last fraction boiled at 98.0 to 98.3°C (760 mm Hg), had a refractive index $n_D^{20} = 1.4425$, $n_D^{25} = 1.4329$, and analyzed by the Carius method 35.50% sulfur (calculated 35.57% sulfur). A weighed amount of this material was dissolved in kerosene, and a 10-cc sample was titrated.

The potentiometric analysis indicated a purity of 93.3% on the original sample before fractionation. The purified sample analyzed 97.6% (average) by the potentiometric method, and 97.1% by the iodometric method of Kimball, Kramer, and Reid (7), given below.

PRECISION. The relative accuracy of individual determinations is governed by two factors—the solubility of the sample in alcohol and the amount of silver nitrate consumed per sample. The end point of titrations can be established with an accuracy of ±0.02 cc. This amount indicates 2×10^{-7} mole of mercaptan in the sample, which should be the error in the absolute amount of mercaptan found in the titration.

The authors of this book have used the method of Tamele and Ryland on ethyl, amyl, and dodecyl mercaptans; cysteine; and 2-mercaptopyridine. Accuracies of ±1% of theoretical were obtained, with precision of ±0.5%.

Adapted from P. K. C. Tseng, and W. F. Gutknechy

[*Anal. Chem.*, **47**, *2316* (*1975*)]

Several mercaptans have been measured in alkaline solution by a direct potentiometric method with a silver sulfide electrode. Nernstian responses were shown by the electrode.

7. J. W. Kimball, R. L. Kramer, and E. E. Reid, *J. Am. Chem. Soc.*, **43,** 1199–1200 (1921).

Table 3. Influence of Dilution

Mercaptan	Present %	Found %	Purity of Sample %
Ethyl	0.3229	0.2706	83.8
	0.3229	0.2761	85.5
	0.0807	0.0684	84.7
	0.0202	0.0176	87.2
Isopropyl	0.3816	0.3220	84.4
	0.3816	0.3180	83.3
	0.3816	0.3160	82.8
	0.0954	0.0789	82.7
	0.0239	0.0195	81.6
n-Butyl (1)	0.4109	0.3895	94.9
	0.1027	0.0968	94.3
	0.0257	0.0243	94.6
n-Butyl (2)	0.2877	0.2681	93.2
	0.2877	0.2681	93.2
	0.2877	0.2681	93.2
	0.2677	0.2674	92.9
n-Butyl (3)	0.1041	0.0971	93.3
	0.01041	0.0097	93.2
	0.001041	0.0010	96.1
Isobutyl	0.3860	0.3680	95.4
	0.3860	0.3690	95.6
	0.0965	0.0924	95.8
	0.0242	0.0231	95.5
n-Amyl	0.4036	0.3945	97.8
	0.1009	0.0984	97.5
	0.0252	0.0246	97.6
	0.0252	0.0250	99.2
n-Heptyl	0.4149	0.4110	99.1
	0.4149	0.4110	99.1
	0.1037	0.1023	98.7
	0.0259	0.0258	99.6
	0.0259	0.0258	99.6
Benzyl	0.4002	0.3898	97.4
o-Thiocresol	0.4013	0.3377	84.2
	0.1004	0.0846	84.3
	0.0250	0.0206	82.4

EXPERIMENTAL

The mercaptide ion selective electrode used was prepared in the laboratory. To prepare a membrane, 0.5 gram of reagent grade silver sulfide was ground with a mortar and pestle and then compressed in a potassium bromide die at 30,000 psi. The resulting membrane had a thickness of 1 mm and was quite rugged. This membrane was attached to a glass tube with epoxy resin. Before the electrode was used, the membrane surface was polished on Microcloth (Buehler Ltd.) with 0.05-μm alumina as the polishing agent. The filling solution of this electrode was 0.004*M* silver nitrate solution and the internal electrode was a silver wire. The reference electrode was a saturated calomel electrode. The connection between the reference electrode and the test solution was made with a fiber-tipped secondary junction filled with 1*M* potassium nitrate solution. The laboratory-made sulfide electrode responded linearly to mercaptan in sodium hydroxide solution over a concentration range of 0.1 to $10^{-6}M$.

Cysteine solutions were 0.1*M* in sodium hydroxide, the thioglycolic acid solutions in 1.0*M* sodium hydroxide and the 2-mercaptoethanol and 3-mercaptopropionic acid in either 1.0 or 5.0*M* sodium hydroxide. These high concentrations of base were found to be necessary to obtain rapid, stable responses. Oxidation of these mercaptans was not a serious problem under the conditions used. In addition, the sodium hydroxide solutions maintained both the pH and ionic strength of the solutions at constant values. Water was purified by deionization followed by distillation.

All the cell emf measurements were made with a Beckman Model SS-3 pH meter. Solutions were stirred during each measurement, and the solution temperatures were ambient at 20 to 21°C. The final potential of each test solution was taken as the point at which the measured potential changed less than 0.5 mV over a 5-minute period. Replicate measurements were reproducible to within ± 1 mV if care was taken to repolish the electrode surface lightly after each measurement.

RESULTS AND DISCUSSION

Linear plots were obtained for the measured potential versus logarithm of the mercaptan concentration for the four compounds, 2-mercaptoethanol, 3-mercaptopropionic acid, thioglycolic acid, and cysteine. The detection limits were all about $10^{-4}M$. The electrode response times were only a few minutes for these and other mercaptans tested.

Freshly prepared solutions of the four mercaptans showed constant potential values during a minimum of 4 hours, indicating no rapid oxidations. The other mercaptans tested, 1,2-dithioethane and thiophenol, showed unstable potentials. White precipitates formed in these two mercaptans on standing. Evidence from infrared, nuclear magnetic resonance, and mass spectrometric measurements supported the conclusion that these precipitates were disulfides formed as oxidation products:

$$4RSH + O_2 \rightarrow 2RSSR + 2H_2O$$

Thioacetic acid hydrolyzes easily and the ion detected was the sulfide ion, S^{2-}.

AMPEROMETRIC TITRATION

One advantage amperometric titration has over the potentiometric titration is a more sensitive end point indication in the low concentration ranges.

Method of I. M. Kolthoff and W. E. Harris

[*Adapted from Ind. Eng. Chem., Anal. Ed.*, **18,** *161–2* (*1946*)]

$$AgNO_3 + RSH \rightarrow RSAg + HNO_3 \text{ (removed by } NH_4OH)$$

A rotating platinum wire electrode is the indicator in this method. The mercaptan is titrated with silver nitrate. When excess of silver ion is added, an increase in current is noted on a microammeter. A plot of current versus milliliters of reagent will show a sharp break at the end point (see Fig. 18.6). If the titration is carried out in ammoniacal solution, chlorides and small amounts of bromides will not interfere.

APPARATUS

Figure 18.4 illustrates the complete set up. The platinum wire electrode (Fig. 18.5) is 6 to 8 mm long and 0.5 mm in diameter. An ordinary stirring motor can be used to rotate the electrode.

The reference electrode *F* has a potential of −0.23 volt (against the saturated calomel electrode). The electrolyte for the reference half-cell is prepared by dissolving 4.2 grams of potassium iodide and 1.3 grams of mercuric iodide in 100 ml of saturated potassium chloride solution. A layer of mercury serves as the

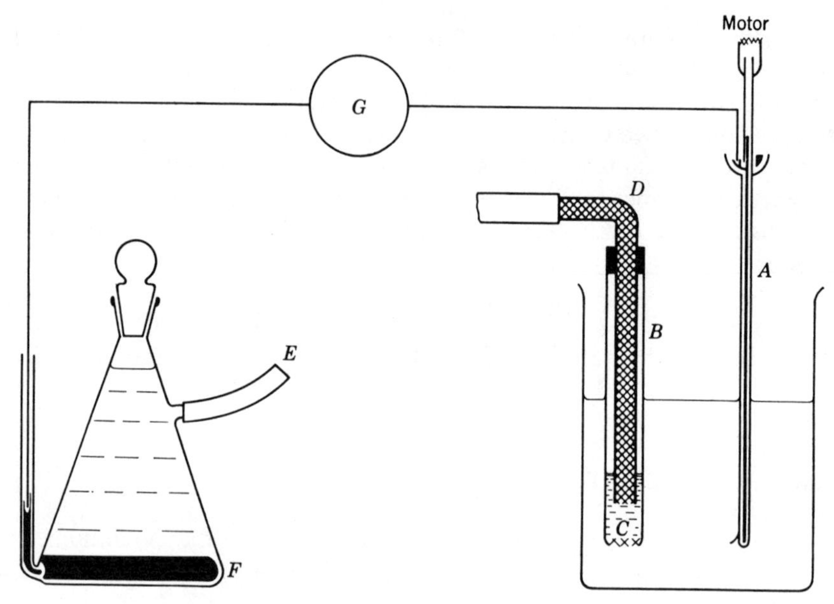

Fig. 18.4. Amperometric titration apparatus.

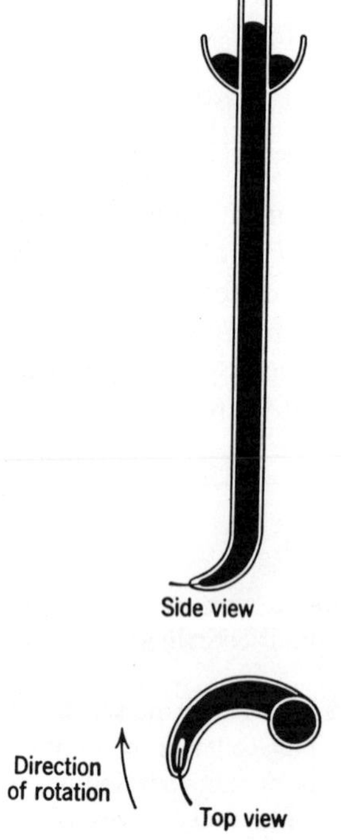

Fig. 18.5. Rotating platinum electrode.

electrode. The salt bridge *E* consists of a rubber tube 2 feet long and 6 mm inside diameter filled with saturated potassium chloride solution. The rubber tubing is connected to a short length of glass tubing *D*, which is filled with a gel of 3% agar and 30% potassium chloride. To minimize contamination, a coarse sintered-glass disk may be inserted. For further protection against contamination with iodide, the tube *B*, having an agar or fine sintered-glass plug, can be inserted over the agar-filled glass tube *D*. The electrolyte solution *C* (saturated potassium chloride) used in the tube *B* can be rinsed out and replaced as soon as it becomes contaminated with iodide. It is essential that all sources of high resistance such as air bubbles be eliminated from the salt bridge.

To complete the circuit, the two half-cells are short circuited through the microammeter *G*. For the experiments used to test this procedure, a Weston (Weston Electrical Instrument Corp., Newark, N.J.), Model 430, microammeter was used. Instead of the microammeter, other current-indicating devices may be used, such as a pointer galvanometer with a sensitivity of 0.25 μA per division on the attached scale.

REAGENTS

Standard silver nitrate, 0.005*N*.
Ethanol, 95%.
Concentrated ammonium hydroxide.
Ammonium nitrate.

PROCEDURE

Place a sample containing about 5 mg of mercaptan sulfur in a 250-ml beaker and dissolve in 100 ml of 95% ethanol. Make the solution about 0.25*M* with respect to ammonium hydroxide and 0.01 to 0.1*M* with ammonium nitrate. Immerse the end of the salt bridge and the rotating platinum electrode in the solution and titrate the solution with 0.005*M* silver nitrate. Read the diffusion current after each addition; the increments of silver nitrate solution should be small when close to the end point. Add a few more small portions of silver nitrate solution when the ammeter shows that the end point has been passed. Then plot the ammeter readings against the milliliters of silver nitrate added. The curve will resemble Fig. 18.6.

Draw a straight line through the points before the end point and another straight line through the points just after the end point. The intersection of these lines will correspond to the end point. There is usually no recordable current before the end point in these titrations.

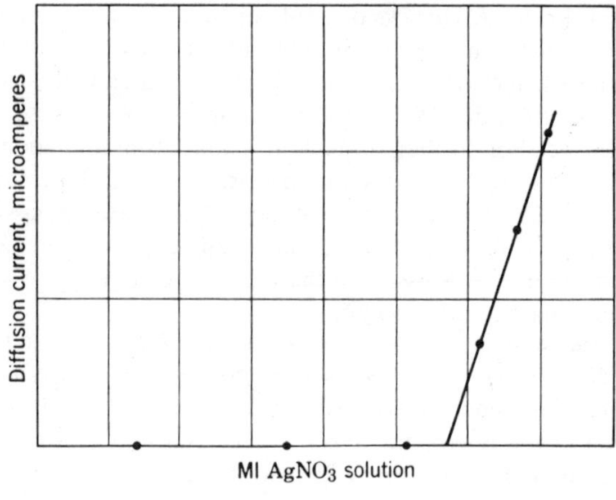

Fig. 18.6

The platinum electrode may become insensitive or erratic after long use or when titrating large amounts of mercaptans. The sensitivity can be restored by wiping the electrode with a cloth.

Some samples have a bad effect on the platinum electrode. As soon as the first excess of silver nitrate is indicated on the ammeter, stop the rotating electrode and wipe it. Add another small increment of silver nitrate, start the rotating again, and take the reading. Repeat this procedure; wipe the electrode after addition of each portion of silver nitrate, and take the reading immediately after the electrode is started.

The electrodes should be thoroughly cleaned with concentrated nitric acid when it is deemed necessary. When a new or freshly cleaned electrode is placed in an ammoniacal silver solution and the cell is short-circuited, a large current of 20 to 30 μA may be observed. The current decreases rapidly and is practically zero after 5 to 10 minutes.

This procedure is very accurate and is simple to carry out. Moreover, the same type of amperomeric titration can be used to determine halides, alone and in the presence of each other, and cyanides. Samples containing more than 2 mg of mercaptan sulfur can be determined with the accuracy and precision of $\pm 0.3\%$. Substances used to test this procedure were ethyl, amyl, *n*-dodecyl, and cyclopentyl mercaptans, and also mercaptans in the C_{12} range whose purity was predetermined by another method.

ACIDIMETRIC SILVER METHOD

The direct titrations with silver ion just shown are quite widely applicable. Interference with halides does not often occur, since the dissociation of the silver mercaptides is small enough to be differentiated by potentiometric methods in ammoniacal systems. Halide interference can occur, however, especially with bromide or iodide and cyanide interference is more pronounced.

There is a method by which excess silver nitrate is added, forming the mercaptide and an equivalent amount of nitric acid. The titration of this acid can be used to determine mercaptans. Indicator end points can be used in the silver approach.

Method of B. Saville

[*Adapted from Analyst, **86**, 29–32 (1961)*]

Methods based on this principle, have been proposed by Sampey and Emmet Reid (8) and by Mapstone (9). The former workers utilized the related reaction involving mercuric chloride and titrated the liberated hydrochloric acid to a methyl red or methyl orange end point; the disadvantage mentioned was that the rather acidic solution at the equivalence point in the titration led to slightly low results. Titration to a higher pH was impossible, since mercuric oxide or hydroxide would have been formed, thereby invalidating the titration. Mapstone used silver sulfate and titrated the liberated sulfuric acid.

In this work, a weighed amount of the thiol is added to an excess of silver nitrate dissolved in aqueous pyridine, the mixture is diluted with water, and the pyridinium nitrate formed is titrated with standard alkali to a phenolphthalein end point. The reaction, involving coordinated silver ions (Py_nAg^+), is

$$Py_nAg^+ + RSH \rightleftharpoons (n-1)Py + PyH^+ + AgSR$$

The use of the aqueous pyridine solvent has several advantages:

1. The thiol is soluble in this medium, so that the mercaptide precipitate does not occlude unreacted thiol.
2. Since the silver ions are coordinated, there is little possibility of the phenolphthalein end point being in error.
3. Decomposition of the silver salts of certain unsaturated thiols to silver sulfide, a reaction occurring in the presence of excess of free silver ions, is avoided.
4. The removal of free protons as PyH^+ ensures that the mercaptide-forming reaction is quantitative and that the reverse reaction is avoided.

8. J. R. Sampey and E. Emmet Reid, *J. Am. Chem. Soc.*, **54,** 3404 (1932).
9. G. E. Mapstone, *Austral. Chem. Inst.. J. Proc.*, **13,** 232, 373 (1946).

REAGENTS

Pyridine, reagent grade.
Silver nitrate, approximately $0.4M$, aqueous.
SODIUM HYDROXIDE, $0.1N$, AQUEOUS. Standardize this solution with pure potassium hydrogen phthalate.

PROCEDURE

To 15 ml of pyridine in a stoppered 250-ml Erlenmeyer flask, add an accurately weighed amount (0.0010–0.0018 mole) of the thiol. As soon as possible, gradually add from a pipet 5 ml of the silver nitrate solution. Insert the stopper and set the flask aside for 5 minutes. Add about 100 ml of distilled water and 3 to 4 drops of phenolphthalein indicator solution, and titrate with $0.1N$ sodium hydroxide to a light pink end point. Note that the end point is often recognized as a change in the color of the yellow silver mercaptide to white, since the pink tint of the supernatant aqueous phase is almost complementary to the yellow of the precipitate.

A simple mixture of the silver nitrate solution and pyridine was found to be neutral to phenolphthalein and sensitive to 1 drop of added $0.1N$ sodium hydroxide; hence no blank correction is necessary.

CALCULATION

10 ml of $0.1N$ sodium hydroxide $\equiv$ 0.001 group equivalent of —SH

The percentage purity of the thiol is given by the expression

$$\frac{\text{Milliliters of NaOH} \times N\text{NaOH} \times \text{equivalent wt. of thiol.} \times 100}{\text{Weight of thiol taken} \times 1000}$$

The results (Table 4) indicate that the method can be applied to the assay of both saturated and unsaturated primary and secondary thiols and to tertiary aliphatic thiols. The validity of the method is shown particularly by the results for 2-methylpentane-2-thiol (10), since this compound was demonstrated to be pure by careful gas-liquid chromatographic examination.

The reproducibility of results by the technique can be judged from the results for but-2-ene-1-thiol and 4-methylpent-3-ene-2-thiol. These two sets of results suggest that a mean deviation of 0.14% and a maximum deviation of 0.37% might be accepted as a provisional description of the consistency of the method.

10. D. F. Lee, B. Saville, and B. R. Trego, *Chem. Ind.*, **1960,** 868.

Table 4. Purities of Various Thiols

Sample	Molecular Weight	Calculated Purity, %
Ethanethiol[a]	62.08	100.4
		101.0
But-2-ene-1-thiol	88.11	96.8
		96.7
		96.7
		96.6
		96.9
2-Methylpropane-2-thiol	90.18	83.8
		82.8
		100.6
		100.0
2-Methylpent-2-ene-1-thiol[b]	116.2	98.5
		98.6
4-Methylpent-3-ene-2-thiol	116.2	100.1
		100.1
		99.5
		100.0
		100.1
2-Methylpentane-2-thiol	118.2	100.2
		100.1
2-Methylpentane-3-thiol	118.2	97.9
Heptane-4-thiol	132.3	97.3
		96.9
3-Methylhexane-3-thiol	132.3	99.3
		99.5
Dodecane-1-thiol	202.4	98.1
		98.3

[a] Results for this thiol refer to the contents of commercially supplied ampoules immediately after being opened.

[b] Other isomeric thiols present in sample.

Oxidation Methods

IODINE METHOD

Adapted from J. W. Kimball, R. L. Kramer, and E. E. Reid

[*J. Am. Chem. Soc.*, ***43***, *1199–1200* (*1921*)]

$$2RSH + I_2 \rightarrow RSSR + 2HI$$

REAGENTS

Standard iodine, 0.1*N*.
Sodium thiosulfate, 0.1*N*.
Starch-indicator solution.

PROCEDURE

Weigh a sample containing 0.002 to 0.003 equivalent of mercaptan in a sealed glass ampoule (pp. 862–4). Place the ampoule in a 250-ml glass-stoppered Erlenmeyer flask with 50 ml of 0.1*N* iodine and some glass beads to aid in breaking the ampoule. Shake the flask vigorously to break the ampoule, and continue shaking to assure complete reaction. Titrate the excess iodine with 0.1*N* standard sodium thiosulfate, using indicator.

CALCULATIONS

$$\frac{C \times N \text{ Thiosulfate} \times \text{mol. wt. of mercaptan} \times 100}{\text{Grams of sample} \times 1000} = \% \text{ mercaptan}$$

where

A = milliliters of thiosulfate needed for 50 ml of the iodine used

B = milliliters of thiosulfate used in titration

$C = A - B$ = milliliters of thiosulfate equivalent to the mercaptan present

The following mercaptans were used to test this procedure: methyl, ethyl, propyl, *n*-butyl, *sec*-butyl, isobutyl, isoamyl, phenyl, *p*-cresyl, 2-naphthyl. Mercaptans determined by other authors were lauryl mercaptan, β-mercaptoethanol, and terpene mercaptan. This procedure is reproducible to ±0.4%. Hydrogen sulfide will interfere, as will any material oxidized by iodine.

CUPRIC ION OXIDATION METHODS

Adapted from the Cupric Alkyl Phthalate Method of E. Turk and E. E. Reid

[*Ind. Eng. Chem., Anal. Ed.*, **17,** *713–4* (*1945*)]

This method involves the oxidation of mercaptans with the cupric ion as follows:

$$2Cu^{2+} + 4RSH \rightarrow 2CuSR + RSSR + 4H^+$$

Cupric butyl phthalate and cupric octyl phthalate are used as the reagents to titrate the mercaptans directly. This method is fast and accurate but somewhat less accurate than the iodine method. This procedure is advantageous because it is effective in the presence of many substances that would be oxidized in the iodine method, thus yielding erroneous results. Unsaturated compounds will also not interfere in this procedure, whereas they usually do in the iodine method. Hydrogen sulfide will interfere in this procedure because of precipitation of copper sulfide. No interference was observed by hydrogen cyanide, organic thiocyanates, organic sulfides, thiocyanoacetates, and terpenes.

REAGENTS

CUPRIC BUTYL PHTHALATE. To 50 ml of *n*-butanol contained in a 500-ml Erlenmeyer flask, add 74 grams of finely divided phthalic anhydride. Heat the mixture with agitation to about 105°C. (The phthalic anhydride should not be allowed to hydrate by exposure to moisture. A sample of anhydride should remain clear when heated to 131°C in a test tube.) Stop the heating while continuing the agitation. The temperature continues to rise to 120°C, and the mixture becomes clear in a few minutes. Then cool the solution and pour it into a solution of 20 grams of sodium hydroxide in 1500 ml of water. If the resulting solution is not acid, it should be made so with acetic acid. Filter the solution into a 4-liter beaker. Slowly pour a solution of 65 grams of cupric sulfate (pentahydrate) in 500 ml of water (filtered to remove any solid particles) into the butyl phthalate mixture, with vigorous agitation. Collect the cupric butyl phthalate that precipitates on a Büchner funnel, wash with water, and dry in air. Pulverize the precipitate in a mortar and dry in a vacuum desiccator. The yield is about 95%.

The purity of the cupric butyl phthalate has to be determined, since this material is used as the standard. Dissolve a 0.3- to 0.5-gram sample in 5 ml of glacial acetic acid, dilute with 50 ml of water, and determine the copper iodometrically. The cupric butyl phthalate is stable for very long periods of time.

CUPRIC OCTYL PHTHALATE. This reagent is prepared exactly as the cupric butyl phthalate except that 68 grams of *n*-octanol is used instead of the *n*-butanol. Cupric octyl phthalate is difficult to filter, and washing by decantation is often necessary. Cupric butyl phthalate is more easily prepared and is more generally applicable. Therefore, there is no need to use the cupric octyl phthalate.

For the standard solutions, use 25.29 grams of the cupric butyl phthalate (or 30.9 grams of cupric octyl phthalate) to prepare a 0.1N solution. If P = purity of the cupric butyl phthalate, $25.29 \times 100/P$ grams is required for a 0.1N solution. Weigh the cupric butyl phthalate into a 1-liter volumetric flask, dissolve in 50 ml of acetic acid, and dilute to 1 liter with Pentasol, a mixture of the isomers of amyl alcohol (butanol, pentanol, or any hydrocarbon solvent can be used). Standardize the solution by the methods of Weatherburn, Weatherburn, and Bayley (11),

11. A. S. Weatherburn, M. W. Weatherburn, and C. H. Bayley, *Ind. Eng. Chem., Anal. Ed.*, **16,** 703 (1944).

which consists of pipetting a sample of solution containing about 50 to 100 mg of copper into a beaker and adding 10 ml of 1:1 hydrochloric acid. Boil the mixture for 2 to 5 minutes, then cool and pour into a separatory funnel. Draw off the aqueous layer, and wash the organic layer three times with 15 ml of water; combine the water washings with the original water layer. If the oil and water layers tend to emulsify, 3 to 5 ml of isopropyl alcohol can be added. Make the water solution alkaline with ammonium hydroxide and then very slightly acidic with glacial acetic acid. Add about 2 grams of potassium iodide, and titrate the liberated iodine with 0.1N thiosulfate, starch indicator being used.

The standardization should be carried out at about the same temperature as that in the procedure given below. A 10° temperature difference causes about a 0.1-ml error per 10-ml titration. The normality of the cupric $\begin{Bmatrix}\text{octyl}\\ \text{butyl}\end{Bmatrix}$ phthalate can be expressed as

$$\frac{\text{Milliliters of thiosulfate} \times N \times 2}{\text{Milliliters of copper solution}} = N \text{ copper solution}$$

(The factor 2 in the numerator is due to the reaction of 2 moles of mercaptan per mole copper reagent.)

The cupric alkyl phthalate solutions are stored in dark bottles to retard peroxide formation. Solutions containing considerable amounts of precipitate should be discarded.

PROCEDURE

Weigh a sample containing 0.1 to 0.3 gram of mercaptan (depending on molecular weight) into a 125-ml Erlenmeyer flask containing 50 ml of Pentasol (a synthetic amyl alcohol, a mixture of five of the isomers of amyl alcohol) or any hydrocarbon solvent. Weigh low-boiling mercaptans in sealed ampoules (pp. 862–864). Titrate the contents of the flask with the standard cupric alkyl phthalate solution. Add the copper solution in increments of about 0.5 ml until the end point is almost reached. The end point is the persistence of the blue-green color of the reagent. The solution may darken during the titration, but it clears before the end point. The end point can be observed more readily when the flask is illuminated and viewed against a white background.

In some cases, particularly with mercaptans of low molecular weight, the yellow cuprous mercaptide precipitates, but in others it remains dissolved. In either case, the end point is readily discernible. A definite amount of copper reagent is necessary to impart a green color to 50 ml of solution. The amount is small but significant. Therefore, run a blank to determine the amount of reagent necessary to produce the color.

CALCULATIONS

1 ml of $1N$ cupric alkyl phthalate = 0.0331 grams of —SH

If V is the volume used in titration, N the normality of copper reagent, and W the sample weight,

$$\frac{(V-\text{blank})\times N\times 0.0331\times 100}{W}=\%\ \text{—SH}$$

This method is quite generally applicable to all types of mercaptan. Two mercaptans that cannot be determined are thioglycolic acid and dithioethylene glycol. Methyl and butyl thioglycolates, however, behave normally.

Mercaptans used to test this procedure were pentadecanethiol; methyl and butyl thioglycolate; terpene mercaptan; *n*-butyl, isobutyl, *tert*-butyl, lauryl, and cetyl mercaptans; 2-mercaptoethanol. This method is accurate and precise to about ±1%.

Adapted from the Method of S. Bose, M. P. Sahasrabuddhey, and K. Verma

[*Talanta,* ***23,*** *725–6* (*1976*)]

In this method the sample dissolved in water, methanol, dimethylformamide, or acetonitrile is treated with a measured excess of copper sulfate. The excess ion is then back-titrated with a standard mercaptoacetic acid solution. This procedure overcomes end point difficulties encountered in the direct titration of some mercaptans and permits the determination on a semimicro scale.

REAGENTS

Standardized $0.1M$ copper sulfate and $0.1M$ mercaptoacetic acid, standardized by treatment with excess of iodine and thiosulfate back-titration.

PROCEDURE

Weigh a sample containing 0.3–1.0 meq. of mercaptan into a 150-ml Erlenmeyer flask and dissolve it in 20 ml of water, methanol, dimethylformamide, or acetonitrile. Add a measured excess of $0.1M$ cupric sulfate solution and swirl for 1 minute. Titrate with $0.1M$ mercaptoacetic acid from a 10-ml buret until the deep violet precipitate first formed changes to a permanent yellow. Near the end point, titrate slowly with vigorous swirling. Run a blank determination on the same volume of cupric sulfate solution.

RESULTS AND DISCUSSION

A wide variety of mercaptans, including primary, secondary, and tertiary compounds, was examined. In all cases the reaction was rapid and complete, and the end point was unaffected. Table 5 compares results obtained by this method and by independent methods.

Table 5. Determination of Mercaptans with Copper(II)

Compound	Purity, %		Comparison Method Technique
	Copper(II) Method[a]	Comparison Method	
1-Dodecanethiol	98.1	98.2	Hg^{2+} Titration[b]
o-Mercaptobenzoic acid	96.6	96.5	Pb^{4+} Titration[c]
2-Mercaptoethanol	99.1	99.3	Pb^{4+} Titration[c]
p-Chlorobenzenethiol	99.3	99.5	Alkalimetry[d]
1-Butanethiol	91.0	90.8	Iodimetry[e]
Toluene-α-thiol	98.6	98.4	Acetylation[f]
Allylthiol	93.5	93.6	Pb^{4+} Titration[c]
3-Mercaptopropionic acid	100.0	99.8	Iodimetry[e]
Mercaptosuccinic acid	96.7	96.5	Hg^{2+} Titration[b]
2-Mercaptopropionic acid	95.9	95.7	Iodimetry[e]
2-Methylpropane-2-thiol	99.0	98.9	Acetylation[f]
2-Naphthalenethiol	98.2	98.0	Alkalimetry[d]
p-Toluenethiol	97.6	97.6	Alkalimetry[d]
2-Diethylaminoethanethiol	98.6	98.4	Iodimetry[e]
2-Mercaptobenzothiazole	99.2	99.0	Alkalimetry[d]
2-Mercaptobenzimidazole	97.8	97.6	Alkalimetry[d]
2-Mercaptobenzoxazole	98.9	98.8	Alkalimetry[d]
2-Butanethiol	97.9	97.8	Hg^{2+} Titration[b]
Cyclohexanethiol	95.3	95.6	Pb^{4+} Titration[c]

[a] Average of 10 determinations; average deviation in the range 0.2 to 0.3%.
[b] J. S. Fritz and T. A. Palmer, *Anal. Chem.*, **33,** 98 (1961).
[c] K. K. Verma and S. Bose, *Anal. Chim. Acta*, **65,** 236 (1973).
[d] K. K. Verma, *Talanta*, **22,** 920 (1975).
[e] J. W. Kimball, R. L. Kramer, and E. E. Reid, *J. Am. Chem. Soc.*, **43,** 1199 (1921).
[f] G. H. Schenk and J. S. Fritz, *Anal. Chem.*, **32,** 987 (1960).

Organic sulfides, disulfides, thiocyanates, unsaturated compounds, chloride, bromide, and sulfite do not interfere. Serious interference is caused by thiourea and ions such as sulfide, thiosulfate, thiocyanate, and cyanide. Ethylenedithiol and *o*-aminobenzenethiol do not reduce copper(II) but form cupric mercaptides.

Mixtures of Mercaptan and Free Sulfur

Davies and Armstrong (12) published a modification of the potentiometric method of Tamele and Ryland (pp. 708–714), which gave separate potentiometric peaks for the titration of the free sulfur and the mercaptan. Karchmer's expanded form of the approach of Davies and Armstrong is given below.

Study of J. H. Karchmer

[*Reprinted in Part from Anal. Chem.*, ***29****, 425–31 (1957)*]

The original method of Tamele and Ryland (13) (pp. 708–714) consists of potentiometrically titrating the sample dissolved in alcohol containing 0.1*N* sodium acetate to buffer the solution, with an alcoholic solution of silver nitrate using a silver indicator electrode and a mercury–sodium acetate half-cell. Later a glass electrode was employed as the reference electrode (14) in place of the mercury–sodium acetate half-cell. External calomel cells have also been successfully used. These cells are electrically connected to the solution by an agar-saturated potassium nitrate bridge to avoid contaminating the solution with chloride ions from the calomel cell. In the presence of hydrogen sulfide and/or elemental sulfur, erroneous results may be obtained unless the titration curve is correctly interpreted. When hydrogen sulfide is present, the initial potential between the silver electrode and the solution is approximately −0.7 volt (Fig. 18.7). As the silver ion is added to the solution, silver sulfide is precipitated, and after all the sulfide ion has reacted, the potential breaks sharply to the voltage characteristic of the particular mercaptan present. This voltage is influenced largely by the solubility product of the silver mercaptide in the solvent. For *n*-butyl mercaptan it is about −0.35 volt. Upon continued addition of silver ion to the solution, silver mercaptide begins to precipitate and finally another sharp "break" in voltage is observed, which corresponds to the end point of the mercaptan titration.

The presence of hydrogen sulfide presents no difficulty because the volume of silver nitrate used to reach the first break in voltage is taken as that required for the sulfide ion, and the volume of silver nitrate from that point to the last break in voltage is taken as that required for the mercaptan sulfur. In calculating the sulfide sulfur content, 0.016 is taken as the milliequivalent weight; 0.032 is taken to calculate the mercaptan

12. E. R. H. Davies and J. W. Armstrong, *J. Inst. Petrol.* **29,** 323 (1943).
13. M. W. Tamele and L. B. Ryland, *Ind. Eng. Chem., Anal. Ed.*, **8,** 16 (1936).
14. Louis Lykken and F. D. Tuemmler, *Ind. Eng. Chem., Anal. Ed.*, **14,** 67 (1942).

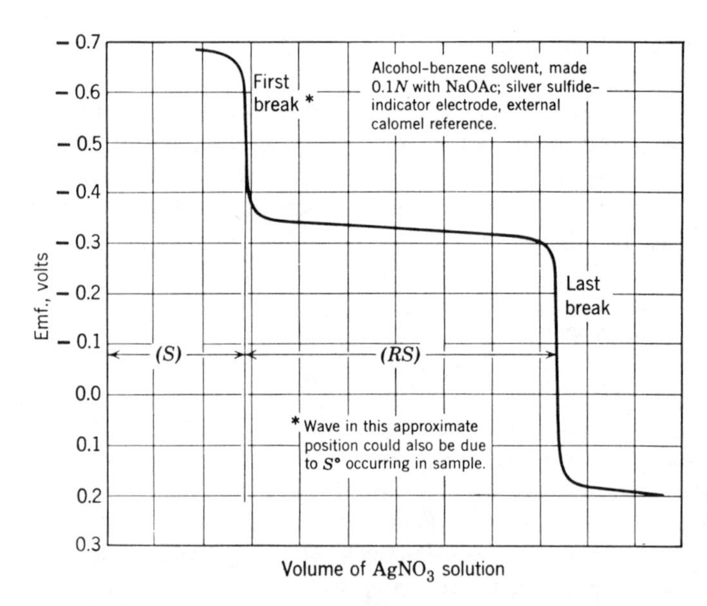

Fig. 18.7. Typical titration curve of hydrogen sulfide and mercaptan with silver nitrate.

sulfur content. When elemental sulfur is present in the sample with a mercaptan, a reaction occurs which is accelerated when the sample is placed in an alkaline titrating solvent. One of the products of the reaction is the sulfide ion or some material that reacts with silver ion to yield a sharp break in potential in the region of the sulfide break. Thus when sulfur is present with the mercaptan, the titration characteristics of the sample are the same as they would be if hydrogen sulfide were present. Since the analyst does not know whether hydrogen sulfide or elemental sulfur is present, he must treat the sample to remove one or the other of these materials before a correct interpretation of the titration curve can be made.

Contact with mercury to remove elemental sulfur is not satisfactory, since this treatment partially removes the mercaptans, as shown by Schindler, Ayers, and Henderson (15). This finding was confirmed in Karchmer's laboratory. The other alternative removal of hydrogen sulfide by washing with acidified cadmium salts (16, 17), has been recommended by Davies and Armstrong (18).

15. Hans Schindler, G. W. Ayers, and L. M. Henderson, *Ind. Eng. Chem., Anal. Ed.*, **13,** 327 (1941).
16. P. Borgstrom and E. E. Reid, *Ind. Eng. Chem., Anal. Ed.*, **1,** 186 (1929).
17. W. F. Faragher, J. C. Morrell, and G. S. Monroe, *Ind. Eng. Chem.*, **19,** 1281 (1927).
18. E. R. H. Davies and J. W. Armstrong, *J. Inst. Petrol.* **29,** 323 (1943).

This leaves elemental sulfur in the sample, which will react with the mercaptan in the alkaline solvent to yield two titration breaks. In calculating the mercaptan content, however, the first break is ignored and the total volume of silver nitrate used in titrating to the second break is employed with 0.032 as the milliequivalent weight.

Davies and Armstrong (18), in presenting the stoichiometric relationship between the elemental sulfur and the mercaptan, postulate the formation of an intermediate, sodium alkyl disulfide, which reacts with silver ion to produce silver sulfide and an organic trisulfide.

$$2S + 2RSNa \rightarrow 2RSSNa \quad (1)$$

$$2RSSNa + 2Ag^{+} \rightarrow Ag_2S + R_2S_3 + 2Na^{+} \quad (2)$$

At the first break, the titration corresponds to the reaction shown in eq. 2, whereas the second break corresponds to the reaction of silver ion with the remainder of the mercaptan sulfur.

$$Ag^{+} + RSH \rightarrow RSAg + H^{+} \quad (3)$$

This interpretation indicates two facts of analytical importance: (1) that the mercaptan content of a sample can be determined even though elemental sulfur is present, and (2) that elemental sulfur itself can be determined by measurement of the first break. In both cases 0.032 is used as the milliequivalent weight.

In using this method with synthetic samples of many types it was found that the mercaptan content in many cases was low, and often the first break did not correlate with the amount of elemental sulfur present.

Consequently, an investigation was undertaken to explain some of the difficulties and to determine the conditions that would yield complete recoveries of mercaptans in various types of refinery samples. The practice of taking the sum of both breaks to obtain the mercaptan content is satisfactory for certain types of samples, but in other types when the ratio of mercaptan sulfur to elemental sulfur is less than 1 : 1, the titration must commence immediately after the sample is placed in the titration solvent; even after 5 minutes, noticeably lower results are obtained (first part of Table 6).

This effect may be largely eliminated by the use of a more acidic type of solvent. In the course of this work there was some experimental evidence to indicate the existence of the monoalkyl disulfide ion, $(RSS)^{-}$, as postulated by Davies and Armstrong (18). This ion is difficult to detect because its silver salt decomposes readily, the ion degrades to yield a sulfide ion on standing even in air-free systems, and the potential produced by this ion versus the silver sulfide electrode is very nearly the

Table 6. Effect of Time and Sulfur Content on Accuracy of Potentiometric Mercaptan Determination

Sample Mercaptan plus Rhombic Sulfur	Solvent	Sulfur Present in Sample, mg.		Approximate Ratio, SH:S	Time, min.	Volume of Titrant[a]			Total Mercaptan Found	
		Mercaptan	Elemental			Apparent Sulfide	Apparent Mercaptan	Total	mg.	%
n-Amyl	Regular	2.47	0	Blank	5	—	7.70	7.70	2.47	100.0
					120	—	7.41	7.41	2.37	96.0
		2.47	5.04	1:2	5	6.32	0	6.32	2.02	82.0
					30	4.21	0	4.21	1.35	54.7
					60	3.00	0	3.00	0.96	38.9
					120	1.69	0	1.69	0.54	21.9
		2.47	2.52	1:1	5	7.17	0	7.17	2.30	93.0
					30	6.55	0	6.55	2.10	85.0
					60	6.12	0	6.12	1.96	79.4
					120	4.59	0	4.59	1.47	59.6
		2.47	0.81	3:1	5	4.47	3.18	7.65	2.45	99.3
					30	3.16	4.33	7.49	2.40	97.3
					60	5.06	2.40	7.46	2.39	96.8
					120	2.67	4.44	7.01	2.28	91.0
	Acidic	2.47	0	Blank	5	0.0	7.70	7.70	2.47	100.0
					120	0.0	7.70	7.70	2.47	100.0
		2.47	4.520	1:2	5	7.72	0.0	7.72	2.48	100.4
					30	7.50	0.0	7.50	2.40	97.2
					60	7.25	0.0	7.25	2.33	94.3
					120	6.50	0.0	6.50	2.09	84.6
		2.47	2.58	1:1	5	7.71	0.0	7.70	2.47	100.0
					30	7.65	0.0	7.65	2.45	99.2
					60	7.35	0.0	7.35	2.36	95.5
					120	6.95	0.0	6.95	2.23	90.3
		2.47	0.774	3:1	5	3.55[b]	4.18	7.73	2.48	100.4
					30	2.51[b]	5.21	7.72	2.48	100.4
					60	2.68[b]	5.04	7.72	2.48	100.4
					120	2.58[b]	4.87	7.45	2.40	97.2
Phenyl	Acidic	3.11	0	Blank	5	—	9.70	9.70	3.11	100.0
					120	—	8.92	8.92	2.86	92.0
		3.11	6.05	1:2	5	9.82	0.0	9.82	3.15	101.3
					30	9.63	0.0	9.63	3.09	99.4
					60	9.04	0.0	9.04	2.90	93.2
					120	8.48	0.0	8.48	2.72	87.5
		3.11	3.03	1:1	5	9.88	0.0	9.88	3.17	101.9
					30	9.54	0.0	9.54	3.06	98.4
					60	9.20	0.0	9.20	2.95	94.9
					120	8.48	0.0	8.48	2.72	87.5
		3.11	1.01	3:1	5	5.10	4.70	9.80	3.14	101.0
					30	4.50	4.90	9.40	3.01	96.8
					60	4.88	4.32	9.20	2.95	94.9
					120	4.0	4.58	8.58	2.75	88.4
tert-Butyl	Acidic	3.25	0	Blank	0	—	10.12	10.13	3.25	100.0
					120	—	10.06	10.09	3.24	99.7
		3.25	6.05	1:2	5	0.0	10.19	10.19	3.27	100.6
					30	7.18	3.01	10.19	3.27	100.6
					60	6.90	3.11	10.05	3.22	99.1
					120	6.50	2.95	9.45	3.03	93.2
		3.25	3.03	1:1	5	0.0	10.40	10.40	3.33	102.5
					30	4.05	6.15	10.20	3.27	100.6
					60	3.85	6.25	10.10	3.24	99.7
					120	3.68	6.26	9.94	3.19	95.7
		3.25	1.01	3:1	5	0.0	10.32	10.32	3.31	101.8
					30	0.0	10.18	10.18	3.26	100.3
					60	0.60	9.60	10.10	3.24	99.7
					120	1.02	8.91	9.93	3.18	97.8
		3.19	12.6	1:4	5	9.77	0	9.77	3.13	98.2
					30	9.71	0	9.71	3.11	97.5
					60	9.65	0	9.65	3.09	96.9
					120	8.44	0	8.44	2.71	84.8

[a] Alcoholic 0.01N $AgNO_3$.
[b] Poor breaks.

same as that produced by the sulfide ion. Consequently, a break between these is not readily apparent. The stoichiometry of the reaction of mercaptide and elemental sulfur is not always that shown in eq. 2, since there is a difference in the degree to which various organic radicals retain sulfur atoms in forming organic polysulfides. Consequently, the practice of determining the sulfur content from the first break should be discouraged, except under highly controlled conditions, when the type of mercaptan is known and the mercaptan–elemental sulfur ratio is high.

METHOD, APPARATUS, AND MATERIALS USED

The procedure followed in carrying out these determinations was essentially that described by Tamele and Ryland (13) (pp. 708–714), with the following exceptions.

A Recordomatic titrator (Precision Scientific Co.) was used. The silver sulfide–glass electrode system described by Lykken and Tuemmler (14) was employed for most of the experimental work; in certain cases an external calomel cell was substituted for the glass electrode. The exceptions are indicated. Calomel and glass cells are equal in performance in their application, although the initial voltages and the magnitude of the voltage breaks may differ slightly.

The mercaptans used were obtained from Eastman Kodak Company (Distillation Products Industries), White Label grade. The synthetic blends were prepared by dissolving the mercaptan in isopropyl alcohol and using the actual mercaptan sulfur content as obtained from a potentiometric determination.

In preparing the titration solvents, a mixture of methanol, isopropyl alcohol, and benzene was substituted for ethyl alcohol, originally recommended by Tamele and Ryland, because this mixture could dissolve larger quantities of hydrocarbon samples. These materials were purified by silica gel percolation. Two titration solvents were used. The "regular" solvent was prepared by dissolving 13.7 grams of sodium acetate trihydrate in 20 ml of distilled water and adding to it a mixture of 400 ml of methanol and 400 ml of isopropyl alcohol. Sufficient benzene was added to bring volume to 1 liter. The "acidic" solvent was prepared by dissolving 13.7 grams of sodium acetate trihydrate and 6 ml of glacial acetic acid in 500 ml of methanol and diluting to 1 liter with benzene. Although isopropyl alcohol and a small amount of water were used in the regular solvent employed in this particular study, these are not essential, and a simple regular solvent of methanol, benzene, and sodium acetate would be equivalent.

TITRATION OF MERCAPTANS IN PRESENCE OF ELEMENTAL SULFUR

The first series of experiments was designed to establish a base case and to demonstrate the difficulties that may arise under extreme conditions with the alkaline (regular) solvent. In this experiment, a series of synthetic samples was prepared containing *n*-amyl mercaptan and elemental sulfur in different mole ratios. A portion of each of these blends was dissolved in the alkaline titration solvent and titrated 5, 30, 60, and 120 minutes after being put into solution (first portion of Table 6). The volumes of silver nitrate required to reach the first and last breaks are shown together with the total mercaptan recovery as calculated by taking the total silver nitrate titer and using 0.032 as the milliequivalent weight. The greater the ratio of elemental sulfur to mercaptan sulfur and the longer the time of standing in the alkaline titration solvent, the poorer are the recoveries. As long as elemental sulfur was equal to or greater than the mercaptan sulfur content, all the titratable material was found at the sulfide break. When mercaptan sulfur was in excess, two breaks were obtained. In the solutions exposed to air, the correlation of the first break with the amount of elemental sulfur appears to be erratic.

After a series of preliminary experiments it was observed that the poor results were related in some way to the alkalinity of the solution. It was then decided to employ a less basic solvent to carry out the titration. It could not be too acidic because of the possibility of losing the mercaptans of low molecular weight or increasing the solubility of the silver mercaptides. The solvent selected was the same as used in the polarographic determination of elemental sulfur and other sulfur compounds (19) whose preparation has been described.

Various blends of the three different types of mercaptans with elemental sulfur were prepared, and the mercaptan sulfur–elemental sulfur ratios varied from 1:4 to 3:1. Portions of each of these blends were placed in the acidic titrating solvent and the titration was started 5, 30, 60, and 120 minutes after the addition of the sample to the solvent. The volumes of $0.01N$ silver nitrate solution required to reach both the first and the last break were recorded. The mercaptans employed in this series were *n*-amyl mercaptan, phenyl mercaptan, and *tert*-butyl mercaptan.

From these data (Table 6) it may be seen that the acidic solvent minimizes the errors obtained, even under the drastic conditions of high elemental sulfur content and exposure to air oxidation. Although the acidic solvent gives better results for the mercaptan content, it is not recommended in determining elemental sulfur.

The foregoing experiments were carried out in open titration vessels, because open vessel titrations are commonly employed in many service laboratories. Initially it was thought that the poor recoveries were entirely due to the oxidation of sodium sulfide to sodium polysulfide and sodium thiosulfate. The latter two materials also titrate with silver nitrate in the

19. J. H. Karchmer and M. T. Walker, *Anal. Chem.*, **26**, 271 (1954).

regular solvent but yield poorly defined curves, which may confuse the interpretation. Furthermore, sodium polysulfide does not react with silver nitrate in the same stoichiometric ratio as does sodium sulfide. This was illustrated by titrating a freshly prepared solution of sodium sulfide with silver nitrate in the regular solvent and comparing the results with those obtained when the same solution with added sulfur (1:1 mole ratio) was titrated 5, 30, 60, and 120 minutes after standing. Table 7 shows that in the presence of elemental sulfur, the sulfide recovery is low and becomes lower as time of standing increases.

Table 7. Reaction of Sodium Sulfide and Elemental Sulfur in Alkaline Solvent

Solvent. Regular
Sample. Sodium sulfide and rhombic sulfur added to 100 ml. of titration solvent

Sample Composition, mg.		Reaction Time, min.	Sulfide Sulfur Found	
Sulfide Sulfur	Elemental Sulfur		mg.	%
2.24	0	5	2.24	100
		120	2.17	96.9
2.24	2.52	5	1.28	57.1
		30	0.70	31.2
		60	0.59	26.3
		120	0.32	14.3

TITRATION IN ABSENCE OF AIR

To determine whether oxidation was the only cause of the poor recoveries a series of experiments was conducted under air-free conditions. As may be seen in Table 8, even in the absence of air, the mercaptan recoveries are low when an excess of sulfur is present and losses become greater as the length of time increases. The losses in air, however, were greater.

Although the elimination of air from the titration did not produce complete recoveries, it minimized the formation of sodium polysulfide and sodium thiosulfate, hence allowed more information to be obtained from the titration curves. Sodium polysulfide and sodium thiosulfate produce a series of breaks in both titration solvents that obscure other end points and make interpretations of the intermediate portions (as reported in Table 6) of the titration curve extremely doubtful.

Table 8. Effects of the Exclusion of Air on the Potentiometric Determination of Mercaptan

Titration solvent. Regular, 50–50 volume % isopropyl alcohol–benzene made 0.1*N* with sodium acetate
Sample. 3.975 mg. of *n*-amyl mercaptan sulfur + 8.028 mg. of rhombic sulfur (approximate ratio SH:S = 1:2)
Solutions blown with nitrogen before mixing and blanketed after mixing and during storage period

Time of Standing, min.	Mercaptan Recovered, %	
	In Air[a]	In Nitrogen
5	82.0	90.0
30	54.7	83.1
60	38.9	75.7
120	21.9	69.7

[a] Data taken from Table 6.

Synthetic solutions of a mercaptan and elemental sulfur (3:1 mole ratio) were prepared in a nitrogen atmosphere and allowed to stand in the regular solvent for 5, 30, 60, and 120 minutes before titration under a nitrogen blanket. The compounds used were phenyl, *n*-butyl, and *tert*-butyl mercaptans (Table 9). Figure 18.8 gives titration curves for only the *tert*-butyl mercaptan mixtures. The titration breaks of the remaining two mercaptans, while showing similar characteristics, were not as well defined as *tert*-butyl mercaptan. The lack of resolution of these curves is probably due to the presence of the polysulfide ion, which can form in this solution even in absence of air.

From Fig. 18.8 it may be seen that the 5-minute titration curve possessed only two breaks: one with an end point in the vicinity of −0.43 volt and the other at −0.05 volt. On standing 30 minutes, an additional break is visible at −0.57 volt; at 60 minutes the material causing the break, which occurs at about −0.57 volt, has increased, and at 120 minutes its concentration has increased still further. Concurrent to this change the concentration of the material causing the second break (approximately −0.43 volt) is decreasing. The same trend may be observed in the data (Table 9) for the other two mercaptans.

To determine whether the sulfide ion was involved in this change, a portion of the *tert*-butyl mercaptan and sulfur mixture that had been allowed to stand for 120 minutes was purged with nitrogen for 10 minutes and hydrogen sulfide was identified in the effluent. The titration curve of

Table 9. Effects of Mercaptan Type and Storage Time on the Accuracy of Potentiometric Mercaptan Determination

Air-free system
Regular titration solvent
Ag_2S–ext. calomel electrodes
Approximate mercaptan–elemental sulfur ratio, 3 to 1

Sample	Length of Standing, min.	Volume of 0.01*N* $AgNO_3$ Ml. to Titrate					Total Sulfur Found	
		1 $(S)^{--}$	2 $(RSS)^{-}$[a]	3 1 + 2	4 $(RS)^{-}$	5 3 + 4	mg.	%
$tC_4SH + S^{\circ}$	5	0.00	3.23	3.23	10.46	13.69	4.39	99.3
(4.42 mg. mercaptan sulfur	30	0.36	2.58	2.94	10.76	13.70	4.39	99.3
1.535 mg. elemental sulfur)	60	0.80	1.60	2.40	10.90	13.30	4.26	96.5
	120	1.02	0.90	1.92	11.03	12.95	4.15	93.9
$nC_4SH + S^{\circ}$	5	0.00	5.78	5.78	9.76	15.54	4.98	99.6
(5.00 mg. mercaptan sulfur	30	2.15[b]	3.67	5.82	9.99	15.81	5.06	101.2
1.616 mg. elemental sulfur)	60	3.30[b]	2.38	5.68	9.27	14.95	4.79	95.9
	120	4.30[b]	1.72	6.02	8.98	15.00	4.81	96.1
$\phi SH + S^{\circ}$	5	[c]	[c]	11.55	7.30	18.85	6.04	97.8
(6.18 mg. mercaptan sulfur	30	[c]	[c]	12.03	5.27	17.30	5.54	89.7
2.046 mg. elemental sulfur)	60	7.52	2.48	10.00	6.24	16.24	5.20	84.2
	120	7.02	2.29	9.31	6.92	16.23	5.20	84.2

[a] Material titrated by intermediate break, postulated to be $(RSS)^{-}$.
[b] Poor break, making values for $(S)^{--}$ and $(RSS)^{-}$ somewhat doubtful.
[c] Very poor break, making values for $(S)^{--}$ and $(RSS)^{-}$ unreliable.

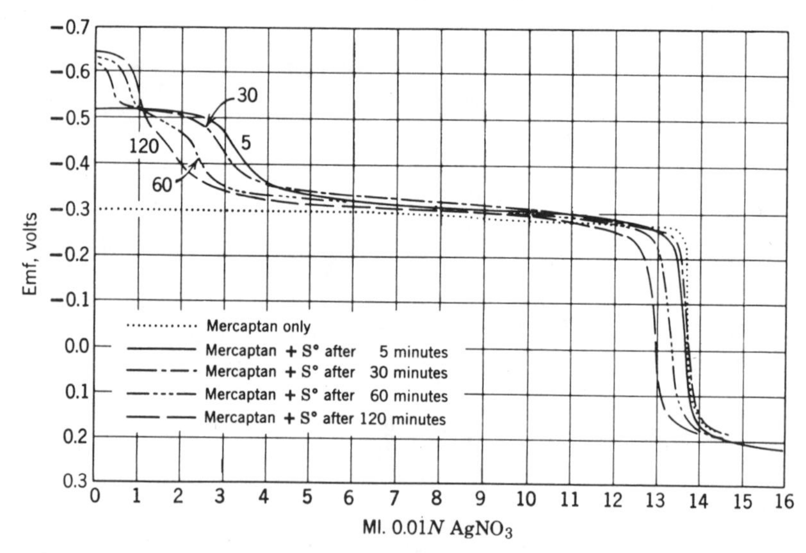

Fig. 18.8. Titration curve of *tert*-butyl mercaptan in the presence of sulfur, but in the absence of air. Regular titration solvent; silver sulfide–calomel electrodes; titrant, 0.001*N* silver nitrate; mercaptan–elemental sulfur ratio, 3:1.

this purged sample reflected a substantial lowering of the titer required to reach the first break. Purging of a portion of the sample that had stood only for 5 minutes revealed no significant loss of hydrogen sulfide and no essential change in the amount of titer to reach the break at −0.43 volt. It thus appears that the first break is caused by the presence of sulfide ion and the intermediate break is caused by some compound that cannot be removed from the solution by 10-minute nitrogen blowing. The sulfide was not present immediately, but required some time to form. These facts tend to support the Davies and Armstrong (18) postulation that sulfur plus a mercaptide yields $(RSS)^-$, and to indicate that this $(RSS)^-$ ion will decompose to a sulfide on standing, even in absence of air.

INTERPRETATION OF EXPERIMENTAL RESULTS

CHEMISTRY INVOLVED. Not too much is definitely known about the reaction of elemental sulfur and mercaptans in an alkaline solution. In a comprehensive review of the chemistry of mercaptans, all that Malisoff, Marks, and Hess (20) have to say about it is that the reaction is "surmised" to be

$$2RSH + S \rightarrow R_2S_2 + H_2S \qquad (4)$$

20. W. M. Malisoff, E. M. Marks, and F. G. Hess, *Chem. Rev.*, **7**, 493 (1930).

Faragher, Morrell, and Monroe (17) note that elemental sulfur, mercaptans, and sodium hydroxide react to form sodium sulfide and an organic disulfide. Holmberg (21) states that ethyl mercaptan dissolved in an alkaline solution produces ethyl disulfide when sulfur is added.

The reaction presented by Davies and Armstrong (eqs. 1 and 2) is of interest because it can serve as a starting point to interpret certain experimental observations herein reported.

Equation 1 may be written in the ionic form

$$(RS)^- + S^\circ \rightarrow (RSS)^- \quad (1')$$

This ion may react with silver ion to form the unstable silver compound

$$(RSS)^- + Ag^+ \rightarrow AgSSR \quad (5)$$

which decomposes to

$$2AgSSR \rightarrow Ag_2S + RSSSR \quad (6)$$

The sum of eqs. 1′, 5, and 6 is

$$2(RS)^- + 2S^\circ + 2Ag^+ \rightarrow Ag_2S + RSSSR \quad (7)$$

which is identical with the sum of eqs. 1 and 2.

However if a solution is allowed to stand (without the silver ion and in absence of air), sulfide ions are produced, as was shown experimentally. This reaction could be represented

$$2(RSS)^- \rightarrow (S)^{--} + RSSSR \quad (8)$$

As shown experimentally (Table 7), this sulfide ion could combine with elemental sulfur

$$(S)^{2-} + S^\circ \rightarrow (SS)^{2-} \quad (9)$$

to form a polysulfide ion (disulfide ion in this case), and if there is a sufficient excess of elemental sulfur, the trisulfide or tetrasulfide ion probably forms. It is not known which of the polysulfide ions, $(SS)^{2-}$, $(SSS)^{2-}$, or $(SSSS)^{2-}$, yields poorer recoveries. Presumably the poorer recoveries are associated with the higher polysulfides, and each ionic species could yield a slightly different titration break with silver ion.

Another mechanism by which the existence of polysulfide ions in the mixtures can be postulated is as follows. The monoalkyl disulfide ion, $(RSS)^-$, may be able to add elemental sulfur to form a monoalkyl trisulfide ion

$$(RSS)^- + S^\circ \rightarrow (RSSS)^- \quad (10)$$

21. Bror Holmberg, *Ann.*, **359,** 81–99 (1908).

This material on standing could decompose

$$(RSSS)^- \rightarrow (SS)^{2-} + (RS)^+ \tag{11}$$

The $(RS)^+$ could combine with some of the negative ions

$$(RS)^+ + (RSS)^- \rightarrow RSSSR \tag{12}$$

These mechanisms are largely speculative, since there is no physical evidence proving the existence of the $(RS)^+$ and $(RSS)^-$ ions. However some experimental data support the validity of eqs. 7 and 8, since polarograms of mercaptan sulfur and elemental sulfur obtained with the acidic solvent indicated the presence of organic polysulfides. A trisulfide was actually identified as the product in eq. 7, using the polarographic techniques described by Karchmer and Walker (19), employing *n*-butyl mercaptan and elemental sulfur.

These equations indicate possible routes by which polysulfide ions could form in the solutions, as it was shown that the presence of polysulfide ions is related to the poor recoveries. In the presence of air the polysulfide would further oxidize to thiosulfate and complicate the titration still more.

Equation 9 is accelerated in presence of a base; hence in the more acidic solvent, we may expect a decrease in the amount of polysulfide ion produced, which is consistent with the fact that better recoveries are obtained in the acidic solvent. By analogy, the reaction shown in eq. 10 could be minimized in the acidic solvent.

To determine whether the existence of polysulfide ions was dependent on the presence of mercaptans, elemental sulfur without any mercaptan was allowed to stand in the regular solvent for 5, 30, 60, and 120 minutes before titration with silver nitrate. Although a discoloration of the solution was observed after the addition of a small amount of silver nitrate, the potential observed in all cases was lower than +0.2 volt, which is more positive than the final potential obtained for the mercaptan titration. Thus in 2 hours the reaction of the elemental sulfur with the regular solvent would not be a significant factor. The results of this experiment were essentially the same when it was repeated in presence of ammonium hydroxide, which was added to minimize the formation of silver oxide.

EFFECT OF ORGANIC SUBSTITUENT OF MERCAPTANS. In this work there was some experimental evidence of differences in the behavior of the various mercaptans. It was suspected that the stoichiometry of the reaction shown in eq. 7 may not be the same for all mercaptan types. For example, Eby (22) showed that tertiary mercaptans could form stable tetrasulfides in

22. L. T. Eby, Division of Organic Chemistry, 118th Meeting of the American Chemical Society, Chicago, September 1950.

presence of an alkaline sulfide solution, whereas *n*-butyl tetrasulfide was less stable. Work by Karchmer and Walker (19) and Birch, Cullum, and Dean (23) has indicated a difference in the degree to which various organic radicals in polysulfides retain sulfur atoms. Thus the end product in eq. 7 would not necessarily be a trisulfide, and the amount of sulfur consumed could be more or less than indicated. When R is a normal alkyl, eq. 7 is essentially correct and the organic trisulfide is formed: however when R is a tertiary alkyl group, such as *tert*-butyl or *tert*-amyl, the reaction could be

$$2(tC_4S)^- + 3S^\circ + 2Ag^+ \rightarrow tC_4SSSStC_4 + Ag_2S \qquad (13)$$

When R is a phenyl group the overall reaction may be

$$2(\phi S)^- + S^\circ + 2Ag^+ \rightarrow \phi SS\phi + Ag_2S \qquad (14)$$

This is supported by data in Table 9, which show that when a 3:1 ratio of *tert*-butyl mercaptan sulfur to elemental sulfur is used, the amount of unreacted mercaptan present (after 5 minutes) equals 76.5% (10.46 ml out of a total of 13.69 ml of silver nitrate used). According to eq. 13, when 9 moles of mercaptan is used with 3 moles of elemental sulfur, only 2 moles of the mercaptan should be converted to a sulfide, leaving $\frac{7}{9}$ or 77.8% of mercaptan unreacted, which is in close agreement with the experimental data. With *n*-amyl mercaptan, which reacts as shown in eq. 7, 1 mole of mercaptan should react with 1 mole of elemental sulfur; thus at a 3:1 mercaptan elemental sulfur ratio, 2 of the 3 moles of mercaptan should be unreacted for a theoretical 66.7%. From Table 9 it may be computed that at the end of 5 minutes, 9.76/15.54 or 62.8% of mercaptans remained. By the same line of reasoning, the amount of excess phenyl mercaptan sulfur remaining when 3:1 mixtures are used should be 1 mole out of 3 or 33.4%, according to eq. 14. The data in Table 9 for phenyl mercaptan indicate that 7.30/18.85 or 38.8% of unreacted mercaptan sulfur was present at the end of 5 minutes. Although the agreements in the latter two cases were not as good as that obtained with the *tert*-butyl mercaptan, the trend is nevertheless apparent.

From the differences in the observed reactions of elemental sulfur with various mercaptans, certain deductions may be made. Since di-*tert*-butyl tetrasulfide is a reasonably stable compound, its formation uses the excess elemental sulfur that otherwise would promote the formation of inorganic polysulfides deleterious to the titration. Since the di-*n*-butyl tetrasulfide is less stable than the di-*n*-butyl trisulfide and the diphenyl trisulfide is less stable than the diphenyl disulfide, increasing amounts of inorganic polysulfides may form with the same mercaptan–elemental sulfur ratio as

23. S. F. Birch, T. V. Cullum, and R. H. Dean, *J. Inst. Petrol.* **39**, 206 (*April* 1953).

one progresses from *tert*-butyl to *n*-butyl to phenyl. It would also follow that the so-called unstable organic polysulfides could react with other mercaptans.

Adapted from the Study of M. W. Tamele, L. Ryland, and R. N. McCoy

[*Anal. Chem.*, ***32***, *1007* (*1960*)]

The quantitative nature of the precipitation reactions was confirmed for aqueous mixtures of low molecular weight mercaptans and hydrogen sulfide in almost all proportions. However coprecipitation occurs in aqueous electrolyte solutions with mixtures of hydrogen sulfide and higher molecular weight mercaptans, resulting in low recoveries of sulfide and correspondingly high values for mercaptans; a reversed coprecipitation effect, which occurs in alcoholic electrolyte, results in high values for sulfide and correspondingly low values for mercaptans. In these cases it is necessary to titrate a separate portion of the sample from which the sulfide has been removed.

The interference of elemental sulfur was also confirmed. The presence of elemental sulfur is evidenced by formation of a yellow color when the sample is added to the electrolyte, by a high potential, and by the formation of a black silver sulfide during titration of a sample from which hydrogen sulfide has been removed. In this case the total titration value to the mercaptan end point, if mercaptan is in excess of sulfur, accurately represents the mercaptan originally present. If elemental sulfur is in excess of the mercaptan, all the mercaptan appears as sulfide, the excess sulfur remaining unreacted.

APPARATUS

The silver sulfide indicating electrode consisted of a straight silver (99.9% Ag) rod about 2 mm in diameter that was immersed a few centimeters in the titrated solution. It was polished to a clean silver surface with a fine emery cloth, then soaked in potassium cyanide solution to clean it further; finally it was coated with silver sulfide, either electrolytically or by use of a preliminary titration of sodium sulfide. After being coated, the electrode was rinsed, wiped with a soft tissue, and burnished lightly with a fine emery cloth. The electrode was soaked in alcohol containing sodium acetate and 0.5% silver nitrate for 5 minutes before use and was stored in the same solution when not in use. Before each titration the electrode was burnished lightly with a soft cloth. This treatment is normally sufficient for a large number of titrations, but if anomalous titration curves are obtained, it is advisable to remove the sulfide coating completely and to repeat the procedure above.

A glass electrode is preferred as a reference electrode.

REAGENTS

All reagent solutions must be free of dissolved air and peroxidic impurities to prevent losses of the easily oxidized sulfur compounds. Percolation of the solvents through activated alumina, removal of dissolved air by bubbling nitrogen for a few minutes, and storage of solvents under inert gas are convenient for maintaining a supply of solvents.

ACID CADMIUM SULFATE SOLUTION. Dissolve 150 grams of cadmium sulfate octahydrate in distilled water, add 10 ml of 45% sulfuric acid, and dilute to 1 liter.

ALCOHOLIC TITRATION SOLVENT. Dissolve 13.6 grams of sodium acetate trihydrate in 25 ml of distilled water and add 975 ml of anhydrous ethyl alcohol (2B or 3A alcohol is satisfactory).

AQUEOUS TITRATION SOLVENT. Dissolve 40 grams of sodium hydroxide in distilled water, add 3.3 ml of ammonium hydroxide (sp. gr., 0.90) and dilute to 1 liter with distilled water.

SILVER NITRATE SOLUTION, STANDARD $0.1N$ AQUEOUS.

SILVER NITRATE SOLUTION, STANDARD $0.1N$ ALCOHOLIC. Dissolve 17 grams of silver nitrate in 170 ml of distilled water and dilute to 1 liter with isopropyl alcohol.

SILVER NITRATE SOLUTION, STANDARD $0.01N$ AQUEOUS. Prepare by exact dilution of the $0.1N$ aqueous silver nitrate solution with distilled water.

SILVER NITRATE SOLUTION, STANDARD $0.01N$ ALCOHOLIC. Prepare by exact dilution of the $0.1N$ alcoholic silver nitrate solution with 92% isopropyl alcohol. Solutions of $0.01N$ silver nitrate should not be stored longer than 24 hours.

PROCEDURE

DETERMINATION OF MERCAPTANS AND HYDROGEN SULFIDE IN AQUEOUS SOLUTION. Measure a quantity of sample into the titration cell that will require preferably from 2 to 10 ml of 0.1 or $0.01N$ silver nitrate solution to titrate the sulfide and mercaptan present. Add 100 ml of aqueous sulfide titration solvent, immediately place the cell on the titration stand, place the tip of the buret below the surface of the solution, and adjust the stirrer to give vigorous stirring without drawing air through the solution. Plot the titration curve, and select the end points for the sulfide and mercaptan present at the bottom of the steep portions of the titration curve. Blank titrations are unnecessary if pure reagents are used.

DETERMINATION OF MERCAPTANS AND HYDROGEN SULFIDE IN PETROLEUM PRODUCTS. Measure a quantity of sample as directed above into a titration cell containing 100 ml of titration solvent. Limit the sample size to a maximum of 25 grams. Samples that are not soluble in alcohol may be dissolved in up to 40 ml of benzene before addition of the titration solvent. It is important to begin the titration as soon as possible and to continue without interruption.

ADDITIONAL TITRATIONS. If the titration curve shows only the presence of mercaptan, both hydrogen sulfide and elemental sulfur absent, no additional titrations are required. If the titration curve shows the presence of hydrogen sulfide and elemental sulfur is not known to be absent, it is necessary to perform an additional titration after the hydrogen sulfide has been extracted.

HYDROGEN SULFIDE EXTRACTION. Measure at least 3 times the quantity of hydrocarbon sample used in the initial titration into a nitrogen-flushed separatory funnel and add, if necessary, enough isooctane to make a total volume of at least 25 ml of hydrocarbon phase. Add an equal volume of acid cadmium sulfate solution and shake vigorously for several minutes. Allow to stand until a good separation is obtained, drain, and discard the aqueous phase containing the yellow cadmium sulfide precipitates. Repeat until no further cadmium sulfide is precipitated. Filter the hydrocarbon layer through a dry filter paper into a volumetric flask. Rinse the separatory funnel with isooctane. Transfer an aliquot to a titration cell containing 100 ml of alcoholic titration solvent and titrate as before. Disregard any apparent hydrogen sulfide end point and calculate the mercaptan content of the sample from the total titration to the mercaptan end point.

Particularly when $0.01N$ titrant is used, stable electrode potentials are reached more rapidly in the vicinity of the end point if the cell is blanketed with an inert gas. A static blanket is preferable to flowing gas because of the volatility of certain mercaptans, even from alkaline solutions. Presumably, this more rapid equilibration was due to prevention of air oxidation of the small amounts of sulfide or mercaptan present in solution as the end point is approached.

RESULTS AND DISCUSSION

Some of the mixtures tested and the recoveries obtained are listed in Table 10. The results are generally satisfactory for the mixtures of sulfide and methyl, ethyl, and *n*- and isopropyl, and *sec*-, *tert*-, and isobutyl mercaptans. The results obtained for *n*-butyl and *n*-amyl mercaptans

Table 10. Titrations of Mixtures of Sulfide and Mercaptans

Mercaptan	Equivalent Ratio, Sulfide to Mercaptan	Recover, %		
		Sulfide	Mercaptan	Total
Sodium sulfide and various mercaptans with $0.1N$ silver nitrate in aqueous $1N$ sodium hydroxide, $0.05N$ ammonium hydroxide				
Methyl	1:1	99.0	101.9	100.3
	1:5	98.0	102.0	101.2
	10:1	99.5	96.2	99.5
Ethyl	1:1	99.0	101.0	100.0
	1:5	98.0	100.0	99.7
	10:1	99.0	105.8	99.6
n-Propyl	1:1	99.0	100.0	99.5
	1:5	99.0	100.0	99.8
	10:1	99.0	100.0	99.1
Isopropyl	1:1	98.0	103.6	100.8
	1:5	90.0	102.0	100.0
	10:1	97.0	91.8	96.5
Isobutyl	1:1	98.0	101.9	100.0
	1:5	89.1	101.9	99.8
	10:1	99.0	104.8	99.6
sec-Butyl	1:1	98.5	102.0	100.3
	1:5	90.0	102.0	100.0
	10:1	99.5	102.0	99.7
tert-Butyl	1:1	99.0	101.1	100.0
	1:5	97.1	104.3	103.0
	10:1	100.0	108.7	100.7
n-Butyl	1:1	97.1	104.0	100.5
	1:5	58.8	109.1	100.5
	10:1	99.0	111.1	100.1
n-Amyl	1:1	79.3	122.8	100.0
	1:5	0.0	122.2	100.2
Hydrogen sulfide[a] and mercaptans[b] in alcoholic sodium acetate				
n-Propyl (5)[c]	1:1	96.2	104.3	101.1
n-Butyl (3)	1:1	105.2	93.2	99.1
n-Amyl (6)	1:1	99.2	100.7	100.0
n-Hexyl (3)	1:1	106.8	92.3	99.7
n-Heptyl (2)	1:1	104.2	94.8	99.3

[a] Added as aqueous solution of sodium sulfide.

[b] Added as isooctane solution.

[c] Figures in parentheses indicate number of determinations averaged.

show a marked coprecipitation effect, even when the sulfide and mercaptan are present in approximately equivalent amounts. This effect is evident even with the lower molecular weight mercaptans and is considered to be caused by adsorption and subsequent coprecipitation because the total titration value obtained for the various mixtures is correct, although the sulfide recovery is low and the mercaptan recovery is high. In this case sulfide ion is not completely precipitated at the selected end point. The data for mixtures tested in alcoholic medium also show that a marked coprecipitation effect occurs, but unlike that observed in aqueous solvent, high values are obtained for sulfide and correspondingly low recoveries for mercaptans. Thus it appears that some silver mercaptide is precipitated in the sulfide portion of the titration curve. Titrations in more dilute solutions with $0.01N$ titrant did not markedly decrease this coprecipitation.

The presence of elemental sulfur in samples complicates the application of the potentiometric procedure and interpretation of the titration curves, particularly for the determination of hydrogen sulfide. In strongly alkaline aqueous solutions, elemental sulfur and mercaptides are generally considered to react according to the following (24)

$$S + 2NaSR \rightarrow Na_2S + R_2S_2$$

This reaction is rapid and complete. Because sulfide is produced in an amount equal to the mercaptan consumed, the potentiometric titration curve will show the presence of sulfide; however the total titration to the mercaptan end point represents the amount of mercaptan initially present, provided the mercaptan is in excess. If the sulfur is in excess of the mercaptan, the titration curve will not show the presence of mercaptan, but the hydrogen sulfide titration will represent the amount of mercaptan originally present. If hydrogen sulfide is also present in the sample, correct interpretation of a single titration curve is impossible.

Adapted from the Study of F. Peter and R. Rosett

[*Anal. Chim. Acta.* ***64,*** *397* (*1973*); ***70,*** *149, 1974*]

Causes of errors encountered in the potentiometric titration of thiols and hydrogen sulfide with silver nitrate solution were studied, and some conclusions were reached on ways to reduce them. Hydrogen sulfide, methanethiol, and ethanethiol are so volatile that titrations should be done in the reverse direction from the normal one; the sulfur compounds

24. B. A. Stagner, *Ind. Eng. Chem.*, **27,** 275 (1935).

to be determined should be contained in a micrometer buret with a tight Teflon plunger and used as the titrant. If the titration is carried out in aqueous solution, silver hydroxide precipitation, which causes adsorption problems, must be avoided. This can be done by operating in a sufficiently acidic medium (pH range 1–2) or by adding gelatin solution to the solution to be titrated, at a concentration of about 1%.

Colorimetric Methods for Trace Quantities of Mercaptans

As indicated earlier (p. 708), the potentiometric titration of mercaptans can often be extended to the range of 0.001 meq. if 0.001*N* silver nitrate is used. The colorimetric methods, however, extend below this range.

Shinohara (25) devised a procedure based on the Folin-Looney (26) phosphototungstic acid reagent. The reduction of the reagent results in a blue complex. The approach was further developed by Danehy and Kreuz and is discussed later.

Turbidimetric methods can be applied to mercaptans because most mercaptans precipitate readily in water when various metal ions such as iron, copper, silver, mercury, lead, and cadmium are added, mercaptides are generally quite insoluble unless the mercaptan contains solubilization groups such as hydroxyl, carboxyl, or sulfonic acids. Although the turbidimetric analyses are sensitive, they are not as applicable as we would like. This is often the case with turbidimetric methods, since the particle size of the mercaptide is not reproducible from determination to determination. The particle size of the dispersed phase directly controls the magnitude of the turbidity; the smaller the particle size, the higher the turbidity, though the amount of mercaptide present is the same.

The colorimetric method using *N*-ethylmaleimide provides a convenient method for determining mercaptans.

N-Ethylmaleimide Method—Adapted from N. M. Alexander

[*Anal. Chem.*, ***30***, *1292–4* (*1958*)]

N-Ethylmaleimide (NEM) reacts rapidly with sulfhydryl compounds at neutral pH (27, 28). The rate of reaction of equimolar amounts of

25. K. Shinohara, *J. Biol. Chem.*, **120**, 743–9 (1937).
26. O. Folin and J. M. Looney, *Ibid.*, **51**, 421 (1922).
27. E. Friedmann, *Biochim. Biophys. Acta*, **9**, 65 (1952).
28. J. D. Gregory, *J. Am. Chem. Soc.*, **77**, 3922 (1955).

N-ethylmaleimide and reduced glutathione has been followed spectrophotometrically by a decrease in absorption of the former at 300 nm (28), but the reaction does not go to completion under these conditions. Roberts and Rouser (29) showed that the change in absorbance at 300 nm is proportional to the concentrations of cysteine and glutathione when *N*-ethylmaleimide is present in excess. They used this method to determine the extent to which bovine serum albumin reacts with *N*-ethylmaleimide.

The present report also shows that when present in excess, *N*-ethylmaleimide reacts stoichiometrically with sulfhydryl compounds. The decrease in absorption of this compound at 300 nm can be used as an assay method for sulfhydryl groups.

MATERIALS

Reduced glutathione (GSH) and *N*-ethylmaleimide were obtained from Schwarz Laboratories, Inc., Mount Vernon, N.Y. Cysteine hydrochloride and mercaptoethanol were purchased from Eastman Organic Chemicals, Rochester, N.Y. The glycolic acid was a product of Evans Chemetics, Inc., New York, and was labeled as 99% pure. Homocysteine and crystalline egg albumin were bought from Nutritional Biochemicals Corp., Cleveland, Ohio. Ergothioneine and thiolhistidine were obtained from the California Foundation for Biochemical Research, Los Angeles. A boiled aqueous extract of rat liver was prepared as described (30). *N*-Ethylmaleamic acid was prepared by adding 0.3 ml of 0.5*N* sodium hydroxide to 1 ml of 0.01*M* *N*-ethylmaleimide and allowing to stand for 5 minutes. Six milliliters of 0.1*M* phosphate buffer (pH 6.8) was added to neutralize the solution before carrying out the reaction of glutathione with *N*-ethylmaleimide.

N-Ethlymaleimide was stored in the refrigerator, and the molar extinction coefficient was determined on a sample that had been twice recrystallized from ethanol. Once this coefficient is known, the highly pure commercial *N*-ethylmaleimide can be used directly, because any slight contaminants in the solid material or the degradation products formed after it is brought into solution do not react with thiol groups.

All the solid thiol compounds were dried to constant weight *in vacuo* (55 mm) at 55°C just prior to use. The liquid compounds were diluted directly to a desired concentration. Prior to reacting with *N*-ethylmaleimide, the concentrations of the thiol solutions were determined by iodometric titration. The titrations showed that glutathione was 99.4% pure and that cysteine hydrochloride contained 101.7% cysteine. The value for cysteine hydrochloride is explained by the fact that it loses hydrochloride (31). Homocysteine was 98.5% pure. Thiolhistidine

29. E. Roberts and G. Rouser, *Anal. Chem.*, **30,** 1291 (1958).
30. N. Alexander, *J. Biol. Chem.*, **227,** 975 (1957).
31. P. D. Boyer, *J. Am. Chem. Soc.*, **76,** 4331 (1954).

and ergothioneine had negligible reducing action in the acid iodometric titration, in agreement with the earlier observation on ergothioneine (32).

PROCEDURE

The reaction is usually carried out in $0.1M$ phosphate buffer (pH 6.8) with $0.001M$ *N*-ethylmaleimide and a sulfhydryl concentration less than $0.0009M$ and greater than $0.0001M$. Prepare a $0.001M$ *N*-ethylmaleimide solution in buffer and read the absorbances of both solutions at 300 nm. Solutions containing everything but *N*-ethylmaleimide serve as blanks. Divide the difference in absorption between the reacted and unreacted *N*-ethylmaleimide solutions by the molar extinction coefficient of the compound; this quotient is equal to the molar sulfhydryl concentration of the sample. Make the spectrophotometric measurements in 1-cm, matched silica cells in a Beckman DU spectrophotometer. The concentration of reactants may be varied as long as the *N*-ethylmaleimide concentration is in 10% excess of the sulfhydryl concentration. The reaction can be carried out at neutral or acid pH (29), although it proceeds somewhat faster at neutrality (28).

RESULTS

The molar extinction coefficient of *N*-ethylmaleimide at 300 nm is 620. This agrees with Gregory's result (28) obtained at 302 nm.

When varying concentrations of glutathione react with *N*-ethylmaleimide, the percentage decrease in absorbance of the latter at 300 nm is exactly equal to the percentage of glutathione concentration in relation to the *N*-ethylmaleimide concentration. Identical results, within experimental error, were obtained for homocysteine, mercaptoethanol, cysteine, and thioglycolic acid. The assay is unaffected by the presence of 18 other amino acids, Duponol (Du Pont), ascorbic acid, glucose, and Versene.

N-Ethylmaleimide spontaneously decomposes in $0.1M$ phosphate (pH 6.8) at a fairly slow rate when stored in the refrigerator, as seen in experiment 3 of Table 11. The decomposition product is *N*-ethylmaleamic acid, which does not absorb light at 300 nm. If *N*-ethylmaleamic acid could react with sulfhydryl groups at an appreciable rate, freshly prepared *N*-ethylmaleimide solutions would have to be used. Table 11 gives data demonstrating that *N*-ethylmaleamic acid does not

32. G. E. Woodward and E. G. Fry, *J. Biol. Chem.*, **97**, 465 (1932).

Table 11. Reaction of *N*-Ethylmaleimide with Glutathione in the Presence of *N*-Ethylmaleamic Acid

Contents	Experiment Number	Absorbance at 300 mμ	Δ Absorbance
0.001*M* Fresh NEM	1	0.618	
+0.0005*M* GSH	2	0.309	0.309
0.001*M* NEM aged for 1 week	3	0.547	
+0.0005*M* GSH	4	0.237	0.310
+0.001*M* N-ethylmaleamic acid	5	0.570	
+0.0005*M* GSH	6	0.262	0.308

react with sulfhydryl groups in the conditions of the assay. The decrease in absorption of *N*-ethylmaleimide in the presence of a constant amount of glutathione is the same with varying amounts of *N*-ethylmaleamic acid (experiments 3–6) as in a freshly prepared *N*-ethylmaleamide solution (experiments 1 and 2).

Table 12 gives the results obtained with this method in a deproteinated rat liver extract. Two milliliters of extract gave a sulfhydryl value double that of 1 ml of extract (experiments 1–3), demonstrating compliance with Beer's law. Pretreatment of the extract for 10 minutes with sodium *p*-chloromercuribenzoate abolishes any reaction with *N*-ethylmaleimide, indicating that it is reacting only with sulfhydryl groups in the extract.

Table 12. Determination of Sulfhydryl Content of Rat Liver Extract

(Total volume of each reaction mixture, 10 ml.)

Flask Number	NEM, 10 μmoles	Liver Extract, 20 mg. ml.[a]	GSH Added, μmoles	Absorbance at 300 mμ[b]	Δ Absorbance	Calculated μmoles Sulfhydryl Groups
1	+	—	—	0.625	—	0
2	+	1	—	0.468	0.157	2.53
3	+	2	—	0.315	0.310	5.00
4	+	—	2.0	0.500	0.125	2.01
5	+	—	4.0	0.378	0.247	3.98
6	+	1	2.0	0.346	0.279	4.50
7	+	1	4.0	0.220	0.405	6.53

[a] Concentration in terms of nonvolatile organic material and determined according to M. J. Johnson [*J. Biol. Chem.* **181,** 707 (1949)], using crystalline bovine serum albumin as standard.

[b] Blank for each determination contained all components except NEM as indicated.

Glutathione can be added to the extract (experiments 6 and 7) and quantitatively determined over and above the amount of sulfhydryl groups initially present in the extract.

Two milliliters of the filtrate from a 1:5 diluted, hemolyzed human blood sample in 5% metaphosphoric acid gave a reaction with *N*-ethylmaleimide equivalent to 0.75 μM of sulfhydryl groups; 4 ml of the same filtrate gave a value of 1.48 μM. As in the case with the liver extract, known amounts of glutathione added to the filtrate quantitatively reacted with *N*-ethylmaleimide. The reaction with blood filtrates was carried out in the presence of 4 ml of 1*M* phosphate buffer (pH 6.8) to neutralize the metaphosphoric acid. This did not affect the extinction coefficient of *N*-ethylmaleimide.

Commercial crystalline egg albumin reacts with *N*-ethylmaleimide only after being denatured with Duponol. By reacting a 1% solution of crystalline egg albumin with 0.001*M* *N*-ethylmaleimide in the presence of 0.5% Duponol for 15 minutes, 2.2 equivalents of sulfhydryl per mole of egg albumin (assuming mol. wt. = 46,000) reacts with *N*-ethylmaleimide. No further reaction occurred after 1 hour. This same egg albumin sample when reacted with *p*-chloromercuribenzoate at pH 4.6 for 15 minutes (31) gave a value equal to about 3 sulfhydryl groups per mole of protein.

DISCUSSION

This procedure is fast, simple, and highly specific for sulfhydryl groups. The principal advantage is that no standard thiol solutions have to be assayed along with the unknown sample for comparison. Since added glutathione was quantitatively recovered in a tissue extract and in a blood filtrate, the method may be of use in determining the thiol concentrations of other complex biological solutions.

The method appears to be limited in determining the total sulfhydryl content of proteins because *p*-chloromercuribenzoate reacted with more thiol groups in the same egg albumin sample. Moreover, Roberts and Rouser (29) found that *N*-ethylmaleimide reacted with only 60% as many sulfhydryl groups in bovine serum albumin as did *p*-chloromercuribenzoate. This assay would, however, obviate the need for using nitroprusside as an external indicator (33) to determine the extent to which *N*-ethylmaleimide reacts with a protein. One per cent solutions with one reactive sulfhydryl group per 100,000 molecular weight would be 0.0001*M* with respect to the sulfhydryl concentration, which is within the limits of sensitivity of the method.

33. T. C. Tsao and K. Bailey, *Biochim. Biophys. Acta*, **11,** 102 (1953).

Benesch et al. (34) have advantageously used *N*-ethylmaleimide to detect thiols and thiol esters as pink spots on paper chromatograms in a strongly alkaline, nonaqueous medium. In contrast to their results, thiol esters and ergothioneine do not react with *N*-ethylmaleimide in neutral, aqueous media.

Phosphotungstic Acid Method–Adapted from J. P. Danehy and J. A. Kreuz

[*J. Am. Chem. Soc.*, ***83***, *1109* (*1961*)]

REAGENTS

PHOSPHOTUNGSTIC ACID REAGENT. To a 1-liter flask equipped with a reflux condenser, add 100 grams of sodium tungstate dihydrate, 200 ml of distilled water, and 50 ml of 85% phosphoric acid. Reflux gently for 1 hour, add 5 drops of bromine to discharge the light blue color that develops, gently reflux for 5 minutes, and then boil vigorously for 20 minutes to expel excess bromine. Allow the solution to cool, and dilute to 1 liter with distilled water.
ACETATE BUFFER, pH 5. Dissolve 600 grams of sodium acetate trihydrate and 50 ml of glacial acetic acid in sufficient distilled water to give 2 liters of solution.

PROCEDURE

Add 2 ml of $1N$ hydrochloric acid solution to about 2 meq. of mercaptan in 2 to 5 ml of water and aspirate for 10 minutes with nitrogen to remove any hydrogen sulfide. Transfer to a 50-ml volumetric flask containing 10 ml of acetate buffer, 4 ml of phosphotungstic acid reagent, and a single drop of $0.002M$ cupric sulfate. Allow the blue color to develop for at least 15 minutes but no more than 30 minutes.

Read the absorption at 660 nm and compare with a graph prepared with known amounts of mercaptan.

The method was used successfully for mercaptoacetic acid, 2-mercaptopropionic acid, and 3-mercaptopropionic acid. Sulfinic acids do not interfere.

34. R. Benesch, R. E. Benesch, M. Gutcho, and L. Laufer, *Science*, **123,** 981 (1956).

19

Dialkyl Disulfides

RSSR

Dialkyl disulfides can be determined via two chemical approaches. By the first approach the disulfides can be reduced to the corresponding mercaptan and the mercaptan measured.

$$RSSR + \xrightarrow{H_2} 2RSH$$

The other approach is the bromine oxidation of the disulfide to the corresponding sulfonic acid (see equation, p. 762).

The reduction method followed by mercaptan titration is preferred because it is the more specific. The oxidation method using bromine suffers from more interferences, since bromine either oxidizes or substitutes with many organic compounds. The oxidation method is the faster and the simplest, however, when it can be used.

There are several reduction methods that can be applied to convert mercaptans to disulfides. Kolthoff et al. used a Jones-type reductor of zinc amalgam (see pp. 760–1). Siggia and Stahl (see below) used sodium borohydride. Veibel and Wrónski used butyllithium (p. 757–8), and Humphrey and Potter used tributylphosphine (p. 758–9). Bell and Agruss (1) and Karchmer and Walker (2) used zinc acetic acid reductions. Hubbard, Haines, and Ball (3) compared the zinc and acetic acid method to a zinc and caustic method. The zinc and acid (or alkali) methods suffer from inconsistency of reaction with different disulfides. Time, temperature, and agitation must be varied and controlled.

Reduction Methods

SODIUM BOROHYDRIDE REDUCTION

Adapted from Method of S. Siggia and C. R. Stahl

[*Reprinted in Part from Anal. Chem.*, ***29,*** *154–5* (*1957*)]

Because of its freedom from interferences, reduction is the better means of determining disulfides, if quantitative reduction can be accomplished in a reasonable length of time. After various means of reducing

1. R. T. Bell and M. S. Agruss, *Ind. Eng. Chem., Anal. Ed.*, **13,** 297 (1941).
2. J. H. Karchmer and M. T. Walker, *Anal. Chem.*, **30,** 85–90 (1958).
3. R. L. Hubbard, W. E. Haines, and J. S. Ball, *Ibid.*, **30,** 91–3 (1958).

disulfides were considered, metal hydrides seemed to be the most promising reducing agents to investigate, since it had been observed that when reductions with hydrides are quantitative, they proceed in a short time. Also, the hydrides present no problem in the analysis of the mercaptan formed on reduction once they were decomposed. Lithium aluminum hydride was first tried, but high results were obtained. This appeared to be due to reduction of part of the disulfide to hydrogen sulfide, since two potential breaks were observed on titration of the reduced samples with silver nitrate. These breaks corresponded to those obtained by titrating mixtures of hydrogen sulfide and mercaptans, and no silver nitrate was consumed in blank titrations of the decomposed reagent.

Reduction with sodium borohydride was attempted when it appeared unlikely that conditions could be adjusted to eliminate the difficulties observed in using lithium aluminum hydride. Brown and Rao found that in the presence of aluminum chloride a diethylene glycol–dimethyl ether solution of sodium borohydride (4) reduced diphenyl disulfide. An estimated reaction completeness of about 90% is reported. A method for quantitatively determining disulfides was developed with this reduction as a basis. After reduction the excess hydride is decomposed, and the mercaptan obtained is titrated potentiometrically in an aqueous solution of sodium and ammonium hydroxides with silver nitrate. The reducing agent will interfere with the titration unless it is completely decomposed.

The decomposition of the hydride presented a problem in the development of the method, because sodium borohydride can readily be decomposed only by the addition of acid, which causes a rapid evolution of gas and loss of mercaptan. Because sodium borohydride and aluminum chloride apparently form aluminum borohydride (4, 5) to some extent, it seemed probable that the addition of dilute sodium hydroxide would destroy most of the hydride in this case, and no mercaptan would be lost from the basic solution; however the decomposition with sodium hydroxide was found to be vigorous, and mercaptan was lost unless the decomposition was carried out in a flask submerged in an ice bath and equipped with a long coiled condenser. In some cases a small amount of sodium borohydride remained; since little gas was evolved, however, it was possible to decompose this with acid without loss of mercaptan. The procedure finally adopted was to add sodium hydroxide and then nitric acid to ensure complete decomposition.

REAGENTS

Sodium borohydride, 2 grams dissolved in 100 ml of diethylene glycol dimethyl ether (Ansul Chemical Co., Marinette, Wis.).

4. H. C. Brown and B. C. Rao, *J. Am. Chem. Soc.*, **78,** 2582 (1956).
5. H. C Brown and B. C. Rao, *Ibid.*, **77,** 3164 (1955).

Anhydrous aluminum chloride.
Sodium hydroxide, $1N$.
Sodium hydroxide, $6N$.
Nitric acid, $3M$.
Concentrated ammonium hydroxide.
Standard $0.1N$ silver nitrate.

PROCEDURE

Accurately weigh a sample containing approximately 0.001 mole of disulfide and place it in a 150-ml round-bottomed flask containing 15 ml of sodium borohydride solution and approximately 0.5 gram of aluminum chloride. Immediately attach the flask to a coil condenser 40 cm long, and allow the reduction to proceed at room temperature for 30 minutes. The order of addition of reagents and sample to the flask is unimportant except in the case of methyl disulfide. When the sodium borohydride solution and the aluminum chloride are mixed, a rapid reaction takes place, and methyl disulfide is lost if it is added to the solution before the aluminum chloride. In the determination of methyl disulfide, mix the sodium borohydride solution and the aluminum chloride and allow to stand in an ice bath for a few minutes while the sample is being weighed. Another approach is to add the aluminum chloride in solution in 5 ml of the diglycol-diether solvent. This solution is added dropwise to the mixture of borohydride solution and sample.

When reduction is complete, submerge the flask in an ice bath, and add 5 ml of $1N$ sodium hydroxide through the condenser. Add the sodium hydroxide a few drops at a time until the initial vigorous reaction subsides, and then add the remainder and rinse the condenser with a few milliliters of distilled water. Allow the solution to stand for 2 or 3 minutes and add 10 ml of $3M$ nitric acid. Remove the ice bath and add 10 ml of $6N$ sodium hydroxide after about 2 minutes. In determining methyl disulfide it is better to allow the solution to stand for 5 minutes in the ice bath after adding the nitric acid and to add the sodium hydroxide before removing the ice bath.

Rinse the condenser with a few milliliters of distilled water and remove the flask. Rinse the contents of the flask into a 400-ml beaker, and add 10 ml of concentrated ammonium hydroxide. Using a pH meter equipped with silver and calomel electrodes, titrate the solution potentiometrically with standard $0.1N$ silver nitrate solution. The break occurs between approximately −325 and −175 mV, although it varies somewhat for the different mercaptans being titrated. The percentage of disulfide is calculated in the following manner:

$$\%\ \text{Disulfide} = \frac{\text{Milliliters} \times N \text{ of } AgNO_3 \times \text{mol. wt.} \times 100}{\text{Weight of sample} \times 2000}$$

RESULTS AND DISCUSSION

Results obtained for the six disulfides used to test the procedure are given in Table 1. The precision and accuracy of the method are within ±1%. Mercaptans and sulfides do not interfere with the procedure. When sulfides are treated in the same manner used to determine disulfides, no

Table 1. Determination of Disulfide by Sodium Borohydride Reduction

Compound	Disulfide Found, %
Phenyl disulfide[a]	99.2
	99.3
	99.8
	99.6
Methyl disulfide[a]	99.4
	98.9
	99.0
	99.2
Ethyl disulfide[a]	97.5
	97.2
	97.8
	97.1
n-Butyl disulfide	98.7
	98.7
	98.6
	99.1
Isoamyl disulfide	95.5
	95.7
	96.3
	96.6
	95.9
	95.4
Benzyl disulfide	99.0
	99.1
	99.4
	99.0

[a] Sulfur values obtained on these samples. Phenyl disulfide found = 29.52%, theory = 29.37%. Methyl disulfide found = 67.80%, theory = 68.08%. Ethyl disulfide found = 50.63%, theory = 52.46%.

silver nitrate is consumed on titration. In determining the disulfide content of a sample that contains mercaptan, the mercaptan must be determined on an unreduced sample and subtracted from the value obtained after reduction. The mercaptan is unchanged by reduction and titrates with the reduced disulfide.

BUTYLLITHIUM REDUCTION

Adapted from the Method of S. Veibel and M. Wrónski

[*Anal. Chem.*, ***38***, *910* (*1966*)]

Butyllithium as the reagent permits the maintenance of an alkaline solution, which eliminates the possibility of introducing errors as a result of evaporation of low-boiling members of the series during acidification.

REAGENTS

BUTYLLITHIUM IN HEXANE. Commercially available.
O-HYDROXYMERCURIBENZOATE (HMB), $0.01N$. Dissolve 3.2 grams of *o*-hydroxymercuribenzoic anhydride in 20 ml of $1N$ sodium hydroxide solution and dilute to 1 liter with 50% ethanol. Calculate the normality from the weight of the anhydride.
INDICATOR. Prepare a solution of thiofluorescein in $0.5N$ ammonia in 50% aqueous ethanol. The blank value of the indicator should not be more than $0.05N$ HMB per mole of the indicator solution.

PROCEDURE

Dissolve a sample of organic disulfide in 10 ml of dry benzene and treat with 0.5 ml of $6N$ butyllithium. Choose the time and temperature depending on the disulfide type (Table 2). Add 15 ml of ethanol cooled in ice, and titrate the solution with $0.01N$ HMB in the presence of 1 ml of the thiofluorescein indicator to the disappearance of the blue color.

DISCUSSION AND RESULTS

The results are summarized in Table 2.

In the reaction, one mole of thiol is produced per mole of disulfide and at the same time one mole of thioether is formed.

$$RSSR + C_4H_9Li \rightarrow RSC_4H_9 + RSLi$$

Table 2. Determination of Organic Disulfides with Butyllithium

Disulfide	Reaction with Butyllithium: Time, sec.	Reaction with Butyllithium: Temp., °C	Sample, mg	Range of Errors, %	Average Recovery, %	Number of Determinations
Methyl	30	25	1–12	−1.1 to +0.6	99.5	5
Ethyl	60	40	2–12	−1.1 to +0.4	99.4	5
Butyl	60	70	5–25	−0.9 to −0.1	99.5	5
Isopropyl	60	70	5–25	−0.8 to +0.4	99.7	5
Phenyl	60	25	5–25	−1.4 to −0.5	99.1	5
Benzyl	10	15	8–40	−0.4 to +0.6	100.2	5
Trimethylene (1,2-dithiolane)	60	70	4–12	−1.7 to −1.2	98.5	5

To some extent a reaction of butyllithium and thioether can take place.

$$C_4H_9Li + RSR \rightarrow C_4H_9R + RSLi$$

This reaction, however, is very slow in comparison with the disulfide reaction, and usually quantitative conversion of disulfide is obtained without significant interference from the subsequent reaction. As an example, after 60 seconds at 70°C, the determinations of methyl ethyl, and benzyl disulfides result in 107, 102, and 120% recovery, respectively. Table 2 reveals that in the estimation of these disulfides, temperatures of 25, 40, and 15°C, respectively, are recommended; the interference is negligible even for benzyl disulfide.

REDUCTION WITH TRIBUTYLPHOSPHINE

Adapted from the Method of R. E. Humphrey and J. L. Potter

[*Anal. Chem.*, **37,** *164* (*1965*)]

Aromatic disulfides are quantitatively reduced to thiols by triphenylphosphine at room temperature in aqueous methanol containing perchloric acid (6). Alkyl disulfides, however, are reduced by only 20 to 30% in 4 hours at room temperature by this procedure and by 80 to 90% after reflux in aqueous methanol for several hours. Some alkyl disulfides and benzyl disulfide are not reduced by triphenylphosphine in benzene at elevated temperatures.

Tributylphosphine, on the other hand, because of its more basic and nucleophilic nature, is more effective than triphenylphosphine in reducing alkyl disulfides to thiols.

6. R. E. Humphrey and J. M. Hawkins, *Anal. Chem.*, **36,** 1812 (1964).

REAGENTS

Tributylphosphine is available from M & T Chemicals and can be used without further purification.

PROCEDURE

Dissolve the weighed sample containing 0.05 mM of disulfide and a weighed amount of tributylphosphine (0.10 mM) in 1 to 2 ml of acetone and dilute to approximately 100 ml with 10% aqueous methanol. Alternatively, use aliquots of millimolar solutions of the disulfide and of the phosphine. Allow the reaction mixture to stand for up to 1 hour. Add 1 to 2 ml of a 0.10*M* sulfur solution in benzene, equivalent to the amount of phosphine introduced, to remove excess phosphine. Then add 1 ml of 70% perchloric acid and titrate the thiol amperometrically with 0.01*M* silver nitrate solution.

DISCUSSION AND RESULTS

The reaction was presumed to proceed according to the following equation:

$$RSSR + (C_4H_9)_3P + H_2O \rightarrow 2RSH + (C_4H_9)_3PO$$

Essentially quantitative reductions of the disulfides investigated within 5 to 60 minutes were realized, as indicated in Table 3.

Table 3. Reduction of Disulfides with Tributylphosphine

Compound	Time,[a] min.	RSSR,[b] mg	Reduction, %
Bis(*o*-nitrophenyl) disulfide	20	13.6	97
4,4′-Dithiodianiline	20	13.9	98
Phenyl disulfide	5	9.5	95
	30	8.9	100
p-Tolyl disulfide	5	10.4	100
	15	9.7	98
Benzyl disulfide	30	18.2	94
	60	18.2	103
Butyl disulfide	30	13.3	88
	60	13.5	98
Propyl disulfide	60	11.3	97

[a] Elemental sulfur added after this period of time to remove remaining phosphine and stop reaction.

[b] Sufficient tributylphosphine added so ratio of phosphine to disulfide was approximately 2:1.

JONES–TYPE REDUCTION

The Jones–type reduction does not yield values as accurate and precise as does the sodium borohydride reduction. The results are generally about 5% low, with a precision of ±2%. But the Jones reduction approach can be applied to solution media to which the borohydride is inapplicable.

Method of I. M. Kolthoff, D. R. May, P. Morgan, H. A. Laitinen, and A. S. O'Brian

[*Ind. Eng. Chem., Anal. Ed.*, ***18,*** *442–4* (*1946*)]

REAGENTS

Sulfuric acid, 0.3*N*, in 90% ethanol.

ZINC AMALGAM (0.02% IN MERCURY). By stirring with 6*N* hydrochloric acid for 1 minute, etch 100 grams of 20-mesh granulated zinc. Next add 27 mg of mercuric chloride, for 1 minute, and wash the zinc with distilled water. The Jones reductor should be of convenient size to hold 100 grams of 20-mesh zinc.

Sulfuric acid, 10*N*.

The reagents for the determination of the mercaptan formed by the reduction are given with the procedures for mercaptans (pp. 717–20).

PROCEDURE

Dissolve a sample containing about 0.0003 mole of disulfide in 90% ethanol that is 0.3*N* in sulfuric acid, heat 25-ml portions of the solution to 50°C, and allow this to flow through the reductor, which is immersed in a bath held at 50°C (Fig. 19.1). The rate of flow is such that the disulfide solution remains in contact with the amalgamated zinc for about 5 minutes. Then rinse the reductor quantitatively with three successive 25-ml portions of 0.3*N* sulfuric acid in 90% ethanol. The solution is analyzed for mercaptan. Kolthoff et al. used the amperometric silver method described previously (pp. 717–20); however any of the other procedures described in Chapter 18 on mercaptans (pp. 707–28), except possibly the iodimetric method, may be used.

Incomplete reduction occurs if the reductor is not completely clean. For cleaning, use a 10*N* sulfuric acid solution. The reductor should be checked with a standard disulfide solution before each series of runs to be sure that it is in working order. If the reductor remains slow even after

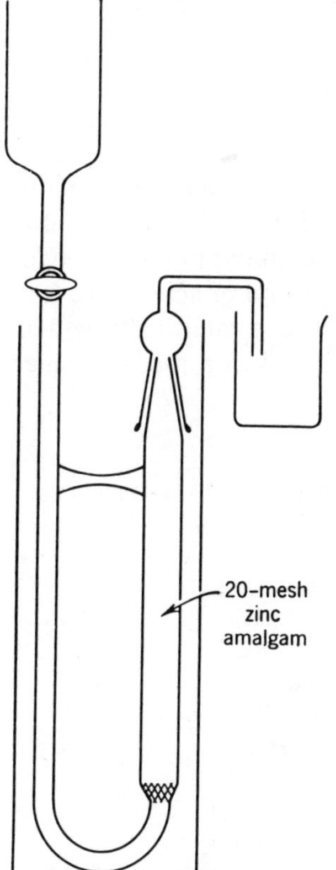

Fig. 19.1. Disulfide reduction apparatus.

rinsing with 10*N* sulfuric acid, it should be refilled with fresh zinc amalgam.

If the sample contains mercaptan, the mercaptan is determined on an unreduced sample. A sample is then reduced, and the total mercaptan is determined. The difference will yield the disulfide content. This method was tested only with *n*-dodecyl disulfide.

Oxidation Method Using Bromine

Adapted from S. Siggia and R. L. Edsberg

[*Anal. Chem.*, **20,** *938–9* (*1948*)]

The bromination procedure as described for alkyl sulfides can also be used to determine disulfides. This method is faster and simpler than the reduction methods described, but more compounds interfere. The reaction is

$$RSSR + 5Br_2 + 4H_2O \rightarrow 2RSO_2Br + 8HBr$$

For most disulfides, more acid is required to achieve complete reaction than is needed for alkyl sulfides. The alkyl sulfides require only 3 ml of concentrated hydrochloric acid. Table 4 shows the amount of acid needed for various disulfides. When in doubt as to the amount of acid to use for a particular disulfide, it is best to use 25 ml of concentrated sulfuric acid, since this amount of acid is great enough to work for the most inert

Table 4

Sample	Milliliters of Acid Used	Percentage Found
Butyl disulfide in 50 ml. of 80% glacial acetic acid and 20% water	3 ml. conc. HCl	87.2
	5 ml. conc. HCl	92.2
	10 ml. conc. HCl	95.8
	25 ml. conc. HCl	96.8
	5 ml. 6*N* H_2SO_4	79.6
	5 ml. 12*N* H_2SO_4	83.0
	10 ml. 12*N* H_2SO_4	86.9
	10 ml. conc. H_2SO_4	97.8
	20 ml. conc. H_2SO_4	98.8
	25 ml. conc. H_2SO_4	98.8
	30 ml. conc. H_2SO_4	98.8
Ethyl disulfide in 50 ml. of 80% glacial acetic acid and 20% water	3 ml. conc. HCl	71.3
	10 ml. conc. HCl	86.4
	20 ml. conc. H_2SO_4	95.6
	25 ml. conc. H_2SO_4	95.3
	30 ml. conc. H_2SO_4	95.3
Phenyl disulfide in 50 ml. of 80% glacial acetic acid 20% water	3 ml. conc. HCl	94.5
		95.0
		94.3
	25 ml. conc. HCl	94.0
l-Cystine in 50 ml. of water	3 ml. conc. HCl	100.0
		99.6
	25 ml. conc. HCl	99.6
	20 ml. conc. H_2SO_4	99.6

disulfides. Disulfides that will react quantitatively with much less acid will also react satisfactorily in the presence of 25 ml of concentrated sulfuric acid. Also, it is best to titrate the solution while it is warm (about 30–50°C). Solutions cooled below these temperatures react too slowly to give a sharp end point.

Mixtures of alkyl sulfides and alkyl disulfides can be determined by first obtaining the total of the two types by the bromination method just discussed, then determining the disulfide by the reduction methods of Siggia and Stahl (pp. 753–61). The difference between the two analyses will yield the amount of the alkyl sulfide.

Samples containing mercaptans can also be determined, provided the conditions are the same as those described for alkyl sulfides.

$$\frac{\text{Milliliters of } BrO_3^- - Br^- \times N \times \text{mol. wt. of compound} \times 100}{\text{Grams of sample} \times 10{,}000} = \%\ \text{compound}$$

Houff and Schuetz (7) carried out the bromine oxidation analyses using a dead-stop electrometric end point. This makes possible the titration of colored samples.

Mixtures of Dialkyl Disulfides and Dialkyl Sulfides

Mixtures of alkyl disulfides and sulfides can be determined by first applying the bromine oxidation, which yields both the disulfide and the sulfide. Then, by using the reduction methods (pp. 775–76), the disulfide alone can be determined; the sulfide is obtained by difference.

Mixtures of Dialkyl Disulfides and Mercaptans

These mixtures can readily be determined by first determining the free mercaptan by any of the methods in Chapter 18. A second sample can then be reduced by any of the reduction methods shown, the total mercaptan can be determined, and the disulfide can then be determined by difference. For the latter it is preferable to use the sodium borohydride for the reduction method, since the accuracy is better than the Jones reductor method and the difference value is less subject to error.

The total mercaptan and disulfide can be determined using the bromination method, provided the mercaptan content is low (pp. 775–8), since

7. W. H. Houff and R. D. Schuetz, *Anal. Chem.*, **25,** 1258 (1953).

mercaptans yield results that are 5% low by bromination. If the mercaptan content is high, this error will accumulate in the value for disulfide.

Another method involves the reaction of the mercaptan with acrylonitrile and determination of the disulfide in the reaction mixture. This is the method of Earle.

Method of T. E. Earle

[*Adapted from Anal. Chem.*, ***25****, 769* (*1953*)]

In seeking a method by which disulfides could be determined directly, consideration was given to methods utilizing precipitation or extraction to eliminate the interfering thiols. Removal of thiols as silver salts (8), or as other insoluble heavy-metal mercaptides, was unattractive because of the gummy nature of the precipitate formed by the mixed thiols present in petroleum distillates. Removal of thiols by extraction with alkali is not quantitative, and air oxidation of thiols to disulfides is accelerated in alkaline media. A specific reaction of the thiols was therefore sought, the reagents and products of which would not interfere in the subsequent reduction of disulfide to thiols or in the titration of thiols with silver nitrate.

Acrylonitrile and other conjugated unsaturated nitriles, esters, and ketones react with thiols in the presence of small amounts of alkaline condensing agents to form thioethers (9–11). This reaction suggested the use of an excess of an unsaturated compound of this type to remove thiols quantitatively from mixtures with disulfides.

DEVELOPMENT OF METHOD

Acrylonitrile and related unsaturated compounds were screened as thiol acceptors, and three alkaline condensing agents were tested. The necessary amount of the selected acceptor was then determined, and an operable procedure for the determination of disulfides was developed.

The three thiol acceptors tested were acrylonitrile, mesityl oxide, and methyl acrylate. Alkalies tested were potassium hydroxide,

8. I. M. Kolthoff, D. R. May, and Perry Morgan, *Ind. Eng. Chem., Anal. Ed.*, **18,** 442 (1946).
9. D. W. Beesing, W. P. Tyler, D. M. Kurtz, and S. A. Harrison, *Anal. Chem.*, **21,** 1073 (1949).
10. L. L. Gershbein and C. D. Hurd, *J. Am. Chem. Soc.*, **69,** 241 (1947).
11. M. W. Harman (to Monsanto Chemical Co.), U.S. Patent 2,413,917 (Jan. 7, 1947).

trimethylbenzylammonium hydroxide, and trimethylphenylammonium hydroxide. Every combination of acceptor and alkali was tested with known solutions of 1-butanethiol (*m*-butyl mercaptan) and 2-methyl-2-propanethiol (*tert*-butyl mercaptan) in *n*-heptane. The absence of thiols was indicated by the absence of lead sulfide formation when the heptane solution was treated with sodium plumbite solution and sulfur (12). Table 5 indicates that thiol-free heptane solutions were obtained in the shortest reaction times when acrylonitrile and potassium hydroxide were used together.

Table 5. Screening of Combinations of Thiol Acceptors and Alkalies

(Minutes required for complete reaction)

Thiol Acceptor[a]	Alkali		
	7% Potassium Hydroxide in Ethanol	40% Aqueous Trimethylbenzyl-Ammonium Hydroxide	22% Aqueous Trimethylphenyl-Ammonium Hydroxide
1-Butanethiol, 0.84 Gram of Thiol Sulfur per Liter			
Acrylonitrile	<10	15	15
Mesityl oxide	15	>20	<20
Methyl acrylate	10	>20	>20
2-Methyl-2-propanethiol, 0.91 Gram of Thiol Sulfur per Liter			
Acrylonitrile	<10	15	<10
Mesityl	15	>20	>20
Methyl acrylate	<10	>20	>20

[a] Molar ratio of acceptor to thiol: 12:1 to 24:1.

Investigation of the necessary excess of acrylonitrile was made in a similar manner, with the results given in Table 6. A 6:1 molar ratio of acrylonitrile to thiol is sufficient to give quantitative reaction with eight alkyl and aryl thiols. Minimum ratios of acrylonitrile to thiol were not investigated.

To determine the behavior of thiols in the presence of acrylonitrile in acid solution, the reaction of eight thiols with acrylonitrile in 2-propanol

12. American Society for Testing and Materials, 1949 *Book of ASTM Standards*, pt. 5, Philadelphia, 1949, p. 923.

Table 6. Molar Ratio of Acrylonitrile to Thiol Giving Complete Reaction in 10 Minutes

(Alkali, KOH, 7% in ethyl alcohol)

Thiol	Thiol Sulfur, G./L.	Molar Ratio
n-Decyl	0.53	3:1
tert-Butyl	0.81	2:1
tert-Octyl	0.44	6:1
Cycloalkyl[a]	0.49	3:1
Phenyl	0.81	3:1
p-Tolyl	0.92	3:1
2-Naphthyl	0.47	3:1
Aryl[b]	0.26	3:1

[a] Mixed cyclohexyl and methylcyclopentyl.
[b] Mixture from petroleum.

and acetic acid was studied. The results are given in Table 7. The alkyl thiols did not react with acrylonitrile, but the aryl thiols reacted to a considerable extent. It is therefore necessary to remove the excess acrylonitrile before converting the disulfides to thiols by reduction in acid solution.

Table 7. Reaction of Thiols with Acrylonitrile in Presence of Acetic Acid

Thiol	% Reacted in 2 Hours
n-Decyl	0
tert-Butyl	0
tert-Octyl	0
Cycloalkyl	0
Phenyl	60
p-Tolyl	70
2-Naphthyl	50
Aryl	30

PROCEDURE

Determination of disulfides in the presence of thiols was achieved by reaction of acrylonitrile with thiols, removal of excess acrylonitrile, reduction of disulfides to thiols, and titration with silver nitrate.

To 50 ml of sample in a 125-ml separatory funnel, add a minimum of 6 moles of acrylonitrile per mole of thiol present, followed by 2 ml of 1*N* alcoholic potassium hydroxide. Allow the mixture to react for 10 minutes, with occasional swirling, and wash it three times with 50 ml of water to remove the excess acrylonitrile. Wash the contents of the separatory funnel into a 250-ml beaker with 100 ml of 99% 2-propanol, and add 50 ml of glacial acetic acid. Pass the solution through a zinc reductor (8), at room temperature for samples boiling below 200°C, or at 60°C for higher-boiling samples. Then wash the reductor successively with 100 ml of 5% acetic acid in 2-propanol and with 100 ml of 2-propanol. Dilute the reduced material and washings to a definite volume with 2-propanol, and take an aliquot for titration of thiol with standard silver nitrate solution (13) (see pp. 708–14).

DISCUSSION

Because only aryl thiols react with excess acrylonitrile in acid solution, the water washing may be omitted if no aryl disulfides are present. With certain types of sample it may be found convenient to remove excess acrylonitrile by reaction for 30 minutes with excess *o*- or *p*-cresol, rather than by water washing; the effectiveness of this technique is illustrated in Table 8.

Table 8. Comparison of Methods of Removal of Acrylonitrile

Added, Grams/Liter		Method of Removal	Disulfide Sulfur Found, Grams/Liter
Disulfide Sulfur	Thiol Sulfur		
	Phenyl Thiol and Diphenyl Disulfide		
0.62	1.00	None	0.40
		Water washing	0.59, 0.60
1.06	0.81	Phenol	0.55
		m-Cresol	0.44
		o-Cresol	1.02
		p-Cresol	1.03, 0.98
	p-Tolyl Thiol and Di-*p*-tolyl Disulfide		
0.29	0.86	Water washing	0.29, 0.30
		o-Cresol	0.27, 0.28
0.57	0.58	None	0.28, 0.19
		Water washing	0.57, 0.56
		o-Cresol	0.57, 0.56

13. M. W. Tamele and L. B. Ryland, *Ind. Eng. Chem., Anal. Ed.*, **8**, 16 (1936).

Table 9 presents the analyses of heptane solutions of mixtures of thiols and the corresponding disulfides by the procedure described. The agreement of the disulfide values with the amounts taken is generally within 5%.

Table 9. Disulfide Analyses of Thiol-Disulfide Mixtures

	Thiol Sulfur, Grams/Liter	Disulfide Sulfur, Grams/Liter	
		Taken	Found
n-Butyl	0.20	0.18	0.18, 0.17, 0.18
n-Butyl	0.60	0.10	0.10, 0.09, 0.10
n-Dodecyl	0.59	0.40	0.40, 0.40
tert-Butyl	0.44	0.38	0.40, 0.36, 0.39
tert-Octyl	0.44	0.43	0.41, 0.41, 0.37
Cycloalkyl	0.24	0.13	0.13, 0.14
Phenyl	1.13	0.62	0.59, 0.60
p-Cresyl	0.86	0.29	0.29, 0.30
p-Cresyl	0.58	0.57	0.57, 0.56

Table 10 compares analyses of three petroleum distillates with the results obtained by the indirect method. The first distillate contained small amounts of olefins; the second, considerable olefins and aromatics; and the third, small amounts of aromatics. When the thiol-disulfide ratio

Table 10. Disulfide Analyses of Petroleum Distillates

Boiling Range, °C	Thiol Sulfur, Grams/Liter	Disulfide Sulfur, Grams/Liter	
		Indirect Method	Present Method
25–70	0.52	−0.05, −0.06, −0.01	0.05, 0.05, 0.05
	0.06	0.10, 0.09, 0.11	0.12, 0.12, 0.12
	0.31	0.10, 0.13, 0.11	0.20, 0.20, 0.21
	0.06	0.12, 0.12, 0.12	0.14, 0.14, 0.15
66–204	0.31	0.02, 0.02, 0.02	0.04, 0.04, 0.04
	0.26	0.04, 0.04, 0.05	0.05, 0.05, 0.05
	0.09	0.12, 0.11, 0.11	0.12, 0.12, 0.12
	0.06	0.18, 0.17, 0.18	0.16, 0.16, 0.17
165–293	0.68	0.02, 0.11, 0.10	0.10, 0.11, 0.10
	0.65	0.04, 0.04, 0.03	0.10, 0.10, 0.10
	0.15	0.51, 0.49, 0.49	0.49, 0.51, 0.52
	0.03	0.56, 0.56, 0.53	0.55, 0.55, 0.54

was less than 10:1 and volatile thiols were not present, good areement was obtained. Low results were given by the indirect method in cases where the thiol-disulfide ratio was high or where volatile thiols were present.

The precision of the method, as indicated by the data in Tables 9 and 10, is on the order of 0.01 gram of disulfide sulfur per liter for normal alkyl and aryl disulfides. The slightly lower precision for tertiary disulfides does not reduce the precision of the method when applied to petroleum distillates, which do not generally contain compounds of this type.

In the alkaline medium of the reaction step, any free sulfur present would be expected to react with mercaptan, resulting in a high value for disulfide content. Polysulfides, which contain loosely bound sulfur, would probably behave similarly. These interfering materials are not present in significant amounts in untreated petroleum distillates—the type of sample for which the method is most useful.

Determination of Traces of Disulfides

Traces of disulfides can be determined by reduction by any of the foregoing approaches and determination of the resulting mercaptan by any of the colorimetric methods shown earlier (pp. 747–52). The Jones-type reduction has an advantage here in that no extraneous materials are added to the sample. The relatively low accuracy is not disturbing on the trace level. The sodium borohydride has an advantage, however, if the resulting mercaptan can be distilled or steam-distilled. In these cases large quantities of sample can be taken and reduced, and the resulting mercaptans can be concentrated by distillation. The distillate can then be analyzed titrimetrically or colorimetrically (see Chapter 18).

20

Sulfides

RSR

Organic sulfides (thioethers) are readily oxidized in general according to the following overall reactions:

$$R_2S + (O) \rightarrow \underset{\text{Sulfoxide}}{R_2SO} \rightarrow \underset{\text{Sulfone}}{R_2SO_2}$$

The reaction to the sulfoxide goes very rapidly; that to the sulfone proceeds more slowly. Therefore when oxidation is used as a quantitative means for the determination of sulfides, the results tend to be high when obtained from reaction to sulfoxide and low when obtained from reaction to sulfone.

The difficulties of overoxidation can be controlled by the use of carefully controlled conditions such as by avoiding an excess of the oxidizing agent or by quenching the oxidation after the appropriate reaction time interval. In the first method to be described, bromine is the oxidizing agent and the titration is made until a slight excess of bromine is detected. In this way there is never a large excess of bromine present to carry the reaction beyond the sulfoxide stage. The bromine color end point does fade because of oxidation to the sulfone, but it fades so slowly that the end point can be detected without difficulty. If the analysis is carried out by adding excess bromine, the excess bromine accelerates the reaction to sulfone, and high results are obtained.

This procedure, with slight modification (pp. 761–3) may also be employed to determine disulfides. A method for determining sulfides and disulfides in the presence of each other is also described (p. 763).

Bromine is involved in the second method as well, but it is present only in catalytic amounts, with the actual titrations made with lead tetraacetate solution. The reactions taking place during the application of this method were formulated as follows:

$$2Br^- \xrightarrow{\text{Oxidant}} Br_2$$

$$Br_2 + \begin{matrix} R \\ \quad S \\ R_1 \end{matrix} \longrightarrow \left[\begin{matrix} R \\ \quad S^+Br \\ R_1 \end{matrix} \right] \cdot Br^-$$

$$\left[\begin{matrix} R \\ \quad S^+Br \\ R_1 \end{matrix} \right] \cdot Br^- \xrightarrow{\text{HOAc–}H_2O} \begin{matrix} R \\ \quad S{=}O \\ R_1 \end{matrix} + 2HBr$$

The hydrobromic acid liberated is continuously oxidized by the oxidizing agent added during the titration and is used up immediately in the second step of the reaction series.

A third method involves oxidation with hydrogen peroxide and potentiometric titration with perchloric acid of the sulfoxide formed.

Adapted from the Method of Siggia and Edsberg

[*Ind. Eng. Chem., Anal. Ed.*, ***20,*** *938–9* (*1948*)]

REAGENTS

BROMATE–BROMIDE, 0.1*N*. Dissolve 2.78 grams of dry potassium bromate and 10 grams of potassium bromide in water and dilute to 1 liter. Standardize the solution by adding 6 ml of 6*N* sulfuric acid to a 50-ml aliquot of solution in a 250-ml iodine flask. Chill the flask in an ice bath, and add 10 ml of 20% potassium iodide by placing the potassium iodide solution in the well and raising the stopper carefully so that the solution flows into the flask without allowing the bromine vapors to escape. Shake the flask for 2 to 3 minutes to allow the bromine vapors to react with the potassium iodide. The iodine liberated is titrated with standard 0.1*N* thiosulfate, starch indicator being used.

Glacial acetic acid.

Concentrated hydrochloric acid.

PROCEDURE

Weigh a sample containing about 0.002 mole of alkyl sulfide into a 250-ml Erlenmeyer flask and dissolve it in 40 ml of glacial acetic acid. To this solution add 10 ml of water (less water can be added if the sample comes out of solution). Next add 3 ml of concentrated hydrochloric acid and titrate the solution with 0.1*N* bromate-bromide solution until the first yellow color due to excess of bromine is detected. The end point is not as sharp as a starch-iodine end point, but it is easily distinguished. Run a blank on the solvents alone to correct for the amount of excess bromine needed to detect the end point. The blank amounts to about 0.25 ml.

CALCULATIONS

$$\frac{\text{Milliliters of } BrO_3^-\text{—}Br^- \times N \times \text{mol. wt. of alkyl sulfide} \times 100}{\text{Grams of sample} \times 2000} = \%\ \text{alkyl sulfide}$$

This procedure is accurate to better than 0.5% if large titrations (about 40 ml) are used. This minimizes end point errors; the end point is reproducible to about 0.1 ml.

Diethyl, di-*n*-butyl, diisobutyl, and dibenzyl sulfides are used to standardize the procedure.

Houff and Schuetz (1) carried out the bromine oxidation analyses using a dead-stop electrometric end point, which makes possible the titration of colored samples.

Adapted from the Method of C. Casalini, G. Cesarano, and G. Mascellani

[*Anal. Chem.*, ***49***, *1002* (*1977*)]

EXPERIMENTAL

A Mettler automatic titrator was used, with platinum and saturated calomel electrodes.

The solution to be titrated contained approximately 2.5×10^{-4} mole of the sulfide and approximately 2.5×10^{-6} mole of potassium bromide or 1.2×10^{-6} mole of bromine in about 70 ml of 70% acetic acid in water. Potassium bromide was preferred to bromine water. Lead tetraacetate (0.1*N*) was used as the titrant. At the beginning, the titrant was added at a rate of 4 ml per minute. In the proximity of the potential jump, the titrator adjusted automatically the delivery rate of the titrant to the shape of the potentiometric curves.

RESULTS AND DISCUSSION

The method described was applied to alkyl sulfides, aryl alkyl sulfides, diphenyl sulfides, and the cyclic sulfides represented by cephalosporins. Tables 1 and 2 give the results.

The bromide-to-bromine oxidation system is established in solution only after all the sulfide has reacted and when an excess of bromide is used. Thus the bromide reaction is seen in the titration curves after completion of the sulfide oxidation process. Therefore traces of potassium bromide acting like a catalyst are sufficient to bring about the oxidation of sulfides to sulfoxides.

In cases where the titration curve does not permit a clear differentiation between the potential jump associated with the oxidation of the sulfide and that associated with the oxidation of the bromide, as in the case of cyclic sulfides, it is acceptable for analytical purposes to carry out a blank

1. W. H. Houff and R. D. Schuetz, *Anal. Chem.*, **25**, 1258 (1953).

Table 1. Results for Sulfides by Lead Tetraacetate Titrations in Presence of Catalytic Amounts of Potassium Bromide

Compound	Found, %	Precision		Found by Other Methods, %
		σ	$\sigma/\sqrt{n}$	
$CH_3CH_2SCH_2CH_3$	97.3	0.6	0.2	97.7
$C_6H_5CH_2S(CH_2)_2NH(CH_2)_2CH(C_6H_5)_2$	96.6	0.6	0.2	97.4
$C_6H_5S(CH_2)_3NH(CH_2)_2CH(C_6H_5)_2$	99.4	0.2	0.1	99.8
$C_6H_5SCH_3$	99.1	0.6	0.3	97.6
p-$CH_3O(C_6H_4)S(CH_2)_3OH$	101.9	0.3	0.2	99.3
$C_6H_5SC_6H_5$	99.2	0.2	0.1	98.0

Table 2. Results for Cyclic Sulfides by Lead Tetraacetate Titrations in Presence of Catalytic Amounts of Potassium Bromide

S, RNH, N, CH_2R_1, O, $COOR_2$

R	R_1	R_2	Found, %	Precision		Found by Other Methods, %
				σ	$\sigma/\sqrt{n}$	
H	H	H	97.0[a]	1.1	0.4	98.6
H	$OCOCH_3$	H	89.2[a]	0.6	0.3	90.2
$C_6H_5CH(NH_2)CO$	H	H	97.7	0.9	0.4	98.8
$CNCH_2CO$	$OCOCH_3$	Na	94.8	0.9	0.4	96.7
H	H	CH_2CCl_3	98.6	0.3	0.1	99.2
$(NH_2)(COOH)CH(CH_2)_3CO$	$OCOCH_3$	Na	97.7	0.3	0.1	95.5
$C_6H_5OCH_2CO$	H	$CH_2C_6H_4NO_2$	99.1	1.0	0.4	98.8

[a] Treated with 5 ml of acetic anhydride and 5 ml of acetic acid.

titration with the solvent containing the appropriate amount of catalyst and to subtract the corresponding volume of titrant from the sample titration volume.

HYDROGEN PEROXIDE OXIDATION

Adapted from the Method of C. B. Puchalsky

[*Anal. Chem.*, ***41,*** *843* (*1969*)]

The bromide-bromate procedure described above gave no discernible end point for the fungicide 5-carboxanilido-6-methyl-2,3-dihydro-1,4-oxathiin (Vitavax, Uniroyal, Inc., trademark). This procedure was developed for this product and was checked for its applicability for other organic sulfides.

The method is based on the room-temperature oxidation of sulfide to sulfoxide by hydrogen peroxide in acetic acid. Depending on the particular molecule, maximum yields of the sulfoxide are generally attained between 15 and 60 minutes. Acetic acid and unused hydrogen peroxide are removed by vacuum topping in a rotary evaporator. The sulfoxide derivative is then titrated potentiometrically in acetic anhydride with perchloric acid.

INTERFERENCES

Sulfoxide impurities in the sample can be titrated directly without peroxide oxidation beforehand because sulfides (and sulfones) exhibit no measurable basicity in acetic anhydride.

In general, any basic substance persisting through hydrogen peroxide oxidation or formed by hydrogen peroxide will interfere. These substances are limited in number and not usually found in organic sulfide samples.

APPARATUS AND REAGENT

The electrode system and titrant are those described elsewhere (p. 646).

PROCEDURE

Accurately transfer about 1 mM of the sample to a 200-ml round-bottomed flask. Insert a small Teflon-covered stirring bar, wash down the interior surface with 5 ml of acetic acid, and mix to attain complete solution. Add another 5 ml of acetic acid and, with stirring, add 0.15 ±0.01 ml of 30% hydrogen peroxide. This addition is done conveniently with a small hypodermic syringe and needle. Stopper the flask, remove it from the stirrer, and roll it at an angle to collect any droplets.

Let the flask stand for the appropriate time. Place the flask on a rotary evaporator and with a large air bleed, begin rotation. Place a bath of room-temperature water under the flask to prevent freezing of the solution. Samples judged to produce volatile sulfoxides should have a condenser interposed between the flask and the rotary evaporator. Gradually decrease the pressure to about 1 to 2 mm. When the solvent is gone (generally about 5 minutes), remove the room-temperature water bath and substitute a bath of water at 100°C, maintaining that temperature with a hot plate. Heat the flask for 10 minutes at 1 to 2 mm to remove the hydrogen peroxide. It is important that this be entirely removed, or low, erratic results will be obtained.

Cool the flask before restoring pressure. Use a total volume of about 80 ml of acetic anhydride to transfer the contents to a 100-ml beaker, along with the stirring bar. Warm the acetic anhydride slightly, if needed, to dissolve solid derivatives.

Insert the electrodes and while stirring, adjust the pH meter to an apparent pH between 10 and 11. Add the perchloric-acetic titrant from a 10-ml buret. Plot milliliters of titrant versus pH and read the end point.

RESULTS

The sulfoxide yield for various compounds as a function of time of oxidation is shown in Table 3.

Table 3. Sulfoxide Yield as a Function of Oxidation Time

		Percentage of Theory			
Sample	Purity	15 min.	30 min.	60 min.	Bromate-Bromide Method
Vitavax	Recrystallized 3 times from cyclohexane (m.p., 97.8–99.0°C)	—	97.9	97.9[a]	[b]
Benzyl sulfide	Eastman White Label	99.1	—	94.4	101.7
Hexyl sulfide	Eastman White Label	96.9	95.8	95.7	102.5 102.9
Dilauryl thiodipropionate	Technical	88.3	93.9	95.9	100.9 100.4
2,2′-Thiodiethanol	169–170°C at 20 mm	—	—	95.1[c]	100.7 100.4

[a] Average of 10 determinations, done over a 4-month period; the standard deviation was 0.20.
[b] No end point.
[c] Poor end point.

MIXTURES OF DIALKYL SULFIDES AND MERCAPTANS

Samples containing mercaptans can be analyzed by the foregoing bromine oxidation, provided they contain no more than 10% mercaptan.

This procedure can be used to determine mercaptans with a precision of about ±0.8% but with an accuracy of only ±5% of the amount contained in the sample. This procedure, then will determine both the sulfide and the mercaptan; the mercaptan alone can be determined by one of the methods described in the chapter on mercaptans. The sulfide content is then determined by difference (Table 4). Since the accuracy of this method for mercaptans is only ±5%, if the sample contains too much mercaptan, a serious error in the sulfide value may be introduced when the mercaptan value is subtracted. In addition to the method just mentioned for handling mixtures of mercaptans and dialkyl sulfides, another was devised by Jaselskis.

Table 4. Sulfides

	Concentrated Acid Used, ml.	%
n-Butyl sulfide in 50 ml. of 80% HAc-20% H_2O	3 HCl	99.5
	13 HCl	99.0
	25 H_2SO_4	99.5
Benzyl sulfide[a] in 50 ml. of 80% HAc-20% H_2O	3 HCl	99.5
		100.3
		100.1
Ethyl sulfide in 100 ml. of 80% HAc-20% H_2O	3 HCl	98.3
		99.0
		99.1
		98.3

[a] Sulfur analysis was run on this sample: % S found = 15.20, % S calculated = 14.95.

Method of B. Jaselskis

[*Reprinted in Part from Anal. Chem.*, ***31,*** *928–31* (*1959*)]

REAGENTS AND APPARATUS

POTASSIUM TRIIODIDE SOLUTION, APPROXIMATELY 0.1*N*. Dissolve iodine in potassium iodide solution and standardize against arsenious oxide.

STANDARD $0.1N$ BROMATE–BROMIDE SOLUTION. Weigh analytical grade potassium bromate and add at least a sevenfold excess of potassium bromide. Check the normality against arsenious oxide.

Acrylonitrile used for the condensation reaction with alkyl mercaptans was obtained from Matheson. It was used without further dilution.

The dialkyl sulfide solutions were reduced in a refluxing flask containing a water-cooled condenser and a cold finger cooled by an ice-salt mixture. The latter attachment is necessary when using ethyl and propyl sulfides.

PROCEDURE

ALKYL MERCAPTAN–DIALKYL SULFIDE MIXTURES. Place two equal aliquots of the alkyl mercaptan–dialkyl sulfide mixture in separate flasks. Dilute the contents with a sufficient amount (30–50 ml) of alcohol to keep the alkyl mercaptan and dialkyl sulfide in solution throughout the titration. Acidify one of the aliquots with approximately 2.0 ml of acetic acid; then titrate mercaptan with the standard iodine solution until the color of free iodine appears. For small amounts of mercaptan use a 10-ml buret to enable the titrant to be measured within ±0.01 ml.

Alkyl mercaptans with higher molecular weights than propyl mercaptan could be titrated in the absence of oxygen in acidic potassium iodide solution with the standard bromate-bromide. Deaerate this mercaptan solution using pieces of dry ice. The results are somewhat lower than in the direct iodometric determination. When this method is used, only one standard solution is necessary for determination of mercaptan and sulfide.

To the contents of the second flask add approximately 3 drops of 10% potassium hydroxide, then add acrylonitrile from a 1-ml graduated pipet. Determine the volume of acrylonitrile to be added by the amount of mercaptan present. Approximately a twofold excess as compared to mercaptan is usually sufficient. However the volume of acrylonitrile is not critical as long as there is an excess, and provided blank determinations are run using the same amount of acrylonitrile.

After the addition of acrylonitrile, stir the solution for about 2 minutes to ensure completion of the reaction. Then acidify with approximately 15 ml of glacial acetic acid containing about 3 ml of concentrated hydrochloric acid and titrate with the standard bromate-bromide solution to the first appearance of free bromine throughout the solution for at least 30 seconds.* The end point can be observed best in a white light against a white background. Under these conditions, the blank volume of $0.1N$ bromate-bromide necessary to impart a yellow coloration to the solution

* See method of Siggia and Edsberg, pp. 771–3.

does not exceed 0.1 ml. In the presence of approximately 0.2 ml of acrylonitrile, the blank requires about 0.15 ml.

The amount of alkyl sulfide present in the mixture is calculated from the two titrations as follows:

$$\text{Alkyl sulfide, grams} = [\text{meq.}_{(BrO_3^- - Br^-)} \text{ minus } 2 \text{ meq.}_{I_3^-}] \times E$$

where milliequivalents of bromine is for the second titration corrected for the blank, milliequivalents of iodine is for the first titration corrected for the blank, and E is the equivalent weight of sulfide.

RESULTS AND DISCUSSION

The determination of alkyl mercaptans and dialkyl sulfides in the presence of alcohol cannot be performed using exhaustive oxidation with bromate-bromide in strongly acid solution because of the oxidation of alcohol. In the absence of alcohol, however, the oxidation proceeds slowly to completion in the following manner:

$$RSH + 3Br_2 + 2H_2O = RSO_2^+ + 6Br^- + 5H^+ \quad (1)$$

$$RSR + Br_2 + H_2O = RSOR + 2Br^- + 2H^+ \quad (2)$$

$$RSOR + Br_2 + H_2O = RSO_2R + 2Br^- + 2H^+ \quad (3)$$

The first and third reactions are slow and incomplete at room temperature, but they can be speeded by heating. In the presence of alcohol, the determination is based on the condensation reaction of alkyl mercaptan with acrylonitrile

$$RSH + CH_2 = CHCN \xrightarrow{(OH)^-} RS\text{—}CH_2\text{—}CH_2\text{—}CN \quad (4)$$

The resulting sulfide is readily oxidized in dilute acid solution by bromine according to eq. 2. It is titrated to the first appearance of free bromine throughout the solution. The double bond in acrylonitrile is not brominated readily. The end point can be determined to within ±0.03 ml.

The average values of five determinations for mixtures of alkyl mercaptans and dialkyl sulfides are summarized in Table 5. The results for alkyl mercaptan in the absence of dialkyl sulfide determined by iodine or by bromate–bromide titrations, after the condensation of mercaptan with acrylonitrile, are in good agreement. The end point can be reproduced readily to ±0.03 ml, yielding reproducibility better than 1% for the 10-ml volume of titrant used.

Table 5. Titration of Alkyl Mercaptan–Dialkyl Sulfide Mixtures

(Results are average of five determinations)

	Millimoles of Alkyl Mercaptan			Millimoles of Dialkyl Sulfide	
		Found by			
Sample	Added	Iodometric Method	Acrylonitrile Condensation, and Bromate Titration	Added	Found by Bromate Titration[a]
Ethyl mercaptan	0.9574	0.956	0.957	—	—
Mixture IA[b]	0.9574	0.956	—	0.2470	0.244
Mixture IB[b]	0.4787	0.477	—	0.6175	0.615
Mixture IC[b]	0.1915	0.190	—	1.2350	1.227
Diethyl sulfide	—	—	—	1.2350	1.236
n-Propyl mercaptan	1.1130	1.109	1.115	—	—
Mixture IIA[c]	1.1130	1.107	—	0.2469	0.245
Mixture IIB[c]	0.5665	0.556	—	0.6172	0.616
Mixture IIC[c]	0.2226	0.221	—	1.2340	1.232
Dipropyl sulfide	—	—	—	1.2340	1.233
n-Butyl mercaptan	1.0640	1.061	1.065	—	—
Mixture IIIA[d]	1.0640	1.059	—	0.1844	0.183
Mixture IIIB[d]	0.5320	0.531	—	0.4611	0.460
Mixture IIIC[d]	0.2128	0.211	—	0.9222	0.921
Dibutyl sulfide	—	—	—	0.9222	0.921
n-Octyl mercaptan	0.9015	0.901	0.901	—	—
Mixture IVA[e]	0.9015	0.901	—	0.1844	0.183
Mixture IVB[e]	0.4507	0.449	—	0.4611	0.460
Mixture IVC[e]	0.1803	0.179	—	0.9222	0.921

[a] Millimoles of dialkyl sulfide obtained from $[(\text{meq.})_{BrO_3^-} - 2(\text{meq.})_{I_3^-}]/2$.
[b] Mixture I, ethyl mercaptan and diethyl sulfide.
[c] Mixture II, *n*-propyl mercaptan and dipropyl sulfide.
[d] Mixture III, *n*-butyl mercaptan and dibutyl sulfide.
[e] Mixture IV, *n*-octyl mercaptan and dibutyl sulfide.

The results for dialkyl sulfide in the presence of mercaptan can be reproduced to approximately 1% or better if small amounts are present. The results are in better agreement with higher molecular weight dialkyl sulfides than with diethyl sulfide.

The presence of acrylonitrile affects the blank correction somewhat. The change of volume of acrylonitrile from 0.2 to 0.5 ml increases the blank correction by approximately 0.1 ml of 0.1*N* bromate. The condensation of acrylonitrile with mercaptan is rapid and complete in alkaline solution at pH 12 or above.

Normal and secondary mercaptans can be titrated with the standard iodine solution. Tertiary mercaptans should be titrated argentimetrically. Hydrogen sulfide and all the substances that can be oxidized by iodine will interfere. Bromate-bromide titrations can be used in the absence of all the substances that can be oxidized by bromine and all unsaturated compounds prone to bromination.

MIXTURES OF DIALKYL SULFIDES WITH DIALKYL DISULFIDES

A procedure to analyze these mixtures is outlined in Chapter 19 in the section concerned with disulfides (p. 763).

21

Sulfoxides

R_2SO

Sulfoxides exhibit weakly basic properties as a result of the tendency to form hydrogen bonds. They can be titrated directly as bases in acetic anhydride solvent with perchloric acid as titrant. Other analytically useful reactions are reduction to sulfides and oxidation to sulfones. Titanous chloride is used as a reductant and potassium dichromate as an oxidant. Other reduction methods such as those applied to disulfides are not effective on sulfoxide. Sulfoxides are quite resistant to reductants weaker than titanous chloride. Chromous ions also should be applicable as a reductant. A strong oxidant is also needed to oxidize these stable compounds.

Titration Method—Adapted from D. C. Wimer

[*Reprinted in Part from Anal. Chem.*, **30,** *2060* (*1958*)]

The present method is an extension of work described earlier (pp. 183–9) on the titration of amides in acetic anhydride. Streuli (1) reported the successful titration of dimethyl sulfoxide in acetic anhydride. No other compounds of this type were cited.

APPARATUS

A Precision-Dow Recordomatic Tirometer, Model K-3-247, was used in all titrations. The modified calomel-glass electrodes used were described in Ref. 2 (see pp. 183–9 on titration of amides).

REAGENTS

Acetic anhydride, ACS reagent grade.

Dioxane, purified by the procedure shown on p. 694 for isocyanate-isothiocyanate analysis

1. C. A. Streuli, *Anal. Chem.*, **30,** 997–1000 (1958).
2. D. C. Wimer, *Ibid.*, **30,** 77–80 (1958).

PERCHLORIC ACID, 70% VACUUM DISTILLED. Available from the G. Frederick Smith Chemical Company, Columbus, Ohio.
PERCHLORIC ACID TITRANT. Prepare a $0.1N$ solution by diluting 9 ml of 70% perchloric acid to 1 liter with purified dioxane. The solution is allowed to stand 24 hours prior to use. The titrant may be standardized either visually or potentiometrically against primary standard potassium acid phthalate dissolved in acetic acid.

PROCEDURE AND RESULTS

Dissolve an approximately 0.001-mole sample, accurately weighed, in 75 ml of acetic anhydride, and titrate potentiometrically with freshly standardized $0.1N$ perchloric acid in dioxane. The end point may be determined by inspection or calculation of maximum $\Delta E/\Delta V$.

All samples were analyzed as received without further purification. The relative degree of basicity may be ascertained by inspection of the first derivative values listed in Table 1. Attempts to resolve mixtures of sulfoxides by differential titration were unsuccessful. Only single inflections were observed. This indicates that the differences in the magnitude of maximum $\Delta E/\Delta V$ probably represent minor differences in actual base strength.

DISCUSSION

Sulfides and sulfones, possible impurities in sulfoxides, in general exhibit no measurable basicity in acetic anhydride when titrated with perchloric acid.

Trimethylphosphine oxide, although not a sulfoxide, has been included for comparison because amine oxides have been found to exhibit very sharp inflections in acetic anhydride. Diethoxy sulfoxide (diethyl sulfite) could not be titrated as a base in acetic anhydride.

The direct titration of sulfoxides in acetic anhydride offers a rapid and reproducible approach to purity determination. The method appears to be limited only by the solubility of the particular sulfoxide in acetic anhydride.

Titanous Chloride Reduction Method—Adapted from D. Barnard and K. R. Hargrave

[*Reprinted in Part from Anal. Chim. Acta*, **5,** *536–42* (*1951*)]

Table 1. Titration of Various Compounds in Acetic Anhydride

Compound	Purity, %	Max. $\Delta E/\Delta V$, mv./ml. (Approximate)
Bis(2-hydroxyethyl) sulfoxide	96.9 96.9	120
Dibenzyl sulfoxide	98.1 98.1	580
Diethoxy sulfoxide (diethyl sulfite)	No titration curve inflection obtained	
Dimethyl sulfoxide	99.9 99.8	1030
Diphenyl sulfoxide	99.0 99.4	130
Di-*p*-tolyl sulfoxide	98.8 98.4	370
p,p'-Dihydroxydiphenyl sulfoxide	55.0[a]	300
Phenothiazine 5-oxide	96.9 97.8	230
Phenoxathin 10-oxide	98.6 98.6	300
Thianthrene 10-oxide	95.9 95.0	100
Thianthrene 5,10-dioxide	95.0 95.3	100
Trimethylphosphine oxide	100.1	1420

[a] Incompletely soluble in acetic anhydride.

SATURATED SULFOXIDES

REAGENTS

TITANOUS CHLORIDE SOLUTION, APPROXIMATELY 0.1*N*. Prepared by diluting a 15% commercial sample (100 ml) with water (850 ml) and hydrochloric acid (sp.

gr., 1.18, 50 ml) both freed from dissolved oxygen. Store solution under nitrogen and dispense by means of the device described in Ref. 3.* The titer of the stock solution remains unchanged for several months.

FERRIC ALUM SOLUTION. Dissolve 200 grams ferric alum per liter in distilled water containing 60 ml of concentrated sulfuric acid.

POTASSIUM DICHROMATE, 0.05*N*. Dissolve 75 ml of orthophosphoric acid and 75 ml of concentrated sulfuric acid in 350 ml of distilled water.

INDICATOR. 0.25% solution of diphenylamine sulfonic acid.

SOLVENT. The acetic acid used was of analytical grade. The nitrogen was of "oxygen-free" standard and was passed through Fieser's solution (4) and a drying train before use.

PROCEDURE A

Dissolve the sample, containing 0.7 to 1.0 meq. of sulfoxide, in acetic acid (10 ml) in a 250-ml Erlenmeyer flask fitted with a standard ground joint and equipped with a cone and tap adaptor held by two spring clips. Then evacuate the flask to 20 mm pressure and fill with nitrogen. Add 15 ml of 0.1*N* titanous chloride solution from a pipet, and at once reevacuate the flask and fill it with nitrogen; repeat this procedure twice. Immerse the flask one-third in a water bath maintained at 80°C and leave for 1 hour. After this period, add a boiling solution consisting of 5 ml of ferric alum solution diluted to 50 ml with water, allow it to stand for 30 seconds, and then cool it rapidly. Add 10 ml of phosphoric acid solution and 15 ml of carbon tetrachloride, shake the mixture vigorously, and titrate the ferrous ion with 0.05*N* dichromate solution, using 6 drops of diphenylaminesulfonic acid as indicator.

Carry out blank determinations on the solvent. They should not differ by more than 0.1 ml of 0.05*N* solution from the titer of the titanous chloride when standardized directly.

$$\% \text{ Sulfoxide estimated} = \frac{(\text{Blank} - \text{titer found}) \times \text{dichromate normality} \times \text{mole. wt. of sulfoxide} \times 100}{2000 \times \text{sample weight}}$$

3. D. Barnard and K. R. Hargrave, *Anal. Chim. Acta*, **5**, 479 (1951). Storage vessel had rubber stopper in which was inserted a loosely fitting pipet. The solution is kept under nitrogen and is dispersed by filling the pipet with a positive nitrogen pressure.
4. L. F. Fieser, *J. Am. Chem. Soc.*, **46** (1924) 2639.

RESULTS

Table 2. Standardization of Procedure Against Pure Sulfoxides

Sulfoxide	% Estimated (mean of 5 determinations)	Standard Deviation
Diphenyl	100.0	0.3
Phenyl cyclohexyl	99.0	0.5
Phenyl methyl	99.4	0.2
Di-*n*-butyl	100.4	0.3
Methyl cyclohexyl	98.8	0.2
Diphenyl selenoxide	100.2	0.3

DISCUSSION

The excess titanous chloride can be satisfactorily estimated in two ways; direct titration with ferric ion using thiocyanate as indicator; or by adding excess ferric iron and then titrating the resulting ferrous iron with standard dichromate solution. The latter procedure was actually adopted, since it gave the more precise results and had the advantage of dispensing with titrations under an inert atmosphere.

INTERFERING SUBSTANCES. Interference with the dichromate titration by thioethers and primary alcohols is discussed in Ref. 3. Thioethers are inevitably present but can be satisfactorily removed from the zone of dichromate oxidation by shaking the mixture with carbon tetrachloride before titration. Failure to do this leads to serious errors. Low molecular weight primary alcohols that do not pass wholly into the organic layer necessitate the direct titration of titanous chloride with ferric iron.

Substances (e.g., nitro compounds, amine oxides, azo and diazo compounds, and hydroperoxides) reduced by titanous chloride must be absent. Aldehydes present in amounts up to 2 meq. are reducible under the experimental conditions only to a relatively small extent (not exceeding 1 ml of 0.05N solution), whereas sulfones, disulfides, ketones, and compounds containing ethylenic double bonds do not interfere at all.

SOLVENTS. Although acetic acid is recommended, *tert*-butyl alcohol and benzene may also be used. The volume of water-immiscible solvent should not greatly exceed 10 ml owing to the unfavorable partition coefficient of the sulfoxide between the organic and aqueous layers.

TIME OF REACTION. At temperatures above 60°C the reduction is very rapid, being complete in 30 minutes at 60°C and 10 to 15 minutes at 80°C. The recommended time of 1 hour can therefore be considerably reduced if desired. No change in the amount estimated or the blank titration occurs when the reaction time is extended to several hours.
SAMPLE SIZE. By using reagents diluted as described in Ref. 3, the sample size can be reduced to 0.2 to 0.3 meq. with only a slight reduction in precision (standard deviation ~2%).

UNSATURATED SULFOXIDES

The analytical procedure established for saturated sulfoxides does not hold when applied to certain allylic sulfoxides. In such cases precise but low estimations of the order 80 to 95% were obtained for crystalline samples whose purity could not be in doubt. This implies that under the estimation conditions, part of the sulfoxide is decomposed by some secondary reaction—possibly acid hydrolysis, since heating in acid solution prior to reduction by titanous chloride produced a further decrease in the percentage estimated. Solutions of titanous chloride buffered with sodium acetate to as low an acidity as is practicable have been used to minimize such hydrolysis; it is known, moreover, that under these conditions the reducing power is increased (5). The accuracy was improved by this means, but always with a loss in experimental precision. However the presence of ammonium thiocyanate during the reduction of unsaturated sulfoxides allows both an accurate and precise analysis of all the compounds detailed below. In this case the procedure already described for the determination of excess titanous chloride cannot now be followed and the alternative method, that is, direct titration with standard ferric iron solution, must be used.

Table 3 compares results from the three methods given in this section.

REAGENTS

In addition to the reagents described in connection with Procedure A, the following are required:

Ferric alum solution, 0.05*N*, standardized against potassium dichromate using the stannous chloride reduction method.

Ammonium thiocyanate solution, 3*M*.

5. P. Butts, W. Meikle, J. Shovers, D. Kouba, and W. Becker, *Anal. Chem.*, **20** (1948) 947.

Table 3. Comparison of Results for the Reduction of Pure Unsaturated Sulfoxides

(1) With titanous chloride alone.
(2) With titanous chloride buffered with sodium acetate.
(3) With titanous chloride in the presence of ammonium thiocyanate.

Sulfoxide	(1)		(2)		(3)	
	%	S.D.	%	S.D.	%	S.D.
n-Propyl cinnamyl	94.0	0.6	100.3	0.9	99.7	0.4
n-Butyl cinnamyl	94.2	0.2	101	0.9	101.0	0.6
Phenyl cinnamyl	100.6	0.3	99.8	0.6	99.2	0.5
Diallyl	91.0	0.6	83	3.5	100.1	0.4
Methyl 2-methylallyl	95.6	0.5	95	2.0	99.6	0.4
n-Propyl 1,3-dimethylallyl	93.5	0.4	98.5	2.0	99.8	0.2
Phenyl allyl	99.3	0.3	100	1.0	99.6	0.5
Cyclohexenyl methyl[a]	80.0	0.4	90	2.0	97.1	0.4
Dibenzyl[a,b]	86.8	0.2	95	1.0	97.5	0.3
Phenyl benzyl[b]	99.1	0.3	99.0	0.6	99.2	0.3

[a] Purest sample obtainable.
[b] Can be regarded as allylic-type sulfoxides.

PROCEDURE

Add 1.5 ml of 3*M* ammonium thiocyanate solution to the sample (0.7–1.0 meq. sulfoxide) dissolved in acetic acid (10 ml) and carry out the titanous chloride reduction exactly as described in Procedure A. After the period of heating, titrate the hot solution with 0.05*N* ferric alum solution, maintaining a rapid stream of nitrogen through the flask and preferably employing some form of magnetic stirring. The end point is indicated by the first permanent ferric thiocyanate color. Blanks are again within 0.1 ml of 0.05*N* solution of the titanous chloride titer when standardized directly. Calculate the percentage of sulfoxide from the formula given in Procedure A.

PROCEDURE USING BUFFERED TITANOUS CHLORIDE SOLUTION

Add 25 ml of sodium acetate solution (40% w/w $CH_3COONa \cdot 3H_2O$) to the sample in acetic acid and carry out the reduction as previously.

After the heating period, add 6.5 ml of hydrochloric acid (sp. gr., 1.18); the excess titanous chloride can then be estimated by either of the ways described in Procedures A and B. Blank determinations in this case are usually lower than the direct titanous chloride titer.

DISCUSSION

MODE OF ACTION OF THIOCYANATE. Ammonium and potassium thiocyanates similarly affect the estimation, indicating that the thiocyanate ion is the significant factor. The addition of thiocyanate to solutions of titanous chloride considerably modifies their normal violet color, suggesting the formation of a complex ion. If the titanic ion were preferentially involved, the reducing power of the system should be increased and the reduction of sulfoxide might then be facilitated relative to any secondary process. Actually, thiocyanate complexes have been reported for both valency states of titanium (6), and indeed the oxidation-reduction potential of the system Ti^{+3}/Ti^{+4} as determined experimentally shows no significant change when thiocyanate is added. In contrast, similar determinations show that buffering with sodium acetate produces a substantial increase in the reducing power of this system. Thus we find that the reduction of nitro compounds by titanous chloride is virtually unaffected by thiocyanate (in fact, a slight retardation is observed) whereas the rate of reduction of dinitrotoluene is considerably increased by the addition of sodium acetate (5). As illustrated for typical examples in Fig. 21.1, however, the rate of reduction for both saturated and unsaturated sulfoxides is significantly increased by the presence of thiocyanate. In these experiments, the lower temperature was used to decrease the normal rates to more easily measured values.

Clearly, the thiocyanate must engage in some specific interaction with sulfoxide groups. In this respect, it is significant that a mixture of titanous chloride and sulfoxide in concentrations as recommended gives rise to complex precipitates when thiocyanate is added. The formation of these precipitates is dependent on the presence of all three components, the concentration of thiocyanate and to some extent the nature of sulfoxide. The addition of 5 ml of 3*M* ammonium thiocyanate to such mixtures produces precipitates in the case of all the sulfoxides listed, both saturated and unsaturated, except for methyl 2-methylallyl and diallyl sulfoxides. These complexes, which redissolve during the course of the analysis, vary from red to green and from semiliquid to flocculent. Their nature has

6. J. Newton Friend, *Textbook of Inorganic Chemistry*, Vol. V, Griffin and Co. Ltd., 1921, p. 254.

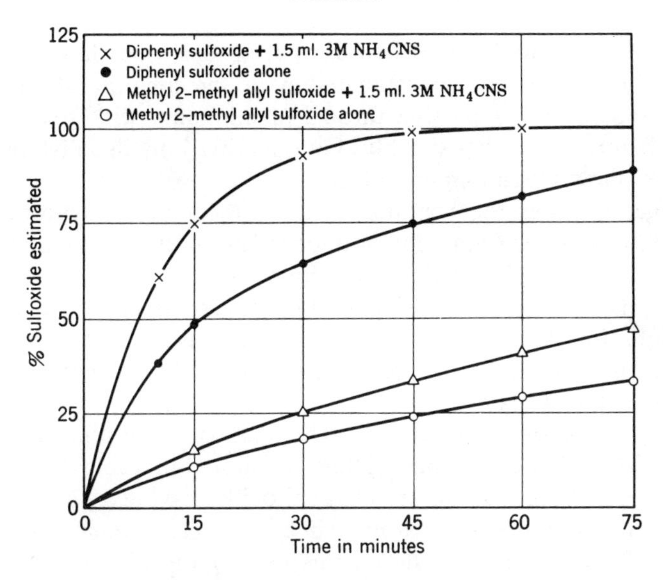

Fig. 21.1. Effect of ammonium thiocyanate on rate of reduction of diphenyl sulfoxide and methyl 2-methallyl sulfoxide by titanous chloride at 40°C.

not been fully investigated, but they appear to be unstable when isolated and of variable composition, although indications were obtained that the titanous and sulfoxide contents were about equimolar. Although these complexes are not always precipitated under the recommended procedure, the inherent tendency toward their formation is undoubtedly closely associated with the catalysis by thiocyanate.

AMOUNT OF THIOCYANATE REQUIRED. The data in Table 4 show that the

Table 4. Variation of Amount of Thiocyanate in Recommended Procedure

Volume of 3*M* Ammonium Thiocyanate Present, ml.	% Estimated	
	n-Propyl Cinnamyl Sulfoxide	Methyl 2-Methylallyl Sulfoxide
0	91.1	96.3
0.01	93.7	92.8
0.02	95.5	91.8
0.05	96.8	92.8
0.10	98.7	96.1
0.20	—	97.9
0.40	—	98.9
1.50	99.7	99.6

sulfoxides appear to fall into two distinct groups, one showing a steady increase in the percentage estimated with the amount of thiocyanate present and the other passing through a minimum with small amounts present before increasing steadily. Only methyl 2-methylallyl and diallyl sulfoxides show this minimum effect.

In general, under the conditions of the recommended procedure, as little as 0.01 ml of 3*M* ammonium thiocyanate solution is found to have a noticeable effect, 0.5 ml being the minimum amount required for complete estimation. Beyond this quantity, no further change is observed. The results for saturated sulfoxides, and the unsaturated sulfoxides not requiring thiocyanate for complete estimation, remain unaffected by the presence of thiocyanate.

EFFECT OF SULFOXIDE STRUCTURE. The examples of phenyl cinnamyl, phenyl allyl, and phenyl benzyl sulfoxides quoted in Table 3, suggest as a general principle that an unsaturated sulfoxide, having a phenyl group as one substituent on the sulfur atom, will be correctly estimated even in the absence of thiocyanate. No doubt this reflects the increased stability of either the sulfoxide molecule or some intermediate in the reduction stage.

Legault and Groves (7) slightly modified the method of Barnard and Hargrave to make it apply to dimethylsulfoxide and 2-hydroxy diethyl sulfoxide.

DICHROMATE OXIDATION OF SULFOXIDES

Adapted from V. V. Savant, J. Gopalakrishnan, and C. C. Patel

[*Z. Anal. Chem.*, ***238***, *273* (*1968*)]

REAGENTS

Ferrous ammonium sulfate, 0.1*N*, in about 1*N* sulfuric acid.
Standard 0.1*N* potassium dichromate solution.
P-Phenyl anthranilic acid indicator in sodium hydroxide solution.

PROCEDURE

To a flask containing about 0.1 gram of sample, add 25 ml of 0.1*N* potassium dichromate and sufficient 18*M* sulfuric acid to give an initial acid concentration of 2*M*. Add a few glass beads, and cover the flask with

7. A. R. Legault and K. Groves, *Anal. Chem.*, **29**, 1495–6 (1957).

a small funnel. Heat the contents of the flask to boiling (105–110°C) on a sand bath and maintain this temperature for 10–12 minutes. Cool the contents and dilute with about 50 ml of water. Titrate the unreacted dichromate with the ferrous ammonium sulfate solution with *N*-phenyl anthranilic acid as the indicator. Run a blank under identical conditions.

CALCULATION

$$\text{Sulfoxide, \% by weight} = \frac{(V_1 - V_2)N \times \text{mol. wt.}}{\text{Grams of sample} \times 20}$$

where

V_1 = milliliters of ferrous solution required for the blank

V_2 = milliliters of ferrous solution corresponding to the excess dichromate

RESULTS

Results for dimethyl and diphenyl sulfoxides are given in Table 5. The method was applied successfully to some copper sulfoxide complexes also and should be generally applicable to sulfoxides in the absence of organic materials that consume dichromate.

Table 5. Determination of Sulfoxides by Dichromate Oxidation

	Sulfoxides		
Compound	Taken, g	Found, g	Recovery, %
Dimethyl sulfoxide	0.04996	0.04995	99.97
	0.05971	0.05963	99.86
	0.07020	0.07000	99.72
	0.07423	0.07406	99.77
	0.08772	0.08760	99.87
	0.10460	0.10440	99.84
Diphenyl sulfoxide	0.08418	0.08440	100.20
	0.09500	0.09506	100.10
	0.10060	0.10060	100.00
	0.10470	0.10450	99.77
	0.10650	0.10620	99.72

22

Sulfonic Acids, Sulfonate Salts, Sulfonamides, and Sulfinic Acids

Sulfonic Acids

Sulfonic acids are very strong acids comparable to sulfuric acid and are readily determined by titration. If the sample is water insoluble, it can usually be dissolved in excess caustic and the excess back-titrated or, it can be readily titrated in alcoholic solvents, glycols or glycol mixtures, dioxane, pyridine, or almost any solvent in which a titration for acids can be run (pp. 000–0) without causing the sulfonic acid to react with the solvent. Sodium hydroxide (aqueous for aqueous titrations or alcoholic for nonaqueous systems) can be used, and indicators work well in their respective solvents. The end points are so sharp for sulfonic acids that an analyst rarely has to resort to potentiometric titration.

Sulfonate Salts

The alkali metal salts of sulfonic acids are neutral and cannot be titrated as acids or bases. The alkaline earth or transition metal sulfonates are generally acidic and should be titratable, but they are generally so insoluble that titration is impossible.

A determination based on ignition to the sulfate is accurate and precise and will operate for most metal sulfonates. It has a severe limitation, however, in that any other inorganic cation in the sample will cause high results for sulfonate.

Aryl sulfonates are converted to sulfite and the corresponding phenol by alkali fusion in an inert atmosphere. Measurements of the phenol or sulfite produced are the bases for general, accurate, and precise methods for these sulfonates.

Precipitation methods with inorganic cations also suffer from lack of specificity, since any sulfate ion in the sample (and sulfonates generally contain sulfate from their synthesis) will also either precipitate or co-precipitate with the sulfonate. The benzidine precipitation is the most specific method for determining sulfonate salts, but it tends toward low results owing to a slight solubility of the precipitate.

Titration with quaternary cations can be applied to alkali metal salts of alkyl or aryl sulfonic acids, where the alkyl or aryl groups are fairly large, to contribute an organic solubility to the molecule.

IGNITION METHOD

PROCEDURE

Weigh about 2 grams of sample into a tared crucible (platinum is preferable but porcelain can be used). Just cover the sample with concentrated sulfuric acid. Heat the crucible with a Bunsen burner, taking care not to heat too vigorously, which will cause spattering of the contents of the crucible. Continue heating until fumes of sulfur trioxide are no longer visible. Then strongly ignite the crucible in a muffle furnace (a Meker type burner can be used) until the contents are white. Cool the ignition residue and moisten well with sulfuric acid, and repeat the heating with a Bunsen burner until all the acid has been driven off. Then ignite the crucible in the muffle oven for 30 minutes, cool, and weigh.

CALCULATIONS

$$\frac{\text{Grams of residue} \times \text{gravimetric factor} \times 100}{\text{Grams of sample}} = \%\ \text{salt}$$

The procedure is applicable only when the sample contains no cation other than that of the salt. If such cations were present, they too would form sulfates and would be weighed in the residue. This is the best method for the determination of sulfonic acid salts, but there is often some inorganic material present that makes this method useless.

ALKALI FUSION METHOD

The conversion of aryl sulfonates proceeds according to the following equation:

$$ArSO_3M + M'OH \rightarrow ArOH + MM'SO_3$$

where M and M′ can be the same or different cations and M can be a hydrogen atom. Two approaches are possible for the quantitative measurement; either the sulfite or the phenol produced can be measured. If the sulfite is measured, the total sulfonate analysis is obtained. If the phenol is measured, it is possible to obtain both the total sulfonate analysis and an analysis for each sulfonate present in the sample by quantitative gas chromatography. Sulfite is measured volumetrically and the procedure is described here.

Method of S. Siggia, L. R. Whitlock, and J. C. Tao

[*Reprinted in Part from Anal. Chem.*, ***41,*** *1387* (*1969*)]

APPARATUS

The equipment is illustrated in Fig. 22.1. The sample and caustic are held in a stainless steel crucible *D* (Scientific Glass Co., Catalog no. 2610). The crucible enclosed in a glass chamber *C* is placed in a sand bath *E* and heated by an open flame. Oxygen-free nitrogen is passed into the chamber through *A*, over the fusion mixture, and vented through *B*. Hot copper mesh (temperature 450–600°C) is used to remove traces of oxygen from the nitrogen.

PROCEDURE

Place approximately 3 mM of sulfonate along with 5 grams of potassium hydroxide pellets in a stainless steel crucible. Place the crucible in the glass chamber and flush the system with oxygen-free nitrogen for at least 5 minutes to displace the air. Bring the temperature up to 400°C and maintain for 30 to 60 minutes. Measure the temperature with a thermometer in the sand directly adjacent to the glass chamber.

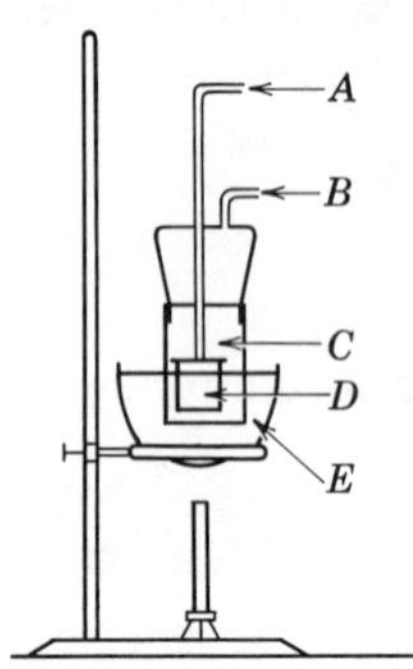

Fig. 22.1. Apparatus for alkali fusion used in the sulfite-measuring method.

After fusion, remove the sand bath to allow the system to cool to room temperature. Maintain the flow of nitrogen during the cooling period. Remove the crucible from the glass chamber and add 30 to 50 ml of distilled water. Stir the mixture gently with a magnetic stirring paddle until dissolution is complete. Transfer the solution quantitatively to a 250-ml beaker. Add 6*N* sulfuric acid dropwise until the pH of the system is 9 to 10, as measured by a glass-calomel electrode system. Then add 0.1*N* sulfuric acid via a buret until the pH reached indicates that the excess potassium hydroxide is just neutralized. This pH, designated P_1 in Fig. 22.2, is approximately 8 but varies slightly with each sulfonate being measured. The exact pH should be determined for accurate work. At this point, add 1 ml of formaldehyde (Certified ACS 36.8%): the pH is observed to rise as hydroxide is liberated by the bisulfite addition reaction.

$$Na_2SO_3 + HCHO + H_2O \rightarrow \underset{\displaystyle SO_3Na}{\underset{|}{CH_2OH}} + NaOH$$

Check the formaldehyde for the presence of formic acid by a blank titration (the blank is generally negligible). Titrate the liberated base potentiometrically with standard 0.1*N* sulfuric acid to the inflection point P_2 in Fig. 22.2. Determine the inflection point from a plot of pH versus milliliters.

CALCULATION

$$\%\ \text{Sulfonate} = \frac{V \times N \times M \times 100}{S \times 1000}$$

where V is volume of standard acid used to titrate the liberated base, N is the normality of the acid, M is the molecular weight of the sulfonate, and S is the weight of the sample in grams.

RESULTS AND DISCUSSION

Iodimetric measurement of the sulfite resulting from the fusion gave erratic and low results. Complete recoveries were obtained when sodium samples were put through the procedure and the formaldehyde treatment was used. Potassium hydroxide gave better fusion than did sodium hydroxide. The minimum temperature is the melting point of potassium hydroxide, 360°C. Temperatures as high as 450°C could be tolerated

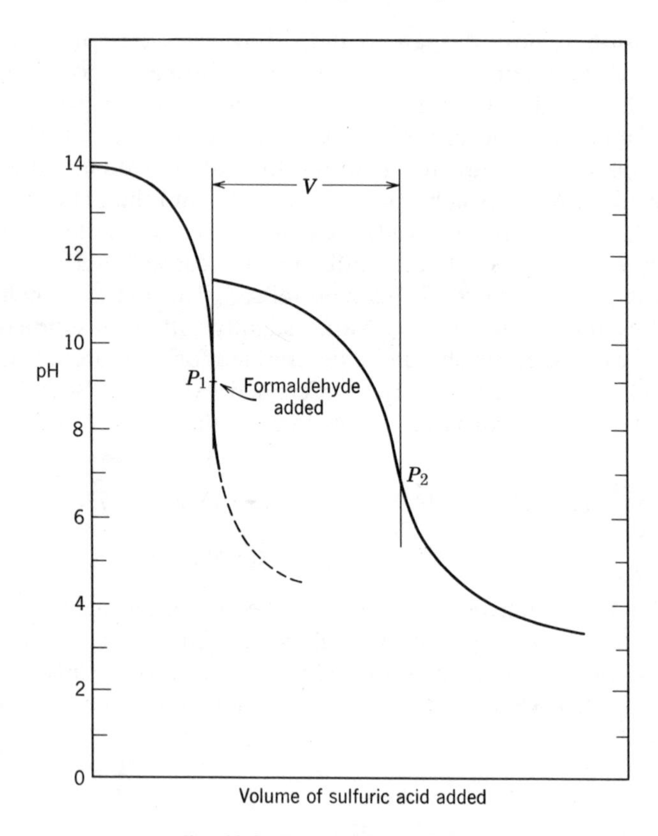

Fig. 22.2. Typical titration curve.

without decomposition in most cases. Fusion times varied from 30 minutes to 36 hours; however, 1 hour was found to be sufficient in most cases.

The fusion apparatus was so arranged that oxygen-free nitrogen was passed over the sample during fusion. This was needed to preserve both the sodium sulfite and the phenols from oxidation. The oxidation of the phenols, particularly the polyhydric phenols, was a more serious problem than was the oxidation of the sulfite. The polyhydric phenols oxidize very rapidly to yield quinoid-type compounds, which react weakly with sulfite, resulting in poor titration end points. With monohydric phenols, purified grade tank nitrogen was satisfactory. However with polyhydric phenols a hot (450–600°C) copper mesh was used to scavenge traces of oxygen from the nitrogen.

The inflection points in the potentiometric titration curves in the case of monosulfonates were sharp, as in Fig. 22.2. However in the case of disulfonated or hydroxysulfonated compounds when polyhydric phenols resulted, somewhat buffered titration curves resulted because of reaction by-products. To circumvent this, excess acid was added to react with the base liberated from the reaction of the sulfite and the formaldehyde. The excess acid was back-titrated with standard base. This procedure yielded somewhat better titration inflection points, probably because the formaldehyde-sulfite reaction is an equilibrium and the excess acid forces it to completion (see method of Siggia and Maxcy, p. 104). An attempt to determine *p*-hydroxybenzenesulfonate, however, was unsuccessful. No inflection was obtained in the titration curve, probably as a result of oxidation of hydroquinone to quinone, as evidenced by the appearance of a colored solution. Quinhydrone and sodium sulfite react.

Table 1 summarizes the results of the analysis of some sulfonic acids and salts by the sulfite method. The method has been applied successfully to both benzenesulfonates and naphthylsulfonates. Attempts on sulfanilamide and *p*-diphenylaminesulfonic acid sodium salts were unsuccessful because the sulfonates decomposed during fusion.

Table 1. Analysis of Sulfonic Acids and Salts by Alkali Fusion and Measurement of Sulfite

Compound	Analysis,[a] wt. %	Standard Deviation
p-Toluenesulfonic acid	99.4 (30)	1.6
p-Toluenesulfonic acid (recrystallized)	100.0 (5)	1.0
Benzenesulfonic acid sodium salt	97.4 (6)	0.4
Benzenesulfonic acid monohydrate[b]	97.9 (5)	6.3
p-Acetylbenzenesulfonic acid sodium salt	97.9 (3)	2.5
p-Acetylbenzenesulfonic acid (recrystallized)	100.8 (3)	0.7
2-Naphthalenesulfonic acid sodium salt	100.1 (4)	3.3
2-Naphthalenesulfonic acid sodium salt (recrystallized)	100.0 (3)	2.3
2-Naphthalenesulfonic acid monohydrate	98.6 (3)	3.8
2-Anthraquinonesulfonic acid[c]	93.3 (2)	4.0

[a] Figures in parentheses indicate number of trials.

[b] Compound is very hydroscopic, which may account for poor end points obtained.

[c] Compound gave poor end points, probably because of reaction of quinone with sulfite.

Some sulfonates give phenols of low volatility that cannot be measured by gas chromatography. These include the polysulfonated compounds, sulfonates containing other very polar functional groups, and certain higher molecular weight sulfonates. To extend the alkali fusion method to include these sulfonates, a method for phenol measurement based on the phenolic group reactivity was developed (see p. 46).

Method of L. R. Whitlock, S. Siggia, and J. E. Smola

[*Reprinted in Part from Anal. Chem.*, ***44***, *532* (*1972*)]

PROCEDURE

Weigh between 3 and 10 mg of sulfonate in a platinum sample boat. Add approximately 30 mg of solid potassium hydroxide along with 5 to 8 mg of sodium acetate to act as a flux. Place the sample boat in the fusion oven (Fig. 22.1), flush the system with helium for several minutes, and increase the oven temperature slowly to 360°C. After completion of the fusion reaction, usually in 15 minutes, remove the sample boat from the fusion apparatus and quickly place into a 50-ml volumetric flask containing water and sufficient hydrochloric acid to neutralize the caustic. Adjust the final pH to 7.0 with either dilute hydrochloric acid or sodium bicarbonate and adjust the solution volume to the mark. Take aliquots of this solution for spectrophotometric measurement following the procedure used for phenols (p. 62).

DISCUSSION AND RESULTS

To test this combined procedure, two sulfonates that had been analyzed by gas chromatography were taken as test samples. Sulfanilic acid was used for the coupling reaction. Data from the analysis of benzenesulfonate and 2-naphthalenesulfonate are given in Table 2. The results demonstrate that this procedure works well for these sulfonates.

An important group of sulfonates whose analysis was not previously attempted because of the low volatility of the phenols produced are the polysulfonates. The results of the analysis of this group of sulfonates are given in Table 2 also. To achieve complete conversion during the fusion of *m*-benzenedisulfonate and 1,3,5-benzenetrisulfonate, one-third longer fusion times were used, 30 to 35 minutes at 380°C. The fusion reaction products of each compound were clear white. An infrared analysis of each

Table 2. Sulfonate Analysis by Alkali Fusion and Measurement of the Phenol by Formation of an Azo Dye[a]

Compound	Analysis,[b] mole %	Standard Deviation
2-Naphthalenesulfonic acid sodium salt	99.2 (4)	1.9
Benzenesulfonic acid sodium salt	98.7 (5)	2.0
m-Benzenedisulfonic acid disodium salt	92.5 (6)	1.8
m-Benzenedisulfonic acid disodium salt (purified)	100.2 (5)	1.7
1,3,5-Benzenetrisulfonic acid trisodium salt	97.3 (5)	3.6
2,7-Naphthalenedisulfonic acid disodium salt	98.4 (6)	2.0

[a] Diazotized sulfanilic acid was used for coupling.
[b] Figures in parentheses represent number of trials.

fusion product showed only resorcinol and phloroglucinol, respectively. The first samples of *m*-benzenedisulfonate were taken as received from the supplier. The average of six determinations was 92.5% as the disulfonate. However when further purified samples were taken (elemental analysis showed 99.4% of the theoretical amount of sulfur), the average of five determinations was 100.4%. The fusion of 2,7-naphthalenedisulfonate was found to go to completion without difficulty at the normal fusion conditions. The average of the standard deviations for all the compounds given in Table 2 is about 2.0% for sample sizes between 3 and 10 mg.

The alkali fusion of *p*-chlorobenzenesulfonate gave 50% conversion to resorcinol rather than the predicted product, *p*-chlorophenol. A nuclear magnetic resonance spectrum of the starting sulfonate showed that it was indeed para substituted. However an infrared identification of the fusion products revealed only the presence of resorcinol. A probable explanation of this result is that both the hydrogen and sulfonate groups of halogenated sulfonates undergo alkaline hydrolysis.

When a high-temperature reaction such as alkali fusion is used, the possibility of thermal degradation of the starting sample as well as that of the reaction product is always present. To help more easily establish first, which sulfonates can be analyzed by fusion, and second, the upper limit of their fusion temperatures that can be used safely, an examination of their thermal stabilities is suggested. The data obtained for a large number of sulfonates and several phenolates are presented in Table 3. This aspect of

Table 3. Decomposition Temperatures[a] of Some Sulfonic Acids, Alkali Metal Sulfonates, and Alkali Metal Phenolates in Air and Helium Atmospheres

Compound	Decomposition Temperature, °C	
	Air	Helium
Potassium phenolate	215	420
Potassium 2-naphtholate	250	480
Sodium *p*-nitrophenolate	—	350**
Benzenesulfonic acid sodium salt	520	520
m-Benzenedisulfonic acid disodium salt	—	570
1,3,5-Benzenetrisulfonic acid trisodium salt	—	560
p-Sulfobenzoic acid monopotassium salt	390	405**
m-Sulfobenzoic acid monosodium salt	—	430**
p-Chlorobenzenesulfonic acid sodium salt	450	450
p-Aminobenzenesulfonic acid sodium salt	—	255
m-Nitrobenzenesulfonic acid sodium salt	—	400**
p-Acetylbenzenesulfonic acid sodium salt	—	350
Dodecylbenzenesulfonic acid sodium salt	400	400
p-Diphenylaminedisulfonic acid sodium salt	420	420
1-Naphthalenesulfonic acid sodium salt	—	480
2-Naphthalenesulfonic acid sodium salt	—	510
2,7-Naphthalenedisulfonic acid disodium salt	—	520
4-Amino-1-naphthalenesulfonic acid sodium salt	285	—
1-Anthraquinonesulfonic acid sodium salt	—	450**
2-Anthraquinonesulfonic acid sodium salt	—	430**
m-Aminobenzenesulfonic acid	—	365*
p-Aminobenzenesulfonic acid	—	260*
p-Toluenesulfonic acid	—	270*

[a] No asterisk, determined by thermogravimetric analysis; one asterisk, determined by differential scanning colorimeter; two asterisks, determined by thermogravimetric analysis, confirmed by differential scanning colorimeter.

the fusion reaction was studied using thermogravimetry in an atmosphere of both air and helium. Most sulfonate salts examined were stable to temperatures well above 400°C, and some were stable above 500°C. Sulfonic acids, however, started to decompose at much lower temperatures than did their corresponding salts. The acids could be analyzed without difficulty, however, because they were quickly neutralized by the alkali, and fusion of the sulfonate salt proceeded normally. The phenolates were found to be much more stable in a helium atmosphere than in air. For this reason, the fusion reaction oven must be continuously purged with a flow of helium.

PRECIPITATION METHODS

Using Inorganic Cations

Many sulfonic salts will be precipitated with such cations as barium, silver, or mercury(I) or (II). Barium is the most general precipitant, but sometimes only silver or mercury ions will work. A pure sample of the salt to be determined should first be tested to see if it precipitates quantitatively with any of the cations mentioned previously. When a convenient cation is found, it must be ascertained whether there are any materials in the actual samples that might interfere; for instance, sulfate ions in the sample would exclude the use of barium, and chloride ions would exclude silver and mercurous ions. If these conditions can be met, the procedure is applicable. It must be remembered that if carboxylic groups are present, they also may react with a metallic ion.

A general procedure cannot be described for this method: each sample must be treated in a way peculiar to it. In most cases, after precipitation is carried out, the precipitate is filtered through a tared Gooch or sintered-glass crucible, then is washed, dried, and weighed. If a precipitated barium salt is waxy, hygroscopic, or in any other way difficult to determine accurately, the precipitate can be soaked with concentrated sulfuric acid and ignited to the sulfate as described in the ignition procedure. This ignition will not work well for silver or mercury precipitates. Free sulfonic acid in a salt sample has to be determined on a separate sample, and the amounted present has to be subtracted from the total acid plus salt as determined by precipitation.

Using Benzidine Salts—Method of D. A. Shiraeff

[*Adapted from S. Dal Nogare, Organic Analysis, Vol. I, Wiley-Interscience, New York, (1953), pp. 382–3, Reprinted in Part*]

Formation of the benzidine salt appears to be quantitative, with two sulfonic acid functions reacting with one molecule of the aromatic diamine. The salt is easily isolated by filtration and is sufficiently stable to permit drying at moderate temperatures. Since the Shiraeff method (1)

1. D. A. Shiraeff, *Am. Dyestuff Rep.*, **36,** 313 (1947); **37,** 411 (1948).

appears to be quite general for high molecular weight sulfonic acids, it is described here.

REAGENTS

BENZIDINE HYDROCHLORIDE SOLUTION. Prepare the solution by dissolving 25.7 grams of benzidine hydrochloride and 50 ml of concentrated hydrochloric acid in enough water to make 1 liter of solution.

Petroleum ether.

Aqueous alcohol. A solution of 200 ml of water and 800 ml of 95% ethyl alcohol is used.

Sodium hydroxide solution, 0.02*N*, standardized.

PROCEDURE

Accurately weigh out enough sample so that 0.1 to 0.4 gram of sulfonic acid is obtained, and dissolve the sample in the minimum volume of distilled water. Remove any insoluble material by filtration. Add 50 ml of the benzidine reagent to the sample solution and allow the solution to stand for at least 3 hours, preferably overnight. Quantitatively collect the precipitated salt on a Whatman No. 41-H filter paper and wash with distilled water until the washings are neutral to blue litmus paper. Transfer the filter paper containing the washed precipitate to a 100-ml beaker and dry it in an oven at 110°C for the minimum time. Remove organic contaminants from the dried precipitate by extraction with 50 ml of petroleum ether added directly to the beaker. Use a stirring rod to obtain good contact of the precipitate with the solvent. Decant the petroleum ether extract through a Whatman No. 40 filter paper and wash the precipitate four more times with 10-ml portions of this solvent. Place the filter paper from the ether washing in the beaker containing the bulk of the precipitate and dissolve the solid by adding 50 ml of aqueous alcohol. Obtain rapid solution by warming the beaker on a hot plate and working the precipitate free of the paper with a stirring rod. After dissolution, filter the hot solution through a Whatman No. 40 filter paper and collect the filtrate in a 100-ml volumetric flask. Wash out the beaker with several 10-ml portions of hot aqueous alcohol, which are filtered and added to the main filtrate. After cooling the solution to room temperature, dilute it to the mark. Pipet exactly 50 ml into a 250-ml Erlenmeyer flask, add 50 ml of distilled water and a few drops of phenolphthalein, and titrate the solution to the pink end point with 0.02*N* standard sodium hydroxide solution. Correct the titration volume for the volume of standard base required to titrate 50 ml of aqueous alcohol and 50 ml of distilled water to the same end point.

CALCULATION

$$\% \text{ Sulfonic acid} = \frac{\text{Milliliters of NaOH} \times N \times \text{E.W.} \times 2}{\text{Grams of sample} \times 1000} \times 100$$

where E.W. is the equivalent weight of the sulfonic acid.

The procedure has been successfully applied to a number of commercial aryl sulfonates, sulfonic acids of fatty acid condensates, and sulfonated fatty alcohols. A certain amount of care should be taken in using the procedure when the composition and components of the sample are unknown. In some instances a higher concentration of alcohol may be required to dissolve the benzidine salt. If large quantities of inorganic sulfate are present, the volume of benzidine hydrochloride solution used may be doubled.

TITRATION METHOD USING QUATERNARY SALTS

This approach is widely used for determining anionic surface active agents. A known quaternary ammonium salt is used as titrant and the sulfonate is titrated directly.

The titration method using quaternary salts involves the reaction in a two-phase water-chloroform system of the sulfonate, with methylene blue forming a chloroform soluble salt. The methylene blue color is then in the chloroform layer. The quaternary, as it is added, reacts with sulfonate, releasing the methylene blue, allowing it to transfer to the water layer. For this method to be operable, the sulfonate salt must have a hydrophobic group of sufficient size to contribute chloroform solubility to the addition compound between it and the quaternary. Hence low molecular weight sulfonates cannot generally be determined using this method.

Although the method does not have as wide a range of applicability as one would like, it is simple to operate and is fairly specific. Alkyl or aryl sulfates and phosphates interfere in the analysis, but inorganic sulfates or other salts do not.

Accuracies and precisions as high as ±1% can be obtained for the more ideal systems, but these values vary with the particular sulfonate and its environment.

Antara Method*

Methylene blue is much less soluble in chloroform than it is in water, but its alkyl or aryl sulfonate salt can be quantitatively extracted from water by chloroform if the alkyl or aryl group is C_8 or greater. Addition of Hyamine† (*p*-diisobutyl phenoxyethoxyethyl dimethyl benzyl ammonium chloride monohydrate) displaces the methylene blue from its organic salt, causing the methylene blue to become redissolved in the aqueous portion. Thus if the amount of methylene blue present is very minute, its extraction back into the aqueous portion may be detected visually, and this becomes the basis for an end point in the reaction of Hyamine with an alkyl or aryl sulfonate.

REAGENTS

PERCHLORIC ACID, 0.1*M*. Add 8.5 ml of 70 to 72% perchloric acid, reagent grade, to 900.0 ml of glacial acetic acid. Add 15.0 ml of reagent grade acetic anhydride and dilute to 1000.0 ml with acetic acid. Let stand overnight to let the acetic anhydride remove all the water.

Glacial acetic acid, reagent grade.

Potassium acid phthalate, primary standard grade.

APPARATUS

Pyrex beaker, 250 ml.

Steam bath.

pH Meter with glass and silver–silver chloride electrode system, indicating millivolts.

Buret, 50 ml.

PROCEDURE

Transfer 150 ml of glacial acetic acid to a 250-ml beaker. Add 0.5 grams of potassium acid phthalate to the nearest 0.0001 gram. Heat on the steam bath to dissolve, if necessary. Cool to room temperature. Place the beaker at the pH meter so that the electrode system is immersed in the solution. Titrate from the buret with 0.1*M* perchloric acid to be standardized, plot millivolts against milliliters, and determine the end point from the curve.

* Antara Chemical Co., 435 Hudson St., New York.

† Hyamine 1622, obtainable from Rohm and Haas.

CALCULATION

$$\text{Molarity of perchloric acid} = \frac{\text{Weight of sample} \times 1000}{\text{Milliliters of perchloric acid} \times 204.15}$$

Hyamine* Solution, 0.004*M*

APPARATUS

Pyrex beaker, 250 ml.
pH Meter with glass and silver–silver chloride electrode system indicating pH.
Buret, 50 ml.
Graduate for measuring, 150 ml.

REAGENTS

Glacial acetic acid, reagent grade.
Mercuric acetate, reagent grade.
Perchloric acid, 0.1*M*.

PROCEDURE

Measure 150 ml of glacial acetic acid into the 250-ml beaker. Weigh 0.90 gram of Hyamine to the nearest 0.0001 gram and dissolve in the acetic acid. Dissolve 1 gram of mercuric acetate in the solution. Place the beaker at the pH meter so that the electrode system is immersed in the solution. Titrate from the buret with the standard 0.1*M* perchloric acid, plotting pH against milliliters. Determine the end point from the curve.

CALCULATIONS

$$\% \text{ Hyamine} = \frac{\text{Milliliters of } HClO_4 \times \text{molarity of } HClO_4 \times 448.1 \times 100}{\text{Weight of sample} \times 1000}$$

* See footnote†, p. 804.

PREPARATION OF EXACTLY 0.00400M SOLUTION OF HYAMINE

Grams of Hyamine required for 1 liter of 0.00400M solution =

$$\frac{179.24}{\%\ \text{purity of Hyamine}}$$

Weigh the amount of Hyamine indicated by the foregoing calculation and transfer to a 1000-ml volumetric flask with distilled water. Dissolve and dilute to the mark. This solution is exactly 0.00400M.

Methylene Blue Indicator Solution

Dissolve 0.030 gram of methylene blue, reagent grade, and 50 grams of anhydrous sodium sulfate, reagent grade, in 500 ml of distilled water. Add 6.5 ml of sulfuric acid (96% reagent grade) and dilute to 1000 ml with distilled water. Shake well.

PROCEDURE FOR DETERMINING SULFONATE SALTS

Weigh a sample containing about 0.1 mole of quaternary salt to the nearest 0.0001 gram. Transfer to a 100-ml volumetric flask with distilled water, dissolve, and dilute to the mark with distilled water. Pipet 10 ml of the sample solution prepared above into a 250-ml glass-stoppered Erlenmeyer flask. Pipet 25 ml of methylene blue indicator solution and 15 ml of chloroform into the flask. Run in 10 ml of Hyamine solution and shake for 30 seconds. Then add 1 to 2 ml at a time, stoppering and shaking the flask for 30 seconds after each addition. At the appearance of signs of the approaching end point (i.e., the more rapid separation of the temporary emulsification and increased cloudiness of the aqueous layer), reduce the additions to 0.5 ml. When the water layer begins to turn blue, add the Hyamine solution dropwise, shaking for 30 seconds after each addition, inverting, and waiting for 60 seconds until the aqueous layer is clear and appears as blue as the chloroform layer. The comparison is made by placing the flask in reflected light and obstructing one's view of the interface with a stirring rod.

CALCULATION

$$\%\ \text{Active ingredient} = \frac{\text{ml}_{\text{Hyamine}} \times N_{\text{Hyamine}} \times \text{mol. wt.}_{\text{quat}} \times 100}{0.1 \times \text{Weight of sample} \times 1000}$$

Sulfonamides

Adapted from the Method of J. S. Fritz and R. T. Keene

[*Reprinted in Part from Anal. Chem.*, **24**, *308–10 (1952)*]

The —SO_2NH— group found in sulfa drugs and other sulfonamides of primary amines is known to be feebly acidic. When sulfa drugs are dissolved in basic organic solvents, the acidic properties of this group are sufficiently enhanced to permit direct titration with a strong base. This is the basis of the method described. This procedure is simple, rapid, and accurate. Because it involves titration of the —SO_2NH— group, this method is applicable to the determination of sulfathalidine, sulfasuxidine, and other sulfonamides which cannot be analyzed by the diazo method.

REAGENTS AND SOLUTIONS

BUTYLAMINE. The commercially available material (Sharples) was used without further purification.

DIMETHYLFORMAMIDE. The commercially available material (Du Pont) was used without further purification.

p-NITROBENZENEAZORESORCINOL (AZO VIOLET). An approximately saturated solution in benzene.

SODIUM METHOXIDE, 0.1*N*. Prepare as described by Fritz and Lisicki (2). Wash about 6 grams of sodium in methanol and dissolve immediately in 100 ml of methanol. Protect the solution from carbon dioxide while the sodium is dissolving; if necessary, cool the solution in cold water to prevent the reaction from becoming too violent. When all of the sodium has reacted, add 150 ml of methanol and 1500 ml of benzene, and store the reagent in borosilicate glassware protected from carbon dioxide. Standardize this solution by titration against benzoic acid dissolved in benzene-methanol. Although the reagent is reasonably stable, it should be restandardized every few days.

SULFA DRUGS, U.S.P. GRADE. The other sulfonamides were prepared in the laboratory of Fritz and Keene according to the procedures of Shriner and Fuson (3) and are of 98 to 100% purity.

THYMOL BLUE. Dissolve 0.3 gram in 100 ml of methanol.

PROCEDURE

Weigh a sample of suitable size into a 50-ml beaker and dissolve in 10 to 20 ml of dimethylformamide or butylamine. Add indicator, cover the

2. Fritz and Lisicki, *Anal. Chem.*, **23**, 589 (1951).
3. R. L. Shriner and R. C. Fuson, *Systematic Identification of Organic Compounds*, 3rd ed., Wiley, New York, 1948.

beaker with a cardboard provided with a small hole for the buret tip, and titrate to the first appearance of a clear blue color. During the titration, agitate the solution by means of a magnetic stirrer.

The solvents employed contain some acidic impurities. This is best corrected for by exactly neutralizing the solvent with sodium methoxide shortly before it is to be used.

Standardize the sodium methoxide against benzoic acid by using dimethylformamide as solvent and thymol blue as indicator. Titration of benzoic acid in butylamine gives erratic results because of gel formation, but no gel forms in dimethylformamide if 20 to 25 ml of solvent is used for each 100 mg of benzoic acid.

SOLVENTS AND INDICATORS

Dimethylformamide is an excellent solvent for sulfonamides. Good end points are obtained for moderately acid sulfonamides using thymol blue indicator. Butylamine enhances the acid strength of sulfonamides more than dimethylformamide and is therefore used for the titration of the more feebly acidic sulfonamides. Azo violet is the preferred indicator for use in butylamine. Butylamine absorbs carbon dioxide more readily than dimethylformamide and is therefore slightly less convenient to use.

SULFA DRUGS

Eight sulfa drugs in common use were analyzed by the foregoing procedure. Sulfaguanidine does not appear to have any acid properties. Sulfanilamide is not acidic to thymol blue in dimethylformamide, but can be accurately titrated in butylamine using azo violet indicator. In the other drugs the —SO_2NH— linkage is sufficiently activated by substituent groups to permit titration in dimethylformamide. Table 4 shows the results of these analyses.

Some differential titrations are possible. Mixtures such as sulfathiazole-sulfanilamide can be analyzed by first titrating sulfathiazole in the presence of sulfanilamide, using dimethylformamide and thymol blue, then titrating both drugs using butylamine and azo violet. The acid sulfa drugs may be accurately determined in the presence of sulfaguanidine. Table 5 gives data for the analysis of some mixtures.

OTHER SULFONAMIDES

In addition to sulfa drugs, the sulfonamides of several primary amines were titrated, to determine the scope of the method and to show the effect of

Table 4. Titration of Sulfa Drugs with Sodium Methoxide

Compound	Sample Weight, mg.	Base, *N*	Base, ml.	Purity, %	Solvent	Indicator
Sulfamethazine	153.7	0.0866	6.36	99.6	DMF	Thymol blue
	157.9		6.55	99.9		
	153.7		6.36	99.6		
Sulfamerazine	152.3	0.0866	6.65	99.8	DMF	Thymol blue
	150.6		6.58	99.9		
	152.2		6.64	99.8		
Sulfanilamide	102.3	0.0866	6.84	99.6	$BuNH_2$	Azo violet
	104.1		6.98	99.9		
	101.1		6.76	99.6		
Sulfadiazine	250.0	0.1783	5.57	99.7	DMF	Thymol blue
	250.9		5.60	99.9		
	251.0		5.59	99.7		
Sulfapyridine	124.5	0.1093	4.56	100.1	DMF	Thymol blue
	150.8		5.53	100.2		
	152.5		5.58	100.0		
Sulfathalidine	126.4	0.1093	5.67	98.8	DMF	Thymol blue
	130.3		5.85	98.8		
	126.9		5.70	98.9		
Sulfathiazole	125.6	0.1093	4.51	100.1	DMF	Thymol blue
	127.7		4.59	100.2		
	125.5		4.50	99.9		

substituents on the acidity of sulfonamides. Results are given in Table 6.

p-Toluenesulfonamide is slightly acid and gives a good end point in butylamine but not in dimethylformamide. Substitution of one hydrogen by a phenyl group ($PhSO_2NHPh$) increases the acidity sufficiently to permit titration in dimethylformamide. This is probably due to the tendency of a phenyl group to attract electrons. The electron-donating properties of an alkyl group, however (as in $PhSO_2NHR$), weaken the acidity so that a poor end point is obtained on titration in butylamine. From the titrations carried out, it appears that the order of acidity is $ArSO_2NHPh$, $ArSO_2NHPyr$, $ArSO_2NHthiazole > ArSO_2NH_2$, $ArSO_2$-$NHnaph > ArSO_2NHCH_2Ph > ArSO_2NHR > NaphSO_2NHR$. A detailed study of acidity of the more acidic sulfonamides has been made by Bell and Roblin (4).

The presence of the phenolic group in *p*-$HOC_6H_4NHSO_2C_6H_4Br$

4. Bell and Roblin, *J. Am. Chem. Soc.*, **64,** 2905 (1942).

Table 5. Titration of Sulfonamide Mixtures

Sulfonamide Taken, mg.		Base, ml.	Base, N	Found, %
Sulfaguanidine (0.2 gram) added as impurity. DMF solvent, thymol blue indicator				
Sulfapyridine	173.5 mg.	4.70	0.1477	100.0
	183.3	4.97	0.1477	100.0
Sulfanilamide (0.2 gram) added as impurity. DMF solvent, thymol blue indicator				
Sulfathiazole	181.7 mg.	4.82	0.1477	99.9

Butylamine solvent, azo violet indicator

Sulfanilamide, me.	Sulfathiazole, me.			Total Taken, me.	Total Found, me.
0.413	0.396	5.57	0.1477	0.810	0.823
0.418	0.399	5.61	0.1477	0.817	0.828
0.412	0.422	5.62	0.1477	0.829	0.829

caused no interference in dimethylformamide, except that a slight green color formed shortly before the end point. In butylamine using azo violet indicator, the acidity of the phenolic group is sufficiently enhanced to interfere with the end point.

DISCUSSION

For the compounds in which a poor end point was observed in butylamine and dimethylformamide, a somewhat sharper end point was obtained in 95 to 100% ethylenediamine. Because of lack of adequate supply of this solvent in the laboratory of Fritz and Keene, further investigation was discontinued.

In the identification of organic amines, sulfonamides are among the commonly used derivatives. It is suggested that titration of the sulfonamides be used in the determination of the equivalent weight of primary amines. By observing the sharpness of the end points in each of the two solvents recommended, knowledge of the basic strength and character of the parent amine may be obtained.

Table 6. Titration of Sulfonamides

Compound	Solvent	Indicator	Purity
$CH_3C_6H_4SO_2NHC_4H_9$	$BuNH_2$	Azo violet	100.7 101.3 100.8
$BrC_6H_4SO_2NHC_7H_{15}$	$BuNH_2$	Azo violet	100.4 100.1 101.9
$C_{10}H_7SO_2NHC_6H_{11}$	$BuNH_2$	Azo violet	99.8 98.2 97.3
$(CH_3C_6H_4SO_2NH)_2C_2H_4$	$BuNH_2$	Azo violet	98.6 98.6 98.9
$CH_3C_6H_4SO_2NHC_6H_4NHSO_2C_6H_4CH_3$	$BuNH_2$ + DMF	Azo violet	96.3 96.7 96.9
$BrC_6H_4SO_2NHC_6H_4OCH_3$	$BuNH_2$	Azo violet	99.6 99.8
$BrC_6H_4SO_2NHC_6H_4Cl$	DMF	Thymol blue	98.1 98.1 98.0
$C_{10}H_7SO_2NHCH_2C_6H_5$	$BuNH_2$	Azo violet	97.8 97.8 97.8
$CH_3C_6H_4SO_2NHCH_2C_6H_5$	$BuNH_2$	Azo violet	99.6 99.6 99.7
$BrC_6H_4SO_2NH-C_5H_4N$	DMF	Thymol blue	100.2 100.0 100.2

Table 6. (contd.)

Compound	Solvent	Indicator	Purity
$Br\text{-}C_6H_4\text{-}SO_2NH\text{-}C_6H_4\text{-}OH$	DMF	Thymol blue	98.7 98.8 98.8
$CH_3\text{-}C_6H_4\text{-}SO_2NH_2$	$BuNH_2$	Azo violet	100.0 100.1 100.0
$Br\text{-}C_6H_4\text{-}SO_2NH\text{-}C_{10}H_7$	$BuNH_2$	Azo violet	100.0 100.0

Methods have been used for the potentiometric titration of weak acids in basic solvents. Electrode systems used in dimethylformamide with sodium methoxide or titrant were described in Chapter 3 (pp. 163–4), and, the potentiometric titration of weak acids was discussed earlier (pp. 50–57); these methods should also be applicable to the titration of sulfonamides.

DETERMINATION OF SULFINIC ACIDS IN THE PRESENCE OF MERCAPTANS

Both mercaptans and sulfinic acids are titrated quantitatively by nitrous acid.

$$2RSO_2H + HONO \rightarrow (RSO_2)_2NOH + H_2O$$

$$RSH + HONO \rightarrow RSNO + H_2O$$

Since the concentration of mercaptan in a mixture of mercaptan and sulfinic acid can be determined without interference from the latter by the method of Danehy and Kreuz (p. 000), the sulfinic acid present can be obtained by difference.

Method of J. P. Danehy and V. J. Elia

[*Anal. Chem.*, ***44***, *1281* (*1972*)]

PROCEDURE

Dissolve an accurately weighed sample containing 0.2 to 0.8 mM total of mercaptan and sulfinic acid in 10 ml of water. Stir magnetically and add 5 ml of 5N sulfuric acid. Titrate the solution with continuous stirring by addition of standard 0.05M sodium nitrite solution until an external end point is obtained; one drop of solution gives an instantaneous blue color when added to a starch–potassium iodide solution.

Determine the mercaptan content of the sample by the method of Danehy and Kreuz and obtain the sulfinic acid content by difference

Table 7. Titration of Mercaptans, Alone and in Mixtures with *p*-Chlorobenzenesulfinic Acid with Sodium Nitrite

Mercaptan				
Name	Amount, mM	Sulfinic Acid mM	Sodium Nitrite, mM	Sulfur accounted for by HONO, %
L-Cysteine	0.244	—	0.240	98
	0.507	—	0.509	100
	0.496	—	0.500	100
	0.814	—	0.810	100
	—	0.465	0.226	98
	0.490	0.472	0.720	99
2-Mercaptoethanol	0.106	—	0.102	96
	0.207	—	0.197	95
	—	0.067	0.032	97
	—	0.133	0.067	100
	—	0.177	0.086	98
	0.063	0.133	0.126	98
	0.063	0.067	0.094	98
	0.106	0.053	0.127	96
Mercaptosuccinic acid	0.085	—	0.084	99
	0.170	—	0.168	99
	—	0.052	0.026	100
	0.085	0.052	0.108	97
	0.085	0.104	0.136	100
2-Mercaptoethyl-ammonium chloride	0.246	—	0.243	99
	0.258	—	0.250	97
	0.412	—	0.407	99
	0.588	—	0.570	97
2,5-Dimercaptoadipic acid	0.56	—	0.394	78
	0.64	—	0.452	71
	0.234	—	0.172	73
p-Chlorothiophenol	0.245	—	0.245	100
	0.737	—	0.744	101
Thiophenol	0.257[a]	—	0.244	95
	0.771	—	0.760	99
m-Mercaptobenzoic acid	0.436[a]	—	0.439	101

[a] Spectrophotometric determination by method of Danehy and Kreuz (p. 752).

based on the relationship

$$\text{Millimoles of HONO} = \text{millimoles of RSH} + \text{millimoles of } \frac{RSO_2H}{2}$$

No one of several disulfides tested consumed any nitrous acid under the titration conditions employed. Table 7 gives the results of the titration of known concentrations of several mercaptans and of *p*-chlorobenzenesulfinic acid, alone and in pairs.

Of the eight mercaptans examined, only 2,5-dimercaptoadipic acid, a dithiol, failed to react quantitatively. If the amount of mercaptan present is large compared to the sulfinic acid, the difference between the millimoles of nitrite used in the titration and the millimoles of mercaptan determined independently is subject to poor accuracy and precision.

23

Techniques and Reasoning in Developing New Analytical Methods or Modifying Existing Methods*

For Compounds Whose Groups Cannot be Determined by Existing Methods

The methods available for functional group determination have been amply described in the preceding chapters. Methods exist for almost all the functional groups, but situations necessitating changes in methods, or sometimes a completely new procedure, often arise. This chapter is concerned with the techniques and reasoning involved in devising a new method of analysis or in making changes in known procedures to make them applicable to the samples to be analyzed.

When a new procedure is necessary, the reactions of the functional group in question must be completely surveyed. The factors involved are the presence of at least one reactant or product that can be measured, the completeness of the reaction, and the effect of impurities.

A reaction in which a base, acid, oxidant, reductant, or any other easily measured material is liberated or consumed is a good one to use for functional group determination.

In devising a new method or altering a known procedure, several difficulties may be encountered that must be overcome before the method can be used. These difficulties are time for complete reaction (including equilibrium), solvent difficulties, and interferences.

TIME OF REACTION

A reaction used for a quantitative measurement has to go to completion in a reasonable length of time; the shorter the time, the better. The time necessary for complete reaction can be shortened by several means.

* Reprinted in part, with permission, from the article of S. Siggia, *Anal. Chem.*, **22,** 378 (1950).

1. Increasing concentration of reagents used to react with the functional group, especially if one of the products of the reaction is being determined. If the excess of reagent is being determined, it will be necessary to determine too large an excess, and the analysis will be based on the difference between two large figures, which is never desirable.
2. Increasing the temperature of the reaction by using higher-boiling solvents than are ordinarily used, so that reflux temperatures will be higher.
3. Use of a more reactive reagent.
4. Resorting to higher pressure reactions in some instances.
5. Catalysts.

Technique 1 is employed in the acetylation of hydroxyl and amino groups (pp. 12–14). At least one part of acetic anhydride to 3 parts of pyridine must be used, to operate in a reasonable length of time. If more pyridine is used, the reaction time increases sharply. In the dechlorination of 1-chloro-2-methoxy-2-phenylethane, if $0.5N$ sodium hydroxide is used, the reaction proceeds to only 12% of completion in 4 hours; if $5N$ sodium hydroxide is used, the reaction is complete in 30 minutes.

The saponification of some esters requires the use of technique 2. Alcoholic (methanol) sodium hydroxide is usually used for saponification of esters. Some esters do not saponify readily, however, and the reflux temperature of the methanol is too low to permit complete reaction in a reasonable time. The use of a higher alcohol overcomes this difficulty (see pp. 167–171).

The dechlorination of methyl-α-chloroacrylate can also be used in the application of technique 2. If ethyl alcohol is used as a solvent, the dechlorination proceeds to about 70% completion in 2 hours; with trimethylbenzene (mixture of various isomers, b.p., 170°C) the reaction goes to completion in 30 minutes.

The dechlorination of methyl-α-chloroacrylate can also be used to demonstrate technique 3. If pyridine is used as the dechlorinating agent, no reaction takes place; with piperidine, which is a much stronger base, the reaction proceeds to completion in 30 minutes. When applied to the dechlorination of 1-chloro-2-methoxyl-2-phenylethane, pyridine yields no reaction in 2 hours, piperidine yields about 75% reaction, and $5N$ sodium hydroxide (a stronger base than piperidine) yields complete reaction.

Technique 4 is applicable only when one of the reactants is a gas. In quantitative hydrogenation, the reaction proceeds at a faster rate, the higher the pressure. In hydrogenation of vinyl alkyl ethers, a 20-gram sample is saturated in about 2 hours at 10 cm Hg, or in about 30 minutes at 50 cm Hg.

Catalysts (technique 5) can sometimes be employed to speed up a reaction. The use of mercuric sulfate will speed up the addition of bromine to an unsaturated linkage (1). 2-Heptyne will brominate completely in 7 minutes in the presence of mercuric sulfate, whereas without the catalyst the results are 23% low under the same conditions. Dichloroethylene will brominate completely in 5 minutes with the catalyst, but without the catalyst it will go only 2% toward completion in 20 minutes.

The completeness of a reaction is often affected by the existence of an equilibrium. There are several ways of eliminating or at least minimizing the effects of an equilibrium. The reaction can be pushed closer to completion by using a higher concentration of reagent, or it can be pulled to completion by removing one of the products from the reaction. The latter method is preferable. The reaction of acetophenone and hydroxylamine hydrochloride is a good example of how equilibrium difficulties can be overcome.

$$\phi\overset{\overset{\displaystyle O}{\|}}{C}\text{—}CH_3 + NH_2OH{\cdot}HCl \rightleftharpoons \phi\overset{\overset{\displaystyle NOH}{\|}}{C}\text{—}CH_3 + H_2O + HCl$$

In an aqueous system, the reaction equilibrium does not permit the reaction to proceed more than about 50% to completion. In a nonaqueous solvent (1:1 ethylene glycol–isopropyl alcohol), the reaction goes to 80% completion. When a weak base, di(β-hydroxyethyl)aniline, is added to the solvent mixture to tie up the hydrogen chloride liberated, the reaction proceeds to 100% completion. If water is added to the system containing even the weak base, the equilibrium is set up again (see pp. 95–96 for additional discussion).

Equilibrium difficulties are also encountered in determining carbonyl compounds using sodium sulfite.

$$Na_2SO_3 + \text{—}CHO + H_2O \rightleftharpoons \underset{\underset{\displaystyle SO_3Na}{|}}{\text{—}CHOH} + NaOH$$

When a carbonyl compound is introduced into a solution of sodium sulfite, the pH of the solution rises, indicating liberation of the hydroxide. On titration of the hydroxide, a few milliliters of acid brings the pH of the solution back to the original pH of the sulfite; but on standing, the pH rises again as more hydroxide is formed. On addition of an excess of standard acid, the reaction can be brought to completion. Titration of the excess of acid indicates the amount of carbonyl compound. However as the excess acid is consumed and the end point is approached, the bisulfite

1. H. J. Lucas and D. Pressman, *Anal. Chem.*, **10**, 140–2 (1938); see also pp. 383–386 of this book.

addition compound breaks up and the equilibrium is again noticed. This action tends to dull the end point. To minimize this effect, a large excess of sodium sulfite is used to keep the reaction essentially at completion as the excess of acid is consumed. When aliphatic aldehydes through the butyraldehyde are determined, on addition of the sample to the sulfite, the hydroxide liberated can be titrated if time is allowed between additions of acid for the reaction to proceed further. The problem here is the low boiling points of some of these aldehydes, which cause loss. On addition of excess acid to the sulfite solution, the reaction is so near completion that no aldehyde odor can be noted, in spite of the low boiling points and strong odors of some aldehydes. On back-titration of the excess acid, as soon as the end point is close the odor of free aldehyde becomes noticeable, showing the presence of an equilibrium.

When aldehydes such as benzaldehyde are determined, the equilibrium is noticeable, and the end point on back-titration of the excess acid is dulled to a significant extent, but results can still be obtained.

With acetone and most ketones, even a large excess of sulfite cannot keep the reaction anywhere near completion. As the acid is consumed on the back-titration, the bisulfite addition compound breaks down and the reaction proceeds in the reverse direction. As a result, no end point is visible (see 105–6 for further discussion).

SOLVENT PROBLEMS

Since nonaqueous titrations began to be used, a number of the problems of the organic analyst have been solved. One of the main problems of the organic analyst has been the solution of his samples. Very often a procedure was found that should have operated well for a certain compound, but the sample failed to dissolve in the solvent used for the analysis. It has now been learned that we can titrate effectively in many organic solvents, so that if the solvent used in a certain method does not dissolve the sample, one can usually be discovered in which the reaction and the final titration can be run. Some such solvents have been mentioned in this book (see Determination of Enols, pp. 46–57; Carboxylic Acids and Salts, pp. 160–4; Amines, pp. 533–8; Sulfonamides, pp. 807–812).

Another solvent problem arises from the reactivity of certain samples such as acid anhydrides (pp. 230–56) and acid chlorides (pp. 223–30). These samples react with many organic solvents. However solvents have been found that are unreactive toward these compounds; although titrations can still be run in them. Dimethylformamide, acetone, and chlorobenzene are such solvents. Also, some solvents affect the analytical reaction to some degree. In the foregoing paragraphs the effect of water on the completion of a reaction was indicated, since it affected the equilibrium of the reaction. A nonaqueous solvent was found that would circumvent this problem, in which the sample and reagents were soluble and titration was possible.

The use of nonaqueous solvents has also enabled the analyst to accelerate stubborn analytical reactions by making possible the use of higher-boiling solvents, which yield a higher temperature for a reaction, yet permit the final analysis.

When poor end points are obtained in water or other solvents, this difficulty can often be resolved by using certain nonaqueous solvents. The titrations of weak acids that show poor or sometimes no end points in titration curves when run in solvents such as acetone, dimethylformamide, ethylene diamine, and butylamine will show very sharp end points (pp. 161–4, 807–12). The same is true with weak bases that show no end point or a very poor end point in water. These materials will usually show very good end points in glacial acetic acid, ethylene glycol–isopropanol, and acetone (pp. 533–5).

INTERFERENCES

In functional group analysis, as in all types of analysis, the problem of interferences arises. A sample containing a component that interferes with the desired analysis can be treated in several ways.

1. By physical means (distillation, adsorption, extraction) to separate the desired component from the interference.
2. Chemically, to alter the interference into a compound that is harmless to the analysis or to alter the desired component and put it in a form in which it is determinable in the presence of the interference. Also, both the desired component and the interference can be chemically altered, but the altered form of the desired component is determinable in the presence of the altered form of the interference.
3. By a method that determines both the desired component and the interference; then the interference alone is determined and the desired component is obtained by difference.

An example of technique 1 is the determination of free phenol in a phenolic resin by bromination. If the bromination procedure is carried out on the sample as is, the analysis for the phenol will be high, owing to substitution of bromine on the resin. If the free phenol is separated from the resin by dissolving the sample in chloroform and extracting the free phenol from the solution using aqueous sodium hydroxide solution, the aqueous extract may be used for the determination.

An example of technique 2 is the determination of diethylaniline, monoethylaniline, aniline, and ethyl alcohol in the presence of each other. All three amines are very basic in glycol–isopropyl alcohol and can be titrated. On reaction of the sample with acetic anhydride, the aniline and monoethylaniline are converted to the corresponding anilides, which are not basic; then the reaction mixture can be titrated directly with standard acid for diethylaniline in the glycol–isopropyl alcohol medium (pp. 569–72). This is an example of reacting the interferences to remove them.

The determination of the aniline is an example of chemically reacting the desired component to make it determinable. In this case the total base content of the mixture is determined. Then salicylaldehyde is added to a separate sample to react with the aniline to form the Schiff base, which is only weakly basic in the glycol-alcohol medium. The remaining basicity is determined in this sample. The difference between the total base and the base content after addition of the salicylaldehyde yields the aniline analysis (pp. 569–72).

The determination of the ethanol is an example of reacting both the desired component and the interference, but the reacted form of the desired component is determinable in the presence of the reacted form of the interference. Ethanol can be determined by acetylation, but the aniline and monoethyl aniline interfere. This sample can be handled by reacting with acetic anhydride. The alcohol is converted to the ester, and the two interfering amines are converted to their anilides. The ester can be saponified quantitatively with caustic, but the anilides are not noticeably affected. The saponification figure then yields the alcohol analysis (pp. 40–1).

Technique 3 can be demonstrated in the analysis of mixtures of acetals and vinyl ethers. The total ether plus acetal can be determined by hydrolyzing the sample and determining the acetaldehyde formed (pp. 510–511). On a separate sample, the vinyl ether can be determined iodimetrically (pp. 515–7). The acetal is obtained by difference.

Another example of technique 3 is the determination of the components of a mixture of $HOCH_2C{\equiv}CCH_2OH$, $HOCH_2CH{=}CHCH_2OH$, and $HOCH_2CH_2CH_2CH_2OH$. A quantitative acetylation picks up all three components. A quantitative bromination picks up the unsaturated components (pp. 383-6). The acetylenic member of the series is obtained using the hydration method (pp. 467-72). By subtracting the analysis for the acetylenic compound from the bromination analysis, the ethylenic compound can be obtained. By subtracting the bromination analysis from the acetylation, the saturated butanediol is obtained.

24

The Role of Quantitative Functional Group Determination in the Identification of Organic Compounds*

In the past, identifications have been handled mainly by running elemental analyses and by preparing solid derivatives. This chapter indicates the usefulness of quantitative functional group determination in the solution of identification problems.

Identification problems lie in two main categories. In the first category, the chemist is rather sure of the identity of his compound, since he carried out a known reaction with known reagents, yet he needs verification of the identity of his compound before he can proceed with his work. In the second category, we have the identification of unknowns.

In the first category are the compounds that are prepared under known conditions and by known reagents, but whose identity has yet to be established. It has been the standard practice for many years for the chemist who has been working on a synthesis to send the product for the usual elemental analyses to verify the identity of the product. It has become recognized as more advantageous to verify the identity of an organic compound by quantitatively determining the functional groups that should be on the molecule. The analysis is more meaningful if we determine that the synthesized compound contains the theoretical amount of nitro group or amino group rather than just the theoretical amount of nitrogen. Along the same lines the identity of a hydroxy acid, for example, is more firmly established by a hydroxyl-group determination and carboxyl-group determination than by a carbon and hydrogen determination.

Functional group analysis also eliminates the uncertainty that exists when the elemental analyses are close to the theoretical but not quite there. If a carbon analysis is 1% low out of a theoretical 50.0% carbon, the compound could be 98% pure but very often is much less pure, since all organic compounds also contain carbon.

Elemental analysis has a definite role in organic analysis, and the foregoing should not be construed to mean that functional group analysis

* The source of this chapter is S. Siggia, *Anal. Chem.*, **23,** 667 (1951).

should replace elemental analysis. However the emphasis is on the greater specificity of the functional group determination.

In the second category is the case of the analyst who is not *verifying* the identity of a compound but actually must establish the identity of an unknown. It is best to illustrate this category with several examples:

It was discovered that in the reaction

$$HC{\equiv}CH + CH_2O \rightarrow HC{\equiv}CCH_2OH$$

several side reactions took place, and the by-products had to be identified. The product of the reaction was distilled, and two materials were isolated that did not have the physical constants of any of the known materials from the reaction mixture. All the functional groups involved in the system were known, and no matter which functional groups were consumed in forming the by-products, there still should have been some unreacted groups that would be determinable. In the case of these two by-products, the following determinations were run: hydroxyl, acetylenic hydrogen, unsaturation, and free and combined carbonyl groups (the combined carbonyl groups being acetal-like compounds). The hydroxyl and free carbonyl values were zero. However definite values were obtained for acetylenic hydrogen, for unsaturation, and for combined carbonyl. From these values, the corresponding "equivalent weight" was calculated. (The term "equivalent weight" is used here in the same sense as neutralization and saponification equivalents are used. Hydroxyl, carbonyl, amide, etc., equivalents are just as useful as the neutralization and saponification equivalents but for some reason have been completely neglected in the analytical field. The "equivalent" is actually the molecular weight of the compound if only one of the determined functional groups is on the molecule. It is one-half the molecular weight if there are two such functional groups on the molecule; it is one third the molecular weight if three groups are present, etc.)

From the "equivalent weights" determined by these analyses and also from the known reactants and conditions present in the original reaction mixture, the identity of the two materials was established as

$$(1)\ HC{\equiv}CCH(OCH_3)_2 \qquad (2)\ H_2C(OCH_2C{\equiv}CH)_2$$

In the case of compound 1, the functional group analysis not only told what groups were present and which of the starting groups were absent but gave the "equivalent weight" according to each group. This indicated that there was one acetylenic hydrogen for each acetal-like group and that there was either one triple bond for each of the groups above or two double bonds. Figuring the molecular weight of the compound as equal to the "equivalent weight," the formula shown for compound 1 was reached.

From familiarity with the system, it was known that propargyl alcohol, the desired product of the reaction, does oxidize on standing to the corresponding aldehyde. Also it was known that methanol was present in the formalin used in the synthesis. The reaction mixture was also slightly acidic, so that all the prerequisites existed to obtain the compound indicated by the functional-group determination.

In the case of compound 2, again the functional group determination indicated the groups present and which of the starting groups were absent. The "equivalent weight" according to each group indicated that two acetylenic hydrogens were present for each acetal-like group and also that there were two triple bonds (or an unlikely four double bonds) for each acetal-like group. It was evident then that the molecular weight was at least the "equivalent weight" as calculated from the acetal analysis with two acetylenic hydrogens and two triple bonds, or that the molecular weight had to be some multiple of that value. From these data and from the knowledge that formaldehyde and propargyl alcohols were present in the original reaction and that they form formals under slightly acid conditions (which were present in the reaction), it was simple to arrive at the identity of compound 2.

Functional group determination does not make an absolute identification. It does, however, yield data that serve to indicate the probable identity of the compound. The absolute identification is made by actually synthesizing, by known methods, the compound indicated by the functional group determination and then comparing the known against the unknown by standard techniques such as comparison of the infrared curves of each; the X-ray diffraction patterns, in the case of crystalline solids; microscopical examination; also, for crystalline solids, crystalline form, refractive indices of the crystals, and other optical behavior can be compared. Other physical properties such as boiling point, freezing point, refractive index, and density can be used to compare the known and the unknown.

The value of the functional group determination lies in the fact that it indicates the type of unknown compound and also gives the molecular weight, or a factor thereof, with the ratios of the functional groups to each other on the molecule, which are very important data in an identification.

Another example of an identification that would be classed in the second category was the identification of a detergent, which was said to be a fatty acid–ethylene oxide compound ($RCOO[CH_2CH_2O]_xH$). The "equivalent weight" in this case was determined by a saponification determination, since the compound is an ester. This value was of such magnitude that either R or x had to be small. Since the sample was water soluble, however, x had to be quite large. A determination was run for combined ethylene oxide (pp. 278–83), and this was found to be quite large; thus contrary to prior information, R could not be a fatty acid. From the "equivalent weight" and from the ethylene oxide content, the size of the R group was computed. This was verified by isolating the sodium salt of the acid, which came out of solution on alcoholic saponification, and determining the sodium carboxylate "equivalent" by the combustion procedure (pp. 168–9). The "equivalent weight" of the acid then calculated from this value compared very closely to that calculated from the "equivalent weight" of the ester, corrected for the ethylene oxide content. Knowing the "equivalent weight" of the acid made it a simple matter to obtain samples of all the known, commercially available acids of approximately that "equivalent weight" and to get the comparisons necessary for an absolute identification.

In summary, then, quantitative determination of the functional groups can be a valuable aid in identifying complete unknowns as well as in verifying the identity of suspected compounds. In the latter type of identification, the functional group determination, because of its specificity, yields a more meaningful identification than an elemental analysis.

The main value of functional group analysis in identifying complete unknowns lies in the determination of the ratio of the groups to one another and the determination of the functional equivalent weight of the compound, which is a factor of the molecular weight.

25

The Use of Differential Reaction Rates to Analyze Mixtures Containing the Same Functional Group

The rate of reaction gives us another parameter that we can use in analysis to handle mixtures. We have already seen some examples. In the bromination analysis for unsaturation (pp. 375–6), we saw that the faster rate of addition of bromine to the unsaturation over the substitution of bromine for the hydrogen atoms on the molecule could make analysis possible. Another example was seen in the determination of hydroxyl compounds in the presence of amines (pp. 40–1), where the acetylation reaction yields esters from the hydroxy compounds and amides from the amines. Saponification of the esters was then often possible without significant effect from the amides, since the esters saponify so much more rapidly than the amides.

The foregoing represent extreme cases, where reaction with one component proceeds so much more rapidly than with the other that the reagent is basically reacting with only one component. However it is now possible to resolve functional groups whose rates of reaction (rate constants) differ by only a factor of 2. In these cases, both components are reacting with the reagent, but one is reacting faster than the other.

Lee and Kolthoff (1) pointed out the potentialities of the kinetic approach. Their approach is empirical, however, and consists of reacting a mixture with a reagent at a fixed temperature for a specified length of time. A calibration curve is prepared indicating the amount of each component reacted per unit time. The composition of the unknown is determined from the curve. The disadvantages of this approach are (1) the compounds being determined must be known, so that rate constants for each component can be obtained at the given temperature and calibration curves plotted; and (2) the experiments must be run at constant temperature and concentrations. Lee and Kolthoff indicated that the theoretical limit of resolution of their method was for compounds

1. T. S. Lee and I. M. Kolthoff, *Ann. N.Y. Acad. Sci.*, **53,** 1093 (1951).
2. H. Etienne, *Ind. Chim. Belge*, **17,** 455 (1952).

whose rates of reaction with a given reagent differed by more than a factor of 4. Their experiments, however, showed no example of analysis of mixtures whose rates of reaction were closer than a factor of 17.

The authors, encountering the difficulties of the Lee and Kolthoff approach, used conventional second-order kinetic considerations and calculations and found that these could be applied very readily with no need for temperature control or knowledge of the rate constants of the compounds involved.

The method of Siggia and Hanna permits a wide range of application, resolving components whose rates differ by only a factor of 2. The method may permit a better resolution than this, but no examples with closer rate constants have been found to try the method further. This approach permits resolution of such closely related compounds as isomers; that is, 1-propanol and 2-propanol, 1-butanol from 2-butanol, 2-pentanone from 3-pentanone. It also permits resolution of homologs such as 1-butanol from 1-pentanol, 1-pentanol from 1-heptanol, *n*-hexylamine from *n*-heptylamine, which differ by only one carbon atom. The approach is so sensitive that it will resolve two of the same functional groups on the same molecule, such as the hydroxyl groups on propylene glycol and glycerine, and the amino groups on phenylene diamine.

Determination of Mixtures of Hydroxy Compounds and Mixtures of Carbonyl Compounds

Method of S. Siggia and J. G. Hanna

[*Reprinted in Part from Anal. Chem.*, **33**, *896–900* (*1961*)]

The approach uses the conventional plotting of second-order reaction data. In a second-order reaction, the rate is dependent on the concentrations of two reactants in the system. Hence if A and B are reacting, their concentrations a and b determine the rate. If x describes the amount of A and B that has reacted at time t, then the rate of reaction dx/dt can be described as

$$\frac{dx}{dt} = k(b-x)(a-x)$$

On integration, taking into account that $x=0$ when $t=0$, and $x=x$ when $t=t$, the equation develops as follows:

$$k = \frac{2.303}{t(b-a)} \log \frac{a(b-x)}{b(a-x)}$$

Plots of $\log(b-x)/(a-x)$ against t in the conventional manner yield straight lines for second-order reactions. When two second-order reactions are proceeding in the same mixture, a curve with two straight-line portions is obtained if the reaction rates of the two reactions are sufficiently different.

Many reactions used to determine organic compounds via their functional groups are second-order reactions. Some of these lend themselves very readily to rate studies, which makes possible the resolution of the components in the mixture. Below are examples of reactions that have been found usable with this rate approach to resolve mixtures containing the same group.

ANALYSIS OF MIXTURES OF ALCOHOLS INCLUDING MIXTURES OF ISOMERIC PRIMARY AND SECONDARY ALCOHOLS

The literature contains no general chemical method for the analysis of mixtures of two homologous alcohols or mixtures of isomeric primary and secondary alcohols. Methods (2, 3) for the specific determination of 2-propanol in mixtures involve oxidation of the 2-propanol to acetone and subsequent determination of the acetone produced. A method (4) has been described for the determination of small amounts of secondary in primary alcohols. This also involves oxidation of the secondary alcohol to a ketone and colorimetric estimation of the ketone.

The method described here is based on the differences in reaction rates of different alcohols with acetic anhydride.

It is necessary to know the total hydroxyl content of the sample used in order to select the sample size and to make the calculations for the analysis. This can be done by an established acetylation procedure (pp. 12–22). The amount of the less reactive component is found by difference.

REAGENTS

Reagent grade pyridine and reagent grade acetic anhydride.

Titration indicator, a 2:1 mixture of 0.1% Nile blue sulfate in 50% ethanol and 1% phenolphthalein in 95% ethanol.

The alcohols used as standards were reagent grade chemicals.

The propanols were dried over calcium chloride and distilled.

3. F. Strache and E. Martienssen, *Z. Lebensm.-Untersuch. Forsch.*, **104,** 339 (1956).
4. F. E. Critchfield and J. A. Hutchinson, *Anal. Chem.*, **32,** 862 (1960).

PROCEDURE

Transfer a sample containing 0.05 mole of hydroxyl to a 250-ml volumetric flask, using pyridine, and dilute it to almost 240 ml with pyridine. Pipet 10 ml of acetic anhydride into the flask and rapidly dilute the mixture to volume with pyridine. Note the time. At intervals, pipet 10-ml aliquots into glass-stoppered flasks, add 5 ml of water to each, and again note the time. Allow each to stand at least 10 minutes and then titrate with 0.1*N* alcoholic potassium hydroxide. Run a blank by pipetting 10 ml of acetic anhydride into a 250-ml volumetric flask and diluting to volume with pyridine. A 10-ml aliquot of this is treated in the same manner as the sample.

Log $(b-x)/(a-x)$ is plotted against t, where x is the decrease in concentration of reactant in time t, and a and b are the initial concentrations of alcohol and anhydride, respectively. If a mixture of two alcohols is indicated, straight lines are drawn, representing the two slopes. The extrapolation procedure for the determination of the more reactive alcohol is illustrated in Fig. 25.1. This plot is the second-order reaction curve for the mixture of 1-butanol and 2-butanol listed in Table 1. The

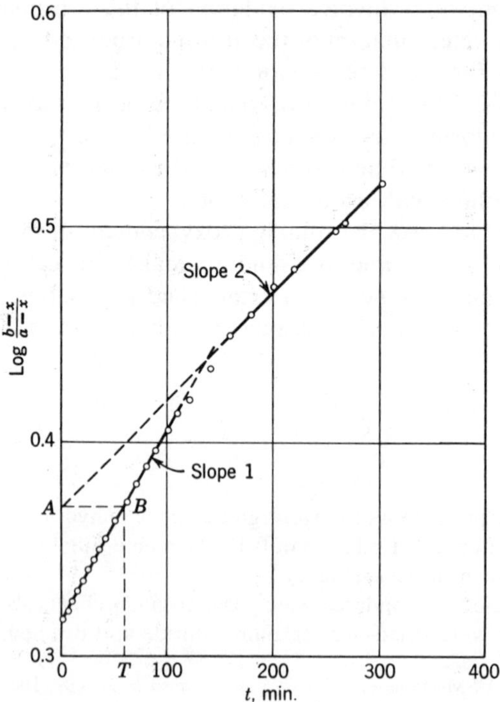

Fig. 25.1. Second-order reaction curve for mixtures of butanols.

Table 1. Alcohol Mixtures

Mixture Number	Alcohols		% Primary	
	Primary	Secondary	Present	Found
1	1-Propanol	2-Propanol	1.01	1.00[a]
2	1-Propanol	2-Propanol	50.5	50.0[b]
3	1-Propanol	2-Propanol	20.3	21.0[b]
4	3-Amino-1-propanol	1-Amino-2-propanol	49.6[c]	49.5
5	1-Butanol	2-Butanol	19.9	19.9
6	1-Octanol	2-Octanol	74.8	64.2[d]
7	1-Pentanol	3-Pentanol	14.9	15.5

[a] Alcohol in excess of acetic anhydride.
[b] Isocyanate reaction.
[c] 99.1% pure by acidimetric titration.
[d] Concentration of primary alcohol is so large that points for portion of curve representing reaction rate of secondary alcohol are near end of reaction, where they are not reliable.

line representing the slope of the less reactive alcohol (slope 2) is extrapolated to point A at zero time. A line AB is drawn parallel to the time axis. The time T at point B of intersection between this line and slope 1 is read. The concentration of the more reactive alcohol is then the concentration of alcohol reacted at this time.

The best plot of $\log (b-x)/(a-x)$ versus time is obtained by first plotting x versus t, which are the experimental data. The points are then taken off this plot to constitute the final plot. The procedure serves two purposes; it averages the errors in the experimental data, and it provides a large number of points for a good plot of $\log (b-x)/(a-x)$ and t. The latter is important to issue enough points to clearly define the two straight-line portions of the curve.

For the example illustrated, T was found to be 59.5 minutes. From a plot of x versus t, the concentration of alcohol reacted in this time was found to be 0.0398 mole per liter. The total mole per liter of butanol present was 0.2000. Then

$$\frac{0.0398}{0.2000} \times 100 = 19.9\% \text{ 1-butanol found}$$

An alternate calculation is as follows.

Let $a = a_1 - a_2$, where a_1 represents the concentration of the more reactive hydroxyl. When slope 2 is extrapolated to zero time,

$$x = a_1 \quad \text{at} \quad t = 0$$

and

$$\log (b-x)/(a-x) = \log (b-a_1)/(a-a_1)$$
$$= \text{value at point } A, \text{ the intercept } (t=0).$$

For Fig. 25.1,

$$b = 0.415$$
$$a = 0.2$$

and

$$\log \frac{(b-a_1)}{(a-a_1)} = 0.369$$

Thus $a_1 = 0.0396$ mole per liter or 19.8% 1-butanol.

DISCUSSION AND RESULTS

If this method of analysis is to be successful, the difference between the specific rates of reaction of the components in the mixture must be large enough so that a separate and distinct linear plot is obtained for each component when the second-order reaction rate plot is made. However no mixture of alcohols was encountered where this difference was not sufficient for this method to be used. From the standpoint of ease of following the reaction and for the best resolution of the two slopes, the most practical molar ratio of anhydride to hydroxyl is about 2:1.

Since the rate of reaction is dependent on the concentration of the two reactants in the system, the rate may be increased by increasing the concentration of the alcohol or anhydride or of both. However if the reaction is too rapid, difficulty is experienced with the resolution of the two portions of the curve. The concentrations of alcohol and anhydride chosen must not be equal, because then $a = b$ and $\log (b-x)/(a-x)$ will always be zero, and two separate slopes will not be obtained. This condition is rarely attained, since an excess of reagent is generally used.

The data in Table 1 indicate the applicability of the method for the determination of primary alcohols in the presence of secondary alcohols. The results in Table 2 show that it is possible to determine an alcohol in

Table 2. Mixtures of Homologous Alcohols

More Reactive Alcohol	Less Reactive Alcohol	% More Reactive Alcohol	
		Present	Found
1-Propanol	1-Octanol	5.00	4.99
1-Butanol	1-Pentanol	29.7	30.1
1-Pentanol	1-Heptanol	24.6	24.4

the presence of its next higher homolog as well as in the presence of one further separated in the series. This is illustrated by the results for 1-butanol in the presence of 1-pentanol. Table 3 contains data for polyhydric alcohols containing both primary and secondary hydroxyl groups. All the reactions were followed for about 300 minutes.

Table 3. Polyhydric Alcohols Containing Both Primary and Secondary Hydroxyl Groups

	% Primary Alcohol	
Alcohol	Present	Found
1,2-Propanediol	50.0	50.2
1,3-Butanediol	50.0	50.5
Glycerol	66.7	66.4

The reaction between isocyanates and alcohols can also be used to analyze mixtures of alcohols. Results for mixtures 2 and 3 in Table 1 were obtained in this way. Triethylenediamine was used to catalyze the reaction. A disadvantage of this system is the difficulty introduced by the presence of water in the reaction mixture. It was necessary to run a rate study on a blank along with the sample, to correct for any water in the reagents and solvents. The reaction with water appears to have a measurable rate. The sample must also be dry.

The presence of water when acetic anhydride is used will cause some of the anhydride to be consumed, changing its concentration in the mixture; but this reaction with small amounts of water (less than 1% of the sample) is rapid and has substantially no effect on the shape of the reaction plot, thus none on the final result. Large quantities of water, however, will destroy sufficient anhydride to affect the esterification of the alcohol.

If more than about 70% of the more reactive alcohol is present in the mixture, difficulty is experienced in plotting. As the reaction approaches completion, the upper portion of the reaction rate plot levels out and becomes unreliable. Then enough reliable points cannot be obtained to make a linear plot for the less reactive component. This difficulty is illustrated by results for mixture 6 in Table 1. To overcome this, a known amount of the less reactive alcohol can be added to the mixture and the final result corrected for the amount added.

To determine small amounts of the more reactive alcohol in a mixture, it was found expedient to use a larger sample and to have the alcohol in excess of the anhydride. The result for mixture 1 in Table 1 was obtained in this manner. The molar ratio of alcohol to anhydride was 5:1.

The method is generally limited to mixtures of two alcohols. However when the reaction rate plot was made for a mixture of ethylene glycol, diethylene glycol, and triethylene glycol, three separate linear parts of the curve were noted. Calculation for the composition of the mixture showed 18.9% found versus 19.3% ethylene glycol present, 33.2% found versus 33.1% diethylene glycol present, and 47.9% found by difference versus 47.6% triethylene glycol present.

An attempt was made to analyze mixtures of amines and mixtures of phenols in the same manner. In both cases, however, the reactions with acetic anhydride were too rapid for the rate to be measured. Phthalic anhydride was substituted for acetic anhydride, but again, the amine reaction was too rapid. No reaction between phthalic anhydride and phenol was detected, even at elevated temperature.

ANALYSIS OF MIXTURES OF ALDEHYDES AND KETONES

Chemical methods are available for the determination of aldehydes in the presence of ketones and for specific carbonyl compounds in a mixture. However none of these are general for the determination of one carbonyl compound in the presence of another. A mixture of an aldehyde and a ketone can be analyzed by first determining the total carbonyl present, using a hydroxylamine hydrochloride method (5), then determining the aldehyde alone, using a bisulfite (6) or an argentimetric method (7, 8) (see pp. 95–107, 115–37). The amount of ketone is obtained by difference, Dimedone and cyanide (9) have been used to analyze mixtures of formaldehyde and propionaldehyde. Deniges (10) used a modified Schiff reagent for the detection of formaldehyde in the presence of higher aldehydes. This has been used along with the standard Schiff reagent to analyze mixtures of formaldehyde and 2-furaldehyde, and formaldehyde and acetone (11). Chromotropic acid has been used also to determine formaldehyde in the presence of higher aldehydes (12).

Methods based on rates of reaction have been proposed. Ionescu and

5. W. M. D. Bryant and D. M. Smith, *J. Am. Chem. Soc.*, **57**, 57–61 (1935).
6. S. Siggia and W. Maxcy, *Ind. Eng. Chem., Anal. Ed.*, **19**, 1023 (1947) (see pp. 104–107, this book).
7. J. Mitchell, Jr. and D. M. Smith, *Anal. Chem*, **22**, 746–50 (1950).
8. S. Siggia and E. Segal, *Anal. Chem.*, **25**, 640 (1953) (see pp. 118–19, this book).
9. G. Hoepe and W. D. Treadwell, *Helv. Chim. Acta*, **25**, 353–61 (1942).
10. G. Denigès, *Compt. Rend.*, **150**, 529–31 (1910).
11. R. I. Veksler, *Zhr. Anal. Khim.*, **4**, 14–20 (1949); **5**, 32–8 (1950).
12. C. E. Bricker and H. R. Johnson, *Ind. Eng. Chem., Anal. Ed.*, **17**, 400–2 (1945).

Slusanchi (13) were able to differentiate between formaldehyde and acetaldehyde by the rates of precipitation of the solid derivatives with dimedone. Lee and Kolthoff (1) have given a procedure based on the difference in reaction rates for the analysis of mixtures of two organic compounds that contain the same functional group. Calibration curves relate the concentration at a chosen time with the original concentration. Analysis of mixtures of aldehydes based on the rate of decomposition of the bisulfite addition compound was mentioned, but no data were given. Methods based on the competing rates of oxime formation have been proposed (14, 15). These methods also use calibration curves to relate the amount of reaction after a period of time to the original concentration. The methods are limited to the determination of aromatic aldehydes in the presence of aromatic ketones, where the difference in the reaction rates is large and the oximation of the aldehyde is complete when a very small amount of the ketone has reacted.

The method proposed here also makes use of the different rates of reaction of carbonyl compounds with hydroxylamine hydrochloride. However with this method, mixtures of carbonyl compounds can be analyzed when the reaction rates are much closer than those of aromatic aldehydes and aromatic ketones. This method will resolve not only the aliphatic aldehydes and ketones, but also mixtures of two aldehydes and mixtures of two ketones.

The mixture is reacted with hydroxylamine hydrochloride, and the amount of reaction is determined at successive time intervals. A second-order reaction rate plot is made. If the carbonyls present react at different rates, the plot will show two straight-line portions. The contribution of the slower reacting component is separated from the slope of the more reactive component by extrapolation, and the more reactive component is then measured.

REAGENTS AND APPARATUS

The methanol used as the solvent was reagent grade, acetone-free material.

The purities of the aldehydes and ketones used were determined by established methods (7, 16) and the synthetic mixtures prepared based on these analyses.

A Beckman Model H pH meter equipped with a glass and calomel electrode system was used for the pH measurements.

13. M. V. Ionescu and H. Slusanchi, *Bull. Soc. Chim. Fr.*, **53** (4), 909–18 (1933).
14. L. Fowler, *Anal. Chem.*, **27,** 1686–8 (1955).
15. L. Fowler, H. R. Kline, and R. S. Mitchell, *Ibid.*, **27,** 1688–90 (1955).
16. L. P. Hammett, *Physical Organic Chemistry*, McGraw-Hill, New York, 1940.

PROCEDURE

Weigh the sample containing 0.004 mole of carbonyl and dilute it to 100 ml with a 4:1 methanol-water solution. To an 800-ml beaker, add 480 ml of 4:1 methanol-water solution. Pipet a 10-ml portion of $0.1N$ hydroxylamine hydrochloride in 4:1 methanol-water into the solution. Place the beaker in an ice bath and lower the temperature to 4°C, using mechanical agitation. For aromatic ketones, allow the reaction to proceed at room temperature. Place the glass and calomel electrodes of a pH meter in the solution and adjust the pH of the solution to 3.5. Pipet a 10-ml aliquot of the sample into the solution and note the time. Maintain the pH of the solution at 3.5 ± 0.02, using standardized $0.02N$ sodium hydroxide solution. Note the milliliters of sodium hydroxide solution added at 5-minute intervals.

The concentration of carbonyl remaining and the concentration of hydroxylamine hydrochloride remaining after time t are calculated. In the second-order rate equation, the original quantities of carbonyl compound can be represented as a and the hydroxylamine reagent as b. The amount of each reacted after time t is represented as x. Then $\log (b-x)/(a-x)$ is plotted against t. If a mixture of two carbonyl compounds is indicated, straight lines are drawn, representing the two slopes. The extrapolation procedure for the determination of the more reactive carbonyl compound is illustrated in Fig. 25.2. This plot is the second-order reaction curve for the mixture of 2-butanone and 3-pentanone.

The concentration of the more reactive carbonyl compound is then obtained by an operation similar to those described for alcohols. The total carbonyl content of the mixture is determined by a regular oximation procedure (5).* The difference is then a measure of the less reactive component of the mixture.

DISCUSSION AND RESULTS

The reaction of carbonyl compounds with hydroxylamine to form the oxime is second-order (16, 17). Separate and distinct linear plots were obtained for each component in the mixtures when the second-order reaction rate plots were made. No mixture of carbonyls was encountered for which the rates of reaction were not different enough for this method to be used.

* See also pp. 95–100 of this text for other oximation methods.

17. G. H. Stempel and G. S. Shaffel, *J. Am. Chem. Soc.*, **66,** 1158 (1944).

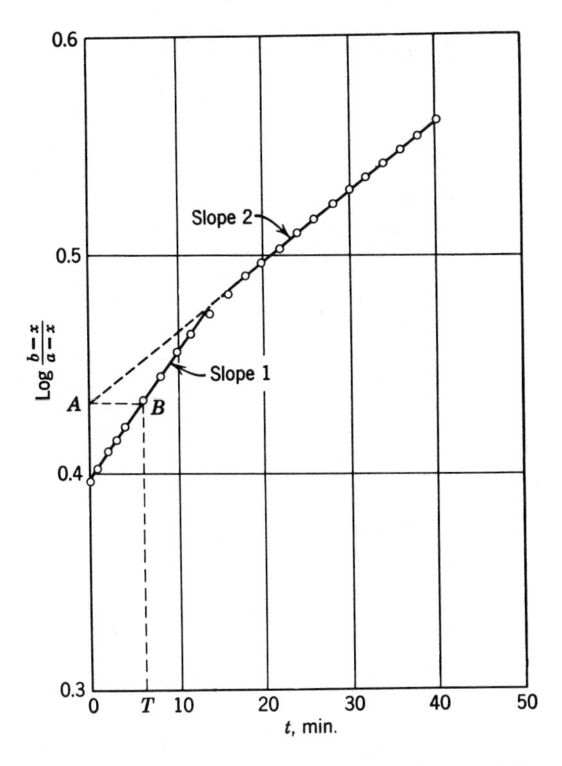

Fig. 25.2. Reaction rate curve for mixture of 2-butanone and 3-pentanone.

The concentration of either the carbonyl or the hydroxylamine hydrochloride can be varied to speed up or slow down the reaction. In no case, however, should the concentrations chosen be equal, because then $a = b$ and $\log (b - x)/(a - x)$ will always be zero, and two separate slopes will not be obtained.

The procedure has been applied to mixtures of aldehydes, mixtures of ketones, and mixtures of aldehydes and ketones. Table 4 shows that it is possible to determine an aldehyde or a ketone in the presence of its next higher homolog, as well as in the presence of one further separated in the series.

Except in the case of the aromatic ketones, the reaction was run at 0 to 5°C. At higher temperatures, the reaction proceeds too rapidly to be practical. For aldehydes and aliphatic ketones, the determination takes less than an hour. The aromatic ketones were reacted at room temperature and required approximately 90 minutes for the determination.

Table 4. Analytical Results

Mixture	%A	
	Present	Found
1. A. Formaldehyde	13.9	14.7
B. Acetaldehyde		
2. A. Acetaldehyde	24.7	24.7
B. Acetone		
3. A. Acetone	7.84	7.80
B. 2-Butanone		
4. A. Propionaldehyde	11.8	11.7
B. Acetone		
5. A. 2-Butanone	17.4	17.0
B. 3-Pentanone	17.4	17.6
6. A. Crotonaldehyde	7.5	7.5
B. 3-Pentanone		
7. A. 2-Pentanone	18.2	17.9
B. 3-Pentanone		
8. A. Cyclopentanone	7.37	7.36
B. 3-Pentanone		
9. A. Cyclohexanone	65.8	65.4
B. Cyclopentanone		
10. A. Crotonaldehyde	18.0	17.4
B. Hexaldehyde		
11. A. Hexaldehyde	15.6	15.6
B. 3-Pentanone		
12. A. Cyclohexanone	84.5	86.1
B. 2-Butanone		
13. A. Benzaldehyde	28.3	28.0
B. Salicylaldehyde		
14. A. Acetophenone	8.4	8.0
B. Benzophenone		

Cyclopentanone reacted more rapidly than 3-pentanone, and cyclohexanone more rapidly than cyclopentanone. For the other mixtures of aldehydes and mixtures of ketones tested, the carbonyl of lower molecular weight reacted more rapidly. For mixtures of aldehydes and ketones, the aldehydes reacted more rapidly.

Acidic or basic impurities in the sample interfere and should be neutralized before the determination is made.

Determination of Mixtures of Amines Using Differential Reaction Rates

Method of J. G. Hanna and S. Siggia

[*Anal. Chem.*, ***34***, *547–9* (*1962*)]

Amine mixtures can be resolved much in the same manner as were the alcohols and carbonyl compounds. For amines, the reaction of the amines with phenyl isothiocyanate was used. This compound reacts with primary and secondary amines as follows:

$$C_6H_5N{=}C{=}S + H_2NR \rightarrow C_6H_5NH\overset{\overset{\displaystyle S}{\|}}{C}NHR$$

$$C_6H_5N{=}C{=}S + HN(R)_2 \rightarrow C_6H_5NH\overset{\overset{\displaystyle S}{\|}}{C}N(R)_2$$

Hence the rate approach can distinguish primary and secondary amines; tertiary amines are also determinable, since they do not react with the isothiocyanate. The tertiary amines can thus be determined by difference, or they can be determined on the reaction mixture after the reaction with isocyanate has proceeded to completion.

The reaction is followed by titrating the residual unreacted amine at each time interval. Secondary amines generally reacted more rapidly than the corresponding primary amine. Tertiary amines do not react at all with the isothiocyanate and are left intact at the end of the reaction. In addition, aromatic amines reacted before aliphatic amines.

Tables 5 and 6 indicate analyses on known mixtures of amines. These samples included mixtures of primary and secondary amines; mixtures of primary, secondary, and tertiary amines; mixtures of two primary amines; and mixtures of two secondary amines. A diamine was also run, indicating that the individual amino groups on the same molecule can be resolved.

Table 5. Amine Mixtures

Mixture	% A	
	Present	Found
1. A. di-*n*-Butylamine	80.8	80.1
B. *n*-Butylamine	43.1	42.7
2. A. di-*n*-Butylamine	22.9	23.1
B. di-*n*-Hexylamine		
3. A. *n*-Hexylamine		
B. *n*-Heptylamine	4.6	4.3
4. A. Aniline		
B. di-*n*-Hexylamine	25.0	24.7
5. A. Aniline		
B. 2-Naphthylamine	10.7	10.7
6. A. *p*-Toluidine		
B. Aniline	14.7	14.7
7. A. Aniline		
B. *m*-Phenylenediamine[a]	10.2	11.0
8. A. *m*-Phenylenediamine[a]	50.0	49.0

[a] The two amino groups in *m*-phenylenediamine reacted at different rates. The faster group reacted at the same rate as the amino group in aniline.

Table 6. Mixture of Primary, Secondary, and Tertiary Amines

	%	
	Present	Found
1. A. *n*-Butylamine	7.2	7.3
B. di-*n*-Butylamine	23.8	23.8
C. tri-*n*-Butylamine	69.0	68.7
2. A. Aniline	48.4	47.8
B. *n*-Propyl aniline	35.1	35.1
C. di-*n*-Propyl aniline	16.5	17.1

The reaction between the amine mixture and the isothiocyanate is carried out in an acidified medium, to slow the reaction sufficiently to make possible the rate study. Isocyanates were tried as a reagent, but these reacted too rapidly for the rate approach to be applied.

REAGENTS AND APPARATUS

The purities of the amines were determined by titration with standard acid and the synthetic mixtures prepared based on these analyses.

The dioxane was dried by distillation from dispersed sodium.

A Beckman Model H pH meter equipped with a glass and calomel electrode system was used for the potentiometric titrations.

Phenyl isothiocyanate, Eastman White Label.

PROCEDURE

Transfer a sample containing 0.025 equivalent of amine to a 250-ml volumetric flask with dioxane. Add acetic acid: 30 to 40 ml for aromatic amines, and 15 to 25 ml for alkyl amines. Add 0.1 mole of phenyl isothiocyanate, rapidly dilute the mixture to volume using dioxane, and note the time. At intervals, pipet 20-ml aliquots into beakers containing 50 ml of acetic acid to stop the reaction, and note the time again. Titrate each potentiometrically with 0.1N perchloric acid in acetic acid (see Fig. 25.3).

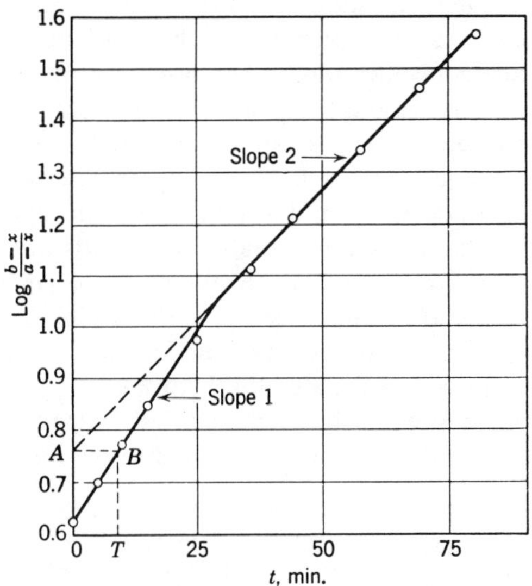

Fig. 25.3. Reaction rate curve for mixture of di-n-butylamine and di-n-hexylamine.

The analyses are calculated as shown above for alcohols and carbonyl compounds.

Determination of Mixtures of Unsaturated Compounds

Method of S. Siggia, J. G. Hanna, and N. M. Serencha

[*Anal. Chem.*, ***35***, *362* (*1963*)]

Rates of addition to olefinic bonds have been used in special cases for the analysis of mixtures of unsaturated compounds. The rates of reaction of perbenzoic acid with olefinic double bonds have been used as a basis of analysis (1, 18). The amount of reaction is measured after a specified time, and this is related to the original concentration by reference to a calibration curve prepared from the analysis of known mixtures under the same conditions. A procedure has been described for the determination of internal and external double bonds in polymers which also uses the rates of reaction with perbenzoic acid (19). The composition of the original mixture is obtained by extrapolation to zero time of the flat portion of the curve obtained by a plot of percentage reacted versus time. Mixtures of ethyl elaidinate and ethyl oleate have been analyzed, using the difference in rates of addition of mercuric acetate (20). Again, reference to a calibration curve relates the amount of reaction at a specified time to the concentrations of compounds in the original mixture. The separation of the amount of substitution obtained while adding bromine to unsubstituted linkages can be measured, based on differences in rates of bromine consumption (21). Several additional determinations are made at increased reaction times in excess of that necessary for complete saturation of the double bonds. The results are plotted against time and extrapolated to zero time to obtain the correct value for the addition reaction. It is sometimes possible to determine one component in a mixture by a selective hydrogenation under such conditions that the desired reaction is obtained without the hydrogenation of other groups.

The rate approach previously described for hydroxyl, carbonyl, and amine compounds (pp. 826–40), when applied to the analysis of mixtures

18. I. M. Kolthoff, T. S. Lee, and M. A. Mairs, *J. Polym. Sci.*, **2,** 199 (1947).
19. A. Saffer and B. L. Johnson, *Ind. Eng. Chem.*, **40,** 538 (1948).
20. T. Connor and G. P. Wright, *J. Am. Chem. Soc.*, **68,** 256 (1946).
21. A. Polgar and J. L. Jungnickel, in *Organic Analysis*, Vol. III, J. Mitchell, Jr., Ed., Wiley-Interscience, New York, 1956, p. 237.

of unsaturated compounds, was found to be more general than the aforementioned methods. Both bromination and hydrogenation can be used.

The bromination reaction follows a second-order rate process. Free bromine is used, and the decrease in bromine content is followed colorimetrically. The standard second-order rate plot shows linear portions for each unsaturated component present. Hydrogenation is performed catalytically, and the decrease in hydrogen content with time is noted. Since the hydrogen is present in a large excess relative to the sample, this reaction can be considered pseudo first order. The standard first-order rate plot then gives linear portions for each unsaturated compound present.

BROMINATION PROCEDURE

A solution containing 0.00125 equivalent per liter of sample in the appropriate solvent (Table 7) is rapidly mixed with an equal volume of a solution containing 0.0025 equivalent of bromine per liter in a colorimeter cell and the time noted. A Bausch & Lomb Spectronic 20 colorimeter was used in this work. At successive time intervals, the absorbance is read at 400 nm. The concentration of bromine at these time intervals is read from a previously prepared calibration curve relating absorbance to concentration of bromine.

$\text{Log}\,(b-x)/(a-x)$ is plotted against t, where x is the decrease in concentration of reactant in time t, and a and b are the initial concentrations of unsaturation and bromine, respectively. If a mixture of two unsaturated compounds is indicated, straight lines are drawn representing the two slopes. The concentration of the more reactive component is then

Table 7. Relative Rates of Bromination in Different Solvents

Solvent	Unsaturated Compound				
	Methyl Butynol	Allyl Esters	Methyl Hexynol	Butyne-Diol	Oleates
Water	Satisfactory	Fast	—	Satisfactory	—
Methanol	Slow	Satisfactory	—	—	—
Acetic Acid	Slow	Slow	Satisfactory	Slow	Fast
Carbon Tetrachloride	Slow	Slow	Slow	—	Satisfactory

obtained by an operation similar to that previously described (pp. 828–30). The total unsaturation is determined by allowing the reaction to proceed until no more bromine is consumed.

HYDROGENATION PROCEDURE

The hydrogenation apparatus and procedure used were those of Clauson-Kaas and Limborg (22) described on pp. 402–6. Ethanol is used as the solvent. The sample size is about 0.0002 equivalent. Platinum oxide catalyst (2–4 mg) is hydrogenated, and the pressure in the system equilibrated. A platinum beaker containing the solid sample is dropped into the solvent and the time noted. If the sample is liquid, it is contained in a sealed glass capillary tube, which is broken and dropped into the solvent. The pressure in the hydrogenation vessel is maintained equal to that in the compensating vessel by the addition of mercury from a buret to the hydrogenation vessel. The amount of mercury added, therefore the amount of hydrogen consumed, is read at successive time intervals.

Because the concentration of hydrogen is large compared to that of the sample, the reaction is treated as pseudo-first order. The integrated form of a first-order reaction expression is

$$kt = 2.303 \log \frac{a}{(a-x)}$$

where a is the original concentration of unsaturation and x is the concentration reacting in time t. A plot of $\log (a-x)$ against t results in a straight line having a slope of $-(k/2.303)$. For two unsaturated bonds reacting at different rates, two slopes are obtained. If the second slope is extrapolated to zero time, then at the point of intersection, y, $x = a_1$, the concentration of the faster reacting component and

$$\log (a - a_1) = y.$$

This equation can be solved for a_1, since a is known from the total hydrogenation value, and y is read at the intercept.

DISCUSSION AND RESULTS

Better plots of $\log (b-x)/(a-x)$ versus t for brominations and $\log (a-x)$ versus t for hydrogenations are obtained by first plotting x versus t,

22. N. Clauson-Kaas and F. Limborg, *Acta Chem. Scand.*, **1,** 884 (1947); see pp. 402–6, this book.

using the experimental data. A smooth curve is drawn, and points taken from this curve are used to construct the final plot. This procedure averages the errors in the experimental data and provides a large number of points to clearly define the straight-line portions of the final reaction rate plot.

In the use of the bromination procedure, the main limiting factor was the ability to adjust the conditions in order to control the reaction at a practical rate. Proper choice of solvent was the primary means used to control the speed of reaction. Some relative rates of reaction in different solvents are indicated in Table 7. The relative rates in different solvents are in the order

$$\text{water} > \text{methanol} > \text{acetic acid} > \text{carbon tetrachloride}$$

The use of the second-order rate equation was justified because straight lines were obtained when a plot was made of the bromination of compounds containing a single double bond.

The data in Table 8 indicate the applicability of the bromination

Table 8. Bromination of Mixtures of Unsaturated Compounds

Mixture	Solvent	Per Cent A	
		Found	Present
1. A. Maleic acid[a] B. Fumaric acid[a]	Water	50.5	49.6
2. A. Methyl oleate B. Ethyl oleate	Carbon tetrachloride	22.9	22.6
3. A. Methyl oleate B. Butyl oleate	Carbon tetrachloride	8.4	8.8
4. A. 5-Methyl-1-hexyne B. 2-Butyne-1,4-diol	Acetic acid-water	23.8[b]	23.6

[a] Sodium salts.
[b] The methyl hexyne was brominated totally faster than the butynediol.

procedure to mixtures of unsaturated compounds. Compounds containing triple bonds or two double bonds consumed one-half the total bromine at a faster rate than the second half, as is indicated by the data in Table 9. Figure 25.4 is a plot of the second-order reaction curve for the mixture of methyl hexyne and butynediol listed in Table 8. The solvent used in this case was a 1:1 mixture of acetic acid and water. In this system the triple

Table 9. Bromination of Compounds Containing Triple Bonds or Two Double Bonds

Compound	Solvent	Per Cent Bromine Added at Faster Rate
1. 2-Butyne-1,4-diol	Water	49, 50
2. 2-Methyl-3-butyne-2-ol	Water	50, 51
3. 5-Methyl-1-hexyne	Acetic acid	49
4. Sorbic acid	Acetic acid	51

bond of methyl hexyne was completely saturated more rapidly than the first mole of bromine was added to the triple bond of butynediol (see Table 7). A close inspection of the first slope of the curve shows a suggestion of two slopes due to the triple bond of methyl hexyne. But the resolution of these two slopes is not definite enough to be used as a basis for differentiation of the components of the mixture.

Maleic and fumaric acids brominated too slowly for practical purposes, but their sodium salts reacted fast enough for the procedure to be applied.

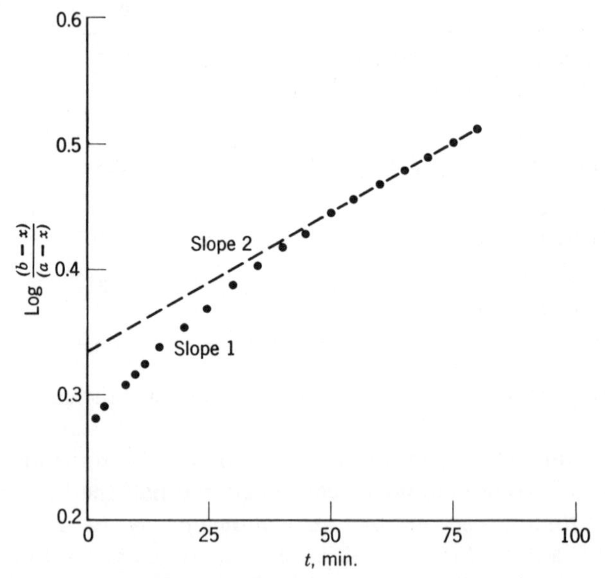

Fig. 25.4. Reaction rate curve for the bromination of a mixture of 5-methyl-1-hexyne and 2-butyne-1,4-diol.

This difference in rate has been observed previously (23). No satisfactory system was found for methyl hexynol (too slow), *p*-mentha-1,5-diene (too slow), and allyl alcohol (too fast). Allyl acetate, allyl formate, and allyl propionate each gave straight-line plots when run separately. When mixtures of these compounds were brominated, definite straight-line portions were not obtained, probably because of ester interchange. It should be pointed out that in no case should the concentrations of bromine and unsaturation be chosen equal, because then $a = b$ and $\log (b - x)/(a - x)$ will always be zero.

Plots containing a single straight line were obtained when the results of the hydrogenation of a single double bond were treated as a first-order process. Compounds containing triple bonds or two double bonds consumed one-half the total hydrogen at a faster rate than the second half, as indicated in Table 10. The data in Table 11 indicate the applicability of

Table 10. Hydrogenation of Compounds Containing Triple Bonds or Multiple Double Bonds

Compound	Per Cent Hydrogen Added at Faster Rate
1. 2-Butyne-1,4-diol	51.3
2. 5-Methyl-1-hexyne	47.2
3. 3-Methyl-1-hexyne-3-ol	47.5
4. 2-Methyl-3-butyne-2-ol	49.2
5. 2,5-Dimethyl-3-hexyne-2,5-diol	50.7
6. 3,6-Dimethyl-4-octyne-3,6-diol	48.4
7. 3-Methyl-1-pentyne-3-ol	49.4
8. *p*-Mentha-1,5-diene	49.2, 49.5

the hydrogenation procedure to mixtures of unsaturated compounds. Figure 25.5 is a plot of the first-order reaction curve for a mixture of methyl and ethyl oleates, as listed in Table 11.

Hydrogenation of mixtures of compounds containing triple bonds in most cases resulted in rate plots showing three or more flat portions. The relationship between each flat portion of the curve and its corresponding unsaturated unit must be established before a calculation of the composition of the mixture can be made. This can be done by running mixtures of known composition.

23. F. E. Critchfield, *Anal. Chem.*, **31**, 1406 (1959).

Table 11. Hydrogenation of Mixtures of Unsaturated Compounds

Mixture	Per Cent A	
	Found	Present
1. A. Crotonic acid	34.7	34.5
B. Fumaric acid	27.1	27.4
2. A. Sorbic acid	67.6	70.1
B. Fumaric acid		
3. A. Sorbic acid	85.9	85.7
B. Maleic acid		
4. A. Methyl oleate	53.1	53.8
B. Ethyl oleate	12.9	13.0
	32.4	30.8
5. A. Methyl oleate	44.1	44.9
B. *n*-Butyl oleate	62.8	61.3
	36.3	36.4
6. A. 2-Methyl-3-butyne-2-ol	27.4	27.1
B. 3-Methyl-1-pentyne-3-ol		
7. A. 2-Methyl-3-butyne-2-ol	52.2	54.8
B. 3-Methyl-1-hexyne-3-ol		
8. A. 3-Methyl-1-pentyne-3-ol	36.3	37.3
B. 3-Methyl-1-hexyne-3-ol		
9. A. 2-Methyl-3-butyne-2-ol	56.9	56.7
B. 2-Butene-1-ol		
10. A. 2-Methyl-3-butyne-2-ol	39.6	41.0
B. 2-Butyne-1,4-diol	79.5	79.1

The hydrogenation procedure appears to be more general than the bromination procedure. Although among the compounds studied, the necessity for a variety of conditions was not encountered, it is conceivable that in some situations different temperatures, concentrations, or pressures would be advantageous. The hydrogenation method is more adaptable to changes in these conditions than is the bromination procedure.

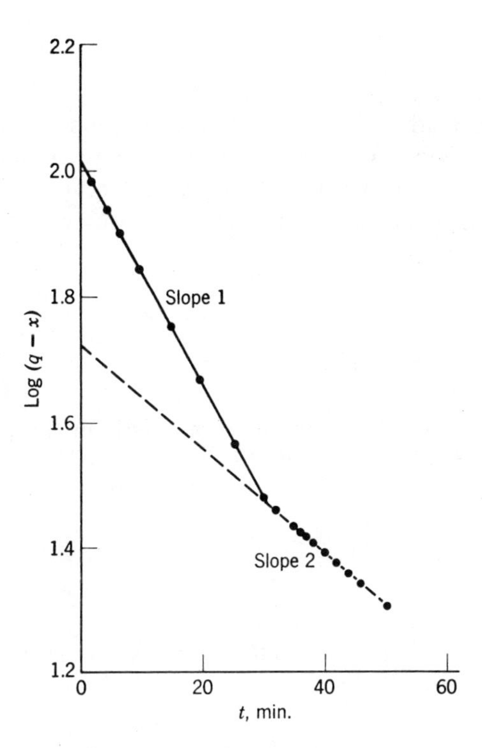

Fig. 25.5. Reaction rate curve for the hydrogenation of a mixture of methyl oleate and ethyl oleate.

Determination of Mixture of Diazonium Compounds

Method of S. Siggia, J. G. Hanna, and N. M. Serencha

[*Anal. Chem.*, ***35***, *575* (*1963*)]

The feasibility of the differential reaction rate technique has been demonstrated previously for the determination of components in mixtures. The work to follow shows the applicability of the kinetic approach to the analysis of mixtures of diazonium compounds by following their first-order rates of decomposition.

PROCEDURE

The apparatus and procedure have been described (pp. 681–4). The only adaptation made is to perform the decomposition at a constant temperature (80–90°C), using an electric heating mantle. The volume of nitrogen evolved is recorded at 1-minute intervals. The timing of the reaction is begun when the sample and cuprous chloride solutions are mixed in the reaction flask.

RESULTS AND DISCUSSION

The decomposition of diazonium compounds in the presence of cuprous chloride results in the quantitative evolution of nitrogen.

$$\left[C_6H_5N \equiv N \right]^+ Cl^- \xrightarrow{CuCl,\ HCl,\ H_2O} C_6H_5OH + HCl + N_2$$

The decomposition follows a first-order rate process. The integrated first-order rate expression is

$$kt = 2.303 \log \frac{a}{a-x}$$

and a plot of $\log(a-x)$ versus t yields a straight line for a single compound. For mixtures of diazonium compounds, a is the total diazonitrogen, as determined by allowing the reaction to go to completion, and it is denoted as 100 for the calculation; x is the percentage of the total diazonitrogen evolved in time t. For a mixture of two diazonium compounds decomposing at different rates, two slopes are obtained. If the second slope is extrapolated to zero time, then at the point of intersection y, $x = a_1$, the percentage of the faster decomposing compound and

$$\log(100 - a_1) = y$$

can be solved for a_1.

Results obtained for mixtures of diazonium compounds are given in Table 12. The rates for the diazonium compounds studied were sufficiently different for mixtures of three compounds to be resolved by this method. Results for three component mixtures are included in the table. Figure 25.6 is the plot obtained for mixture 1 in the table.

Table 12

Mixture	Components and Relative Rates of Decomposition	Present	Found
1	$A > B$	40.9% A	40.9% A
2	$A > B$	63.3% A	63.3% A
3	$A > B$	11.7% A	12.3% A
4	$A > C$	28.5% A	28.7% A
5	$A > C$	57.7% A	57.9% A
6	$A > C$	15.3% A	14.9% A
7	$B > C$	38.3% B	38.8% B
8	$B > C$	58.1% B	57.9% B
9	$B > C$	12.7% B	13.1% B
4	$A > B > C$	22.4% A	21.5% A
		40.2% B	41.8% B
		37.4% C	36.7% C
5	$A > B > C$	6.9% A	6.7% A
		29.8% B	29.9% B
		63.3% C	63.4% C

Compound A Sodium *p*-ethoxybenzenediazosulfonate

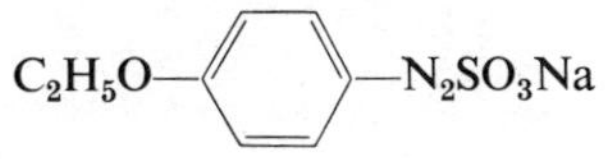

C_2H_5O—⟨benzene ring⟩—N_2SO_3Na

Compound B *p*-Amino-N-ethyl-2-hydroxyethylbenzenediazoniumchlorozincate

HOC_2H_4 / C_2H_5 —N—⟨benzene ring⟩—$N_2Cl\cdot\frac{1}{2}ZnCl_2$

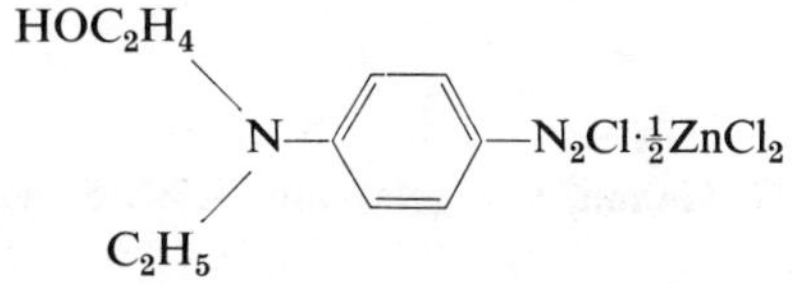

Compound C 2,5-Dibutoxy-4-morpholinobenzenediazoniumchlorozincate

O⟨morpholine ring⟩N—⟨benzene ring, C_4H_9O above, OC_4H_9 below⟩—$N_2Cl\cdot\frac{1}{2}ZnCl_2$

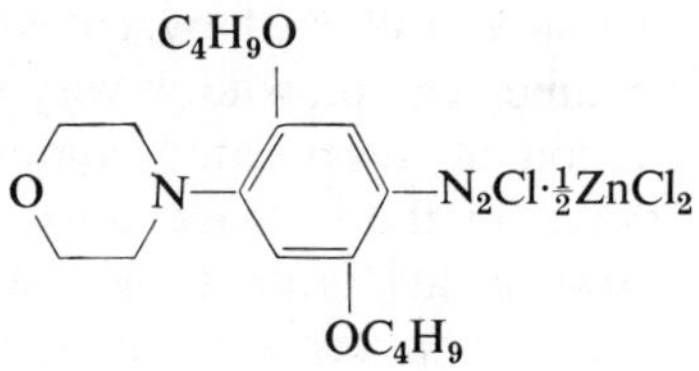

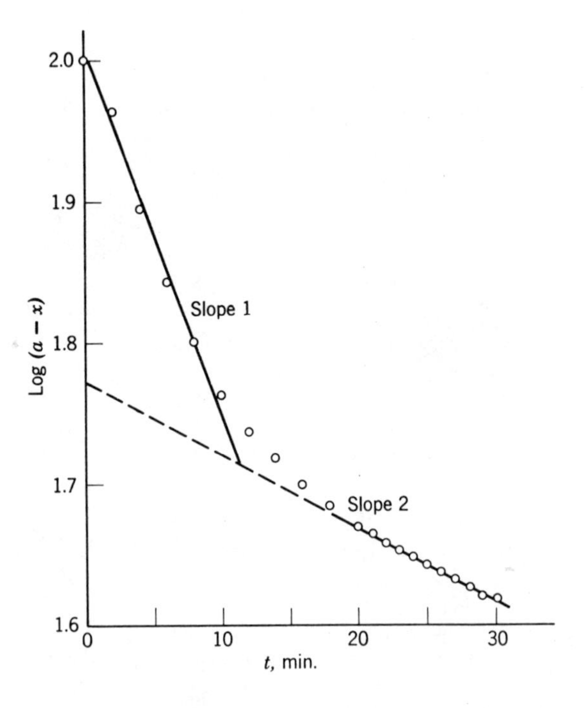

Fig. 25.6. Reaction rate curve for mixture 1 of Table 12.

Determinations of Mixtures of Amides and Mixtures of Nitriles

Method of J. G. Hanna, S. Siggia, and N. M. Serencha

[*Anal. Chem.*, **36**, *227* (*1964*)]

The hydrolysis of amides and nitriles follows pseudo-first-order kinetics. The reaction is followed by collecting and titrating the liberated ammonia with time. The plots look very much like those on page 847 for the hydrogenation of unsaturated compounds, and the calculations are identical to those for the hydrogenation.

Table 13 shows results for mixtures of amides. Table 14 shows results for mixtures of nitriles and also for mixtures of nitriles and amides.

Table 13. Mixtures of Amides

Mixture	Per Cent *A*	
	Found	Present
A. Acetamide	29.7	29.2
B. Benzamide	57.4	58.7
A. Acrylamide	16.3	16.3
B. Benzamide		
A. Propionamide	13.3	14.3
B. Benzamide		
A. Butyramide	21.7	22.8
B. Benzamide		
A. Acetamide	60.0	60.6
B. Acrylamide		
A. Acetamide	54.5	52.9
B. Propionamide		

Table 14. Mixtures of Nitriles and Mixtures of Amides and Nitriles

Mixture	Percent *A*	
	Found	Present
A. Butyronitrile	22.4	22.9
B. Valeronitrile	53.6	54.2
A. Butyronitrile	21.3	22.7
B. Benzonitrile	51.9	51.0
A. Valeronitrile	27.2	25.6
B. Benzonitrile	53.0	53.4
A. Butyronitrile	24.7	24.3
B. Anisonitrile	58.2	51.9
A. Benzonitrile	28.0	27.2
B. Anisonitrile	22.9	22.4
A. Acetamide	24.5	24.4
B. Acetonitrile	52.5	52.0
A. Butyramide	75.0	75.7
	30.5	30.1
B. Butyronitrile	53.0, 55.5	54.9
A. Benzamide	24.7	26.1
B. Benzonitrile	52.5	52.5

APPARATUS AND REAGENTS

Kjeldahl distillation apparatus.
Tetrahydrofuran distilled from sodium.
Boric acid solution, 4%.

PROCEDURE

A sample containing 0.0016 equivalent of amides or nitriles is placed in the distillation flask along with a 5:1 mixture of tetrahydrofuran-water for amides or a 20:1 mixture of tetrahydrofuran-water for nitriles and for binary mixtures of nitriles and amides. Approximately 10 ml of 50% sodium hydroxide is added. The system is assembled for distillation with a nitrogen sweep of about 2 bubbles per second. The reaction mixture is maintained at 60°C for amides or 70°C for nitriles. The distillate is collected in boric acid, which is adjusted at the start to pH 4.00, using a pH meter. The first indication of change in pH of the boric acid solution is taken as zero time. The acid solution is maintained at pH 4.00 by adding 0.04*N* hydrochloric acid. The amount of normal acid added is read at successive time intervals, and the equivalents of the mixture decomposed at these intervals are calculated.

CALCULATIONS

The composition of the original mixture is calculated from the first-order rate plot in the manner shown earlier (pp. 842–3) for the hydrogenation of unsaturated compounds.

DETERMINATIONS OF MIXTURES OF ESTERS

Method of T. I. Munnelly

[*Adapted from Anal. Chem.*, **40,** *1495* (*1968*)]

The second-order extrapolation approach was used for the determination of three- to five-component mixtures of esters. A binary temperature form of differential analysis described presents a general approach for studying mixtures in which the various individual reactions may be partitioned conveniently into reasonable time periods.

EXPERIMENTAL

Saponification of the samples was monitored by removal of suitable aliquots; these aliquots were quenched by addition of a known excess of standard acid and titrated with standard sodium hydroxide solution with phenolphthalein as the indicator. Specifically, a solution of each component was prepared and aliquots of each were mixed together. For mixtures of four and five components, the resulting solution was then divided into two portions, each to be studied at one of the operating temperatures. When reactants were combined, the concentration of the individual acetates were on the order of 0.001 to 0.002*M*. For phenyl acetate, it was necessary to add alcohol in a preparative step for dissolution; therefore in mixtures containing phenyl acetate a reaction medium composed of 98% water and 2% alcohol was established. A solution of 0.02*M* sodium hydroxide was the titrant.

Temperature was controlled to ±0.1°C at both 15.0 and 30.0°C by a Blue M Magic Whirl temperature bath with an Automatic Dual Microtol. The latter device made it possible to switch quickly from one temperature level to the other preselected setting.

RESULTS AND DISCUSSION

THREE–COMPONENT SOLUTIONS. Ten acetate mixtures were studied, and each was examined at either 15 or 30°C. When the experimental data were plotted, as illustrated in Fig. 25.7 for a mixture comprised of

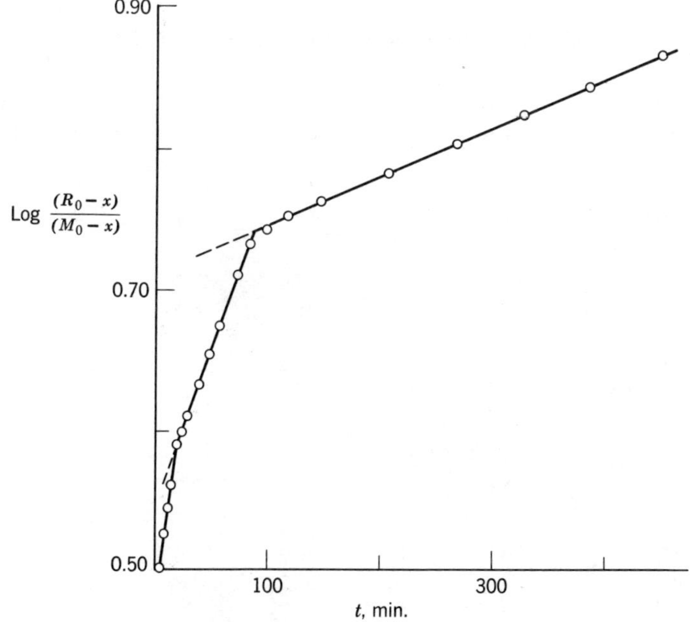

Fig. 25.7. Saponification of *n*-propyl, isopropyl, and *tert*-butyl acetate mixture.

Table 15. Three-Component Ester Analysis at 30°C

Acetate[a] Mixture[a]	Component A, %		Component B, %		Component C, %	
	Found	Present	Found	Present	Found	Present
A. Methyl	19.6	20.0	40.3	39.8	40.1	40.2
B. Isopropyl	44.1	43.4	22.3	21.7	33.6	34.9
C. *tert*-Butyl						
A. Ethyl	24.7	25.1	26.1	25.0	49.2	49.9
B. Isopropyl	35.9	35.7	27.7	28.6	36.4	35.7
C. *tert*-Butyl						
A. *n*-Propyl	26.3	27.8	29.0	27.8	44.7	44.4
B. Isopropyl	38.9	39.2	29.9	29.4	31.2	31.4
C. *tert*-Butyl						
A. Phenyl	39.5	40.1	24.0	23.9	36.5	36.0
B. Isopropyl	24.6	23.8	38.8	40.6	36.6	35.6
C. *tert*-Butyl						
A. 2-Hydroxyethyl	33.6	33.0	36.7	38.3	29.7	28.7
B. *n*-Butyl	24.0	22.3	30.2	33.2	45.8	44.5
C. *tert*-Butyl						

[a] Relative saponification rates are in the order A>B>C.

Table 16. Three-Component Ester Analysis at 15°C

Acetate Mixture[a]	Component A, %		Component B, %		Component C, %	
	Found	Present	Found	Present	Found	Present
A. Phenyl	22.3	23.2	34.9	35.2	42.8	41.6
B. 2-Hydroxyethyl	42.7	42.1	28.5	31.0	28.8	26.9
C. Isopropyl						
A. Phenyl	32.7	33.4	35.0	33.5	32.3	33.1
B. 2-Hydroxyethyl	23.7	25.0	48.7	48.4	27.6	26.6
C. *n*-Propyl						
A. Phenyl	22.8	23.1	35.2	33.3	42.0	43.6
B. 2-Hydroxyethyl	36.7	37.4	36.4	36.3	26.9	26.3
C. Ethyl						
A. 2-Hydroxyethyl	22.5	24.1	27.4	26.4	50.1	49.5
B. Ethyl	35.2	35.7	26.3	27.2	38.5	37.1
C. Isopropyl						
A. Phenyl	40.7	41.7	20.3	21.0	39.0	37.3
B. Methyl	42.4	43.5	33.7	33.7	23.9	22.8
C. Isoamyl						

[a] Relative saponification rates are in the order A>B>C.

n-propyl, isopropyl, and *tert*-butyl acetate, the overall curve obtained consisted of three straight lines of varying slope, corresponding to the relative reactivity of each component. Upon graphical extrapolation of the intermediate line to zero time, the determined intercept y was equated to the total concentration of ester M_0 (evaluated by allowing the rate reaction to go to completion) and hydroxide R_0 by

$$\log \frac{R_0 - x}{M_0 - x} = y$$

Evaluation of x yielded the concentration of the fastest acting component. Similar extrapolation and calculation involving the final line gave the total concentration of the constituents A and B; on taking differences, the amount of the intermediate and slowest reacting compounds present was determined. Results for various analyzed mixtures are listed in Tables 15 and 16.

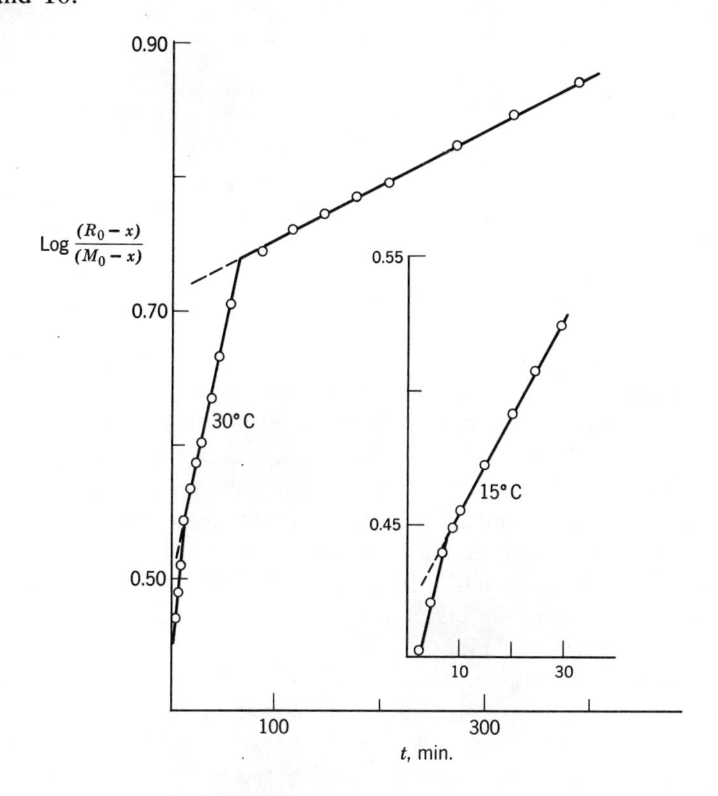

Fig. 25.8. Saponification of 2-hydroxyethyl, ethyl, isopropyl, and *tert*-butyl acetate mixture.

Table 17. Four-Component Ester Analysis

Acetate Mixture[a]	Component A, %		Component B, %		Component C, %		Component D, %	
	Found	Present	Found	Present	Found	Present	Found	Present
A. Methyl	16.0	17.3	22.0	22.5	27.6	27.7	34.4	32.5
B. Isoamyl	27.6	29.7	26.7	26.5	24.7	23.5	21.0	20.3
C. Isopropyl								
D. *tert*-Butyl								
A. 2-Hydroxyethyl	29.2	30.7	32.2	31.4	19.0	18.8	19.6	19.1
B. Ethyl	16.0	17.6	18.3	18.6	32.5	31.9	33.2	31.9
C. Isopropyl								
D. *tert*-Butyl								
A. 2-Hydroxyethyl	29.9	30.8	21.8	22.1	15.5	15.6	32.8	31.5
B. *n*-Propyl	18.1	19.6	22.5	20.2	18.6	20.3	40.8	39.9
C. Isopropyl								
D. *tert*-Butyl								
A. Phenyl	37.8	39.7	20.6	20.0	22.7	20.3	18.9	20.0
B. Methyl	16.5	17.4	32.8	34.3	32.7	31.0	18.0	17.3
C. *n*-Amyl								
D. Isopropyl								
A. Phenyl	30.2	28.4	27.7	28.6	28.0	28.5	14.1	14.5
B. 2-Hydroxyethyl	15.8	14.3	29.8	28.6	25.1	28.7	29.3	28.4
C. Ethyl								
D. Isopropyl								
A. Phenyl	27.9	28.4	19.1	18.2	17.3	17.9	35.7	35.5
B. Ethyl	17.0	16.6	34.3	32.6	21.4	22.9	27.3	27.9
C. Isopropyl								
D. *tert*-Butyl								

[a] Relative saponification rates are in the order A>B>C>D.

Selection of the temperature level to be used for each mixture was necessarily prescribed by the composition. The presence of the reactivity extremes for these esters, phenyl–hydroxethyl acetate on the one hand and isopropyl–*tert*-butyl acetate on the other, requires measuring temperatures of 15 and 30°C, respectively, for effective analysis. One mixture containing both phenyl and *tert*-butyl, encompassing a reactivity differential of about a thousandfold, was analyzed at 30°C. By working at this temperature, no data are collected on the individual reaction undergone by phenyl acetate; only the other two reactions are of measurable length, showing up in a two-line plot.

FOUR–COMPONENT SOLUTIONS. Introduction of another component into a ternary mixture, such as 2-hydroxyethyl acetate to a mixture of ethyl, isopropyl, and *tert*-butyl acetates, creates a system in which the range of reactivities has increased noticeably. This mixture could be analyzed completely at 15°C but would require an extremely long time. By operating at two temperatures, the reactions were separated into two stages. Low-temperature measurements yielded mainly the reaction sequence of 2-hydroxyethyl and ethyl acetate up to a relatively short time of 30 minutes, whereas at the higher temperature, saponification of isopropyl and *tert*-butyl acetates is essentially taking place (Fig. 25.8). After the data had been plotted to ascertain the respective linear portions, the points in each linear series were treated by a least-squares method for determining intercepts, to minimize errors in correlating data from two runs. As previously, successive concentration values were determined and differences taken. The calculated and experimentally determined values are compared in Table 17.

FIVE–COMPONENT SOLUTIONS. Extension of the dual-temperature method to mixtures of five esters, again of sufficient reactivity differences, resulted in the successful separation of the extremely fast reacting constituents from the slow ones. As illustrated in Fig. 25.9 for the acetates of phenyl, 2-hydroxyethyl, *n*-propyl, isopropyl, and *tert*-butyl alcohols, measurements showed adequate reaction monitoring at 15°C for three components and at 30°C for the remaining two.

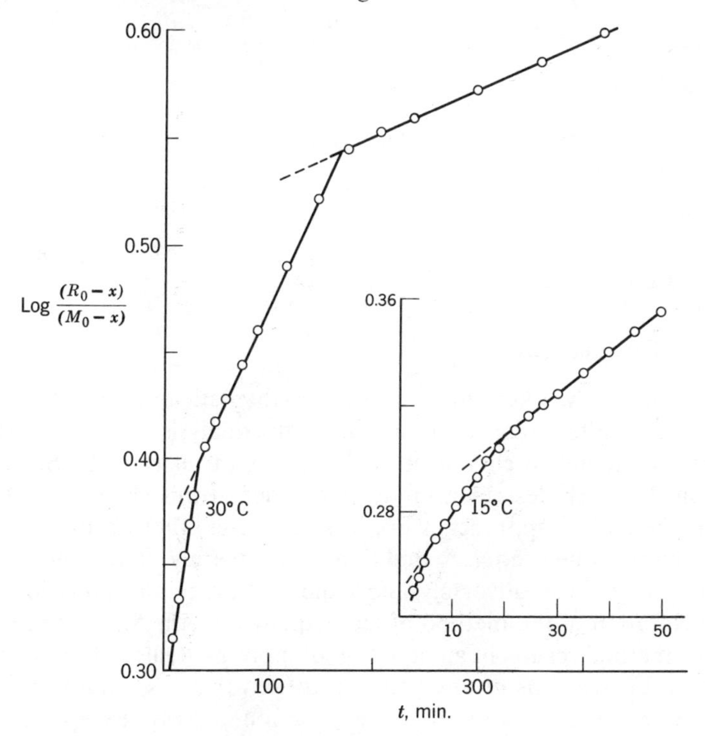

Fig. 25.9. Saponification of phenyl, 2-hydroxyethyl, *n*-propyl, isopropyl, and *tert*-butyl acetate mixture.

Table 18 shows the results obtained for five-component mixtures calculated in an analogous manner to four-component mixtures but with the additional numerical extrapolation carried out. Although the graphs show a total of six linear segments, only four (the more comprehensive) are actually applied to the composition determination.

Table 18. Five-Component Ester Analysis

Acetate Mixture[a]	Component A, %		Component B, %		Component C, %		Component D, %		Component E, %	
	Found	Present	Found	Present	Found	Present	Found	Present	Found	Present
A. Phenyl	15.1	14.2	12.4	14.3	23.0	21.5	20.1	21.4	29.4	28.6
B. 2-Hydroxyethyl	26.1	27.3	13.9	15.0	16.9	15.1	14.7	15.3	28.4	27.3
C. *n*-Propyl										
D. Isopropyl										
E. *tert*-Butyl										
A. Phenyl	19.0	18.5	24.6	26.1	13.8	15.5	17.9	16.9	24.7	23.0
B. Methyl	24.1	22.8	16.1	15.0	15.3	16.7	21.8	24.3	22.7	21.2
C. Isoamyl										
D. Isopropyl										
E. *tert*-Butyl										
A. Phenyl	21.6	20.0	17.1	20.0	21.8	20.0	18.9	20.0	20.6	20.0
B. 2-Hydroxyethyl	19.5	21.4	15.3	14.3	31.0	28.6	11.5	14.2	22.7	21.5
C. *n*-Butyl										
D. Isopropyl										
E. *tert*-Butyl										
A. Phenyl	14.9	14.3	14.7	17.2	21.2	19.9	21.9	22.9	27.3	25.7
B. 2-Hydroxyethyl	27.2	25.8	21.9	25.7	16.3	14.2	15.5	14.3	19.1	20.0
C. Ethyl										
D. Isopropyl										
E. *tert*-Butyl										

[a] Relative saponification rates are in the order A>B>C>D>E.

General Discussion

Reilley and co-workers have devised mathematical approaches to the application of differential reaction rates to the analysis of mixtures. These approaches attempt to circumvent the use of plots as done by Siggia and Hanna in the work described in the remainder of this chapter. Although these mathematical approaches do, in some cases, shorten the time of a kinetic analysis, they require conditions that severely limit their range of applicability and also adversely affect their accuracy and precision.

An adaptation of the method of least squares for the Siggia and Hanna graphical method results in an equation of more general usefulness. It can be applied to reactions of any order, provided there is a linear relationship between a concentration term and time. Also, except that the equation is indeterminate if the value of only one point is used, it is not limited otherwise to two experimental points. The analyst can use his discretion in choosing the number of points, based on his confidence in the accuracy of the experimental values and on the accuracy needed. Accuracy comparable to the graphical method is to be expected if the number of points used is the same as the number of points describing the final linear portion of the graphical method plot.

The final linear portion of the reaction rate curve for a mixture can be represented by

$$j = m + pt \tag{1}$$

where m and p are constants and j is, for example, $\log(a-x)$ for a first-order or pseudo-first-order reaction, or $\log[(b-x)/(a-x)]$ for a second-order reaction. The designations a, b, and x have been defined (p. 826). To fit the data, a straight line is chosen, using equations from the method of least squares (24). For the line represented by eq. 1, these are

$$\Sigma j = nm + p\Sigma t \tag{2}$$

$$\Sigma jt = m\Sigma t + p\Sigma t^2 \tag{3}$$

where n is the number of observations.

Solving eqs. 2 and 3 simultaneously for m gives

$$m = \frac{\Sigma t^2 \Sigma j - \Sigma t \Sigma jt}{n\Sigma t^2 - (\Sigma t)^2} \tag{4}$$

Then, for the line extrapolated to $t = 0$, eq. 1 becomes for the intercept

$$j = m \tag{5}$$

and combining eqs. 4 and 5 results in the following general equation:

$$j = \frac{\Sigma t^2 \Sigma j - \Sigma t \Sigma jt}{n\Sigma t^2 - (\Sigma t)^2} \tag{6}$$

For a first-order reaction, $j = \log(a - a_1)$ at $t = 0$, and eq. 6 becomes

$$\log(a - a_1) = \frac{\Sigma t^2 \Sigma[\log(a-x)] - \Sigma t \Sigma\{t \log(a-x)\}}{n\Sigma t^2 - (\Sigma t)^2} \tag{7}$$

where a_1 is the initial concentration of the faster reacting species, and for a second-order reaction

$$j = \log\frac{(b-a_1)}{(a-a_1)} \qquad \text{at} \quad t = 0$$

$$\log\frac{(b-a_1)}{(a-a_1)} = \frac{\Sigma t^2 \Sigma\{\log[(b-x)/(a-x)]\} - \Sigma t \Sigma\{t \log[(b-x)/(a-x)]\}}{n\Sigma t^2 - (\Sigma t)^2} \tag{8}$$

24. A. G. Worthing and J. Geffner, *Treatment of Experimental Data*, Wiley, New York, 1943, p. 240.

Reilley and Pappa (25) devised a kinetic analytical approach to systems where a and b are equal in the kinetic equation shown on page 826. This simplifies the mathematics required but severely limits applicability, since in most analytical methods, an excess of reagent is required for the analysis to proceed at a satisfactory rate.

Garman and Reilley (26) devised a method of proportional equations for first-order and pseudo-first-order systems.

The graphical approach of Siggia and Hanna as depicted earlier (pp. 825–859) has several advantages over the more mathematical approaches of Lee and Kolthoff (1) and those of Reilley and co-workers (25, 26, 30). These are as follows:

VERSATILITY

1. Approach applies to first-, pseudo-first-, and second-order reaction systems. The same reasoning applies to all types.

2. No rate constants are required for the calculations, hence rigidly controlled reaction conditions are not required. In fact, one set of conditions can be used for one sample and another set of conditions for a second sample. Each analysis is independent of the others. This makes the method applicable to systems widely divergent in reactivity or in concentration of the diverse components. Optimum experimental conditions can be chosen for each case. The examples in Tables 1 to 14 show this versatility.

ACCURACY AND PRECISION

1. The value used to calculate the analysis in the graphical method is an extrapolation from a number of experimental points. This results in an "averaging" of the data and a decreasing of the effects of experimental error. The mathematical approaches are based on either one or two points; hence all errors attendant to those one or two points end up in the analysis. In mixtures, nonjudicious selection of the one or two points can result in serious errors.

2. The use of rate constants in some of the mathematical approaches tends toward error because any deviation from the fixed condition results in a different rate, which then affects the analysis. Also, rate constants are

25. C. N. Reilley and L. J. Pappa, *Anal. Chem.*, **34,** 801–4 (1962).
26. R. G. Garman and C. N. Reilley, *Anal. Chem.*, **34,** 600–6 (1962).
27. S. Siggia, J. G. Hanna and N. M. Serencha, *Anal. Chem.*, **35,** 365 (1963).
28. J. Block, E. Morgan, and S. Siggia, *Anal. Chem.*, **35,** 573 (1963).

determined on pure samples of the various components. When the components are mixed, their rate constants are not always identical to those when each component occurred alone (29).

The graphical approach has now been applied to some 80 mixtures with 96% success. The same approach has also been used to analyze mixtures using physical reactions rather than chemical ones (27) and has been used in combustion analysis (28).

29. S. Siggia and J. G. Hanna, *Anal. Chem.*, **36,** 228 (1964).
30. L. J. Pappa, H. B. Mark, and C. N. Reilley, *Anal. Chem.*, **34,** 1513 (1962).

26

Weighing of Volatile or Corrosive Liquids

Need often arises for the determination of liquid samples that are highly volatile. These liquids cannot be weighed even in glass-stoppered weighing bottles without serious losses. Some liquids give off corrosive vapors or gases, which may cause injury to the balance being used.

A technique has been devised whereby the sample is sealed in a glass ampoule so that there can be no loss of vapors. Samples boiling as low as 0°C can be successfully weighed by this technique.

The preparation of the ampoule is described in Fig. 26.1. The bulbs are blown as shown in steps 1 to 4. Then the arm of the bulb is bent, as in step 5, and the constriction is made, as in step 6.

The design of the bulb makes possible the removal of a sample from a bottle (step 7) without awkward tilting of the sample bottles. The constriction makes possible a very quick seal (step 8) without significant vaporization of sample. Vaporization of more volatile components of a mixture can result in an unrepresentative sample.

To take a sample, proceed as follows. Warm the bulb of the ampoule gently in a flame to drive out some of the air by expansion (step 6). Then quickly insert the tip of the capillary below the surface of the sample. Immerse the bulb of the ampoule in a beaker containing a mixture of dry ice and methanol or acetone. The contraction of the air in the bulb will draw the sample into the bulb. As soon as the flow of sample into the bulb has diminished, but has not yet ceased, withdraw the tip of the capillary from the sample bottle and allow the sample in the capillary side arm to be drawn past the constriction. Keep the bulb of the ampoule in the dry ice bath while the ampoule is sealed.

The sealing is accomplished by placing the constriction in the arm of the ampoule in the flame of a micro burner or in the flame produced by a Bunsen burner with the chimney removed. The flame can be brought to the edge of the beaker of dry ice and solvent without the contents of the beaker catching fire. This is probably because of the extremely low temperature of the solvent and the atmosphere of carbon dioxide over its surface. To be safe, hold the beaker with a pair of tongs. This procedure is more comfortable, since the beaker is extremely cold. The liquid

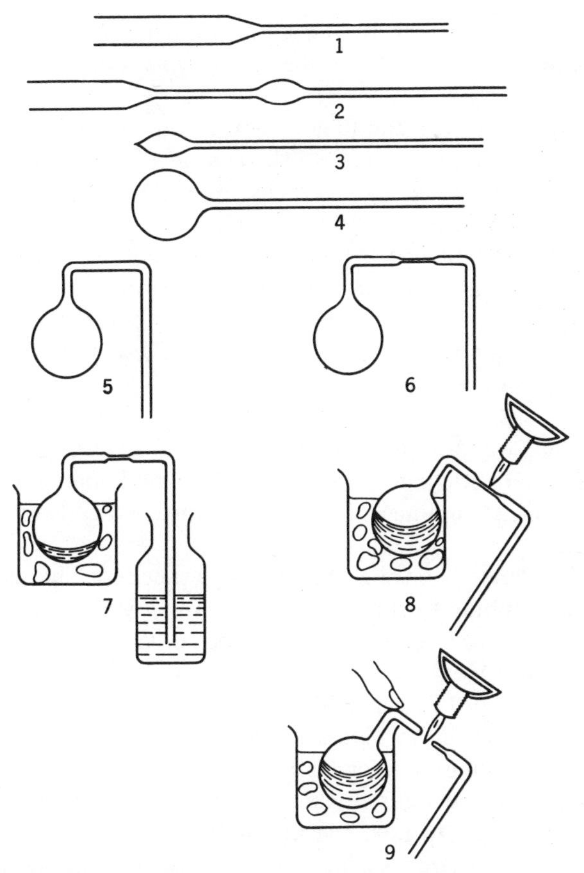

Fig. 26.1. Ampoules for weighing volatile liquids.

contained in the capillary side arm, which now is separated from the bulb (step 9), is driven out by holding the capillary in a flame with a pair of tweezers. This liquid should be entirely removed, since the capillary side arm has to be weighed with the full ampoule because it was included in the weight of the empty ampoule.

A few precautions in sampling liquids are as follows.

1. No vapor should be lost from the bulb at any time. The sample should be drawn up in one operation. Do not try to draw up additional sample by warming the sample already in the bulb and then recooling the bulb. This would result in driving off the lower boiling portion of the sample and obtaining an unrepresentative sample.

In the same respect, error in sampling can result if the bulb has been brought to a temperature too high for that particular sample. The liquid first striking the hot bulb can be partially vaporized off; this is indicated by bubbles escaping from the tip of the capillary as soon as the first liquid hits the bulb.

2. Support the bulb with the fingers as the constriction is being sealed so that the bulb will not fall over into the cold alcohol or acetone bath. If the hot tip strikes the cold bath, cracks often develop in the seal and portions of the samples can be lost through these cracks.

3. Do not fill the bulb more than two-thirds full when the sample is very volatile. The bulb may burst as it comes to room temperature.

4. The face of the worker should be shielded when handling sealed ampoules that contain liquids boiling below 20°C, since faulty ampoules may not be able to bear the pressure.

This technique was found very satisfactory in analyzing methyl vinyl ether (b.p. 5–6°C), ethylene oxide (10.7°C), ethyl amine (17°C), and acetaldehyde (21°C). Fuming sulfuric acid samples were also measured by this technique.

The bulb size depends on the size of sample needed. The use of 5-mm outside diameter tubing in stage 1 (Fig. 26.1) will yield bulbs about $\frac{1}{8}$ to $\frac{3}{16}$ in. in diameter; these bulbs will hold about 0.05 to 0.2 gram. The use of 7-mm tubing will yield bulbs that range from $\frac{1}{4}$ to $\frac{1}{2}$ in. in diameter and will hold about 0.2 to 0.7 gram. When 9- and 12-mm tubing is used, the bulbs will be $\frac{3}{4}$ in. to 1 in. in diameter and will hold about 1 to 3 grams. For the very low-boiling samples, 15°C or below, the ampoules should be blown small so that the walls remain rather heavy. The ampoule should be tested for its ability to hold the pressure. A simple test is to tap the bulb sharply on a wooden surface, when it is in stage 4, holding the ampoule by the end of the stem. If the bulb withstands this test, it will hold a sample boiling down to 0°C. (During the test, the bulb should be turned slightly between taps to make sure there are no weak spots.)

When gases such as ethylene oxide, butadiene, or methyl vinyl ether are sampled at room temperature, the cylinder containing the sample should be cooled well below the boiling point of the sample. The cylinder is then inverted and the sample is *poured* in the liquid state into a tall bottle that is immersed about half its length in crushed dry ice. The bottle should be capped while it is cooling in the ice to prevent condensation of water on the inside. Also, it should be capped immediately after the sample is in, to prevent carbon dioxide from dissolving in the sample (this is sometimes very significant). The sampling of the material in the liquid state ensures obtaining a representative sample from the cylinder. The samples are then drawn from the bottle into ampoules, as described earlier.

Index